# Biofilms in the food and beverage industries

**Related titles:**

*Foodborne pathogens: Hazards, risk analysis and control Second edition*
(ISBN 978-1-84569-362-6)
Effective control of pathogens continues to be of great importance to the food industry. The first edition of *Foodborne pathogens* quickly established itself as an essential guide for all those involved in the management of microbiological hazards at any stage in the food production chain. This major new edition strengthens that reputation, with extensively revised and expanded coverage, including more than ten new chapters. Chapters in Part I cover pathogen detection, microbial modelling, the risk assessment procedure, pathogen control in primary production, hygienic design and sanitation, among other topics. Parts II and III then review the management of key bacterial and non-bacterial foodborne pathogens.

*Handbook of hygiene control in the food industry*
(ISBN 978-1-85573-957-4)
Complementing the highly successful *Hygiene in food processing*, this book reviews recent research on improving hygiene in food processing. Part I considers recent research on contamination risks such as biofilms and how they can be assessed. Part II reviews ways of improving hygienic design of both buildings and equipment, including clean room technology. The final part of the book discusses ways of improving hygiene practice and management.

*Emerging foodborne pathogens*
(ISBN 978-1-85573-963-5)
Developments such as the increasing globalisation of the food industry, constant innovations in technologies and products, and changes in the susceptibility of populations to disease, have all highlighted the problem of emerging pathogens. Pathogens may be defined as emerging in a number of ways. They can be newly discovered (e.g., through more sensitive analytical methods), linked for the first time to disease in humans, or first associated with a particular food. A pathogen may also be defined as 'emerging' when significant new strains emerge from an existing pathogen, or if the incidence of a pathogen increases. This collection discusses ways of identifying emerging pathogens and includes chapters on individual pathogens, their epidemiology, methods of detection and means of control.

Details of these books and a complete list of Woodhead's titles can be obtained by:

- visiting our web site at www.woodheadpublishing.com
- contacting Customer Services (e-mail: sales@woodheadpublishing.com; fax: +44 (0) 1223 893694; tel.: +44 (0) 1223 891358 ext. 130; address: Woodhead Publishing Limited, Abington Hall, Granta Park, Great Abington, Cambridge CB21 6AH, UK)

# Biofilms in the food and beverage industries

Edited by
Pina M. Fratamico, Bassam A. Annous and Nereus W. Gunther IV

CRC Press
Boca Raton Boston New York Washington, DC

WOODHEAD PUBLISHING LIMITED
Oxford Cambridge New Delhi

Published by Woodhead Publishing Limited, Abington Hall, Granta Park,
Great Abington, Cambridge CB21 6AH, UK
www.woodheadpublishing.com

Woodhead Publishing India Private Limited, G-2, Vardaan House, 7/28 Ansari Road,
Daryaganj New Delhi – 110002, India
www.woodheadpublishingindia.com

Published in North America by CRC Press LLC, 6000 Broken Sound Parkway, NW,
Suite 300, Boca Raton, FL 33487, USA

First published 2009, Woodhead Publishing Limited and CRC Press LLC


British Library Cataloguing in Publication Data
A catalogue record for this book is available from the British Library.

Library of Congress Cataloging in Publication Data
A catalog record for this book is available from the Library of Congress.

Woodhead Publishing ISBN 978-1-84569-477-7 (book)
Woodhead Publishing ISBN 978-1-84569-716-7 (e-book)
CRC Press ISBN 978-1-4200-9484-8
CRC Press order number: WP9484

The publishers' policy is to use permanent paper from mills that operate a sustainable forestry policy, and which has been manufactured from pulp which is processed using acid-free and elemental chlorine-free practices. Furthermore, the publishers ensure that the text paper and cover board used have met acceptable environmental accreditation standards.

Typeset by SNP Best-set Typesetter Ltd., Hong Kong
Printed by TJ International Limited, Padstow, Cornwall, UK

# Contents

# Contributor contact details

(* = main contact)


**Editors**

Dr Pina. M. Fratamico*, Bassam A. Annous and Nereus W. Gunther IV
United States Department of Agriculture
Agricultural Research Service
Eastern Regional Research Center
600 E. Mermaid Lane
Wyndmoor, PA 19038
USA

E-mail: Pina.Fratamico@ars.usda.gov
bassam.annous@ars.usda.gov
jack.gunther@ars.usda.gov

**Chapter 1**

Professor T. E. Cloete*, I. Molobela, A. Van Der Merwe and M. Richards
Stellenbosch University
South Africa

E-mail: eugenecloete@son.ac.za

**Chapter 2**

Dr James L. Smith*, Dr Pina M. Fratamico and Dr Gaylen Uhlich
United States Department of Agriculture
Agricultural Research Service
Eastern Regional Research Center
600 E. Mermaid Lane
Wyndmoor, PA 19038
USA

E-mail: james.smith@ars.usda.gov
pina.fratamico@ars.usda.gov
gaylen.uhlich@ars.usda.gov

**Chapter 3**

Professor Z. Lewandowski*
Center for Biofilm Engineering
Department of Civil Engineering
Montana State University
Bozeman, MT 59717-3980
USA

E-mail: zl@erc.montana.edu

H. Beyenal
Gene and Linda Voiland School of Chemical Engineering and Bioengineering
Center for Environmental, Sediment, Aquatic Research
Washington State University
Pullman, WA 99163-2710
USA

E-mail: beyenal@wsu.edu

**Chapter 4**

Ana Pereira and Luis F. Melo*
LEPAE/Department of Chemical Engineering
Faculty of Engineering
University of Porto
Rua Dr. Roberto Frias, s/n
4200-465 Porto
Portugal

E-mail: lmelo@fe.up.pt

**Chapter 5**

Dr Xianghe Yan* and Dr Pina M. Fratamico
United States Department of Agriculture
Microbial Food Safety Research Unit
Eastern Regional Research Center
United States Department of Agriculture
600 E. Mermaid Lane
Wyndmoor, PA 19038
USA

E-mail: xianghe.yan@ars.usda.gov
Pina.Fratamico@ars.usda.gov

Jin Gui
College of Management and Technology
Walden University
155 Fifth Avenue South
Minneapolis, MN 55401
USA

**Chapter 6**

Dr D. R. Korber*, A. K. Mangalappalli-Illathu and S. Vidović
Food and Bioproduct Sciences
51 Campus Drive
University of Saskatchewan
Saskatoon
Saskatchewan
Canada

E-mail: darren.korber@usask.ca

**Chapter 7**

Dr A. Rodríguez-Lozano
Campden BRI
Station Road
Chipping Campden
Gloucestershire GL55 6LD
UK

Dr L. McLandsborough*
Department of Food Science
University of Massachusetts
Chenoweth Laboratory
100 Holdsworth Way
Amherst, MA 01003
USA

E-mail: lm@foodsci.umass.edu

**Chapter 8**

Dr C. Gamazo*
Department of Microbiology
University of Navarra
Irunlarrea, 1
31008-Pamplona
Spain

E-mail: cgamazo@unav.es

Dr C. Solano and Dr I. Lasa
Instituto de Agrobiotecnología y Recursos Naturales
Departamento de Producción Agraria
Universidad Pública de Navarra
Pamplona
Spain

E-mail: ilasa@unavarra.es

**Chapter 9**

Dr S. Langsrud
Nofima Mat. Norwegian Food Research Institute
Oslov. 1
N-1430 Ås
Norway

E-mail: solveig.langsrud@nofima.no

**Chapter 10**

Dr D. Lindsay*
Analytical Development (Microbiology)
Fonterra Co-operative Group Limited
Dairy Farm Road
Palmerston North 4442
New Zealand

E-mail: denise.lindsay@fonterra.com

Associate Professor S. Flint
Institute of Food Nutrition and Human Health
Massey University
Private Bag 11222
Palmerston North 4442
New Zealand

E-mail: S.H.Flint@massey.ac.nz

**Chapter 11**

Dr C. Faille*
INRA-UR638
369 rue J. Guesde
BP 20039
F-59651 Villeneuve d'Ascq cedex
France

E-mail: faille@lille.inra.fr

Dr B. Carpentier
AFSSA
Laboratoire d'études et de recherches sur la qualité des aliments et des procédés agroalimentaires
23 avenue du Général de Gaulle
F-94706 Maisons-Alfort cedex
France

E-mail: b.carpentier@afssa.fr

**Chapter 12**

Dr D.A. Grinstead
JohnsonDiversey, Inc
8310 16th Street
Sturtevant, WI 53177
USA

E-mail: dale.grinstead@johnsondiversey.com

**Chapter 13**

Professor S. E. Martin*
Department of Food Science and Human Nutrition
University of Illinois
486 Animal Sciences Building
1207 West Gregory Drive
Urbana, IL 61801
USA

E-mail: semartn@illinois.edu

Dr H. Feng
Department of Food Science and Human Nutrition
University of Illinois
382B Agricultural Engineering Sciences Building
1304 West Pennsylvania Avenue
Urbana, IL 61801
USA

E-mail: haofeng@illinois.edu

**Chapter 14**

Dr B. Carpentier
Agence française de sécurité sanitaire des aliments
Laboratoire d'études et de recherches sur la qualité des aliments et sur les procédés agroalimentaires
23 avenue du Général de Gaulle
F-94706 Maisons-Alfort cedex
France

E-mail: b.carpentier@afssa.fr

**Chapter 15**

Associate Professor P. Bremer* and Mr B. Seale
Department of Food Science
University of Otago
P O Box 56
Dunedin 9054
New Zealand

E-mail: phil.bremer@otago.ac.nz

Associate Professor S. Flint and Dr J. Palmer
Institute of Food Nutrition and Human Health
Massey University
Private Bag 11222
Palmerston North 4442
New Zealand

**Chapter 16**

Dr E. Storgårds* and Dr O. Priha
VTT Technical Research Centre of Finland
P.O. Box 1000
FI-02044 VTT
Finland

E-mail: erna.storgards@vtt.fi

**Chapter 17**

Dr J. W. Arnold
United States Department of Agriculture, Agricultural Research Service
Russell Research Center
950 College Station Road
Athens, GA 30605
USA

E-mail: Judy.Arnold@ars.usda.gov

**Chapter 18**

Nasib Qureshi
United States Department of Agriculture
Agricultural Research Service
National Center for Agricultural Utilization Research
1815 N. University Street
Peoria, IL 61604
USA

E-mail: Nasib.Qureshi@ars.usda.gov

**Chapter 19**

Dr Kathleen T. Rajkowski
United States Department of Agriculture
Agricultural Research Service
Eastern Regional Research Center
600 E. Mermaid Lane
Wyndmoor, PA 19038
USA

E-mail: kathleen.rajkowski@ars.usda.gov

**Chapter 20**

Dr Bassam A. Annous,* Dr James L. Smith and Dr Pina M Fratamico
United States Department of Agriculture
Agricultural Research Service
Eastern Regional Research Center
600 E. Mermaid Lane
Wyndmoor, PA 19038
USA

E-mail: bassam.annous@ars.usda.gov
pina.fratamico@ars.usda.gov

Dr Ethan B. Solomon
DuPont Chemical Solutions Enterprise
Experimental Station Laboratory
Wilmington, DE 19880-0402
USA

E-mail: Ethan.B.Solomon@usa.dupont.com

**Chapter 21**

Dr David E. Nivens* and Bonnie M. Co
Food Science Department
Purdue University
West Lafayette, IN
USA

E-mail: dnivens@purdue.edu

Dr Michael J. Franklin
Montana State University
Department of Microbiology
The Center for Biofilm Engineering
Bozeman, MT
USA

# Preface

Normally, the presence of biofilms on food and on food contact surfaces negatively impacts food quality and safety. Compared to non-attached cells, biofilm-associated bacteria are more resistant to inactivation by antimicrobial compounds and biocides, as well as to physical and chemical stresses. Through the use of sophisticated imaging techniques and molecular analysis methods, great advances have been made in understanding the dynamics of biofilm formation and the influence of environmental factors and cell-to-cell signaling processes in modulating biofilm development. A clearer understanding of the conditions and factors that promote biofilm formation by spoilage and pathogenic bacteria will help in the development of novel control strategies to enhance food safety and quality.

The purpose of this book is to provide a comprehensive reference covering a variety of aspects of microbial biofilms. The chapters in Part I of the book present an introduction on biofilm formation and related problems for the food and beverage industry. Other chapters focus on the ecology and characteristics of biofilms, molecular mechanisms of biofilm formation, and methods for imaging, sampling, and quantifying biofilms. There is also a chapter describing a centralized database that will assist researchers who are studying biofilms and the role of quorum sensing, as well as methods to control biofilms by modulating quorum sensing processes. Part II has chapters on biofilm formation by specific organisms, including spoilage microorganisms, spore formers, *Listeria*, *Salmonella*, and Gram-positive bacteria. The third part has chapters on prevention of biofilms and on methods for their inactivation and removal. The final section focuses on issues related to biofilms in various food commodities, including red meats, dairy products, produce, fish, poultry, and beer and also includes a chapter on industrial applications of biofilms.

We, the editors, hope that this book will serve as a valuable reference source for research scientists in the food industry, academia, and government, graduate students, regulatory agencies, and individuals interested in learning more about biofilms. We also anticipate that the information presented in the various chapters written by a distinguished international group of scientists will stimulate ideas for interdisciplinary research efforts.

We gratefully acknowledge the assistance of Dr James L. Smith in reviewing the chapters, and we also extend heartfelt thanks to the authors of the various chapters in this book for contributing their time, knowledge, and expertise to this endeavor. It has been a pleasure working with such a fine group of professionals.

*Pina M. Fratamico, Bassam A. Annous, and Nereus W. Gunther IV*

# Part I

# Biofilms in the food and beverage industries

# 1

# Biofilms in the food and beverage industries: an introduction

**E. Cloete, Stellenbosch University, South Africa and I. Molobela, A. Van Der Merwe and M. Richards, University of Pretoria, South Africa**

**Abstract**: Microbes, (bacteria, fungi, yeasts, moulds, viruses, etc.) are primarily found in two forms, either floating freely in water (planktonic), or attached to a surface (sessile) where they congregate in large numbers to form a film or slime layer known as a biofilm. Although planktonic microbes have been studied for many decades, it is only in the past two decades that industry has begun to understand and appreciate the complexity of a biofilm as a living organism and the huge problems biofilm formation causes in industry. It is now acknowledged that biofilms have huge cost implications to industry and can negatively affect a company's profitability. Biofilm is extremely difficult to remove and control. It is common knowledge that many chemical biocides are ineffective against biofilm and, at best, merely control planktonic microbes – leaving the biofilm intact to continue to recontaminate, corrode and to build resistance against antimicrobial compounds. The added danger then is when a particularly undesired pathogen (e.g., *E. coli* in food processing, *Legionella* in cooling towers, *Pseudomonas* in poultry rearing) enters a system and becomes attached to the biofilm, it becomes extremely difficult to remove and control, causing enormous problems that often result in production and plant shut-downs – costing companies millions in lost revenue.

**Key words**: biofilms, extracellular polymers, biocides, bacteria, resistance.

## 1.1 Introduction

A biofilm can be defined as the sessile community of microbes characterized by cells that are irreversibly associated with a surface, embedded in a matrix of extracellular polymeric substances (EPS), and display an altered phenotype with respect to gene expression, protein production and growth. Biofilms can form on a wide variety of surfaces and the biofilm thickness varies from a single cell layer to a thick community of cells embedded in

EPS. Materials like corrosion particles, clay, slit and mineral crystals, depending on the environment in which the biofilm has formed, may also be found in the biofilm matrix (Donlan, 2002).

Microorganisms in biofilms display some particular features that are not shared with the same microorganisms in suspended form. In biofilms, the cells are embedded in a polymer matrix of their own origin that mainly consists of polysaccharides and proteins. Biofilms contain mixed populations of bacteria, fungi, protozoa and if conditions allow, they can host even higher organisms in the food chain such as nematodes and larvae.

Biofilms represent a very complex form of microbial life that is mainly characterized by a high degree of interaction between different types of organisms and by a more or less immobilized form of life. This allows the formation of stable aggregates in which synergistic effects can develop. In biofilms, the cells can tolerate much higher concentrations of biocides in suspension (Flemming, 1998). Another characteristic of biofims is their heterogeneity; for example, aerobic microorganisms in aerobic system consume oxygen resulting in an anaerobic zone within the biofilm (Flemming, 1998). This process provides habitats for anaerobes that could not proliferate under aerobic conditions.

A biofilm is mainly composed of water (80–90%), extracellular polymer substances (EPS) that contribute 85–98% of the organic matter, the microorganisms, entrapped organic and inorganic particles, substances sorbed to EPS, cells or particles and substances dissolved in the interstitial water (Flemming, 1998). Microorganisms are characterized as freely suspended cells (planktonic) and are described on the basis of their growth characterizations in nutritionally rich culture media.

In a natural aquatic ecosystem, surface associated microorganisms consist of different organisms in suspension. The ability for bacteria to colonize surfaces is advantageous from an ecological point because specialized bacteria are preferentially targeted to specific locations, encouraging symbiotic relationships. Examples of these relationships are taking place in nature, e.g. prokaryotic diazotrophs that colonize the roots of the legumes (Pearce *et al.*, 1995) and the diverse residential microbial flora inhabiting the digestive tract of ruminants that promotes the degradation and recycling of insoluble materials (Cheng *et al.*, 1995).

It is advantageous for bacteria to form biofilms since this mode of growth provides the bacteria with elevated levels of protection (Allison *et al.*, 2000; Mara and Horan, 2003) against fluctuating environmental conditions (pH, nutrients, oxygen and shear forces), disinfectants, antibiotics and host immune responses (Jefferson, 2004). All bacteria within biofilms live together and depend on other microorganisms for energy, carbon and other nutrients. Thus, most of the bacteria live in micro ecosystems filled with hundreds of other microorganisms and form biofilms. When in this form, bacteria can be up to a 1000 times more resistant to antimicrobials than their planktonic counterparts (Mah and O'Toole, 2001; Drenkard,

2003). Biofilm bacteria attach and reside in environments that are favorable and, when exposed to unfavorable conditions, cells can detach and move to new locations.

The metabolism of the cells in a biofilm is reduced due to its heterogeneous nature. Heterogeneity arises due to different diffusion rates of nutrients and oxygen within the biofilm, creating oxygen, substrate and pH gradients. This in turn allows bacteria at different locations in the same biofilm to carry out different metabolic tasks and produce different gene products (Jefferson, 2004). Kreft (2004) points out another advantage of communal growth by confirming altruistic behavior in biofilms. Altruism can be defined as the unselfish behavior of one group in order to enhance the fitness of the entire group. An example of altruistic behavior is where cells exhibit a low growth rate in order to decrease the use of nutrient sources and increase the growth yield and strength of the group (Kreft, 2004).

Gene transfer between biofilm bacteria is believed to be favored due to the extended contact time of bacteria within biofilms and the steadiness of cells on the surface (Allison *et al.*, 2000). Gene transfer increases the genetic diversity of bacteria and supplies the bacteria with antibiotic resistance genes and other genes of interest (Jefferson, 2004). Ghigo (2001) demonstrated the role of conjugative plasmids in the development of biofilms. These plasmids encode pilli that assist in the attachment of *E. coli* and other Gram-negative cells to each other and to surfaces and therefore lead to the conversion of planktonic cells to biofilm state. In this manner the plasmids ensure their spread to other bacterial cells (Ghigo, 2001).

The switch from planktonic to a biofilm mode of growth is an intricate process, which occurs in response to changes in environmental conditions (O'Toole *et al.*, 2000). For this switch to take place certain genes are up- or down-regulated (Kuchma and O'Toole, 2000). Biofilm bacteria differ from their planktonic counterparts in genes they express (Prigent-Combaret *et al.*, 1999; Becker *et al.*, 2001; Sauer and Camper, 2001; Whiteley *et al.*, 2001; Oosthuizen *et al.*, 2002; Sauer *et al.*, 2002; Schembri *et al.*, 2003; Beloin *et al.*, 2004; Beenken *et al.*, 2004; Ren *et al.*, 2004); the rate at which the cells grow (Huang *et al.*, 1995; Cloete, 2003a,b) and their resistance to biocides and antibiotics (Anderl *et al.*, 2000; Brooun *et al.*, 2000; Norwood and Gilmour, 2000; Ryu and Beuchat, 2005).

Biofilms can be formed on all types of surfaces in most ecosystems where nutrients are available. Biofilms can be divided into two groups according to their growth phase, namely the planktonic or sessile. In the planktonic stage, microbes live as individual free floating organisms, whereas in the sessile phase, they are attached to a surface and function as a closely integrated community (Poulsen, 1999). The water-system biofilm is highly complex, containing non-cellular materials such as mineral crystal, corrosion particles, clay or silt particles, freshwater diatoms and filamentous bacteria, while biofilms on medical devices, on the other hand, appear to be composed of a single, coccoid organism or blood components (Prakash *et al.*, 2003a,b).

The ability for bacteria to become surface bound is ubiquitous in diverse ecosystems that have a strong survival or selective advantage for surface dwellers over their free ranging counterparts. Bacterial adhesion may occur as a result of nutrients in an aqueous environment that tend to concentrate near a solid surface. Many specialized structures and complex ligand interactions, that evolved in prokaryotes designed specifically for surface recognition and biofilm formation, play an important role in microbial adhesion (Dunne, 2002).

From an evolutionary standpoint, the selective advantage of bacterial adhesion has been postulated to favor the localization of surface bound bacterial populations in nutritionally favorable, non-hostile environments and at the same time provide some level of protection from external predation. When the environment ceases to support the bacterial load, the equilibrium is shifted to favor dissociation of individual cells from the biofilm to seek a more favorable habitat.

## 1.2 Biofilm formation

The pattern of biofilm development involves the initial attachment of the bacteria to a solid surface, the formation of micro-colonies on the surface and the differentiation of these micro-colonies into mature biofilms encased in exopolysaccharides.

Bacteria sense certain environmental cues that trigger the transition from planktonic growth to biofilm formation (Nyvad and Kilian, 1990; O'Toole *et al.*, 2000; O'Toole and Kolter, 1998a,b). These environmental signals, including pH, temperature, oxygen tension and iron availability, vary among different microorganisms (Nyvad and Kilian, 1990; O'Toole and Kolter, 1998a,b).

The process of bacterial adhesion is controlled by a number of variables. These include the species of bacteria, environmental factors, essential gene products and the surface composition (Carpentier and Cerf, 1993).

Bacterial adhesion can be divided into two stages: the primary or docking stage and the secondary or locking phase (An *et al.*, 2000; Pearce *et al.*, 1995). An additional stage, surface conditioning, can also be included to describe the interaction of the substratum with its environment (Gristina, 1987). Conditioning occurs when a film of organic molecules form on clean surface in fluid surroundings (van Loodsrecht *et al.*, 1990). The properties of a conditioned surface are permanently altered and may cause a difference in an organism's affinity for the native and conditioned surface (Dunne, 2002).

## 1.3 Stages involved during attachment

### 1.3.1 The primary stage (reversible)

During the process of attachment, the organism must be brought into close approximation of the surface, propelled either randomly or in a directed fashion via chemotaxis and motility. Once an organism reaches critical proximity to a surface, the final determination of adhesion depends on the net sum of attractive or repulsive forces generated between the two surfaces. These forces include electrostatic and hydrophobic interactions (Melo and Bott, 1997). Initial attachment of the bacteria to the conditioned surface is facilitated by weak forces acting over lengthy distances, such as electrostatic and van der Waal's attractions (Denyer *et al.*, 1993). This attachment is unstable and reversible and if the environment is not favorable for microbial attachment, cells can detach from the surface (Ghannoum and O'Toole, 2004).

The primary adhesion stage constitutes the beneficial meeting between a conditioned surface and planktonic microorganisms. This stage is reversible and it is characterized by a number of physiochemical variables that define the interaction between the bacterial cell surface and the conditioned surface of interest (An *et al.*, 2000; Liu *et al.*, 2004; Singh *et al.*, 2002).

During this stage of attachment, planktonic microorganisms can also stick to each other or different species of surface-bound organisms forming aggregates on the substratum. The presence of mono species of microorganisms on a surface can promote the adhesion of another species. All bacteria produce multiple adhesions and some are regulated at the transcriptional level depending on the genes encoded, permitting organisms to switch from sessile to planktonic forms under different environmental influences. An example of this is in the case with *S. epidermidis*, which produces a polysaccharide intracellular adhesin (PIA) that is essential to cell-to-cell adhesion and biofilm formation (Dunne, 2002).

The acidophilic iron (II) ion oxiding bacterium *Acidithiobacillus (A) ferrooxidans* is one of the most important mesophiles for the extraction of metals from sulphidic ores by bioleaching. Attachment of these bacteria to the mineral surfaces seems to enhance bioleaching of pyrite and other minerals (Kinzler *et al.*, 2003).

### 1.3.2 The secondary stage (irreversible)

This is the anchoring or locking stage and employs molecularly mediated binding between the specific adhesins and the surface. In this process, loosely bound organisms gather together by producing exopolysaccharides that complex with surface materials and/or receptor specific ligands located on the pilli, fimbria and fibrillae or even both (An *et al.*, 2000; Rachid *et al.*, 2000).

If conditions are favorable, the bacterial cells can switch from reversible attachment to a more stable irreversible attachment (Denyer *et al.*, 1993; Dreeszen, 2003). Forces responsible for this type of attachment act over short ranges and are known as dipole, ionic, hydrogen or hydrophobic interactions (Charaklis and Marshall, 1990; Denyer *et al.*, 1993; Allison *et al.*, 2000). Firm attachment of bacterial cells to the surface is assisted by bacterial motility structures (flagella, pili), bacterial surface structures (proteins, LPS) and exopolymers produced by bacteria (Allison *et al.*, 2000).

Once the bacteria have attached irreversibly to the surface they undergo a range genotypic and phenotypic changes to ensure the development and maturation of the biofilm. These changes result in the production of increased amounts of EPS, increased resistance to antibiotics, increased UV resistance, gene exchange events occur more frequently and higher amounts of secondary metabolites are produced (O'Toole *et al.*, 2000).

Various structures such as flagella, fimbriae, outer membrane proteins (OMPs), curli (a proteinaceous surface structure) and extracellular polymers structure (EPS) are involved in biofilm formation. These structures have distinct roles in different species and under different environmental conditions (Giaouris and Nychas, 2006). Flagella motility is important to overcome the forces that repel bacteria from reaching many abiotic materials. Once they reach the surface, the nonflagella appendages other than those involved in transfer of viral or bacterial nucleic acids (called pili), OMPs and curli are then required to achieve stable cell-to-cell and cell-to-surface attachments. Flagella apparently play an important role in the early stages of bacterial attachments by overcoming the repulsive forces associated with the substratum (Giaouris and Nychas, 2006).

Korber *et al.* used motile and nonmotile strains of *Pseudomonas fluorescens* to show that motile cells attach in greater number and against the flow more rapidly than do nonmotile strains. Nonmotile strains also do not recognize the substratum as evenly as motile strains, resulting in slower biofilm formation by the nonmotile organisms. A number of aquatic bacteria possess fimbriae, which have also been shown to be involved in bacterial attachment to animal cells (Prakash *et al.*, 2003a,b, Giaouris and Nychas, 2006, Meyer, 2003, http://www.medcape.com/viewarticle/441355_4).

At the conclusion of the second stage, adhesion becomes irreversible in the absence of physical or chemical intervention, and the organisms are attached firmly to the surface. With certain organisms, several distinct adhesins might be used for surface attachment depending on the state of the environment. For examplc, in the case of *Vibrio cholerae* EI Tor, a toxin-coregulated pilus is used as an attachment and colonizes intestinal epithelium during the process of human infection. In addition, a mannose-sensitive hemagglutinin is the primary adhesin used to anchor to abiotic surfaces in an aquatic environment (Watnick *et al.*, 1999).

## 1.4 Formation of three-dimensional structures

During the attachment phase of biofilm development, the transcription of specific genes takes place. These are required for the synthesis of EPS. Attachment itself can initiate synthesis of the extracellular matrix in which the sessile bacteria are embedded, followed by formation of water-filled channels in the circulatory systems that help in delivering nutrients to and removing waste products from the communities of the cell in the micro colonies (Prakash *et al.*, 2003a,b).

## 1.5 Microcolony formation

After the adherence of bacteria to the inert surface, the association becomes stable for micro-colony formation (Palmer and White, 1997). The bacteria begin to multiply while sending out chemical signals that intercommunicate among the bacterial cells. Once the signal intensity exceeds a certain threshold level, the genetic mechanisms underlying exopolysaccharide production are activated. In this way, the bacteria multiply within the embedded exopolysaccharide matrix, thus giving rise to formation of a micro-colony (Prakash *et al.*, 2003a,b).

Micro-colonies further develop into macrocolonies that are divided by fluid-filled channel and enclosed in an extacellular polysaccharide matrix (Allison, 2003). Macro-colonies, compared to micro-colonies, are composed of a larger amount of cells, are enclosed in an extracellular matrix and have a higher metabolic and physiological heterogeneity (Ghannoum and O'Toole, 2004). The number of cells within the colonies increases by three methods: (i) translocation; where cells on the surface exploit twitching motility or other modes of transport to get into connection with already formed microcolonies, (ii) the direct attachment of planktonic cells to microcolonies, (iii) clonal growth through cell division of cells within micro-colonies (Ghannoum and O'Toole, 2004). The final macro-colony structure of Pseudomonads is mushroom-like pillars. These structures are produced by two subpopulations; motile and non-motile bacteria. The non-motile bacteria grow and form the mushroom stalk while the motile bacteria make use of type-IV pili to move up the stalk and collect at the top to form the mushroom cap (Klausen *et al.*, 2003).

Micro-colony development occurs by several mechanisms. In one of the mechanisms attached cells are redistributed by surface motility. Examples of this include the flagella and type IV pili-mediated twitching motility used by *Pseudomonas aeruginose* and the fimbriae, flagella and type I pili used by *Escherichia coli* (O'Toole and Kolter, 1998a,b). Motility appears to assist in surface colonization by Gram-negative organisms, but it is not a prerequisite for biofilm formation as many non-motile bacteria have the ability to form biofilms. In the non-motile *Staphylococcus epidermidis,* polysaccharide and protein adhesins have been linked to adherence, while a novel biofilm-associated protein was found to be involved in attachment and intercellular adhesion of *S. aureus*. In another mechanism, binary fusion of attached cells results in biofilm formation. From the attached point the cells divide and daughter cells spread outward and upward to form micro-colonies (Tolker-Nielsen and Molin, 2000).

## 1.6 Biofilm maturation

Once bacteria have irreversibly attached to a surface, the process of biofilm maturation begins. The overall density and complexity of the biofilm increase as surface-bound organisms begin to actively replicate and extra cellular components generated by attached bacteria interact with organic and inorganic molecules in the immediate environment to create the glycocalyx (Carpentier and Cerf, 1993). The availability of nutrients in the immediate environment, the pouring of those nutrients to cells within the biofilm and the removal of waste, limit the growth potential of any bacterial biofilm. In addition, there is an existence of an optimum hydrodynamic flow across the biofilm that favors growth and pouring rather that erosion of the outmost layers (Carpentier and Cerf, 1993). Other factors that control biofilm maturation include the internal pH, oxygen, carbon source osmolarity, temperature, electrolyte concentration and the flux of materials and surface types. The surface types can be either (O'Toole and Kolter, 1998a,b):

- High surface energy materials are negatively charged; hydrophilic materials such as glass, metals or minerals.
- Low surface energy materials are either low positively or low negatively charged; hydrophobic materials such as plastic made up of organic polymers (Kumar and Prasad, 2006).

At some point, the biofilm reaches a critical mass and a dynamic equilibrium is reached at which the outermost layers of growth begin to generate planktonic organisms. These organisms are free to escape the biofilm and colonize other surfaces. Cells nearest the surface become inactive or die due to a lack of nutrients, decrease in pH, pO2 or an accumulation of toxic metabolic by-products (Dunne, 2002).

## 1.7 Detachment and dispersal of cells from biofilms

As the biofilm ages, to survive and colonize new niches, the attached cells must be able to detach and disperse from the biofilm. This detachment can be due to various factors including fluid dynamics and shear effects of the bulk fluid (Brugnoni *et al.*, 2007). As the thickness of the EPS increases, anaerobic conditions develop within the biofilm (Spiers *et al.*, 2003). Because of the film thickness and the activity of anaerobic species, the film detaches and sloughs off from the surface of the substrate. Polysaccharides enzymes specific for the EPS of different organisms may be produced during different phases of biofilm growth of the organisms and contribute to detachment. It has been suggested that escape of *P. auriginosa* cells from the biofilms matrix involved the action of an enzyme that digest alginate (Prakash *et al.*, 2003a,b).

This process happens for mechanical reason because some bacteria are shed from the colony, some stop producing EPS and are released into the surrounding environment. Biofilm cells may be dispersed either by shedding of daughter cells from actively growing cells or detachment as result of nutrient levels (Spiers *et al.*, 2003). The released microorganisms may be transported to newer locations and restart the biofilm process (Prakash *et al.*, 2003a,b).

*P. fluorescens* and *P. aeruginosa* produce enzymes, known as lyases, which can degrade their exopolysaccharides and lead to the detachment of the cells from the surface (Boyd and Chakrabarty, 1994; Allison *et al.*, 1998). Allison *et al.* (1998) observed that *P. fluorescens* biofilms, grown for longer that 50 h, detached from the surface. This could be ascribed to the presence of alginate lyase in the culture supernatant. Lyase activity was only detected in older culture supernatants (20 h) suggesting that its production was triggered by nutrient limitations (Allison *et al.*, 1998). This is not the only study that supports the link between nutrient deprivation and biofilm detachment. Sawyer and Hermanowicz (1998) as well as Hunt *et al.* (2004) demonstrated that the exhaustion of nutrient sources led to increased detachment of *Aeromonas hydrophila* and *Pseudomonas aeruginosa*, respectively. Host-specific bacteriophages produce enzymes, known as polysaccharide depolymerases, capable of degrading EPS and releasing biofilms from the surface (Hughes *et al.*, 1998a; 1998b).

Five different categories of detachment have been identified: erosion, sloughing, human intervention, grazing and abrasion. Spontaneous detachment of cells from the biofilm has been divided into two processes: erosion and sloughing. Erosion can be defined as the continual detachment of small portions of the biofilm or single cells, while sloughing can be defined as the rapid loss of large portions of the biofilm. Abrasion is the loss of biofilm biomass due to collisions between biofilm carriers, while human intervention and grazing are more physical means of biofilm removal (Characklis, 1990). Some of the important factors in biofilm detachment include EPS-degrading enzymes, microbially generated gas bubbles, availability of multivalent cross-linking cations, the levels of nutrients and the growth status of the microbes, fluid shear stress, the activation of lytic bacteriophages and quorum-sensing signals. Although some of these mechanisms are purely physical, others are predominantly biological (Hunt *et al.*, 2004).

## 1.8 Heterogeneity of biofilm matrix and structures involved in biofilm formation

### 1.8.1 Quorum sensing (QS)

The development of biofilms on surfaces is mediated by density-dependent chemical signals released by bacterial cells densely packed within an EPS matrix. These conditions facilitate or may even represent the results of chemical communication between microbial cells. These signals are called auto inducers and will effect physiological changes in activities of cells when they reach a threshold concentration. In the quorum sensing process, microbial cells act as coordinator units rather than opportunistic individuals. This form of intercellular communication serves to coordinate gene expression and structures morphological differentiation and development responses of bacterial cells (Faqua *et al.*, 1996; Dunlap, 1997).

Quorum sensing plays a part in the dispersal of cells (Puskas *et al.*, 1997; Rice *et al.*, 2005). *Rhodobacter sphaeroides* produce an acyl-homoserine lactone (acyl-HSL) quorum sensing signal that is required for the dispersal of its cells. This novel acyl-HSL is produced by a homologue of *luxI*, known as *cerI*. A *cerI* mutant showed clumping in liquid media but as soon as the homoserine lactone signal was added, the clumps dispersed (Puskas *et al.*, 1997). Davey *et al.* (2003) illustrated that rhamnolipids, produced by *P. aeruginosa*, were capable of interfering with the cell-to-cell interactions as well as the attachment of cells to the surface, causing the detachment of *P. aeruginosa* biofilms.

Quorum-sensing signals can control biofilm detachment by the accumulation of the signal molecules (excreted by bacteria) to a threshold concentration that will eventually trigger the dispersion of the biofilm (Hentzer *et al.*, 2002). When the quorum-sensing mechanism was investigated with a computer model, the results indicated that bacteria in the centre of cell clusters excreted the highest concentration of the signal molecules. In this section of the cell cluster, diffusive egress is most restricted and it is also the part of the biofilm that is most likely to be deficient in nutrients. This led to an alternative explanation for biofilm detachment based on a starvation response. Experiments with pure culture *Pseudomonas aeruginosa* biofilms in continuous-flow conditions indicated that these biofilms detached after flow was stopped. When these biofilms were starved of nutrients under continuous-flow conditions, detachment also occurred. These observations suggest that starvation, and not the accumulation of signal/metabolic products, was responsible for triggering the detachment of these biofilms (Hunt *et al.*, 2004).

### 1.8.2 Conditioning film

As previously explained, organic matter can settle on surfaces that come into contact with an aqueous medium and these deposits form a conditioning film (Donlan, 2002). Components like polysaccharides, proteins, lipids, humic acids, nucleic acids and aromatic amino acids in the surrounding medium can deposit onto the surface and lead to a change in the chemical properties of the surface (Siboni *et al.*, 2007). These changes can affect the number of cells attaching to the surface (Barnes *et al.*, 1999; Carlén *et al.*, 2001; Murga *et al.*, 2001; Pasmore *et al.*, 2001).

A material surface exposed in an aqueous medium will become conditioned or coated by polymers from that medium and the resulting chemical modification will affect the rate and extend of microbial attachment (Prakash *et al.*, 2003a,b). Surfaces cannot be colonized by biofilms unless they have been exposed to organic material from the surrounding environment (Allison *et al.*, 2000). However the effect of surface characteristics like charge, hydrophobicity, roughness and elasticity on microbial attachment cannot be ignored (Bakker *et al.*, 2004).

Most surfaces have intense charges and free energy that can hinder microbial attachment. Thus, before any microorganism can come into close contact with a surface, these surface barriers must be eliminated (Gottenbos *et al.*, 1999; Dreeszen, 2003). This can be achieved through settling of organic matter onto solid surfaces, thereby ridding surfaces of high charges and energy and allowing microbial attachment to occur (Dreeszen, 2003). This layer is known as a conditioning film and consists mostly of polysaccharides, proteins, lipids, humic acids, nucleic acids and aromatic amino acids (Siboni *et al.*, 2007) depending on the surrounding environment. Moreover, this layer of organic matter can promote bacterial attachment (Murga *et al.*, 2001) and provide the microorganisms with nutrients (Beveridge *et al.*, 1997; Dreeszen, 2003; Siboni *et al.*, 2007).

Studies based on the films were made on surfaces exposed in seawater and results found showed that films were organic in nature and they form when expose to surfaces and continue to grow for several hours. The nature of conditioning films may be quite different for surfaces expose in the human host (http://www.medcape.com/viewarticle/441355_4).

Other studies in medical research based on microbial films were involved and results found showed that a number of hosts produce conditioning film such as blood, tears, urine, saliva, intravascular fluid and respiratory secretion influenced the attachment of bacteria to biomaterials and that, the surface energy of the suspending medium may affect hydrodynamic interaction of microbial cells with surfaces by altering the substratum characteristics (http://www.medcape.com/viewarticle/441355_4).

## 1.9 Regulation of biofilm formation

### 1.9.1 Surface

The characteristics of a surface can influence cellular attachment and two properties play an important role in the number of cells that will attach to a surface. These properties comprise the roughness and the physicochemical nature of the surface (Donlan, 2002; Prakash *et al.*, 2003a,b; Ghannoum and O'Toole, 2004). Rough surfaces present a higher surface area to which microbial cells can adhere and, when attached to these surfaces, they experience less friction with surrounding water bodies (Characklis and Marshall, 1990; Donlan, 2002; Prakash *et al.*, 2003a,b; Ghannoum and O'Toole, 2004). When biofilm formation was compared on smooth vs. rough surfaces it was apparent that rough surfaces are favored for biofilm formation (Percival *et al.*, 1999; Carlén *et al.*, 2001; Pasmore *et al.*, 2001; Morgan and Wilson, 2001) and the removal of biofilms from rough surfaces was shown to be more difficult than from smooth surfaces (Pasmore *et al.*, 2002). It was, however, demonstrated that a rough surface might only assist in the attachment of the first cells to the surface, but that it had no further effect on later attachment events (Keevil *et al.*, 1999).

It has been shown that hydrophobic surfaces (like Teflon and other plastics) are easier to colonize than hydrophilic surfaces (like glass and metals) (Fletcher and Loeb, 1979; Pringle and Fletcher, 1983; Bendinger *et al.*, 1993; Pasmore *et al.*, 2001; 2002). A possible reason for this is that the bacterial and colonization surface experience a hydrophobic interaction, allowing the cell to resist repulsive forces and attach reversibly to the surface (Donlan, 2002; Prakash *et al.*, 2003a,b). Absolom *et al.* (1983), however, showed that when bacteria had a higher hydrophobicity than its surrounding environment they attached with greater ease to hydrophobic surfaces and *vice versa*. Surface charge also affects the attachment of bacteria to surfaces (Carlén *et al.*, 2001; Pasmore *et al.*, 2001; 2002). Pasmore *et al.* (2002) demonstrated that surfaces with neutral or small negative charges allowed for easy removal of biofilms while surfaces with high charges (positive or negative) contained biofilms that were not easy to remove.

The surface could be a dead or living tissue or any inert surface. The attachment of microorganisms to surfaces is a complex process with many variables affecting the outcome. Attachment will occur most readily on surfaces that are rougher, more hydrophobic and coated by surface conditioning films (Dunne, 2002). Furthermore, growth requires complex developmental pathways that are regulated in response to environmental and bacterial derived signals. Studies based on the effect of substratum were made and results found showed that the extent of microbial colonization appears to increase as the surface roughness increases (Prakash *et al.*, 2003a,b).

The surface may have several characteristics that are important in the attachment process. An increase in flow velocity, water temperature or nutrient concentrations may also exert to increased attachment if these factors do not exceed critical levels (http://www.medscape.com/viewarticle/441355_4). Properties of the cell surface, specifically the presence of fimbriae, flagella and surface associated polysaccharides or proteins, are also important and may provide a competitive advantage for one organism where a mixed community is involved (http://www.medscape.com/viewarticle/441355_4).

The solid surface may have several characteristics that are important in the attachment process. The physicochemical properties of the surface may also exert a strong influence on the rate an extent of attachment. Most studies concerning these have been made and results found indicated that microorganisms attach more rapidly to hydrophobic, non-polar surfaces than to hydrophilic materials (http://www.medcape.com/viewarticle/441355_4).

### 1.9.2 Gene regulation

Studies based on gene regulation of microbial biofilms were made and results showed that 22% of the genes were up regulated and 16% down regulated in biofilm forming of *Pseudomonas aeruginosa* (Steyn *et al.*, 2001). Other studies demonstrated that *algC* up regulation occurs within minutes of attachment to a surface in a flow cell system. Gene encoding for enzymes involved in glycolysis or fermentation such as phosphoglycerate mutase, triosephosphate isomerase and alcohol dehydrogenase are up regulated in biofilm forming by *Staphylococcus aureus*. A recent genetic study also showed that *algD, algU, rpoS* and genes controlling polyphosphokinase synthesis were up regulated in biofilm formation of *Pseudomonas aeruginosa* (Prakash *et al.*, 2003a,b).

### 1.9.3 Properties of the cells

Cells surface hydrophobicity, presence of fimbriae and flagella, and productions of EPS influence the rate and extent of attachment of microbial cells. The hydrophobicity of the cell surface is important in adhesion because hydrophobic interactions tend to increase with an increasing non polar nature of one or both surfaces involved (i.e., the microbial cell surface and the substratum surface) (http://en.wikipedia.org/wiki/biofilm). Most bacteria are negatively charged but still contain hydrophobic surface components (An *et al.*, 2000; http://www.medcape.com/viewarticle/441355_4).

Fimbrae contribute to cell surface hydrophobicity. Most fimbriae that have been examined contained high proportion of hydrophobic amino acid residues. They play a role in cell surface hydrophobicity and attachment, probably by overcoming the initial electrostatic repulsion barrier that exists between the cell and substratum. A number of aquatic bacteria possess fimbriae which have also been shown to be involved in bacterial attachment to animal cell (Spiers *et al.*, 2003). Several studies have shown that treatment of adsorbed cells with proteolytic enzymes caused a marked release of attached bacteria, providing evidence for the role of protein in attachment (http://wwwmedscape.com/viewarticle/441355_4).

Studies made by several researchers found that mycolic acid containing organisms such as *Corynebacterium, Norcadia* and *Mycobacterium* were more hydrophobic than the non mycolic acid containing bacteria and an increase in mycolic acid chain length coincided with an increase in hydrophobicity. For most bacterial strains tested, adhesion was greater on hydrophobic materials. The O antigen component of lipopolysaccharide (LPS) has also been shown to contribute hydrophilic properties to Gram negative bacteria (Spiers *et al.*, 2003).

Other studies showed that mutants of *P. fluorescens* lacking the O antigen adhered in greater numbers to hydrophobic materials (Spiers *et al.*, 2003). Fletcher and Loeb (1979) found that treatment of attached fresh water bacteria with cations resulted in contraction of the initial adhesives (decrease in the cell distance from the substratum), indicating that the material was an anionic polymer. Glucosidase and N-acetylglucosaminidase (NAG) reduced attachment for *P. fluorescens*, while NAG only reduced attachment for *Desulphovibrio desulforicans*. Lactins preferentially bind to polysaccharides on the cell surface or to the EPS. Binding of lactins by the cells would minimize the attachment sites and affect cell attachment if polysaccharide were involve in attachment. An example of this was made with *P. fragie*, which confirmed the role of polysaccharides in attachment (Dunne, 2002).

With all these findings, it has been shown that cell surface structures such as fimbriae, other proteins, LPS, EPS and flagella play an important role in the attachment process (Dunne, 2002). Cell surface polymers with non-polar sites such as fimbriae, other proteins and components of certain Gram-positive bacteria (mycolic acids) appear to dominate attachment to hydrophobic substrata while EPS and LPS are more important in attachment to hydrophilic material (http://www.medscape.com/viewarticle/441355_4).

## 1.10 Biofilm physiology

### 1.10.1 Factors affecting the growth and development of biofilms

*Nutrients*

Biofilms can form under diverse nutrient concentrations, ranging from high to almost non-detectable (Prakash *et al.*, 2003a,b). Biofilms are, however, more abundant, densely packed and thicker in environments with high nutrient levels (Allison *et al.*, 2000; Prakash *et al.*, 2003a,b; Rochex and Lebeault, 2007). High nutrient concentrations promote the transition of bacterial cells from the planktonic to biofilm state (O'Toole *et al.*, 2000) while depletion of these nutrients has shown to cause detachment of biofilm cells from surfaces (Allison *et al.*, 1998; Hunt *et al.*, 2004; Rochex and Lebeault, 2007).

In an open reticulating system there is an abundance of nutrients derived from water, particularly in cooling towers and spray bond. Close systems (i.e., not exposed to the atmosphere) are ideal systems in that the problem

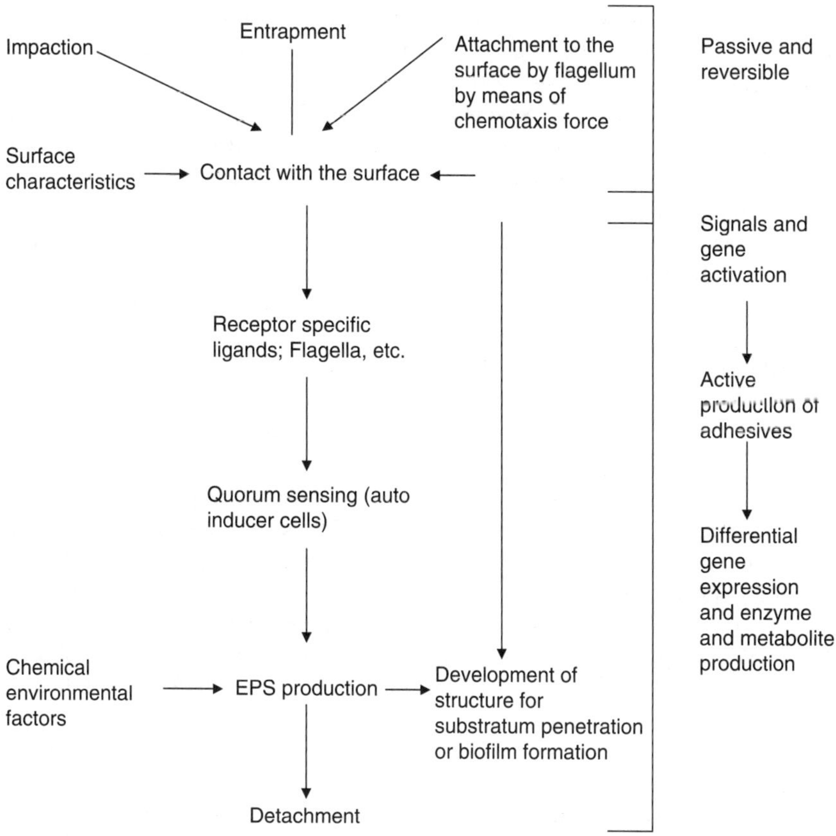


**Fig. 1.1** Summarises biofilm formation in the form of a diagram.

of biofouling is either unlikely to be encountered or reduced (Melo and Bott, 1997). High levels of nutrients appear to produce an open structure in the biofilm, whereas lower concentrations tend to give a more compact structure. The structure of the biofilms has an effect on the availability of nutrients to the constituent cells. An open structure facilitates the diffusion of nutrients to the bacteria in the form of aqueous phase. For aerobic bacteria, the availability of oxygen is necessary unless the particular microorganism can exist under oxygen-starved conditions (Melo and Bott, 1997).

An increase in nutrient concentrations correlated with an increase in the number of attached bacterial cells (Cowan *et al.*, 1991; Prakash *et al.*, 2003a,b; http://www.medcape.com/viewarticle/441355_4; Dunne, 2002). However, nutrient concentrations too low to measure are still sufficient for biofilm growth. Biofilm bacteria acquire nutrients by concentrating trace organics on surfaces by the extracellular polymer, using the waste products from previously attached cells and secondary colonizers and also by pooling the biochemical resources with different enzymes to break down food supplies (Cowan *et al.*, 1991).

Sauer *et al.* (2004) and Rochex and Lebeault (2007), however, showed that high nutrient levels can lead to the detachment of *P. aeruginosa* and *P. putida* biofilms. In contrast, Oh *et al.* (2007) demonstrated by atomic force microscopy (AFM) that *Escherichia coli* O157:H7 biofilms formed faster and higher amounts of cells attached to a glass surface under low nutrient conditions than under high nutrient conditions. Under low nutrient conditions, *Serratia marcescens* form biofilms consisting of single micro-colonies while under high nutrients conditions these micro-colonies can revert to filamentous biofilms (Rice *et al.*, 2005).

### 1.10.2 Temperature

For many bacteria found in cooling water systems, the optimum temperature for maximum growth is about 40°C, which is the level of temperature likely to be encountered in industrial water coolers. At this temperature small changes in temperature are likely to produce substantial changes in biofilm growth, because microbial activity is very sensitive to temperature. For instance, studies have shown that biofilm thickness of *Escherichia coli* increased by 80% by raising the temperature from 30°C to 35°C. (Melo and Bott, 1997).

### 1.10.3 Surface condition

It has been demonstrated that surface condition (e.g., whether rough or smooth) affect the ability of bacteria to adhere to a surface. A material surface exposed in an aqueous medium will inevitably and become conditioned or coated by polymers from that medium, and the resulting chemical modification will affect the rate and extent of microbial attachment (Prakash *et al.*, 2003a,b).

Studies based on the films were made on surfaces exposed in seawater and results showed that films were organic in nature and they formed within minutes of exposure and continued to grow for several hours (Prakash *et al.*, 2003a,b).

### 1.10.4 Velocity and turbulence

Thc area from the surface where no turbulent flow is experienced is known as the boundary layer. Within this area the flow velocity has been shown to be insufficient to remove biofilms (Dreeszen, 2003). The area outside this layer is characterized by high levels of turbulent flow and has an influence on the attachment of cells to the surface (Donlan, 2002). The size of the boundary layer is dependent on the flow velocity of the water. At high velocities, the boundary layer decreases in size and the cells are exposed to high turbulence levels (Donlan, 2002). Hydrodynamic conditions can influence the formation, structure, EPS production, thickness, mass and metabolic activities of biofilms (Stoodley *et al.*, 2002; Liu and Tay, 2001; Simões *et al.*, 2007).

Biofilms formed under turbulent flow can be described as 'streamers' and these are typically formed by filamentous bacteria. The micro-colonies formed under these conditions are stretched out in the direction of the current. Biofilms formed under low shear conditions (laminar flow conditions) are characterized by spherical micro-colonies divided by water channels (Stoodley *et al.*, 2002; Purevdorj *et al.*, 2002). Simões *et al.* (2007) determined the differences between three strains of *P. fluorescens* biofilms (one type strain and two strains isolated from an industrial processing plant) grown under turbulent and laminar flow. All biofilms grown under turbulent flow were denser, had a higher mass, were more active, produced similar amounts of matrix proteins and the same (isolated strains) and higher (type strain) amounts of extracellular polysaccharides (Simões *et al.*, 2007). It has been demonstrated that biofilms formed under higher detachment forces produced higher amounts of extracellular polysaccharides in order to stabilize the biofilm structure (Trinet *et al.*, 1991; Ohashi and Harada, 1994; Chen *et al.*, 1998).

Studies showed that an increase in water flow velocity resulted in an increased bacterial number in biofilms, especially if the amount of biodegradable organic matter is low. This is attributable to better mass transfer of growth-limiting nutrients at the higher flow velocity of water. Biofilms rely on defensive mechanisms to resist detachment by the higher fluid shear (Lehtola *et al.*, 2006). It has been proposed that the viscoelasticity of biofilms allows them to resist detachment as has been found in *Staphylococcus aureus* biofilms. Changes in water flow can lead to resuspension of the sediments in the distribution systems and thus to a deterioration in the quality of drinking water (Lehtola *et al.*, 2006).

### 1.10.5 Hydrodynamics

Studies concerning the hydrodynamics of an aqueous medium demonstrated that the flow velocity adjacent to the substratum/liquid interface is negligible. This zone of negligible flow is termed the hydrodynamic boundary layer. Its thickness is dependent on linear velocity, the higher the velocity, the thinner the boundary layer. The region outside the boundary layer is characterized by substantial mixing or turbulence. For flow regimes characterized as laminar or minimally turbulent, the hydrodynamic boundary layer may affect cell substratum interactions (Kumar and Prasad, 2006).

Cells behave as particles in a liquid, and the rate of settling and association with a submerged surface will depend largely on the velocity characteristics of the liquid. Under very low linear velocities, the cells must traverse the sizeable hydrodynamic boundary layer, and association with the surface will depend on cell size and cell mortality (http://www.medcape.com/viewarticle/441355_4). As the velocity increases, the boundary layer decreases and cells will be subjected to increasingly greater turbulence and mixing. Higher linear velocities would therefore be expected to equate to more rapid association with the surface, at least until velocities become high enough to exert substantial shear forces on the attaching cells resulting in detachment of these cells (http://www.medcape.com/viewarticle/441355_4).

The hydrophobic/hydrophilic nature of the bacterial cell surface and the attachment surface play an important role in the formation of biofilms. Lipopolysaccharides are significant determinants of the hydrophobic/hydrophilic nature of bacterial surfaces. Alterations in lipopolysaccharides (LPS) can lead to changes in microbial attachment. *P. aeruginosa* produce A- and B-band LPS and studies have shown that mutation of B-band LPS resulted in cells having a higher affinity for hydrophobic than hydrophilic surfaces. In contrast, mutation of A-band LPS did not drastically influence attachment to either surface, indicating that B-band LPS are responsible for changes in cell surface hydrophobicity/hydrophilicity (Makin and Beveridge, 1996). Al-Tahhan *et al.* (2000) showed that treatment of *P. aeruginosa* with a biosurfactant (rhamnolipid) caused the removal of LPS and subsequently gave rise to a hydrophobic cell surface. These cells could easily attach to and therefore degrade hydrophobic substrates (Al-Tahhan *et al.*, 2000).

### 1.10.6 Effects of particles

Biofouling of industrial equipment occurs together with other kinds of fouling; the most common of all being the simultaneous deposition of small particles that are transported with the incoming water; or are formed in the plant as a consequence of metal corrosion (Melo and Bott, 1997). When the particles are organic in nature, they act as substrates for microorganisms and are being degraded by them contributing to the growth of the biomass. In most cases, however, the biological matrix incorporates inorganic particles that are relatively inert from a strict metabolic standpoint but may cause changes in the structure and activity of the biofilm. The adhesion between particles and microorganisms could be facilitated by the electropositive charges developed at the surface of some particles depending on the pH of the environment. Metabolic inhibitors and toxic metallic ions could be adsorbed on the particle surface favoring biomass formation and microbial respiration stimulation in the presence of particles (Melo and Bott, 1997).

## 1.11 Microbial extracellular polysaccharides (EPS) and their role in biofilms

### 1.11.1 Defining EPS

Geesey (1982) describes EPS as 'extracellular polymeric substances of biological origin that participate in the formation of microbial aggregates'. Although the term implies that EPS is dominated by polysaccharides, it can also contain other molecules (nucleic acids, proteins, lipids and humic material) in high concentrations. These molecules can even in some cases prevail over polysaccharides (Allison *et al.*, 2000). Biofilms are characterized by high levels of EPS and as much as 50–90% of the biofilms' total organic matter consists of EPS (Denyer *et al.*, 1993 and Allison *et al.*, 2000). EPS is highly heterogeneous even among the same bacterial species and therefore its composition and function within the biofilm will differ (Allison *et al.*, 2000). The basic composition and role of EPS in biofilm dynamics will be discussed in the following sections.

### 1.11.2 Composition of EPS

The biofilm matrix mainly consists of polysaccharides and proteins (Wingender *et al.*, 1999; Tsuneda *et al.*, 2003; Keevil *et al.*, 1999). Tsuneda *et al.* (2003) demonstrated that these two components in several studied bacterial strains accounted for 75–89% of the EPS composition (Wingender *et al.*, 1999). The prevalence of polysaccharides in EPS is believed to be the highest (Ghannoum and O'Toole, 2004); however, in certain cases EPS from wastewater biofilms (activated sludge) were shown to have a bigger proportion of proteins than polysaccharides (Bura *et al.*, 1998; Dignac *et al.*, 1998; Jahn and Nielsen, 1998).

The type and quantity of exopolysaccharides synthesized is reliant on the species and environmental factors experienced by the specific microorganism (Ghannoum and O'Toole, 2004). Polysacharides can be categorized into groups based on whether the exopolysaccharides are attached to the microbial cell (known as capsular exopolysaccharides) or liberated in the surrounding environment (known as slime), and whether the structure is linear, branched or cyclic (Ghannoum and O'Toole, 2004). The main configuration of the polysaccharides present in EPS is β and rarely α with a high prevalence of 1,3 or 1,4 linkages in the backbone of the polymer. The number of different polysaccharide compositions in EPS is countless. Compositions can range from very basic polysaccharides, containing only one type of linkage (homopolysaccharides), to more intricate polysaccharides composed of more than two different monosaccharides. Neutral polysaccharides are scarce while positively charged (due to presence of amino sugars) and negatively charged (due to presence of carboxyl, sulfate or phosphate groups) polymers are more abundant (Allison *et al.*, 2000). A variety of acyl groups can also be attached to polysaccharides, which add to their intricacy (Wingender *et al.*, 1999). Polysaccharides can also interact with other molecules like proteins, lectins and lipids to form a complex structure (Sutherland, 2001b).

EPS obtain proteins from living and dead cells and these proteins can be absorbed due to the polyanionic properties of the EPS carbohydrates (Ghannoum and O'Toole, 2004). Proteins are typically responsible for the anionic and hydrophobic nature of EPS (Allison *et al.*, 2000). Proteins present in activated sludge were shown to add to the electrostatic and hydrophobic forces due to the presence of hydrophobic and acidic amino acids in their structure (Dignac *et al.*, 1998). The hydrophobic portion of activated sludge EPS was composed of proteins, and no polysaccharides were present in this section, showing that the polysaccharides do not contribute to the hydrophobic interactions. Proteins assist in the attachment of bacteria to negatively charged surfaces although some of the proteins can have a surplus negative charge. Proteins are alleged to aid in the attachment of ions to hydrophobic surfaces (Characklis and Marshall, 1990). The proteins in the EPS do not only play a structural role, but some of these proteins might also act as enzymes (Characklis and Marshall, 1990). Two types of biofilm involved proteins are prominent in EPS and they are known as lectins and polysaccharases. Lectins can bind to certain carbohydrates, play a role in the pathogenesis of bacteria and assist in the attachment of the cell to its host and to other bacterial cells. Polysaccharases are also prominent in EPS (Ghannoum and O'Toole, 2004) and these enzymes are responsible for the degradation of the EPS itself and other components in the surrounding environment, supplying the biofilm cells with nutrients (Wingender *et al.*, 1999).

Other components, present in smaller amounts, include nucleic acid (DNA and RNA), lipids and other products of cells lysis (Wingender *et al.*, 1999; Sutherland, 2001b; Allison, 2003; Tsuneda *et al.*, 2003). Cellular material makes up as little as 2–15% of the matrix; however, the amounts of these components can fluctuate (Sutherland, 2001b). Research showed that nucleic acids might be released into the EPS via membrane vesicles (Whitchurch *et al.*, 2002) or due to programmed cell lysis (Webb *et al.*, 2003) and as much as 20% of *P. aeruginosa* PAO1 EPS consists of DNA. When biofilms were treated with DNAse I they were removed from the surface, suggesting the possible role of DNA in biofilm stability (Whitchurch *et al.*, 2002). Another explanation for the presence of nucleic acids in the EPS could be due to horizontal gene transfer taking place between tightly packed biofilm cells (Christensen *et al.*, 1998, Hausner and Wuertz, 1999). Soil and water biofilms have been shown to contain humic acids in their matrix (Wingender *et al.*, 1999), while biofilms present in wastewaters and activated sludge plants can contain humic substances trapped within their matrix (Wingender *et al.*, 1999; Jahn and Nielsen, 1998).

Table 1.1 shows the general composition of bacterial EPS.

**Table 1.1** General composition of bacterial EPS (Jahn and Nielsen, 1998)

| EPS | Principal component (subunits, precursors) | Main type of linkage between subunits | Structure of polymer backbone | Substituents (examples) |
|---|---|---|---|---|
| Polysaccharides | Monosaccharides<br>Uronic acids<br>Amino sugars | Glycosidic bonds | Linear, branched | **Organic**: O-acetyl, N-acetyl, succinyl, pyruvyl<br>**Inorganic**: sulfate, phosphate |
| Proteins (polypeptides) | Amino acids | Peptide bonds | Linear | Oligosaccharides (glycoproteins), fatty acids (lipoproteins) |
| Nucleic acids | Nucleotides | Phosphodiester bonds | Linear | – |
| (Phospho)lipids | Fatty acids<br>Glycerol<br>Phosphate<br>Ethanolamine<br>Serine<br>Choline<br>Sugars | Ester bonds | Side-chains | – |
| *Humic substances | Phenolic compounds<br>Simple sugars<br>Amino acids | Ether bonds, C-C bonds, peptide bonds | Cross-linked | – |

* Humic substances are occasionally seen as being part of the EPS matrix.

Up to 98% of the biofilm matrix can consist of water (Keevil *et al.*, 1999; Sutherland, 2001a; 2001b). Polysaccharides, however, vary in the amount of water they can bind. Hyaluronic acid can bind large quantities of water (1 kg water/g polysaccharide), while cellulose, curdlan and mutan are hydrophobic in nature and repel water (Sutherland, 2001a). The water can either be attached in the capsules of microbial cells or can be located around the cells and act as a solvent in which compounds can dissolve. Water also plays an important role in the diffusion of material.

In some cases it is believed that all other compounds of the biofilm matrix, except for carbohydrates, have their origin from cell lysates and surrounding water/environments. Nevertheless, when the EPS of pure cultures were examined it was observed that some of the non-carbohydrate compounds were produced by the bacteria (Wingender *et al.*, 1999).

### 1.11.3 Production and secretion of EPS

The formation of EPS is dependent upon the polymerization of comparable or identical organic compounds. These organic compounds can be organized as repeating units within the polymer structure. The structure and physicochemical properties of EPS can, however, be changed by the presence of non-polymeric organic (acetyl, pyruvyl and succinyl) and inorganic (sulphate and phosphate) compounds. Proteins and lipids can also form part of the EPS. The glycosylation of proteins with oligosaccharides form glycoproteins and the oligosaccharides can be replaced with fatty acids to form lipoproteins (Wingender *et al.*, 1999).

EPS are situated on or in close proximity to the outer surface of biofilm cells. It can be released into the surrounding environment by active secretion, during the lysis of bacterial cells, or by the detachment of materials from the cell surface. Certain EPS components can, however, be taken up from the surrounding environment. Several methods to actively secrete EPS proteins (Filloux *et al.*, 1998; Hueck, 1998) and polysaccharides (Rehm and Valla, 1997; Becker *et al.*, 1998; Jonas and Farah, 1998) are known, but it is not yet clear whether extacellular DNA is excreted actively or passively (Lorenz and Wackernagel, 1994). Some important cellular components, such as lipopolysaccharides (LPS) of Gram-negative bacteria, can be released from the outer membrane by surface blebbing. This process takes place during the normal growth phase of the cells and its occurrence is possible due to metabolic turnover processes. Other components released in this manner include periplasmic compounds like nucleic acids and enzymes as well as membrane components such as phospholipids (Wingender *et al.*, 1999). DNA of several Gram-negative bacteria has been shown to be released into the surrounding environment by the formation of blebs (membrane-derived vesicles) (Lorenz and Wackernagel, 1994 Dorward and Garon, 1990). During cell lysis the content of the cell is released and captured within the biofilm matrix and therefore the EPS may also contain substances originally located inside the cell (Wingender *et al.*, 1999).

### 1.11.4 Roles that EPS play in biofilms

Energy is a valuable component of microbial survival and therefore any process taking up some of this energy is seen as being essential to the microorganism. An example of such a process is the production of extracellular polysaccharides (Wingender *et al.*, 1999). The function of EPS has only been partially revealed up to now due to its high heterogeneity among different bacterial strains (Tsuneda *et al.*, 2003). EPS is, however, known to play a role in the structural stability (Yildiz and Schoolnik, 1999; Danese *et al.*, 2000; Hentzer *et al.*, 2001; Nivens *et al.*, 2001; Bomchil *et al.*, 2003), attachment (Mack *et al.*, 1996; Danese *et al.*, 2000), protection (Vrany *et al.*, 1997; Yildiz and Schoolnik, 1999; Anderl *et al.*, 2000; Hentzer *et al.*, 2001; Leid *et al.*, 2002; Zheng and Stewart, 2002) and infection abilities of biofilms (Gilligan, 1991; Bereswill and Geider, 1997; Hentzer *et al.*, 2001).

The stability of the biofilm structure can be ascribed to the presence of non-covalent interactions (London forces, electrostatic interactions and hydrogen bonds) between the matrix components (Keevil *et al.*, 1999; Wingender *et al.*, 1999; Allison, 2003). These interactions can link the polysaccharide chains directly or indirectly by making use of multivalent cation bridges (Wingender *et al.*, 1999; Allison, 2003). Interactions between proteins via multivalent cations are proposed to be more dominant than those between polysaccharides and the proteins have also been implicated in hydrophobic interactions within the biofilm matrix (Wingender *et al.*, 1999). Hentzer *et al.* (2001) pointed out that a *P. aeruginosa* strain overproducing alginate formed a biofilm that had a more stable structure than biofilms formed by a non-mucoid strain.

EPS are also involved in the attachment of cells to a surface and in holding the biofilm cells together (Wingender *et al.*, 1999; Tsuneda *et al.*, 2003; Allison, 2003; Sutherland, 2001a). EPS can change the physicochemical properties of the attachment surface and in some cases make it easier for cells to attach (Tsuneda *et al.*, 2003). Tsuneda *et al.* (2003) concluded that when low amounts of EPS were produced, electrostatic interactions reduced bacterial attachment, while the presence of high amounts of EPS increased adhesion due to interactions formed between the polymeric substances. In certain cases EPS was found to not influence microbial attachment as indicated by Hanna and colleagues. *E. coli* produces a capsular polysaccharide known as colanic acid. Hanna *et al.* (2003) showed that this polysaccharide obstructs both specific binding and time-dependent interactions which take place between bacteria and their attachment surfaces.

The resistance of biofilms to a wide range of biocides and antibiotics can partly be attributed to EPS (Wingender *et al.*, 1999; Sutherland, 2001a; Allison, 2003). EPS act as a protective layer against antibiotics

(Anderl, 2000; Kostenko *et al.*, 2007), biocides (de Beer *et al.*, 1994; Chen and Stewart, 1996; Stewart and Costerton, 2001; Grobe *et al.*, 2002), heavy metals (Teitzel and Parsek, 2003), host immune responses and UV irradiation (Elasri and Miller, 1999), reducing their penetration into the cells or by deactivating/neutralizing certain of these antimicrobials (Ghannoum and O'Toole, 2004). Another function of this substance is the protection of biofilm cells against dehydration under water-limited conditions (Ophir and Gutnick, 1994; Schnider-Keel *et al.*, 2001) and other environmental conditions such as temperature fluctuations (Mao *et al.*, 2001) and osmotic shock (Schnider-Keel *et al.*, 2001). EPS can bind high amounts of water and in some cases water comprises as much as 97% of the matrix content. The outer layer of EPS can dry out under water-deficient conditions and form a hard, protective layer, preventing dehydration of the inner cells (Sutherland, 2001a).

Bereswill and Geider (1997) showed that the production of acidic EPS (amylovoran) by *Erwinia amylovora* is linked to the virulence of this strain. A strain that was incapable of producing amylovoran was avirulent on its host plants (Bereswill and Geider, 1997). Several studies concluded that mucoid *P. aeruginosa* initiate chronic lung infection due to the high production of EPS by this strain (Gilligan, 1991; Hentzer *et al.*, 2001). This 'mucoid exopolysaccharide' (MEP) is therefore seen as a virulence factor in cystic fibrosis infections (Gilligan, 1991).

## 1.12 Biofilm applications and problems

Biofilm formation can be on a wide spectrum of surfaces. These include human tissues (e.g., lung and skin) (Costerton *et al.*, 1999), medical devices (e.g., urinary catheter) (Habash and Reid, 1999), natural aquatic systems and industrial pipelines (Donlan, 2002), to name a few.

In industry, biofilms can be both beneficial and detrimental. The three properties of biofilms that can affect industrial processes are: physical (structural), chemical (metabolic) and biological (living) (Ludensky, 2003).

For example, industry can benefit from biofilms by using their metabolic activities (Gilbert *et al.*, 2003). Metabolic activities can be used to manage wastewaters (Manem and Rittmann, 1992; Gilbert *et al.*, 2003), to treat sewage (Yu *et al.*, 1999; Gilbert *et al.*, 2003) and also in biotechnological processes in a variety of solid-state fermentations (Kirimura *et al.*, 1999; Gilbert *et al.*, 2003).

In contrast, biofilms also cause a lot of damage in various industries, which amounts to billions of dollars annually. Biofilm formation has serious implications in industrial, environmental, public health and medical situations because they express biofilms-specific properties such as increased resistance to antibiotics, UV light and chemical biocides, increased rates of genetic exchange, altered biodegradability and increased secondary metabolite production (Giaouris and Nychas 2006, Bourne *et al.*, 2006). The major problems caused by biofilms associated with industry are: huge losses in efficiency, down-time of processes, damage of equipment, obstruction of pipelines and lowering the competence of heat exchangers (Characklis, 1990; Gilbert *et al.*, 2003). The occurrence of biofilms in food processing environments can cause post-processing contamination leading to lower shelf life of products and transmission of diseases (Poulsen, 1999). EPS is an example of a structural property of biofilms which can cause havoc in cooling water systems (Ludensky, 2003). EPS causes energy loses by increasing frictional resistance of fluid and heat transfer resistance (Ludensky, 2003).

### 1.12.1 Microbial biofilms negatively affecting different areas

*Biofilm related problems in drinking water*

Biofilms create economic and health problems in many areas. For example, in water systems they cause fouling of industrial equipment such as heat exchangers, resulting in heat transfer, energy loss, increased fluid fractional resistance and accelerated corrosion. Biofilm formation in water distributions decreases water quality and increases health risks. Microorganisms in water distributions differentiate themselves from other industrial water contaminants by their ability to utilize available nutrient resources, reproduce and generate intra and extracellular organic and inorganic substances in water (Stoodley *et al.*, 2002).

Drinking water is an important resource all around the globe. Nevertheless, research has focused on identifying bacteria in water distribution systems. The bacteria present in wastewater treatment plants are well characterized by advanced molecular techniques. The monitoring and identification of bacteria present in drinking water have been achieved by plating and isolation strategy. Studies showed that the majority of these culturable bacteria described in drinking water networks are in the phylum *Proteobacteria* and to a lesser extent in the phyla *Actinobacter*, *firmicutes* and *Bacteriodes*. Some of the commonly detected genera include *Pseudomnas*, *Caulobactor*, *Aeromonas*, *Acinetobacter* and *Bacillus* (Martiny *et al.*, 2005).

Aqueous environments are easily colonized by biofilms and can lead to energy losses, a reduction in heat transfer, an increase in the resistance against water flow and the corrosion of surfaces in water systems, pipelines and cooling towers (Melo and Bott, 1997; Johansen *et al.*, 1997; Cloete *et al.*, 1998; Ludensky, 2003). Conditions in cooling water towers and other industrial systems favor the formation of biofilms due to the high nutrient and oxygen content of the water, suitable temperatures (Melo and Bott, 1997; Keevil *et al.*, 1999; Ludensky, 2003), the presence of light and large surface areas for attachment. Biofilms that form on condensers and heat exchangers result in increased energy losses and resistance against heat transfer (Keevil *et al.*, 1999; Ludensky, 2003). The amount of resistance against heat transfer is almost the same for biofilms than for water. However, the gel-like matrix of biofilms causes the heat transfer to be diffusive instead of convective. This, in turn, leads to higher energy expenditures, friction resistance and running costs (Flemming, 2002). Water systems differ in the amount of utilizable organic and inorganic nutrients that are available, but are always capable of allowing microbial growth (Cloete *et al.*, 1998).

Financial losses are experienced due to microbial-induced corrosion (MIC) of water systems; the equipment needs to be replaced and in some cases cooling water systems need to be stopped in order to clean colonized surfaces and reduce or eliminate health risks caused by certain biofilm microorganisms (Ludensky, 2003). Biofouling of heat exchangers has been shown to amount in losses of £200–400 million per year in the UK, while chemical treatment of biofilms in a moderate cooling water system could range from £20–30,000 per year (Keevil *et al.*, 1999).

Water transport and storage systems are also subject to biofilm growth. Owing to the high surface-to-volume ratio in water transport systems, more than 95% of the biological material present in the water is attached to the surface, while the remainder is dissolved in the water. Biofilms can cause great damage to these systems but it has been shown that they rarely contain pathogenic microorganisms. In fact, biofilms in these environments have been suggested to reduce the prevalence of pathogenic microorganisms and assist in the cleaning of water by utilizing organic matter present. Problems can, however, occur if surfaces are capable of sustaining the growth of microorganisms, increasing the possibility that pathogenic microorganisms may be present (Flemming, 2002). If nutrients are provided and a biofilm exists, coliforms can incorporate into the biofilm and multiply. Kilb *et al.* (2003) confirmed the absence of biofilms on pipelines but the formation of biofilms harboring coliforms on rubber-coated valves. Substances escaping from the rubber were suggested to sustain its growth. It is, however, still not certain if coliforms attach due to these substances or due to secondary metabolites produced by the already existing biofilm bacteria (Kilb *et al.*, 2003).

There is still some controversy about the effect of surface materials on biofilm development. In the past, the majority of pipelines in drinking water networks were made of iron-based materials. More recently, polymeric materials have been preferred, mainly polyvinyl chloride (PVC) and PE, because they are easier to handle and implement (Manuel *et al.*, 2007). Studies by Kerr *et al.* (1999), Niquette *et al.* (2000), and Momba and Keleni (2002) demonstrated that drinking water biofilms grew less on polymeric material (PE, PVC, Teflon) than on iron matrices. This fact was attributed to iron corrosion products that favor biofilm protection from the effect of flow rate and of disinfectants. Different results were found in studies made by Cloete *et al.* (1998), where biofilm formation was higher on the PVC surfaces than on steel piping. On the contrary, van der Kooij and Veenendaal (2001) observed that PE supports biofilm formation to a higher degree that

PVC. Studies of Wingender *et al.* (1999), and Zacheus *et al.* (2000), concluded that there was no significant difference in the colonization of stainless steel, PVC and PE.

The metabolic state of the cells in suspension and in biofilms can be a measure of their growing potential and therefore of the risk of water contamination (Manuel *et al.*, 2007). Wingender *et al.* (1999), observed that the cultivable cells were at most 3.5% of the total cells in biofilms aging from 2 years while Boe-Hansen *et al.* (2002) reported that the highest percentages up to 24% were observed in the youngest biofilm in an assay of 522 days. The biofilm densities of cultivable bacteria on these different materials were less than the total cell number and 35–38% of the bacteria were metabolically active except for copper. In addition, Schaule *et al.* (1993) found that the percentage of metabolically active bacteria in biofilms sampled within the distribution system was about 30–35%.

*Biofilm-related problems in the food industry*

Food processing environments are susceptible to biofilm formation, and biofilms can result in losses due to corrosion of equipment, reduction in heat transfer, obstruction of pipelines, spoilage of food products and the loss of time when these systems need to be stopped to remove biofilms (Bower *et al.*, 1996; Kumar and Anand, 1998; Chmielewski and Frank, 2003; Trachoo, 2003). Contaminants in the food environment originate from food surfaces, equipment used during food processing or from the surrounding air (Kumar and Anand, 1998). Failure to clean food processing surfaces can lead to the accumulation of soil on the surface and if water is used during food processing, microbial growth may be enhanced. These surfaces can contain high amounts of food deposits and minerals derived from the food products which assist in biofilm formation on the surface and also shield the biofilm cells from harmful environmental factors (Chmielewski and Frank, 2003). Pathogenic microorganisms present in biofilms are difficult to control due the increased resistance of biofilm bacteria to disinfectants and can therefore present a potential health risk (Trachoo, 2003). Food-borne pathogens capable of forming biofilms in food processing environments include: *Listeria monocytogenes*, *Pseudomonas* species, *Bacillus* species and *Salmonella* species (Chmielewski and Frank, 2003).

In nature food systems, microorganisms become attached to solid surfaces conditioned with nutrients, ions and other organic materials that are sufficient for their viability and growth. If these organisms are not removed, they will accumulate in the pipes, resulting in the formation of biofilms. (Poulsen, 1999).

Several food spoilage and pathogenic bacteria have been reported to attach and form biofilms *in vitro* and on food contact surfaces such as stainless steel, polystyrene or rubber. The persistence of microorganisms in biofilms is a serious hygienic problem in food industries, causing processing and post-processing cross-contamination leading to reduced product shelf life and effectiveness of sanitizing treatments as well as potentially affecting consumer health (Lapidot *et al.*, 2006).

Studies have shown that even with cleaning and sanitation procedures consistent with good manufacturing practices, microorganisms can remain on equipment surfaces. These organisms may survive for prolonged periods depending on the environmental conditions. Moreover, microbial attachment greatly reduces the heat transfer and operating efficiency of the processing equipment. For example, biofilm formation greatly reduces the permeability of the filtration membranes (Brugnoni *et al.*, 2007).

The incidence of human pathogens on fresh produce is a serious concern in industrialized countries. *Salmonella* is among the most common isolated pathogens associated with fresh fruit and vegetables. Outbreaks of salmonellosis have been linked to a wide variety of fresh produce including alfalfa sprouts, lettuce, fennel, cilantro, cantaloupes, unpasteurized orange juice, tomatoes, melons, mango, celery and parsley. Contaminated fresh parsley has also been linked to outbreaks of *Shigella sonnei*, *Escherichia coli* and verotoxinogenic *Citrobacter freundii* (Lapidot *et al.*, 2006).

*Listeria monocytogens* also represents an important food-borne pathogen responsible for major outbreaks associated with dairy and other food products (Hentzer *et al.*, 2007). The organism is ubiquitous throughout nature and is frequently isolated from the food processing environments. Post-process contamination of food with *L. monocytogens* represents a serious problem as it is able to survive and grow at refrigeration temperature (Kalmokoff *et al.*, 2001; Timothy and Hansen, 2006).

*E. coli* O157:H7 was also identified as a pathogen and is now recognized as an important cause of food-borne disease (Lapidot *et al.*, 2006; Emtiazi *et al.*, 2004). The illness caused by this organism can be manifested as hemorrhagic colitis, hemolytic uremic syndrome (HUS) and thrombocytopenic purpura (TTP). Hemorrhagic colitis is the most common syndrome and is known by severe abdominal pain and glossy bloody diarrhea (Dewanti and Wong, 1995). HUS is the leading cause of renal failure in children, and patients often require dialysis and blood transfusions. Symptoms of TTP are similar but more severe than HUS. Death may results from HUS or TTP. Outbreaks due to *E. coli* O157:H7 have been associated primarily with consumption of undercooked beef. Other modes of transmission are unpasteurized milk, roast beef, apple cider and person-to-person (Dewanti and Wong, 1995). *Staphylococcus aureus* is also one of the moist frequent food-borne pathogens being found in food processing plants in which it adhere to surfaces and forms biofilms. It can enhance the contamination of processed food products (Hentzer *et al.*, 2007).

Contamination of fruits and vegetables may occur at various stages during production, harvest, processing and transport. Attached microorganisms (pathogens and spoilage bacteria) are not easily removed by washing with water or antimicrobial agents (Lapidot *et al.*, 2006). Chemical treatments such as calcium or sodium hypochlorite, hydrogen peroxide, ethanol and a variety of detergents partially reduced the populations of the pathogens. At present, chlorine at a concentration of 50–200 mg/L is the primary post-harvest sanitizing agent in routine use in the produce industry. This concentration of chlorine is usually ineffective in eliminating pathogens from leafy vegetables (Lapidot *et al.*, 2006).

According to the National Institute of Health, biofilms are responsible for more than 60% of microbial infections (Ghannoum and O'Toole, 2004). Biofilms are associated with a wide range of diseases with the most prominent being periodontitis, chronic lung infections (as encountered in cystic fibrosis patients) and a range of nosocomial infections linked to biofilm-invaded medical devices. Biofilms can also colonize teeth surfaces (Wecke *et al.*, 2000; Noiri *et al.*, 2002), prosthetic devices, medical implants (Schierholz and Beuth, 2001) and contact lenses (McLaughlin-Borlace *et al.*, 1998; Bruisma *et al.*, 2001; Mah and O'Toole, 2001; Stewart and Costerton, 2001; Donlan, 2002). Approximately 65% of nosocomial infections involve biofilms. Treatment of these infections results in costs of more than $1 billion per annum (Mah and O'Toole, 2001), whereas infections due to colonized prosthetic devices can amount to £100 million per annum (Allison *et al.*, 2000). When bacteria are in the biofilm state they are much more resistant to antimicrobial treatment and attacks from the host immune system. Some biofilms (*P. aeruginosa* and *S. epidermidis*) can cause persistent infections, which affect the well-being of the patient but are seldom life-threatening. Treatment of these infections with antibiotics is effective against the planktonic counterparts, but reoccurrence of the disease after treatment can take place due to the presence of persistent biofilms (Stewart and Costerton, 2001). It is, however, very difficult to prove conclusively that an infection or disease is caused by biofilm cells and not by their planktonic counterparts. This is due to the fact that in most diseases and infections both forms of growth are present and it is sometimes impossible or very difficult to obtain samples from infected areas. If samples can be obtained and cultured, it may not represent the true form of the causative organism(s), since bacteria are capable of losing their phenotypic traits when cultured. Infectious kidney stones, bacterial endocarditis and cystic fibrosis airway infections are, however, examples of diseases where biofilms play a definite role (Parsek and Singh, 2003).

## 1.13 References and further reading

ABSOLOM, D.R., LAMBERTI, F.V., POLICOVA, Z., ZINGG, W., VAN OSS, C.J. and NEUMANN, A.W. (1983). Surface thermodynamics of bacterial adhesion. *Applied Environmental Microbiology*, **46**: 90–97.

ALLISON, D.G. (2003). The biofilm matrix. *Biofouling*, **19**(2): 139–150.

ALLISON, D.G., RUIZ, B., SANJOSE, C., JASPE, A. and GILBERT, P. (1998). Extracellular products as mediators of the formation and detachment of *Pseudomonas fluorescens* biofilms. *FEMS Microbiology Letters*, **167**: 179–184.

ALLISON, D.G., GILBERT, P., LAPPIN-SCOTT, H.M. and WILSON, M. (2000). Community structure and co-operation in biofilms. Cambridge University Press, Cambridge.

AL-TAHHAN, R.A., SANDRIN, T.R., BODOUR, A.A. and MAIER, R.M. (2000). Rhamnolipid-induced removal of lipopolysaccharide from *Pseudomonas aeruginosa*: Effect on cell surface properties and interaction with hydrophobic substrates. *Applied and Environmental Microbiology,* **66**(8): 3262–3268.

AN, Y.H., DICKISON, R. and DOYLE, R.J. (2000). Mechanisms of bacterial adhesion and pathogenesis of implant and tissue infections. P. 1–27: In An Y.H. and Friedman R.J. Ed. *Handbook of bacterial adhesion: Principles, methods and applications.* Humana Press, Totowa, N.J.

ANDERL, J.N. (2000). Role of antibiotic penetration limitation in *Klebsiella pneumoniae* biofilm resistance to ampicillin and ciprofloxacin. *J. Antimicrob. Agents Chemother.*, **44**: 1818–1824.

BARNES, L.-M., LO, M.F., ADAMS, M.R. and CHAMBERLAIN, A.H.L. (1999). Effect of milk proteins on adhesion of bacteria to stainless steel surfaces. *Applied and Environmental Microbiology*, **65**(10): 4543–4548.

BAKKER, D.P., BUSSCHER, H.J., VAN ZANTEN, J., DE VRIES, J., KLIJNSTRA, J.W. and VAN DER MEI, H.C. (2004). Multiple linear regression analysis of bacterial deposition to polyurethane coatings after conditioning film formation in the marine environment. *Microbiology*, **150**: 1779–1784.

BECKER, A., KATZEN, F., PÜHLER, A. and IELPI, L. (1998). Xanthan gum biosynthesis and application: a biochemical/genetic perspective. *Applied Microbiology and Biotechnology*, **50**: 145–152.

BECKER, P., HUFNAGLE, W., PETERS, G. and HERRMANN, M. (2001). Detection of differential gene expression in biofilm-forming versus planktonic populations of *Staphylococcus aureus* using micro-representational-difference analysis. *Applied and Environmental Microbiology*, **67**(7): 2958–2965.

BEENKEN, K.E., DUNMAN, P.M., MCALEESE, F., MACAPAGAL, D., MURPHY, E., PROJAN, S.J., BLEVINS, J.S. and SMELTZER, M.S. (2004). Global gene expression in *Staphylococcus aureus* biofilms. *Journal of Bacteriology*, **186**(14): 4665–4684.

BELOIN, C., VALLE, J., LATOUR-LAMBERT, P., FAURE, P., KZREMINSKI, M., BALESTRINO, D., HAAGENSEN, J.A.J., MOLIN, S., PRENSIER, G., ARBEILLE, B. and GHIGO, J-M. (2004). Global impact of mature biofilm lifestyle on *Escherichia coli* K-12 gene expression. *Molecular Microbiology*, **51**(3): 659–674.

BENDINGER, B., RIJNAARTS, H.H.M., ALTENDORF, K. and ZEHNDER, A.B.J. (1993). Physicochemical cell surface and adhesive properties of Coryneform bacteria related to the presence and chain length of mycolic acids. *Applied and Environmental Microbiology*, **59**(11): 3973–3977.

BERESWILL, S. and GEIDER, K. (1997). Characterization of the *rcsB* gene from *Erwinia amylovora* and its influence on exopolysaccharide synthesis and virulence of the fire blight pathogen. *Journal of Bacteriology*, **179**(4): 1354–1361.

BEVERIDGE, T.J., MAKIN, S.A., KADURUGAMUWA, J.L. and LI, Z. (1997). Interactions between biofilms and the environment. *FEMS Microbiology Reviews*, **20**: 291–303.

BISHOP, P.L. (2007). The role of biofilms in water reclamation and reuse. *Water Science and Technology*, **55**: 19–26.

BOE-HANSEN, R., ALBRECHTSEN, H.J., ARVIN, E. and JORGESEN, C. (2002). Dynamics of biofilm formation in a model drinking water distribution systems. *J. Wat. Supply Res. Technol-Aqua.*, **51**: 399–406.

BOMCHIL, N., WATNICK, P. and KOLTER, R. (2003). Identification and characterization of a *Vibrio cholerae* gene, *mbaA*, involved in maintenance of biofilm architechture. *Journal of Bacteriology*, **185**(4): 1384–1390.

BOURNE, D.G., HOJ, L., WEBSTER, N.S., SWAN, J. and HALL, M.R. (2006). Biofilm development within a larval rearing tank of tropical rock lobster, *Panulirus ornatus*. *J. Aquacul.*, **260**: 27–38.

BOWER, C.K., MCGUIRE, J. and DAESCHEL, M.A. (1996). The adhesion and detachment of bacteria and spores on food-contact surfaces. *Trends in Food Science and Technology*, **7**: 152–157.

BOYD, A. and CHAKRABARTY, A.M. (1994). Role of alginate lyase in cell detachment of *Pseudomonas aeruginosa. Applied Environmental Microbiology*, **60**: 2355–2359.

BROOUN, A., LIU, S. and LEWIS, K. (2000). A dose-response study of antibiotic resistance in *Pseudomonas aeruginosa* biofilms. *Antimicrobial Agents and Chemotherapy*, **44**(3): 640–646.

BRUGNONI, L.I., LOZANO, J.E and CUBITTO, M.A. (2007). Potential of yeast isolated from apple juice to adhere to stainless steel surfaces in the apple juice processing industry. *Inter J. Food Res.*, **40**: 332–340.

BRUISMA, G.M., RUSTEMA-ABBING, M., VAN DER MEI, H.C. and BUSSCHER, H.J. (2001). Effects of cell surface damage on surface properties and adhesion of *Pseudomonas aeruginosa. Journal of Microbiological Methods*, **45**(2): 95–101.

BURA, R., CHEUNG, M., LIAO, B.Q., FINLAYSON, J., LEE, B.C., DROPPO, I.G., LEPPARD, G.G. and LISS, S.N. (1998). Composition of extracellular polymeric substances in the activated sludge floc matrix. *Water Science and Technology*, **37**: 325–334.

CARLÉN, A., NIKDEL, K., WENNERBERG, A., HOLMBERG, K. and OLSSON, J. (2001). Surface characteristics and in vitro biofilm formation on glass ionomer and composite resin. *Biomaterials*, **22**: 481–487.

CARPENTIER, B. and CERF, O. (1993). Biofilms and their consequencies with particular reference to hygiene in the food industry. *J. Appl. Bacteriol.*, **75**: 499–511.

CHARACKLIS, W.G. (1990). Biofilm process. In: Characklis, W.G., Marshall, K.C. (Eds) *Biofilms*. New York: Wiley. 195–231.

CHARACKLIS, W.G. and MARSHALL, K.C. (1990). *Biofilms*. Wiley-Interscience Publication, Canada.

CHARACKLIS, W.G., MCFETERS, G.S. and MARSHALL, K.C. (1990). Physiological ecology in biofilm systems. In: Characklis, W.G., Marshall, K.C. (Eds) *Biofilms*. New York: Wiley. 341–349.

CHEN, M.J., ZHANG, Z. and BOTT, T.R. (1998). Direct measurement of the adhesive strength of biofilms in pipes by micromanipulation. *Biotechnological Techniques*, **12**(12): 875–880.

CHEN, X. and STEWART, P.S. (1996). Chlorine penetration into artificial biofilm is limited by a reaction-diffusion interaction. *Environmental Science and Technology*, **30**: 2078–2083.

CHENG, K.J., MCALLISTER, T.A. and COSTERTON, J.W. (1995). *Biofims of the ruminant digestive tract. Microbial biofilms*, 1st ed, Cambridge University Press, New York. 221–232.

CHMIELEWSKI, R.A.N. and FRANK, J.F. (2003). Biofilm formation and control in food processing facilities. *Comprehensive Reviews in Food Science and Food Safety*, **2**: 22–32.

CHRISTENSEN, B.B., STERNBERG, C., ANDERSEN, J.B., EBERL, L., MØLLER, S. and GIVSKOV, M. (1998). Establishment of new genetic traits in a microbial biofilm community. *Applied and Environmental Microbiology*, **64**(6): 2247–2255.

CLOETE, T.E. (2003a). Biofouling control in industrial water systems: what we know and what we need to know. *Materials and Corrosion*, **54**: 520–526.

CLOETE, T.E. (2003b). Resistance mechanisms of bacteria to antimicrobial compounds. *International Biodeterioration and Biodegradation*, **51**: 277–282.

CLOETE, T.E., JACOBS, L. and BRÖZEL, V.S. (1998). The chemical control of biofouling in industrial water systems. *Biodegradation*, **9**: 23–37.

COSTERTON, J.W., STEWART, P.S. and GRENBERG, E.P. (1999). Bacterial biofilms: a common cause of persistent infections. *Science*, **248**: 1318–1322.

COWAN, M.M., WARREN, T.M. and FLETCHER, M. (1991). Biofouling. *J. App. Environ. Microbiol.*, **3**: 23–34.

DANESE, P.N., PRATT, L.A. and KOLTER, R. (2000). Exopolysaccharide production is required for development of *Escherichia coli* K-12 biofilm architechture. *Journal of Bacteriology*, **182**(12): 3593–3596.

DAVEY, M.E., CAIAZZA, N.C. and O'TOOLE, G.A. (2003). Rhamnolipid surfactant production affects biofilm architecture in *Pseudomonas aeruginosa* PAO1. *Journal of Bacteriology*, **185**(3): 1027–1036.

DE BEER, D., SRINIVASAN, R. and STEWART, P.S. (1994). Direct measurement of chlorine penetration into biofilms during disinfection. *Applied and Environmental Microbiology*, **60**(12): 4339–4344.

DENYER, S.P., GORMAN, S.P. and SUSSMAN, M. (1993). *Microbial biofilms: Formation and control.* Blackwell Scientific Publications.

DEWANTI, R. and WONG, A.C.L. (1995). Influence of culture conditions on biofilm formation by Escherichia coli O157:H7. *Inter. J. Food Microbiol.*, **26**: 147–164.

DIGNAC, M.-F., URBAIN, V., RYBACKI, D., BRUCHET, A., SNIDARO, D. and SCRIBE, P. (1998). Chemical description of extracellular polymers: implication on activated sludge floc structure. *Water Science and Technology*, **38**(8): 45–53.

DONLAN, R. (2002). Biofilm: Microbial life on surfaces. *Emerging Infectious Diseases*, **8**(9): 881–890.

DORWARD, D.W. and GARON, C.F. (1990). DNA is packaged within membrane-derived vesicles of Gram-negative but not Gram-positive bacteria. *Applied and Environmental Microbiology*, **56**(6): 1960–1962.

DREESZEN, P.H. (2003). *The key to understanding and controlling bacterial growth in automated drinking water systems.* Second edition. Edstrom Industries.

DRENKARD, E. (2003). Antimicrobial resistance of *Pseudomonas aeruginosa* biofilms. *Microbes and Infection*, **5**: 1213–1219.

DUNLAP, P.V. (1997). N-acyl-L-homoserine lactone autoinducers in bacteria: unity and diversity. In Shapiro, J.A. and Dworkin, M. (eds), *Bacteria as Multicellular Organisms*, New York University Press, pp. 69–106.

DUNNE, W.M. (2002). Bacterial adhesion: Seen any good biofilms lately? *J. Clin. Microbiol.*, **15**: 155–166.

ELASRI, M.O. and MILLER, R.V. (1999). Study of the response of a biofilm bacterial communit to UV radiation. *Applied and Environmental Microbiology*, **65**(6): 2025–2031.

EMTIAZI, F., SCHWARTZ, T., MARTEN, S.M., KROLLA-SIDENSTEIN, P. and OBST, U. (2004). Investigation of natural biofilms formed during the production of drinking water from surface water embankment filtration. *Wat Res.*, **38**: 1197–1206.

FAQUA, C., WINANS, S.C. and GREENBERG, E.P. (1996). Census and consensus in bacterial ecosystem: The Lux R-Lux I family of quorum sensing transcriptional regulators. *An. Rev. Microbiol.*, **50**: 727–751.

FILLOUX, A., MICHEL, G. and BALLY, M. (1998). GSP-dependent protein secreation in Gram-negative bacteria: the Xcp system of *Pseudomonas aeruginosa. FEMS Microbiology Reviews*, **22**: 177–198.

FLEMMING, H.C. (1998). Relevance of biofilms for the biodeterioration of surfaces of polymeric materials. *J. Pol. Degrad. Stab.*, **59**: 309–315.

FLEMMING, H.-C. (2002). Biofouling in water systems – cases, causes and countermeasures. *Applied Microbiology and Biotechnology*, **59**: 629–640.

FLETCHER, M. and LOEB, G.I. (1979). Influence of substratum characteristics on the attachment of a marine Pseudomonas to solid surfaces. *Applied and Environmental Microbiology*, **37**(1): 67–72.

GEESEY, G.G. (1982). Microbial exopolymers: Ecological and economic considerations. *ASM American Society for Microbiology News*, **48**(1): 9–14.

GHANNOUM, M. and O'TOOLE, G.A. (2004). *Microbial biofilms.* ASM Press, USA.

GHIGO, J-M. (2001). Natural conjugative plasmids induce bacterial biofilm development. *Nature*, **412**: 442–445.

GIAOURIS, E.D. and NYCHAS, G.J.E. (2006). The adherence of *Salmonella Enteritidis* PT4 to stainless steel: The importance of the air-liquid interface and nutrient availability. *J. Food Microbiol.*, **23**: 747–752.

GILBERT, P., MCBAIN, A.J. and RICKHARD, A.H. (2003). Formation of microbial biofilm in hygienic situations: a problem of control. *International Biodeterioration and Biodegradation*, **51**: 245–248.

GILLIGAN, P.H. (1991). Microbiology of airway disease in patients with cystic fibrosis. *Clinical Microbiology Reviews*, **4**(1): 35–51.

GOTTENBOS, B., VAN DER MEI, H.C. and BUSSCHER, H.J. (1999). Models for studying initial adhesion and surface growth in biofilm formation on surfaces. *Methods in Enzymology*, **310**: 523–534.

GRISTINA, A.G., ORGA, M., WEBB, L.X. and HOBGOOD, C.D. (1987). Adherent bacterial colonization in the pathogenesis of osteomyelitis. *Science*, **228**: 990–993.

GROBE, K., ZAHLER, J. and STEWART, P. (2002). Role of dose concentration in biocide efficacy against *Pseudomonas aeruginosa* biofilms. *Journal of Industrial Microbiology and Biotechnology*, **29**: 10–15.

HABASH, M. and REID, G. (1999). Microbial biofilms: their development and significance for medical device-related infections. *The Journal of Clinical Pharmacology*, **39**: 887–898.

HANNA, A., BERG, M., STOUT, V. and RAZATOS, A. (2003). Role of capsular colanic acid in adhesion of Uropathogenic *Escherichia coli. Applied and Environmental Microbiology*, **69**(8): 4474–4481.

HAUSNER, M. and WUERTZ, S. (1999). High rates of conjugation in bacterial biofilms as determined by quantitative in situ analysis. *Applied and Environmental Microbiology*, **65**(8): 3710–3713.

HENTZER, M., TEITZEL, G.M., BALZER, G.J., HEYDORN, A., MOLIN, S., GIVSKOV, M. and PARSEK, M.R. (2001). Alginate overproduction affects *Pseudomonas aeruginosa* biofilm structure and function. *Journal of Bacteriology*, **183**(18): 5395–5401.

HENTZER, M., RIEDEL, K., RASMUSSEN, T.B., HEYDORN, A., ANDERSEN, J.B., PARSEK, M.R., RICE, S.A., EBERL, L., MOLIN, S., HØIBY, N., KJELLEBERG, S. and GIVSKOV, M. (2002). Inhibition of quorum sensing in *Pseudomonas aeruginosa* biofilm bacteria by a halogenated furanone compound. *Microbiology*, **148**: 87–102.

HENTZER, M., WU, H., ANDERSEN, J.B., RIEDEL, K., RASMUSSEN, T.B., BAGGE, N., KUMAR, N., SCHEMBRI, M.A., SONG, Z., KRISTOFFERSEN, P., MANEFIELD, M., COSTERTON, J.W., HERRERA, J.J.R., CABO, M.L., GONZALEC, A., PAZOS, I. and PASTORIZA, L. (2007). Adhesion and detachment kinetics of several strains of *Staphylococcus aureus* subsp. Aureus under three different experimental conditions. Food Microbiol., **24**: 585–591. http://www.medcape.com/viewarticle/441355_4.

HUANG, C-T., YU, F.P., MCFETERS, G.A. and STEWART, P.S. (1995). Nonuniform spatial patterns of respiratory activity within biofilms during disinfection. *Applied and Environmental Microbiology*, **61**(6): 2252–2256.

HUECK, C.J. (1998). Type III protein secretion systems in bacterial pathogens of animals and plants. *Microbiology and Molecular Biology*, **62**(2): 379–433.

HUGHES, K.A., SUTHERLAND, I.W. and JONES, M.V. (1998a). Biofilm susceptibility to bacteriophage attack: the role of phage-borne polysaccharide depolymerase. *Microbiology*, **144**: 3039–3047.

HUGHES, K.A., SUTHERLAND, I.W., CLARK, J. and JONES, M.V. (1998b). Bacteriophage and associated polysaccharide depolymerases – novel tools for study of bacterial biofilms. *Journal of Applied Microbiology*, **85**: 583–590.

HUNT, S.M., WERNER, E.M., HUANG, B., HAMILTON, M.A. and STEWART, P.S. (2004). Hypothesis for the role of nutrient starvation in biofilm detachment. *Applied and Environmental Microbiology*, **70**(12): 7418–7425.

JAHN, A. and NIELSEN, P.H. (1998). Cell biomass and exopolymer composition in sewer biofilms. *Water Science and Technology*, **37**(1): 17–24.

JEFFERSON, K.K. (2004). What drives bacteria to produce a biofilm? *FEMS Microbiology Letters*, **236**: 163–173.

JOHANSEN, C., FALHOLT, P. and GRAM, L. (1997). Enzymatic removal and disinfection of bacterial biofilms. *Applied and Environmental Microbiology*, **63**(9): 3724–3728.

JONAS, R. and FARAH, L.F. (1998). Production and application of microbial cellulose. *Polymer Degradation and Stability*, **59**(1): 101–106.

KALMOKOFF, M.L., AUSTIN, J.W., WAN, X.D., SANDERS, G., BANERJEE, S. and FARBER, J.M. (2001). Adsorption, attachement and biofilm formation among isolates of *Listeria monocytogenes* using model conditions. *J. Appl. Microbiol.*, **91**: 725.

KEEVIL, C.W., GODFREE, A., HOLT, D. and DOW, C. (1999). *Biofilms in the aquatic environment.* The Royal Society of Chemistry, UK.

KERR, C.J., OSBORN, K.S., ROBOSON, G.D. and HANDLEY, P.S. (1999). The relationship between pipe material and biofilm formation in a laboratory model system. *J. Appl. Microbiol.*, **85**: 29–38.

KILB, B., LANGE, B., SCHAULE, G., FLEMMING, H.-C. and WINGENDER, J. (2003). Contamination of drinking water by coliforms from biofilms grown on rubber-coated valves. *International Journal of Hygiene and Environmental Health*, **206**: 563–573.

KINZLER, K., GEHRKE, T., TELEGDI, J. and SAND, W. (2003). Bioleaching – a result of interfacial processes caused by extracelluar polymeric substances (EPS). *J. Hydromet.*, **71**: 83–88.

KIRIMURA, K., YUSA, S., RUGSASEEL, S., NAKAGAWA, H., OSUMI, M. and USAMI, S. (1999). Amylose-like polysaccharide accumulation and hyphal cell-surface structure in relation to citric acid production by *Aspergillus niger* in shake culture. *Applied Microbiology and Biotechnology*, **52**: 421–428.

KLAUSEN, M., HEYDORN, A., RAGAS, P.C., LAMBERTSEN, L., AAES-JØRGENSEN, A., MOLIN, S. and TOLKER-NIELSEN, T. (2003). Biofilm formation by *Pseudomonas aeruginosa* wild type, flagella, and type IV pili mutants. *Molecular Microbiology*, **48**: 1511–1524.

KOSTENKO, V., CERI, H. and MARTINUZZI, R.J. (2007). Increased tolerance of *Staphylococcus aureus* to vancomycin in viscous media. *FEMS Immunol Med Microbiol*, **51**: 277–288.

KREFT, J.-U. (2004). Biofilms promote altruism. *Microbiology*, **150**: 2751–2760.

KUCHMA, S.L. and O'TOOLE, G.A. (2000). Surface induced and biofilms induced changes in gene expression. *Curr. Opin. Biotechnol.*, **11**: 429–433.

KUMAR, A. and PRASAD, R. (2006). *Biofilms.* Dep. Microbiology. Himalayan Institute of Medical Sciences, Dehradun, U.A. **8**: 14–17.

KUMAR, C.G. and ANAND, S.K. (1998). Significance of microbial biofilms in food industry: a review. *International Journal of Food Microbiology*, **42**: 9–27.

LAPIDOT, A., ROMLING, U. and YARON, S. (2006). Biofilm formation and the survival of *Salmonella typhimurium* on parsley. *Int. J. Food Microbiol.*, **109**: 229–233.

LEHTOLA, M.J., LAXANDER, M., MIETTINEN, I.T., HIRVONEN, T.V. and MARTIKAINEN, P. (2006). The effects of changing water flow velocity on the formation of biofils and water quality in pilot distribution system consisting of copper or polyethelene pipes. *Wat. Res.*, **40**: 2151–2160.

LEID, J.G., SHIRTLIFF, M.E., COSTERTON, J.W. and STOODLEY, P. (2002). Human leukocytes adhere to, penetrate, and respond to *Staphylococcus aureus* biofilms. *Infection and Immunity*, **70**(11): 6339–6345.

LIU, Y. and TAY, J.H. (2001). Metabolic response of biofilm to shear stress in fixed-film culture. *Journal of Applied Microbiology*, **90**: 337–342.

LIU, Y., YANG, S.F., LI, Y., XU, H., QIN, L. and TAY, J.H. (2004). The influence of cell substratum surface hydrophobicities on microbial attachment. *J. Biotechnol.*, **110**: 251–256.

LORENZ, M. and WACKERNAGEL, W. (1994). Bacterial gene transfer by natural genetic transformation in the environment. *Microbiological Reviews*, **85**(3): 563–602.

LUDENSKY, M. (2003). Control and monitoring of biofilms in industrial applications. *International Biodeterioration and Biodegradation*, **51**: 255–263.

MACK, D., FISCHER, W., KROKOTSCH, A., LEOPOLD, K., HARTMANN, R., EGGE, H. and LAUFS, R. (1996). The intercellular adhesin involved in biofilm accumulation of *Staphylococcus epidermidis* is a linear β-1,6-linked glucosaminoglycan: purification and structural analysis. *Journal of Bacteriology*, **178**(1): 175–183.

MAH, T.-F. and O'TOOLE, G.A. (2001). Mechanisms of biofilm resistance to antimicrobial agents. *Trends in Microbiology*, **9**(1): 34–39.

MAKIN, S.A. and BEVERIDGE, T.J. (1996). The influence of A-band and B-band lipopolysaccharides on the surface characteristics and adhesion of *Pseudomonas aeruginosa* to surfaces. *Microbiology*, **142**: 299–307.

MANEM, J.A. and RITTMANN, B.E. (1992). Removing trace level organic pollutants in a biological filter. *Journal of American Water Works Association*, **84**: 152–158.

MANUEL, C.M., NUNES, O.C. and MELO, L.F. (2007). Dymanics of drinking water biofilm inflow/non-flow conditions. *Wat. Res.*, **41**: 551–562.

MAO, Y., DOYLE, M.P. and CHEN, J. (2001). Insertion mutagenesis of *wca* reduces acid and heat tolerance of enterohemorrhagic *Escherichia coli* O157:H7. *Journal of Bacteriology*, **183**(12): 3811–3815.

MARA, D. and HORAN, N. (2003). *The Handbook of Water and Wastewater Microbiology.* Academic Press, UK.

MARTINY, A.C., ALBRECHTSEN, H.J., ARVIN, E. and MOLIN, S. (2005). Identification of bacteria in biofilm and bulk water samples from a nonchlorinated model drinking water distribution system: Detection of a large nitrite-oxidizing population associated with *Nitrospira* spp. *J. Appl. Environ. Microbiol.*, **71**: 8611–8617.

MCLAUGHLIN-BORLACE, L., STAPLETON, F., MATHESON, M. and DART, J.K.G. (1998). Bacterial biofilm on contact lenses and lens storage cases in wearers with microbial keratitis. *Journal of Applied Microbiology*, **84**: 827–838.

MELO, L.F. and BOTT, T.R. (1997). Biofouling in water systems. *J. Exp. Therm. Fluid Sci.*, **14**: 375–381.

MEYER, B. (2003). Approaches to prevention, removal and killing of biofilms. *Inter. J. Biodeter. Biodegrad.*, **51**: 249–253.

MOMBA, M.N.B. and KELENI, P. (2002). Regrowth and survival of indicator microorganisms on the surfaces of household containers used for the storage of drinking water in rural communities of South Africa. *Wat. Res.*, **36**: 3023–3028.

MORGAN, T.D. and WILSON, M. (2001). The effects of surface roughness and type of denture acrylic on biofilm formation by *Streptococcus oralis* in a constant depth film fermentor. *Journal of Applied Microbiology*, **91**: 47–53.

MURGA, R., MILLER, J.M. and DONLAN, R.M. (2001). Biofilm formation by Gram-negative bacterial on central venous catheter connectors: Effect of conditioning films in a laboratory model. *Journal of Clinical Microbiology*, **39**(6): 2294–2297.

NIQUETTE, P. SERVALS, P. and SAVOIR, R. (2000). Impacts of pipematerials on densities of fixed bacterial biomass in a drinking water distribution system. *Wat. Res.*, **34**: 1952–1956.

NIVENS, D.E., OHMAN, D.E., WILLIAMS, J. and FRANKLIN, M.J. (2001). Role of alginate and its O acetylation in formation of *Pseudomonas aeruginosa* microcolonies and biofilms. *Journal of Bacteriology*, **183**(3): 1047–1057.

NOIRI, Y., EHARA, A., KAWAHARA, T., TAKEMURA, N. and EBISU, S. (2002). Participation of bacterial biofilms in refractory and chronic periapical periodontitis. *Journal of Endodontics*, **28**: 679–683.

NORWOOD, D.E. and GILMOUR, A. (2000). The growth and resistance to sodium hypochlorite of *Listeria monocytogenes* in a steady-state multispecies biofilm. *Journal of Applied Microbiology*, **88**: 512–520.

NYVAD, B. and KILIAN, M. (1990). Microbial associated with experimental root surface caries in humans. *Infect Immmun.*, **58**: 1628–1633.

O'TOOLE, G., KAPLAN, H.B. and KOLTER, R. (2000). Biofilm formation as microbial development. *Annual Reviews in Microbiology*, **54**: 49–79.

O'TOOLE, G.A. and KOLTER, R. (1998a). Flagellar and twitching motility are necessary for *Pseudomonas aeruginosa* biofilm development. *Molecular Microbiology*, **30**(2): 295–304.

O'TOOLE, G.A. and KOLTER, R. (1998b). Initiation of biofilm formation in *Pseudomonas fluorescens* WCS365 proceeds via multiple, convergent, signaling pathways: A genetic analysis. *J. Mol. Microbiol.*, **28**: 449–461.

OH, Y.J., JO, W., YANG, Y. and PARK, S. (2007). Influence of culture conditions on *Escherichia coli* O157:H7 biofilm formation by atomic force microscopy. *Ultramicroscopy*, **107**: 869–874.

OHASHI, A. and HARADA, H. (1994). Adhesion strength of biofilm development in an attached-growth reactor. *Water Science and Technology*, **29**(10–11): 281–288.

OOSTHUIZEN, M.C., STEYN, B., THERON, J., COSETTE, P., LINDSAY, D., VON HOLY, A. and BRÖZEL, V.S. (2002). Proteomic analysis reveals differential protein expression by *Bacillus cereus* during biofilm formation. *Applied and Environmental Microbiology*, **68**(6): 2770–2780.

OPHIR, T. and GUTNICK, D.L. (1994). A role for exopolysaccharides in the protection of microorganisms from desiccation. *Applied and Environmental Microbiology*, **60**(2): 740–745.

PALMER, R. JR. and WHITE, D.C. (1997). Developmental biology of biofilms: implications for treatment and control. *Trends in Microbiology*, **5**(1): 435–440.

PARSEK, M.R. and SINGH, P.K. (2003). Bacterial biofilms: An emerging link to disease pathogenesis. *Annual Reviews in Microbiology*, **57**: 677–701.

PASMORE, M., TODD, P., SMITH, S., BAKER, D., SILVERSTEIN, J., COONS, D. and BOWMAN, C.N. (2001). Effects of ultrafiltration membrane surface properties on *Pseudomonas aeruginosa* biofilm initiation for the purpose of reducing biofouling. *Journal of Membrane Science*, **194**: 15–32.

PASMORE, M., TODD, P., PFIEFER, B., RHODES, M. and BOWMAN, C. (2002). Effect of polymer surface properties on the reversibility of attachment of *Pseudomonas aeruginosa* in the early stages of biofilm development. *Biofouling*, **18**(1): 65–71.

PEARCE, D., BASIN, M.J. and LYNCH, J.M. (1995). The rhizosphere as a biofilm. *Microbial biofilms*, 1st ed. Cambridge University Press, New York, N.Y. 207–220.

PERCIVAL, S.L., KNAPP, J.S., WALES, D.S. and EDYVEAN, R.G.J. (1999). The effect of turbulent flow and surface roughness on biofilm formation in drinking water. *Journal of Industrial Microbiology and Biotechnology*, **22**: 152–159.

POULSEN, L.V. (1999). Microbial biofilm in food processing. *Lebensm.Wiss. J. Technol.*, **32**: 321–326.

PRAKASH, B., VEEREGOWDA, B.M. and KRISHNAPPA, G. (2003a). Biofilms: A survival strategy of bacteria. *J. Curr. Sci.*, **85**: 9–10.

PRAKASH, B., VEEREGOWDA, B.M. and KRISHNAPPA, G. (2003b). Biofilms: A survival strategy of bacteria. *Current Science*, **85**(9): 1299–1307.

PRIGENT-COMBARET, C., VIDAL, O., DOREL, C. and LEJEUNE, P. (1999). Abiotic surface sensing and biofilm-dependent regulation of gene expression in *Escherichia coli*. *Journal of Bacteriology*, **181**(19): 5993–6002.

PRINGLE, J.H. and FLETCHER, M. (1983). Influence of substratum wettability on attachment of freshwater bacteria to solid surfaces. *Applied and Environmental Microbiology*, **45**(3): 811–817.

PUREVDORJ, B., COSTERTON, J.W. and STOODLEY, P. (2002). Influence of hydrodynamics and cell signaling on the structure and behavior of *Pseudomonas aeruginosa* biofilms. *Applied and Environmental Microbiology*, **68**(9): 4457–4464.

PUSKAS, A., GREENBERG, E.P., KAPLAN, S. and SCHAEFER, A.L. (1997). A quorum-sensing system in the free-living photosynthetic bacterium *Rhodobacter sphaeroides. Journal of Bacteriology*, **179**(23): 7530–7537.

RACHID, S., OHLSEN, K., WITTE, W., HACKER, J. and ZIEBUHR, W. (2000). Effect of subinhibitory antibiotic concentrations on polysaccharide intercellular adhesion expression in biofilm forming *Staphylococcus epidermidis. J. Antimicrob. Agents Chemother.*, **44**: 3357–3363.

REHM, B.H. and VALLA, S. (1997). Bacterial alginates: biosynthesis and applications. *Applied Microbiology and Biotechnology*, **48**: 281–288.

REN, D., BEDZYK, L.A., THOMAS, S.M., YE, R.W. and WOOD, T.K. (2004). Gene expression in *Escherichia coli* biofilms. *Applied Microbiology and Biotechnology*, **64**: 515–524.

RICE, S.A., KOH, K.S., QUECK, S.Y., LABBATE, M., LAM, K.W. and KJELLERBERG, S. (2005). Biofilm formation and sloughing in *Serratia marcescens* are controlled by quorum sensing and nutrient cues. *Journal of Bacteriology*, **187**(10): 3477–3485.

ROCHEX, A. and LEBEAULT, J.-M. (2007). Effects of nutrients on biofilm formation and detachment of a *Pseudomonas putida* strain isolated from a paper machine. *Water Research*, **41**: 2885–2892.

RYU, J.-H. and BEUCHAT, L.R. (2005). Biofilm formation by *Escherichia coli* O157: H7 on stainless steel: Effect of exopolysaccharide and curli production on its resistance to chlorine. *Applied and Environmental Microbiology*, **71**(1): 247–254.

SAUER, K. and CAMPER, A.K. (2001). Characterization of phenotypic changes in *Pseudomonas putida* in response to surface-associated growth. *Journal of Bacteriology*, **183**(22): 6579–6589.

SAUER, K., CAMPER, A.K., EHRLICH, G.D., COSTERTON, J.W. and DAVIES, D.G. (2002). *Pseudomonas aeruginosa* displays multiple phenotypes during development as a biofilm. *Journal of Bacteriology*, **184**(4): 1140–1154.

SAUER, K., CULLEN, M.C., RICKARD, A.H., ZEEF, L.A.H., DAVIES, D.G. and GILBERT, P. (2004). Characterization of nutrient-induced dispersion in *Pseudomonas aeruginosa* PAO1 biofilm. *Journal of Bacteriology*, **186**(21): 7312–7326.

SAWYER, L.K. and HERMANOWICZ, S.W. (1998). Detachment of biofilm bacteria due to variations in nutrient supply. *Water Science and Technology*, **37**: 211–214.

SCHAULE, G., FLEMMING, H.C. and RIDGWAY, H.F. (1993). Use of 5-cyano-2.3 ditolyl tetrazolium chloride for quantifying planktonic and sessile respiring bacteria in drinking water. *J. Appl. Bacteriol.*, **59**: 3850–3857.

SCHEMBRI, M.A., KJAERGAARD, K. and KLEMM, P. (2003). Global gene expression in *Escherichia coli* biofilms. *Molecular Microbiology*, **48**(1): 253–267.

SCHIERHOLZ, J.M. and BEUTH, J. (2001). Implant infections: a haven for opportunistic bacteria. *Journal of Hospital Infection*, **49**: 87–93.

SCHNIDER-KEEL, U., LEJBØLLE, K.B., BAEHLER, E., HAAS, D. and KEEL, C. (2001). The sigma factor AlgU (AlgT) controls exopolysaccharide production and tolerance towards desiccation and osmotic stress in the biocontrol agent *Pseudomonas fluorescens* CHA0. *Applied and Environmental Microbiology*, **67**(12): 5683–5693.

SIBONI, N., LIDOR, M., KRAMARSKY-WINTER, E. and KUSHMARO, A. (2007). Conditioning film and initial biofilm formation on ceramics tiles in the marine environment. *FEMS Microbiology Letters*, **274**: 24–29.

SIMÕES, M., PEREIRA, M.O., SILANKORVA, S., AZEREDO, J. and VIEIRA, M.J. (2007). The effect of hydrodynamic conditions on the phenotype of *Pseudomonas fluorescens* biofilms. *Biofouling*, **23**(3/4): 249–258.

SINGH, P.K., PARSEK, M.R., GREENBERG, E.P. and WELSH, M.J. (2002). A component of innate immunity prevents bacterial biofilm development. *Nature*, **417**: 552–555.

SPIERS, A.J., BOHANNON, J., GEHRIG, S.M. and RAINEY, P.B. (2003). Biofilm formation at the ari-liquid interface by the *Pseudomonas Fluorescens* SBW25 wrinkly spreader requires an acetylated form of cellulose. *J. Mol. Microbiol.*, **50**: 15–27.

STEWART, P.S. and COSTERTON, J.W. (2001). Antibiotic resistance of bacteria in biofilms. *The Lancet*, **358**: 135–138.

STEYN, B., OOSTHUIZEN, M.C., MACDONALD, R., THERON, J. and BROZEL, V.S. (2001). The use of glass wool as an attachment surface for studying phenotypic changes in *Pseudomonas aeruginosa* biofilms by two dimensional gel elecrophoresis. *J. Proteo.*, **1**: 871–879.

STOODLEY, P., SAUER, K., DAVIES, D.G. and COSTERTON, J.W. (2002). Biofilms as complex differentiated communities. *Annual Reviews of Microbiology*, **56**: 187–209.

SUTHERLAND, I.W. (2001a). Biofilm exopolysaccharides: a strong and sticky framework. *Microbiology*, **147**: 3–9.

SUTHERLAND, I.W. (2001b). The biofilm matrix – an immobilized but dynamic microbial environment. *Trends in Microbiology*, **9**(5): 222–227.

TEITZEL, G.M. and PARSEK, M.R. (2003). Heavy metal resistance of biofilm and planktonic *Pseudomonas aeruginosa. Applied and Environmental Microbiology*, **69**(4): 2313–2320.

TIMOTHY, C.E. and HANSEN, L.T. (2006). Strain and growth temperature influence *Listeria* spp attachment to intact and cut cabbage. *Int. J. Food Microbiol.*, **111**: 34–42.

TOKLER-NEILSEN, T. and MOLIN, S. (2000). Spatial organization of microbial biofilm communities. *Microbial Ecology*, **40**: 75–84.

TRACHOO, N. (2003). Biofilms and the food industry. *Songklanakarin Journal of Science and Technology*, **25**(6): 807–815.

TRINET, F., HEIM, R., AMAR, D. and CHANG, H. (1991). Study of biofilm and fluidization of bioparticles in a three-phase liquid fluidized-bed reactor. *Water Science and Technology*, **23**: 1347–1354.

TSUNEDA, S., AIKAWA, H., HAYASHI, H., YUASA, A. and HIRATA, A. (2003). Extracellular polymeric substances responsible for bacterial adhesion onto solid surface. *FEMS Microbiology Letters*, **223**: 287–292.

VAN DER KOOIJ, H. and VEENENDAAL, H.R. (2001). Biomass production potential of materials in contact with drinking water: Method and practical importance. *Wat. Sci. Technol.*, **3**: 39–45.

VAN LOOSDRECHT, M.C.M., LYKLEMA, J., NORDE, W. and ZEHNDER, A.J.B. (1990). Influence of interfaces on microbial activity. *Microbiological Reviews*, **54**(1): 75–87.

VRANY, J.D., STEWART, P.S. and SUCI, P.A. (1997). Comparison of recalcitrance of ciprofloxacin and levofloxacin exhibited by *Pseudomonas aeruginosa* biofilms displaying rapid-transport characteristics. *Antimicrobial Agents and Chemotherapy*, **41**(6): 1352–1358.

WATNICK, P.I., FULLNER, K.J. and KOLTER, R.E. (1999). *Staphylococcus epidermidis* adhesion to hydrophobic biomedical polymer is mediated by platelets. *J. Infec. Dis.*, **167**: 329–336.

WEBB, J.S., THOMPSON, L.S., JAMES, S., CHARLTON, T., TOLKER-NIELSEN, T., KOCH, B., GIVSKOV, M. and KJELLEBERG, S. (2003). Cell death in *Pseudomonas aeruginosa* biofilm development. *Journal of Bacteriology*, **185**(15): 4585–4592.

WECKE, J., KERSTEN, T., MADELA, K., MOTER, A., GÖBEL, U.B., FRIEDMANN, A. and BERNIMOULIN, J.-P. (2000). A novel technique for monitoring the development of bacterial biofilms in human periodontal pockets. *FEMS Microbiology Letters*, **191**: 95–101.

WHITCHURCH, C.B., TOLKER-NIELSEN, T., RAGAS, P.C. and MATTICK, J.S. (2002). Extracellular DNA required for bacterial biofilm formation. *Nature*, **295**: 1487.

WHITELEY, M., BANGERA, M.G., BUMGARNER, R.E., PARSEK, M.R., TEITZEL, G.M., LORY, S. and GREENBERG, E.P. (2001). Gene expression in *Pseudomonas aeruginosa* biofilms. *Nature*, **413**: 860–864.

WINGENDER, J., NUE, T.R. and FLEMMING, H.-C. (1999). *Microbial extracellular polymeric substances: Characterization, structure and function.* Springer, Gemany.

YILDIZ, F.H. and SCHOOLNIK, G.K. (1999). *Vibrio cholerae* O1 El Tor: Identification of a gene cluster required for the rugose colony type, exopolysaccharide production, chlorine resistance, and biofilm formation. *Proc. Natl. Acad. Sci.*, **96**: 4028–4033.

YU, J.A., JI, M. and YUE, P.L. (1999). A three-phase fluidized bed reactor in the combined anaerobic/aerobic treatment of wastewater. *Journal of Chemical Technology and Biotechnology*, **74**: 61.

ZACHEUS, O.M., LIVANAINEN, E.K., NISSINEN, T.K., LEHTOLA, M.J. and MARTIKAINEN, P.J. (2000). Bacterial biofilm formation on polyvinyl chloride, polyethylene and stainless steel exposed to ozonated water. *Wat. Res.*, **1**: 63–70.

ZHENG, Z. and STEWART, P.S. (2002). Penetration of rifampin through *Staphylococcus epidermidis* biofilms. *Antimicrobial Agents and Chemotherapy*, **46**(3): 900–903.

# 2

# Molecular mechanisms involved in biofilm formation by food-associated bacteria

**J. Smith, P. M. Fratamico and G. Uhlich, United States Department of Agriculture, USA**

**Abstract**: This chapter discusses the role of cell-to-cell signaling processes, known as quorum sensing, on biofilm formation in food-borne pathogens. When bacterial populations reach the 'quorum' level, signaling molecules bind to appropriate transcription regulators resulting in the activation or repression of target genes. In many bacterial species, the effect of quorum sensing on gene expression is closely linked with biofilm formation, which may be either negatively or positively regulated by quorum sensing. A potential strategy to control biofilm formation is by modulating quorum sensing through the use of signal analogs or by enzymatic degradation of the signaling molecules.

**Key words**: biofilms, quorum sensing, bacteria, mechanisms, food, pathogens.

## 2.1 Introduction: overview of biofilm formation and quorum sensing

Bacteria in the natural environment form complexes which are closely associated with abiotic and biotic surfaces of solids and liquids. These bacterial communities are adherent to a surface, an interface or to each other and are known as biofilms. Biofilms on surfaces have a characteristic construction consisting of tower- and mushroom-shaped microcolonies enclosed in a hydrated matrix of extracellular polymeric material consisting of nucleic acid, proteins and polysaccharides produced by the microorganisms present in the biofilm. Usually, biofilms consist of a mixed bacterial population but they may be composed of a single bacterial species (Donlan, 2002; Stoodley *et al.*, 2002). The stages involved in the development of a biofilm include (Stoodley *et al.*, 2002):

Stage 1: reversible attachment of bacteria to a surface.
Stage 2: irreversible attachment mediated by the formation of exopolymeric material.
Stage 3: formation of microcolonies and the beginning of biofilm maturation.
Stage 4: formation of mature biofilm with a three-dimensional structure containing cells packed in clusters with water channels between the clusters allowing for nutrient transport and waste removal.
Stage 5: detachment and dispersion of cells from the biofilm and initiation of new biofilm formation; dispersed cells are more similar to planktonic cells than to mature biofilm cells.

It is of great interest that the microorganisms present in biofilms behave differently when compared to planktonic (i.e., non-adherent) cells. Biofilm organisms are significantly more resistant to microbially deleterious substances than planktonic cells and they are more resistant to other environmental stresses. In addition, biofilms present on medical devices or on infected tissues are more resistant to host immune responses than the corresponding planktonic cells (Donlan, 2002; Sauer, 2003; Stoodley *et al.*, 2002). Genomics and proteomics studies show that the gene expression patterns and protein synthesis of bacteria in biofilms differ from those of planktonic cells indicating that there is a quite different physiology between the two types of cells (Sauer, 2003).

There are many advantages to the biofilm mode of growth. Biofilms represent an environment where bacteria are protected from the inhibitory effects of antimicrobial compounds and biocides as well as from physical and chemical stresses. The increased binding of water by the polymeric matrix decreases the likelihood of dehydration of the bacterial cells, a stress that planktonic cells are subject to. Due to the close proximity of the microorganisms in biofilms, nutrients, metabolites, and genetic material are readily exchanged (Davey and O'Toole, 2000; Donlon, 2002; Trachoo, 2003). In a mature biofilm, cell division is uncommon; rather, energy is used to produce exopolysaccharides, which can be used by the biofilm cells when nutrients are scarce (Watnick and Kolter, 2000). Stewart and Franklin (2008) have reviewed the physiological heterogeneity of cells in biofilms; the cell populations have a diversity of genotypes and phenotypes with divergent metabolism, stress responses and other biological activities. Jefferson (2004) stated that biofilms are the default mode of bacterial growth, whereas, planktonic growth is an *in vitro* artifact.

Quorum sensing (QS) is a cell density-dependent signaling system which regulates gene expression and is mediated by chemical autoinducer molecules produced by bacteria. When the bacterial population reaches the 'quorum' level, the autoinducer molecules bind to appropriate transcription regulators followed by the activation or repression of target genes. QS allows the microorganisms to display a unified response beneficial to survival of the population (Smith *et al.*, 2004). There are several QS autoinducer mechanisms in bacteria. The QS system in Gram-negative bacteria depends on homologues of the LuxI-LuxR regulatory proteins originally identified in *Vibrio fischeri* (Miller and Bassler, 2001). The LuxI-like proteins are responsible for the biosynthesis of the autoinducer compounds known as acylated homoserine lactones (AHLs) consisting of a homoserine lactone ring with a variable length acyl side chain. The AHLs are produced inside of the cell and then secreted outside of the cell. The concentration of AHLs increases as the bacterial numbers increase. At a critical threshold level, the AHLs re-enter the bacterial cell and bind to the N-terminal domain of the LuxR-like protein receptor. The LuxR-AHL complexes activate or repress target gene transcription (Miller and Bassler, 2001).

Gram-positive bacteria, similar to Gram-negative, also use QS to regulate a number of cellular processes. However, Gram-positive bacteria utilize secreted peptides as autoinducer compounds. The concentration of peptides secreted to the external environment increases as the bacterial population increases (Miller and Bassler, 2001). Two-component sensor kinases detect the peptide signals and initiate a series of phosphorylation/dephosphorylation reactions which eventually leads to the phosphorylation of the response regulator. Phosphorylation activates the response regulator which then binds DNA to alter transcription of the QS-controlled target gene (Miller and Bassler, 2001).

A large number of bacteria, both Gram-negative and Gram-positive, have a common QS system, mediated by autoinducer 2 (AI-2), which is a product of the enzyme, LuxS (Smith *et al.*, 2004; Vendeville *et al.*, 2005). LuxS is also a key component of the activated methyl cycle (AMC) and generates the major methyl donor, S-adenosylmethionine. Toxic S-ribosylhomocysteine is produced as part of the AMC pathway. LuxS detoxifies S-ribosylhomocysteine by forming 4,5-dihydroxy-2,3-pentanedione (DPD) and homocysteine. The DPD cyclizes with boron to form AI-2; thus, AI-2 is a by-product of the AMC cycle (McDougald *et al.*, 2007; Vendeville *et al.*, 2005). LuxS is, therefore, implicated in quorum sensing as well as in cellular metabolism. AI-2 in involved in bioluminescence by *Vibrio harveyi* and binds to LuxP as a furanosyl-borate-diester. LuxP is only found in *Vibrio* species (McDougald *et al.*, 2007; Vendeville *et al.*, 2005). AI-2 in *Salmonella* Typhimurium does not contain boron (Vendeville *et al.*, 2005). In *Salmonella* Typhimurium and *Escherichia coli*, AI-2 regulates the *lsrACDBFGE* operon which is involved in the transport of extracellular AI-2 into the cell (Ahmer, 2004; Taga *et al.*, 2003; Walters and Sperandio, 2006b; Xavier and Bassler, 2005). In *S.* Typhimurium and *E. coli*, AI-2 binds to LsrB which activates transcription of the *lsr* operon (Hardie and Heurlier, 2008). The binding proteins for AI-2 in other bacterial species have not been identified. Since LuxS has a dual role, it is necessary to separate the metabolic role of the enzyme from the quorum sensing activity of the LuxS product, AI-2 (Doherty *et al.*, 2006). Some of the phenotypes seen with *luxS* may be due to a metabolic defect due to the loss of LuxS function in the activated methyl cycle rather than due to a signaling defect. Therefore, the proper study of the effects of LuxS and AI-2 indicates that experiments must include complementation with both *luxS* (to discount second-site mutation effects) and purified AI-2 in order to separate the effects of AI-2 as a QS compound from metabolic effects under the control of the *luxS* gene (Hardie *et al.*, 2003; McDougald *et al.*, 2007). The only two phenotypes known to be AI-2 dependent (purified or in vitro synthesized) are the expression of the *lsr* operon in *S.* Typhimurium (and probably *E. coli*) and luminescence in *V. harveyi.*

The autoinducer-3/epinephrine/norepinephrine (AI-3/epi/norepi) signaling system is found in a number of Gram-negative bacterial species, as well as in food-borne enteric pathogens including *E. coli, Shigella, Salmonella*, and *Yersinia* (Walters and Sperandio, 2006b). The AI-3/epi/norepi system has been shown to have an important role in the virulence of *E. coli* O157:H7 (Sperandio *et al.*, 2003; Walters and Sperandio, 2006b).

In a recent review, Ammor *et al.* (2008) examined the potential role of QS in food spoilage. They discuss the various QS mechanisms and their detection and quantification. The spoilage of a variety of food products, including foods of animal origin, seafood and fish, fruits, and vegetables are covered. It is probable that food spoilage involves the formation of biofilms. In addition, the inhibition of QS as a means of preventing food spoilage is discussed (Ammor *et al.*, 2008).

Quorum sensing and biofilm formation are closely intertwined in many bacterial species. The formation of biofilms may be negatively or positively regulated by QS. In the present review, the relationship between QS and and the various stages of biofilm formation are examined in important food- and water-borne pathogens, including *L. monocytogenes*, enterohemorrhagic *E. coli, Salmonella* spp., *Campylobacter jejuni, Staphyococcus aureus*, and *Shigella* spp., and others. Research on other genera, including *Aeromonas hydrophila, Klebsiella pneumonia, Serratia marcescens*, and others will also be reviewed since the potential role of quorum sensing in biofilm formation in these bacteria has been reported.

## 2.2 Biofilm formation and quorum sensing in bacteria

### 2.2.1 *Salmonella enterica*

*Salmonella enterica* consists of a large number of serovars; they are facultative anaerobic Gram-negative rod-shaped bacteria. The salmonellae are associated with food- and water-borne gastroenteritis. Solomon *et al.* (2005) found that biofilms were formed by *Salmonella enterica* on polystyrene microtiter plates by 15 clinical isolates (6 serovars), 26 produce-related isolates (19 serovars), and 31 meat-related stains (15 serovars). Other studies indicate that *S.* Enteritidis formed biofilms at the air/liquid interface on stainless steel coupons (Giaouris and Nychas, 2006) and glass slides (Niemira and Solomon, 2005). Latasa *et al.* (2005) and Scher *et al.* (2005) observed that *S.* Enteritidis formed pellicles at the air/liquid interface in glass tubes. *Salmonella* Typhimurium DT104 formed biofilms on polyvinyl chloride microplates, stainless steel and glass surfaces (Kim and Wei, 2007), thermanox (polyolefin) cover slips (Ledeboer *et al.*, 2006), and polystyrene microtiter plates (Ngwai *et al.*, 2006). Ledeboer *et al.* (2006) demonstrated *S.* Typhimurium biofilms on HEp-2 cells and chicken intestinal epithelium. In the presence of bile, *S.* Typhi and *S.* Typhimurium formed biofilms on the surface of gall stones (Prouty *et al.*, 2002). Thus, a cursory survey suggests that biofilm formation is a common trait in different *Salmonella* serotypes.

The LuxS/AI-2 and the AI-3/epi/norepi systems are present in *Salmonella* (Walters and Sperandio, 2006b); however, information about the AI-3/epi/norepi system in *Salmonella* is lacking. In addition, *Salmonella* species have a LuxR homolog, SdiA; however, they do not have the LuxI homolog and thus do not synthesize AHLs (Walters and Sperandio, 2006b). In addition SdiA of *Salmonella* can detect the AHLs produced by other bacteria (Walters and Sperandio, 2006b) but has only been shown to regulate a few genes such as the chromosomal *srgE* gene (*sdiA* regulated gene), which has no known function, and the *rck* (resistant to complement killing) operon, which is located on the *Salmonella* virulence plasmid (Ahmer, 2004). The regulation of these genes by SdiA is AHL-dependent. Genes in the *rck* operon play roles in adhesion to host tissues and resistance to killing by complement. The *rck* operon and the *srgE* gene are not present in *E. coli* (Ahmer, 2004). In the mouse, chicken, and bovine models for salmonellosis, mutation in *sdiA* had no effect on virulence (Ahmer, 2004). Thus, the exact function of SdiA in *Salmonella* is not clear.

Instead of a furanosyl borate diester AI-2, the AI-2 of *Salmonella* lacks boron (Miller *et al.*, 2004). In *Salmonella*, LuxS and AI-2 regulate the expression of an operon encoding an ABC transport system (homologous to the ribose transporter), the *lsr* (*luxS* regulated) operon. LuxS is involved in the synthesis of AI-2, and AI-2 accumulates extracellularly. Basal expression of the Lsr transporter allows enough AI-2 to enter the cytoplasm to be phosphorylated by LsrK. The phosphorylated AI-2 then inactivates LsrR (the repressor of the *lsr* operon) leading to the transcription of the *lsr* operon with ultimate increased uptake and phosphorylation of AI-2. Phosphorylated AI-2 appears to be further processed and/or degraded by *lsr*-encoded enzymes (Taga *et al.*, 2003).

A type III secretion system, encoded by genes located on the *Salmonella* pathogenicity island-1 (SPI-1), is necessary for virulence of *S.* Typhimurium during infection of the GI tract (Lostroh and Lee, 2001). The transcription regulator protein, InvF (encoded by *invF*) is required for expression of several genes encoding proteins of the SPI-1 type III secretion system (Darwin and Miller, 1999). Choi *et al.* (2007) demonstrated that the induction of the *invF* gene was eliminated in a *luxS*-negative mutant strain of *S.* Typhimurium. In addition, the *luxS* mutant showed a 3-fold decrease in the ability to invade HEp-2 cells. The expression of SPI-1 genes and epithelial cell invasion was restored to the *luxS* mutant by the addition of DPD or by inserting a *luxS*-containing plasmid. De Keersmaecker *et al.* (2005) had previously demonstrated that DPD was active in the luminescence assay of *V. harveyi*, thereby, indicating that it had AI-2-like activity. The work of Choi *et al.* (2007) indicates that QS via AI-2 is involved in the expression of the the InvF regulator of genes present on SPI-1 and thus, has an important role in the invasive ability of *S.* Typhimurium.

On soft agars, *S.* Typhimurium differentiates into hyperflagellated swarmer cells. Flagella and lipopolysaccharide are essential to swarming (Toguchi *et al.*, 2000; Wang *et al.*, 2006). Toguchi *et al.* (2000) suggested that the O antigen provided a surfactant function during swarming. Mutations in *luxS* or *sdiA* do not inhibit swarming and while externally added AHLs activated *sdiA,* there was no effect on swarming (Toguchi *et al.*, 2000). A mutation in *flhE* (located in a flagellar locus but not involved in swimming motility) abolished swarming in *S.* Typhimurium; swarming could be restored with a plasmid containing the *flhE* gene (Stafford and Hughes, 2007). The *flhE* mutant showed a 5-fold increase in biofilm production on the surface of polyvinyl chloride microtiter plates as compared to the wild type. Even though sdiA and AI-2 are not involved in swarming, Kim and Surette (2006) found that *S.* Typhimurium swarmer cells produced elevated levels of AI-2 and upregulated the expression of the *rck* operon in an SdiA-dependent manner in the presence of exogenous AHLs. The *lsr* operon was not expressed indicating that the enzymes encoded by the *rck* operon did not act on AI-2 (Kim and Surette, 2006). Quorum sensing does not appear to be involved in swarming in *S.* Typhimurium.

Flagellar phase variation is the process by which serovars of *S. enterica* express two different flagellar filament proteins, FljB and FliC (Bonifield and Hughes, 2003). Utilizing *S.* Typhimurium, Karavolos *et al.* (2008) demonstrated that FliC was predominant in a LuxS mutant whereas FljB was predominant in the parent wild type strain. Complementation of the mutant strain with the LuxS gene restored the parental phenotype; however, addition of parental cell-free supernatant to the mutant had no effect, suggesting that AI-2 is not involved in flagellar phase variation. Karavolos *et al.* (2008), unfortunately, did not add purified AI-2 or DPD to the mutant strain. In *S.* Typhimurium, *luxS* has a role in flagellar phase variation and triggers the expression of the less immunogenic FljB and represses the expression of the highly immunogenic FliC independently of AI-2 signaling. The expression of the less immunogenic flagellar protein may allow *S.* Typhimurium to evade an immune response when the intestine is invaded by salmonellae.

Prouty *et al.* (2002) demonstrated that a *S.* Typhimurium *luxS* mutant was unable to form a biofilm on the surface of gallstones; however, they did not add AI-2 to determine if biofilm formation was restored in the mutant. A *luxS* mutant of *S.* Typhimurium did not form a mature biofilm on polystyrene (De Keersmaecker *et al.*, 2005). However, biofilm formation similar to that of the wild-type strain resulted when the mutant was complemented by a functional *luxS* gene; however, biofilm formation was not restored with addition of DPD (De Keersmaecker *et al.*, 2005). Biofilm formation was similar in a AI-2-producing strain and an isogenic AI-2-negative strain of *S.* Typhimurium (Yoon and Sofos, 2008). Thus, the relationship between biofilm formation and the presence of an active LuxS system and AI-2 in *Salmonella* is not clear.

The QS response involving AI-3, epinephrine (epi), and norepinephrine (norepi) in *E. coli* is modulated by the QseBC two-component system, QseC acts as a histidine sensor kinase, and QseB is the response regulator that modulates motility of *E. coli* in the presence of AI-3, epi, and norepi (Walters and Sperandio, 2006b). Bearson and Bearson (2007) demonstrated that the motility genes of wild-type *S.* Typhimurium were upregulated in the presence of norepi, and motility was enhanced by 2- to 3-fold. The motility-enhancing activity of norepi was abolished by the addition of phentolamine, an adrenergic antagonist. The *flhDC* genes (encode for the master regulators for motility, FlhDC) are regulated by the QseBC system. While wild-type *S.* Typhimurium showed enhanced motility on exposure to norepi, a *qseC* mutant did not (Bearson and Bearson, 2007). The data indicated that the QseBC system is necessary for norepi enhancement of motility in *S.* Typhimurium. The *qseC* gene is important in the colonization of swine. There was decreased colonization of swine tonsils and ceca by the *qseC* mutant as compared to wild-type *S.* Thyphimurium (Bearson and Bearson, 2007). In addition, there was decreased fecal shedding of the mutant as compared to the wild type.

Malcova *et al.* (2008) isolated 94 *S.* Typhimurium strains from human, animal, and environment sources and determined their ability to form biofilms on polystyrene microtiter plates. The most efficient biofilm-forming strains were phage type DT104 *S.* Typhimurium carrying SGI1 (*Salmonella* genomic island-1; responsible for multiple drug resistance). Strains lacking SGI1 demonstrated decreased biofilm formation (Malcova *et al.* 2008). The SGI1 increased both antibiotic resistance and biofilm formation.

### 2.2.2 *Escherichia coli*

*Escherichia coli* is a facultative anaerobic Gram-negative rod-shaped bacterium; pathogenic strains are associated with food- and water-borne gastroenteritis or extraintestinal infections. Strains of *E. coli* form biofilms on the surfaces of glass, stainless steel, high density polyethylene, polyamide-6, polyvinyl chloride, and Teflon® coupons (Faille *et al.*, 2002). Other surfaces on which *E. coli* have been shown to form biofilms include glass wool (Ren *et al.*, 2004a), polystyrene microtiter plates, and glass coverslips (Hancock and Klemm, 2007). Uropathogenic strains produce biofilms inside bladder epithelial cells (Anderson *et al.*, 2003) and on silicone urinary catheters (Ferrières *et al.*, 2007; Jones *et al.*, 2006), while Ferrières *et al.* (2007) found that uropathogenic strains of *E. coli* produced very little biofilm on polystyrene plates and glass tubes. An important enterohemorrhagic strain, *E. coli* O157:H7, can form biofilms on stainless steel and buna S rubber coupons, polystyrene tissue culture plates, glass vials (Pawar *et al.*, 2005), polycarbonate membrane filters (Pérez-Conesa *et al.*, 2006), and glass slides (Niemira, 2007). Cooper *et al.* (2007) demonstrated that *E. coli* O157 formed biofilms on flint shingles (gravel-like stones) from freshwater streams located in agricultural areas.

The SdiA quorum sensing system which can detect AHLs is present in *E. coli*; however, the organism lacks the ability to produce AHLs (Ahmer, 2004; Walters and Sperandio, 2006b). It has been shown that the *Salmonella* SdiA, in the presence of AHLs, regulates the *rck* operon and the *srgE* gene, however, no *E. coli* genes have been shown to be regulated by a single chromosomal copy of *sdiA* (Ahmer, 2004; Walters and Sperandio, 2006b). A cloned *sdiA* gene on a multi-copy plasmid upregulated *ftsQAZ* (cell division genes) in *E. coli*; however, a *sdiA* mutant was not defective in cell division (Wang *et al.*, 1991). The expression of *sdiA* from a high-copy-number plasmid led to abnormal cell division, reduced adherence to epithelial cells, and reduced expression of proteins encoded by the locus of enterocyte effacement (LEE) pathogenicity island in *E. coli* O157:H7 (Kanamaru *et al.*, 2000). A *sdiA* mutant was not studied, and the effects seen when a high-copy-number plasmid was used may be artifacts introduced by the use of abnormally high levels of SdiA (Walters and Sperandio, 2006b).

The LuxS/AI-2 system is present in *E. coli* (Ahmer, 2004; Walters and Sperandio, 2006a). The *lsr* operon of *E. coli* functions in an analogous fashion to the *lsr* operon of *S.* Typhimurium in the uptake and further processing of AI-2 (Vendeville *et al.*, 2005: Xavier and Bassler, 2005). In *E. coli*, both the synthesis and uptake of AI-2 are influenced by sugar catabolite repression by the cyclic AMP receptor protein complex (CRP). In the presence of glucose, the level of the cAMP-CRP complex is low, and there is little transcription of the *lsr* operon leading to an increase in external AI-2 since LuxS is not repressed (Wang *et al.*, 2005). In the absence of glucose, the cAMP-CRP complex stimulates the expression of *lsr* with AI-2 uptake and phosphorylation and represses LuxS expression.

The AI-3/epi/norepi system regulates the expression of the locus of enterocyte effacement (LEE) genes and the flagella regulon in *E. coli* O157:H7 (Sperandio *et al.*, 2003; Walters and Sperandio, 2006a). In *E. coli*, a mutation in *luxS* leads to a decrease in AI-3 production. AI-3 is not synthesized by LuxS; however, AI-3 synthesis is affected by a mutation in *luxS* due to changes in cell metabolism (Walters *et al.*, 2006). Addition of purified AI-3 to the *E. coli* O157:H7 *luxS* mutant restored activation of LEE gene expression (Sperandio *et al.*, 2003). Epinephrine can substitute for AI-3 in activating the transcription of LEE genes (Sperandio *et al.*, 2003), and Walters and Sperandio (2006a) showed that there is a synergistic relationship between AI-3 and epi in activation of LEE gene expression. Thus, the combined effects of the bacterial signal, AI-3, and the host signals, epi/norepi, lead to increased LEE expression and to more efficient infection. Studies by Kendall *et al.* (2007) showed that a mutation in *LuxS* mainly affects central metabolic genes in both enterohemorrhagic and non-pathogenic strains of *E. coli;* however, addition of exogenous AI-2 (as DPD) did not completely restore gene expression in the *luxS*-deficient strains. Addition of epi or AI-3 but not AI-2 increased expression of the LEE regulon (Kendall *et al.*, 2007).

During infection of the gastrointestinal tract, *E. coli* O157:H7 will be exposed to the hormones, epi and norepi, as well as to indole. Bansal *et al.* (2007) found that exposure of the bacteria to the hormones led to migration of the cells toward the hormones whereas the bacteria migrated away from indole. Similarly, the hormones stimulated bacterial motility and biofilm formation, while indole inhibited motility and biofilm formation. There was several-fold increased attachment of *E. coli* O157:H7 to HeLa cells in the presence of epi and norepi but there was a 50% decrease in attachment in the presence of indole (Bansal *et al.*, 2007). In addition, Bansal *et al.* (2007) demonstrated that the *lsr* operon involved in uptake and further metabolism of AI-2 was down-regulated by the hormones whereas, indole had no effect. Thus, hormones such as epi and norepi and the bacterial product, indole, which are present in the intestine, may have a profound effect on an intestinal infection induced by *E. coli* O157:H7.

The LEE pathogenicity island genes of enteropathogenic and enterohemorrhagic *E. coli* are regulated via the AI-3/epi/norepi QS system (Sperandio *et al.*, 2003). The LEE genes encode a type 3 secretion system, intimin adhesin (encoded by *eae*), translocation intimin receptor (encoded by *tir*), proteins secreted by the type III secretion system (encoded by *espABD*), and an H-NS-like protein (Ler, encoded by *ler*) that activates the LEE genes (Sperandio *et al.*, 2002). The AI-3/epi/norepi QS system activates transcription of the regulator QseA (QS *E. coli* regulator A), which then activates the transcription of *ler* leading to the transcription of the LEE genes (Russell *et al.*, 2007; Sharp and Sperandio, 2007; Sircili *et al.*, 2004; Sperandio *et al.*, 2002).

The AI-3/epi/norepi QS system also controls the flagellar regulon through QseBC, which activates the transcription of *flhDC*, the master flagellar activator. QseB/QseC is a two-component system; QseC is the sensor kinase and QseB is the response regulator (Clarke *et al.*, 2006; Clarke and Sperandio, 2005). An enterohemorrhagic *E. coli* mutant unable to produce AI-3 (i.e., a *luxS* mutant) can activate flagella/motility genes if AI-3 or epi is added exogenously; however, these additions did not restore motility in a QseC mutant (Clarke *et al.*, 2006; Sperandio *et al.*, 2003).

Autophosphorylation of QseC occurs when QseC senses AI-3, epi, or norepi; AI-2 did not induce autophosphorylation. Phosphorylated QseC passes its phosphate to QseB, which activates the transcription of *flhDC* (Clarke *et al.*, 2006). Recently, in an excellent review, Hughes and Sperandio (2008) examined signaling between bacteria and their hosts.

Utilizing the rabbit enteropathogenic *E. coli* (REPEC) strain (a model for enterohemorrhagic *E. coli*), Clarke *et al.* (2006) showed that inoculation of rabbits with wild-type REPEC produced diarrhea in 6/7 rabbits whereas inoculation of rabbits with a *qseC* mutant produced diarrhea in 2/8 rabbits suggesting that QseC plays a role in bacterial pathogenesis.

The two-component regulatory system, QseEF, in *E. coli* O157:H7 is part of the AI-3/epi/norepi QS signaling system involved in the activation of the transcription of *espFu* to drive actin polymerization during attaching and effacing pedestal formation (Reading *et al.*, 2007). QseE is a sensor kinase, and QseF is the response regulator. Wild-type *E. coli* O157:H7 formed pedestals whereas a *qseF* mutant did not. Reading *et al.* (2007) demonstrated that QseE phosphorylated QseF. EspFu (encoded by *espFu*) is a non-LEE encoded effector protein required for pedestal formation in *E. coli* O157:H7. A *qseF* mutant did not transcribe *espFu*; however, transcription was at the level of the wild type when the mutant was complemented with an active copy of *qseF* (Reading *et al.*, 2007). The QseEF proteins regulate the remodeling of the infected host's actin to form enterocyte pedestals in *E. coli* O157:H7 via activation of *espFu*.

Mutations that lead to decreased extracellular and intracellular levels of indole in *E. coli* K-12 result in increased motility and biofilm formation indicating that indole is an inhibitor of biofilm formation (Domka *et al.*, 2006; Lee *et al.*, 2007b). Addition of indole to the wild type or to the mutants resulted in a decrease in motility and biofilm formation. In microarray studies, the addition of 600 μM indole induced *sdiA* almost 3-fold. Deletion of *sdiA* enhanced both motility and biofilm formation; therefore, SdiA is a repressor of motility and biofilm formation. Addition of indole to to a *sdiA* mutant had little effect on biofilm formation but did decrease motility (Lee *et al.*, 2007b). In wild-type *E. coli* K-12, addition of short-chain AHLs (in the presence of low levels of indole) led to the reduction in biofilm formation; however, the *sdiA* mutant did not respond to AHLs. Lee *et al.* (2007b) suggested that indole inhibits biofilm formation by binding to SdiA. Hydroxyindoles such as 5- and 7-hydroxyindoles, like indole, are potent inhibitors of biofilm formation in *E. coli* O157:H7 and K-12 but do not appear to work through *sdiA* (Lee *et al.*, 2007a). However, isatin, indole-2,3-dione, increased biofilm formation in *E. coli* O157:H7 but not K-12. Isatin inhibited the *lsr* operon but induced flagellar gene systems, *flg* and *fli* (Lee *et al.*, 2007a).

The addition of enzymatically synthesized AI-2 increased motility and stimulated the formation of biofilms in wild type *E. coli* K-12 strains. There was no stimulation of biofilm formation when a strain with a mutation in *lsrK* was treated with AI-2; LsrK is a kinase necessary for the uptake of external AI-2 (González Barrios *et al.*, 2006). The failure of the mutant to internalize the added AI-2 indicates that intracellular AI-2 is necessary for biofilm formation. Using a *luxS* mutant, González Barrios *et al.* (2006) found that the addition of enzymatically synthesized AI-2 to the mutant strain led to the induction of a number of motility genes including *qseB, flhD, fliA, fliC*, and *motA*. QseBC regulates the transcription of the master regulon *flhDC* required for the expression of the genes in the flagellar regulon, *fliA* encodes the sigma factor $\sigma^{28}$, *fliC* encodes flagellin, and *motA* encodes the proton exchange conductor for the movement of the flagellum. Deleting the *mqsR* gene led to a several-fold reduction in the transcription of *qseB*, decreased biofilm formation, and loss of motility (González Barrios *et al.*, 2006). González Barrios *et al.* (2006) suggested that AI-2 stimulates biofilm formation in *E. coli* by stimulating the expression of *mqsR*. MqsR then induces the expression of *qseBC*, and then QseBC promotes motility via the master motility regulon *flhDC* leading eventually to biofilm formation. However, Yoon and Sofos (2008) found that an AI-2 producing strain of *E. coli* O157:H7 and its isogenic non-AI-2-producing strain behaved similarly in terms of biofilm synthesis on solid surfaces.

There was increased swimming motility and biofilm formation when the *tqsA* gene of *E. coli* K-12 was deleted; complementation of the mutant with an active gene led to the wild type phenotype (Herzberg *et al.*, 2006). There was a decreased extracellular and increased intracellular level of AI-2 in the *tqsA* mutant indicating that TqsA inhibits the importation of extracellular AI-2 so that there is little AI-2 inside the cell. The results obtained by Herzberg *et al.* (2006) indicate that TqsA inhibits biofilm formation in *E. coli* by interfering with AI-2 import.

Curli fimbriae have important roles in biofilm formation in *E. coli* K-12. Flagellar motility is not necessary for biofilm formation in curli-producing strains. Curli have a role in initial attachment to an abiotic surface and cell-to-cell interactions essential to microcolony formation. Curli also form interbacterial bundles, permitting cohesive and stable association of cells, thereby stabilizing the three-dimensional biofilm structure (Prigent-Combaret *et al.*, 2000). The formation of a three-dimensional mature biofilm with characteristic water channels and bacterial pillars is dependent on the presence of curli (Kikuchi *et al.*, 2005). Using both curli-deficient and curli-producing strains of *E. coli* O157:H7, Ryu *et al.* (2004) found that both strains attached to stainless steel coupons but biofilms were formed only by the curli – producing strains. Curli play a role in biofilm formation on abiotic surfaces such as polystyrene, glass, stainless steel, rubber, or teflon by various strains of Shiga-toxin producing *E. coli* (Pawar *et al.*, 2005; Uhlich *et al.*, 2006).

The viscous capsular polysaccharide of *E. coli* K-12, colonic acid, does not appear necessary for bacterial attachment and further biofilm formation but does contribute to three-dimensional biofilm architecture and formation of voluminous biofilms (Danese *et al.*, 2000b; Prigent-Combaret *et al.*, 2000). The adhesin, Ag43 mediates the autoaggregration of cells in *E. coli* through Ag43–Ag43 interaction and it promotes biofilm formation by inducing microcolony formation (Danese *et al.*, 2000a). A non-organelle-type adhesin, AIDA is responsible for cell aggregation in diarrheagenic *E. coli* through AIDA-AIDA interaction and enhances biofilm formation on abiotic surfaces. AIDA can interact with Ag43 and thereby further enchances cellular aggregration (Sherlock *et al.*, 2004). TibA is a bacterial adhesion present in enterotoxigenic *E. coli*, which mediates attachment to human cells and promotes invasion of human cells (Sherlock *et al.*, 2005). TibA-TibA interaction induces cellular aggregation and enhances biofilm formation on abiotic surfaces (Sherlock *et al.*, 2005). Microcolony formation and cell aggregration in enteropathogenic *E. coli* is an important aspect of biofilm formation and is mediated by several adhesins including BFP (bundle-forming pilus) and EspA filament protein (Moreira *et al.*, 2006). The expression of *espA* is regulated through AI-3/epi/norepi QS and QseA.

The RNA binding global regulatory protein CsrA (carbon storage regulator) of *E. coli* K-12 acts as both a repressor of biofilm formation and as an activator of biofilm dispersal (Jackson *et al.*, 2002). Biofilm formation is affected by CsrA through regulation of intracellular glycogen biosynthesis and catabolism.

Biofilm formation is promoted in *E. coli* K-12 strains carrying conjugative plasmids. These plasmids promote biofilm formation by promoting cell aggregation (Gonzáles Barrios *et al.*, 2005; Reisner *et al.*, 2003; Van Houdt and Michiels, 2005). Flagella, type 1 fimbriae, Ag43 (surface protein adhesion), and curli are not required for conjugative plasmid-promoted biofilm formation (Reisner *et al.*, 2003). Hoffman *et al.* (2005) found that the addition of subinhibitory concentrations of aminoglycoside antibiotics such as tobramycin induced biofilm formation in *E. coli*, but the mechanism by which the antibiotic acts is not clear.

### 2.2.3 *Campylobacter jejuni*

*Campylobacter jejuni* is a curved, rod-shaped Gram-negative microaerophilic bacterium commonly associated with food-borne gastroenteritis. The production of AHLs has not been documented in *C. jejuni* (Smith *et al.*, 2004). The enzyme, LuxS is present in *C. jejuni* (Elvers and Park, 2002; Jeon *et al.*, 2005; Reeser *et al.*, 2007), and an AI-2-like product has been demonstrated (Cloak *et al.*, 2002; Elvers and Park, 2002; Jeon *et al.*, 2003); however, chemical characterization of the *C. jejuni* AI-2-like compound has not been done. The production of AI-2 activity in milk and chicken broth by both *C. jejuni* and *C. coli* has been demonstrated, and the *luxS* gene has been sequenced in both organisms (Cloak *et al.*, 2002).

*Campylobacter jejuni* exist as three types of monospecies biofilms in liquid culture: attached to glass or other surfaces, unattached aggregates or flocs, or pellicles at the liquid-gas interface (Joshua *et al.*, 2006). The bacterium produces biofilms on abiotic surfaces such as coupons of acryl-nitrile-butadiene-styrene plastic, polyvinyl chloride plastic, stainless steel, copper (Kalmokoff *et al.*, 2006; Reeser *et al.*, 2007), nitrocellulose membranes, glass fiber filters (Kalmokoff *et al.*, 2006), and glass beads (Dykes *et al.*, 2003). *Campylobacter* jejuni showed enhanced attachment and survival in preformed biofilms on polyvinyl chloride coupons colonized by chicken house isolates of Gram-positive bacteria or a *Pseudomonas* species. There was very little attachment of *C. jejuni* to coupons without preformed biofilms (Trachoo and Frank, 2002; Trachoo *et al.*, 2002). The presence of preformed biofilms on polyvinyl chloride coupons improved the attachment and survival of *C. jejuni* as compared to attachment to surfaces without preformed biofilms (Trachoo and Frank, 2002). When mixtures of *C. jejuni* and rinses from chicken carcasses were incubated in the presence of stainless steel coupons, Sanders *et al.* (2007) found that the surface covered by *C. jejuni* increased significantly as compared to *C. jejuni* incubated without the carcass rinses. Buswell *et al.* (1998) found that the survival of *C. jejuni* and *C. coli* in water was improved when they attached to glass coupons with preformed biofilms of autochthonous microflora from tapwater.

Biofilm formation by *C. jejuni* on solid surfaces was decreased approximately 3-fold with a change of atmosphere from 10% $CO_2$ to aerobic conditions and decreased approximately 13-fold with reduction of temperature from 37°C to 25°C (Reeser *et al.*, 2007). There was approximately 50% decrease of biofilm formation when the NaCl level increased from 0.00 M to 0.05 M or when the sucrose level increased from 0.0% to 0.5%. There was approximately 60% inhibition of biofilm formation on solid surfaces when the glucose level increased from 0.0% to 0.5% (Reeser *et al.*, 2007).

The *maf5* aflagellated non-motile mutant of *C. jejuni* does not form a pellicle nor attach to a solid matrix and the *fliS* aflagellated non-motile mutant has a reduced ability to form a pellicle and does not attach to a solid suface (Joshua *et al.*, 2006). The aflagellated non-motile *flhA* mutant does not attach to a solid surface and does not form a pellicle (Kalmokoff *et al.*, 2006). Thus, flagella and flagellar motility plays a role in the initial attachment of *C. jejuni* to solid surfaces during biofilm formation, which is necessary for the cell-to-cell interaction involved in pellicle formation.

While *C. jejuni* has adjacent and similar flagellin genes, *flaA* and *flaB*, the *flaA* gene is mainly responsible for flagellin synthesis and for motility (Nuijten *et al.*, 1990). A null mutation in the *luxS* gene of *C. jejuni* abolished approximately 43% of the transcription of *flaA* (Jeon *et al.*, 2003). The *luxS* mutant was unable to produce AI-2; however, it is not clear that the lack of AI-2 is involved in the decrease of *flaA* transcription since Nuijten and his co-workers did not add exogenous AI-2 to the *luxS* mutant. A *luxS* mutant of *C. jejuni* M129 showed a 3.2-fold decrease in biofilm formation as compared to the wild type and there was a 2.5-fold decrease in biofilm formation by a *flaAB* mutant of a *luxS*-positive of M129 (Reeser *et al.*, 2007). The addition of filtered bacterial culture (24 h) supernatant fluids of *C. jejuni* M129, *Pseudomonas aeruginosa, P. fluorescens* or *Arcanobacterium pyogenes* to the *C. jejuni luxS* mutant led to an increase in biofilm development, whereas the addition of culture supernatants from *Clostridium perfringens* or *Chromobacterium violaceum* had no effect on biofilm formation. Unfortunately, Reeser *et al.* (2007) did not show that these culture supernatants actually had AI-2-like activity; therefore, it is not clear why the culture supernatants stimulated biofilm formation in the *luxS* mutant. In a comparison between an isogenic *luxS* mutant and an AI-2-producing wild-type *C. jejuni,* He *et al.* (2008) found that growth rate and swarming behavior were reduced in the mutant at 37°C but growth rate and swarming were similar in both strains at 42°C. At 42°C, the *luxS* mutant was more sensitive to $H_2O_2$ and cumene hydroperoxide as compared to the wild type; this sensitivity suggests that the mutant is less able to cope with stress (He *et al.*, 2008). The *luxS* mutation data obtained with *C. jejuni* indicate that the LuxS enzyme plays a role in biofilm formation, rate of growth, swarming, and stress but it is not clear that AI-2 has a role.

### 2.2.4 *Listeria monocytogenes*

*Listeria monocytogenes* is a facultative intracellular pathogenic Gram-positive rod-shaped bacterium that can grow at low temperatures (2 to 3°C), and it causes food-borne illness. *Listeria monocytogenes* has a LuxS/AI-2 system. The maximal amount of AI-2 is produced during the mid-exponential phase of growth, and the level decreases during the stationary phase. A *luxS* mutant did not show AI-2 activity (Challan Belval *et al.*, 2006; Sela *et al.*, 2006). Little is known about possible peptide QS compounds (common in Gram-positive bacteria) in *L. monocytogenes*. The virulence genes of *L. monocytogenes* are controlled by the transcriptional regulator PrfA. Ermolaeva *et al.* (2004) have suggested that a 'peptide' mediates a quorum sensing mechanism which restricts the expression of the PrfA virulence regulon. Recent data indicate that *L. monocytogenes* has an *agr* system. There are 4 genes, *agrB, agrD, agrC* and *agrD,* in the *agr* locus organized as an operon (Rieu *et al.*, 2007). Thus, limited data indicates that QS in *L. monocytogenes* is mediated by a peptide.

Biofilm formation by *L, monocytogenes* has been demonstrated on glass fiber filters (Lebert *et al.*, 2007), polyvinyl chloride microtiter plates (Lemon *et al.*, 2007), conveyer belt materials (polypropylene, acetal, stainless steel) (Tolvanen *et al.*, 2007), stainless steel, polyethylene, and teflon coupons (Pan *et al.*, 2006; Rodríguez and McLandsborough, 2007), glass slides (Chae *et al.*, 2006), and floor drains of food processing facilities (Zhao *et al.*, 2004). Lemon *et al.* (2007) demonstrated the importance of motility for biofilm formation on abiotic surfaces (polyvinyl chloride, polystyrene, polypropylene, glass, and stainless steel) in *L. monocytogenes.* At ≤30°C, wild-type *L.*

*monocytogenes* was motile and formed peritrichious flagella; however, at 37°C, the cells were not motile, and very few flagella were present. A *flaA* mutant (flagella-minus) and a paralyzed flagellar mutant were non-motile at all temperatures whereas the wild type and a non-glycosylated flagellar mutant (due to a mutation in glycosyltransferease responsible for flagellin glycosylation) were motile at temperatures ≤30°C (Lemon *et al.*, 2007). The *flaA* and paralyzed flagellar mutants did not form biofilms at 30°C, whereas the wild type and non-glycosylated flagellar mutants did form biofilms on the various abiotic surfaces. Flagella-mediated motility is necessary for early attachment to abiotic surfaces and for biofilm formation (Lemon *et al.*, 2007).

Of interest to food microbiologists, Rodríguez and McLandborough (2007) studied the transfer of *L. monocytogenes* present as attached cells (formed by 5 min contact time) or biofilms (formed by 48 h contact time) on stainless steel or polyethylene slides onto bologna or American cheese slices. More bacteria were transferred to food from stainless steel than from polyethylene. There was a greater transfer of *L. monocytogenes* to foods from biofilms than from cells attached to a surface, and there was a better transfer of bacterial cells to bologna than to cheese. Interestingly, there was more transfer of the pathogen to foods from a dried biofilm than from a moist biofilm (Rodríguez and McLandborough, 2007). Further work by Rodríguez *et al.*, (2007) demonstrated that biofilms dried on the surface of stainless steel transferred to hard salami and bologna better than from a moist biofilm and that more cells from the dried biofilm transferred to bologna than to hard salami.

*agrA* or *agrD* deletion mutants of *L. monocytogenes* demonstrated an approximately 62% decrease in the number of cells attached to glass slides as compared to the wild type (Rieu *et al.*, 2007). In addition, the amount of biofilm formed on polystyrene during the first 24 h was decreased by approximately 33% in the deletion mutants as compared to the wild type; however, at 48 and 72 h, the amount of biofilm formed by the mutants and wild type was approximately the same. Rieu *et al.*, (2007) suggested that the early delay in biofilm formation was due to decreased adhesion of the cells to the plastic surface.

The presence of the *luxS* gene in *L. monocytogenes* represses its attachment to polystyrene and the bacteria's ability to form biofilm (Sela *et al.*, 2006). A mutation in *luxS* derepressed biofilm formation leading to a 4-fold thicker biofilm than in the wild type. Addition of wild type AI-2 positive culture supernatants did not induce repression of biofilm formation in the *luxS* mutant (Sela *et al.*, 2006). Challan Belval *et al.* (2006) also demonstrated that a *luxS* mutant of *L. monocytogenes* produced thicker biofilms than the wild type and that the addition of in vitro synthesized AI-2 to cultures of the mutant did not repress biofilm formation. There was an almost 2-fold increase in the levels of the activated methyl cycle intermediates, S-adenosyl homocysteine (SAH) and S-ribosyl homocysteine (SRH), accumulated by the *luxS* mutant as compared to the wild-type *L. monocytogenes* (Challan Belval *et al.*, 2006). The increased accumulation would be expected since LuxS is absent in the mutant. The results obtained by Sela *et al.* (2006) and Challan Belval *et al.*, (2006) do not indicate a quorum sensing role for AI-2 in *L. monocytogenes* biofilm formation.

### 2.2.5 *Yersinia pestis, Y. pseudotuberculosis,* and *Y. enterocolitica*

The yersiniae are facultative anaerobic Gram-negative short rods. *Yersinia pestis* is the cause of plague, whereas *Y. enterocolitica* and *Y. pseudotuberculosis* cause enterocolitis and may be associated with food-borne illness. *Yersinia pestis* and *Y. pseudotuberculosis* strains form biofilms on glass or polystyrene (Joshua *et al.*, 2003; Patel *et al.*, 2006), as well as on the head of the nematode, *Caenorhabditis elegans* (Joshua *et al.*, 2003; Tan and Darby, 2004). Flea infection by *Y. pestis* is also manifested as a biofilm (Jarrett *et al.*, 2004). Camper *et al.* (1985) demonstrated that *Y. enterocolitica* formed biofilms on granular activated carbon columns but reports of biofilm formation on other abiotic surfaces by *Y. enterocolitica* were not found. Biofilms were not produced by *Y. enterocolitica* when the *Caenorhabditis elegans* model was utilized (Joshua *et al.*, 2003).

There are two *luxI/luxR*-like genes in *Y. pestis*, *yspI/yspR* and *ypeI/ypeR* (Kirwan *et al.*, 2006). The gene *yspI* encodes the AHL synthase YspI which synthesizes mainly 3-oxo-octanoyl-L-homoserine lactone (3-oxo-$C_8$-HSL) and N-3-oxo-hexanoyl-L-homoserine lactone (3-oxo-$C_6$-HSL) in an approximately 1:1 ratio (Kirwan *et al.*, 2006). The *ypeI/ypeR* system has not been characterized. Similarly, *Y. pseudotuberculosis* has two LuxI synthase genes: *ypsI* and *ytbI* (Ortori *et al.*, 2007). Wild-type *Y. pseudotuberculosis* produced at least 24 different AHLs. However, the most abundant AHLs were the 3-oxo-substituted $C_6$, $C_7$, and $C_8$ compounds and the unsubstituted $C_6$ and $C_8$ compounds (Ortori *et al.*, 2007). Studies with mutants indicated that YpsI is responsible for the synthesis of 3-oxo-$C_6$-HSL, 3-oxo-$C_7$-HSL and $C_6$-HSL; Ytb1 is responsible for 3-oxo-$C_8$-HSL and $C_8$-HSL production. A double mutant did not produce AHLs (Ortori *et al.*, 2007). In *Y. enterocolitica*, the synthesis of the quorum sensing compounds, 3-oxo-$C_6$-HSL and $C_6$-HSL, are directed by YenI; the two compounds are found in an approximately 1:1 ratio (Atkinson *et al.*, 2006). In addition, YenI directed the synthesis of smaller amounts of 3-oxo-$C_{10}$-HSL, 3-oxo-$C_{12}$-HSL, and 3-oxo-$C_{14}$-HSL. Medina-Martinez *et al.* (2006, 2007) demonstrated the production of AHLs in liquid beef, fish, and pork extracts and milk by *Y. enterocolitica*; however, there was no production of AHLs in liquid extracts of mixed lettuce, cucumber, or soy bean. Jarrett *et al.* (2004) demonstrated the presence of the *luxS* gene in *Y. pestis* and the gene is probably present in *Y. pseudotuberculosis* and *Y. enterocolitica*. Walters and Sperandio (2006b) stated that the AI-3/epi/norepi signaling system is also present in *Yersinia* species.

Biofilm formation in fleas due to *Y. pestis* infection requires the bacterial *hms* genes involved in hemin storage (Jarrett *et al.*, 2004); these genes are also necessary for the formation of biofilms on glass and for the production of the extracellular matrix (Jarrett *et al.*, 2004; Kirillina *et al.*, 2004). When Jarrett *et al.*, (2004) tested a triple mutant (inactive *yspI, ypeI,* and *luxS* genes) of *Y. pestis*, they found that the mutant produced biofilms in vitro as well as the wild type strain. The extent of biofilm formation by *Y. pestis* depended on the level of the polyamine, putrescine (Patel *et al.*, 2006). It is not known if putrescine synthesis is dependent on QS.

Swarming motility is a type of translocation in bacteria which is a flagella-dependent movement in the presence of extracellular slime, allowing bacterial spreading over a surface. It is distinct from swimming motility. Swarming has been implicated in the formation of biofilms (Daniels *et al.*, 2004; Harshey, 2003). Flagellin production, swimming motility and swarming motility are under the control of the motility master regulatory operon, *flhDC* in *Y. enterocolitica* (Young *et al.*, 1999). Mutations within the motility master regulatory locus led to complete loss of flagellin protein production with concomitant loss of swimming and swarming activities. Atkinson *et al.* (2006) demonstrated that a mutation in *yenI* down-regulated *flhDC*; there was a delay in the initiation of swimming motility; however, swarming motility was abolished completely. The synthase, YenI, produces both short chain AHLs (3-oxo-$C_6$-HSL and $C_6$-HSL) and long-chain AHLs (3-oxo-$C_{10}$-HSL, 3-oxo-$C_{12}$-HSL, and 3-oxo-$C_{14}$-HSL). The addition of these AHLs to a *yenI* mutant did not restore swimming or swarming motility (Atkinson *et al.*, 2006). The failure of the exogenous addition of AHLs to restore swimming and swarming motility was not due to failure of *Y. enterocolitica* to take up the AHLs. However, swimming and swarming motility were restored to the parental level when a *yenI* mutant was complemented with a plasmid-borne copy of *yenI* (Atkinson *et al.*, 2006). While *yenI* is necessary for swimming and swarming and probably biofilm formation in *Y. enterocolitica*, the role of AHLs is not clear.

### 2.2.6 *Staphylococcus aureus* and *S. epidermidis*

*Staphylococcus aureus* and *S. epidermidis* are facultative anaerobic Gram-positive cocci that cause a wide variety of infections. *Staphylococcus aureus* has been associated with food-borne illness. *Staphylococcus aureus* strains form biofilms on polystyrene and glass microtiter plates (Gross *et al.*, 2001). In addition, biofilms are formed on Teflon catheters and other medical devices (Beenken *et al.*, 2004; Götz, 2002). Human and animal infections are associated with biofilm formation by *S. aureus* and include endocarditis, osteomyelitis, skin infections, as well as other types of infections (Yarwood and Schlievert, 2003). Biofilm-forming strains of *S. aureus* are also involved in bovine mastitis (Melchior *et al.*, 2006; Vasudevan *et al.*, 2003). Intravascular catheters removed from patients have been shown to be positive for biofilms of *S. aureus* (Ekkelenkamp *et al.*, 2008). *Staphylococcus epidermidis* strains formed biofilms in coverglass cell-culture chambers (Qin *et al.*, 2007), on polystyrene microtiter plates (Vuong *et al.*, 2003), and on tygon® silicone tubing inserted into rabbits (Vuong *et al.*, 2004). Biofilms were formed by *S. epidermidis* on platelet aggregates and on the inner surfaces of polyvinyl chloride platelet storage bags under simulated platelet storage conditions (Greco *et al.*, 2007). Both *S. aureus* and *S. epidermidis* form biofilms on various medical devices (Davey and O'Toole, 2000). Biofilms of *S. aureus* and *S. epidermidis* consist of different physiological cell types: cells growing aerobically, cells growing fermentatively, viable dormant cells (approximately 80% of the biofilm), and dead cells (approximately 10%) (Rani *et al.*, 2007). Staphylococcal biofilm formation consists of two stages: attachment of the cells to a surface followed by an accumulation stage involving cell multiplication and formation of a mature, multi-layered, structured biofilm (Yarwood and Schlievert, 2003).

In *S. aureus* and *S. epidermidis*, QS is based on the secretion of a short peptide attached to a five-membered thiolactone ring (autoinducing peptide, AIP) encoded by the *agr* (accessory gene regulator) locus. The secreted AIP attaches to the AIP cell surface receptor and initiates a series of phosphorylation reactions leading to the activation of the transcription regulator, RNA-III. Activated RNA-III then initiates transcription of the targeted genes (Otto, 2001, 2004). Expression of *agr* is important in controlling virulence in staphylocci and is involved in the the expression of several secreted virulence factors (Kong *et al.*, 2006; Yarwood and Schlievert, 2003). *S. aureus* strains belong to one of four AIP subgroups; the AIP of one subgroup inhibits expression of the *agr* regulon of the other subgroups but does not inhibit growth. Each subgroup AIP activates the virulence response in strains of its own subgroup and inhibits the *agr*-mediated virulence response in other AIP subgroups (Otto, 2001; Smith *et al.*, 2004).

The LuxS/AI-2 system is present in *S. aureus* (Doherty *et al.*, 2006; Kong *et al.*, 2006) and *S. epidermidis* (Kong *et al.*, 2006; Xu *et al.*, 2006). AI-2 accumulates during the logarithmic phase of growth in *S. epidermidis* and decreases during the stationary phase. A *luxS* mutant of *S. epidermidis* did not produce AI-2; however, when complemented with a vector containing *luxS*, the mutant regained almost 70% of its ability to produce AI-2 (Xu *et al.*, 2006). AI-2 production by *S. aureus* occurred under both aerobic and anaerobic conditions; however, aerobic production of AI-2 was 2 to 3 times greater than under anaerobic conditions (Doherty *et al.*, 2006). AI-2 decreased during the stationary stage of growth in the absence of glucose, but there was little or no decrease when glucose was present. Complementation of the *luxS* mutant with a plasmid *luxS* gene restored AI-2 production. Doherty *et al.* (2006) determined that inactivation of the *luxS* gene in *S. aureus* had no effect on the *agr*-dependent AIP signaling system and vice versa by studying the effect of AI-2 and AIP on the *S. aureus agr* P3 and *luxS* reporter constructs. The addition of AI-2 had no effect on transcription from either the *luxS* or *agr* P3 promoters, whereas, the addition of AIP stimulated the *agr* P3 promoter but not the *luxS* promoter.

The formation of biofilms is considered to be the main virulence factor in *S. epidermidis* (Xu *et al.*, 2006). Using plastic microtiter plates, Xu *et al.* (2006) demonstrated that a *luxS* mutant produced biofilms with a biomass 1.7-times that of the wild type. The biofilm produced by the mutant was more compact and thicker than that of the wild type. Addition of cell-free supernatants from exponential-growth phase cultures of wild-type *S. epidermidis* to the *luxS* mutant resulted in a decrease in biofilm formation with no decrease in staphylococcal growth (Xu *et al.* 2006). However, the exogenous addition of synthetic AI-2 was not done. There was an infection rate in a rat model of 61.5 ± 7.8% with the *luxS* mutant as compared to a rate of 27.5 ± 2.8% for the wild type, thereby indicating that the mutant of *S. epidermidis* was more virulent than the wild type (Xu *et al.*, 2006). Production of the major biofilm exopolysaccharide PIA (polysaccharide intercellular adhesin) produced by *S. epidermidis* is controlled by products of the *ica* (intercellular adhesion) operon. There was an approximately 4-fold increase in *ica* transcription and an approximately 3-fold increase in PIA production in the *luxS* mutant as compared to the wild type (Xu *et al.*, 2006). Xu *et al.* (2006) stated that LuxS-dependent gene regulation has a negative impact on biofilm formation, production of PIA, and virulence. However, in a later publication, these investigators stated that the results obtained by Xu *et al.* (2006) may not be due to AI-2 acting as a QS compound but may be due to LuxS acting in the activated methyl cycle (Kong *et al.*, 2006).

Comparison of gene expression in a *S. epidermidis luxS* mutant and in the mutant supplemented with synthesized AI-2 indicated that AI-2 controlled genes involved in the metabolism of sugars, nucleotides, amino acids and nitrate (Li *et al.*, 2008). In addition, AI-2 regulated genes encoded potential virulence factors such as lipase, phenol-soluble modulins (pro-inflammatory peptides), and the bacterial apoptosis protein, LrgB. The data obtained by Li *et al.* (2008) suggest that AI-2 acts as a QS compound in *S. epidermidis*.

Doherty *et al.* (2006) found that a mutation in the *luxS* gene of *S. aureus* did not affect virulence-associated traits such as the production of hemolysins, extracellullar proteases, biofilms (on polystyrene microtiter plates), or the *agr* signaling system. They stated that the gene had a role in metabolism of the cell but was not involved in QS.

An *agr*-negative mutant of *S. epidermidis* demonstrated enhanced biofilm formation on a polystyrene surface as compared to the wild type (Vuong *et al.*, 2003). The *agr* mutant (the entire agr system was deleted) produced high levels of the autolysin, AtlE, which is necessary for primary attachment of the cells to a polystyrene surface (Vuong *et al.*, 2003). The mutant also produced no δ-toxin, which has detergent-like activity resulting in decreased attachment of *S. epidermidis* cells during later stages of biofilm formation. The expression of the δ-toxin peptide is encoded within the part of the *agr* locus which codes for RNAIII and therefore, its expression is linked to *agr* activity. Thus, deletion of *agr* results in elimination of δ-toxin formation (Vuong *et al.*, 2003). Addition of exogenous purified δ-toxin decreased biofilm formation in *S. epidermidis* (Vuong *et al.*, 2000). The main determinant of the second stage of staphylococcal biofilm formation, the accumulation stage, is the exopolysaccharide, PIA, which is encoded by the *ica* locus. The synthesis of PIA was similar in both the wild type and *agr* mutant; therefore, transcription of the *ica* genes is not regulated by *agr* in *S. epidermidis* (Vuong *et al.*, 2003).

The ClpP protease is necessary for attachment to polystyrene and biofilm formation in *S. epidermidis*. An isogenic *clpP* mutant showed approximately 22-fold less biofilm formation compared to the wild-type strain; biofilm formation was restored by addition of a plasmid containing the *clpP* gene (Wang *et al.*, 2007). The level of *clpP* expression was decreased approximately 2-fold in the wild-type *S. epidermidis* as compared to an *agr* mutant. Thus, *agr* down-regulates the expression of the *clpP* gene (Wang *et al.*, 2007). While the ClpP proteins of *S. epidermidis* and *S. aureus* are similar (amino acid identity of 98%) (Wang *et al.*, 2007), a *clpP*-deficient mutant of *S. aureus* demonstrated enhanced biofilm formation (Frees *et al.*, 2004) in contrast to the biofilm formation behavior of a *clpP*-deficient mutant of *S. epidermidis*. It is not clear if *agr* regulates *clpP* in *S. aureus*.

If the *ica* locus is deleted in *S. aureus* or *S. epidermidis*, biofilm formation and PIA synthesis are abolished (Crampton *et al.*, 1999; Heilmann *et al.*, 1996). Complementation of the knockout mutants with the wild-type *ica* locus led to restoration of the ability to form biofilms and to produce PIA. While the *ica* locus is necessary for biofilm formation in *S. aureus* and *S. epidermidis*, not all *ica*-harboring strains form biofilms (Crampton *et al.*, 1999; Grinholc *et al.*, 2007; Heilmann *et al.*, 1996).

Vuong *et al.* (2000) studied 100 *S. aureus* strains and found that 78% of strains lacking the *agr* locus (21/27) produced biofilms on polystyrene whereas only 6% (5/73) of $agr^+$ strains produced biofilms. Addition of *S. epidermidis* AIP to the *agr* mutant of *S. aureus* led to a 5-fold decrease in biofilm formation. AtlE and PIA are necessary for biofilm formation in staphylococci and Vuong *et al.* (2000) demonstrated that these activities were not under the control of *agr* in *S. aureus*. The addition of δ-toxin to the *agr* mutant of *S. aureus* decreased biofilm formation (Vuong *et al.*, 2000). Since the gene for δ-toxin synthesis is located in the *agr* region, δ-toxin is not produced by the *agr* mutant.

The hemolytic cytotoxin, α-toxin, encoded by the *hla* gene, is necessary for biofilm formation on polystyrene (Caiazza and O'Toole, 2003) and for teflon device-related infections by *S. aureus* (Goerke *et al.*, 2001). Caizza and O'Toole (2003) found that there was an approximately 5-fold decrease in biofilm formation on polystyrene in a *hla* mutant as compared to the wild type. Under in vitro conditions, synthesis of the α-toxin is regulated by three loci: *agr, sarA*, and *sae* (Goerke *et al.*, 2001). However, under in vivo conditions (during teflon device initiated infection of guinea pigs), mutations in *agr* or *sarA* had no effect on the expression of *hla;* whereas, a mutation in the *sae* gene resulted in downregulation of *hla* expression both in vivo as well as *in vitro* (Goerke *et al.*, 2001).

Turning off the *agr* system to limit staphylococcal virulence expression may not be a good idea especially in terms of biofilm formation both *in vitro* and *in vivo*. In *S. epidermidis*, *agr* mutants produce increased amounts of AtlE and decreased levels of δ-toxin, which leads to increased biofilm formation. Mutations in *agr* eliminate δ-toxin production in *S. aureus*. The toxin has been shown to inhibit biofilm formation and therefore, an *agr* mutant demonstrates increased biofilm formation as compared to the wild type.

### 2.2.7 *Bacillus cereus*

*Bacillus cereus* is a rod-shaped facultative aerobic Gram-positive endospore-forming bacterium. It has been associated with food-borne illness. Biofilm formation has been demonstrated in *B. cereus*. Surfaces that support biofilm formation include glass wool (Lindsay *et al.*, 2006; Oosthuizen *et al.*, 2002), stainless steel (Kreske *et al.*, 2006; Peng *et al.*, 2002), and polystyrene microtiter plates (Hsueh *et al.*, 2006) as well as glass, high density polyethylene, polyamide-6, polyvinyl chloride, and teflon (Faille *et al.*, 2002). In *B.* cereus, biofilm formation is most efficient at the air/liquid interface; less biofilm formation is seen in submerged systems (Wijman *et al.*, 2007). Intravenous catheters colonized with *B. cereus* led to bacteremia in patients (Chiu *et al.*, 1996; Hernaiz *et al.*, 2003; Koch and Arvand, 2005); thus, medical devices can act as sites for biofilm formation by *B. cereus*. Sporulation of *B. cereus* takes place in biofilms and the subsequent release of spores in a food processing environment can lead to contamination of food and thereby affect food quality and safety (Lindsay *et al.*, 2006; Wijman *et al.*, 2007).

The PlcR-PapR QS system controls the expression of extracellular virulence factors in various species in the *B. cereus* group. The *plcR* gene encodes a transcriptional regulator, and the *papR* gene encodes a cell-cell signaling peptide (Slamti and Lereclus, 2002, 2005). The PapR polypeptide is secreted outside the cell and then is reimported and processed to a pentapeptide, which activates the PlcR regulon, allowing PlcR to bind to its DNA target promoters (Slamti and Lereclus, 2002, 2005). A LuxS/AI-2 system is also present in *B. cereus* (Auger *et al.*, 2006).

In comparison to its isogenic wild-type parent, a *plcR* mutant of *B. cereus* produced approximately 4-times more biofilm on polystyrene (Hsueh *et al.*, 2006). The mutant also produced large quantities of a biosurfacant, necessary for biofilm formation, in comparison to the wild type. Precoating of polystyrene plates with surfactin (a biosurfacant produced by *B. subtilis*) improved biofilm formation by the wild-type strain of *B. cereus* (Hsueh *et al.*, 2006).

Addition of culture supernatants from *B. cereus* to a Δ*luxS* mutant of *Photohabdus luminescens* restored luminescence to the organism, thereby indicating that the AI-2 produced by *B. cereus* was active (Auger *et al.*, 2006). Exogenous AI-2 had an inhibitory effect on *B. cereus* biofilm formation in a dose-dependent manner. At 24 h, an approximate 12-fold decrease in biofilm formation was induced by 3 μM AI-2 (Auger *et al.*, 2006). Addition of AI-2 to 24 h preformed biofilms decreased biofilm density by releasing cells from the biofilm. Thus, AI-2 inhibits biofilm formationby *B. cereus* and promotes the release of cells from a preformed biofilm. The precursor of AI-2, DPD and its bis-(O)-acetylated derivative, also inhibited biofilm formation by *B. cereus* (Frezza *et al.*, 2007).

*lsrACDB*-like genes are present in *B. cereus,* as well as a *lsrR*-like gene encoding the regulator of the *lsr* operon and *lsrK* and *lsrF*-like genes encoding products, which process internalized AI-2 (Auger *et al.*, 2006). Thus, similar to *S.* Typhimurium and *E. coli*, *B. cereus* has a LsR-like system responsible for AI-2 uptake and processing.

Several species of *Bacillus*, including *B. cereus*, have the interesting property of being able to deactivate AHLs by the action of AHL lactonase (Dong and Zhang, 2005; Roche *et al.*, 2004). The lactonase, AiiA, is encoded by the *aiiA* gene, and the enzyme acts by hydrolyzing the ester bond of the lactone ring with the production of an acylhomoserine. There were little or no AHLs detected when a co-culture of AHL-producing *Y. enterocolitica* and *aiiA*-positive *B. cereus* were grown together in liquid medium or in pork extract as compared to control cultures containing only *Y. enterocolitica* (Medina-Martínez *et al.*, 2007). There was no loss of AHLs in co-cultures with *Y. enterocolitica* and an *aiiA*-negative *B. cereus* strain.

### 2.2.8 *Clostridium difficile, C. perfringens,* and *C. botulinum*

The *Clostridia* are anaerobic spore-forming rod-shaped bacteria associated with disease. Both *C. botulinum* and *C. perfringens* are important food-borne pathogens, and *C. difficile* induces pseudomembranous colitis. The LuxS/AI-2 system is present in *C. perfringens, C. botulinum,* and *C. difficile* (Lee and Song, 2005; Ohtani *et al.*, 2002; Zhao *et al.*, 2006). The production of AI-2 was maximal during the mid-to-late exponential phase of growth for both *C. perfringens* and *C. difficile,* and the level of AI-2 declined during the stationary phase (Lee and Song, 2005; Ohtani *et al.*, 2002). Ohtani *et al.* (2002) demonstrated that a *luxS* mutant did not produce AI-2. Do these *Clostridium* species have a peptide-mediated QS similar to that seen in *Bacillus* or *Staphylococcus*? There is no mention of peptide-mediated QS in the published literature. There appears to be a deficiency of literature concerning biofilm formation in these pathogenic species of *Clostridium*; however, Leung *et al.* (2000) reported the presence of *C. perfringens* in biofilms involved in biliary stent blockage.

A *luxS* mutant of *C. perfringens* produced reduced levels of alpha-, kappa-, and theta-toxins. The addition of culture supernatants from wild-type *C. perfringens* to the mutant stimulated the production of the toxins and activated the expression of the *pfoA* gene (encodes theta-toxin) (Ohtani *et al.*, 2002). The transcript levels of the *tcdA* (encodes toxin A) and *tcdB* (encodes toxin B) genes were upregulated in *C. difficile* when growth medium was supplemented with *C. difficile* culture supernatant (containing AI-2) but did not increase the production of toxin A (Lee and Song, 2005). In contrast to Lee and Song (2005), Carter *et al.* (2005), using a similar system, showed that the addition of culture supernatant (containing AI-2) or the AI-2 inhibitor, 4-hydroxy-5-methyl-3(2H) furanone, had no effect on the production of A or B toxins in *C. difficile*. The results obtained by Carter *et al.* (2005) and Lee and Song (2005) are not clear since *luxS* mutants were not used. The results obtained with the *Clostridium* species may not really represent QS but may be dependent on the role of LuxS in the activated methyl cycle.

### 2.2.9 *Vibrio cholerae, V. prahemolyticus* and *V. vulnificus*

Members of the genus *Vibrio* are faculatative anaerobic Gram-negative curved rod-shaped bacteria. *Vibrio cholerae* is the cause of cholera, *V. parahaemolyticus* is associated with food-borne illness, and *V. vulnificus* has been associated with enteritis, bacteremia, wound infections, and death in immunocompromised individuals.

*Vibrio cholerae*

Cholerae autoinducer-1, CAI-1, is synthesized by the CqsA synthase in *V. cholerae*. The autoinducer has recently been identified as (S)-3-hydroxytridecan-4-one (Higgins *et al.*, 2007). The sensor (receptor) for CAI-1 is CqsS (Miller *et al.*, 2002). The LuxS/AI-2 system is also present in the organism. LuxP in conjunction with LuxQ detects AI-2. Information from these two signals, CAI-1 and AI-2 is conducted through LuxU to LuxO (Miller *et al.*, 2002). Miller *et al.* (2002) suggested that there is a third sensory circuit, 'System 3', which acts in parallel to systems 1 and 2.

In *V. cholerae* El Tor, QS is controlled by the two central regulators, LuxO and HapR. When the cell density and the level of autoinducers, CAI-1 and AI-2, are low, CqsS and LuxQ act as kinases and transfer phosphate to LuxU. LuxU then transfers the phosphate to Lux O. Phosphorylated LuxO together with $\sigma^{54}$ leads to the transcription of small regulatory RNAs that hybridize with and inhibits the translation of the mRNA encoding the QS master regulator, HapR. Since the concentration of the HapR protein is low at low cell density, target genes that are normally repressed by HapR are expressed (Lenz and Bassler, 2007; Miller *et al.*, 2002; Vendeville *et al.*, 2005). These expressed genes include *aphA*, whose gene products activate virulence regulons (Lin *et al.*, 2007) and the *vps* (*Vibrio* polysaccharide synthesis) operon, which is involved in biofilm formation (Hammer and Bassler, 2003; Zhu and Mekalanos, 2003). Genes that require HapR for activation are not expressed such as *hapA*, which encodes the hemagglutinin/protease (HA/protease) necessary for the detachment of *V. cholerae* from intestinal epithelium (Finkelstein *et al.*, 1992; Joelsson *et al.*, 2006). At high cell concentrations and high autoinducer concentrations, the autoinducers bind to their respective sensors, CqsS and LuxQ, and the sensors change from kinases to phosphatases. LuxO is dephosphorylated, its activity decreases, and *hapR* expression is enhanced leading to production of the protein HapR. Target genes that require HapR for activation (for example, *hapA*) are expressed; however, genes that are repressed by HapR (*aphA, vps*) are not expressed (Lenz and Bassler, 2007; Miller *et al.*, 2002). Thus, in terms of biofilm formation, at low levels of AI-2 and CAI-1, HapR is repressed and biofilm formation is enhanced whereas, at high levels of autoinducers, biofilm formation is reduced. High levels of 3′,5′cyclic diguanylic acid (c-di-GMP) enhance biofilm formation in *V. cholerae* whereas low levels repress biofilm formation. HapR controls the transcription of genes encoding proteins that synthesize and degrade c-di-GMP, thus at high cell concentrations, HapR reduces the level of cellular c-di-GMP with a decrease in biofilm formation (Waters *et al.*, 2008). Therefore, at high cell concentrations, HapR represses biofilm formation by two mechanisms: binding and repressing the biofilm transcriptional activator, *vpsT* and reducing the level of c-di-GMP. Unlike *V. cholerae* El Tor, classical *V. cholerae* has a non-functional HapR but exhibits QS-controlled gene expression via a HapR-independent pathway in which small regulatory RNAs (sRNAs) hypbridize with a new target mRNA and activate translation (Hammer and Bassler, 2007).

*Vibrio cholerae* formed biofilms at the air/surface interface of glass beads (Zhu and Mekalanos, 2003), of glass tubes (Joelsson *et al.*, 2006), on glass coverslips (Hammer and Bassler, 2003), and polyvinyl chloride microtiter plates (Fong and Yildiz, 2007). Zooplankton and phytoplankton provide biotic surfaces for biofilm formation by *V. cholerae* (Tamplin *et al.*, 1990). Islam *et al.* (2007) found that *V. cholerae* associated with algae formed biofilms on an abiotic surface such as plexiglass discs.

*Vibrio cholerae* exists as two colonial morphologies: rugose colonies, which are wrinkled and opaque and form floating pellicles at the air/liquid interface; and smooth colonies which are translucent and grow beneath the liquid surface as a suspension of single cells (Yildiz and Schoolnik, 1999). Rugosity is associated with augmented resistance to osmotic, acid, and oxidative stresses and shows enhanced production of biofilms. Rugose colonies are associated with the production of exopolysaccharide (EPS) encoded by the chromosomal *vps* loci; cells of the rugose type up-regulate the *vps* operons (Fong *et al.*, 2006). The EPS mediates cell survival in sodium hypochlorite and biofilm formation on the walls of glass test tubes and polyvinyl chloride microtiter plates (Yildiz and Schoolnik, 1999). The gene *rbmA* (rugosity and biofilm structure modulator A) encodes a protein, RbmA, which is necessary for rugose colony formation and biofilm integrity in *V. cholerae*. RbmA allows cells to form more structured, detergent-resistant biofilms (Fong *et al.*, 2006). The *vps* genes are organized into two regions on the chromosome, *vps-I* and *vps-II*, which are separated by an intergenic region; the *rbm* genes A-F are located in this intergenic region. *rbmA-F* encode proteins that modulate rugosity and biofilm matrix production and maintenance of biofilm structure and stability (Fong and Yildiz, 2007).

Protozoon flagellate grazing eliminated planktonic cells of *V. cholerae* but biofilm cells were resistant to grazing, and there was a decline in the number of protozoa when they were associated with biofilms. Mutants defective in HapR produced biofilms that were eliminated by grazing protozoa whereas biofilms produced by the wild-type rugose strains were unaffected (Matz *et al.*, 2005).

Biofilm formation by *V. cholerae* is enchanced in the presence of bile acids, and biofilm cells are more resistant to bile acids as compared to planktonic cells. The induction of biofilms by bile acids is dependent on the *vps* genes and their transcriptional activator, *vpsR* (Hung *et al.*, 2006). This interesting in vitro effect may be important in *V. cholerae* infections since the organism will incur the presence of bile in the gastrointestinal tract.

Planktonic cells of *V. cholerae* are 1000-fold more sensitive to acid pH (4.5, 30 min) than biofilm cells from the wild type or from the QS-deficient *hapR* mutant indicating that biofilms should traverse the stomach readily without harm (Zhu and Mekalanos, 2003). There was an approximately 16-fold increase in the detachment of cells from established wild-type biofilms as compared to that of established mutant *hapR* biofilms indicating that QS promotes cellular detachment from biofilms after the biofilms pass through the stomach (Zhu and Mekalanos, 2003). Using an infant mouse assay, Zhu and Mekalanos (2003) demonstrated that there was a 10-fold decrease in the ability of the *hapR* mutant biofilms to colonize the intestines of infant mice. The poor detachment of cells from *hapR* biofilms may explain the decreased colonization of infant mice intestines. Detachment of cells from the biofilm is necessary for bacterial infection of the intestinal epithelial cells (Zhu and Mekalanos, 2003). Thus, the biofilm structure of *V. cholerae* protects the organism from the acid stomach. On reaching the intestine, individual bacterial cells detach from the biofilm and colonize intestinal cells. As the number of vibrios increase in the intestine, there is activation of HA/protease production allowing the bacterial cells to detach from the intestinal epithelium to find new foci of infection or to exit the host (Zhu and Mekalanos, 2003).

HapR also has a role in stress responses in *V. cholerae*. The wild-type strain was >100-fold more resistant to $H_2O_2$ than a *hapR*-negative strain

(Joelsson *et al.*, 2007). Complementation of the *hapR* mutant with the active gene led to $H_2O_2$ resistance similar to the wild type. The expression of *rpoS* (encodes the RNA polymerase sigma factor RpoS) was approximately 3.5-fold higher in the wild-type *V. cholerae* as compared to the *hapR* mutant. By upregulating the expression of RpoS, HapR aids in protecting *V. cholerae* against stress effects (Joelsson *et al.*, 2007).

*Vibrio parahaemolyticus*

In *V. parahaemolyticus*, AHLs are produced by the LuxM synthase, and a two-component sensor, LuxN, senses the AHLs (Henke and Bassler, 2004; McCarter, 1998). The LuxS/AI-2 system is also present (Defoirdt *et al.*, 2006; Henke and Bassler, 2004). Biofilms are formed on glass coverslips (Shime-Hattori *et al.*, 2006) and polystyrene microtiter plates (Enos-Berlage *et al.*, 2005; Güvener and McCarter, 2003). In addition, *V. parahaemolyticus* formed pellicles at air/liquid interfaces (Enos-Beriage *et al.*, 2005; Güvener and McCarter, 2003).

Colonial morphology in *V. parahaemolyticus* can be variable, switching between opaque (OP) and translucent (TR) colony types. The OP type is under the control of the *opaR* gene (has a similar role as the *hapR* gene in *V. cholerae*). In TR colony types, *opaR* is not expressed. In an OP background, disruption of *opaR* leads to a permanent TR colony type (McCarter, 1998). Both OP and TR types (as well as the *opaR* mutant) produce pellicles at the air-liquid interface when grown in glass tubes, even with high speed shaking (Güvener and McCarter, 2003).

The OP colonial type produces more capsular polysaccharide (CPS) than the TR type. OpaR activates the expression of the *cpsA* gene which encodes CPS synthesis (Güvener and McCarter, 2003). *cps* mutants fail to produce CPS and are deficient in pellicle formation, attachment to polystyrene surfaces, and in 3-dimensional biofilm architecture. There was a 43.4% decrease in attachment to polystyrene by the *cpsA* mutant, 34.0% decrease by the *opaR* mutant and 81.1% decrease in attachment by the *cpsA:opaR* double mutant (Güvener and McCarter, 2003). Excellent biofilm formation occured in TR strains despite the fact that TR strains produced less CPS than OP strains. The increased levels of CPS in OP strains may give a biofilm architecture that differs from that of the TR strains (Enos-Berlage *et al.*, 2005).

In addition, OpaR regulates swarming by repressing the *laf* gene encoding the lateral flagellar system (Jacques and McCarter, 2006). The lateral flagella are adapted for movement over moist or viscous surfaces and therefore mediate swarming activity. The cells express lateral flagella only when a surface is sensed. Since OpaR is a repressor of the *laf* gene, OP strains produce few lateral flagella and are poor swarmers as compared to TR strains. Swarming is an important aspect of biofilm formation (Daniels *et al.*, 2004; Harshey, 2003).

*Vibrio parahaemolyticus* has two types of IV-A pilus genes: *pilA* and *mshA* that are involved in biofilm formation. Shime-Hattori *et al.* (2006), utilizing glass coverslips, found that *mshA* mutants formed cellular aggregates but the attachment of bacteria to the coverslips were reduced in comparison to the wild type. This result indicated that the *mshA* gene is involved in bacterial attachment during biofilm formation. *pilA* mutants attached to the coverslips but did not form aggregates indicating that the *pilA* gene contributes to bacterial agglutination during formation of biofilms (Shime-Hattori *et al.*, 2006). Thus, both pili types are necessary for biofilm formation. It is not known if QS has an effect on the role of pili in biofilm formation.

*Vibrio vulnificus*

The presence of the LuxS/AI-2 system has been demonstrated in *Vibrio vulnificus* (Kim *et al.*, 2003; McDougald *et al.*, 2001), as well as the LuxR homologue, SmcR (McDougald *et al.*, 2001; Shao and Hor, 2001). A number of AHLs have been isolated from cultures of *V. vulnificus* including $C_4$-, $C_6$-, 3-oxo-$C_8$-, 3-oxo-$C_{10}$-, 3-oxo-$C_{12}$-, and 3-oxo-$C_{14}$-HSL (Morin *et al.*, 2003). *V. vulnificus* must have a synthase similar to Lux M of *V. harveyi* (Henke and Bassler, 2004) responsible for the synthesis of the AHLs. Biofilms in *V. vulnificus* form on glass tubes (Joseph and Wright, 2004; Paranjpye and Strom, 2005), glass cover slips (McDougald *et al.*, 2006), and polystyrene (Lee *et al.*, 2004) surfaces. Serovar E of *V. vulnificus* formed biofilms on the skin of eels (Marco-Noales *et al.*, 2001). As a wound pathogen, the organism probably forms biofilms in human tissue.

Motility and swarming are necessary for biofilm formation. A *flgE*-deleted mutant of *V. vulnificus* lacks flagella and is not motile (Lee *et al.*, 2004). At 72 h, there was an approximately 5-fold increase in biofilm formation on polystyrene by the wild-type strain as compared to the non-motile mutant, and there was an approximately 45-fold increase in adherence of the wild-type strain to INT-407 (human embryronic intestinal) cells as compared to the mutant strain (Lee *et al.*, 2004). Thus, motility is necessary for adherence and biofilm formation in *V. vulnlificus* as compared to the wild type.

A mutation in the *V. vulnificus pilA* gene (encodes type IV pilin) decreased biofilm formation by approximately 25%, decreased attachment of bacterial cells to HEp-2 (human laryngeal epithelial) cells by approximately 50%, and decreased *V. vulnificus* lethality for mice (Paranjpye and Strom, 2005).

McDougald *et al.* (2001, 2006) found that a *smcR* (a *luxR* homologue) mutant produced approximately 5-fold increase of biofilm on polystyrene as compared to the wild type. Mutation in the *luxS* gene of *V. vulnificus* eliminated AI-2 production, decreased the production of metalloprotease (encoded by *vvpE*), increased hemolysin (encoded by *vvhA*), decreased cytotoxicity to HeLa (human cervix epithelial) cells, and decreased mouse lethality (Kim *et al.*, 2003). The addition of wild-type culture supernatant to the *luxS* mutant led to a delay in hemolysin production and an increase of protease production to a level similar to the wild type (Kim *et al.*, 2003).

A mutation in *vvpE* abolished swarming in *V. vulnificus* (Kim *et al.*, 2003), whereas, a mutation in *vvhA* had no effect on swarming (Kim *et al.*, 2003). Since swarming is involved in biofilm formation, LuxS, through its effect on the *vvpE* gene, plays a role in biofilm formation, probably due to quorum sensing by AI-2. It would have been helpful if Kim *et al.* (2003) had added synthetic AI-2 to cultures of the *luxS* mutant in addition to wild type culture supernatants.

A mutation in *smcR* led to decreased production of the metalloprotease and an increase in the production of the hemolysin similar to that seen with a *luxS* mutant (Shao and Hor, 2001). Insertion of a plasmid containing *smcR* into the mutant restored the protease levels to that of the wild type. However, the complemented strain had a greatly reduced hemolysin activity during the early log phase but not in the late log phase, and there was a slight decline of activity during the stationary phase; thus, the complemented strain did not act quite like the wild-type strain (Shao and Hor 2001). Shao and Hor (2001) did not add AHLs to the cultures of the *smcR* mutant to determine if the wild-type production of protease and hemolysin was restored.

### 2.2.10 *Shigella* species

Members of the genus *Shigella* species are strict pathogens of primates and consist of facultative anaerobic Gram negative rod-shaped bacteria. These organisms are associated with water and food-borne dysentery or gastroenteritis. The LuxR analog, SdiA, is present in *Shigella* species, but a LuxI analog is not (Case *et al.*, 2008). However, there is no information available concerning the role of SdiA. All four species of *Shigella* were positive for AI-2 production (Day and Maurelli, 2001). Day and Maurelli (2001) found that a *luxS* mutant of *S. flexneri* 2a was as invasive toward L2 (rat lung fibroblast) cells as the wild type and behaved similarly to the wild type in the Sereny test for virulence. Culture supernatants containing AI-2 activity from wild type *S. flexneri* or from enteroinvasive *E. coli* enhanced the expression of *virB* (a transcription factor that regulates the *Shigella* invasive operon) by a *luxS* mutant of *S. flexneri* (Day and Maurelli, 2001). The effect of purified AI-2 was not examined. Since the *luxS* mutant of *S. flexneri* was as invasive and virulent as the wild type, the role of LuxS and AI-2 is not clear. The AI-3/epi/norepi system is present in *Shigella* (Walters and Sperandio, 2006b). Systematic studies on biofilm formation by *Shigella* species do not appear to have been done.

### 2.2.11 *Enterobacter sakazakii*

*Enterobacter sakazakii* is a Gram-negative rod-shaped bacterium that can cause an invasive infection in neonates and is associated with powdered infant formula. *Enterobacter sakazakii* formed biofilms on stainless steel coupons suspended in reconstituted infant formula (Kim *et al.*, 2007) and on latex, silicone, and polycarbonate pieces suspended in infant milk formula (Iversen *et al.*, 2004). Kim *et al.* (2006) demonstrated biofilm formation by *E. sakazakii* on pieces of polyvinyl chloride enteral feeding tubes suspended in infant formula broth. *Enterobacter sakazakii* formed biofilms on polyvinyl chloride microtiter plates at the air/solid interface (Lehner *et al.*, 2005), pellicles on glass tubes at the air/liquid interface, and aggregates (flocs) at the glass/liquid interface of glass growth tubes (Lehner *et al.*, 2005). Biofilm formation on glass, pellicle formation on glass tubes, and clumping by *E. sakazakii* was associated with the expression of cellulose production and formation of curli fimbriae (Zogaj *et al.*, 2003). Lehner *et al.* (2005) found that *E. sakazakii* produced AHLs including 3-oxo-$C_6$-HSL and/or 3-oxo-$C_8$-HSL; however, they did not determine if these compounds had a role in biofilm formation nor did they determine if LuxI/LuxR homologs were present. It is not known if AHLs are involved in the regulation of biofilm or cellulose production and formation of curli fimbriae in *E. sakazakii*.

### 2.2.12 *Aeromonas hydrophila*

*Aeromonas hydrophila*, present in fresh and brackish waters, is a facultative anaerobic Gram-negative rod which causes opportunistic infections, as well as occasional cases of food- and water-borne illness. Biofilms were produced by *Aeromonas hydrophila* on stainless steel coupons (Lynch *et al.*, 2002) and on polystyrene microtiter plates (Canals *et al.*, 2006a). Since *A. hydrophila* is an opportunistic pathogen in humans causing extraintestinal and intestinal infections and is an animal pathogen leading to infections in fish, amphibians, and mammals (Farmer *et al.*, 1991), they undoubtedly form biotic biofilms. Quorum sensing in *A. hydrophila* is regulated by LuxR/LuxI homologs. *ahyI* encodes the AHL synthase and *ahlR* encodes AhyR, the transcription regulator (Swift *et al.*, 1997). The major AHL produced by *A. hydrophila* is butanoyl-L homoserine lactone ($C_4$-HSL); a minor component is $C_6$-HSL (Swift *et al.*, 1997). Maximal expression of *ahyI* requires the presence of $C_4$-HSL. AhyR influences the production of AhyI but is not absolutely required (Kirke *et al.*, 2004). AhyI can be detected in the exponential phase of growth but is degraded during the stationary phase; in an *ahyR* mutant; however, the AhyI protein is produced only during the stationary phase, indicating that AhyR controls the timing and turnover of AhyI (Kirke *et al.*, 2004). Insertion inactivation of the *ahyR* gene eliminated a number of virulence determinants of *A. hydrophila* and eliminated lethality of the organism in a fish model (Bi *et al.*, 2007).

There is constitutive production of a single polar unsheathed flagellum in *Aeromonas hydrophila*,which allows the organism to swim in liquid media; however, growth on a solid medium or in highly viscous media induces the expression of peritrichous lateral flagella, which is associated with swarming motility and colonization on surfaces (Altarriba *et al.*, 2003). Mutations in the *A. hydrophila* lateral flagella loci reduced the adherence of the organism to HEp-2 (human laryngeal epithelial) cells and the ability to form biofilms on polystyrene microtiter plates (Gavin *et al.*, 2002). Mutants lacking polar flagella but which produce lateral flagella demonstrated an approximately 86% decrease in adhesion to HEp-2 cells and a 55% reduction in biofilm formation on polystyrene microtiter plates (Canals *et al.*, 2006b). Mutants lacking the ability to produce lateral flagella but positive for the polar flagellum showed an approximately 80% reduction in adhesion to HEp-2 cells and a 62% reduction in biofilm formation as compared to the wild type (Canals *et al.*, 2006a). In addition, Canals *et al.* (2006a) found that a $\sigma^{54}$ (*rpoN*) mutant failed to produce both the polar flagellum and lateral flagella and showed approximately 89% reduction in adhesion to HEp-2 cells and an approximately 75% reduction in the ability to form biofilms. Thus, both the *A. hydrophila* lateral flagella and polar flagellum are necessary for maximum biofilm formation.

The formation of mature biofilms on stainless steel coupons by *A. hydrophila* required the production of $C_4$-HSL since an *ahyI* mutant did not produce a mature biofilm. A mutation in the *ahyR* gene has no effect on biofilm formation (Lynch *et al.*, 2002) suggesting that there must be another receptor for the AHLs produced by AhyI. Exogenous addition of $C_4$-HSL to the *ahyI* mutant only partially restored mature biofilm formation. Thus, Lynch *et al.* (2002) have shown that quorum sensing regulates biofilm formation and a number of workers have shown that mutations in polar and lateral flagella formation decrease biofilm formation (Altarriba *et al.*, 2003; Canals *et al.*, 2006a, b; Gavin *et al.*, 2002). While studies have not been reported, it is probable that flagella formation in *A. hydrophila* is regulated via quorum sensing.

### 2.2.13 *Helicobacter pylori*

*Helicobacter pylori* is a microaerophilic helical shaped Gram-negative bacterium present in the stomach and duodenum. It has been associated with pyloric ulcers and stomach cancer. *H. pylori* can withstand low acid conditions and can be found in food and water, suggesting that both food and water can be vehicles of human infection. A gene encoding AHL synthase has not been demonstrated in *Helicobacter pylori* (Forsyth and Cover, 2000); however, the LuxS/AI-2 system is present. Maximal levels of AI-2 are produced during early- and mid-exponential phase of growth and decreases 8- to 9-fold during the stationary phase (Forsyth and Cover, 2000; Joyce *et al.*, 2000). LuxS persists during the stationary phase even though the extracellular AI-2 decreases (Hardie *et al.*, 2003). Disruption of the *luxS* gene resulted in the loss of AI-2 production; complementation of the mutant with a vector containing *luxS* resulted in the recovery of AI-2 production (Forsyth and Cover, 2000; Joyce *et al.*, 2000).

*Helicobacter pylori* formed biofilms at the air/liquid interface on the glass wall of culture vessels during growth (Stark *et al.*, 1999), on surfaces present in water distribution systems (Watson *et al.*, 2004), and at the air/liquid interface on glass coverslips; however, *H. pylori* did not consistently form biofilms on polystyrene microtiter trays (Cole *et al.*, 2004). Park *et al.* (2001) demonstrated the presence of *Helicobacter* sp. biofilms on cast iron water main distribution pipes. Using scanning electron microscopy of biopsies from human gastric mucosa, Coticchia *et al.* (2006) established that 97.3% of the gastric mucosa surfaces of peptic ulcer patients who were urease-positive (indicating the presence of *H. pylori*) was covered by dense mature biofilms, whereas, only 1.6% of the mucosa surface was covered with biofilms in control individuals who were urease-negative.

In *H. pylori*, *flaA* (encodes flagellin) expression increases as the culture density increases, and *luxS* may play an important role in regulating growth-dependent *flaA* expression since *luxS* mutants did not show growth-dependent regulation of *flaA* (Loh *et al.*, 2004). While bacterial-free supernatants from wild type *H. pylori* cultures had AI-2 activity, the addition of the culture supernatants to *luxS* mutants had only a slight effect on *flaA* transcription. Therefore, *flaA* expression is dependent on *luxS* but the role of AI-2 is not clear. A *luxS* mutant of *H. pylori* showed decreased motility as compared to the wild type; however, the mutant and wild-type strains had similar flagellar morphology (Osaki *et al.*, 2006). In addition, infection of Mongolian gerbils with wild type and *luxS*-negative strains indicated that only infection with the wild type led to gastric inflammation and increased antibody titers. While Osaki *et al.*, (2006) demonstrated the production of AI-2 by their strain of *H. pylori*, they did not study the role of AI-2 in motility and infectivity.

Rader *et al.* (2007) demonstrated that a mutation in *luxS* led to a decrease in motility in *H. pylori*. In addition, they found that there was a marked decrease in *flhA* transcription. The *flhA* gene is an important membrane component of the flagellar export machinery. A deletion in *flhA* in either a wild type or *luxS* mutant background resulted in loss of motility, flagella, and flagellar gene expression (Rader *et al.*, 2007). Addition of DPD (demonstrated to have AI-2 activity) to the *luxS* mutant restored *flhA* transcription to that of the wild type; addition of DPD to wild type *H. pylori* increased the transcription of *flhA* approximately 4-fold. Thus, Rader *et al.* (2007) have demonstrated a role for AI-2 in motility for *H. pylori*.

A mutation in *luxS* increased *H. pylori* biofilm formation on glass surfaces by approximately 3-fold as compared to the wild-type strains (Cole *et al.*, 2004). Mutations in the *cagE* gene (responsible for the assembly of the type IV secretory apparatus) also led to an approximately 3-fold increase in the formation of biofilms by *H. pylori* (Cole *et al.*, 2004). The genes, *luxS* and *cagE,* apparently exert negative control on biofilm formation in *H. pylori.* The role of AI-2 in biofilm formation was not determined by Cole and her coworkers (2004). While Rader *et al.* (2007) presented data indicating that AI-2 affects motility in *H. pylori*, it is not clear if AI-2 has an effect on biofilm formation or virulence.

### 2.2.14 *Klebsiella pneumoniae*

*Klebsiella pneumoniae* is a faculatative anaerobic Gram-negative rod-shaped bacterium commonly associated with opportunistic infections. *Klebsiella pneumoniae* forms biofilms on polystyrene and polyvinyl chloride microtiter plates (Balestrino *et al.*, 2005; Boddicker *et al.*, 2006). The organism also forms biofilms on urinary catheters (Burton *et al.*, 2006), and it is probable that biofilms are formed on other implanted medical devices. *Klebsiella pneumoniae* does not produce AHLs (Balestrino *et al.*, 2005; Boşgelmez-Tinaz *et al.*, 2005) but does possess an AHL receptor, SdiA, similar to *Salmonella* and *E. coli* (Smith and Ahmer, 2003), as well as the LuxS/AI-2 system (Balestrino *et al.*, 2005). The maximum production of AI-2 occurs during the late-exponential and early stationary growth phases (Balestrino *et al.*, 2005). Balestrino *et al.* (2005) found that a *luxS* mutant of *K. peneumoniae* was able to form a mature biofilm but had a reduced capacity to form microcolonies in the early steps of biofilm formation. The authors suggested that AI-2 plays a role in early stages of biofilm formation. However, they did not add exogenous AI-2 to the *luxS* mutant to determine its effect on biofilm formation.

### 2.2.15 *Serratia marcescens*

*Serratia marcescens* is a facultative anaerobic Gram-negative rod-shaped bacterium associated with nosocomial infections, particularly urinary tract and wound infections. *Serratia marcescens* forms biofilms on glass (Koh *et al.*, 2007; Rice *et al.*, 2005), polystyrene microtiter plates (Labbate *et al.*, 2004), and silicone urinary catheters (Jones *et al.*, 2006). The LuxS/AI-2 system has been demonstrated in *S. marcescens* strains. The production of AI-2 was maximal at the late exponential phase in strain 274 and was maximal during transition into the stationary phase in strain 39006. The AI-2 activity decreased several fold during the stationary phase (Coulthurst *et al.*, 2004). *luxS* mutants did not produce AI-2. The LuxI/LuxR homologs in *S. marcescens* MG1 are SwrI (the AHL synthase) and SwrR (AHL receptor). The *swrI* gene is responsible for directing the synthesis of $C_4$-HSL and $C_6$-HSL (Eberl *et al.*, 1996). Similarly, *S.* marcescens strain 12 has the homologs SmaI (synthetase) and SmaR (receptor), and $C_4$-HSL and $C_6$-HSL are produced (Coulthurst *et al.*, 2006). There are two forms of flagella-driven motility in *S. marcescens*: swimming (motility in a liquid environment) and swarming (motility on a solid surface). Flagellar expression is governed by the *flhDC* master operon (Givskov *et al.*, 1998). The *flhDC* and *swrI* genes are necessary for swarming activity in the MG1 strain (Givskov *et al.*, 1998).

Swarming behavior is abolished in a *swrI* mutant of *S. marcescens* MG1; however, the activity is restored upon addition of $C_4$-HSL; a 6-fold increased level of $C_6$-HSL was necessary to restore swarming activity to the mutant strain in comparison to $C_4$-HSL (Eberl *et al.*, 1996). A *swrI* mutant produced a biofilm that is distinct from the wild type by being thinner and nonmature; addition of $C_4$-HSL restored biofilm formation by the mutant to that of the wild type (Labbate *et al.*, 2004). Two additional quorum sensing-regulated genes, *bsmA* and *bsmB*, were necessary for proper biofilm development. A mutation in *smaI* reduced biofilm formation in *S. marcescens* strain 12 by approximately 50%; addition of $C_4$-HSL to the mutant restored biofilm formation (Coulthurst *et al.*, 2006). Labbate *et al.* (2007) demonstrated that quorum sensing was necessary for attachment of *S. marcescens* MG1 to polystyrene. A *swrI* mutant did not attach to the plastic abiotic surface, and the addition of $C_4$-HSL restored the ability of the mutant to attach. However, quorum sensing was not necessary for the attachment of the MG1 strain to human corneal epithelial cells (Labbate *et al.*, 2007). A mature biofilm of *S. marcescens* MG1 detaches or sloughs off after 75 to 80 h. Rice *et al.* (2005) found that both biofilm formation and sloughing of cells from the biofilm is dependent on SwrI/SwrR quorum sensing. Deletion of the *spnI* gene (AHL synthetase) in *S. marcescens* AS-1 eliminated prodigiosin production; however, the addition of $C_6$-HSL restored its production (Morohoshi *et al.*, 2007). The *spnI* mutant lacked swarming activity, which was restored by the addition of $C_6$-HSL. There was approximately 55% decrease in biofilm formation by the mutant, and addition of $C_6$-HSL partially restored biofilm formation. Studies on the effect of AI-2 on biofilm formation in *S. marcescens* have not been done.

*Serratia marcescens* MG1 produces a lipopeptide biosurfactant, serrawettin W2, necessary for swarming activity. A *swrI* mutant did not produce serrawettin and did not swarm, thereby indicating that production of surfactant was dependent on QS (Lindum *et al.*, 1998). However, inactivation of the *flhD* locus had no effect of the production of the surfactant. The *swrA* gene encodes the production of serrawettin W2 (Lindum *et al.*, 1998), and the expression of the gene is controlled by QS. *swrI* and *swrI/swrA* mutants lack swarming motility; however, the activity could be restored by the addition of serrawettin W2 suggesting that the QS-controlled gene involved in swarming is *swrA* (Lindum *et al.*, 1998). Thus, QS in necessary for swarming motility and biofilm formation in *S. marcescens*.

## 2.3 Inactivation of quorum sensing molecules and inhibition of quorum sensing

At pH values above 7, there is nonenzymatic degradation of AHLs in bacterial cultures (Byers *et al.*, 2002; Flodgaard *et al.*, 2003; Yates *et al.*, 2002). At alkaline pH values, the degradation of AHLs is due to lactonolysis, i.e., opening up of the lactone ring by hydrolyzing the ester bond of the ring to give an acylhomoserine (Yates *et al.*, 2002). AHLs also can be degraded enzymatically. Lactonases are produced by a number of bacteria, which induce lactonolysis of AHLs (Dong and Zhang, 2005; Rasmussen and Givskov, 2006; Roche *et al.*, 2004). A few bacteria have been shown to produce acylases, which degrade AHLs by cleaving the amide bond connecting the lactone ring to the acyl chain with the production of homoserine lactone and a fatty acid (Dong and Zhang, 2005; Liu *et al.*, 2007; Rasmussen and Givskov, 2006; Roche *et al.*, 2004). In addition to microbial sources of AHL degrading enzymes, AHL lactonases have been demonstrated in mammalian sera (Yang *et al.*, 2005) and an AHL acylase, acylase I, has been identified in porcine kidney. (Xu *et al.*, 2003). Cytochrome P450 isolated from *B. megatherium* can oxidize AHLs and acyhomoserines (produced by lactonase action) (Chowdhary *et al.*, 2007). Dong *et al.* (2001) inserted the AHL lactonase gene from a *Bacillus* sp. into tobacco and potato plants and demonstrated protection of plants against the plant pathogen, *Erwinia carotovora*. Thus, Dong *et al.* (2001) established the feasibility of the use of AHL lactonases as antimicrobial agents.

The marine alga, *Delisea pulchra* produces halogenated furanones (de Nys *et al.*, 1993), and Givskov *et al.* (1996) found that these halogenated furanones were antagonistic towards AHL-controlled processes in bacteria. The binding of the halogenated furanones by the AHL receptor protein, LuxR, destabilizes the protein and leads to its destruction (Manefield *et al.*, 2002). Once the inhibitor is bound, the destruction of the protein is not reversible by addition of AHLs. However, the antagonistic action of brominated furanones is not limited to AHLs. The brominated furanone, (5Z)-4-bromo-5-(bromomethylene)-3-butyl-2(5H)-furanone was shown to inhibit the QS action of both AHL (AI-1) and AI-2 (Ren *et al.*, 2001). The inhibitor probably inhibits AHL action by destabilizing LuxR; however, it is not known if the brominated furanone inhibits AI-2 action by binding to AI-2 receptors.

AHLs have been found in the sputum of cystic fibrosis patients infected with *P. aeruginosa* and *Burkholderia cepacia*, suggesting that AHLs may have a role in the disease (Middleton *et al.*, 2002). Using a mouse model of cystic fibrosis, Wu *et al.* (2004) found that the intravenous administration of brominated furanones to mice inhibited 3-oxo-$C_6$-HSL dependent gene expression by *Pseudomonas aeruginosa*. The brominated furanones increased the survival time of mice infected by *P. aeruginosa*, promoted bacterial clearance from the mouse lung, and reduced lung pathology (Wu *et al.*, 2004). The use of brominated furanones as therapeutic agents could be a feasible option for treatment of cystic fibrosis and other diseases. AHL signaling is involved in biofilm formation by *P. aeruginosa* since a *lasI* (gene that encodes synthesis of AHLs) mutant does not produce a mature biofilm (Davies *et al.*, 1998). Hentzer *et al.* (2002) demonstrated that brominated furanones had an adverse effect on the architecture (i.e., interference with maturation) of *P. aeruginosa* biofilms and enhanced the detachment of bacteria from the biofilm (Hentzer *et al.*, 2002).

AI-2 promotes biofilm and swimming motility in *E. coli* (González Barrios *et al.*, 2006). Swarming activity and biofilm formation by *E. coli* was inhibited by (5Z)-4-bromo-5-(bromomethylene)-3-butyl-2-(5H)-furanone at concentration levels that had no effect on bacterial growth; however, swimming was not inhibited (Ren *et al.*, 2001, 2004b). The concentration of AI-2 was decreased by the brominated furanone; however, there was no effect on *luxS* and *pfs*, which encode the proteins for AI-2 production (Ren *et al.*, 2004b).

Mutations in the AI-2 QS system (including mutations in *luxS, luxP, lux0)* attenuated the virulence of of *V. harveyi* towards gnotobiotically cultured brine shrimp, *Artemia franciscana* (Defoirdt *et al.*, 2005); however, mutations in the AHL system (including mutations in *luxM, luxN*) had no effect on virulence of *V. harveyi*. The QS-disrupting brominated furanones, (5Z)-4-bromo-5-(bromomethylene)-3-butyl-2(5H)-furanone and (5Z)-4-bromo-5-(bromomethylene)-2(5H)-furanone, protected the brine shrimp against infection by *V. harveyi, V. campbellii,* and *V. parahaemolyticus* at levels that had no effect on growth of the vibrios (Defoirdt *et al.*, 2006). The AHL analogue, N-nonanoyl-cyclopentyamide, prevented prodigiosin synthesis, swarming, and biofilm formation in *S. marcescens* AS-1 by antagonizing quorum sensing (Morohoshi *et al.*, 2007).

Furocoumarins (dihydroxybergamottin and bergamottin) isolated from grapefruit juice inhibited the AHL- and AI-2-associated activities in *V. harveyi* by almost 100%. In addition, the furocoumarins inhibited biofilm formation on polystyrene by *E. coli* O157:H7 by >50% but was less effective in inhibiting biofilm formation in *S.* Typhimurium (Girennavar *et al.*, 2008). The concentrations of furocoumarins used had little effect on bacterial growth. González Barrios *et al.* (2006) have demonstrated that the addition of enzymatically synthesized AI-2 to wild type K-12 or to a *luxS* mutant led to a several fold increase in biofilm formation. A mutation in *sdiA* increased biofilm formation in *E. coli* K-12 several fold, whereas, the addition of AHLs to the wildtype decreased formation. It is probable that the furocoumarins inhibited biofilm formation in the *E. coli* strains by inhibiting the action of AI-2. Under the conditions that were used by Girennavar *et al.* (2008), AHLs were not present and biofilm stimulation was not an AHL-associated event.

Medellin-Peña *et al.* (2007), using size exclusion chromatography to isolate fractions from cell-free spent medium of *Lactobacillus acidophilus* La-5 that significantly reduced the extracellular AI-2 production by *E. coli* O157:H7. In addition, these fractions inhibited the expression of the LEE operons; however, the production of Shiga toxin 2 was not inhibited (Medellin-Peña *et al.*, 2007). While the mechanism(s) of inhibition is not clear, the data does suggest that probiotics may be useful in antagonizing QS.

Purified LuxS produces AI-2 by an aldose-ketose isomerization of S-ribosylhomocysteine followed by a β-elimination. The analogues, S-anhydroribosyl-L-homocysteine and S-homoribosyl-L-cysteine inhibited aldose-ketose isomerization and β-elimination, respectively (Alfaro *et al.*, 2004), thereby inhibiting the formation of AI-2 by LuxS. Studies on the therapeutic potential of these LuxS inhibitors have not been reported.

AI-2 is a boric acid complex, and Ni *et al.* (2008) found that a number of phenylboronic acids inhibited AI-2 mediated bioluminescence by *V. harveyi*. It is probable that phenylboronic acids can inhibit AI-2 action in other microorganisms.

The few examples given above indicate that QS compounds can either be destroyed or their action inhibited under real life situations. Rasmussen and Givskov (2006) suggested that targeting the QS system of a pathogen as a disease control strategy since it would inhibit the expression of QS-regulated virulence factors. QS is not directly implicated in essential growth processes and does not affect viability, and therefore, interference with QS is less likely to induce a strong selective pressure for the development of resistance, unlike the case for antibiotics.

An immunopharmacotherapeutic approach to the inhibition of QS compounds may hold promise as a means of controlling QS-mediated virulence in microorganisms. A monoclonal antibody against *S. aureus agr*-regulated QS peptide, AIP-type 4, inhibited *S. aureus*-induced abscess formation in SKH1 euthymic hairless mice and protected the mice against a lethal challenge of *S. aureus* (Park *et al.*, 2007). It would be interesting to know if the AIP-4 hapten could act as a vaccine. Immunization of BALB/c mice with a 3-oxo-$C_{12}$-HSL-protein conjugate protected the mice against a lethal *P. aeruginosa* lung infection (Miyairi *et al.*, 2006). Thus, vaccination may be a way of targeting QS compounds to protect against infections by pathogens.

Macrolides such as azithromycin inhibit the AHL synthases of *P. aeruginosa,* thereby decreasing the AHL levels with reduction in biofilm formation. Administration of azithromycin to cystic fibrosis patients led to significant improvement in pulmonary function (Martin *et al.*, 2008). Despite its effectiveness in interfering with AHL communication, excess use of a macrolide as a therapeutic agent can lead to microbial resistance.

## 2.4 Genetic transfer in biofilms

It seems logical to think that gene transfer can take place readily in biofilms and the frequency of such transfer in biofilms should be quite high since bacteria are in very close contact. The frequency of gene transfer in planktonic cells is probably lower than that seen in cells located in biofilms (Roberts *et al.*, 2001). A number of studies indicate that transfer of genes is a common phenomenon in biofilms.

In single species biofilms on glass beads, a strain of donor *E. coli* harboring three different plasmids transferred (probably by conjugation) the plasmids to an *E. coli* strain present in the biofilm (Lebaron *et al.*, 1997). When *Pseudomonas putida* carrying pBF1 (confers mercury resistance) was mixed with either an unspecified *Vibrio* strain, *V. fischeri*, *Deleya marina*, or *P. putida* (lacking pBF1) and inoculated onto polycarbonate filters floating on artificial sea water, Dahlberg *et al.* (1998) found that pBF1 from the donor strain was transferred by conjugation to the recipient strains in the biofilm formed on the filters. Roberts *et al.* (1999) added a strain of *B. subtilis* containing the conjugative plasmid Tn*5387* (confers tetracycline resistance) to 21 day biofilms on bovine enamel disks which had been inoculated with human saliva. The biofilm population consisted of species of *Streptococcus, Lactobacillus, Actinomyces*, and *Veillonella*. Within six hours of adding the donor *B. subtilis* strain to the biofilm, Roberts *et al.* (1999) were able to show that *S. acidominimus*, isolated from the biofilm, was resistant to tetracycline. A rifampicin-resistant strain of *E. coli* (recipient) was allowed to form biofilms on glass; at 8 days, a donor strain of *E. coli* carrying the plasmid R1*drd*19 (confers resistance to chloramphenicol and ampicillin) was added to the biofilm. Within 24 hours, rifampicin- resistant transconjugants with resistance to chloramphenicol and ampicillin were isolated (Licht *et al.*, 1999).

Horizontal gene transfer by transformation was demonstrated in strains of *S. mutans* by Li *et al.* (2001). Using DNA from a plasmid with the *gfp* (green fluorescent protein) gene, Li *et al.* (2001) demonstrated the gene transferred to *S. mutans* biofilms on the surface of polystyrene microtiter plates. Biofilm-grown cells were transformed at frequencies several-fold higher than that seen with planktonic-grown cells. In addition, Li *et al.* (2001) found that dead cells of an erythromycin resistant strain of *S. mutans* in biofilms acted as DNA donors of resistance to erythromycin in non-resistant strains of *S. mutans*. Using a monoculture biofilm of an *Acinetobacter* strain, Hendrickx *et al.* (2003) added DNA from a plasmid with the gene for *gfp* (green fluorescent protein) and found that gene transfer occurred in young and actively growing biofilms. Maeda *et al.* (2006) used two F-minus (non-conjugative) *E. coli* strains, one with a chromosomal tetracycline resistance gene and the other strain with a non-conjugative plasmid with a kanamycin resistance gene to study gene transfer in biofilms. In mixed biofilms of the two *E. coli* strains on nylon membrane filters,

Maeda *et al.* (2006) detected cells with resistance to both antibiotics. At 25°C, gene transfer in biofilm cells was approximately one log higher than that of planktonic cells. The mechanism of gene transfer was probably transformation (Maeda *et al.*, 2006). Using *V. cholerae* O139 DNA added to a chitin biofilm of *V. cholerae* O1 El Tor suspended in artificial sea water, Blokesch and Schoolnik (2007) found transformants with O139 strain characteristics. In a mixed biofilm of O1 El Tor (rifampicin resistant and streptomycin sensitive) and O139 (rifampicin sensitive and streptomycin resistant), transformants with O139 characteristics and rifampicin resistance and streptomycin sensitivity were isolated. In the presence of DNase, no transformants were found, thereby indicating that the mechanism of gene transfer was transformation.

Antibiotic resistance gene transfer is common in the biofilm environment. This is important from the point of view concerning disease in humans. For example, some infections such as periodonitis or endocarditis involve biofilms (Fux *et al.*, 2005), biofilms are found in wounds resulting from diabetic foot ulcers, pressure ulcers, or venous leg ulcers (James *et al.*, 2008), and biofilms are often present on inserted medical devices (Fux *et al.*, 2005). Owing to the peculiar physiology of biofilm cells, they are able to resist killing by antibiotics and as a consequence, long term antimicrobial treatment is administered. Long term treatment increases the selection pressure for resistance and the high cell density in the biofilm increases the likelihood of gene transfer (Fux *et al.*, 2005).

Thus, a number of workers have shown that gene transfer occurs readily in biofilms by either conjugation or transformation. In addition, genetic exchange occurs at a higher frequency in biofilm cells than in planktonic cells, and gene transfer takes place in mixed-genera biofilms.

## 2.5 Genomics and proteomics of biofilm formation

Genomics (DNA microarrays) and proteomics (gel electrophoresis of proteins) can have a role in microbial food safety by providing tools for the rapid detection and identification of pathogens present in food or the food environment, and they can provide information concerning the cellular diversity and responses to stress by the pathogen. Genomics and proteomics may be useful in determining metabolic pathways that contribute to growth and survival of the pathogen in foods, the food processing environment, and in the human host (Abee *et al.*, 2004). However, these techniques have not revealed a common gene and/or protein expression pattern for biofilm formation in microorganisms (Ghigo, 2003; Sauer, 2003). Sauer (2003) stated that biofilm cells and planktonic cells differ in their patterns of gene expression (and probably protein expression).

Using an extraintestinal pathogenic *E. coli* strain that formed urinary catheter biofilms, Orme *et al.* (2006) demonstrated that the urinary catheter strain upregulated protein OmpA (outer membrane protein A) as compared to a laboratory non-infectious strain of *E. coli*. Orme *et al.* (2006) suggested that OmpA upregulation may be a common feature of urinary tract-infectious *E. coli* and therefore, OmpA could be a potential target for the inhibition of biofilm formation on urinary catheters. Trémoulet *et al.* (2002) did a proteomic analysis of planktonic and biofilm cells of *E. coli* O157:H7 and determined that the level of 14 proteins increased, whereas the level of three proteins decreased in biofilm cells compared to planktonic cells. Biofilm cell up-regulated proteins included the DNA protection during starvation protein, D-ribose-binding periplasmic protein, D-galactose-binding protein, malate dehydrogenase, amino acid ABC transporter-binding protein, thiamine-phosphate pyrophosphorylase, and DNA-binding protein H-NS; however, not all of the proteins were identified (Trémoulet *et al.*, 2002).

Wen and Burne (2002) did a genomic study of *S. mutans*, an obligate biofilm-forming organism. They determined that a strain carrying an insertionally inactivated *brpA* (biofilm regulatory protein) gene were unable to form biofilms. Insertional inactivation of the *luxS* gene had no effect on the formation of biofilms by *S. mutans*. In addition, Wen and Burne (2002) found that a *ccpA*-deficient mutant (*ccpA* encodes the carbon catabolite control protein) gave an approximately 60% decrease in biofilm formation. Thus, BrpA and CcpA proteins regulate genes necessary for biofilm formation by *S. mutans*, whereas LuxS does not have a role.

A proteomic analysis of biofilm-forming *C. jejuni* indicated that proteins involved in motility are up-regulated in comparison with planktonic cells (Kalmokoff *et al.*, 2006). Proteins involved in motility include the flagellins (FlaA, FlaB), the filament cap (FliD), the basal body (FlgG, GlgG2), and the chemotactic protein (CheA), proteins involved in general (GroEL, GroES), and oxidative (Tpx, Ahp) stress responses, adhesins (Peb1, FlaC), and proteins involved in biosynthesis, energy generation, and catabolite functions are up-regulated in biofilm-forming cells. Kalmokoff *et al.* (2006) also studied another type of biofilm in *C. jejuni*, pellicle formation at the air/liquid interface. They found that insertional inactivation of the flagellar genes, *flaA, flab, flaC, fliA,* or *flhA,* led to a delay in pellicle formation, whereas a mutation in *flaG* resulted in the loss of pellicle formation by *C. jejuni*.

When compared to planktonic cells, biofilm-forming cells of *S. aureus* expressed increased levels of proteins linked to peptidoglycan synthesis and cell attachment; in addition, there was an increase in fibrinogen-binding proteins (Resch *et al.*, 2006). Enzymes involved in formate and pyruvate metabolism were also up-regulated. Biofilm cells expressed increased levels of SarA (staphylococcal accessory regulator A protein) which regulates the *ica* operon and biofilm formation. Decreased levels of proteases, immunodominant antigen A and SsaA (staphylococcal secretory antigen) were found in the biofilm cells as compared to the planktonic cells of *S. aureus* (Resch *et al.*, 2006).

In a proteomic comparison of *L. monocytogenes* (serotype 1/2a) cells grown planktonically and in biofilms formed on nitrocellulose membranes, Hefford *et al.* (2005) found that 19 proteins were up-regulated in biofilm-grown cells. These upregulated proteins included glyceraldehyde 3-phosphate dehydrogenase, triose phosphate isomerase, phosphoglycerase mutase, chaperon GroEL (a stress protein), and elongation factor EF-Tu (involved in the transportation of aminoacyl-tRNAs to the A site of the ribosome during elongation) (Hefford *et al.*, 2005). Thus, a short survey of the genomics and proteomics of biofilm formation does not give an indication of a common pathway or a common gene(s) involved in biofilm formation by members of the microbial world.

## 2.6 Research needed

There is a limited amount of information concerning the role of QS in foods and the food processing environment. Studies are needed to answer the following questions:

1. What is the role of QS in food spoilage?
2. Does QS have a role in biofilm formation in foods (both raw and processed) and in food processing plants?
3. What is the effect of QS on the growth of spoilage or pathogenic microorganisms in foods?
4. Does QS play a role in toxin production by pathogens present in foods?
5. Does the food environment have a role in the initiation or inhibition of QS?
6. What influence does food or the food environment exert on bacterial gene transfer?
7. What genomic and proteomic studies are needed to understand, at the molecular level, the mechanism of biofilm formation and the role of quorum sensing?

## 2.7 Sources of further information and advice

The following review articles will be a useful source of information on biofilms and QS, and in addition, may suggest useful ideas for research:

Ammor, M S, Michaelidis, C and Nychas, G-J (2008), Insights into the role of quorum sensing in food spoilage, *J Food Prot*, **71**, 1510–1525.

Danhorn, T and Fuqua, C (2007), Biofilm formation by plant-associated bacteria, *Annu Rev Microbiol*, **61**, 401–422.

Ghandhi, M and Chikindas, M L, (2007), *Listeria*: A foodborne pathogen that knows how to survive, *Int. J. Food Microbiol*, **113**, 1–15.

González, J E and Keshavan, N D (2006), Messing with bacterial quorum sensing, *Microbiol Molec Biol Rev*, **70**, 859–875.

Ni, N, Li, M, Wang, J and Wang B, (2009), Inhibitors and antagonists of bacterial quorum sensing, *Med Res Rev*, **29**, 65–124.

Parsek, M R and Greenberg, E P (2005), Sociomicrobiology: the connections between quorum sensing and biofilms, *Trends Microbiol*, **13**, 27–33.

Puttamreddy, S, Carruthers, M F, Madsen, M L and Minion, F C, (2008), Transcriptome analysis of organisms with food safety relevance, *Foodborne Pathog Dis*, **5**, 517–529.

Waters, C M and Bassler, B L (2005), Quorum sensing: cell-to-cell communication in bacteria, *Annu Rev Cell Dev Biol*, **21**, 319–346.

West, S A, Griffin, A S, Gardner, A and Diggle, S P (2006), Social evolution theory for microorganisms, *Nat Rev Microbiol*, **4**, 597–607.

## 2.8 References

ABEE, T, VAN SCHAIK, W and SIEZEN, R J (2004), Impact of genomics on microbial food safety, *Trends Biotechnol*, **22**, 653–660.

AHMER, B M M (2004), Cell-to-cell signaling in *Escherichia coli* and *Salmonella enterica*, *Mol Microbiol*, **52**, 933–945.

ALFARO, J F, ZHANG, T, WYNN, D P, KARSCHNER, E L and ZHOU, Z S (2004), Synthesis of LuxS inhibitors targeting bacterial cell-cell communication, *Org Lett* **6**, 3043–3046.

ALTARRIBA, M, MERINO, S, GAVÍN, R, CANALS, R, RABAAN, A, SHAW, J G and TOMÁS, J M (2003), A polar flagella operon *(flg) of Aeromonas hydrophila* contains genes required for lateral flagella expression, *Microb Pathogen* **34**, 249–259.

AMMOR, M S, C MICHAELIDIS and NYCHAS G-J E, (2008), Insights into the role of quorum sensing in food spoilage, *J Food Prot*, **71**, 1510–1525.

ANDERSON, G G, PALERMO, J L, SCHILLING, J D, ROTH, R, HEUSER, J and HULTGREN, S J (2003), Intracellular bacterial biofilm-like pods in urinary tract infections, *Science*, **301**, 105–107.

ATKINSON, S, CHANG, C-Y, SOCKETT, R E, CÁMARA, M and WILLIAMS, P (2006), Quorum sensing in *Yersinia enterocolitica* controls swimming and swarming motility. *J Bacteriol*, **188**, 1451–1461.

AUGER, S, KRIN, E, AYMERICH, S and GOHAR, M (2006), Autoinducer 2 affects biofilm formation by *Bacillus cereus*, *Appl Environ Microbiol*, **72**, 937–941.

BALESTRINO, D, HAAGENSEN, J A J, RICH, C and FORESTIER, C (2005), Characterization of type 2 Quorum sensing in *Klebsiella pneumoniae* and relationship with biofilm formation, *J Bacteriol*, **187**, 2870–2880.

BANSAL, T, ENGLERT, D, LEE, J, HEGDE, M, WOOD, T K and JAYAARAMAN, A (2007), Differential effects of epinephrine, norepinephrine, and indole on *Escherichia coli* O157:H7 chemotaxis, colonization and gene expression, *Infect Immun*, **75**, 4597–4607.

BEARSON, B L and BEARSON, S M D (2007), The role of the QseC quorum sensor kinase in colonization and norepinephrine-enchanced motility of *Salmonella enterica* serovar Typhimurium, *Microb Pathogen*, **44**, 271–273.

BEENKEN, K E, DUNMAN, P M, MCALEESE, F, MACAPAGAL, D, MURPHY, E, PROJAN, S J, BLEVINS, J S and SMELTZER, M S (2004), Global gene expression in *Staphyloccus aureus* biofilms, *J Bacteriol*, **186**, 4665–4684.

BI, Z. X., Y. J. LIU and C. P. LU (2007), Contribution of AhyR to virulence of *Aeromonas hydrophila* J-1, *Res Vet Sci*, **83**, 159–156.

BLOKESCH, M and SCHOOLNIK, G R (2007), Serogroup conversion of *Vibrio cholerae* in aquatic reservoirs, *PLoS Pathog*, **3 (6)**, e81.

BODDICKER, J D, ANDERSON, R A, JAGNOW, J and CLEGG, S (2006), Signature-tagged mutagenesis of *Klebsiella pneumoniae* to identify genes that influence biofilm formation on extracellular matrix material, *Infect Immun*, **74**, 4590–4597.

BONIFIELD, H R and HUGHES, K T (2003), Flagellar phase variation in *Salmonella enterica* is mediated by a posttranscriptional control mechanism. *J. Bacteriol.* **185**, 3567–3574.

BOŞGELMEZ-TINAZ, G, S. ULUSORY, B. ARIDOĞAN, F. EROĞLU and S. KAYA (2005), N-butanoyl-L-homoserine lactone (BHL) deficient *Pseudomonas aeruginosa* isolates from an intensive care unit, *Microbiol Res*, **160**, 399–403.

BURTON, E, GAWANDE, P V, YAKANDAWALA, N, LO VETRI, K, ZHANEL, G G, ROMEO, T, FRIESEN, A D and MADHYASTHA, S (2006), Antibiofilm activity of GlmU enzyme inhibitors against catheter-associated uropathogens, *Antimicrob Agents Chemother*, **50**, 1835–1840.

BUSWELL, C M, HERLIHY, Y M, LAWRENCE, L M, MCGUIGGAN, J T M, MARSH, P D, KEEVIL, C W and LEACH, S A (1998), Extended survival and persistence of *Campylobacter* spp. in water and aquatic biofilms and their detection by immunofluorescent-antibody and rRNA staining, *Appl Environ Microbiol*, **64**, 733–741.

BYERS, J T, LUCAS, C., SALMOND, G P C and WELCH, M (2002), Nonenzymatic turnover of an *Erwinia carotovora* quorum-sensing signaling molecule, *J Bacteriol*, **184**, 1163–1171.

CAIAZZA, N C and O'TOOLE, G A (2003), Alpha-toxin is required for biofilm formation by *Staphylococcus aureus*, *J Bacteriol*, **185**, 3214–3217.

CAMPER, A K, LECHEVALLIER, M W, BROADWAY, S C and MCFETERS, G A (1985), Growth and persistence of pathogens on granular activated carbon filters, *Appl Environ Microbiol* **50**, 1378–1382.

CANALS, R, ALTARRIBA, M, VILCHES, S, HORSBURGH, G, SHAW, J G, TOLMÁS, J M and MERINO, S (2006a), Analysis of the lateral flagellar gene system of *Aeromonas hydrophila* AH-3. *J Bacteriol* **188**, 852–862.

CANALS, R, RAMIREZ, S, VILCHES, S, HORBURGH, G, SHAW, J G, TOMÁS, J M and MERINO, S (2006b), Polar flagellum biogenesis in *Aeromonas hydrophila*, *J Bacteriol*, **188**, 542–555.

CARTER, G P, PURDY, D, WILLIAMS, P and MINTON, N P (2005), Quorum sensing in *Clostridium difficile*: analysis of a *luxS*-type signaling system, *J Med Microbiol*, **54**, 119–127.

CASE, R J, LABBATE, M and KJELLEBERG, S (2008), AHL-driven quorum-sensing circuits: their frequency and function among the Proteobacteria, *ISME J*, **2**, 345–349.

CHAE, M S, SCHRAFT, H, HANSEN, L T and MACKERETH, R (2006), Effects of physicochemical surface characteristics of *Listeria monocytogenes* strains on attachment to glass, *Food Microbiol*, **23**, 250–259.

CHALLAN BELVAL, S, GAL, L, MARGIEWES, S, GARMYN, D, PIVETEAU, P and GUZZO, J (2006), Assessment of the roles of LuxS, S-ribosyl homocysteine, and autoinducer 2 in cell attachment during biofilm formation by *Listeria monocytogenes* EGD-e, *Appl Environ Microbiol*, **72**, 2644–2650.

CHIU, S Y, HA, S Y, CHAN, C F, WONG, S S Y, YUEN, K Y, CHAN, K L and LAU, Y L (1996), The use of central venous catheter in paediatric haematology/oncology unit, *HK J Paediatr (New Series)*, **1**, 64–66.

CHOI, J, SHIN, D and RYU, S (2007), Implication of quorum sensing in *Salmonella* virulence: the *luxS* gene is necessary for expression of genes in pathogenicity island 1, *Infect Immun* **75**, 4885–4890.

CHOWDHARY, P K, KESHAVAN, N, NGUYEN, H Q, PETERSON, J A, GONZÁLEZ, J E and HAINES, D C (2007), *Bacillus megaterium* CYP102A1 oxidation of acyl homoserine lactones and acyl homoserines, *Biochemistry*, **46**, 14429–14437.

CLARKE, M B and SPERANDIO, V (2005), Transcriptional regulation of *flhDC* by QseBC and $\sigma^{28}$ (FliA) in enterohaemorrhagic *Escherichia coli*, *Mol Microbiol*, **57**, 1734–1749.

CLARKE, M B, HUGHES, D T, ZHU, C, BOEDEKER, E C and SPERANDIO, V (2006), The QseC sensor kinase: a bacterial adrenergic receptor, *Proc Natl Acad Sci USA* **103**, 10420–10425.

CLOAK, O M, SOLOW, B T, BRIGGS, C E, CHEN, C-Y and FRATAMICO, P M (2002), Quorum sensing and production of autoinducer-2 in *Campylobacter* spp., *Escherichia coli* O157:H7 and *Salmonella enterica* serovar Typhimurium in foods, *Appl Environ Microbiol*, **68**, 4666–4671.

COLE, S P, HARWOOD, J, LEE, R, SHE, R and GUINEY, D G (2004), Characterization of monospecies biofilm formation by *Helicobacter pylori*, *J Bacteriol*, **186**, 3124–3132.

COOPER, I R, TAYLOR, H D and HANLON, G W (2007), Virulence traits associated with verocytotoxigenic *Escherichia coli* O157 recovered from freshwater films, *J Appl Microbiol*, **102**, 1293–1299.

COTICCHIA, J M., SUGAWA, C, TRAN, V R, GURROLA, J, KALWALSKI, E and CARRON, M A (2006), Presence and density of *Helicobacter pylori* biofilms in human gastric mucosa in patients with peptic ulcer disease, *J Gastrointest Surg*, **10**, 883–889.

COULTHURST, S J, KURZ, C L and SALMOND, G P C (2004), *luxS* mutants of *Serratia* defective in autoinducer-2-dependent 'quorum sensing' show strain-dependent impacts on virulence and production of carbapenem and prodigiosin, *Microbiology*, **150**, 1901–1910.

COULTHURST, S J, WILLIAMSON, N R, HARRIS, A K P, SPRING, D R and SALMOND, G P C (2006), Metabolic and regulatory engineering of *Serratia marcescens*: mimicking phage-mediated horizontal acquisition of antibiotic biosynthesis and quorum-sensing capacities, *Microbiology*, **152**, 1899–1911.

CRAMPTON, S E, GERKE, C, SCHNELL, N F, NICHOLS, W W, and GÖTZ, F (1999), The intracellular adhesion (ica) locus is present in *Staphylococcus aureus* and is required for biofilm formation, *Infect Immun*, **67**, 5427–5433.

DAHLBERG, C, BERGSTRÖM, M and HERMANSSON, M (1998), In situ detection of high levels of horizontal plasmid transfer in marine bacterial communities, *Appl Environ Microbiol*, **64**, 2670–2675.

DANESE, P N, PRATT, L A, DOVE, S L and KOLTER, R (2000a), The outer membrane protein, Antigen 43, mediates cell-to-cell interactions within *Escherichia coli* biofilms, *Mol Microbiol*, **37**, 424–432.

DANESE, P N, PRATT, L A and KOLTER, R (2000b), Exopolysaccharide production is required for development of *Escherichia coli* K-12 biofilm architecture, *J Bacteriol*, **182**, 3593–3596.

DANIELS, R, VANDERLEYDEN, J and MICHIELS, J (2004), Quorum sensing and swarming migration in bacteria, *FEMS Microbiol Revs*, **28**, 261–289.

DARWIN, K H and MILLER, V L (1999), InvF is required for expression of genes encoding proteins secreted by the SPI1 type III secretion apparatus in *Salmonella typhimurium*. *J Bacteriol*, **181**, 4949–4954.

DAVEY, M E and O'TOOLE, G A (2000), Microbial biofilms: from ecology to molecular genetics, *Microbiol Mol Biol Rev*, **64**, 847–867.

DAVIES, D G, PARSEK, M R, PEARSON, J P., IGLEWSKI, B H, COSTERTON, J W and GREENBERG, E P (1998), The involvement of cell-to-cell signals in the development of a bacterial biofilm, *Science*, **280**, 295–298.

DAY, W A and MAURELLI, A T (2001), *Shigella flexneri* LuxS quorum-sensing system modulates *virB* expression but is not essential for virulence, *Infect Immun*, **69**, 15–23.

DEFOIRDT, T, BOSSIER, P, SORGELOOS, P and VERSTRAETE, W (2005), The impact of mutations in the quorum sensing systems of *Aeromonas hydrophila*, *Vibrio anguillarum* and *Vibrio harveyi* on their virulence towards gnotobiotically cultured *Artemia franciscana*, *Environ Microbiol*, **7**, 1239–1247.

DEFOIRDT, T, CRAB, R, WOODS, T K, SORGELOOS, P, VERSTRAETE, W and BOSSIER, P (2006), Quorum sensing-disrupting brominated furanones protect the gnotobiotic brine shrimp *Artemia franciscana* from pathogenic *Vibrio harveyi*, *Vibrio campbellii* and *Vibrio parahaemolyticus* isolates, *Appl Environ Microbiol*, **72**, 6419–6423.

DE KEERSMAEACKER, S C J, VARSZEGI, C, VAN BOXEL, N, HABEL, L W, METZGER, K, R. DANIELS, R, MARCHAL, K, DE VOS, D and VANDERLELYDEN, J (2005), Chemical synthesis of (S)-4,5-dihydroxy-2,3-pentanedione, a bacterial signal molecule precursor, and validation of its activity in *Salmonella typhimurium*, *J Biol Chem*, **280**, 19563–19568.

DE NYS, R, WRIGHT, A D, KÖNIG, G M and STICHER, O (1993), New halogenated furanones from the marine alga *Delisea pulchra* (cf. *fimbriata*), *Tetrahedron*, **49**, 11213–11220.

DOHERTY, N, HOLDEN, M T G, QAZI, S N, WILLIAMS, P and WINZER, K (2006), Functional analysis of *luxS* in *Staphylococcus aureus* reveals a role in metabolism but not quorum sensing, *J Bacteriol*, **188**, 2885–2897.

DOMKA, J, LEE, J and WOOD, T K (2006), YlliH (BssR) and YceP (BssS) regulate *Escherichia coli* K-12 biofilm formation by influencing cell signaling, *Appl Environ Microbiol*, **72**, 2449–2459.

DONG, Y-H and ZHANG, L-H (2005), Quorum sensing and quorum-quenching enzymes, *J Microbiol*, **43** (special issue), 101–109.

DONG, Y-H, WANG, L-H, XU, J-L, ZHANG, H-B, ZHANG, X-F and ZHANG, L H (2001), Quenching quorum-sensing-dependent bacterial infection by an N-acyl homoserine lactonase, *Nature*, **411**, 813–817.

DONLAN, R M (2002), Biofilms: microbial life on surfaces, *Emerg Infect Dis* **8**, 881–890.

DYKES, G A, SAMPATHKUMAR, B and KORBER, D R (2003), Planktonic or biofilm growth affects survival, hydrophobicity and protein expression patterns of a pathogenic *Campylobacter jejuni* strain, *Int J Food Microbiol*, **89**, 1–10.

EBERL, L, WINSON,M K, STERNBERG, C, STEWART, G S A B, CHRISTIANSEN, G, CHHABRA, S R, BYCROFT, B, WILLIAMS, P, MOLIN, S and GIVSKOV, M (1996), Involvement of N-acyl-L-homoserine lactone autoinducers in controlling the multicellular behavior of *Serratia liquefaciens*, *Mol Microbiol*, **20**, 127–136.

EKKELENKAMP, M B, VAN DER BRUGGEN, T, VAN DE VIJVER, D A, WOLFS, T F and BONTEN, M J (2008), Bacteremic complications of intravascular catheters colonized with *Staphylococcus aureus*, *Clin Infect Dis*, **46**, 114–118.

ELVERS, K T and PARK, S F (2002), Quorum sensing in *Campylobacter jejuni*: detection of a *luxS* encoded signaling molecule, *Microbiology*, **148**, 1475–1481.

ENOS-BERLAGE, J L, GUVENER, Z T, KEENAN, C E and MCCARTER, L L (2005), Genetic determinants of biofilm development of opaque and translucent *Vibrio parahaemolyticus*, *Mol Microbiol*, **55**, 1160–1182.

ERMOLAEVA, S, NOVELLA, S, VEGA, Y, RIPLO, M-T, SCORTII, M and VÁZQUEZ-BOLAND, J A (2004), Negative control of *Listeria monocytogenes* virulence genes by a diffusible autorepressor, *Mol Microbiol*, **52**, 601–611.

FAILLE, C, JULLIEN, C, FONTAINE, F, BELLON-FONTAINE, M-N, SLOMIANNY, C and BENEZECH, T (2002), Adhesion of *Bacillus* spores amd *Escherichia coli* cells to inert surfaces: role of surface hydrophrobicity, *Can J Microbiol*, **48**, 728–738.

FARMER, J J, M. ARDUINO, J and HICKMAN-BRENNER, F W (1991), The genera *Aeromonas* and *Plesiomonas*, p. 3012–3045. In *The Prokaryotes*, Vol. III, 2nd Ed., edited by A. Balows, H. G. Trüper, M. Dworkin, W. Harder and K-H. Schleifer, Springer-Verlag New York, Inc.

FERRIÈRES, L, HANCOCK, V and KLEMM, P (2007), Specific selection for virulent urinary tract infectious *Escherichia coli* strains during catheter-associated biofilm formation, *FEMS Immunol Med Microbiol*, **51**, 212–219.

FINKELSTEIN, R A, BOESMAN-FINKELSTEIN, M, CHANG, Y and HÄSE, C C (1992), *Vibrio cholerae* hemagglutinin/protease, colonial variation, virulence, and detachment, *Infect Immun*, **60**, 472–478.

FLODGAARD, L R, CHRISTENSEN, A B, MOLIN, S, GIVSKOV, M and GRAM, L (2003), Influence of food preservation parameters and associated microbiota on production rate, profile and stability of acylated homoserine lactones from food-derived Enterobacteriaceae, *Int J Food Microbiol*, **84**, 145–156.

FONG, J C N and YILDIZ, F H (2007), the *rbmBCDEF* gene cluster modulates development of rugose colony morphology and biofilm formation in *Vibrio cholerae*, *J Bacteriol*, **189**, 2319–2330.

FONG, J C N, KARPLUS, K, SCHOOLNIK, G K, YILDIZ, F H (2006), Identification and characterization of RbmA, a novel protein required for the development of rugose colony morphology and biofilm structure in *Vibrio cholerae*, *J Bacteriol*, **188**, 1049–1059.

FORSYTH, M H and COVER, T L (2000), Intracellular communication in *Helicobacter pylori*: *luxS* is essential for the production of an extracellular signaling molelcule, *Infect Immun*, **68**, 3193–3199.

FREES, D, CHASTANET, A, QAZI, S, SØRENSEN, K, HILL, P, MSADEK, T and INGMER, H (2004), Clp ATPases are required for stress tolerance, intracellular replication and biofilm formation in *Staphylococcus aureus*, *Mol Microbiol*, **54**, 1445–1462.

FREZZA, M, SOULÈRE, L, BALELSTLRINO, D, GOHAR, M, DESHAYES, D, QUENEAU, Y, FORESTIER, C and DOUTHEAU, A (2007), $Ac_2$-DPD, the bis-(O)-acetylated derivative of 4,5-dihydroxy-2,3-pentanedione (DPD) is a convenient stable precursor of bacterial quorum sensing autoinducer AI-2, *Biorg Med Chem Lett*, **17**, 1428–1431.

FUX, C A, COSTERTON, J W, STEWART, P S and STOODLEY, P (2005), Survival strategies of infectious biofilms, *Trends Microbiol* **13**, 34–40.

GAVIN, R, RABAAN, A A, MERINO, S, TOMÁS, J M, GRYLLOS, I and SHAW, J G (2002), Lateral flagella of *Aeromonas* species are essential for epithelial cell adherence and biofilm formation, *Mol Microbiol*, **43**, 383–397.

GHIGO, J-M (2003), Are there biofilm-specific physiological pathways beyond a reasonable doubt? *Res Microbiol*, **154**, 1–8.

GIAOURIS, E D and NYCHAS, G-J E (2006), The adherence of *Salmonella* Enteritidis PT4 to stainless steel: the importance of the air-liquid interface and nutrient availability, *Food Microbiol*, **23**, 747–752.

GIRENNAVAR, B, CEPEDA, M L, SONI, K A, VIKRAM, A, JESUDHASAN, P, JAYAPRAKASHA, G K, PILLAI, S D and PATIL, B S (2008), Grapefruit juice and its furocoumarins inhibits autoinducer signaling and biofilm formation in bacteria, *Int J Food Microbiol*, **125**, 204–208.

GIVSKOV, M, DE NYS, R, MANEFIELD, M, GRAM, L, MAXIMILIEN, R, EBERL, E, MOLIN, S, STEINBERG, P D and KJELLEBERG, S (1996), Eukaryotic interference with homoserine lactone-mediated prokaryotic signaling, *J Bacteriol*, **178**, 6618–6622.

GIVSKOV, M, ÖSTLING, J, EBERL, L, LINDUM, P W, CHRISTENSEN, A B, CHRISTIANSEN, G, MOLIN, S and KJELLEBERG, S (1998), Two separate regulatory systems participate in control of swarming motility of *Serratia liquefacians* MG1, *J Bacteriol*, **180**, 742–745.

GOERKE, C, FLUCKIGER, U, STEINHUBER, A, ZIMMERLI, W and WOLZ, C (2001), Impact of the regulatory loci *agr*, *sarA* amd *sae* of *Staphylococcus aureus* on the induction of α-toxin during device-related infection resolved by direct quantitative transcript analysis, *Mol Microbiol*, **40**, 1439–1447.

GONZÁLEZ BARRIOS, A F, ZUO, R, REN, D and WOOD, T K (2005), Hha, YbaJ and OmpA regulate *E. coli* K12 biofilm formation and conjugation plasmids abolish motility, *Biotechnol Bioeng*, **93**, 188–200.

GONZÁLEZ BARRIOS, A F, ZUO, R, HASHIMOTO, Y,YANG, L, BENTLEYH, W E and WOOD, T K (2006), Autoinducer 2 controls biofilm formation in *E. coli* through a novel motility quorum-sensing regulator (MqsR, B3022), *J Bacteriol*, **188**, 305–316.

GÖTZ, F (2002), *Staphylococcus* and biofilms, *Mol Microbiol*, **43**, 1367–1378.

GRECO, C, MARTINCIC, I, GUSINJAC, A, KALAB, M, YANG, A-F and RAMÍREZ-ARCOS, S (2007), *Staphylococcus epidermidis* forms biofilms under simulated platelet storage conditions, *Transfusion*, **47**, 1143–1153.

GRINHOLC, M, WEGRZYN, G and KURLENDA, J (2007), Evaluation of biofilm production and prevalence of the *icaD* gene in methicillin-resistant and methicillin-susceptible *Stapylococcus aureus* strains isolated from patients with nosocomial infections and carriers, *FEMS Immunol Med Microbiol*, **5**, 375–379.

GROSS, M, CRAMTON, S E, GÖTZ, F and PESCHEL, A (2001), Key role of teichoic acid net charge in *Staphylococcus aureus* colonization of artificial surfaces, *Infect Immun*, **69**, 3423–3426.

GÜVENER, A T and MCCARTER, L L (2003), Multiple regulators control capsular polysaccharide production in *Vibrio parahaemolyticus, J Bacteriol*, **185**, 5431–5441.

HAMMER, B K and BASSLER, B L (2003), Quorum sensing controls biofilm formation in *Vibrio cholerae*, *Mol Microbiol*, **50**, 101–114.

HAMMER, B K and BASSLER, B L (2007), Regulatory small RNAs circumvent the conventional quorum sensing pathway in pandemic *Vibrio cholerae, Proc Natl Acad Sci USA*, **104**, 11145–11149.

HANCOCK, V and KLEMM, P (2007), Global gene expression profiling of asymptomatic bacteriuria *Escherichia coli* during biofilm growth in human urine, *Infect Immun*, **75**, 966–976.

HARDIE, K R, COOKSLEY, C, GREEN, A D and WINZER, K (2003), Autoinducer 2 activity in *Escherichia coli* culture supernatants can be actively reduced despite maintenance of an active synthase, LuxS, *Microbiology*, **149**, 715–728.

HARDIE, K R and HEURLIER, K (2008), Establishing bacterial communities by 'word of mouth': LuxS and autoinducer 2 in biofilm development, *Nat Rev Microbiol*, **6**, 635–648.

HARSHEY, R M (2003), Bacterial motility on a surface: many ways to a common goal, *Annu Rev Microbiol*, **57**, 249–273.

HE, Y, FRYE, J G, STROBAUGH, T P and CHEN, C-Y (2008), Analysis of AI-2/LuxS-dependent transcription in *Campylobacter jejuni* strain 81–176, *Foodborne Pathog Dis*, **5**, 399–415.

HEFFORD, M A, D'AOUST, S, CYR, T D, AUSTIN, J W, SANDERS, G, KHERADPIR, E and KALMOKOFF, M L (2005), Proteomic and microscopic analysis of biofilms formed by *Listeria monocytogenes* 568, *Can J Microbiol*, **51**, 197–208.

HEILMANN, C, SCHWEITZER, O, GERKE, C, VANITTANAKOM, N, MACK, D and GÖTZ, F (1996), Molecular basis of intercellular adhesion in the biofilm-forming *Staphylococcus epidermidis*, *Mol Micrrobiol*, **20**, 1083–1091.

HENDRICKX, L, HAUSNER, M and WUERTZ, S (2003), Natural genetic transformation in monoculture *Acinetobacter* sp. strain BD413 biofilms, *Appl Environ Microbiol*, **69**, 1721–1727.

HENKE, J M and BASSLER, B L (2004), Quorum sensing regulates type III secretion in *Vibrio harveyi* and *Vibrio parahaemolyticus*, *J Bacteriol*, **186**, 3794–3805.

HENTZER, M, RIEDEL, K, RASMUSSEN, T B, HEYDORN, A, ANDERSEN, J B, PARSEK, M R, RICE, S A, EBERT, L, MOLIN, S, HØIBY, N, KJELLEBERG, S and GIVSKOV, M (2002), Inhibition of quorum sensing in *Pseudomonas aeruginosa* biofilm bacteria by a halogenated furanone compound, *Microbiology*, **148**, 87–102.

HERNAIZ, C, PICARDO, A, ALOS, J I and GOMEZ-GARCES, J L (2003), Nosocomial bacteremia and catheter infection by *Bacillus cereus* in an immunocompetent patient, *Clin Microbial Infect*, **9**, 973–975.

HERZBERG, M, KAYE, I K, PETI, W and WOOD, T K (2006), YdgG (TqsA) controls biofilm formation in *Escherichia coli* K-12 through autoinducer 2 transport, *J Bacteriol*, **188**, 587–598.

HIGGINS, D A, POMIANEK, M E, KRAML, C M, TAYLOR, R K, SEMMELHACK, M F and BASSLER, B L (2007), The major *Vibrio cholerae* autoinducer and its role in virulence factor production, *Nature*, **450**, 883–886.

HOFFMAN, L R, D'ARGENIO, D A, MACCOSS, M J, ZHANG, Z, JONES, R A and MILLER, S I (2005), Aminoglycoside antibiotics induce bacterial biofilm formation, *Nature*, **436**, 1171–1175.

HSUEH, Y-H, SOMERS, E B, LERECLUS, D and WONG, A C L (2006), Biofilm formation by *Bacillus cereus* is influence by PlcR, a pleiotropic regulator, *Appl Environ Microbiol*, **72**, 5089–5092.

HUGHES, D T and SPERANDIO, V (2008), Inter-kingdom signaling: communication between bacteria and their hosts, *Nat Rev Microbiol*, **6**, 111–120.

HUNG, D T, ZHU, J, STURTEVANT, D and MEKALANOS, J J (2006), Bile acids stimulate biofilm formation in *Vibrio cholerae*, *Mol Microbiol*, **59**, 193–201.

ISLAM, M S, JAHID, M I K, RAHMAN, M M, RAHMAN, M Z, ISLAM, M S, KABIR, M S, SACK, D A and SCHOOLNHIK, G K (2007), Biofilm acts as a microenvironment for plankton-associated *Vibrio cholerae* in the aquatic environment of Bangladesh, *Microbiol Immunol*, **51**, 369–379.

IVERSEN, C, LANE, M and FORSYTHE, S J (2004), The growth profile, thermotolerance and biofilm formation of *Enterobacter sakazakii* grown in infant formula milk, *Lett Appl Microbiol*, **38**, 378–382.

JACKSON, D W, SUZUKI, K, OAKFORD, L, SIMECKA, J W, HART, M E and ROMEO, T (2002), Biofilm formation and dispersal under the influence of the global regulator CsrA of *Escherichia coli*, *J Bacteriol*, **184**, 290–301.

JACQUES, S and MCCARTER, L L (2006), Three new regulators of swarming in *Vibrio parahaemolyticus*, *J Bacteriol*, **188**, 2625–2635.

JAMES, G A, SWOGGER, E, WOLCOTT, R, PULCINI, E, SECOR, P, SESTRICH, J, COSTERTON, J W and STEWART, P S (2008), Biofilmls in chronic wounds, *Wound Repair Regen*, **16**, 37–44.

JARRETT, C O, DEAK, E, ISHERWOOD, K E, OYSTON, P C, FISHER, E R, WHITNEY, A R, KOBAYASHI, S D, DELEO, F R and HINNEBUSCH, B J (2004), Transmission of *Yersinia pestis* from an infectious biofilm in the flea vector, *J Infect Dis*, **190**, 783–792.

JEFFERSON, K J (2004), What drives bacteria to produce a biofilm?, *FEMS Microbiol Lett*, **236**, 163–173.

JEON, B, ITOH, K, MISAWA, N and RYU, S (2003), Effect of quorum sensing on flaA transcription and autoagglutination in *Campylobacter jejuni*, *Microbiol Immunol*, **47**, 833–839.

JEON, B, ITOH, K and RYU, S (2005), Promoter analysis of cytolethal distending toxin genes (*cdtA*, *B* and *C*) and effect of a *luxS* mutation on CDT production in *Campylobacter jejuni*, *Microbiol Immunol*, **49**, 599–603.

JOELSSON, A, LIU, Z and ZHU, J (2006), Genetic and phenotypic diversity of quorum-sensing systems in clinical and environmental isolates of *Vibrio cholerae*, *Infect Immun*, **74**, 1141–1147.

JOELSSON, A, KAN, B and ZHU, J (2007), Quorum sensing enhances the stress response in *Vibrio cholerae*, *Appl Environ Microbiol*, **73**, 3742–3746.

JONES, G L, MULLER, C T, O'REILLY, M and STICKLER, D J (2006), Effect of triclosan on the development of bacterial biofilms by urinary tract pathogens on urinary catheters, *J Antimicrob Chemotherap*, **57**, 266–272.

JOSEPH, L A and WRIGHT, A C (2004), Expression of *Vibrio vulnificus* polysaccharide inhibits biofilm formation, *J Bacteriol*, **186**, 889–893.

JOSHUA, G W P, KARLYSHEV, A V, SMITH, M P, ISHERWOOD, K E, TITBALL, R W and WREN, B W (2003), A *Caenorhabditis elegans* model of *Yersinia* infection: biofilm formation on a biotic surface, *Microbiology*, **149**, 3221–3229.

JOSHUA, G W P, GUTHRIE-IRONS, C, KARLYSHEV, A V and WREN, B W (2006), Biofilm formation in *Campylobacter jejuni*, *Microbiology*, **152**, 387–396.

JOYCE, E A, BASSLER, B L and WRIGHT, A (2000), Evidence for a signaling system in *Helicobacter pylori*: detection a a *luxS*-encoded autoinducer, *J Bacteriol*, **182**, 3638–3643.

KALMOKOFF, M, LANTHIER, P, TREMBLAY, T-L, FOSS, M, LAU, P C, SANDERS, G, AUSTIN, J, KELLY, J and SZYMANSKI, C M (2006), Proteomic analysis of *Campylobacter jejuni* 11168 biofilms reveals a role for the motility complex in biofilm formation, *J Bacteriol*, **188**, 4312–4320.

KANAMARU, K, KANAMARU, K, TATSUNO, I, TOBE, T and SASAKAWA, C (2000), SdiA, an *Escherichia coli* homologue of quorum sensing regulators, controls the expression of virulence factors in enterohaemorrhagic *Escherichia coli* O157:H7, *Mol Microbiol*, **38**, 805–816.

KARAVOLOS, M H, BULMER, D M, WINZER, K, WILSON, M, MASTROENI, P, WILLIAMS, P and KHAN, C M A (2008), LuxS affects flagellar phase variation independently of quorum sensing in *Salmonella enterica* sv. Typhimurium, *J Bacteriol*, **190**, 769–771.

KENDALL, M M, RASKO, D A and SPLERANDIO, V (2007), Global effects of the cell-to-cell signaling molecules autoinducer-2, autoinducer-3, and epinephrine in a luxS mutant of enterohemorrhagic *Escherichia coli*, *Infect Immun*, **75**, 4875–4884.

KIKUCHI, T, MIZUNOE, Y, TAKADE, A, NAITO, S and YOSHIDA, S (2005), Curli fibers are required for development of biofilm architecture in *Escherichia coli* K-12 and enhance bacterial adherence to human uroepithelial cells, *Microbiol Immunol*, **49**, 875–884.

KIM, H, RYU, J-H and BEUCHAT, L R (2006), Attachment of and biofilm formation by *Enterobacter sakazakii* on stainless steel and enteral feeding tubes, *Appl Environ Microbiol*, **72**, 5846–5856.

KIM, H, RYU, J-H and BEUCHAT, L R (2007), Effectiveness of disinfectants in killing *Enterobacter sakazakii* in suspension, dried on the surface of stainless steel, and in a biofilm, *Appl Environ Microbiol*, **73**, 1256–1265.

KIM, S-H and WEI, C-I (2007), Biofilm formation by multidrug-resistant *Salmonella enterica* serotype Typhimurium phage type DT104 and other pathogens, *J Food Prot*, **70**, 22–29.

KIM, S Y, LEE, S E, KIM, Y R, KIM, C M, RYU, P Y, CHOY, H E, CHUNG, S S and RHEE, J H (2003), Regulation of *Vibrio vulnificus* virulence by the LuxS quorum-sensing system, *Mol Microbiol*, **48**, 1647–1664.

KIM, W and SURETTE, M G (2006), Coordinated regulation of two independent cell-cell signaling systems and swarmer differentiation in *Salmonella enterica* serovar Typhimurium, *J Bacteriol*, **188**, 431–440.

KIRILLINA, O, FETHERSTON, J D, BOBROV, A G, ABNEY, J and PERRY, R D (2004), HmsP, a putative phosphodiesterase, and HmsT, a putative diguanylate cyclase, control Hms-dependent biofilm formation in *Yersinia pestis*, *Mol Microbiol*, **54**, 75–88.

KIRKE, D F, SWIFT, S, LYNCH, M J and WILLIAMS, P (2004), The *Aeromonas hydrophila* LuxR homologue AhyR regulates the N-acyl homoserine lactone synthase, AhyI positively and negatively in a growth phase-dependent manner, *Microbiol Lett*, **241**, 109–117.

KIRWAN, J P, GOULD, T A, SCHWEIZER, H P, BEARDEN, S W, MURPHY, R C and CHURCHILL, M E A (2006), Quorum-sensing signal synthesis by the *Yersinia pestis* acyl-homoserine lactone synthase YspI, *J Bacteriol*, **188**, 784–788.

KOCH, A and ARVAND, M (2005), Recurrent bacteraemia by 2 different *Bacillus cereus* strains related to 2 different central venous catheters, *Scand J Infect Dis*, **37**, 772–774.

KOH, K S, LAM, K W, ALHEDE, M, QUECK, S Y, LABBATE, M, KJELLEBERG, S and RICE, S A (2007), Phenotypic diversification and adaptation of *Serratia marcescens* MG1 biofilm-derived morphotypes, *J Bacteriol*, **189**, 119–130.

KONG, K-F, VUONG, C and OTTO, M (2006), *Staphylococcus* quorum sensing in biofilm formation and infection, *Int J Med Microbiol*, **296**, 133–139.

KRESKE, A C, RYU, J-H, PETTIGREW, C A and BEUCHAT, L R (2006), Lethality of chlorine, chlorine dioxide and a commercial produce sanitizer to *Bacillus cereus* and *Pseudomonas* in a liquid detergent, on stainless steel, and in biofilm, *J Food Prot*. **69**, 2621–2634.

LABBATE, M, QUECK, S Y, KOH, K S, RICE, S A, GIVSKOV, M and KJELLEBERG, S (2004), Quorum sensing-controlled biofilm development in *Serratia marcescens* MG1, *J Bacteriol*, **186**, 692–698.

LABBATE, M, ZHU, H, THUNG, L, BANDARA, R, LARSEN, M R, WILLCOX, M D P, GIVSKOV, M, RICE, S A and KJELLEBERG, S (2007), Quorum-sensing regulation of adhesion in *Serratia marcescens* MG1 is surface dependent, *J Bacteriol*, **189**, 2702–2711.

LATASA, C, ROUX, A, TOLEDO-ARANA, A, GHIGO, J-M, GAMAZO, C, PENADÉS, J R and LASA, I (2005), BapA, a large secreted protein required for biofilm formation and host colonization of *Salmonella enterica* serovar Enteritidis, *Mol Microbiol*, **58**, 1322–1339.

LEBARON, P, BAUDE, P, LETT, M C, DUVAL-IFLAH, Y, SIMONET, P, JACQ, E, FRANK, N, ROUX, B, BALEUX, B, FAURIE, G, HUBERT, J C, NORMAND, P, PRIEUR, D, SCHMIDT, S and BLOCK, J C (1997), Recombinant plasmid mobilization between *E. coli* strains in seven sterile microcosms, *Can J Microbiol*, **43**, 534–540.

LEBERT, I, LEROY, S and TALON, R (2007), Effect of industrial and natural biocides on spoilage, pathogenic and technological strains grown in biofilm, *Food Microbiol*, **24**, 281–287.

LEDEBOER, N A, FRYE, J G, MCCLELLAND, M and JONES, B D (2006), *Salmonella enterica* serovar Typhimurium requires the lpf, pef, and tafi fimbriae for biofilm formation on HEp-2 tissue culture cells and chicken intestinal epithelium, *Infect Immun*, **74**, 3156–3169.

LEE, A S Y and SONG, K P (2005), LuxS/autoinducer-2 quorum sensing molecule regulates transcriptional virulence gene expression in *Clostridium difficile*, *Biochem Biophys Res Comm*, **335**, 659–666.

LEE, J, BANSAL, T, JAYARAMAN, A, BENTLY, W E and WOOD, T K (2007a), Enterohemorrhagic *Escherichia coli* biofilms are inhibited by 7-hydroxyindole and stimulated by isatin, *Appl Environ Microbiol*, **73**, 4100–4109.

LEE, J, JAYARAMAN, A and WOOD, T K (2007b), Indole is an inter-species biofilm signal mediated by SdiA, *BMC Microbiol*, **7**, 42.

LEE, J-H, RHO, J B, PARK, K-J, KIM, C B, HAN, Y-S, CHOI, S H, LEE, K-H and PARK, S-J (2004), Role of flagellum and motility in pathogenesis of *Vibrio vulnificus*, *Infect Immun*, **72**, 4905–4910.

LEHNER, A, RIEDEL, K, EBERL, L, BREEUWER, P, DIEP, B and STEPHAN, R (2005), Biofilm formation, Extracellular polysaccharide production, and cell-to-cell signaling in various *Enterobacter sakazakii* strains: aspects promoting environmental persistence, *J Food Prot*, **68**, 2287–2294.

LEMON, K ., HIGGINS, D E and KOLTER, R (2007), Flagellar motility is critical for *Listeria monocytogenes* biofilm formation, *J Bacteriol*, **189**, 4418–4424.

LENZ, D H and BASSLER, B L (2007), The small nucleoid protein Fis is involved in *Vibrio cholerae* quorum sensing, *Mol Microbiol*, **63**, 859–871.

LEUNG, J W, LIU, Y, CHAN, R C Y, TANG, Y, MINA, Y, CHENG, A F and SILVA, J (2000), Early attachment of anaerobic bacteria may play an important role in biliary stent blockage, *Gastrointest Endosc*, **52**, 725–729.

LI, M, VILLARUZ, A E, VADYVALOO, V, STURDEVANT, D E and OTTO, M (2008), AI-2-dependent gene regulation in *Stapylococcus epidermidis*, *BMC Microbiol*, **8**, 4.

LI, Y-H, LAU, P C Y, LEE, J H, ELLEN, R P and CVITKOVITCH, D G (2001), Natural genetic transformation of *Streptococcus mutans* growing in biofilms, *J Bacteriol*, **183**, 897–908.

LICHT, T R, CHRISTENHSEN, B B, KROGFELT, K A and MOLIN, S (1999), Plasmid transfer in the animal intestine and other dynamic bacterial populations: the role of community structure and environment, *Microbiology*, **145**, 2615–2622.

LIN, W, KOVACIKOVA, G and SKORUPSKI, K (2007), The quorum sensing regulator HapR downregulates the expressiojn of the virulence gene transcription factor AphA in *Vibrio cholerae* by antagonizing Lrp- and VpsR-mediated activation, *Mol Microbiol*, **64**, 953–967.

LINDSAY, D, BRÖZEL, V S and VON HOLY, A (2006), Biofilm-spore resonse in *Bacillus cereus* and *Bacillus subtilis* during nutrient limitation, *J Food Prot*, **69**, 1168–1172.

LINDUM, P W, ANTHONI, U, CHRISTOPHERSEN, C, EBERL, L, MOLIN, S and GIVSKOV, M (1998), N-acyl-L-homoserine lactone autoinducers control production of an extracellular lipopeptide biosurfactant required for swarming motility of *Serratia liquefaciens* MG1, *J Bacteriol*, **180**, 6384–6388.

LIU, D, THOMAS, P W, MOMB, J, HOANG, Q Q, PETSKO, G A, RINGE, D and FAST, W (2007), Structure and specificity of a quorum-quenching lactonase (AiiB) from *Agrobacterium tumefaciens*, *Biochemistry*, **46**, 11789–11799.

LOH, J T, FORSYTH, M H and COVER, T L (2004), Growth phase regulation of *flaA* expression in *Helicobacter pylori* is *luxS* dependent, *Infect Immun*, **72**, 5506–5510.

LOSTROH, C P and LEE, C A (2001), The *Salmonella* pathogenicity isaland-1 type III secretion system, *Microb Infect*, **3**, 1281–1291.

LYNCH, M J, SWIFT, S, KIRKE, D F, KEEVIL, G W, DODD, C E R and WILLIAMS, P (2002), The regulation of biofilm development by quorum sensing in *Aeromonas hydrophila*, *Environ Microbiol*, **4**, 18–28.

MAEDA, S, ITO, M, ANDO, T, ISHIMOTO, Y, FUJISAWA, Y, TAKAHASHI, H, MATSUDA, A, SAWAMURA, A and KATO, S (2006), Horizontal transfer of nonconjugative plasmids in a colony biofilm of *Escherichia coli*, *FEMS Microbiol Lett*, **255**, 115–120,

MALCOVA, M, HRADECKA, H, KARPISKOVA, R and RYCHLIK, I (2008), Biofilm formation in field strains of *Salmonella enterica* serovar Typhimurium: identification of a new colony morphology type and the role of SGI1 in biofilm formation, *Vet Microbiol*, **129**, 360–364.

MANEFIELD, M, RASMUSSEN, T B, HENZTER, M, ANDERSEN, J B, STEINBERG, P, KJELLBERG, S and GIVSKOV, M (2002), Halogenated furanones inhibit quorum sensing through accelerated LuxR turnover, *Microbiology*, **148**, 1119–1127.

MARCO-NOALES, E, MILÁN, M, FOUZ, B, SANJUÁN, E and AMARO, C (2001), Transmission to eels, portals of entry, and putative reservoirs of *Vibrio vulnificus* serovar E (biotype 2), *Appl Environ Microbiol*, **67**, 4717–4725.

MARTIN, C A, HOVEN, A D and COOK, A M (2008), Therapeutic frontiers: preventing and treating infectious diseases by inhibiting bacterial quorum sensing, *Eur J Clin Microbiol Infect Dis*, **27**, 635–642.

MATZ, C, MCDOUGALD, D, MORENO, A M, YUNG, P Y, YILDIZ, F H and KJELLEBERG, S (2005), Biofilm formation and phenotypic variation enhance predation-driven persistence of *Vibrio cholerae*, *Proc Natl Acad Sci USA*, **102**, 16819–16824.

MCCARTER, L L (1998), OpaR, a homolog of *Vibrio harveyi* LuxR, controls opacity of *Vibrio parahaemolyticus*, *J Bacteriol*, **180**, 3166–3173.

MCDOUGALD, D, RICE, S A and KJELLEBERG, S (2001), SmcR-dependent regulation of adaptive phenotypes in *Vibrio vulnificus*, *J Bacteriol*, **183**, 758–762.

MCDOUGALD, D, LIN, W H, RICE, S A and KJELLEBERG, S (2006), The role of quorum sensing and the effect of environmental conditions on biofilm formation by strains of *Vibrio vulnificus*, *Biofouling*, **22**, 133–144.

MCDOUGALD, D, RICE, S A and KJELLEBERG, S (2007), Bacterial quorum sensing and interference by naturally occurring biomimics, *Anal Bioanal Chem*, **387**, 445–453.

MEDELLIN-PEÑA, M J, WANG, H, JOHNSON, R, ANAND, S and GRIFFITHS, M W (2007), Probiotics affect virulence-related gene expression of *Escherichia coli*, *Appl Environ Microbiol*, **73**, 4259–4267.

MEDINA-MARTÍNEZ, M S, UYTTENDAELE, M, MEIREMAN, S and DEBEVERE, J (2006), Relevance of N-acyl-L-homoserine lactone production by *Yersinia enterocolitica* in fresh foods, *J Appl Microbiol*, **102**, 1150–1158.

MEDINA-MARTÍNEZ, M S, UYTTENDAELE, M, RAJKOVIC, A, NADAL, P and DEBEVERE, J (2007), Degradation of N-acyl-L-homoserine-lactones by *Bacillus cereus* in culture media and pork extract, *Appl Environ Microbiol*, **73**, 2329–2332.

MELCHIOR, M B, VAARKAMP, H and FINK-GREMMELS, J (2006), Biofilms: a role in recurrent mastitis infections?, *Vet J*, **171**, 398–407.

MIDDLETON, B, RODGERS, H C, CÁMARA, M, KNOX, A J, WILLIAMS, P and HARDMAN, A (2002), Direct detection of N-acylmonoserine lactones in cystic fibrosis sputum, *FEMS Microbiol Lett*, **207**, 1–7.

MILLER, M B and BASSLER, B L (2001), Quorum sensing in bacteria, *Annu Rev Microbiol*, **55**, 165–199.

MILLER, M B, SKORUPSKI, K, LENZ, D H, TAYLOR, R K and BASSLER, B L (2002), Parallel quorum sensing systems converge to regulate virulence in *Vibrio cholerae*, *Cell*, **110**, 303–314.

MILLER, S T, XAVIER, K B, CAMPAGNA, S R, TAGA, M E, SEMMELHACK, M F, BASSLER, B L and HUGHSON, F M (2004), *Salmonella typhimurium* recognizes a chemically distinct form of the bacterial quorum-sensing signal AI-2, *Mol Cell*, **15**, 677–687.

MIYAIRI, S, TATEDA, K, FUSE, E T, UEDA, C, SAITO, H, TAKABATAKE, T, ISHII, Y, HORIKAWA, M, ISHIGURO, M, STANDIFORD, T J and YAMAGUCHI, K (2006), Immunization with 3-oxododecanoyl-L-homoserine lactone–protein conjugate protects mice from lethal *Pseudomonas aeruginosa* lung infection, *J Med Microbiol*, **55**, 1381–1387.

MOREIRA, C G, PALMER, K, WHITELY, M, SIRCILI, M P, TRABULSI, L R, CASTRO, A F P and SPERANDIO, V (2006), Bundle-forming pili and EspA are involved in biofilm formation by enteropathogenic *Escherichia coli*, *J Bacteriol*, **188**, 3952–3961.

MORIN, D, GRASLAND, B, VALLÉE-RÉHEL, K, DUFAU, C and HARAS, D (2003), On-line high-performance liquid chromatography-mass spectrometric detection and quantification of N-acylhomoserine lactones, quorum sensing signal molecules, in the presence of biological matrices, *J Chromatogr A*, **1002**, 79–92.

MOROHOSHI, T, SHIONO, T, TAKIDOUCHI, K, KATO, M, KATO, N, KATO, J and IKEDA, T (2007), Inhibition of quorum sensing in *Serratia marcescens* AS-1 byh the synthetic analogs of N-acylhomoserine lactone, *Appl Environ Microbiol*, **73**, 6339–6344.

NGWAI, Y B, ADACHI, Y, OGAWA, Y and H. HARA, H (2006), Characterization of biofilm-forming abilities of antibiotic-resistant *Salmonella typhimurium* DT104 on hydrophobic abiotic surfaces, *J Microbiol Immunol Infect*, **39**, 278–291.

NI, N, CHOU, H-T, WANG, J, LI, M, LU,, C-D, TAI, P C and WANG, B (2008), Identification of boronic acids as antagonists of bacterial quorum sensing in *Vibrio harveyi*, *Biochem Biophys Res Comm*, **369**, 590–594.

NIEMIRA, B A (2007), Irradiation sensitivity of planktonic and biofilm-associated *Escherichia coli* O157:H7 isolates is influenced by culture conditions, *Appl Environ Microbiol*, **73**, 3239–3244.

NIEMIRA, B A and SOLOMON, E B (2005), Sensitivity of planktonic and biofilm-associated *Salmonella* spp. to ionizing radiation, *Appl Environ Microbiol*, **71**, 2732–2736.

NUIJTEN, P J M, VAN ASSTEN, F J A M, GAASTRA, W and VAN DER ZEIJST, B A M (1990), Structural and functional analysis of two *Campylobacter jejuni* flagellin genes, *J Biol Chem*, **265**, 17798–17804.

OHTANI, K, HAYASHI, H and SHIMIZU, T (2002), The *luxS* gene is involved in cell-cell signaling for toxin production in *Clostridium perfringens*, *Mol Microbiol*, **44**, 171–179.

OOSTHUIZEN, M C, STEYN, B, THERON, J, COSETTE, P, LINDSAY, D, VON HOLY, A and BRÖZEL, V S (2002), Proteomic analysis reveals differential protein expression by *Bacillus cereus* during biofilm formation, *Appl Environ Microbiol*, **68**, 2770–2780.

ORME, R, DOUGLAS, C W, RIMMER, S and WEBB, M (2006), Proteomic analysis of *Escherichia coli* biofilms reveals the overexpression of the outer membrane protein OmpA, *Proteomics*, **6**, 4269–4277.

ORTORI, C A, ATKINSON, S, CHHABRA, S R, CÁMARA, M, WILLIAMS, P and BARRETT, D A (2007), Comprehensive profiling of N-acylhomoserine lactones produced by *Yersinia pseudotuberculosis* using liquid chromatography coupled to hybrid quadrupole-linear ion trap mass spectrometry, *Anal Bioanal Chem*, **387**, 497–511.

OSAKI, T, HANAWA, T, MANZOKU, T, FUKUDA, M, KAWAKAMI, H, SUZUKI, H, YAMAAGUCHI, H, YAN, X, TAGUCHI, H, KURATA, S and KAMIYA, S (2006), Mutation of *luxS* affects motility and infectivity of *Helicobacter pylori* in gastric mucosa of a Mongolian gerbil model, *J Med Microbiol*, **55**, 1477–1485.

OTTO, M (2001), *Staphylococcus aureus* and *Staphylococcus epidermidis* peptide pheromones produced by the accessory gene regulator *agr* system, *Peptides*, **22**, 1603–1608.

OTTO, M (2004), Quorum-sensing control in staphylococci – a target for antimicrobial drug therapy?, *FEMS Microbiol Lett*, **241**, 135–141.

PAN, Y, BREIDT, F and KATHARIOU, S (2006), Resistance of *Listeria monocytogenes* biofilms to sanitizing agents in a stimulated food processing environment, *Appl Environ Microbiol*, **72**, 7711–7717.

PARANJPYE, R N and STROM, M S (2005), A *Vibrio vulnificus* type IV pilin contributes to biofilm formation, adherence to epithelial cells and virulence, *Infect Immun*, **73**, 1411–1422.

PARK, J, JAGASIA, R, KAUFMANN, G F, MATHISON, J C, RUIZ, D I, MOSS, J A, MEIJLER, M M, ULEVICH, R J and JANDA, K D (2007), Infection control by antibody disruption of bacterial quorum sensing signaling, *Chem Biol*, **14**, 1119–1127.

PARK, S R, MACKAY, W G and REID, D C (2001), *Helicobacter* sp. recovered from drinking water biofilm sampled from a water distribution system, *Wat Res*, **35**, 1624–1626.

PATEL, C N, WORTHAM, B W, LINES, J L, FETHERSTON, J D, PERRY, R D and OLIVEIRA, M A (2006), Polyamines are essential for the formation of plague biofilm, *J Bacteriol*, **188**, 2355–2363.

PAWAR, D M, ROSSMAN, M L and CHEN, J (2005), Role of curli fimbriae in mediating the cells of enterohaemorrhagic *Escherichia coli* to attach to abiotic surfaces, *J Appl Microbiol*, **99**, 418–425.

PENG, J-S, TSAI, W-C and CHOU, C-C (2002), Inactivation and removal of *Bacillus cereus* by sanitizer and detergent, *Int J Food Microbiol*, **77**, 11–18.

PÉREZ-CONESA, D, MCLANDSBOROUGH, L and WEISS, J (2006), Inhibition and inactivation of *Listeria monocytogenes* and *Escherichia coli* O157:H7 colony biofilms by micellar-encapsulated eugenol and carvacrol, *J Food Prot*, **69**, 2947–2954.

PRIGENT-COMBARET, C, PRENSIER, G, LE THI, T T, VIDAL, O, LEJEUNE, P and DOREL, C (2000), Developmental pathway for biofilm formation in curli-producing *Escherichia coli* strains: role of flagella, curli and colonic acid, *Environ Microbiol*, **2**, 450–464.

PROUTY, A M, SCHWESINGER, W H and GUNN, J S (2002), Biofilm formation and interaction with the surfaces of gallstones by *Salmonella* spp, *Infect Immun*, **70**, 2640–2649.

QIN, Z, YANG, S, YANG, L, JIANG, J, OU, Y, MOLIN, S and QU, D (2007), Formation and properties of *in vitro* biofilms of *ica*-negative *Staphylococcus epidermidis* clinical isolates, *J Med Microbiol*, **56**, 83–93.

RADER, B A, CAMPAGNA, S R, SEMMELHACK, M F, BASSLER, B L and GUILLLEMIN, K (2007), The quorum sensing molecule AI-2 regulates motility and flagellar morphogenesis in *Helicobacter pylori*, *J Bacteriol*, **189**, 6109–6117.

RANI, S A, PITTS, B, BEYENAL, H, VELUCHAMPY, R A, LEWANDOWSKI, Z, DAVISON, W M, BUCKINGHAM-MEYER, K and STEWART, P S (2007), Spatial patterns of DNA replication, protein synthesis, and oxygen concentration within bacterial biofilms reveal diverse physiological states, *J Bacteriol*, **189**, 4223–4233.

RASMUSSEN, T B and GIVSKOV, M (2006), Quorum sensing inhibitors: a bargain of effects, *Microbiology*, **152**, 895–904.

READING, N, TORRES, A G, KENDALL, M, HUGHES, D T, YAMAMOTO, K and SPERANDIO, V (2007), A novel two-component signaling system that activates transcription of an enterohemorrhagic *E. coli* (EHEC) effector involved in remodeling of host actin, *J Bacteriol*, **189**, 2468–2476.

REESER, R J, MEDLER, R T, BILLINGTON, S J, JOST, B H and JOENS, L A (2007), Characterization of *Campylobacter jejuni* biofilms under defined growth conditions, *Appl Environ Microbiol*, **73**, 1908–1913.

REISNER, A, HAAGENSEN, J A J, SCHEMBRI, M A, ZECHNER, E L and MOLIN, S (2003), Development and maturation of *Escherichia coli* K-12 biofilms, *Mol Microbiol*, **48**, 933–946.

REN, D, SIMS, J J and WOOD, T K (2001), Inhibition of biofilm formation and swarming of *Escherichia coli* by (5Z)-4-bromo-5-(bromomethylene)-3-butyl-2(5H)-furanone, *Environ Microbiol*, **3**, 731–736.

REN, D, BEDZYK, L A, THOMAS, S M, YE, R W and WOOD, T K (2004a), Gene expression in *Escherichia coli* biofilms, *Appl Microbiol Biotechnol*, **64**, 515–524.

REN, D, BEDZYK, L A, YE, R W, THOMAS, S M and WOOD, T K (2004b), Differential gene expression shows natural brominated furanones interfere with autoinducer-2 bacterial signaling system of *Escherichia coli*, *Biotechnol Bioeng*, **88**, 639–642.

RESCH, A, LEICHT, S, SARIC, M, PÁSZTOR, L, JAKOB, A, GÖTZ, F and NORDHEIM, A (2006), Comparative proteome analysis of *Staphylococcus aureus* biofilm and planktonic cells and correlation with transcriptome profiling, *Proteomics*, **6**, 1867–1877.

RICE, S A, KOH, K S, QUECK, S Y, LABBATE, M, LAM, K W and KJELLEBERG, S (2005), Biofilm formation and sloughing in *Serratia marcescens* are controlled by quorum sensing and nutrient cues, *J Bacteriol*, **187**, 3477–3485.

RIEU, A, WEIDMANN, S, GARMYN, D, PIVETEAU, P and GUZZO, J (2007), *agr* system of *Listeria monocytogenes* EGD-e: role in adherence and differential expression pattern, *Appl Environ Microbiol*, **73**, 6125–6133.

ROBERTS, A, MULLANY, P and WILSON, M (2001), Gene transfer in bacterial biofilms, p. 60–65. In R. J. Doyle (ed.), *Methods in Enzymology*, vol. **336**, Academic Press, New York.

ROBERTS, A P, PRATTEN, J, WILSON, M and MULLANY, P (1999), Transfer of a conjugative transposon, Tn*5397* in a model oral biofilm, *FEMS Microbiol Lett*, **177**, 63–66.

ROCHE, D M, BYERS, J T, SMITH, D S, GLANSDORP, F G, SPRING, D R and WELCH, M (2004), Communications blackout? Do N-acylhomoserinelactone-degrading enzymes have any role in quorum sensing?, *Microbiology*, **150**, 2023–2028.

RODRÍGUEZ, A and MCLANDSBOROUGH, L A (2007), Evaluation of the transfer of *Listeria monocytogenes* from stainless steel and high-density polyethylene to bologna and American cheese, *J Food Prot*, **70**, 600–606.

RODRÍGUEZ, A, AUTIO, W R and MCLANDSBOROUGH, L A (2007), Effect of biofilm dryness on the transfer of *Listeria monocytogenes* biofilms grown on stainless steel to bologna and hard salami, *J Food Prot*, **70**, 2480–2484.

RUSSELL, R M, SHARP, F C, RASKO, D A and SPERANDIO, V (2007), QseA and GrflR/GrlA regulation of the locus of enterocyte effacement genes in enterohemorrhagic *Escherichia coli*, *J Bacteriol*, **189**, 5387–5392.

RYU, J-H, KIM, H, FRANK, J F and BEUCHAT, L R (2004), Attachment and biofilm formation on stainless steel by *Escherichia coli* O157:H7 as affected by curli production, *Lett Appl Microbiol*, **39**, 359–362.

SANDERS, S Q, BOOTHE, D H, FRANK, J F and ARNOLD, J W (2007), Culture and detection of *Campylobacter jejuni* within mixed microbial populations of biofilms on stainless steel, *J Food Prot*, **70**, 1379–1385.

SAUER, K (2003), The genomics and proteomics of biofilm formation, *Genome Biol*, **4**, 219.

SCHER, K, ROMLING, U and YARON, S (2005), Effect of heat, acidification, and chlorination on *Salmonella enterica* serovar Typhimurium cells in a biofilm formed at the air-liquid interface, *Appl Environ Microbiol*, **71**, 1163–1168.

SELA, S, FRANK, S, BELAUSOV, E and PINTO, R (2006), A mutation in the *luxS* gene influences *Listeria monocytogenes* biofilm formation, *Appl Environ Microbiol*, **72**, 5653–5658.

SHAO, C-P and HOR, L-I (2001), Regulation of metalloprotease gene expression in *Vibrio vulnificus* by a *Vibrio harveyi* LuxR homologue, *J Bacteriol*, **183**, 1369–1375.

SHARP, F C and SPERANDIO, V (2007), QseA directly activates transcription of LEE1 in enterohemorrhagic *Escherichia coli*, *Infect Immun*, **75**, 2432–2440.

SHERLOCK, O, SCHEMBRI, M A, REISNER, A and KLEMM, P (2004), Novel roles for the AIDA adhesion from diarrheagenic *Escherichia coli*: cell aggregration and biofilm formation, *J Bacteriol*, **186**, 8058–8065.

SHERLOCK, O, VEJBORG, R M and KLEMM, P (2005), The TibA adhesin/invasin from enterotoxigenic *Escherichia coli* is self recognizing and induces bacterial aggregation and biofilm formation, *Infect Immun*, **73**, 1954–1963.

SHIME-HATORI, A, IIDA, T, ARITA, M, PARK, K-S, KODAMA, T and HONDA, T (2006), Two type IV pili of *Vibrio parahaemolyticus* play different roles in biofilm formation, *FEMS Microbiol Lett*, **264**, 89–97.

SIRCILI, M P, WALTERS, M, TRABULSI, L R and SPERANDIO, V (2004), Modulation of enteropathogenic *Escherichia coli* virulence by quorum sensing, *Infect Immun*, **72**, 2329–2337.

SLAMTI, L and LERECLUS, D (2002), A cell-cell signaling peptide activates the PlcR virulence regulon in bacteria of the *Bacillus cereus* group, *EMBO J*, **21**, 4550–4559.

SLAMTI, L and LERECLUS, D (2005), Specificity and polymorphism of the PlcR-PapR quorum- sensing system in the *Bacillus cereus* group, *J Bacteriol*, **187**, 1182–1187.

SMITH, J L, FRATAMICO, P M and NOVAK, J S (2004), Quorum sensing: a primer for food microbiologists, *J Food Prot*, **67**, 1053–1070.

SMITH, J N and AHMER, M M (2003), Detection of other microbial species by *Salmonella*: expression of the SdiA regulon, *J Bacteriol*, **185**, 1357–1366.

SOLOMON, E B, NIEMIRA, B A, SAPERS, G M and ANNOUS, B A (2005), Biofilm formation, cellulose production, and curli biosynthesis by *Salmonella* originating from produce, animal and clinical sources, *J Food Prot*, **68**, 906–912.

SPERANDIO, V, LI, C C and KAPER, J B (2002), Quorum-sensing *Escherichia coli* regulator A: a regulator of the LysR family involved in the regulation of the locus of enterocyte effacement pathogenicity island in enterohemorrhagic *E. coli*, *Infect Immun*, **70**, 3085–3093.

SPERANDIO, V, TORRES, A G, JARVIS, B, NATARO, J P and KAPER, J B (2003), Bacteria-host communication: the language of hormones, *Proc Natl Acad Sci USA*, **100**, 8951–8956.

STAFFORD, G P and HUGHES, C (2007), *Salmonella typhimurium flhE*, a conserved flagellar regulon gene required for swarming, *Microbiology*, **153**, 541–547.

STARK, R M, GERWIG, G J, PITMAN, R S, POTTS, L F, WILLIAMS, N A, GREENMAN, J, WEINZWEIG, I P, HIRST, T R and MILLER, M R (1999), Biofilm formation by *Helicobacter pylori*, *Lett Appl Microbiol*, **28**, 121–126.

STEWART, P S and FRANKLIN, M J (2008), Physiological heterogeneity in biofilms, *Nat Rev Microbiol*, **6**, 199–210.

STOODLEY, P, SAUER, K, DAVIES, D G and COSTERTON, J W (2002), Biofilms as complex differentiated communities, *Annu Rev Microbiol*, **56**, 187–209.

SWIFT, S, KARLYSHEV, A V, FISH, L, DURANT, E L, WINSON, M K, CHHABRA, S R, WILLIAMS, P, MACINTYRE, S and STEWART, G S A B (1997), Quorum sensing in *Aeromonas hydrophila and Aeromonas salmonicida*: identification of the LuxRI homologs AhyRI and AsaRI and their cognate N-acylhomoserine lactone signal molecules, *J Bacteriol*, **179**, 5271–5281.

TAGA, M E, MILLER, S T and BASSLER, B L (2003), Lsr-mediated transport and processing of AI-2 in *Salmonella typhimurium*, *Mol Microbiol*, **50**, 1411–1427.

TAMPLIN, M L, GAUZENS, A L, HUQ, A, SACK, D A and COLWELL, R R (1990), Attachment of *Vibrio cholerae* serogroup O1 to zooplankton and phytoplankton of Bangladesh waters, *Appl Environ Microbiol*, **56**, 1977–1980.

TAN, L and DARBY, C (2004), A movable surface: formation of *Yersinia* sp. biofilms on motile *Caenorhabditis elegans*, *J Bacteriol*, **186**, 5087–5092.

TOGUCHI, A, SIANO, M, BURKART, M and HARSHEY, R M (2000), Genetics of swarming motility in *Salmonella enterica* serovar Typhimurium: critical role for liposaccharide, *J Bacteriol*, **182**, 6308–6321.

TOLVANEN, R, LUNDÉN, J, KORKEALA, H and WIRTANEN, G (2007), Ultrasonic cleaning of conveyor belt materials using *Listeria monocytogenes* as a model organism, *J Food Prot*, **70**, 758–761.

TRACHOO, N (2003), Biofilms and the food industry, *Songklanakarin J Sci Technol*, **25**, 807–815.

TRACHOO, N and FRANK, J F (2002), Effectiveness of chemical sanitizers against *Campylobacter jejuni*-containing biofilms *J Food Prot*, **65**, 1117–1121.

TRACHOO, N, FRANK, J F and STERN, N J (2002), Survival of *Campylobacter jejuni* in biofilms isolated from chicken houses, *J Food Prot*, **65**, 1110–1116.

TRÉMOULET, F, DUCHÉ, O, NAMANE, A, MARTINIE, B and LABADIE, J-C (2002), A proteomic study of *Escherichia coli* O157:H7 NCTC 12900 cultivated in biofilm or in planktonic growth mode, *FEMS Microbiol Lett*, **215**, 7–14.

UHLICH, G A, COOKE, P H and SOLOMON, E B (2006), Analyses of the red-dry-rough phenotype of an *Escherichia coli* O157:H7 strain and its role in biofilm formation and resistance to antibacterial agents, *Appl Environ Microbiol*, **72**, 2564–2572.

VAN HOUDT, R and MICHIELS, C W (2005), Role of bacterial cell surface structures in *Escherichia* coli biofilm formation, *Res Microbiol*, **156**, 626–633

VASUDEVAN, P, KUMAR, M, NAIR, M, ANNAMALAI, T and VENKITANARAYANAN, K S (2003), Phenotypic and genotypic characterization of bovine mastitis isolates of *Staphylococcus aureus* for biofilm formation, *Vet Microbiol*, **92**, 179–185.

VENDEVILLE, A, WINZER, K, HEURLIER, K, TANG, C M and HARDIE, K R (2005), Making 'sense' of metabolism: autoinducer-2, LuxS, and pathogenic bacteria, *Nat Rev Microbiol*, **3**, 383–396.

VUONG, C, SAENZ, H L, GÖTZ, F and OTTO, M (2000), Impact of the *agr* quorum-sensing system on adherence to polystyrene in *Staphylococcus aureus*, *J Infect Dis*, **182**, 1688–1693.

VUONG, C, GERKE, C, SOMERVILLE, G A, FISCHER, E R and OTTO, M (2003), Quorum-sensing control of biofilm factors in *Staphylococcus epidemidis*, *J Infect Dis*, **188**, 706–718.

VUONG, C, KOCIANOVA, S, YAO, Y, CARMODY, A B and OTTO, M (2004), Increased colonization of indwelling medical devices by quorum-sensing mutants of *Staphylococcus epidermidis* in vivo, *J Infect Dis*, **190**, 1498–1505.

WALTERS, M and SPERANDIO, V (2006a), Autoinducer 3 and epinephrine signaling in the kinetics of locus of enterocyte effacement gene expression in enterohemorrhagic *Escherichia coli*, *Infect Immun*, **74**, 5445–5455.

WALTERS, M and SPERANDIO, V (2006b), Quorum sensing in *Escherichia coli* and *Salmonella*, *Int J Med Microbiol*, **296**, 125–131.

WALTERS, M, SIRCILI, M P and SPERANDIO, V (2006), AI-3 synthesis is not dependent on *luxS* in *Escherichia coli*, *J Bacteriol*, **188**, 5668–5681.

WANG, C, LI, M, DONG, D, WANG, J, REN, J, OTTO, M and GAO, Q (2007), Role of ClpP in biofilm formation and virulence of *Staphylococcus epidermidis*, *Microbes Infect*, **9**, 1376–1389.

WANG, L, HASHIMOTO, Y, TSAO, C-Y, VALDES, J J and BENTLEY, W E (2005), Cyclic AMP (cAMP) and CAMP Receptor protein influence both synthesis and uptake of extracellular autoinducer 2 in *Escherichia coli*, *J Bacteriol*, **187**, 2066–2076.

WANG, Q, MARICONDA, S, SUZUKI, A, MCCLELLAND, M and HARSHEY, R M (2006), Uncovering a large set of genes that affect surface motility in *Salmonella enterica* serovar Typhimurium, *J Bacteriol*, **188**, 7981–7984.

WANG, X, DE BOER, P A J and ROTHFIELD, L I (1991), A factor that positively regulates cell division by activating transcription of the major cluster of essential cell division genes of *Escherichia coli*, *EMBO J*, **10**, 3363–3372.

WATERS, C M, LU, W, RABINOWITZ, J D and BASSLER, B L (2008), Quorum sensing controls biofilm formation in *Vibrio cholerae* through modulation of cyclic di-GMP levels and repression of vpsT, *J Bacteriol*, **190**, 2527–2536.

WATNICK, P and KOLTER, R (2000), Biofilm, city of microbes, *J Bacteriol*, **182**, 2675–2679.

WATSON, C L, OWEN, R J, SAID, B, LAI, S, LEE, J V, SURMAN-LEE, S and NICHOLS, G (2004), Detection of *Helicobacter pylori* by PCR but not culture in water and biofilm samples from drinking water distribution systems in England, *J Appl Microbiol*, **97**, 690–698.

WEN, Z T and BURNE, R A (2002), Functional genomics approach to identifying genes required for biofilm development by *Streptococcus mutans*, *Appl Environ Microbiol*, **68**, 1196–1203.

WIJMAN, J G E, DE LEEUW, P P L A, MOEZELAAR, R and ZWIETERING, M H (2007), Air–liquid interface biofilms of *Bacillus cereus*: formation, sporulation and dispersion, *Appl Environ Microbiol*, **73**, 1481–1488.

WU, H, SONG, Z, HENTZER, M, ANDERSEN, J B, MOLIN, S, GIVSKOV, M and HØIBY, N (2004), Synthetic furanones inhibit quorum-sensing and enhance bacterial clearance in *Pseudomonas aeruginosa* lung infection in mice, *J Antimicrob Chemother*, **53**, 1054–1061.

XAVIER, K B and BASSLER, B L (2005), Regulation of uptake and processing of the quorum-sensing autoinducer AI-2 in *Escherichia coli*, *J Bacteriol*, **187**, 238–248.

XU, F, BYUN, T, DUSSEN, H-J and DUKE, K R (2003), Degradation of N-acylhomoserine lactone, the bacterial quorum-sensing molecules, by acylase, *J Biotechnol*, **101**, 89–96.

XU, L, LI, H, VUONG, C, VADYVALOO, A, WANG, J, YAO, Y, OTTO, M and GAO, Q (2006), Role of the *luxS* quorum-sensing system in biofilm formation and virulence of *Staphyloccus epidemidis*, *Infect Immun*, **74**, 488–496.

YANG, F, WANG, L-H, WANG, J, DONG, Y-H, HU, J Y and ZHANG, L-H (2005), Quroum quenching enzyme activity is widely conserved in the sera of mammalian species, *FEBS Lett*, **579**, 3713–3717.

YARWOOD, J M and SCHLIEVERT, P M (2003), Quorum sensing in *Staphyloccus* infections, *J Clin Invest*, **112**, 1620–1625.

YATES, E A, PHILLIPP, B, BUCKLEY, C, ATKINSON, S, CHHABRA, S R, SOCKETT, R E, GOLDNER, M, DESSAUX, Y, CÁMARA, M, SMITH, H and WILLIAMS, P (2002), N-Acylhomoserine lactones undergo lactonolysis in a pH-, temperature-, and acyl chain length-dependent manner during growth of *Yersinia pseudotuberculosis* and *Pseudomonas aeruginosa*, *Infect Immun*, **70**, 5635–5646.

YILDIZ, F H and SCHOOLNIK, G K (1999), *Vibrio cholerae* O1 El Tor: identification of a gene cluster required for the rugose colony type, exopolysaccharide production, chlorine resistance, and biofilm formation, *Proc Natl Acad Sci USA*, **96**, 4028–4033.

YOON, Y and SOFOS, J N (2008), Autoinducer-2 activity of gram-negative foodborne pathogenic baceria and its influence on biofilm formation, *J Food Sci*, **73**, M140-M147.

YOUNG, G M, SMITH, M J, MINNICH, S A and MILLER, V L (1999), The *Yersinia enterocolitica* motility master regulatory operon, *flhDC*, is required for flagellin production, swimming motility, and swarming motility, *J Bacteriol*, **181**, 2823–2833.

ZHAO, L, MONTVILLE, T J and SCHAFFNER, D W (2006), Evidence for quorum sensing in *Clostridium botulinum* 56A, *Lett Appl Microbiol*, **42**, 54–58.

ZHAO, T, DOYLE, M P and ZHAO, P (2004), Control of *Listeria monocytogenes* in a biofilm by competitive-exclusion microorganisms, *Appl Environ Microbiol*, **70**, 3996–4003.

ZHU, J and MEKALANOS, J J (2003), Quorum sensing-dependent biofilms enhance colonization in *Vibrio cholerae*, *Develop Cell*, **5**, 647–656.

ZOGAJ, X, BOKRANZ, W, NIMTZ, M and RÖMLING, U (2003), Production of cellulose and curli fimbriae by members of the family Enterobacteriaceae isolated from the human gastrointestinal tract, *Infect Immun*, **71**, 4151–4158.

# 3

# Methods for imaging and quantifying the structure of biofilms in food processing and other environments

**Z. Lewandowski, Montana State University, USA and H. Beyenal, Washington State University, USA**

**Abstract**: This chapter, as the title indicates, is about imaging and quantifying biofilm structure. The term biofilm structure was conceived in reference to the distribution of biomass in the space occupied by a biofilm, after it became obvious that this distribution was nonuniform. This is the original definition of biofilm structure and we have adhered to this definition in most parts of this text. However, we cannot ignore the fact that the definition of biofilm structure has expanded. Since the time when the original meaning of the term was proposed, nonuniform distribution of other components of biofilm has been demonstrated, in particular the nonuniform distribution of various physiological groups of microorganisms and the nonuniform distribution of various polymeric substances, collectively called extracellular polymeric substances (EPS). The distribution of physiological groups in biofilms has become a well-studied branch of biofilm research – known as microbial community structure analysis, and quantification of the distribution of various polymers constituting EPS has made progress as well. The tools used to quantify the nonuniform distribution of various components in biofilms have much in common. The smallest dimension of the biofilm, the thickness, is conveniently expressed in microns, and it determines that the resolution of tools used to study the distribution of the components in biofilms must provide spatial resolution better than the thickness of the biofilm. This clearly includes several types of microscopy. In this chapter, we discuss these tools and where appropriate we illustrate their use with examples. Finally, the distribution of the components in biofilms needs to be quantified, and we describe software packages that can be used for this purpose. The concept of biofilm structure is dynamic and flexible, and able to accommodate the distributions of the components that have been described as well as the distribution of components that will be described in the future.

**Key words**: biofilms, biofilm structure, quantifying biofilm structure, biofilm imaging, fluorescent microscopy, confocal scanning laser microscopy.

## 3.1 Introduction

Two main biofilm components are microorganisms and extracellular polymers (EPS), and each of these components is distributed in the space occupied by the biofilm in a way that is related to the structure and function of the biofilm and to the way the biofilm accumulates. This chapter gives examples of the relations between the distribution of these components and the function of the biofilm and explains the efforts devoted to quantifying the distribution of these components.

The distribution of microorganisms in biofilms is not random; instead, microorganisms in biofilms are distributed in a way that affects, and perhaps even optimizes, their functions. An example of a biofilm process in which the particular spatial distribution of microorganisms determines the function of the biofilm is the well-known mechanism of microbially influenced corrosion, in which microorganisms deposit manganese oxides on the surface of stainless steel. These oxides serve as the recyclable cathodic reactant in the corrosion process, and their effect on corrosion processes can be evaluated using electrochemical techniques. However, their deleterious effect on material performance takes place only when the manganese-oxidizing microorganisms reside in proximity to the metal surface. If the environmental conditions change, and the motile manganese-oxidizing microorganisms move away from the surface, they deposit manganese oxides farther from the surface of the metal, where they are not in electrical contact with the metal surface. As a consequence, the effect of biomineralized manganese oxides on the corrosion process disappears, which puzzled early investigators. Plate I (between pages 108 and 109) shows the distribution of manganese-oxidizing bacteria, *Leptothrix discophora*, in a mixed population biofilm deposited on the surface of 316L stainless steel.

Since the biofilm was exposed to a solution rich in organics, which favored the growth of facultative heterotrophic bacteria in the biofilm, most of the motile aerobic manganese-oxidizing bacteria (green probe in Plate I (between pages 108 and 109)) moved toward the surface of the biofilm, closer to the locations with higher concentration of oxygen. Consequently, very few manganese-oxidizing bacteria were found near the bottom of the biofilm, where there was little or no oxygen. The effect of this biofilm on the electrochemical behavior of stainless steels could not be predicted just by determining the presence of manganese-oxidizing microorganisms and the presence of manganese oxides in the biofilm. These results were only interpreted when the location of the manganese-oxidizing microorganisms had been precisely determined with respect to the surface of the metal using FISH probes (Campbell *et al.* 2004). This example is not unusual: to understand the function of many types of biofilms it is imperative not only to determine the presence of various physiological groups of the microorganisms residing in the biofilm but also to specify their spatial relations with respect to each other and with respect to the surface.

Plate I (between pages 108 and 109) demonstrates that the location of specific physiological groups of microorganisms in a biofilm – not only the presence of these microorganisms – may determine the function of the biofilm. Similarly, the amount, the composition and the distribution of the extracellular polymers in a biofilm are important factors that determine various functions of the biofilm. In some instances biofilms are used to create barriers preventing the movement of contaminants in contaminated soils, and it is important and beneficial that the microorganisms used in such applications produce copious amounts of extracellular polymers. On the other hand, if biofilms are deposited on filtration membranes, then obviously the amount and the distribution of the extracellular polymers negatively affect the performance of the membrane by increasing the transmembrane pressure drop and decreasing the flux of the permeate. When the biofilm on a filtration membrane is distributed evenly and the layer of extracellular polymers is continuous, the deleterious effect of the biofilm on membrane filtration is more pronounced than when the biofilm is patchy. Not all biofilms look the same. Some microorganisms produce continuous biofilms and some produce patchy biofilms (Siebel and Characklis 1991). This may be caused by the type of the microorganisms residing in the biofilm, but also the environmental conditions, such as the availability of nutrients and the shear stress exerted by flowing water, can affect the structure of a biofilm. In many branches of biofilm research, to understand the effect of biofilm presence on technological processes, biofilm structure must be quantified. Figure 3.1 shows a model of biofilm structure often used as a point of departure in various types of biofilm research in which the structure of the biofilm may play a role.

To understand the effect of biofilm components and their distribution on biofilm function, these components need to be imaged and, if possible,

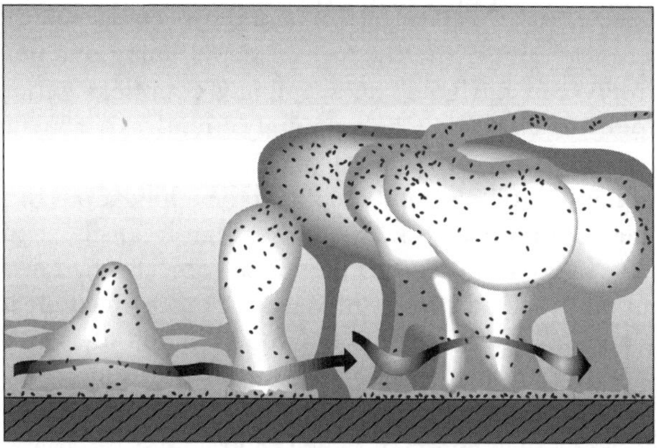

**Fig. 3.1** Model of biofilm structure often used as a point of departure in various types of biofilm research.

their distribution needs to be quantified. Imaging microorganisms distributed in a biofilm is somewhat more involved than imaging microorganisms distributed in suspension; in suspension the aggregates of microorganisms can be disrupted, while in biofilms the structure of the aggregates may determine the function of the biofilm, so it is important to image the microorganisms at the locations they occupy in the intact and fully hydrated biofilm. Although biofilms can be seen with an unaided eye, quantifying their structure in terms of the distribution of extracellular polymers and physiological groups of microorganisms requires the use of several types of microscopy techniques combined with various image acquisition devices. The favorite types of microscopy among biofilm researchers are those that allow the examination of living and fully hydrated biofilms. Such imaging, however, can rarely be achieved, and many techniques for imaging biofilm components affect the viability of the microorganisms and the structure of the extracellular polymers. Such techniques as fluorescent in situ hybridization (FISH) and fluorescent proteins (FP) used as reporter genes have revolutionized microbial ecology and had a pronounced impact on biofilm studies. Using these techniques in conjunction with a suitable microscopy technique, biofilm researchers can detect the presence of the selected physiological groups of microorganisms in the biofilm, their position in the biofilm with respect to other microorganisms and to the surface, and even their physiological state – dead, injured or alive. The application of these probes makes possible the study of population dynamics in biofilms. However, it is important to realize that the procedures used to image biofilm components may introduce various types of artifacts, and an overview of what can and what cannot be accomplished using the state-of-the-art techniques available to biofilm researchers is useful.

## 3.2 Microscopy techniques useful in biofilm studies

Recent progress in optical microscopy is overwhelming, and only those who follow it closely can navigate the seemingly countless variations of the technique, followed closely by the seemingly countless acronyms associated with them. Much of the progress in optical microscopy has followed the advances in quantum optics, combined with the applications of the wonders of modern technology, such as the use of high-intensity monochromatic light sources – tunable lasers – to illuminate objects, digital imaging, and modern computers and advanced software packages to handle image analysis. Optical microscopy is one of the tools commonly used by biofilm researchers, and many of the recent advances are quite relevant to, and some even resulted from, studying biofilms.

Effective use of such complex tools requires that the users understand the principles of the technologies that make them possible. However, understanding the optoelectronic technology behind modern optical microscopy can easily pose a challenge to the users of optical microscopes. The goal of this chapter is to provide an overview of those applications of optical microscopy that are useful in biofilm research, especially in quantifying biofilm structure. It will explain the terminology, discuss the physical principles and, where possible, give examples of the application of optical microscopy to the study and quantification of biofilm structure.

Typical uses of optical microscopy in biofilm research are (1) to characterize the dynamics of the initial attachment of microorganisms to surfaces, (2) to characterize the distribution of microorganisms in a matrix of extracellular polymers, (3) to characterize the distribution of the extracellular polymers – the biofilm matrix, and (4) to characterize the effects of environmental factors on biofilm structure. Some of these applications require the imaging of intact and fully hydrated biofilms, and some require intensive manipulations of the biofilm matrix, such as freezing and slicing, before the images are collected. Many of these applications are made possible by imaging of the distribution of specific fluorophores attached to various components of the biofilm. The images of the components are further analyzed using various forms of morphometry to quantify factors characterizing the size and shape of the microorganisms, microcolonies or EPS. In association with appropriate software, microcopy is often used to produce time-lapse photographic images of the components under study. An example of such an application is the study of the dynamics of microbial attachment to surfaces. Sequences of images showing the temporal distribution of microorganisms on surfaces are further used to compute the rates of such processes as microbial attachment and detachment and to generate kinetic expressions quantifying these processes.

### 3.2.1 Optical microscopy

Optical microscopy usually refers to microscopy that uses the visible part of the spectrum of electromagnetic radiation to illuminate the examined objects. This is the type of microscopy that has been most closely associated with the development of microbiology. By modern standards, the principles that govern optical microscopy are straightforward. The technique has been popular but from its inception it has been known for having two limitations: (1) the resolving power of the optical microscope determines the minimum size of the objects, and (2) out-of-focus objects degrade the images of thicker specimens. These limitations have been addressed by recent advances, and the resulting improvements have reached levels that just a few decades ago were considered impossible because of the physical limitations of optical systems. Much of this progress enhances the use of optical microscopy in biofilm research.

Improving the resolution of optical microscopes has been the central problem occupying microscopists since the invention of the microscope. In the initial stages of developing microscopy, the quality of the lenses used to build the microscope was the limiting factor determining the resolving power of the microscope. Later, as optical glass formulations improved and computers were used to design lenses, the various possible optical aberrations were identified and eventually mitigated. The optical resolution of modern lenses has approached the theoretical limits specified by geometric optics. In an ideal optical system, devoid of all optical aberrations, the resolution of modern lenses is limited by diffraction, which is a limitation imposed by natural laws: it can be manipulated but it cannot be removed entirely.

The theoretical resolution of a microscope is defined as the smallest distance between two objects that can be separated in the field of view, and it is quantified as:

$$d = \lambda / 2A_{\mathrm{N}} \qquad 3.1$$

where $d$ is the distance between the objects, $\lambda$ is the length of the light wave used to illuminate the objects and $A_{\mathrm{N}}$ is the numerical aperture of the microscope, which characterizes the medium and the angle over which the lens can accept light reflected from the illuminated object. Numerical aperture is defined as:

$$A_{\mathrm{N}} = \mathrm{n} \times \sin(\theta) \qquad 3.2$$

where $n$ is the refractive index of the medium, equal to 1 in air, and $\theta$ is one-half of the angle at which the lens can accept light reflected from the object. The importance of determining the possible resolution of optical microscopy warrants a brief discussion of the limitations imposed on the factors in Equation 3.1. For example, this somewhat peculiar reference to the half-angle of acceptance, instead of to the entire angle of acceptance, has practical reasons. The total angle at which a lens accepts light cannot exceed 180°, and half of that angle is 90°, which translates to $\sin(\theta) = 1$. If the entire angle of acceptance were used, then a lens with light acceptance equal to 180° would have a numerical aperture described by a function of $\sin(\theta)$ that reaches a maximum at 90° and decreases to zero at 180°, as $\sin(180°) = 0$. Using the half-angle of acceptance bypasses this problem and the function reaches a maximum at $\sin(\theta) = 1$. To reach the extreme angle of accepting the light reflected from an object, the lens would have to touch the object, making the distance between the lens and the object equal to zero. Lenses cannot focus so close and, therefore, $\sin(\theta)$ is always smaller than 1. The refractive index of the medium refers to the substance between the lens and the object under the microscope. Typically this is air, and the refractive index of air equals 1, by definition. In some techniques this medium is water or a specially formulated liquid that has a refractive index higher than one, which increases the possible resolution of the microscope.

To determine the highest theoretically possible resolution of optical microscopes using air as the medium we can take the extreme values for all the parameters in Equation 3.1 and calculate the smallest distance $d$ between two objects that can be resolved. As predicted by Equation 3.1, optical resolution can be improved by using short wavelengths to illuminate the object and by using an objective with the highest possible numerical aperture, $A_N$. To estimate the best possible resolution we will use violet light, with $\lambda = 400$ nm; a half-angle of incidence $\theta = 90°$, which gives $\sin(\theta) = 1$; and air as the medium, which has a refractive index equal to 1. Introducing all these factors into Equation 3.1 gives the smallest distance between two objects that can be resolved under this set of conditions, equal to $d = 200$ nanometers, or 0.2 micrometers, which is commonly considered to be the theoretical best resolution of the optical microscope. This distance is smaller than the size of many, but not all, bacteria.

The limit on optical resolution determined from Equation 3.1 is known as the Abbes limit, or the diffraction limit. Diffraction refers to the bending of light waves at the edges of opaque objects. Images in optical microscopy are formed by various proportions of diffracted and non-diffracted light reflected from an object. The smaller the object is, the higher the contribution of the diffracted light, which decreases the quality of the images. The diffraction of light is followed by the interference of the light waves: the waves of light bend at the edges of opaque objects (diffraction) and overlap each other, forming alternating bands of bright and dark zones (interference). This effect obviously degrades the quality of images, and the smaller the objects are and the closer they are to each other, the larger the contribution of the diffracted light is to the process of image formation and the more degraded the image quality is. If two small objects are placed in close proximity to each other, and each of them diffracts the light waves that illuminate them, the light waves deflected at the edges of one object interfere with the light deflected at the edges of the other object, degrading the image quality. If these two objects are small enough and placed close enough to each other, they cannot be separated in the image. Hence the name: resolution limited by diffraction. The diffraction limit is quantified in terms of the wavelengths of the interfering light waves. This is often referred to as the Rayleigh criterion, which says that the resolution of a lens is considered diffraction-limited when the minimum of the light wave diffracted from one source point coincides with the maximum of the light wave diffracted from the other source point. Equation 3.1 quantifies this effect in the most concise way: it says that under the most favorable conditions the images of two small objects under a microscope can be resolved if the distance between them is less than one half of the wavelength of the light that illuminates them, which is just another way of stating the Rayleigh criterion. Although the diffraction limit is imposed by natural laws, and therefore appears to be inviolable, it has been continually challenged by those attempting to improve the resolution of their optical tools. Improving the resolution beyond the limit imposed by geometric optics was finally accomplished with the help of quantum optics.

Some of the improvements of resolution in optical microscopy are relevant to imaging biofilms, and it is worthwhile to look at them more closely. When the factors affecting the diffraction limit were first quantified, microscopists increased the resolution of optical microscopes by using immersion liquids with refractive indices higher than one to increase the numerical aperture of the lens. The object under the microscope was covered by the immersion liquid and the lens was immersed in the liquid, which eliminated the air from the space between the lens and the object. Because the immersion liquids had refractive indices larger than one, the numerical aperture in Equation 3.1 increased and *d*, the smallest distance between two objects that can be resolved, decreased. Immersion liquids are still in use and those currently available have refractive indices of about 1.5. Using immersion lenses improves the resolving power of the microscope, but it does not really defy the limitation by diffraction. It just manipulates one of the factors in Equation 3.1 to increase the resolving power of the microscope; the resolving power of the microscope is still determined by the diffraction. Conquering the diffraction limit required more sophisticated approaches, and was finally accomplished with help from quantum mechanics and computer science. As a result, modern optical microscopy does not recognize the diffraction limitation as the ultimate factor determining the resolving power of the optical microscope. The fact that better resolutions than those predicted by the Abbes limit have been accomplished by microscopists does not invalidate the law: researchers just found ways to work around the limitations imposed by the law. In its early stages of development, optical microscopy was exclusively based on illuminating the field of view using external light sources, and objects under the microscope were made visible by the differences in the intensity of the light that was reflected and absorbed by the objects and their backgrounds. In the early stages of optical microscopy, the light illuminating the objects under the microscope was always coming from the outside, and the discussion of the consequences and limitations imposed by Equation 3.1 assumed that both objects were illuminated by the same light source. The limits placed on optical resolution by diffraction were overcome by using another type of illumination of the objects – fluorescence. In fluorescent microscopy, the external light is used to stimulate the secondary illumination, which is emitted by the object itself. In fluorescent microscopy the image of the object is formed by the light emitted by the object under the microscope.

### 3.2.2 Fluorescent microscopy

Fluorescence is the property of a molecule to emit light of a longer wavelength when it is irradiated by light of a shorter wavelength. Typical excitation and emission spectra of a fluorescing molecule are shown in Fig. 3.2. An example of a substance with such excitation and emission spectra is the common stain fluorescein.

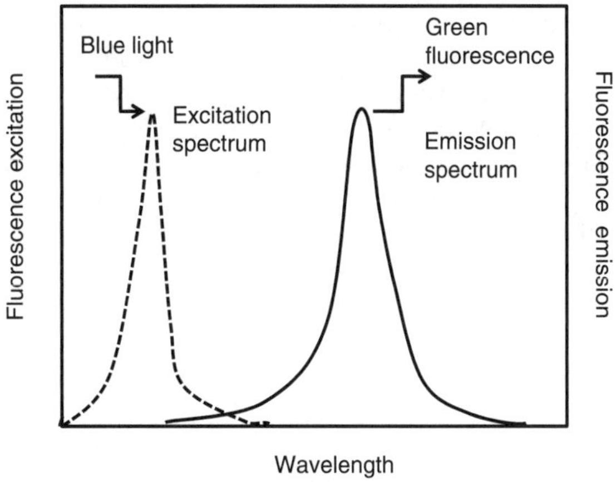


**Fig. 3.2** Typical excitation and emission spectra of a fluorescing molecule.

Imaging fluorescence opened new frontiers in microscopy and allowed the diffraction limit to be overcome. In this type of optical microscopy, objects under the microscope are illuminated with a characteristic wavelength that can be absorbed by the fluorescent markers associated with the objects. Depending on how the sample is illuminated, optical microscopy is often classed into two categories: (1) wide-field microscopy, in which the entire sample is illuminated; and (2) scanning microscopy, in which only part of the sample is illuminated at a time and the beam of illuminating light travels across the sample. In fluorescent microscopy, wide-field microscopy is executed by using white light from mercury lamps, and fluorescent microscopes are equipped with filters to select the desired part of the spectrum from the white light. If the illuminating light is delivered through the objective, the name epi-fluorescent microscopy is often used. In scanning microscopy, the light is delivered by lasers. Lasers are beneficial for many reasons, such as their delivery of high-intensity monochromatic light, but they also limit the ability of the scanning microscopy to the wavelengths that can be delivered by the existing lasers, while wide-field illumination may use a variety of available filters and practically any narrow band of wavelength selected from white light. This disadvantage of using lasers is, however, slowly being mitigated by progress in laser technology and the use of tunable lasers that can deliver more than one wavelength of high-intensity monochromatic light. Also, some applications of optical microscopy, such as photobleaching and multiphoton microscopy, depend on the ability to illuminate objects with a narrow beam of monochromatic light and these applications would be difficult or impossible without lasers.

The physical principles of fluorescent microscopy do not depend on the type of illumination, of course. Whether the illumination is wide-field or scanning, electrons of the fluorescing molecules absorb the photons from the illuminating light and increase their energy above that characteristic for their ground state. These excited molecules quickly lose their excess energy and return to their ground state. In the process most, but not all, of the energy gained by absorbing photons from the illuminating light is released by emitting photons of somewhat lower energy than those initially absorbed. The energy of the incoming photons is related to the wavelength of the light used to illuminate the object by the Planck equation:

$$E = hc/\lambda \qquad 3.3$$

where $h$ is Planck's constant, $c$ is the speed of light, and $\lambda$ is the wavelength. Since the emitted photons have a lower energy than the photons that stimulated the fluorophore, they generate light with a longer wavelength than that used to illuminate the fluorophore. The distance between the peaks of the adsorbed and emitted light spectra is called the Stokes shift.

The concept of optical resolution, as defined by Equation 3.1, needs to be adjusted in fluorescent microscopy by acknowledging that optical resolution depends not only on the distance between two points but also on the contrast difference between the images of the two points, and fluorescent microscopy can improve the latter. The Rayleigh criterion says that the resolution is diffraction-limited if the minimum of the light wave diffracted from one source point coincides with the maximum of the light wave diffracted from the other, and it works well if both objects are evenly illuminated by an external light source. In fluorescent microscopy, the object under the microscope is the source of light. The emitted light is stimulated by an external illuminating light, but the light that is used to image the object is emitted by the object. In variations of fluorescent microscopy that seemingly defy the Rayleigh criterion it is important to note that if the light illuminates mostly one of two objects positioned closer to each other than half the wavelength of the illuminating light, while the other object is illuminated to a much lesser extent, the two objects can be distinguished because of the contrast difference, without violating the Rayleigh criterion. The procedures that allow a resolution to be reached that is better than the diffraction limit using fluorescent microscopy rely on light quenching, in which additional pulses follow the excitation pulse and modify the energy delivered to the fluorophore.

Several procedures have been developed that can improve the resolving power of fluorescent microscopes. One example of such a procedure is stimulated emission depletion (STED) microscopy, which can lower the limit of resolution to about 70 nanometers, which is about two thirds less than the diffraction limit for optical microscopy – 200 manometers – according to the computation shown at the beginning of this chapter. STED microscopy uses blue light to stimulate the fluorophore and orange light to quench the blue light. Light of a given color can be quenched

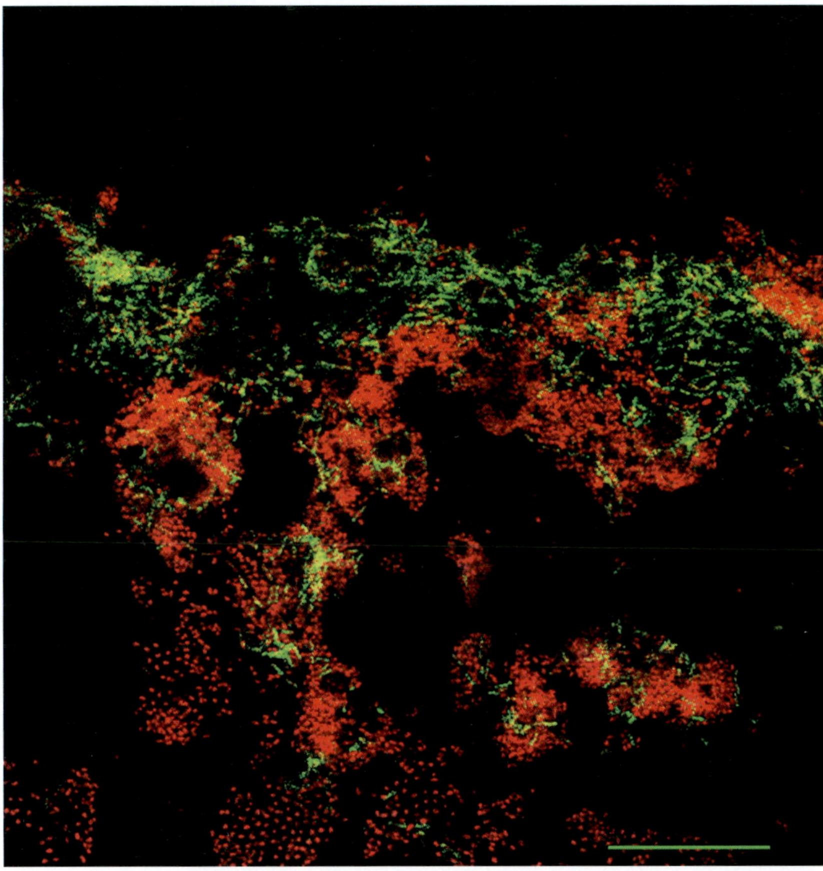

**Plate 1** Distribution of manganese-oxidizing bacteria, *Leptothrix discophora*, in a mixed population biofilm deposited on the surface of 316L stainless steel.

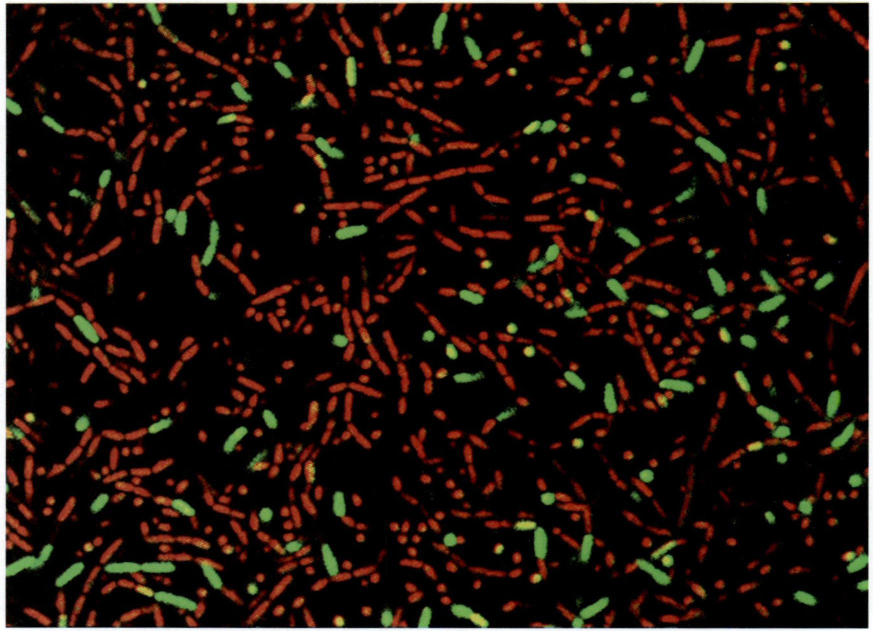

**Plate 2** Confocal laser microscope dual-channel image showing an optical thin section obtained from near the base (substratum) of a thick *Brochothrix thermosphacta* biofilm treated with 0.5% trisodium phosphate (TSP, a foodgrade antimicrobial) and then stained with a fluorescent viability stain. Surviving bacteria are represented by the green fluorescing cells, whereas dead cells are indicated by those fluorescing in the red wavelengths.

using light of the complementary color, the one on the other side of the color wheel. According to color theory, if two complementary colors are mixed in equal proportions, they cancel each other, forming a shade of gray. Blue light is delivered in the form of a narrow beam to illuminate a small spot on the specimen. The orange light is delivered in the form of a doughnut around the small spot; it quenches the light around the spot illuminated by blue light. The emitted fluorescent light is green. In effect, the orange light shrinks the location that is illuminated by the blue light that can stimulate fluorescence, and the fluorescent light is emitted from a spot that is smaller than that illuminated by the blue light. As a result, an illuminated object can be imaged even if it resides in close proximity to another object, because the other object is not illuminated. The resolving power of this technique is better than that predicted by Equation 3.1. By scanning the blue excitation spot over the sample and recording the emitted fluorescent light, an image of the sample is formed. The smaller the excitation spot is, the higher the resolution of the microscope. Because the STED beam is doughnut-shaped and centered over the excitation spot, it is possible to preferentially quench the outer edges of the excitation spot and stimulate fluorescence at the center only.

In another variation of fluorescent microscopy that provides resolution better than that limited by diffraction, one fluorophore can transfer energy to another fluorophore. This type of interaction is called fluorescent resonance energy transfer (FRET), and it is possible only if the donor and the acceptor fluorophore are close to each other, separated by a distance characteristic of molecular interactions – a few nanometers. In this type of interaction, the donor fluorophore is stimulated using light of an appropriate wavelength, as in traditional fluorescent microscopy. However, when the donor fluorophore dissipates the energy, some of the energy, instead of being dissipated by the emission of photons, is transferred to another fluorophore, the acceptor, which in its turn emits photons. Since the light produced by the two fluorophores has different wavelengths, the fluorophores can be resolved in an image. Specialized FRET microscopes are available, but the techniques depend more on the selection of the fluorophores than on specialized microscopy and can be executed using confocal microscopes. Multiphoton microscopy has also been successfully tried with FRET, to make the method even more versatile.

### 3.2.3 Two-photon microscopy

As discussed above, fluorophores are stimulated using light with a short wavelength and, once stimulated, they emit light of a longer wavelength. Two-photon microscopy reverses this process and seemingly achieves the impossible: the light emitted by the fluorophore has a shorter wavelength than the illuminating light, which at first glance could be interpreted as violating natural laws by gaining energy from nothing. This is equivalent to the negative Stokes shift, and the trick can be accomplished with help from quantum mechanics. To stimulate electrons in a fluorophore, a precise amount of energy needs to be delivered, and the amount of energy that is released as photons is always lower than the amount of energy that stimulates the emission: this cannot be violated. In two-photon microscopy, the object is illuminated with photons delivered at short intervals using pulsating light. Two photons can be absorbed by the fluorophore at a single location and, because the time between the deliveries of the photons is very short, on the order of femtoseconds, the total energy delivered to this location is the sum of the energies of these two photons. Before the energy of the fluorophore, increased by its absorption of the first photon from the illuminating light, is dissipated by the emission of a photon of lower energy, the next photon is delivered by the illuminating light and increases the energy level of the fluorophore even further. Consequently, the energy of the two consecutive photons is accumulated by the fluorophore, and the combined energies from two excitation photons dissipate at the same time. As part of this process, a single photon is emitted which has an energy higher than the energy of either of the two photons absorbed from the illuminating light. Since the emitted photon has an energy higher than that of the photons stimulating the emission, it projects itself as light of a shorter wavelength. For example, if the illuminating light is in the infrared range of the spectrum, the emitted photon will have the wavelength of green light. Two-photon microscopy does not violate any physical laws, even though it appears to do just that.

There are advantages to using two-photon microscopy instead of one-photon microscopy. Because the fluorophore in the sample is excited at a single point, a two-photon microscope has an extremely high resolution and does not require the pinhole used in traditional confocal microscopy to reject out-of-focus fluorescent light. There are no out-of-focus objects: the object that is illuminated is the one that is in focus. Also, the light used in two-photon microscopy can penetrate deeper into the specimen than the light used in one-photon fluorescent microscopy. The ultraviolet radiation typically used in one-photon microscopy can cause a host of damages to living cells, while the infrared radiation typically used in two-photon microscopy is considerably less harmful. Therefore, infrared radiation used in two-photon microscopy can penetrate deep into the sample without damaging it, and the depth of penetration is important in biofilm research. The typical light source used in single-photon fluorescent microscopy is an ultraviolet laser, and the typical light source used in two-photon microscopy is an infrared laser, such as the titanium sapphire laser. The main disadvantage of two-photon microscopy is the cost of the instrument: the infrared pulsed laser and the necessity of using optics compatible with infrared light make the application of the technique expensive.

### 3.2.4 Scanning laser confocal microscopy

The first limitation of optical microscopy was related to the lateral resolution: how close two objects can be to each other and still be resolved in an image. In the preceding sections, we tacitly assumed that these objects were positioned on the same plane with respect to the lens. Much changes when these objects are located on different planes with respect to the lens, and the extent to which such objects can be resolved in optical microscopy depends on the depth of field (DOF) of the lens. The depth of field is the range in distance from the lens in which the objects appear to be sharply rendered. In principle, a lens can focus only on a single plane but to the human eye there is a zone which has acceptable focus, and this zone is identified as the depth of field. A large DOF can render sharp images of objects positioned on different planes, even if these planes are separated by relatively large distances. However, DOF depends on many factors, including the numerical aperture of the objective and the distance between the object and the lens. When the objects are very close to the lens and the lens has a high numerical aperture, which are the conditions common in high-resolution optical microscopy, DOF is very thin; in this case, if the imaged objects are located on different planes, only one of them can be in focus. Objects under the microscope that are thicker than the DOF pose difficulties in traditional optical microscopy because parts of the image are out of focus.

This is the physical limitation characteristic of the optical system and nothing can be done to change it; much, however, can be done to bypass this limitation. The solution to the problem is based on optical sectioning of the sample using a lens with a very small DOF. Using traditional light microscopy, a series of images is collected, one focal plane at a time. However, in each such image only one focal plane is sharply in focus, while the remaining focal planes contribute images of the objects that are out of focus and degrade the overall image. Confocal microscopy uses a trick to mitigate this degrading effect due to out-of-focus images. The trick is based on an old invention that preceded the invention of the optical microscope and was known to Aristotle: camera obscura, the dark chamber. The device relies on an optical phenomenon which occurs when light passes through a small opening in a thin material – a pinhole. Part of the light does not scatter but passes undisturbed and forms an image of the object from which it was reflected. This manner of forming an image is different from the manner in which images are formed using lenses. A lens collects light reflected from objects positioned at various distances from the lens and then focuses the light. Because the objects are at different distances from the lens and are at different locations, the light reflected from them passes through different parts of the lens and not all the light can be focused at the single point which defines the plane of focus. This explains why only one distance from the lens can be in sharp focus. When the light passes through a very small opening, many objects that are located at various distances from the focal plane are not imaged because the light reflected from them was not bent enough to find its way to the opening. As a result most out-of-focus objects do not show in the image. This effect, combined with scanning of the sample using various focal lengths, are the bases for confocal microscopy.

Confocal microscopy is not inherently fluorescent microscopy, but samples are often tagged with fluorescent markers to enhance the imaging of various components. This process, called staining, has received much attention, and stains have been developed that can bind specifically to selected parts of specimens and enhance the visibility of various components of biological samples, including biofilm components.

## 3.3 Enhancing the images of microorganisms in biofilms

Objects in fluorescent microscopy may exhibit natural fluorescence or, more often, microscopists may manipulate them into exhibiting fluorescence by attaching fluorescent stains. Most fluorescent stains have been developed with a particular application in mind, and they come with their specific excitation and emission wavelengths clearly specified. Their application is then linked to the ability of the user to generate the conditions under which they can be used, light with suitable wavelengths for stimulating the fluorescent stains and the ability to record the emitted light at its characteristic wavelengths. Among many possible ways of imaging biofilm components, three are very popular among biofilm researchers: 1) live vs. dead stains, 2) fluorescent in situ hybridization probes (FISH) and 3) fluorescent proteins used as reporter genes. These techniques have revolutionized microbial ecology and had a pronounced impact on biofilm studies. Using these probes in conjunction with suitable microscopy, the researcher can detect not only the presence of the selected microorganisms in the biofilm but also their position in the biofilm with respect to other microorganisms and their physiological activity. The application of FISH probes and the use of fluorescent proteins to identify microorganisms in biofilms make possible the study of population dynamics in biofilms. From a practical point of view, the main difference between these two techniques is that FISH staining is done on dead microorganisms while fluorescent proteins are expressed by living microorganisms. Both techniques are powerful, but the ability to monitor living microorganisms gives an advantage in studying biofilm mechanisms that is difficult to overestimate.

### 3.3.1 Fluorescent *in situ* hybridization probes

FISH is a general technique popular outside of biofilm research; for example, it is used in medical diagnostics to detect chromosomal abnormalities, such as Down's syndrome, with various structures of the fluorescently tagged DNA probe. The *in vitro* hybridization techniques require that DNA or

RNA be isolated from the sample and separated on a gel, and that the fluorescent probes then be added to the sample. The *in situ* variety of the technique, which is extensively used in biofilm research, does not require isolating DNA or RNA; instead, the probes are hybridized to the respective nucleotide sequences inside the cells (Biesterfeld *et al.* 2001; Delong *et al.* 1999; Ito *et al.* 2002; Jang *et al.* 2005; Manz *et al.* 1999). *In situ* hybridization uses fluorescence-labeled complementary DNA or RNA probes, often derived from fragments of DNA that have been isolated, purified, and amplified. In microbial ecology, ribosomal RNA in bacterial cells is targeted by fluorescence-labeled oligonucleotide probes. The large number of ribosomes in bacterial cells helps to achieve high signal intensity. However, the accessibility of the targeted site varies among strains of bacteria and the protocols of hybridization need to be modified to account for these differences.

To successfully apply *in situ* hybridization, it is essential that the probe first enter the cell and hybridize with the nucleotides for which it was designed. The excess of the probe is then removed from the cell so that only the copies of the probe that hybridized to the nucleotides remain in the cell. To accomplish this, cell membranes must be permeabilized using chemical agents, such as detergents, which make membranes porous by extracting lipids. The probe enters the cell and hybridizes to the nucleotide, and the excess of the probe that did not hybridize is washed out. Bacterial FISH probes are primers for the 16S RNA region, and in biofilms they are typically used to identify the microorganisms residing in the biofilm matrix, in what are often referred to as microbial community structure studies.

Using FISH probes in biofilm studies requires that the steps of known procedures be modified and adjusted to the existing conditions on a case-by-case basis. This usually calls for a tedious and time-consuming tweaking of the existing protocols. The results can be rewarding, though: not only can the microorganisms in the matrix be identified but, when confocal microscopy is used, the images show the locations of various physiological groups of microorganisms in the matrix of extracellular polymers. Examples of applications are numerous. They include the studies of Biesterfeld *et al.* (2001), who used FISH to quantify ammonia-oxidizing populations within intact biofilm samples collected from a full-scale nitrifying trickling filter and identified the substrate uptake patterns of sulfate-reducing bacteria inhabiting a sewer biofilm with oxygen, nitrate, or sulfate as an electron acceptor by combining microautoradiography and FISH with family- and genus-specific 16S rRNA probes. In another study, Jang *et al.* (2005) used FISH to measure the growth rate of ammonia-oxidizing bacteria in a partially submerged rotating biological contactor. Manz *et al.* (1999) investigated the phylogenetic composition, three-dimensional structure and dynamics of bacterial communities in river biofilms generated in a rotating annular reactor system using FISH and CSLM.

### 3.3.2 Fluorescent proteins

The use of fluorescent proteins (FP) from jellyfish *Aequorea victoria* was initiated in the late 1960s and 1970s (Prendergast 1978), and the gfp protein was cloned and expressed in the 1990s (Chalfie *et al.* 1994). Since then, several other FPs have been discovered and isolated from various organisms, such as red fluorescent protein from *Aequorea victoria* (Matz *et al.* 1999). Many FPs have been optimized for use in fluorescent microscopy by genetic manipulations of the organisms that originally expressed them. When stimulated with light of the appropriate wavelength, the proteins emit fluorescent light of different colors, such as cyan, green, orange, yellow and red, depending on the structure of the protein. Using fluorescent proteins has an advantage over using fluorescent probes to tag microorganisms because FPs can be genetically encoded and tracked in living cells.

Although the use of FPs to study the molecular biology of suspended microorganisms is widespread, their use in biofilm research is inhibited by some of their properties. When the genetic manipulation concerns a single species of microorganisms grown in suspension, the use of FPs is relatively straightforward. To make sure that the microbes in the culture carry the plasmid, an antibiotic resistance gene is inserted into the same plasmid. The microorganisms are grown using a growth medium with the antibiotic to which the microorganisms are resistant. The cells that lose the plasmid cannot grow in this medium, and, consequently, all cells in the growing culture can express the fluorescent protein. This strategy works well in the case of single-species microbial cultures. When a mixed population of microorganisms is used and the biofilms are mixed population cultures, introducing an antibiotic into the growth medium would inhibit all microorganisms except those that carry the antibiotic resistance gene. It is possible to use all microorganisms with antibiotic resistance inserts, but this is not very practical.

To properly enfold, fluorescent proteins need oxygen (Hansen *et al.* 2001). Again, in the culture of suspended microorganisms it is possible to keep oxygen at a constant level by aerating the growth medium. This solution does not work in biofilms, because aerating the bulk solution does not ensure that oxygen is available in deeper layers of the biofilm (Hansen *et al.* 2001).

## 3.4 Staining biofilm components

Although the term 'biofilm components' may encompass many substances in the biofilm, it usually refers to two components: microorganisms and extracellular polymers, and the following sections will describe procedures used in staining these components. There are also general stains such as fluorescein isothiocyanide (FITC), which can stain organic material in general and can be used to calculate biofilm coverage (Larsen *et al.* 2008; Schmid *et al.* 2003).

There are many suppliers of fluorescent dyes, and their catalogs offer impressive numbers of choices, e.g. Moleculer Probes, Inc. (Eugene OR, USA, http://www.invitrogen.com/site/us/en/home/brands/Molecular-Probes.reg.us.html). Once the decision has been made which biofilm component is to be stained, to select a suitable dye, the following factors need to be considered:

1. *Excitation wavelength*: The lasers available with a specific CSLM limit the dye selection. If a dye requires a 400-nm excitation wavelength and the CSLM does not support a 400-nm excitation wavelength, the dye cannot be used.
2. *Optical filters available with the CSLM*: Older CSLMs use removable optical filters to select specific wavelengths of the emitted light. Newer CSLMs use tunable filters.
3. *Emission efficiency*: To improve detection, the dye must fluoresce brightly and the fluorescent light intensity must be significantly higher than the auto-fluorescence of the biofilms.
4. *Emission spectrum*: The CSLM must have filters for emitted light that coincide with the emission spectra of the dyes; using dyes with narrow emission spectra is preferable. When using multiple dyes, their emission spectra should not overlap.
5. *Photobleaching*: Some fluorophores are susceptible to photobleaching, e.g. fluorescein. Avoiding longer image collection times may alleviate this problem. Selecting a dye that is not susceptible to photobleaching addresses this problem directly.
6. *Interference with metabolic processes*: Most of the dyes disrupt metabolic processes within microbial cells. Many dyes are also sensitive to pH and ionic strength, and they may or may not be suitable for a particular application, depending on the pH and ionic strength of the solution.
7. *Dye specificity*: Highly specific dyes that stain the target but do not stain other components of the biofilm are preferable.

### 3.4.1 Staining biofilms to image the microorganisms

Traditionally, the number of live cells is quantified by culturing them in an appropriate growth medium. However, such techniques cannot be used to assess the viability of bacterial cells *in situ* in biofilms, and they do not give any information about the number or location of the dead cells in biofilms. In many biofilm studies in which bacteriocidal or bacteriostatic agents are used, it is important to monitor the effect of these agents on the microbial population in the biofilm. In such studies, live/dead stains are used. The mechanism by which these stains act relies on the condition of the cell membrane. If the cell membrane has been compromised, as in a dead cell, the fluorescent stain penetrates the cell and stains the entire volume of the cell – or its specific component, depending on the type of stain used. This results in a much stronger fluorescent signal from dead cells than that from living cells, which can only absorb the fluorescent marker on their surfaces. In such studies, cells that exclude the stain are considered viable.

One example of a stain that does not penetrate the membranes of living cells but does stain dead cells is a nucleic acid stain, propidium iodide, which is quite popular with biofilm researchers (GrayMerod *et al.* 2005; McNamara *et al.* 2003; Nancharaiah *et al.* 2005). Since the live/dead stain is more effective on dead cells than it is on living cells, other stains are used in tandem which can penetrate both living and dead cells, such as Thiazole orange (Biosciences Catalog No. 349483 or 349480), which is a permeant dye: it enters all cells, live and dead. The combination of two such stains enhances the visibility of all the microorganisms in the biofilm and provides a fast and reliable method for distinguishing the live and dead bacterial cells. Although live/dead staining remains one of the most powerful staining techniques in biofilm research, the reproducibility of the results obtained using this technique raises some questions. For example, GrayMerod *et al.* (2005) used live/dead stains to study the development of *Acinetobacter* biofilm structure. They concluded that monitoring biofilm development using live/dead stains did not produce reproducible results, which limited the use of these stains for that purpose (GrayMerod *et al.* 2005).

Among the many stains that are available for staining microorganisms in biofilms, the following three are quite popular (see also Table 3.1).

*Acridine orange* (3,6-dimethylaminoacridine) is a weak base. It is soluble in water and permeant to live and dead cells. It emits green fluorescence

**Table 3.1** Selected dyes and their applications

| Dye name | Invitrogen catalog number* | Excitation (nm) | Emission (nm) | Applications |
|---|---|---|---|---|
| SYTO 9 | S-34854 | 485 (DNA)<br>486 (RNA) | 498 (DNA)<br>501 (RNA) | Binds to DNA and RNA |
| Acridine orange | A1301 | 500 (DNA)<br>460 (RNA) | 526 (DNA)<br>650 (RNA) | Permeant to live and dead cells, binds DNA and RNA. Used for RNA/DNA discrimination |
| Propidium iodide | P1304MP | 485<br>490** | 617<br>635** | Not permeant to live cells. Stains dead cells. |

* The same products can be bought from other vendors.
** As a component of the live/dead BacLight Bacterial Viability Kit (catalog #L7012).

when bound to compact chromosomal DNA and red fluorescence when bound to the non-intact DNA of dead cells. Acridine orange is used as a general cellular fluorescent stain. As an example of the use of acridine orange, Wirtanen *et al.* (2001) used it and CTC-DAPI to find the effectiveness of selected biocides on biofilms grown on stainless steel surfaces (Wirtanen *et al.* 2001). In another application, Xia *et al.* (1998) used acridine orange to image dead and live cells in biofilms (Xia *et al.* 1998).

*Propidium iodide* is a popular red-fluorescent nuclear and chromosomal counter stain. Since propidium iodide is not permeant to live cells, it is also used to detect dead cells in a population. This stain is one of the components of live/dead staining. As an example, Seo and Frank (1999) used propidium iodide to determine the locations of dead cells in their biofilms (Seo and Frank 1999).

SYTO® dyes are cell-permeant dyes. They are green-fluorescent nucleic acid stains that show a large fluorescence enhancement upon binding nucleic acids. The SYTO® dycs can be used to stain RNA and DNA in both live and dead eukaryotic cells, as well as in both Gram-positive and Gram-negative bacteria. The SYTO 9 kit from Invitrogen®, for example, which stains both live and dead bacteria, comes with instructions on how to use it and is one of the most commonly used dyes in biofilm research.

### 3.4.2 Staining biofilms to image the extracellular polymeric substances

Extracellular polymers form the matrix that holds the biofilm together and in which the microorganisms are imbedded. However, it is still not entirely clear what holds the EPS together. Initially, many biofilm researchers believed that the main components of EPS were alginates, which are linear polysaccharides composed of two different sugar residues: 1,4-linked β-D-mannuronic acid (M) and its 5-epimer, 1,4-linked α-L-guluronic acid (G). Bacterial alginates, especially those produced by *Pseudomonas* sp., differ considerably from the well-known algal alginates (obtained from brown algae such as *Laminaria hyperborea* and *Macrocystis pyrifer*) that are used industrially as gelling agents. The polymer chains of algal alginates contain numerous blocks (>5 consecutive residues) of L-guluronic acid (G-blocks), which enable intermolecular cross-linking via the selective binding of $Ca^{++}$ ions to form gels. In contrast, *Pseudomonas* alginates do not contain G-blocks. Instead, their G residues (0–40%) occur as single residues (..MGM.. type), and their alginates may not form intermolecular junction zones with $Ca^{++}$. A commonly accepted mechanism of polymer matrix formation in biofilms is based on the assumption that calcium ions bridge the alginates; this is called the 'egg-box model' (Skjakbraek *et al.* 1986a; 1986b). However, because microbial alginates do not have adequate numbers of G-blocks (see Table 3.2), using this model to explain processes in bacterial biofilms is risky. Table 3.2 shows the composition of alginate

**Table 3.2** Characteristics of purified alginate from *P. aeruginosa* strain 8830 (Abrahamson *et al.* 1996)

| Property | Native | De-O-acetylated |
|---|---|---|
| Molecular weight ($M_W$) (g/moles) | 1.8 ($10^6$) | 1.5 ($10^6$) |
| Intrinsic viscosity ($\eta$ mL/g) | 29.0 | – |
| O-acetyl content (moles per uronic acid residue) | 0.69 | – |
| Fraction of D-mannuronic acid ($F_M$) | n.d. | 0.95 |
| Fraction of L-guluronic acid ($F_G$) | n.d. | 0.05 |
| Fraction of G units adjacent to a second G ($F_{GG}$) | n.d. | 0.00 |
| Fraction of G units adjacent to a M ($F_{GM+MG}$) | n.d. | 0.05 |

n.d. – not determined.

that was recovered from *Pseudomonas aeruginosa* by precipitation with isopropanol and purified (Abrahamson *et al.* 1996).

According to the current point of view, extracellular polymeric substances consist of polysaccharides, proteins, nucleic acids, and lipids. In contrast to the large number of fluorescent stains available for microorganisms in biofilms, there are no general stains for EPS. There are fluorescent stains for some types of polysaccharides, but not for all those present in biofilms. This should not come as a surprise, considering that EPS comprise a mixture of many substances. It may be difficult to find a single stain that stains them all and at the same time does not stain anything else. Some biofilm researchers use Alcian blue to stain EPS, but it targets anionic molecules only (Wetzel *et al.* 1997). Lipid fractions have been stained using Nile red (Fowler and Greenspan 1985; Greenspan *et al.* 1985; Greenspan and Fowler 1985; Larsen *et al.* 2008).

The recent use of lectins, nonenzymatic proteins exhibiting a high affinity for polysaccharides, expands the availability of stains for EPS (Table 3.3) (Neu and Lawrence 1999). More recently, Peltola *et al.* (2008) used lectins to study the architecture of biofilms growing on glass and on stainless steel and imaged their locations using CSLM and nanobeads as in situ probes (Peltola *et al.* 2008). The overall conclusion is that components of the biofilm can be stained individually or using multiple stains. For example, Chen *et al.* (2007) used multiple stains to image total cells, dead cells, proteins, lipids, and alpha- and beta-polysaccharides in bioaggregates (Chen *et al.* 2007). Larsen *et al.* (2008) used multiple stains to study the efficiency of removing unwanted biofilm from surfaces in industrial water systems (Larsen *et al.* 2008).

### 3.4.3 Staining biofilms to image their structure

Biofilm structure is defined as the distribution of biomass in the space occupied by a biofilm. It has been hypothesized that biofilm structure

**Table 3.3** Selected dyes used to stain EPS

| Dye | Invitrogen catalog number* | Excitation (nm) | Emission (nm) | Specificity |
|---|---|---|---|---|
| Tetramethylrhodamine Concanavalin A (Lectin) | C860 | 555 | 580 | Selectively binds to α-mannopyranosyl and α-glucopyranosyl sugars |
| Lectin PNA from *Arachis hypogaea* (peanut), Alexa Fluor® 488 conjugate (Lectin) | L21409 | 552 | 656 | Lectin PNA is specific for terminal β-galactose |
| Nile red | N-1142 | 625 | 700 | Used to localize and quantify lipids |
| $C_1$-BODIPY 500/510 $C_{12}$ (fluorescent fatty acid analog) | D-3823 | 500 | 510 | Used for lipid trafficking studies |

*Similar dyes can be obtained from other sources.

reflects certain functions of the biofilm and that a biofilm can optimize its structure to control the rate of nutrient delivery to the deeper layers of the biofilm. This and similar hypotheses based on the visual analysis of various biofilms generated the need to quantify biofilm structure. Quantifying the structure of biofilms is an important step towards understanding and describing biofilm systems. Zhang and Bishop determined the densities, porosities, specific surface areas, and mean pore radii of biofilms (Zhang *et al.* 1994; Zhang and Bishop 1994a,b). Hermanowicz *et al.* (1995) used CSLM images to estimate fractal dimension from microscope images and biofilm morphology (Hermanowicz *et al.* 1995). Picioreanu *et al.* (1998) defined both two- and three-dimensional parameters, such as surface enlargement, surface roughness, fractal dimension, porosity, and compactness, for the purpose of describing the structure of computer-simulated biofilms (Picioreanu *et al.* 1998). We conjecture that there exist a finite number of features, and associated quantitative measures, that describe the structure of a biofilm and contain enough information either to reflect variations in the growth dynamics or to predict the functional characteristics of the biofilm. Biofilms are often stained to enhance their visibility and make possible the quantifying of biofilm structure. It is important to use stains and procedures that do not introduce artifacts and do not change the structure of the biofilm. The following section describes some procedures we use to image biofilm structure and to quantify parameters characterizing it.

### 3.4.4 An example of a staining procedure used to image biofilm structure

The first step toward obtaining good-quality images of biofilms is to make sure that the biofilms are grown on a surface that is suitable for imaging. Among many conditions for satisfying this requirement, the surface on which the biofilms are grown should not interact with the stain: the best surface for imaging biofilms is glass. This may sound trivial, but before imaging it is wise to make sure that the sample fits under the CSLM objective. The staining procedure needs to be optimized depending on the actual conditions of the test. Sometimes biofilms are stained before the sample is placed under the microscope. Figure 3.3 shows a procedure we use for biofilm staining using SYTO 9 and propidium iodide. The biofilm was grown in a capillary flow cell and the reactor was operated in a continuous flow mode (500 μl/min). After a mature biofilm was established (approximately three days after inoculation), the flow of the feed nutrient solution was stopped and we gently injected a mixture of SYTO 9 and propidium iodide. This mixture was prepared by adding 3 μL of each component to 1 mL of DI water. The 1-mL solution volume was larger than the volume of the flow cell and ensured that the entire flow cell was filled with the staining solution. We waited for 20 minutes to allow the stains to penetrate the biofilm. During staining, the reactor was covered with aluminum foil to prevent light penetration because most of the dyes are light-sensitive. Then, we washed out the excess stain by pumping the phosphate buffer through the reactor

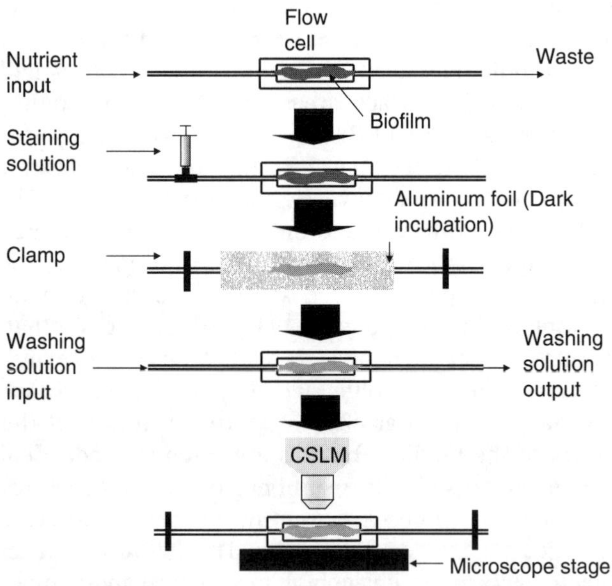


**Fig. 3.3** Procedure used for biofilm staining using SYTO 9 and propidium iodide.

for 20 minutes. To stain our biofilm, we used 20 min for staining and 20 min for washing, but the times for fluorochrome penetration and removal should be optimized on a case-by-case basis.

When selecting a dye for this type of staining, it is important to make sure that the dye does not change the structure of the biofilm. It is important that the biofilm be protected from abrupt changes in the environment while it is imaged, such as excessive temperature and pH changes. It is best if the biofilm can be imaged at the same temperature and in the same solution in which it was grown. In addition, depending on the nature of the study, it may be important to consider whether the differences between the growth medium in the reactor and the solution used for imaging in concentrations of oxygen and carbon dioxide and ionic strength of the medium affect the structure of the biofilm and the results of the test. Obviously, staining and imaging the biofilm in the reactor in which it was grown, as shown in Fig. 3.3, simplifies the procedures and limits the number of manipulations that can affect the structure of the biofilm.

If it is impossible to place the entire biofilm reactor under the microscope (as shown in Fig. 3.3), it may be possible to cut pieces of the biofilm, stain them and then place them under the microscope. The study shown in Fig. 3.4 is an example of such an approach. In this study, the biofilms were grown on a hollow fiber membrane and the hollow fiber was 16 cm long, much too long to be placed under the microscope: we had to cut the hollow fiber into 1-cm pieces. To image the biofilm structure, the 1-cm-long fiber was initially placed in a mixture of SYTO 9 and propidium iodide until the stains penetrated the biofilm. To remove the excess stain, the hollow fiber with the biofilm was immersed several times in a pH buffer solution, the composition of which was adjusted by adding the nutrients to make its composition identical to that of the growth medium. We used the phosphate buffer because it does not interfere with live/dead stains: many other pH buffers interfere with the stains and may cause spurious images. The staining procedure lasted 15 minutes and the hollow fiber with the excess stain was removed from the biofilm by immersing in the pH buffer five times, for 3 minutes each time. Finally, the hollow fiber was placed in a Petri dish, flooded with the pH buffer, and placed under the objective of the microscope, and the structure of the biofilm deposited on the fiber was imaged.

Since the structure of the biofilm needs to be visualized and quantified from the images using appropriate image analysis software, it is important that the images of the biofilm be crisp. The biofilm images can be visualized using appropriate image analysis software such as Imaris® (http://www.bitplane.com), ImageJ NIH (http://rsb.info.nih.gov/ij/), or velocity® (http://www.improvision.com/). Although some commercial software packages do offer limited image analysis capability, biofilm structure is usually quantified using other types of software packages, described later in this text. Figure 3.5 shows examples of acceptable and unacceptable images. Figure 3.5(a) shows individual cells stained by SYTO 9, while Fig. 3.5(b)

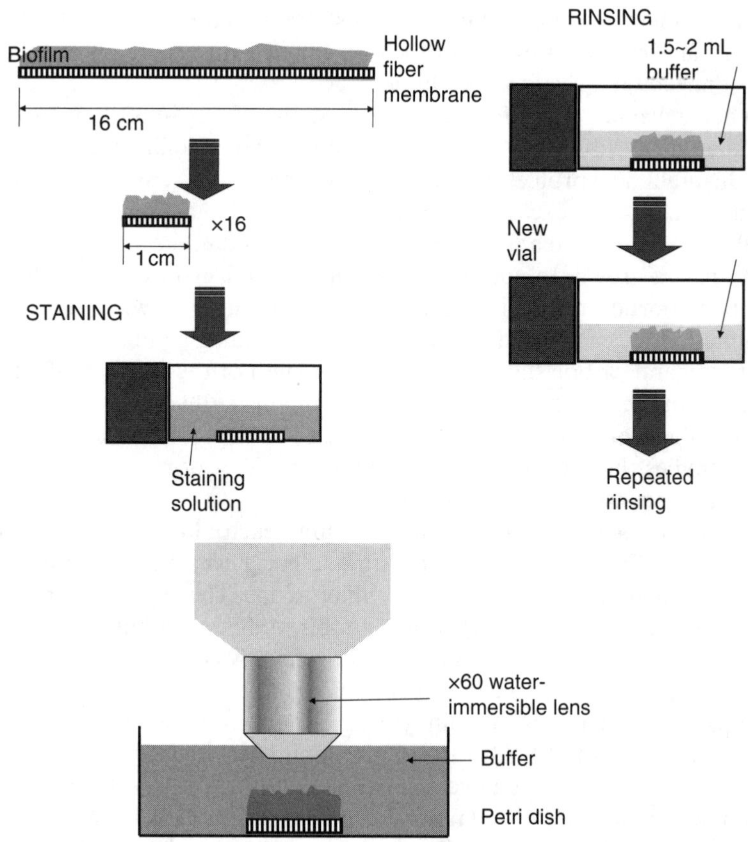


**Fig. 3.4** Staining and rinsing of the biofilm were performed in the dark.

shows only background noise detected by the CSLM. Under higher magnification, the image in Fig. 3.5(a) shows objects with the characteristic shape of bacterial cells while the image in Fig. 3.5(b) shows randomly distributed shapeless clusters of pixels.

## 3.5 Quantifying biofilm structure

The reasons for structural heterogeneity are largely unknown, but they undoubtedly involve both physical and biological factors. Van Loosdrecht *et al.* (1995) discuss how substrate loading rate, shear stress, and growth rate may influence biofilm structure (van Loosdrecht *et al.* 1995). It may be that biological mechanisms create a structure that maximizes mass transport to deep layers within the biofilm. It has been well documented that structural heterogeneity influences the mass transport mechanism and rates

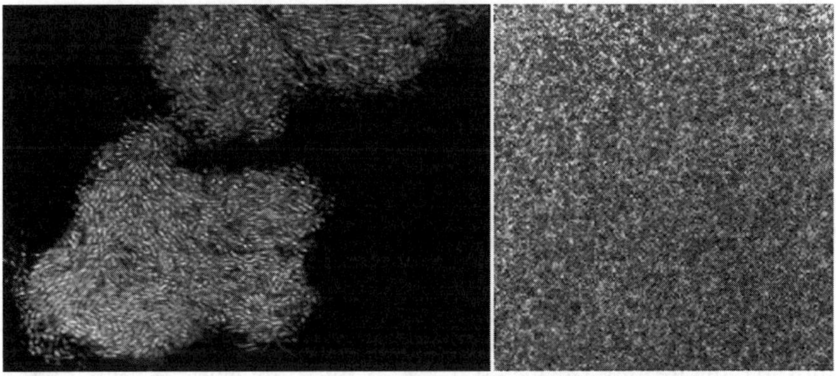

(a) (b)

**Fig. 3.5** Illustration of the difference between properly and improperly stained biofilms. (a) Properly stained biofilm shows a crisp CSLM image of the cells. (b) Improperly stained biofilm shows a distinctly noisy signal in a CSLM image. (Both images were taken at the Center for Biofilm Engineering by Deepak Sharma.)

near and within biofilms (Bishop and Rittmann 1995). For the structural features of biofilms to be related to their function, the structural features need to be quantified.

Images of stained biofilms are stored digitally and retrieved later to quantify parameters characterizing biofilm structure. These parameters by themselves do not necessarily reflect any fundamental biofilm processes. They are values of mathematical functions characterizing the distribution of pixels in images. Some of these functions are conceptually simple, such as the ratio of the number of pixels representing biomass to the number of pixels representing voids, and some are complex, such as the probability that the gray scale intensity of two neighboring pixels is higher than a certain value. Descriptions of these parameters can be found in many publications devoted to this subject (Beyenal *et al.* 2004b; 2004a; Daims *et al.* 2006; Heydorn *et al.* 2000; Lewandowski and Beyenal 2007; Mueller *et al.* 2006; Yang *et al.* 2000). Parameters describing biofilms can be quantified from two-dimensional or three-dimensional images. A two-dimensional image represents a single layer in a biofilm. Three-dimensional images are reconstructed from stacks of two-dimensional images. Most commonly two classes of parameters are computed from such images: 1) textural and 2) areal (for two-dimensional images)/volumetric (for three-dimensional images). Although these parameters quantify biofilm structure, in most instances it is not clear how they are related to the fundamental biofilm processes of attachment, detachment, and growth. Except for a few parameters, such as areal porosity and fractal dimension, it is not immediately clear how these parameters are related to specific biofilm processes or, therefore, how they are relevant.

### 3.5.1 Textural parameters

Texture, described as a repeating pattern of local variations in image intensity (Yang *et al.* 2000), provides information about the spatial distribution of intensity levels in a neighborhood. Textural parameters are calculated from gray level images, without thresholding. In computer science literature, texture is often used to partition images into regions of interest and classify those regions. Textural parameters can be calculated for 2D or 3D images. They quantify biofilm structure by comparing the intensity, position, and/or orientation of the pixels. Each textural parameter measures the character of the cell cluster and interstitial spaces based on the likelihood that pixels of similar or dissimilar types will be neighbors. Textural parameters are calculated in two or three dimensions using a gray level co-occurrence matrix (GLCM), which can be calculated for horizontal and vertical dependence matrices for 2D images and for X, Y and Z direction dependence matrices for 3D images. The GLCM represents the distribution of changes in the gray level values of neighboring pixels in the calculated directions.

### 3.5.2 Areal/volumetric parameters

Areal and volumetric parameters can be calculated from thresholded 2D and 3D biofilm images, respectively. This can be accomplished using one of the software packages developed specifically for the analysis of biofilm images, specified later in this chapter. Areal/volumetric parameters describe the morphology of biofilms: they are concerned with the size, shape and orientation of the constituent parts. Each parameter measures a unique characteristic feature of an imaged component of a biofilm or the interstitial space in the biofilm. Because of this, the first action that is performed in image analysis is to separate cell clusters from interstitial voids. This can be done by a procedure of segmenting an image known as thresholding, in which the values of all pixels with gray levels below a particular value (the threshold) are changed to zero, and all those above the threshold are changed to the maximum pixel value. The resulting black-and-white image has two visible components: the component imaged and void. Examples of areal parameters are the average run lengths (in the horizontal and vertical directions), average and maximum diffusion distances, fractal dimension, and areal porosity. Examples of volumetric parameters are the average run lengths (in the X, Y and Z directions), aspect ratio, average and maximum diffusion distances and fractal dimension.

### 3.5.3 Automatic analysis of biofilm images

Many studies have been devoted to calculating structural parameters of biofilms from their images, and there are software packages, developed by various research groups, for quantifying these parameters. Although commercially available software for image analysis frequently computes the same parameters, many biofilm researchers find it beneficial to develop their own software to quantify parameters from biofilm images. The reason for this is the nature of biofilm images: it is quite different from the nature of the images typically quantified by the commercially available software for image analysis. Commercial packages are excellent for analyzing crisp images of objects with easily identified edges, such as the edges of fluorescent beads, and computing parameters characterizing the structure of such images. Unfortunately, biofilm images do not have well-defined edges and the commercial software packages have limited use in evaluating structural parameters from biofilm images. However, they are very handy when the user needs to calculate the distance between two locations in a biofilm, such as the maximum distance in a single cluster, or the fluoresecent intensity at a selected location (Klayman *et al.* 2008).

Some of the software packages developed by groups of researchers to evaluate biofilm images are:

1. ISA, developed by our research group (Beyenal *et al.* 2004b; 2004a; Yang *et al.* 2000; 2001);
2. COMSTAT, developed by Heydorn *et al.* (2000);
3. PHLIP (Mueller *et al.* 2006); and
4. Daime (Daims *et al.* 2006).

Table 3.4 compares the available software packages for quantifying parameters from biofilm images. The number of parameters that are computed from biofilm images is certainly impressive. However, these parameters are just mathematical functions that describe the distribution of the pixels in the digitized images, and we still lack an understanding of how these parameters are related to the underlying biofilm processes.

## 3.6 Conclusions and directions of future research

The last several years have seen an exponentially increasing interest in imaging components of biofilms and in quantifying biofilm structure from biofilm images. These two efforts are connected: the better the images of biofilms are, the more accurate the quantification of biofilm structure can be. Biofilm research benefits greatly from the progress in molecular biology and microbial ecology, and many biofilm researchers are fluent in using various molecular techniques to image biofilm components. Many fascinating microscopy applications have been used to generate impressive images of the biofilm's inner space. Sadly, the progress in imaging biofilm components has not yet been fully matched by progress in understanding biofilm processes. It appears that satisfying the following two conditions would accelerate the progress toward this understanding:

**Table 3.4** Currently available software packages for quantifying biofilm structure

| Parameter | ISA (Beyenal *et al.* 2004b; 2004a) | COMSTAT (Heydorn *et al.* 2000) | PHLIP (Mueller *et al.* 2006) | DAIME (Daims *et al.* 2006) |
|---|---|---|---|---|
| Biovolume | + | + | + | + |
| Volume to surface area ratio (or surface to volume ratio) | + | + | – | + |
| Area occupied by each layer | + | + | – | – |
| Porosity | + | +* | +* | – |
| Average diffusion distance | + | Only 2D distribution | – | – |
| Average maximum diffusion distance | + | Only 2D distribution | – | – |
| Surface area | + | + | + | + |
| Fractal dimension | + | Only 2D distribution | – | – |
| Average X-run length | + | – | – | – |
| Average Y-run length | + | – | – | – |
| Average Z-run length | + | – | – | – |
| Aspect ratio in 3D | + | – | – | – |
| Spatial spreading | – | – | + | – |
| Co-localization (2D and 3D) | – | – | + | |
| Spatial arrangement | – | – | – | + |
| Fluorescence intensity | – | – | – | + |
| Textural entropy | + | – | – | |
| Energy | + | – | – | |
| Homogeneity | + | – | – | |
| Mean thickness | + | + | – | |

| | | | | |
|---|---|---|---|---|
| Maximum thickness | + additionally can compute percentage of maximums | + | – | – |
| Roughness coefficient | + | + | – | – |
| Identification and areal distribution of microcolonies at the substratum | – | + | – | – |
| Volume of the microcolonies identified at the substratum | + | + | – | – |
| Options | | | | |
| Automatic thresholding | + | – | + | + |
| Interpolation | + | – | – | – |
| Automatic image reversing | + | – | – | + |
| Filtering | + | – | – | + |
| Background reduction | – | – | – | + |
| Three-dimensional imaging | + | – | – | + |
| Operating systems | Windows | Linux, Windows | Linux, OSX, Windows | Linux, Windows |
| Requirements | None | Matlab license and tool boxes | Matlab license only | Free Linux installation |
| Availability | Compiled version costs ~$1500 MATLAB codes are published in a book (Lewandowski and Beyenal 2007) | Free upon request | Free downloadable | Free downloadable |

* The parameter can be computed by the user from other parameters reported by the software.

1. Relating parameters quantifying biofilm structure from biofilm images to the underlying biofilm processes and
2. Including parameters quantified from biofilm images in mathematical models of biofilm activity.

Finally, it is important to reiterate that the imaging of biofilm components is limited by the availability of specific probes that can selectively target the components. Many probes are available to aid in imaging microorganisms in biofilms. However, the existing probes for imaging selected components of the extracellular polymeric substances in the biofilm matrix provide limited information, which in some cases can be inadequate. To improve our understanding of the inner structure of biofilms, the development of probes targeting various components of EPS in intact and fully hydrated biofilms needs to be advanced.

## 3.7 Sources of further information and advice

Protocols for sampling biofilm reactors and the computations we use to characterize biofilm structure are described in our book: Lewandowski, Z. and Beyenal, H. 2007. *Fundamentals of Biofilm Research.* CRC.

## 3.8 References

ABRAHAMSON, M., LEWANDOWSKI, Z., GEESEY, G., SKJAKBRAEK, G., STRAND, W. and CHRISTENSEN, B. E. Development of an artificial biofilm to study the effects of a single microcolony on mass transport. 1996. *Journal of Microbiological Methods* **26**: 161–169.

BEYENAL, H., DONOVAN, C., LEWANDOWSKI, Z. and HARKIN, G. Three-dimensional biofilm structure quantification. 2004a. *Journal of Microbiological Methods* **59**: 395–413.

BEYENAL, H., LEWANDOWSKI, Z. and HARKIN, G. Quantifying biofilm structure: Facts and fiction. 2004b. *Biofouling* **20**: 1–23.

BIESTERFELD, S., FIGUEROA, L., HERNANDEZ, M. and RUSSELL, P. Quantification of nitrifying bacterial populations in a full-scale nitrifying trickling filter using fluorescent in situ hybridization. 2001. *Water Environment Research* **73**: 329–338.

BISHOP, P. L. and RITTMANN, B. E. Modelling heterogeneity in biofilms: Report of the discussion session. 1995. *Water Science and Technology* **32**: 263–265.

CAMPBELL, S., GEESEY, G., LEWANDOWSKI, Z. and JACKSON, G. Influence of the distribution of the manganese-oxidizing bacterium, Leptothrix discophora, on ennoblement of Type 316L stainless steel. 2004. *Corrosion* **60**: 670–680.

CHALFIE, M., TU, Y., EUSKIRCHEN, G., WARD, W. W. and PRASHER, D. C. Green Fluorescent Protein As A Marker for Gene-Expression. 1994. *Science* **263**: 802–805.

CHEN, M. Y., LEE, D. J., TAY, J. H. and SHOW, K. Y. Staining of extracellular polymeric substances and cells in bioaggregates. 2007. *Applied Microbiology and Biotechnology* **75**: 467–474.

DAIMS, H., LUCKER, S. and WAGNER, M. Daime, a novel image analysis program for microbial ecology and biofilm research. 2006. *Environmental Microbiology* **8**: 200–213.

DELONG, E. F., TAYLOR, L. T., MARSH, T. L. and PRESTON, C. M. Visualization and enumeration of marine planktonic archaea and bacteria by using polyribonucleotide probes and fluorescent in situ hybridization. 1999. *Applied and Environmental Microbiology* **65**: 5554–5563.

FOWLER, S. D. and GREENSPAN, P. Application of Nile Red, A Fluorescent Hydrophobic Probe, for the Detection of Neutral Lipid Deposits in Tissue-Sections – Comparison with Oil Red O. 1985. *Journal of Histochemistry & Cytochemistry* **33**: 833–836.

GRAYMEROD, R., HENDRICKX, L., MUELLER, L. N., XAVIER, J. B. and WUERTZ, S. Effect of nucleic acid stain Syto9 on nascent biofilm architecture of Acinetobacter sp BD413. 2005. *Water Science and Technology* **52**: 195–202.

GREENSPAN, P. and FOWLER, S. D. Spectrofluorometric Studies of the Lipid Probe, Nile Red. 1985. *Journal of Lipid Research* **26**: 781–789.

GREENSPAN, P., MAYER, E. P. and FOWLER, S. D. Nile Red – A Selective Fluorescent Stain for Intracellular Lipid Droplets. 1985. *Journal of Cell Biology* **100**: 965–973.

HANSEN, M. C., PALMER, R. J., UDSEN, C., WHITE, D. C. and MOLIN, S. Assessment of GFP fluorescence in cells of *Streptococcus gordonii* under conditions of low pH and low oxygen concentration. 2001. *Microbiology-Sgm* **147**: 1383–1391.

HERMANOWICZ, S. W., SCHINDLER, U. and WILDERER, P. Fractal structure of biofilms: New tools for investigation of morphology. 1995. *Water Science and Technology* **32**: 99–105.

HEYDORN, A., NIELSEN, A. T., HENTZER, M., STERNBERG, C., GIVSKOV, M., ERSBOLL, B. K. and MOLIN, S. Quantification of biofilm structures by the novel computer program COMSTAT. 2000. *Microbiology-UK* **146**: 2395–2407.

ITO, T., NIELSEN, J. L., OKABE, S., WATANABE, Y. and NIELSEN, P. H. Phylogenetic identification and substrate uptake patterns of sulfate-reducing bacteria inhabiting an oxic-anoxic sewer biofilm determined by combining microautoradiography and fluorescent in situ hybridization. 2002. *Applied and Environmental Microbiology* **68**: 356–364.

JANG, A., OKABE, S., WATANABE, Y., KIM, I. S. and BISHOP, P. L. Measurement of growth rate of ammonia oxidizing bacteria in partially submerged rotating biological contactor by fluorescent in situ hybridization (FISH). 2005. *Journal of Environmental Engineering and Science* **4**: 413–420.

KLAYMAN, B. J., KLAPPER, I., STEWART, P. S. and CAMPER, A. K. Measurements of accumulation and displacement at the single cell cluster level in *Pseudomonas aeruginosa* biofilms. 2008. *Environmental Microbiology* **10**: 2344–2354.

LARSEN, P., OLESEN, B. H., NIELSEN, P. H. and NIELSEN, J. L. Quantification of lipids and protein in thin biofilms by fluorescence staining. 2008. *Biofouling* **24**: 241–250.

LEWANDOWSKI, Z. and BEYENAL, H. 2007. *Fundamentals of biofilm research.* CRC.

MANZ, W., WENDT-POTTHOFF, K., NEU, T. R., SZEWZYK, U. and LAWRENCE, J. R. Phylogenetic composition, spatial structure, and dynamics of lotic bacterial biofilms investigated by fluorescent in situ hybridization and confocal laser scanning microscopy. 1999. *Microbial Ecology* **37**: 225–237.

MATZ, M. V., FRADKOV, A. F., LABAS, Y. A., SAVITSKY, A. P., ZARAISKY, A. G., MARKELOV, M. L. and LUKYANOV, S. A. Fluorescent proteins from nonbioluminescent Anthozoa species. 1999. *Nature Biotechnology* **17**: 969–973.

MCNAMARA, C. J., LEMKE, M. J. and LEFF, L. G. Underestimation of bacterial numbers in starvation-survival mode using the nucleic acid stain DAPI. 2003. *Archiv fur Hydrobiologie* **157**: 309–319.

MUELLER, L. N., DE BROUWER, J. F. C., ALMEIDA, J. S., STAL, L. J. and XAVIER, J. B. Analysis of a marine phototrophic biofilm by confocal laser scanning microscopy using the new image quantification software PHLIP. 2006. *BMC Ecology* **6**.

NANCHARAIAH, Y. V., VENUOPALAN, V. P., WUERTZ, S., WILDERER, P. A. and HAUSNER, M. Compatibility of the green fluorescent protein and a general nucleic acid stain for quantitative description of a Pseudomonas putida biofilm. 2005. *Journal of Microbiological Methods* **60**: 179–187.

NEU, T. R. and LAWRENCE, J. R. Lectin-binding analysis in biofilm systems. 1999. *Biofilms* **310**: 145–152.

PELTOLA, M., NEU, T. R., RAULIO, M., KOLARI, M. and SALKINOJA-SALONEN, M. S. Architecture of Deinococcus geothermalis biofilms on glass and steel: a lectin study. 2008. *Environmental Microbiology* **10**: 1752–1759.

PICIOREANU, C., VAN LOOSDRECHT, M. C. M. and HEIJNEN, J. J. Mathematical modeling of biofilm structure with a hybrid differential-discrete cellular automaton approach. 1998. *Biotechnology and Bioengineering* **58**: 101–116.

PRENDERGAST, F. Chemical and physical-properties of aequorin and green fluorescent protein isolated from aequorea-forskalea. 1978. *Biochemistry* **17**: 3448.

SCHMID, T., PANNE, U., HAISCH, C. and NIESSNER, R. Biofilm monitoring by photoacoustic spectroscopy. 2003. *Water Science and Technology* **47**: 25–29.

SEO, K. H. and FRANK, J. F. Attachment of Escherichia coli O157:H7 to lettuce leaf surface and bacterial viability in response to chlorine treatment as demonstrated by using confocal scanning laser microscopy. 1999. *Journal of Food Protection* **62**: 3–9.

SIEBEL, M. A. and CHARACKLIS, W. G. Observations of Binary Population Biofilms. 1991. *Biotechnology and Bioengineering* **37**: 778–789.

SKJAKBRAEK, G., GRASDALEN, H. and LARSEN, B. Monomer Sequence and Acetylation Pattern in Some Bacterial Alginates. 1986a. *Carbohydrate Research* **154**: 239–250.

SKJAKBRAEK, G., SMIDSROD, O. and LARSEN, B. Tailoring of Alginates by Enzymatic Modification Invitro. 1986b. *International Journal of Biological Macromolecules* **8**: 330–336.

VAN LOOSDRECHT, M. C. M., EIKELBOOM, D., GJALTEMA, A., MULDER, A., TIJHUIS, L. and HEIJNEN, J. J. Biofilm structures. 1995. *Water Science and Technology* **32**: 35–43.

WETZEL, R. G., WARD, A. K. and STOCK, M. Effects of natural dissolved organic matter on mucilaginous matrices of biofilm communities. 1997. *Archiv fur Hydrobiologie* **139**: 289–299.

WIRTANEN, G., SALO, S., HELANDER, I. M. and MATTILA-SANDHOLM, T. Microbiological methods for testing disinfectant efficiency on Pseudomonas biofilm. 2001. *Colloids and Surfaces B-Biointerfaces* **20**: 37–50.

XIA, F. H., BEYENAL, H. and LEWANDOWSKI, Z. An electrochemical technique to measure local flow velocity in biofilms. 1998. *Water Research* **32**: 3631–3636.

YANG, X. M., BEYENAL, H., HARKIN, G. and LEWANDOWSKI, Z. Quantifying biofilm structure using image analysis. 2000. *Journal of Microbiological Methods* **39**: 109–119.

YANG, X. M., BEYENAL, H., HARKIN, G. and LEWANDOWSKI, Z. Evaluation of biofilm image thresholding methods. 2001. *Water Research* **35**: 1149–1158.

ZHANG, T. C. and BISHOP, P. L. Evaluation of tortuosity factors and effective diffusivities in biofilms. 1994a. *Water Research* **28**: 2279–2287.

ZHANG, T. C. and BISHOP, P. L. Experimental Determination of the Dissolved-Oxygen Boundary-Layer and Mass-Transfer Resistance Near the Fluid-Biofilm Interface. 1994b. *Water Science and Technology* **30**: 47–58.

ZHANG, T. C., FU, Y. C. and BISHOP, P. L. Competition in Biofilms. 1994. *Water Science and Technology* **29**: 263–270.

# 4

# Monitoring of biofilms in the food and beverage industries

**A. Pereira and L. F. Melo, University of Porto, Portugal**

**Abstract**: Monitoring biofilm formation in the food and beverage industries is an important tool not only to detect and evaluate the surface, product and process contamination, but also to minimize the effects of such contamination, to apply timely countermeasures and to assess their efficacy. The present chapter reviews some of the most used monitoring techniques both in laboratorial and in industrial processes. These methodologies are based on different physical principles, which can range from the traditional visual inspection and classical microbiological techniques to more recent approaches based on the propagation of vibration or optical signals. A brief description of those methods will be provided, as well as a discussion about their (laboratorial vs. industrial) applicability. When relevant, the commercially available devices will also be indicated.

**Key words**: biofilm monitoring, food industry, biofouling.

## 4.1 Introduction

The build up of biofilms in industrial production lines is a problem that has attracted the attention of the scientific and industrial community for many decades. Heat exchangers, membranes, cooling water systems and drinking water storage and transportation systems are examples of equipment typically affected by biofouling (i.e., the unwanted formation of biofilms). The effect that this phenomenon has on the overall process economy is highly dependent on the specific process, but is often perceived as being related to material corrosion and reduction of the thermal and hydrodynamic performance across a given production line, resulting in the increase of the heat transfer resistance, of the pressure drop and of the maintenance costs.

However, in addition to the energetic and corrosion constraints, there are several industrial situations where the presence of microorganisms/

biofilms on surfaces can lead to serious hygiene-related health problems. It should be stressed that the main non-aqueous components of the biofilm matrix, the EPS (extracellular polymeric substances), have a crucial role in protecting the microbial species that thrive within the biological layer from external attacks, by providing pools of nutrients and by sequestering metals and toxins. In fact, the biofilm mode of living itself promotes mechanisms of biological adjustment (based on quorum sensing, phenotypic changes, etc.) to resist the efficacy of preventive or remediation measures.

The food industry and drinking water distributions networks are two examples where biofouling may result in product or water contamination. In the food industry, if the cleaning and disinfection steps are not properly conducted (disinfection is often performed after cleaning and aims at minimizing the number of microorganisms within the industrial plant), the safety and quality of the product will be affected (Hasting, 2005). *Listeria* spp., *Bacillus* spp., *Salmonella*, *Clostridium botulinum* are only a few examples of pathogenic micro-organisms that affect food products. These microbes may grow on the walls of the processing equipment, incorporate in the liquid or solid foodstuff and eventually colonize package surfaces. Concerning potable water systems, the 'biofouling' layers (which usually incorporate abiotic material or are embedded in abiotic deposits) that are found along water distribution networks are preferential shelters for the growth of pathogens (e.g., *Pseudomonas aeruginosa, Legionella, Helicobater pylori*, *Escherichia coli*, etc). The public health is at risk when portions of those biofouling layers (often quite thick) detach and reach the consumer's tap. A similar effect is observed in the food industry where an improper disinfection procedure causes release of microorganisms to the food product during processing or storage. The presence of biofilms can therefore lead to microbial contamination which in the last instances may induce product spoilage and foodborne illnesses (Verran, 2002).

## 4.2 Monitoring approach as a strategic anti-fouling methodology

Several approaches have been studied in order to minimize biofilm formation and/or to maximize its removal through the use of biocides and dispersants (Bott, 2005, Frank, 2000), the optimization of the operating conditions such as flow velocity and temperature gradients (Vieira and Melo, 1999, Pereira *et al.*, 2002, Chen *et al.*, 2005) or the proper design of the equipment, namely the use of anti-fouling surfaces (Zhao *et al.*, 2005) and the elimination of hydrodynamically 'dead zones' in fluid circulation. The chemical approach, although still predominant as an immediate and direct counterattack to the buildup of undesired biofilms, will tend to lose its prime position due to the environmental burden it brings. Instead, a toolbox of eco-friendly preventive methodologies will gradually take up more space in anti-fouling strategies, and this includes not only innovation in equipment and process design, but also the increasing development and generalized application of efficient and reliable on-line monitoring techniques.

The conventional techniques to evaluate the existence of deposits in a given process are based on indirect methods such as the assessment of process performance or product quality (Flemming, 2003) or on the collection of samples for off-line chemical, physical and biological studies. An important step on the microbial characterization of biofilms in the last decade was the development of molecular biology techniques that have improved the accuracy of microbiological analyses and reduced the time needed for the identification and quantification of the microbial species (Warner *et al.*, 2008). However, such indirect measurements are carried out off-line, demand quite specialized technicians to perform them and are time-consuming, delaying the reception and treatment of the information, and can be misguiding because they do not provide real-time reliable information about the extent and location of the possible contaminating layer. Additionally, since the efficacy of a given countermeasure is also evaluated by the same indirect methods, those countermeasures may become highly costly and quite inefficient, because the presence of biofouling layers is only perceived at a late phase (sometimes, too late). A similar idea is shared by (Hasting, 2002) who states that the monitoring of current industrial processes is frequently very poor and the knowledge of the process is typically based on limited and retrospective data. This author reiterates the idea that a problem caused by an inefficient cleaning procedure is typically not detected at the start of the production cycle, when it would be easier to stop and re-clean the line.

This emphasizes the importance of developing online monitoring methods to follow the adhesion, growth and/or removal of deposits from surfaces as a complementary approach to mitigate the effects of biofilm formation, reduce the costs of the cleaning operations and minimize production breaks for maintenance and disinfection.

Milk processing is one of the industrial food processes that better demonstrates the necessity to develop and implement accurate online monitoring techniques, due to the extensive fouling events it suffers (mainly in the pasteurization and sterilization units) and to the regular cleaning procedures (often on a daily basis) that it demands. So, this particular case will be used to exemplify in more detail the problems associated with the lack of timely information about the deposit build up and removal, including both the amount and the type of deposit that forms on the equipment walls. Biofilms may form either in the process side, since the foodstuff contains all the essential nutrients for microbes to develop, or in water circuits that are used to clean the equipment or to cool the product. Particular attention should be given to bacteria that, following a relatively inefficient disinfection phase, were left on the equipment surfaces and can be subsequently released to the food product when production re-starts.

The importance of CIP (cleaning-in-place) practices is totally recognized nowadays in the food processing industry. These automatic procedures typically involve (Fryer *et al.*, 2006):

i) *pre-rinse* step with circulation of water
ii) *detergent cycle* step with the circulation of chemical solutions (alkali and/or acid)
iii) *post-rinse* step with circulation of water
iv) *sanitization* step with the use of disinfectants and surface conditioning
v) *final rinse* step with circulation of water (the step prior to the beginning of production).

The main problem regarding CIP procedures is the inexistence of real time factual evidence, not only about the amount but also about the type of deposit that is being formed on the surface and, additionally, the lack of information about the efficiency of the removal process (Wilson, 2005). The CIP procedure is established based on the assumption that the nature of fouling is already identified and is consistently the same during the process. However, most of the time this is not true, since the nature of fouling depends on many factors, such as the seasonal variation of the raw material and on the processing of different products (in multi-product plants).

Taking into account that these cleaning cycles are currently implemented on a semi-empirical basis, using a fixed time or a fixed cleaning solution concentration and/or volume, independently of the nature of the fouling that is formed on the surface, it becomes clear that the economic and environmental costs associated with such deposit removal are higher than necessary (Fryer *et al.*, 2006). The CIP procedures are designed in order to warrant that in a given plant the biofouling layer is fully removed in spite of the actual nature and amount of deposit formed on the equipment walls. So, the cleaning commodities and cycles tend to be overdesigned – excessive chemicals and water are used, and the CIP operations are sometimes performed more often than needed. Different chemical solutions are applied to remove different deposits (e.g., proteineous layers are removed with alkaline solutions, while mineral deposits are removed with acid solutions), but the CIP procedure does not evaluate *per se* which chemical solution should be applied according to the deposit characteristics.

The previous discussion emphasized the need of monitoring techniques able to provide different levels of information in the food and beverage industries. Among this information it is necessary to determine the microbiological characteristics of the deposit or product in order to guarantee its safety, as well as to follow on-line the amount of biofilm that is being formed. If an effective control of the biofouling extent formed on the surfaces is performed, the production and cleaning cycles can be optimized, the impact of such deposits can be minimized and timely countermeasures can be applied.

The implementation of monitoring systems that provide on-line, in-real time, non-destructive and in-situ information has become an important strategy to minimize biofouling (Hasting, 2005). In fact, this approach can be used to simultaneously gather different types of information:

1 *to monitor the biofouling extent* – this is quite important since it enables the application of timely countermeasures when the problem starts to appear and avoids unnecessarily lengthy CIP procedures.
2 *to evaluate the efficacy of the applied countermeasures* – several countermeasures are typically applied when biofouling is detected. For example, the monitoring of the efficiency of a given cleaning procedure or the evaluation of the effect that a given biocide has on the removal of the attached layer are some examples of the information that a monitoring system should be able to provide. As pointed out before, from the economic point of view, another very relevant issue is the on-line rapid detection of the cleaning end-point.
3 *to assess the information about the microbiological characteristics of the deposit*, both to evaluate the efficacy of disinfection and the product quality control.

## 4.3 Requirements for a suitable monitoring device

A good monitoring system should fulfill some basic requirements:

1 It should be able to provide *on-line*, *real-time*, *in situ* information about the deposit adhesion and removal. The sampling rate at which the information is given should be several times higher than the adhesion/removal rates of the deposit. These features enable the operator to take timely countermeasures and they are strictly related to the additional requirement of being *simple to use*.
2 It should not interfere with the process and so it should be *non-intrusive*. In fact, intrusive sensors are points of possible contamination (which is a major concern in the food and beverage industry) and in some cases they can create fragilities in the piping structure.
3 It should be *robust* and *reliable*. It should be able to operate and give accurate information under realistic industrial conditions, including broad ranges of temperature, pressure and flow velocity. It is also consensual that in many cases the devices should be prepared to resist severe chemical 'attack' (which may occur during cleaning/disinfection procedures).
4 The information provided by the device should be an *integrated value of the deposit* characteristics over a given area. The area of fouling considered for the measurement should be high enough to avoid biasing problems (the deposits are usually not uniformly distributed) but still keep the needed accuracy.

5 It should be possible to apply it to *different types of surface materials*, since – depending on the particular industrial case under study – the equipment may be built with different materials.

One important concern regarding the implementation of monitoring devices is to find out the location of the sensors, taking into account that they should be placed where fouling is most likely to occur with an amount or effect that is representative of what happens in the overall plant. (Withers, 1996). Several devices can be implemented on critical points along the plant so that the information gathered at these locations may be used to infer information about the fouling elsewhere in the plant. A cost/benefit analysis of the implementation of more than one device should be made in order to evaluate the number of devices to be placed along a given processing line. This stresses the need to develop inexpensive sensors.

## 4.4 Biofouling monitoring techniques

The biofouling monitoring systems can provide different levels of information according to their specific design (Flemming, 2003). For example, some are able to assess the biofilm dynamics, attachment/detachment events, but cannot differentiate between the constituents of such layers (e.g., biotic/abiotic). More specific monitoring devices are able to characterize the chemical/biological composition of a given fouling layer, although they are too sophisticated and costly to be operated in an industrial environment. Several reviews (such as Nivens *et al.*, 1995, Flemming, 2003, Janknecht and Melo, 2003, Hasting, 2005) were published that describe numerous monitoring techniques to detect biofouling formation. Such literature discusses in more or less detail both the traditional and the more recent techniques. The reader is invited to look for specific aspects of many techniques in these reviews and in the bibliography listed in the present chapter. In the following paragraphs, the physical principles behind the most well-known/applied biofouling monitoring techniques will be provided, with emphasis on on-line methods with proven or potential capabilities to be implemented in industrial environments.

The practical development of most biofouling on-line monitoring methods is generically based on a simple procedure:

1 A perturbation is introduced on the system, either by an external input signal (light emission, mechanical vibration, etc.) or directly by the presence of the biofilm itself (when the latter produce changes in the overall thermal resistance of a heat transfer process or in the mechanical energy needed for fluid flow in a pipe).
2 The response of the system (given by the modification of some output measured signal – temperature, light absorption, vibration frequency, etc.) is captured by a suitable device/sensor and, if needed, amplified.

3 The output signal is mathematically processed in order to yield a calibrated value representative of the attached biofilm.

### 4.4.1 Visual inspection

The judgment of the degree of cleanliness inside the food industry equipment is often based on visual assessments. A portion of the equipment is typically dismantled and is inspected in order to evaluate the severity of biofouling. In several cases the information provided by visual inspection is complemented with the information gathered from other techniques (such as adenosine triphosphate (ATP) measurements) in order to characterize/quantify bacteria/biofilms present in the system. The visual inspection can be important to confirm the information provided by other monitoring techniques, although it only provides qualitative information often dependent on the subjective evaluation of the operator. Furthermore, the information gathered is only established for visible contaminations. Generally, these methods are time-consuming, retrospective, intrusive and qualitative.

### 4.4.2 Direct measurement of the deposit microbiological characteristics

The hazards that can arise from a deficient detection of microorganisms within the food industry have already been mentioned and are a huge concern to this industry. The application of microbial characterization procedures often requires the sampling of a piece of deposit attached to the surface (off-line measurements) or the swabbing of the surface. Among such techniques it is possible to find microscopic observations and molecular biology analyses, quantification of the metabolic products (e.g., ATP measurements) or chemical analysis (e.g. pH or chlorine measurements). A deeper insight in the subject can be found in Chapter 21 of this book.

*Classical microbiological methodologies*

The standard methodology for microbial quantification is the heterotrophic plate count, which is time-consuming, and requires a skilled operator. Typically, it may take 3–5 days to obtain the results of these tests, which can be quite problematic in cases of product contamination. In some cases, in order to overcome this problem the product stays within the industrial facilities until the results of the test are available. Owing to their low cost and microbiological information provided, the culture methods are applied in almost every food plant, to assess both the cleanliness of the system and the quality of the final product. An alternative and/or complementary approach to these methods is the microscopic inspection of samples (with or without the use of dyes). All these methods are time-consuming, retrospective, require specialized operators, and are invasive.

*ATP bioluminescence*

The ATP bioluminescence determinations are related to the detection of the chemical compound that stores the energy in living cells – the ATP. There are several commercially available portable ATP measuring systems which include a luminometer that measures the light emitted from the conversion of ATP energy into light when catalyzed by an enzyme called luciferase (Hawronskyj and Holah, 1997). The observed light is proportional to the amount of ATP present in the sample, which, through proper calibration, can relate the ATP amount (measured as relative light units – RLU) with cell biomass. These procedures are very quick since the kit already includes the necessary reagents and the measurement can be performed in a few minutes. However, biofilm samples have to be removed from the surface (often by swabbing, which does not capture some portions of the biofilm, particularly on rough surfaces) in order to apply this technique (Chmielewski and Frank, 2003). Several examples of the use of ATP bioluminescence measurements can be found in the literature, such as the monitoring of the cleanliness of stainless steel surfaces in contact with milk (Costa *et al.*, 2006) or of the presence of bacterial contamination in potable water systems (Frundzhyan and Ugarova, 2007). The ATP bioluminescence is only able to provide information about the total amount of contamination observed on the surface rather than giving information about the presence or absence of pathogens (Hawronskyj and Holah, 1997). Some of the commercially available ATP measuring systems can be found in: kaivac (www.kaivac.com), biotracer (www.biotracer.co.uk) or Turner Biosystems (www.luminometer.com). The ATP bioluminescence system is not an online method, it requires a previous calibration for each case. Although the measurements are fast to perform, they are still intrusive, retrospective and localized.

### 4.4.3 Heat transfer and pressure drop measurements

The biofilm adhesion and its removal/detachment affect the pressure and the temperature distributions along and across the overall plant. Heat transfer and/or pressure drop measurements are not only applied industrially, but also widely used in laboratorial research for biofilm growth or cleaning quantification (Bott, 2005). Pressure drop techniques are very easy to implement and commonly applied to monitor heat exchanger fouling, requiring only a differential pressure sensor connected to the inlet and outlet of the equipment. The heat transfer rate can be calculated by an enthalpy balance, once the flow rates and inlet and outlet fluid temperatures are measured. During fouling/biofouling growth, the enthalpy change declines over time, while pressure drop increases. However, these indicators are relatively insensitive to the formation of the initial deposit layer and some of the sensors required here are intrusive, such as temperature and pressure probes. Contrary to the methods previously reported, the heat transfer and/

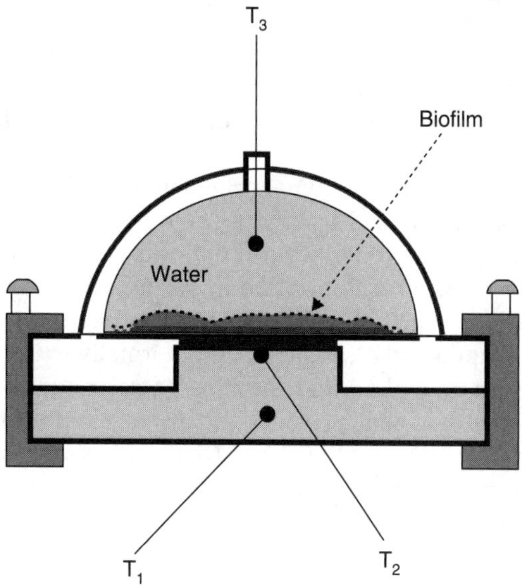


**Fig. 4.1** Schematic representation of a heat transfer monitoring device: by measuring two temperatures in the wall ($T_1$ and $T_2$) and the fluid temperature ($T_3$), the heat transfer rate and the fouling thermal resistance can be calculated (see Vieira *et al.*, 1993).

or pressure drop measurements described above are real-time techniques and produce integrated information about biofouling in the overall equipment or unit.

A more precise technique, although more difficult to implement (impracticable in many heat exchangers), is to measure the heat flux by inserting temperature probes in the wall and in the fluid (Vieira *et al.*, 1993), Fig. 4.1. This technique is better suited for measurements in a side-stream monitor that has to be duly instrumented and it gives a localized measure of the attached deposit (not an overall assessment of the fouled equipment). It should be highlighted that all these techniques are not specific for biofilm detection, but rather give information about the amount of attached deposit, independently of its nature.

There are also some commercially available on-line monitoring sensors specially designed for food and beverage processes based on thermal pulse analysis (www.neo-sens.com), which is a patented technique that measures the local thermal conductivity and the heat transfer variations due to biofouling formation. The temperature of the probe is temporarily increased by the application of a periodic electrical pulse. It is therefore possible to determine the amount of deposit attached to the probe by measuring the temperature before and during the thermal pulse, since as the thickness of the deposit on the probe increases, the transfer of thermal energy to the liquid becomes more and more difficult. The main differences of this type of sensor as regards the traditional heat transfer and pressure drop measurements rely on increased accuracy, which is able to detect deposits only a few microns thick and on the fact that the measurement is localized rather than integrated.

Another technique, the fluid dynamic gauging (FDG) based on pressure measurements, has been used in laboratory research with good results to assess the thickness and strength of soft deposits (Chew *et al.*, 2004) and to study (Chew *et al.*, 2007) the deposition in porous membranes (like macro- and micro-filtration processes). This technique is based on a nozzle that is inserted into the liquid and placed close and normally to the fouled surface. A suction pressure is applied and the flow passes into the nozzle. This is quite a sensitive method that provides the distance between the tip of the nozzle and the top of the fouling layer (or in other words, it is a way to provide the thickness of the deposit), although it does not seem fit to be applicable in most industrial equipment.

### 4.4.4 Measurement of optical signals

There are some methodologies based on the detection of light signals (e.g., visible, infrared, ultraviolet) which rely on the fact that light is emitted, absorbed or scattered by matter. The ATP bioluminescence previously discussed is an example of the measurement of optical signals. One of the monitoring techniques that can be found within this group is the detection of the weak light signals that some organisms are capable of spontaneously emitting and are known as *bioluminescence*. It was successfully applied to measure the number of attached bacteria to different polymeric supports (Ludwicka *et al.*, 1985). In other cases, if this bioluminescence can be genetically manipulated and 'introduced into the biofilm', the biocide/antibiotic effect on=biofilm removal can be successfully monitored (Kadurugamuwa *et al.*, 2004). Associated with this, there is also the detection of *autofluorescence* of biomolecules such as amino acids. Certain molecules have characteristic wavelengths at which they absorb or emit radiation and this absorption or emission is related to the quantum energy levels associated with the vibration of the atoms. So, when a given sample is exposed to a continuum radiation it is possible to measure the absorbed radiation by *spectroscopy* methods and the emitted radiation by *fluorometry* methods (Janknecht and Melo, 2003).

One of the most well-known (commercially available) systems is the FTIR spectrometer (Fourier transform infrared), which combines the measurement of the infrared spectroscopy and the use of Fourier transform techniques. The use of the Fourier techniques in comparison to the dispersive ones increases the sensitivity of the infrared analysis. This methodology was positively applied to detect antimicrobial agents within a biofilm structure in order to understand how the former act within the biofilm (Suci *et al.*, 1998). These techniques are better suited to measure biomolecule signals and, in many cases, to distinguish between biotic and abiotic material, rather than to quantify the amount of deposit, and they are typically not suited for industrial application. Furthermore, the measurements obtained with these techniques should be carefully interpreted, since they can often suffer from interference from external factors (for example, temperature, pH, lack of discrimination between weak and scattering responses). The techniques discussed above are quite important for microbiological and chemical studies, particularly in laboratorial research, although their industrial application should be regarded with some skepticism due to their sophistication, cost and lack of robustness.

Regarding the measurement of the amount of deposited layer, other devices based on light propagation were developed in the last decade. One of them is the *fiber optical sensor* (FOS) which assesses the thickness of the deposit or the number of bacteria that adhere to the tip of an optical fiber inserted into the wall of the pipe through backscattered light (Tamachkiarow and Flemming, 2003) – see Fig. 4.2.

When the measurement is based on the difference of two turbidity values, the device is named *differential turbidity device* (DTM). The DTM compares the turbidity of two systems, one that is measuring the adhesion process and the other that is being continuously cleaned. The FOS and the DTM devices have already been industrially tested to detect the formation of biomass, respectively in brewery water pipelines (Tamachkiarow and Flemming, 2003) and water systems of a paper mill processing unit (Klahre and Flemming, 2000). Regarding the industrial application, the main problem associated with these techniques is the fact that they are not suited for thick deposits – above a given thickness thresh old the optical signal

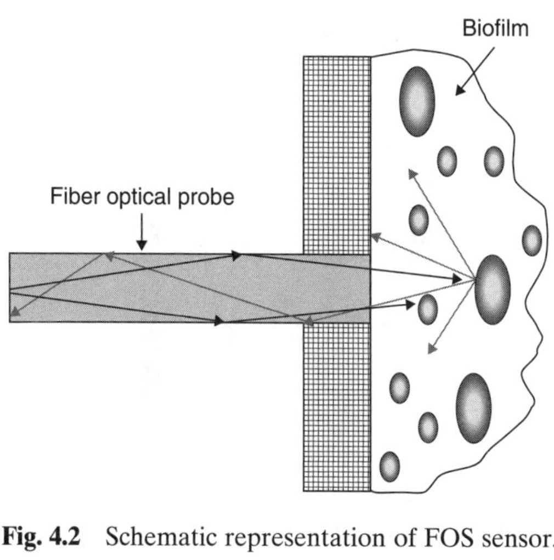


**Fig. 4.2** Schematic representation of FOS sensor.

gets saturated. The DTM requires that one of the turbidity devices is constantly cleaned (e.g., by water jets or by using mechanical wipes), which is a constraint for practical implementation in most food processing units.

Another method, the *photoacoustic spectroscopy sensor* (PAS), is based on the propagation of both light and accoustic waves. When a given sample absorbs energy from a laser source, it physically expands (absorbed energy is transformed into heat) and creates an acoustic wave that propagates through the deposit and can be measured by a vibration sensoring element, such as a piezoelectric film – Fig. 4.3. The sample is excited with a short radiation impulse (electromagnetic radiation, such as laser light) and the measurement principle is based on the difference between the propagation velocities of this impulse (light velocity) and of the acoustic response. The PAS has been applied to investigate the effect of soluble and colloidal substances on the structure and the stability of biofilms, as well as to determine the efficacy of anti-fouling strategies (Schmid *et al.*, 2004, 2003). It is quite an interesting development in monitoring devices mainly for research applications (it is too costly and sophisticated for industrial situations), being able to distinguish between different types of deposits and to detect very small amounts of attached layers.

An infrared monitor has been reliably tested to follow the build-up and detachment of biofilms in the laboratory, as well as to control the growth of microorganisms in a side stream of an industrial cooling plant (Tinham and Bott, 2003). It requires a transparent section through which infrared radiation is emitted (approximately at 950 nm) on one side of

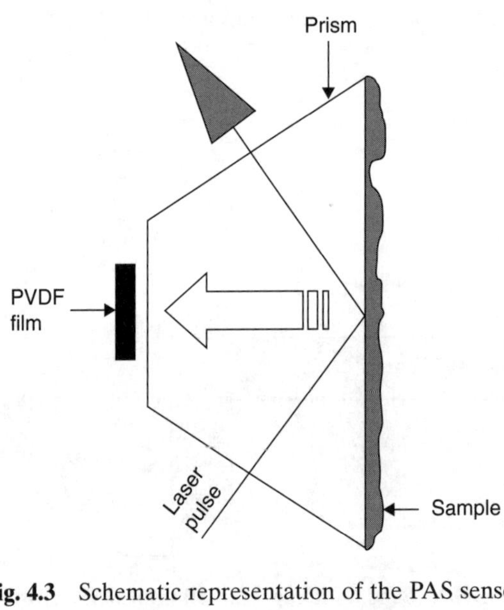


**Fig. 4.3** Schematic representation of the PAS sensor.

the duct surface and collected on the opposite wall. The emitted radiation crosses the transparent section, the fluid and the biofilm (that is attached to the surface) and is absorbed in the biofilm (Tinham and Bott, 2003). It is possible to relate the absorbance of the infrared radiation with the biofilm accumulation on the inside surface of the system, through proper calibration.

### 4.4.5 Measurement of electric signals

The measurement of electric signals is quite popular among the reported monitoring techniques and it includes *electrochemical* measurements and *capacitance/impedance* determinations. The electrochemical measuring devices typically have two or more electrodes (in contact with the aqueous phase) to which a given voltage/current signal is applied. Those devices are based on the effect caused by the deposit build-up on the electric conductivity of the surface/electrode and were applied to monitor, for example, the bio(fouling) formation in nuclear power plants (Licina *et al.*, 1993). Electrochemical measurements have also been applied to study the formation of scaling products on heat transfer surfaces (Tlili *et al.*, 2008). One of the most well-known techniques within this group is the MIC (microbiologically influenced corrosion) monitor, which detects the effect of the biocorrosion caused by microorganisms (due to their metabolic activity) on the electrodes. There are some commercially available MIC monitoring devices, such as the BIOX® (Mollica and Cristiani, 2003), which measures the coupling current between stainless steel and zinc electrodes connected by a resistance. The BIOX has also been used to optimize dosage of biocides in power station condensers (Cristiani, 2005). The BIOGEORGE is another commercial electrochemical sensor developed to monitor biofilm activity (more details can be read at www.structint.com/products/biogeorge). This sensor is a probe with two electrodes, each with a series of stainless steel or titanium discs, which are subjected to intermittent polarization. As the biofilm attaches to the probe, the applied current has to be increased in order to achieve a fixed potential in the system. Additionally, it is possible to follow the attachment/detachment of the biofilm when no potential is applied, since the biofilm generates an electrical current that can be measured.

Concerning the capacitive/impedance devices, they are based on the fact that microorganisms (which contain liquid plasma with dielectric properties) can behave as electric capacitors. This means that microorganisms can store charges when exposed to electric fields and, as such, the movement of the electrical current between two electrodes can be used to detect the presence of capacitance. A capacitive device typically includes two parallel plates (to which a voltage is applied to) separated by a non-conducting material. This methodology was applied to the detection of biofilm layers (Maurício *et al.*, 2006) and the electric capacitance showed to be inversely related to the thickness of such deposit. It was also applied to follow both the formation and the degradation of biofilms in platinum and gold electrodes (Muñoz-Berbel *et al.*, 2008). An example of a capacitive sensor commercially available and devised specifically for food and beverage industries to monitor biomass formation or fermentation processes can be found in www.aber-instruments.co.uk.

### 4.4.6 Measurement of vibration signals

Ultrasound and acoustic techniques have been applied to monitor fouling layers. Two different approaches are typically found in the literature: i) the adhesion of the deposit occurs directly on the transducer (sensor) and then it is related with the sensor's vibration characteristics; ii) the transducer is placed on the outer part of the tubing/equipment and the amount of deposit is related to the acoustic properties of the wave that travels across the equipment surfaces, the bulk liquid and the deposit.

*The deposit adheres directly to a crystal surface*

One of the more often reported devices is the *Quartz crystal microbalance* (QCM), which analyzes the effect that the adhesion of deposits to a quartz crystal causes on the vibration (frequency) of such a surface (Nivens *et al.*, 1995). The sensor consists of thin quartz layers sandwiched between a pair of electrodes. Thanks to the piezoelectric properties of the quartz crystal, the structure can be excited to mechanically vibrate by applying an electric voltage to the electrodes. Because of these characteristics, it becomes quite reliable to relate the resonant frequency of the crystal to the deposit layer adhered to its surface. As the deposit attaches to the crystal surface, its vibration frequency decreases. If the attached layer is thin and rigid, the frequency variation becomes proportional to the mass of such deposit. In those specific cases it is possible to determine the mass of deposit by applying the Sauerbrey relation (Nivens *et al.*, 1993) – the mass is a function of some intrinsic properties of the crystal and of the frequency shift. However, in most practical situations the adhered layer is not rigid but rather viscoelastic, and the former relation is not valid. A viscoelastic deposit dampens the crystal oscillation, which, by means of a proper vibration analysis, can give information about the viscoelasticity and the mass of the attached layer. The QCM has also been experimentally applied to determine the amount of attached bacteria (*P. cepacia*) and its detection limit was found to be $3 \times 10^{-5}$ cells/cm$^2$, although a number of complications regarding temperature and flow velocity were observed (Nivens *et al.*, 1993). It was also applied to detect biofilm formation in quartz polished electrode surfaces indicating that it was able to monitor biofilm adhesion (Helle *et al.*, 2000), but the flow and the nutrients in the bulk seemed to affect the frequency values. There are many commercial versions of the QCM device (e.g., www.Q-Sense.com) including some to detect the initial adhesion of

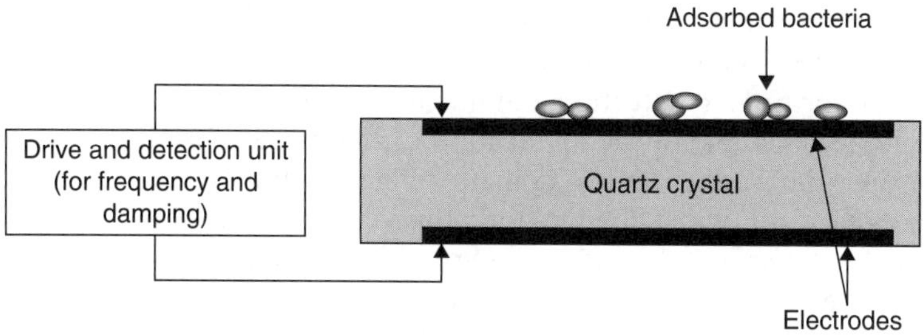


**Fig. 4.4** Schematic representation of the QCM configuration.

bacteria to stainless steel surfaces. In fact, this technique was shown to be quite efficient for the identification of bacteria adhesion (in early initial periods) – see Fig. 4.4 – and the detection of, for example, proteins. The above described QCM technique requires that the deposit is formed directly on the crystal surface and can only measure thin deposits. In its more advanced versions, it is quite an expensive device (tens of thousands of dollars). Because of these intrinsic characteristics, the QCM has wide application in research, but not in industrial processes.

*The transducers are not in contact with the deposit*
An alternative approach relies on the generation of an acoustic pulse that is transmitted through the bulk liquid and the deposit and measured on the wall of the pipe under study. The measurement procedures implemented to detect/monitor fouling can assume different configurations, such as the *pulse-echo* and the *transmission* techniques (Withers, 1996). The main differences between these two configurations are the use of one or more transducers and the type of analysis of the output signal.

In the pulse-echo technique only one transducer (that acts as an exciter and as a sensor) is placed on the outer surface of the tubing and simultaneously transmits an ultrasound pulse to the piping and measures the echoed signals. At each boundary (wall/deposit, deposit/bulk liquid), the wave gives rise to a reflection (echo) and thus the time that the reflection takes to return to the transducer can be related to the amount of fouling, different values being obtained according to the thickness of the deposit. This technique, in its fundamental application (normal beam), was able to provide information about biofouling severity rather than biofouling amount (Withers, 1996). In order to improve the methodology previously presented, some authors (Lohr and Rose, 2003, Hay and Rose, 2003) discussed the implementation of guided waves (which result from the interference of longitudinal and shear waves). The guided waves can be created by placing the transducer(s) within an oblique wedge. The use of the guided waves was shown to be a good indicator of the medium viscosity, since an increase in this physical parameter corresponds to an increase in energy loss. In spite of these still preliminary results (fouling was simulated by using only a viscous film on the pipe inner surface), this technique seems to have great potential for fouling detection and quantification.

Another example of the application of the pulse-echo technique is the ultrasonic time or frequency domain reflectometry. Several studies have been carried out with the ultrasonic time-domain reflectometry (UTDR) and they suggest that this methodology is able to detect the initial adhesion of different foulant layers such as calcium sulphate on membranes of reverse osmosis systems (Mairal *et al.*, 2000) or mixed deposits on ultrafiltration membranes (Li *et al.*, 2003). The amplitude of the reflected wave was shown to be the parameter that can be related to the amount of such deposits. The methodology for data analysis has been improved with the introduction of frequency domain (ultrasonic frequency-domain reflectometry) determinations, since it allows measuring more mature biofilms and their detachment from the polymeric matrix (Kujundzic *et al.*, 2007, Fonseca *et al.*, 2002). These techniques can provide real-time information about the deposit, but they are intrusive methodologies since an acoustic transducer (typically between 5 and 30 MHz) is placed on a polycarbonate coupon surface (contact area of the acoustic reflections: $120 \times 10^3$ $\mu m^2$) (Kujundzic *et al.*, 2007).

The alternative approach, the *transmission technique*, uses one transducer as an exciter and another one as a receiver. When a pipe section is considered, the transducers are placed on opposite sides of the pipe walls so the traveling wave has to pass through the deposit and through the bulk liquid. In comparison to the pulse-echo approach, this technique introduces an advantage since the measured signal does not depend on the deposit characteristics for its reflection. The guided waves approach has also been applied to the transmission setup leading to similar results to the ones already mentioned (Lohr and Rose, 2003, Hay and Rose, 2003) for the pulse-echo approach.

Different acoustic techniques based on low-frequency (0–20 kHz, which use unexpensive actuators and sensors) were designed. The *Mechatronic Surface Sensor* (MSS) has two transducers attached to the monitored surface: one acts as an exciter/actuator and the other one (the sensor) measures the effect that the adhesion/removal of deposits has on the properties of the propagated wave – Fig. 4.5. The MSS was shown to be able not only to detect different build-up rates of biofilm deposits but also to distinguish between these biological layers and mineral ones (silicate deposit) (Pereira *et al.*, 2008). It was also demonstrated that this technology is able to provide information about the amount and relative nature (e.g., proteineous vs. mineral) of different milk components fouling layers (Pereira *et al.*, 2006).

Furthermore, it can be applied to practically all types of industrial surfaces (stainless steel, copper, PVC, glass, etc.) and it is not an intrusive device, since it is placed on the outer surface of the tube or equipment being monitored. A similar methodology was more recently introduced by Merheb

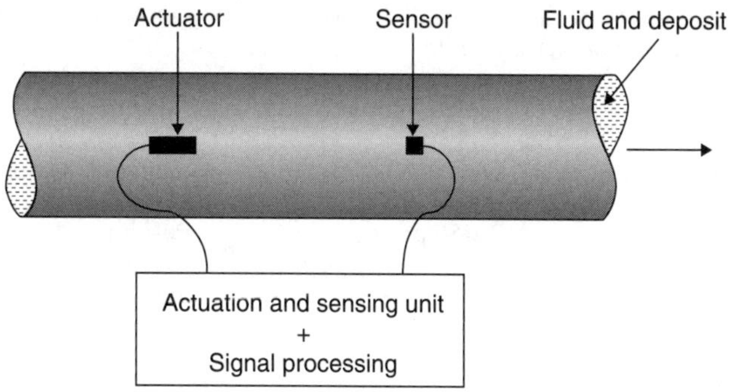


**Fig. 4.5** Schematic representation of the MSS configuration.

*et al.* (2007) to detect fouling in plate heat exchangers. An acoustic transducer is placed on the upper section and excites the heat exchanger plate while three sensors are attached to the lower part of the equipment (at different zones of the heat exchanger). The signal responses are gathered by the three sensors and the acoustic power (integration of the output wave over time) and the delay of the acoustic signals are determined for each sensor. The results seem to indicate good sensitivity and it seems plausible that with further improvements this system can be used to monitor fouling. The comparison of the evolution of the acoustic parameters for the three sensors gives an idea of the amount of fouling on each zone of the heat exchanger.

## 4.5 Conclusions

Biofilm monitoring is always a technical challenge in many industries affected by biofouling problems. However, the food and beverage industry has an important particularity: the hygienic relevance of the contamination not only of the equipment surfaces but, as a consequence, of the fluids in contact with such surfaces. The contamination is obviously favored by the high availability of nutrients in such industrial processes and can be aggravated by the introduction of measuring probes. Being so, a biofilm monitoring system devised for the food and beverage industries should necessarily take into account the hygienic and sterility requirements demanded by such processes. Therefore, whenever possible, the devices should be non-intrusive. Another issue that should be considered is the fact that the control/prevention of deposit formation can be continuously carried out (as happens, for example, in some water treatment programs, where biocide and dispersants can be continuously injected into the system). An efficient monitoring tool should be able to provide on-line, real-time information about the severity and contamination of the system, in order to enable the

**Table 4.1** Summary of the most well-known biofouling monitoring techniques and their respective detection limits

| Technique | Detection limit | Information source |
|---|---|---|
| ATP bioluminescense | 26 pM ATP | Hawronskyj and Holah (1997) |
| Heat transfer | 30–50 μm | Pereira and Melo (authors of present paper) |
| Fiber optical sensor (FOS) | $10^5$ cells/cm$^2$ | Tamachkiarow and Flemming (2003) |
| Differential turbidity device (DTM) | 10 μm | Klahre and Flemming (2000) |
| Photoacoustic spectroscopy sensor (PAS) | Resolution of 10 μm for biological samples | Schmid *et al.* (2003) |
| Electrical capacitance technique | 30 μm | Maurício *et al.* (2006) |
| Infrared absorption (simple method) | 30 μm | Tinham and Bott (2003) |
| Quartz crystal microbalance (QCM) | $3 \times 10^5$ cells/cm$^2$ | Nivens *et al.* (1993) |
| Mechatronic surface sensor (MSS) | 100 μ/cm$^2$ | Pereira and Melo (authors of present paper) |

application of timely counter-measures and to understand how the process variables are affecting biofilm build-up. The latter point contributes to improve the operation management during production and cleaning cycles.

Another important conclusion that can also be made is about the quantitative differences between the monitoring devices described in literature and the ones that are actually applied to industrial processes. Indeed, the fact that some of the monitoring techniques are suitable for laboratorial applications but not for industrial purposes is often related to their high sophistication, high specificity and high cost. As regards the quality of the data obtained with the different techniques, the published literature does not always report values of the respective resolution/accuracy or detection limits. Table 4.1 is a compilation of the available data collected in the literature.

Additionally, it is also well-known that in most of the cases there is a gap between the results obtained with devices operated in research laboratories and their subsequent industrial implementation. In order to offer alternative and viable solutions for biofilm monitoring in the food and beverage industries, it becomes increasingly important to develop and test in industrial environments a higher number of such monitoring techniques.

Ideally, those devices should be able to give information about the microbiological contamination and its location, as well as to indicate the extent and severity of the biofouling phenomenon and its cleaning/removal. Due to the diversity and intrinsic characteristics (e.g., intrusive, localized, cost, non-specificity for biofilms, robustness, etc.) of the reported techniques, it is not possible at the moment to elect the one(s) that best fulfill(s) the requirements of biofilm monitoring. The best solution will probably include the implementation of different monitoring techniques that are able to provide different types of information about biofilm development during both production and cleaning cycles.

## 4.6 References

BOTT, T. R. (2005) Biological growth on heat exchanger surfaces. *Fouling of Heat Exchangers*. Amsterdam, Elsevier.

CHEN, M. J., ZHANG, Z. & BOTT, T. R. (2005) Effects of operating conditions on the adhesive strength of Pseudomonas fluorescens biofilms in tubes. *Colloids and Surfaces B: Biointerfaces*, 43, 61–71.

CHEW, J. Y. M., PATERSON, W. R. & WILSON, D. I. (2004) Fluid dynamic gauging for measuring the strength of soft deposits. *Journal of Food Engineering*, 65, 175–187.

CHEW, Y. M. J., PATERSON, W. R. & WILSON, D. I. (2007) Fluid dynamic gauging: A new tool to study deposition on porous surfaces. *Journal of Membrane Science*, 296, 29–41.

CHMIELEWSKI, R. A. N. & FRANK, J. F. (2003) Biofilm formation and control in food processing facilities. *Comprehensive Reviews in Food Science and Food Safety*, 2, 22–32.

COSTA, P. D., ANDRADE, N. J., BRANDÃO, S. C. C., PASSOS, F. J. V. & SOARES, N. D. F. F. (2006) ATP-bioluminescence assay as an alternative for hygiene-monitoring procedures of stainless steel milk contact surfaces. *Brazilian Journal of Microbiology*, 37, 345–349.

CRISTIANI, P. (2005) Solutions to fouling in power station condensers. *Applied Thermal Engineering*, 25, 2630–2640.

FLEMMING, H.-C. L. (2003) Role and levels of real-time monitoring for successful anti-fouling strategies – An overview. *Water Science and Technology*, 47, 1–8.

FONSECA, A. C., GREENBERG, A. R. & HERNANDEZ, M. (2002) Real time biofilm detection using ultrasonic frequency-domain reflectometry (UFDR). *International Specialised Conference on Biofilm Monitoring*. Porto.

FRANK, J. F. (ed.) (2000) *Control of biofilm in the food and beverage industry*, Chicester, N.Y, John Wiley & Sons Ltd.

FRUNDZHYAN, V. & UGAROVA, N. (2007) Bioluminescent assay of total bacterial contamination of drinking water. *Luminescence*, 22, 241–244.

FRYER, P. J., CHRISTIAN, G. K. & LIU, W. (2006) How hygiene happens: Physics and chemistry of cleaning. *International Journal of Dairy Technology*, 59, 76–84.

HASTING, A. P. M. (2002) Industrial experience of monitoring fouling and cleaning systems. *Fouling, cleaning and disinfection in food processing*. 3–5 April, University of Cambridge, UK.

HASTING, A. P. M. (2005) Improving the monitoring of fouling, cleaning and disinfection in closed process plant. In Lelieveld, H. L. M., Mostert, M. A. & Holah, J. (Eds.) *Handbook of hygiene control in the food industry*. Cambridge, UK, Woodhead Publishing Limited.

HAWRONSKYJ, J. M. & HOLAH, J. (1997) ATP: A universal hygiene monitor. *Trends in Food Science and Technology*, 8, 79–84.

HAY, T. R. & ROSE, J. L. (2003) Fouling detection in the food industry using ultrasonic guided waves. *Food Control*, 14, 481–488.

HELLE, H., VUORIRANTA, P., VALIMAKI, H., LEKKALA, J. & AALTONEN, V. (2000) Monitoring of biofilm growth with thickness-shear mode quartz resonators in different flow and nutrition conditions. *Sensors and Actuators B-Chemical*, 71, 47–54.

JANKNECHT, P. & MELO, L. F. (2003) Online Biofilm Monitoring. *Reviews in Environmental Science and Biotechnology*, 2, 269.

KADURUGAMUWA, J. L., SIN, L. V., YU, J., FRANCIS, K. P., PURCHIO, T. F. & CONTAG, P. R. (2004) Noninvasive optical imaging method to evaluate postantibiotic effects on biofilm infection in vivo. *Antimicrobial Agents and Chemotherapy*, 48, 2283–2287.

KLAHRE, J. & FLEMMING, H. C. L. (2000) Monitoring of biofouling in papermill process waters. *Water Research*, 34, 3657–3665.

KUJUNDZIC, E., CRISTINA FONSECA, A., EVANS, E. A., PETERSON, M., GREENBERG, A. R. & HERNANDEZ, M. (2007) Ultrasonic monitoring of early-stage biofilm growth on polymeric surfaces. *Journal of Microbiological Methods*, 68, 458–467.

LI, J., HALLBAUER, D. K. & SANDERSON, R. D. (2003) Direct monitoring of membrane fouling and cleaning during ultrafiltration using a non-invasive ultrasonic technique. *Journal of Membrane Science*, 215, 33–52.

LICINA, G. J., NEKOKSA, G., WARD, G. L., HOWARD, R. L. & CUBICCIOTTI, D. L. (1993) Monitoring corrosion and biofilm formation in nuclear power plants using electrochemical methods. *Proceedings of the 6th International Symposium on Environmental Degradation of Materials in Nuclear Power Systems-Water Reactors, Aug 1–5 1993*. San Diego, CA, USA, Publ by Minerals, Metals & Materials Soc (TMS), Warrendale, PA, USA.

LOHR, K. R. & ROSE, J. L. (2003) Ultrasonic guided wave and acoustic impact methods for pipe fouling detection. *Journal of Food Engineering*, 56, 315–324.

LUDWICKA, A., SWITALSKI, L. M. & LUNDIN, A. (1985) Bioluminescent assay for measurement of bacterial attachment to polyethylene. *Journal of Microbiological Methods*, 4, 169–177.

MAIRAL, A. P., GREENBERG, A. R. & KRANTZ, W. B. (2000) Investigation of membrane fouling and cleaning using ultrasonic time-domain reflectometry. *Desalination*, 130, 45–60.

MAURÍCIO, R., DIAS, C. J. & SANTANA, F. (2006) Monitoring biofilm thickness using a non-destructive, on-line, electrical capacitance technique. *Environmental Monitoring and Assessment*, 119, 599–607.

MERHEB, B., NASSAR, G., NONGAILLARD, B., DELAPLACE, G. & LEULIET, J. C. (2007) Design and performance of a low-frequency non-intrusive acoustic technique for monitoring fouling in plate heat exchangers. *Journal of Food Engineering*, 82, 518–527.

MOLLICA, A. & CRISTIANI, P. L. (2003) On-line biofilm monitoring by 'BIOX' electrochemical probe. *Water Science and Technology*, 47, 45–49.

MUÑOZ-BERBEL, X., GARCIA-ALJARO, C. & MUÑOZ, F. J. (2008) Impedimetric approach for monitoring the formation of biofilms on metallic surfaces and the subsequent application to the detection of bacteriophages. *Electrochimica Acta*, 53, 5739–5744.

NIVENS, D. E., CHAMBERS, J. Q., ANDERSON, T. R. & WHITE, D. C. L. (1993) Long-term, online monitoring of microbial biofilms using a quartz crystal microbalance. *Analytical Chemistry*, 65, 65–69.

NIVENS, D. E., PALMER, R. J. & WHITE, D. C. L. (1995) Continuous nondestructive monitoring of microbial biofilms – a review of analytical techniques. *Journal of Industrial Microbiology*, 15, 263–276.

PEREIRA, A., ROSMANINHO, R., MENDES, J. & MELO, L. F. (2006) Monitoring deposit build-up using a novel mechatronic surface sensor (MSS). *Food and Bioproducts Processing*, 84, 366.

PEREIRA, A., MENDES, J. & MELO, L. F. (2008) Using nanovibrations to monitor biofouling. *Biotechnology and Bioengineering*, 99, 1407–1415.

PEREIRA, M. O., KUEHN, M., WUERTZ, S., NEU, T. & MELO, L. F. (2002) Effect of flow regime on the architecture of a *Pseudomonas fluorescens* biofilm. *Biotechnology and Bioengineering*, 78, 164–171.

SCHMID, T., HELMBRECHT, C., PANNE, U., HAISCH, C. & NIESSNER, R. L. (2003) Process analysis of biofilms by photoacoustic spectroscopy. *Analytical and Bioanalytical Chemistry*, 375, 1124–1129.

SCHMID, T., PANNE, U., ADAMS, J. & NIESSNER, R. L. (2004) Investigation of biocide efficacy by photoacoustic biofilm monitoring. *Water Research*, 38, 1189–1196.

SUCI, P. A., VRANY, J. D. & MITTELMAN, M. W. (1998) Investigation of interactions between antimicrobial agents and bacterial biofilms using attenuated total reflection Fourier transform infrared spectroscopy. *Biomaterials*, 19, 327–339.

TAMACHKIAROW, A. & FLEMMING, H. C. L. (2003) On-line monitoring of biofilm formation in a brewery water pipeline system with a fibre optical device. *Water Science and Technology*, 47, 19–24.

TINHAM, P. & BOTT, T. R. (2003) Biofouling assessment using an infrared monitor. *Water Science and Technology*, 47, 39–43.

TLILI, M. M., ROUSSEAU, P., BEN AMOR, M. & GABRIELLI, C. (2008) An electrochemical method to study scaling by calcium sulphate of a heat transfer surface. *Chemical Engineering Science*, 63, 559–566.

VERRAN, J. (2002) Biofouling in food processing: Biofilm or biotransfer potential? *Food and Bioproducts Processing: Transactions of the Institution of of Chemical Engineers, Part C*, 80, 292–298.

VIEIRA, M. J. & MELO, L. F. (1999) Intrinsic kinetics of biofilms formed under turbulent flow and low substrate concentrations. *Bioprocess Engineering*, 20, 369–375.

VIEIRA, M. J., MELO, L. F. & PINHEIRO, M. M. (1993) Biofilm formation – hydrodynamic effects on internal diffusion and structure. *Biofouling*, 7, 67–80.

WARNER, J. C., ROTHWELL, S. D. & KEEVIL, C. W. (2008) Use of episcopic differential interference contrast microscopy to identify bacterial biofilms on salad leaves and track colonization by Salmonella Thompson. *Environmental Microbiology*, 10, 918–925.

WILSON, D. I. (2005) Challenges in cleaning: Recent developments and future prospects. *Heat Transfer Engineering*, 26, 51–59.

WITHERS, P. M. (1996) Ultrasonic, acoustic and optical techniques for the non-invasive detection of fouling in food processing equipment. *Trends in Food Science and Technology*, 7, 293–298.

ZHAO, Q., LIU, Y., WANG, C., WANG, S. & MULLER-STEINHAGEN, H. (2005) Effect of surface free energy on the adhesion of biofouling and crystalline fouling. *Chemical Engineering Science*, 60, 4858–4865.

# 5

# A centralized database for use in studying bacterial biofilms and quorum sensing in food processing and other environments: MicroBQs

**X. Yan and P. M. Fratamico, United States Department of Agriculture, USA and Jin Gui, Walden University, USA**

**Abstract**: Biofilm formation in many bacterial species may be negatively or positively regulated by cell-to-cell signaling systems referred to as quorum sensing (QS). This chapter discusses the construction of a comprehensive, centralized database, known as MicroBQs, which will help in understanding and facilitating research related to biofilms, QS, and the role of QS in biofilm formation, ultimately enhancing food safety. Database contents include genomic and proteomic information, research tools and methods, mechanisms and pathways, as well as relevant literature references. Regular updating of information on the MicroBQs website obtained from internet sites, the scientific literature, and in-house research will be performed. The MicroBQs database can be accessed at http://microbqs.arserrc.gov.

**Key words**: biofilms, quorum sensing, website, database, MicroBQs.

## 5.1 Introduction

Bacteria form complex bacterial communities, known as biofilms, that adhere to a surface, an interface, or to each other encased in an exopolysaccharide matrix. Biofilms may cause persistent low-level contamination of foods, which is a food safety concern. Biofilms also account for over 80 percent of microbial infections in the human body (Lewis 2007, 2008). The function of biofilms is dependent on a complex web of symbiotic interactions, which affect the coexistence of the different component species and the structure of the biofilm. Biofilms also enable pathogens, including foodborne bacteria, to resist removal from surfaces, survive disinfection, and elude detection (Costerton *et al.* 1978).

Quorum sensing (QS) is the ability of bacteria to regulate gene expression in response to population density mediated by chemical 'autoinducer' molecules produced by the bacteria (Waters and Bassler 2005). There is evidence that different bacterial species use quorum sensing to negatively or positively regulate biofilm formation under various conditions (Parsek and Greenberg 2005; Chapter 2). Recent advances in genomics, proteomics and high throughput molecular approaches, including microarray technology, have provided information and tools to study bacterial virulence, the mechanism of biofilm formation in different bacteria, how resistance to interventions may arise, and the connections between quorum sensing and biofilm formation (Morales *et al.* 2005; Janssens *et al.* 2008).

There has been considerable interest in research to understand biofilms, quorum sensing, and the role of quorum sensing in biofilm formation. However, these are very complex systems, and knowing which research informatics and tools (genomics, software, strains, plasmids, reagents, synthetic compounds, polymerase chain reaction (PCR) primers, pathway builders, etc.) to use and where and how to obtain them has presented some challenges. This chapter will describe the construction and use of a centralized database/website known as MicroBQs (Microbial Biofilms and Quorum Sensing), which will serve as a comprehensive source of information for the research community. It will consist of the following:

1 All the relevant references in the literature.
2 Genomic and proteomic information, including genes and proteins involved in quorum sensing and biofilm formation, DNA and protein sequences, and publicly available microarray data.
3 Research tools, including bioinformatics tools, reporter strains, plasmids, and compounds needed to conduct research on biofilm formation and quorum sensing.
4 Methods used to study biofilms and quorum sensing.
5 Embedded tools for functional predictions and pathway re-engineering.
6 Mechanisms and genes, proteins, and small molecules involved in quorum sensing and biofilm formation.

This database includes information on bacterial genomes, genes, proteins, compounds/signal molecules, mutants, domains, pathways, and biological processes, which are all related to biofilm formation and quorum sensing. Additional information will be added to the database as necessary and as it becomes available, and this includes identification of candidate genes by genome-comparisons, biofilm and quorum sensing computer modeling based on literature mining, and quorum quenching/inhibition methods to prevent biofilm formation. Text-mining and data-mining software will be embedded in the database website, and scripts will be generated to mine the information by linking to public databases, including but not limited to the DNA DataBank of Japan (DDBJ), the European Molecular Biology

Laboratory (EMBL), PubMed and GenBank at the National Center for Biotechnology Information (NCBI). Information in the database will be automatically updated regularly through the use of these scripts. A similar type of database can be found at https://cabig.nci.nih.gov/ (a website/database serving the Cancer Research Community – an 'initiative to accelerate research discoveries and improve patient outcomes by linking researchers, physicians, and patients throughout the cancer community').

## 5.2 Goal of MicroBQs

The ultimate goal of this bioinformatics/database is to develop and maintain a comprehensive informatics resource covering a range of scientific publications/data providing information on genomics, pathways, biochemistry, functional prediction, pathology/diseases, food safety, methods for detection/analysis of biofilms and quorum sensing, and interventions, including techniques to interfere with quorum sensing. Collected information on biofilm and quorum sensing research will be organized and stored on the database and will be available for widespread public access. This MicroBQs database can be accessed at http://microbqs.arserrc.gov.

## 5.3 Data presentation and source of information for MicroBQs

Data are presented in a variety ways to satisfy scientific community needs, as well as for the purpose of easy maintenance. Most of the database contents including references, genome sequences, and gene and protein context will be linked to public databases, such as NCBI, and some will be tabulated and summarized after a thoroughly reviewed annotation process.

## 5.4 Database contents

The database contents are shown in Fig. 5.1. Some of the information available on the MicroBQs website is described below.

### 5.4.1 Genomics

In understanding biofilm formation and quorum sensing processes, several key elements must be considered:

1 Gene and protein sequences involved in signal molecule biosynthesis, small molecule transporters, lipid and polysaccharide biosynthesis, and signal transduction pathways in bacteria that could potentially relate to

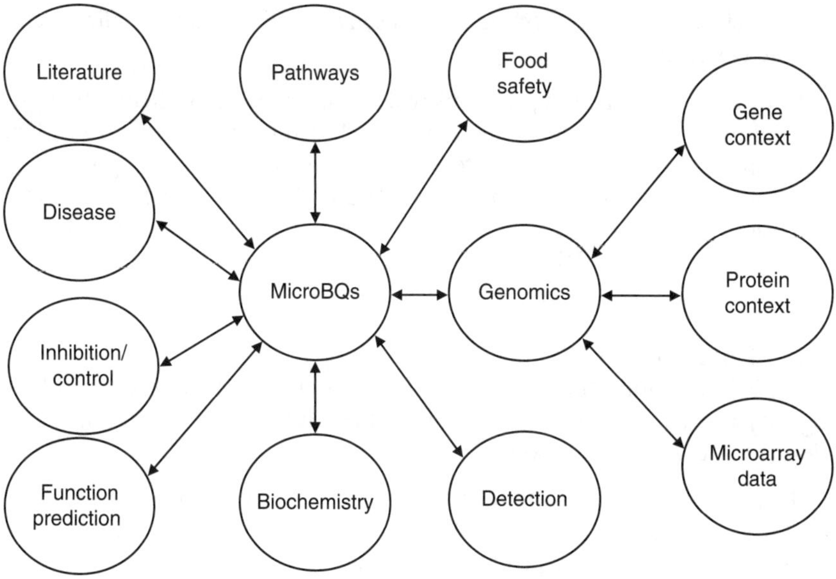


**Fig. 5.1** MicroBQs database contents.

biofilm and quorum sensing will be included in MicroBQs. Also our understanding of gene function has grown with the use of strains with mutations in key genes, which then display different functionalities during biofilm and quorum sensing processes. In a recent morphological study of *Bacillus subtillis* using a large collection of mutant strains, it was demonstrated that certain genes and transcriptional regulators are responsible for switching a cell from a unicellular state to a multicellular communities (Branda *et al.* 2004; Kobayashi 2007).

2 Collection and analysis the structure of proteins involved in biofilm formation and quorum sensing will help us to understand protein–protein interaction networks of these processes.

3 DNA microarray data will be collected and analyzed. This information is especially useful in characterization of gene expression profile of biofilms and non-biofilm cells and of cell-to-cell signaling processes. Stanley *et al.* (2003) identified 519 genes that were differentially expressed at one or more time points as cells transitioned into the biofilm state. The analysis of microarray data will help to understand the unique patterns of gene expression and which genes may be useful targets for interventions to control biofilms or modulate quorum sensing processes. The technology can also be utilized in clinical applications to characterize the genetic background of specific bacteria. Using a mixed whole genome microarray of *E. faecium*, a hospital clade was identified

by hybridization, confirming the presence of 437 clade specific inserts (Leavis *et al.* 2007).

Several gene prediction programs will be run based on particular biofilm- and quorum sensing-related genomic DNA sequences to identify novel functionally related genes or proteins. This collection will be categorized by functional groups, biological processes, and pathways. The whole genome sequence data and the specific DNA and protein sequences will be linked to the appropriate NCBI databases.

### 5.4.2 Pathways

There is information on biofilms and other quorum sensing-regulated biological pathways/processes found in the MicroBQs database. Several examples of these are the following:

- Bioluminescence (*lux* operon) (Engebrecht and Silverman 1984).
- Surface attachment (Dunne 2002).
- Extracellular polymer production (Sakuragi and Kolter 2007).
- Biosurfactant synthesis and biofilm development (Pamp and Tolker 2007).
- Sporulation (Grossman 1995).
- Secretion of nutrient-sequestering compounds.
- Virulence factors and biofilm formation (Rijavec *et al.* 2008).
- Genetic competence and biofilm formation (Wen *et al.* 2005).
- Bacterial motility (O'Toole and Kolter 1998).
- Biosynthesis of bacteriocins (van der Ploeg 2005).
- Biofilm/antibiotic synthesis and drug resistance (Lewis 2008).
- Biofilm formation and quorum sensing related gene expression profile analyses.

### 5.4.3 Function prediction

Several approaches can be used to identify genes responsible for biofilm formation and quorum sensing, and the most common are the following:

- Protein to protein interactions using bacterial two hybrid systems.
- Pathway analysis and functional group categorization coupled with mutant generation.
- Disruption of biofilms by inhibition of quorum sensing.
- Genomic and proteomic approaches, including DNA microarrays, quantitative PCR methods, and function prediction by amino acid sequence analysis or 'homology search'.

Genomics approaches have become a common starting point for experimental designs related to food safety research in the post-genome era. In MicroBQs, functional domains will be collected by literature mining, and the novel functional domains can be identified by comparing the PSI-BLAST sequence alignment profiles and the Pfam protein families database. The well-characterized GGDEF domain can be used as an initial target for assisting in the identification of novel biofilm and quorum sensing related functional domains.

### 5.4.4 Biochemistry

To understand the biochemistry of biofilm formation and quorum sensing is an enormous challenge. Through the careful analysis of genetic data, gene expression profile data, and information on pathways involved in extracellular polymer production, biosurfactant synthesis and other pathways, a greater understanding of the biochemistry of biofilm formation and quorum sensing will be achieved. There are a number of signaling molecules involved in quorum sensing and biofilm formation that have been identified (Camilli and Bassler 2006). In Gram-negative bacteria, N-acyl homoserine lactones (AHL) are mainly involved in intra-species communication. In Gram-positive bacteria, post-translationally modified oligopeptides are involved in quorum sensing, and autoinducer 2 (AI-2) is a signaling molecule that is involved in quorum sensing in both Gram-positive and Gram-negative bacteria. Others signaling molecules include AI-3, indole, PQS, butyrolactones, 3OH PAME, cyclic depeptides, and others. These are described in more detail in Chapter 2.

### 5.4.5 Disease

Human infections may involve various areas of the body, including:

1 oral soft tissues (Foster and Kolenbrander 2004);
2 teeth and dental implants (Quirynen *et al.* 2005);
3 middle ear (Post *et al.* 2007);
4 airway/lung tissue;
5 eyes (Zegans *et al.* 2005);
6 gastrointestinal tract (Macfarlane 2008);
7 urogenital tract/urinary tract prostheses (Kunin 1989).

Biofilms play a role in these infections, and therefore, the bacteria are more likely to be resistant to antimicrobials, and it may be difficult for the host immune system to render an appropriate response (Lewis 2008).

Staphylococcal food poisoning outbreaks are often caused by contamination of food by preparers who are nasal carriers or that have staphylococcal lesions (Bustan *et al.* 1996). Hospital outbreaks of *Staphylococcus aureus* infection can be self-perpetuating and persistent with the forming of biofilms and the development of drug resistance (Smith and Hunter 2008). The initial stage of colonization by the bacteria is mediated by surface proteins binding to host elements on tissues, in body fluids, or on foreign bodies, such as catheters. Spontaneous resolution of the boil caused by *S. aureus* by granulation and fibrosis develops in the skin while the process may continue to form satellite foci and involve other organs.

Clinical evidence indicated that group A streptococci have the capacity to be highly invasive through adherence to mucosal surfaces (Cunningham 2008). Colonization by group A streptococci can cause diseases such as impetigo when combined with minor trauma. Other illnesses caused by group A streptococci include acute infections, acute rheumatic fever, and acute glomerulonephritis.

Highly pathogenic Gram-positive bacteria, including *L. monocytogenes*, can cause human diseases. *Bacillus anthracis*, the cause of anthrax, can form biofilms in the environment, which is part of the ecology associated with its life cycle (Lee *et al.* 2007). Non-toxigenic strains of *Corynebacterium* can cause pharyngitis, which is a disease typically accompanied by a plaque-like pseudomembrane in the throat and trachea (Batty and Wren 2005).

Anaerobes are able to colonize in the respiratory tract (Marik and Careau 1999). *Bacteroides*, *Fusobacterium*, and peptostreptococci, alone or with other facultative anaerobes such as *Escherichia coli* or obligate anaerobes are responsible for the overwhelming majority of localized abscesses in different areas of the body, including sebaceous glands of the skin, the gingival crevices of the gums, the lymphoid tissue of the throat, and the lumina of the intestinal and urogenital tracts, They also play a role in chronic sinusitis, chronic otitis media, aspiration pneumonia, bronchiectatis, septic arthritis, osteomyelitis, and decubitus ulcers (Murdoch 1998).

Enterobacteriaceae colonize the lower gastrointestinal tract of humans and animals (Freter 1981). In nature, many species survive and live wherever there are water and minimal energy sources. In the human body, they are the major facultative components of the colonic bacterial flora and transient colonizers of the skin. Mucosal or skin trauma can allow access of these bacteria to soft tissues and by aspiration to the lung from sites colonized with *Enterobacteriaceae,* including *Salmonella, Shigella, Yersinia enterocolitica,* and certain strains of *E. coli*. These bacteria are able to produce disease in the intestinal tract (Goosney *et al.* 1999). Enteric fever caused by *Salmonella* Typhi is the prototype intestinal infection (Goldberg and Rubin 1988).

The Gram-negative opportunistic pathogen, *P. aeruginosa,* has been isolated from the throat and stool of healthy persons. This bacterium produces infection at a wide range of sites, including the lung, the urinary tract, and soft tissue sites, much like Enterobacteriaceae. *P. aeruginosa* is the most persistent infectious agent that complicates the course of cystic fibrosis (Govan and Deretic 1996). Initial colonization is easier due to the fact that cells from cystic fibrosis patients are less highly sialylated than normal epithelial cells. Once the bronchi are colonized, the organisms remain, forming a biofilm containing microcolonies of bacteria, which become almost impossible to eradicate. This infection is a leading cause of morbidity and death of these patients (Wagner and Iglewski 2008). The role of quorum sensing in the formation of biofilms in persistent and chronic infections warrants further investigation.

### 5.4.6 Food safety

Pathogens may contaminate food and food processing equipment in various ways, and improper sanitation combined with suitable contact surfaces and environmental conditions may lead to biofilm formation. Bacteria within biofilms, which may consist of both spoilage and pathogenic bacteria, are more resistant to disinfectants typically used in food processing operations. In a review by Smith *et al.* (2004) the potential role of quorum sensing in the formation of biofilms in food and food processing environments is discussed, as well as the possibility of controlling biofilms by modulating quorum sensing-related processes.

*Aeromonas hydrophila* is considered as an emerging pathogen in drinking water in the United States, and the organism can produce biofilms on various types of surfaces. AHL-dependent quorum sensing was shown to play a role in *A. hydrophila* biofilm development (Lynch *et al.* 2002). Reeser *et al.* (2007) examined the nutritional and environmental conditions required for biofilm formation in *C. jejuni.* They showed that flagella and quorum sensing were required for maximal biofilm development. Biofilm formation in a *luxS* (gene that encodes the final enzyme in the biosynthetic pathway for AI-2 production) mutant strain was decreased compared to the wild type. Another study found 'AI-2-like' activity on the surfaces of various fruits and vegetables and that tomato surface rinses caused increases (1.8–3.6-fold) in biofilm formation by *E. coli luxS* mutant strains (Lu *et al.* 2005).

### 5.4.7 Methods

Numerous methods have been described for analyzing biofilm development and structure and for detecting biofilms, for monitoring genes expression during biofilm development, as well as for control of biofilms in food and in the environment (Doyle, 1999). All of this information will be available in the MicroBQs website. Some of these techniques have also been described in detail in various chapters of this book. Examples of methods for detection of biofilms include: (1) monitoring biofilm production by the use of capillary tubes and flow through devices; (2) light and fluorescence microscopy, differential interference contrast microscopy, transmission and scanning electron microscopy, and atomic force microscopy; (3) microtiter plate biofilm assays using crystal violet; and (4) rRNA targeted oligonucleotide probing.

A number of methods have been described for detection and identification of quorum sensing molecules and for measuring their activity (Swift *et al.* 1999; Brelles-Mariño and Bednar 2001). Methods for detecting AHLs include cell-based assays using AHL-specific bacteria biosensors, thin layer chromatography, and gas chromatography/mass spectrometry (Ravn *et al.* 2001; Morin *et al.* 2003; Steindler and Venturi 2007). A common method for detection of AI-2 in cell-free culture fluids involves the use of a mutant strain of *V. harveyi*, BB170 (sensor 1–, sensor 2+, luxN::Tn5), which responds to AI-2 but not to AHL (AI-1) signal molecules. Emitted light is proportional to levels of AI-2 (Bassler *et al.* 1994). Another assay method used for real-time quantitative measurement of levels of AI-2 employs AI-2 receptor proteins modified with environmentally sensitive dyes, resulting in measureable fluorescence with binding of AI-2 to the sensors (Zhu and Pei 2008). Because quorum sensing peptides produced by Gram-positive bacteria are usually present in low amounts in complex mixtures, it is a challenge to detect and identify these peptide signaling molecules. A method involving a novel matrix-assisted laser desorption/ionization-quadrupole ion trap mass spectrometer was applied for detection and characterization of trace amounts of quorum sensing-related and other peptides secreted by microorganisms (Kalkum *et al.* 2003). Autoinducing peptides (AIP) from supernatants of *Staphylococcus aureus* were identified, and it was determined that the native group III-AIP was a nonapeptide (RIPTSTGFF) with a lactone ring.

### 5.4.8 Inhibition/control

Various types of inhibitors and antagonists of cell-to-cell signaling have been explored in an attempt to reduce the ability of cells to form biofilms, attenuate virulence, and modulate other processes influenced by quorum sensing. Inhibition of quorum sensing pathways can be accomplished in several ways, including: (1) inhibition of signal synthesis; (2) enzymatic degradation of the signal molecule; and (3) blocking signal transport, secretion, or binding (Ni *et al.* 2009).

AHLs can be degraded enzymatically by lactonases produced by a number of bacteria, including members of the genus *Bacillus*, which produce lactonases that hydrolyze the ester bond of the homoserine lactone ring. In addition, a few bacteria, including *Variovorax paradoxus* and *Pseudomonas aeruginosa*, have been shown to metabolize AHLs using them as a carbon source. By amino acylases that are produced by the bacteria, the amide bond connecting the lactone ring to the acyl chain is cleaved, releasing the homoserine lactone and fatty acid (Rasmussen and Givskov 2006).

Halogenated furanones produced by the marine alga, *Delisea pulchra*, have structural similarity to AHLs and are antagonistic toward AHL-controlled processes (Rasmussen and Givskov 2006). The brominated furanone, (5Z)-4-bromo-5-(bromomethylene)-3-butyl-2(5H)-furanone of *Delisea pulchra*, inhibited swarming motility and decreased biofilm thickness in *E. coli.* (Ren *et al.* 2001). The furanone inhibited quorum sensing mediated by both AHLs and AI-2. Using green fluorescent protein reporter technology, it was shown that a synthetic furanone compound blocked quorum sensing in *Pseudomonas* biofilms, affecting biofilm architecture and enhancing bacterial detachment (Hentzer *et al.* 2002).

Others compounds produced by eukaryotes capable of interfering with bacterial quorum sensing systems, including L-canavanine, an arginine analog found in the seeds of legumes and penicillic acid and patulin, produced by fungi belonging to the genus *Penicillium* have been described (González and Keshavan 2006). Various compounds that interfere with quorum sensing by signal receptor blocking have been designed, and methods for screening and identification of novel compounds that inhibit quorum sensing have also been reported (Kjelleberg *et al.* 2008)

### 5.4.9 Literature

References related to biofilm formation, quorum sensing, and the role of quorum sensing in biofilm formation are included in MicroBQs. References focusing on genomics, pathways, functional prediction, disease, food safety, methods, and inhibition/control will be collected, categorized as comprehensively as possible, and regularly updated.

## 5.5 Web interface

The current version of this web interface was created by a series of static html flat files since the contents of these pages are solely dependent on data collection and annotation. A screen shot of this MicroBQs database front page is shown in Fig. 5.2. In the future, a dynamic server- and client-side Web interface will be developed to allow user-friendly interaction. The content will depend not only on data collection and computational analyses by the database creators, but also on the data entered by each individual researcher who accesses the site.

## 5.6 Conclusions and future work

Currently, MicroBQs is a manually-annotated biological database consisting of publicly available information. In our next version, we plan to synchronize the database information to achieve real-time linking to public databases, including but not limited to NCBI and EMBL (European Molecular Biology Laboratory) to more readily coordinate our research and development efforts with the scientific research community. Important objectives of this database are to assist researchers in conducting studies on biofilms and quorum sensing and to facilitate collaborations among scientists throughout the world. Regular updating of information on the MicroBQs website obtained from internet sites, the scientific literature,

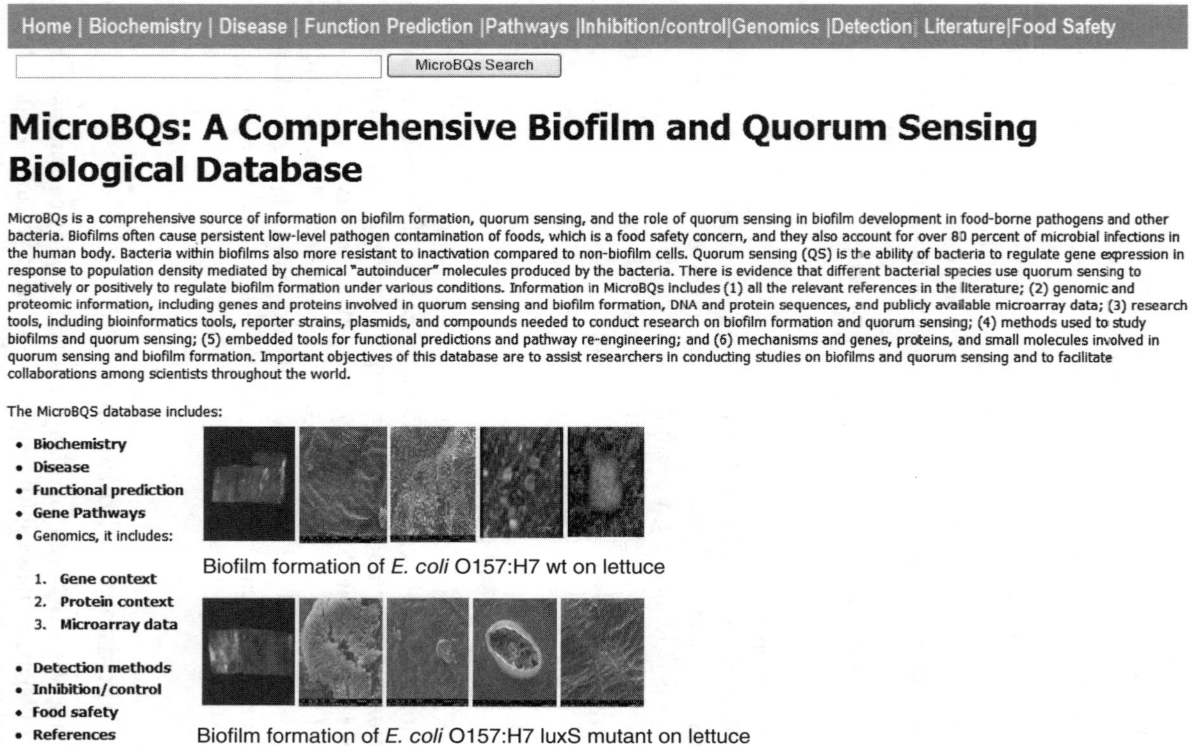


**Fig. 5.2** A screen shot of this MicroBQs database website front page.

in-house research, and through collaborations and joint efforts will be performed, and this effort will enhance our understanding of biofilms and quorum sensing, facilitate research in these areas, ultimately enhancing food safety.

## 5.7 Sources for further information and advice

Additional resources can be found by accessing the following links:
http://www.biofilmsonline.com/cgi-bin/biofilmsonline/index.html
http://www.biosignal.com.au
http://www.bionewsonline.com/u/d/matthew_r_parsek_2004_4427.htm.

The book, *Chemical Communication among Bacteria* (2008), edited by S.C. Winans and B.L. Bassler, ASM Press, Washington, DC, is also a helpful resource.

## 5.8 References

BASSLER BL, WRIGHT M, SILVERMAN MR (1994), 'Multiple signaling systems controlling expression of luminescence in *Vibrio harveyi*: sequence and function of genes encoding a second sensory pathway', *Mol Micribiol* 9:773–786.

BATTY A, WREN MW (2005), 'Prevalence of *Fusobacterium necrophorum* and other upper respiratory tract pathogens isolated from throat swabs', *Br. J. Biomed Sci* 62(2):66–70.

BRANDA SS, GONZALEZ-PASTOR JE, DERVYN E, EHRLICH SD, LOSICK R, KOLTER R (2004), 'Genes involved in formation of structured multicellular communities by *Bacillus subtilis*', *J Bacteriol* 186:3970–3979.

BRELLES-MARIÑO G, BEDNAR EJ (2001), 'Detection, purification, and characterization of quorum-sensing signal molecules in plant-associated bacteria', *J Biotechnol* 91:197–209.

BUSTAN MA, UDO EE, CHUGH TD (1996), 'Nasal carriage of enterotoxin-producing *Staphylococcus aureus* among restaurant workers in Kuwait City', *Epidemiol Infect* 116(3):319–322.

CAMILLI A, BASSLER BL (2006), 'Bacterial small-molecule signaling pathways', *Science* 311:1113–1116.

COSTERTON JW, GEESEY GG, CHENG KJ (1978), 'How bacteria stick', *Sci Am.* 238(1):86–95.

CUNNINGHAM MW (2008), 'Pathogenesis of group A streptoccal infecions and their sequelae', *Adv Exp Med Biol* 609:29–42.

DOYLE RJ (1999), *'Biofilms', A Volume of Methods in Enzymology*: Vol. 310, Academic Press, Orlando.

DUNNE M (2002), 'Bacterial adhesion: seen any good biofilms lately?', *Clin Microbiol Rev* 15:155–166.

ENGEBRECHT J, SILVERMAN M (1984), 'Identification of genes and gene products necessary for bacterial bioluminescence', *Proc Natl Acad Sci USA* 81:4154–4158.

FOSTER JS, KOLENBRANDER PE (2004), 'Development of a multispecies oral bacterial community in a saliva-conditioned flow cell', *Appl Environ Microbiol* 70(7):4340–4348.

FRETER R (1981), 'Mechanisms of association of bacteria with mucosal surfaces', *Ciba Found Symp* 80:36–55.

GOLDBERG MB, RUBIN RH (1988), 'The spectrum of *Salmonella* infection', *Infect Dis Clin North Am* 2(3):571–598.

GONZÁLEZ JE, KESHAVAN ND (2006), 'Messing with bacterial quorum sensing', *Microbiol Mol Biol Rev* 70(4):859–875.

GOOSNEY DL, KNOECHEL DG, FINLAY BB (1999), 'Enteropathogenic *E. coli*, *Salmonella*, and *Shigella*: masters of host cell cytoskeketal exploitation', *Emerg Infect Dis* 5(2):216–223.

GOVAN JR, DERETIC V (1996), 'Microbial pathogenesis in Cystic fibrosis: mucoid *Pseudomonas aeruginosa* and *Burkholderia cepacia*', *Microbiol Rev* 60(3):539–574.

GROSSMAN AD (1995), 'Genetic networks controlling the initiation of sporulation and the development of genetic competence in *Bacillus subtilis*', *Annu Rev Genet* 29:477–508.

HENTZER M, RIEDEL K, RASMUSSEN TB, HEYDORN A, ANDERSEN JB, PARSEK MR, RICE SA, EBERL L, MOLIN S, HØIBY N, KJELLEBERG S, GIVSKOV M (2002), 'Inhibition of quorum sensing in *Pseudomonas aeruginosa* biofilms bacteria by a halogenated furanone compound', *Microbiology* 148:87–102.

JANSSENS JC, STEENACKERS H, ROBIJNS S, GELLENS E, LEVIN J, ZHAO H, HERMANS K, DE COSTER D, VERHOEVEN TL, MARCHAL K, VANDERLEYDEN J, DE VOS DE, DE KEERSMAECKER SC (2008), 'Brominated furanones inhibit biofilm formation by *Salmonella enterica* serovar Typhimurium', *Appl Environ Microbiol* 74(21):6639–6648.

KALKUM M, LYON GJ, CHAIT BT (2003), 'Detection of secreted peptides using hypothesis-driven multistage mass spectrometry', *Proc Natl Acad Sci USA* 100:2795–2800.

KJELLEBERG S, MCDOUGALD D, RASMUSSEN TB, GIVSKOV M (2008), 'Quorum-sensing inhibition', p. 393–416. In *Chemical Communication among Bacteria.* SC Winans and B. Bassler (ed.). ASM Press, Washington, DC.

KOBAYASHI K (2007), '*Bacillus subtilis* pellicle formation proceeds through genetically defined morphological changes', *J Bacteriol* 189:4920–4931.

KUNIN CM (1989), 'Blockage of urinary catheters: role of microorganisms and constituents of the urine on formation of encrustations', *J Clin Epidermiol* 42:835–842.

LEAVIS HL, WILLIEMS RJ, VAN WAMEL WJ, SCHUREN FH, CASPERS MP, BONTEN MJ (2007), 'Insertion sequence-driven diversification creates a globally dispersed emerging multiresistant subspecies of *E. faecium*', *PLoS. Pathog*: 3:e7.

LEE K, COSTERTON JW, RAVEL J, AUERBACH RK, WAGNER DM, KEIM P, LEID JG (2007), 'Phenotypic and functional characterization of *Bacillus anthracis* biofilms', *Microbiology* 153(6):1693–1701.

LEWIS K (2007), 'Persister cells, dormancy and infectious disease', *Nat Rev Microbiol* 5(1):48–56.

LEWIS K (2008), 'Multidrug tolerance of biofilms and persister cells', *Curr Top Microbiol Immunol* 322:107–131.

LU L, HUME ME, PILLAI SD (2005), 'Autoinducer-2-like activity on vegetable produce and its potential involvement in bacterial biofilm formation on tomatoes', *Foodborne Path Dis* 2(3):242–249.

LYNCH MJ, SWIFT S, KIRKE DF, KEEVIL CW, DODD CE, WILLIAMS P (2002), 'The regulation of biofilms development by quorum sensing in *Aeromonas hydrophila*', *Environ Microbiol* 4(1):18–28.

MACFARLANE S (2008), 'Microbial biofilm communities in the gastrointestinal tract', *J. Clin. Gastroenterol* 42(suppl 3 pt 1):S142–143.

MARIK PE, CAREAU P (1999), 'The role of anaerobes in patients with ventilator-associated pneumonia and aspiration pneumonia: a prospective study', *Chest* 115(1):178–183.

MORALES CA, PORWOLLIK S, FRYE JG, KINDE H, MCCLELLAND M, GUARD-BOULDIN J (2005), 'Correlation of phenotype with the genotype of egg-contaminating *Salmonella enterica* serovar Enteritidis', *Appl Environ Microbiol* 71(8):4388–4399.

MORIN D, GRASLAND B, VALLÉE-RÉHEL K, DUFAU C, HARAS D (2003), 'On-line high-performance liquid chromatography-mass spectrometric detection and quantification of N-acylhomoserine lactones, quorum sensing signal molecules, in the presence of biological matrices', *J Chromatogr A* 1002:79–92.

MURDOCH DA (1998), 'Gram-positive anaerobic cocci', *Clin Microbiol Rev* 11(1):81–120.

NI N, LI M, WANG J, WANG B (2009), 'Inhibitors and antagonists of bacterial quorum sensing', *Med Res Rev* 29(1):65–124.

O'TOOLE GA, KOLTER R (1998), 'Flagellar and twitching motility are necessary for *Pseudomonas aeruginosa* biofilm development', *Mol. Microbiol* 30:295–304.

PAMP SJ, TOLKER-NIELSEN T (2007), 'Multiple roles of biosurfactants in structural biofilm development by *Pseudomonas aeruginosa*', *J Bacteriol* 189:2531–2539.

PARSEK M, GREENBERG EP (2005), 'Sociomicrobiology: the connections between quorum sensing and biofilms', *Trends Microbiol* 13:27–33.

POST JC, HILLER NL, NISTICO L, STOODLEY P, EHRLICH GD (2007), 'The role biofilms in otolaryngologic infections', *Curr Opin Otalaryngol Head Neck Surg* 15(5):347–351.

QUIRYNEN M, VOGELS R, PAUWELS M, HAFFAJEE AD, SOCRANSKY SS, UZEL NG, VAN STEENBERGHE D (2005), 'Initial subgingival colonization of "pristine" pockets', *J Dent Res* 84(4):340–344.

RASMUSSEN TB, GIVSKOV M (2006), 'Quorum sensing inhibitors: a bargain of effects', *Microbiology* 152:895–904.

RAVN L, CHRISTENSEN AB, MOLIN S, GIVSKOV M, GRAM L (2001), 'Methods for detecting acylated homoserine lactones produced by Gram-negative bacteria and their application in studies of AHL-production kinetics', *J Microbiol Methods* 44:239–251.

REESER RJ, MEDLER RT, BILLINGTON SJ, JOST BH, JOENS LA (2007), 'Characterization of *Campylobacter jejuni* biofilms under defined growth conditions', *Appl Environ Microbiol* 73(6):1908–1913.

REN D, SIMS JJ, WOOD TK (2001), 'Inhibition of biofilm formation and swarming of *Escherichia coli* by (5Z)-4-bromo-5-(bromomethylene)-3-butyl-2(5H)-furanone', *Environ Microbiol* 3(11):731–736.

RIJAVEC M, MÜLLER-PREMRU M, ZAKOTNIK B, ZGUR-BERTOK D (2008), 'Virulence factors and biofilm production among *Escherichia coli* strains causing bacteraemia of urinary tract origin', *J Med Microbiol* 57:1329–1334.

SAKURAGI Y, KOLTER R (2007), 'Quorum-sensing regulation of the biofilm matrix genes (pel) of *Pseudomonas aeruginosa*', *J Bacteriol* 189:5383–5386.

SMITH K, HUNTER IS (2008), 'Efficacy of common hospital biocides with biofilms of multi-drug resistant clinical isolates', *J Med Microbiol* 57(8):966–973.

SMITH JL, FRATAMICO PM, NOVAK JS (2004), 'Quorum sensing: a primer for food microbiologists', *J Food Prot* 67(5):1053–1070.

STANLEY NR, BRITTON RA, GROSSMAN AD, LAZAZZERA BA (2003), 'Identification of catabolite repression as a physiological regulator of biofilm formation by *Bacillus subtilis* by use of DNA microarrays', *J Bacteriol* 185:1951–1957.

STEINDLER L, VENTURI V (2007), 'Detection of quorum-sensing N-acyl homoserine lactone signal molecules by bacterial biosensors', *FEMS Microbiol Lett* 266: 1–9.

SWIFT S, WILLIAMS P, STEWART GSAB (1999), 'N-acylhomoserine lactones and quorum sensing in proteobacteria', p. 291–313. In GM Dunny and SC Winanas (ed.), *Cell-Cell Signaling in Bacteria*, ASM Press, Washington, DC.

VAN DER PLOEG JR (2005), 'Regulation of bacteriocin production in *Streptococcus mutans* by the quorum-sensing system required for development of genetic competence', *J Bacteriol* 187:3980–3989.

WAGNER VE, IGLEWSKI BH (2008), '*P. aeruginosa* biofilms in CF infection', *Clin Rev Allergy Immunol* 35(3):124–134.

WATERS CM, BASSLER BL (2005), 'Quorum sensing: cell-to-cell communication in bacteria', *Annu Rev Cell Dev Biol* 21:319–346.

WEN Z, SUNTHARALIGHAM P, CVITKOVITCH DG, BURNEL R (2005), 'Trigger factor in *Streptococcus mutans* is involved in stress tolerance, competence development, and biofilm formation', *Infect Immun* 73:219–225.

ZEGANS ME, SHANKS RM, O'TOOLE GA (2005), 'Bacterial biofilms and ocular infections'. *Ocul Surf* 3(2):73–80.

ZHU J, PEI D (2008), 'A LuxP-based fluorescent sensor for bacterial autoinducer II', *ACS Chem Biol* 3:110–119.

# Part II

# Microorganisms and their metabolites in biofilms

# 6

# Biofilm formation by food spoilage microorganisms in food processing environments

**D. R. Korber, A. K. Mangalappalli-Illathu and S. Vidović, University of Saskatchewan, Canada**

**Abstract**: Food processing environments are associated with high nutrient and moisture concentrations, conditions that favor the growth of surface-associated microorganisms and their polymers, known as biofilms. Biofilm bacteria are quite resistant to control strategies and biocides, a feature that contributes to rapid biofilm re-growth. In food processing environments, various spoilage microorganisms often proliferate together as a complex community, within which human pathogens may also take refuge. The possibility of outbreaks of foodborne illness following biofilm-food cross-contamination is a distinct concern, together with the considerable costs associated with food spoilage and biofilm control, emphasizes the ongoing need for the development of effective strategies for biofilm control. This chapter examines: biofilm formation by food spoilage organisms, mechanisms of biofilm antimicrobial resistance, and strategies for biofilm control in food production settings.

**Key words**: food spoilage biofilm, biofilm control, food antimicrobial agents, sanitation, resistance mechanisms.

## 6.1 Introduction

Over the past 20 years, there has been increasing emphasis on improving sanitation and safety of all elements of the food production chain. While long known to be a problem during the manufacture of food, biofilms continue to challenge the food and beverage processing sectors, contributing to millions of dollars in losses each year due to spoiled products and investments in biofilm control strategies (Lewis and Gilmour, 1987). In complex settings like food processing environments, biofilms usually consist of assemblages of different species of microorganisms living together in a complex community. The pioneer colonizing species potentially encourage the colonization of species that are physiologically compatible (i.e., mutually beneficial species), while inhibiting the colonization of the competing species (Kumar and Anand, 1998). One consequence of such complex interrelationships of relevance to the food industry is that these biofilms may harbor pathogenic microorganisms such as *Escherichia coli* O157:H7, *Salmonella* spp., *Listeria monocytogenes*, etc. (Carpentier and Chassaing, 2004), raising the possibility of outbreaks of foodborne illness following biofilm-food cross-contamination.

Foods and food contact surfaces often have relatively high nutrient concentrations, and together with suitable environmental conditions such as temperature, pH, etc., provide an ideal setting for biofilm formation (Chmielewski and Frank, 2003). Biofilms may affect the manufacture and processing of many different solid, semi-solid and liquid foods; meat products or any other food having a high water activity, as well as solid–liquid interfaces, present a particular concern. Owing to the sensitive nature of most foods, the application of chemicals and detergents for biofilm control are restricted to those that have food grade approval. The increasing demand for organic foods precludes the use of certain antimicrobials, placing increased pressure on those approved for use to completely control microbial contamination. It has become common knowledge that microbial biofilms are difficult to control or eradicate because of their inherent resistance to the biocides, a feature that leads to rapid biofilm re-growth following plant cleaning and sanitation (Costerton *et al.*, 1995; Korber and Lawrence, 2004; Mangalappalli-Illathu *et al.*, 2008b).

There is now an abundance of evidence demonstrating that microbial biofilms are distinct from their planktonic counterparts. Biofilm formation is associated with global changes in the cells following attachment to surfaces, including altered gene expression, production of enzymes and acylhomoserine lactone (AHLs), exopolymeric substance (EPS) production, and biofilm physiological and architectural changes resulting in enhanced resistance to antimicrobial agents (Kumar and Anand, 1998; Mittelman, 1998; Sauer and Camper, 2001; Sauer *et al.*, 2002; Mangalappalli-Illathu and Korber, 2006). Quorum sensing (QS) is the complex mechanism of cellular communication between members of a biofilm community using AHLs as the signaling molecules; interspecies and intraspecies interactions such as QS have been reported among the members of bacterial biofilms. Furthermore, biofilms are highly capable of modifying their microenvironment both physically and chemically, so as to be quite distinct from their macroenvironment (Korber and Lawrence, 2004). Thus, various factors contribute to resistance of biofilms to both physical and chemical strategies employed for their control on food processing surfaces (Mittelman, 1998; Korber *et al.*, 2002).

## 6.2 Important physiological and molecular events in biofilm formation

### 6.2.1 Stages of biofilm formation

The transition of microorganisms from the planktonic to the sessile (biofilm) state is often described as occurring in a series of steps or phases (Kumar and Anand, 1998). The stages of biofilm formation include: 1) the formation of a conditioning film; 2) cellular attachment; 3) the formation of microcolonies which eventually merge to become mature biofilms; and 4) biofilm dispersion and recolonization (Korber *et al.*, 1995; Korber and Lawrence, 2004) (Fig. 6.1). The available literature does not suggest that there is any significant difference in the process of biofilm formation of food spoilage organisms and food pathogens.

*Formation of surface conditioning films*

The conditioning of food contact surfaces involves the accumulation of organic and inorganic components or molecules (proteins, lipids, nucleic acids, and other biomolecules) released from foods such as milk, meat, and fruits, resulting in an increased concentration of nutrients on formerly 'clean' surfaces. Sorption of these molecules alters the surface's physicochemical properties (surface free energy, hydrophobicity, and electrostatic charge) (Bryers, 1987; Sjollema *et al.*, 1988; 1990) and plays a key role during the initial attachment of individual bacterial cells that precedes the more serious colonial biofilm formation (Kumar and Anand, 1998). Microorganisms may be transported to these conditioned surfaces along with the

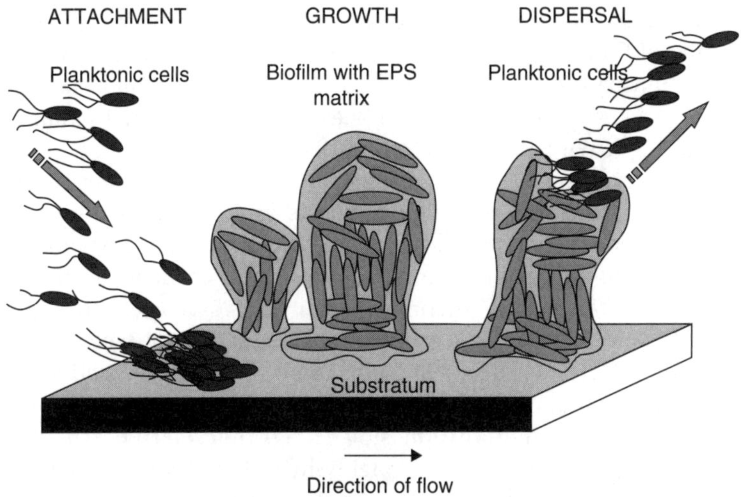


**Fig. 6.1** Stages of bacterial biofilm formation. Dark gray and light gray shades represent cells with planktonic and biofilm phenotypes, respectively.

nutrients via simple diffusion gradients, advective mechanisms, or by directed (chemotactic) motility where the organic film on the surface acts as a 'attractant' for bacterial chemoreceptors (Lawrence *et al.*, 1996). If left unchecked, a biofilm composed of living and dead cells, bacterial polymers and non-bacterial organic exudates, and inorganic compounds ultimately accumulates (Korber and Lawrence, 2004; Sampathkumar *et al.*, 2004), often causing prolonged sanitation issues via cross-contamination. Thus, it is often important that food contact surfaces be cleaned and sanitized following each shift to remove conditioning films along with the pioneer colonizing cells to avoid formation of a persistent biofilm (*see* section on practical considerations below).

*Attachment of cells*

Once on a surface, bacteria may undergo reversible or irreversible adhesion, or attachment. Reversible adhesion is typified by initial weak interactions between the cells and the substratum, where cells often continue to exhibit Brownian or flagellar movement while remaining closely associated with the surface (e.g., vibrations or rotations); such cells may eventually detach. Two theories have been employed to explain reversible adhesion of bacteria: i) the DLVO (Derjaguin, Landau, Verwey, and Overbeek) theory (Van Loosdrecht *et al.*, 1987) that integrates the effect of electrostatic and van der Waals forces on the attraction and adhesion of particles such as microbial cells, and ii) the Free Energy theory, that describes the process in terms of interfacial free energies (or surface hydrophobicity) of bacteria and surfaces (Korber *et al.*, 1995). It is noteworthy that microbial structures like pili, flagella and fimbriae help alter the diameter-dependent repulsion experienced by microbial cells as they approach a surface (Busscher *et al.*, 1990). It is also significant that the composition of the outer membrane of Gram-negative cells (i.e., the O-antigen and other variations in LPS chemistry) influence cell surface hydrophobicity and may impact adhesion to various substrata (sand, glass, polystyrene, etc.) (Makin and Beveridge, 1996; Williams and Fletcher, 1996).

Various interactions operate during irreversible (long-term) adhesion of bacteria to surfaces, including dipole–dipole interactions, hydrogen, ionic and covalent bonding, as well as hydrophobic interactions. Notably, long-term attachment of bacteria cannot entirely be explained by molecular interaction or free energy theories. Other microbiological mechanisms, including the production and secretion of bacterial EPS and receptor-based attachment (Lawrence *et al.*, 1996), along with the timely induction of specific sets of genes that cause the expression of biofilm-specific proteins, also play key roles in the long-term adhesion. EPS (often referred to as glycocalyx, slime, capsule, or sheath) are highly-hydrated polymers excreted by bacterial cells and comprised mainly of polysaccharides, but also containing varying amounts of nucleic (RNA and DNA) and amino acids, various glycoproteins and phosphoproteins, sugars, phospholipids, uronic acids, as well as phenolic compounds (Böckelmann *et al.*, 2006; Wingender *et al.*, 1999). Polymeric EPS fibrils form bridges between the microbial cells and the surface, facilitating irreversible attachment. Bacterial 'holdfast' structures, which have been described for environmental strains of *Hyphomicrobium* and *Caulobacter* spp., are EPS exudates that 'cement' bacteria by one cell-pole to surfaces (Langille and Weiner, 1988; Lawrence *et al.*, 1996). Bacterial EPS not only enhances bacterial adhesion and biofilm stability, but also plays putative roles in nutrient storage, resistance to antimicrobial agents and desiccation, maintenance of the biofilm microenvironment, and predator avoidance (Wolfaardt *et al.*, 1999). Specific receptor-based long-term attachment of bacterial cells to the substratum involves cellular surface structures of bacteria (e.g., the outer membrane and appendages) and various chemical components (e.g., glycoproteins) (Korber *et al.*, 1995; Starkey *et al.*, 2004). One of the better-characterized examples of this is the surface protein adhesins of *Staphylococcus aureus* (Foster and Höök, 1999), which are typically covalently bound to the outer layers of peptidoglycan, and have been designated as microbial surface components recognizing adhesive matrix molecules (MSCRAMM). From a practical standpoint, the removal of irreversibly attached cells from food processing surfaces involves deliberate application of forceful scrubbing and scraping (Korber and Lawrence, 2004), followed by chemical cleaning and disinfection.

*Microcolony formation and biofilm development*

Following irreversible attachment, bacterial cells grow and multiply into microcolonies, then enlarge and coalesce to form a layer of cells covering the surface that further develop into fully formed biofilms. While microcolonies are often described as *maturing* into biofilms, this difference is largely semantic: once attached to surfaces, bacterial cells shift from a planktonic to biofilm mode of growth that includes changes in the expression of genes and proteins (Beloin and Ghigo, 2004; Pringent-Combaret *et al.*, 1999; Ren *et al.*, 2004). Differential expression of proteins during biofilm formation is also stage-specific (Sauer *et al.*, 2002; 2003); proteins that have been shown to be up-regulated in the biofilm cells relative to planktonic cells include membrane proteins, QS proteins, adaptation and protection proteins, and proteins associated with metabolic cycles (Korber and Lawrence, 2004; Mangalappalli-Illathu *et al.*, 2008a). Generally, the genes associated with the biosynthesis of EPS components and QS molecules are induced during biofilm formation in different bacterial species. Likewise, the enzymes associated with stress response, especially oxidative stress, are found to be up-regulated in the biofilm cells as a result of nutrient and $O_2$ limitation in the depth of the biofilms (Ren *et al.*, 2004). Genes associated with the flagellar biosynthesis and other motility structures are usually repressed in biofilm cells (Pringent-Combaret *et al.*, 1999; Mangalappalli-Illathu *et al.*, 2008a). More information continues to point to clear differences between biofilm and planktonic cells; however, these differences are modulated considerably by the conditions of growth and the cells involved. The ability to have a large variety of adaptive and programmed responses to changing conditions has predisposed many bacteria for reproductive success on surfaces (Mangalappalli-Illathu *et al.*, 2008a,b).

*Biofilm redistribution (biofilm dispersion and recolonization)*

As biofilms mature, clumps of cells or individual daughter cells are released into the liquid phase. This phenomenon has been collectively termed sloughing, dispersion, redistribution, seeding, or recolonization. Released cells may be transported to new locations with sufficient quantities of nutrients and space are available, where they recolonize and initiate new foci of biofilm formation (Korber *et al.*, 1989). Biofilm dispersion is controlled by molecular mechanisms by stimulating the production of lytic enzymes that hydrolyze the exopolymeric material (EPS) that cements the individual cells together (Allison *et al.*, 1998; Decho *et al.*, 2000). Daughter cells that are ready to be released from biofilms have already adopted the planktonic cell phenotype, with the up-regulation of motility appendages such as flagella. Proteomic examination of cells released from mature biofilms indicates that these cells are morphologically and physiologically more similar to planktonic cells than to biofilm cells (Korber and Lawrence, 2004). Changes in the concentration of nutrients and $O_2$, along with the accumulation of waste products, hasten the process of biofilm dispersion. Flow is most certainly involved in these processes; Korber and Lawrence (unpublished data) have observed the dispersion of thick biofilms of *P. fluorescens* and other bacteria following the removal of flow stimuli, with the cells returning to the planktonic state within time frames as short as 30 minutes. It is noteworthy that the dispersion of individual cells from biofilms is different from the shearing and erosion of the biofilm biomass induced by fluid dynamics.

### 6.2.2 Intrinsic and extrinsic factors influencing biofilm formation

It is predictable that intrinsic microbial factors such as growth phase, type of flagellation and presence of other surface structures, motility, origin of the bacterial strains (e.g., food isolate or laboratory stain), prior exposure to various stress factors (e.g., high or low temperature and antimicrobial compounds), and mutations in genes encoding biofilm-associated proteins, all potentially contribute to the ability of an organism to attach to surfaces and form a biofilm. Similarly, extrinsic conditions such as ambient temperature, shear force, viscosity and pH of the liquid phase, medium composition (e.g., serum, blood, fruit exudates, and milk), atmosphere, presence of conditioning films, cell surface hydrophobicity and topography, and whether the surface has been previously coated with potential biofilm-inhibiting substances, also have demonstrable affects on biofilm formation. For a more in-depth discussion of the many factors shown to influence biofilm formation, readers are directed to the following excellent books on the topic: *Microbial Biofilms* by Ghannoum and O'Toole (2004); *Biofilms: Methods in Enzymology*, edited by Doyle (1999).

## 6.3 Organisms involved in food industry biofilms

Biofilms in the food industry are often formed by a wide range of different organisms, and are not what one would describe as 'mono-culture' or pure culture systems; however, the following organisms are known to play important roles under certain conditions, and thus are handled in more detail below.

### 6.3.1 *Pseudomonas* spp.

One of the most common and best-studied biofilm-forming genera are the pseudomonads. Pseudomonads are Gram-negative straight or slightly curved rods that are motile by one or several polar flagella. *Pseudomonas* spp. have a strictly-respiratory metabolism, with oxygen as the primary terminal electron acceptor, although nitrate can alternatively be utilized during anaerobic growth (Palleroni, 2001). Reasons that pseudomonas biofilms are ubiquitous is that these organisms are: i) highly-chemisynthetic, and thus do not require organic growth factors, and ii) are often psychrotrophic, and thus can continue to grow on foods under refrigeration. Thus, pseudomonads and other similar Gram-negative organisms introduced into foods via post-processing contamination tend to predominantly be associated with proteinaceous foods stored aerobically at chill temperatures. Widely distributed in nature and able to utilize a wide range of carbon substrates for growth, pseudomonads (typically isolates of *Pseudomonas fluorescens*) are the most commonly documented food spoilage organism, and are problematic in dairy products (Craven and Macauly, 1992), red meat (Dainty, 1996), fish (Gram and Huss, 1996) and poultry (Regez *et al.*, 1988). Spoilage in foods of animal origin by these organisms occurs first by utilization of the non-protein nitrogen fraction, followed by the production of lipases and proteases which liberate fatty and amino acid from the substrate commonly causing off-odors, off-flavors and rancidity (Jos and Huis, 1996). Advanced stages of food spoilage by pseudomonads is characterized by pigmentation and the production of extra-cellular slime (Dainty, 1996). Notably, pseudomonads and other Gram-negative organisms are relatively intolerant to low pH conditions; thus, foods that contain organic acids (e.g., benzoate), or that are naturally low-pH, tend to become spoiled by other organisms such as the lactic acid bacteria or yeasts (Gram *et al.*, 2002).

Many *Pseudomonas* isolates are highly-effective surface colonizers. Biofilms of *Pseudomonas* spp. develop at solid-liquid interfaces exposed to continuous flow of nutrients, and have been reported to produce thick layers of biomass organized in mushroom- and pillar-like structures separated by filled-water spaces (Davies *et al.*, 1998). However, *Pseudomonas* spp. biofilms do not universally adopt this architecture, especially on food processing surfaces; nutrient type and concentration, fluid flow and shear stresses, $O_2$ concentration, and strain differences may all influence the three-dimensional arrangement of biofilm cells on surfaces (Korber *et al.*, 1995). Regularized cleaning schedules can also help to prevent such advanced structures from developing. *Pseudomonas* biofilms consist primarily of hydrated glycocalyx, an extra-cellular polysaccharide matrix in which the cells are embedded (Costerton *et al.*, 1995). Singh and coworkers (2002) found that iron served as an essential signal for the initiation of *Pseudomonas aeruginosa* biofilm formation and development of mushroom-like structures. After cell division in the absence of lactoferrin (a mammalian iron chelator), two daughter cells remained near each other, initiating the formation of a microcolony. In the presence of lactoferrin, the daughter cells moved away from the point of cell division. Singh and coworkers (2002) concluded that increased surface motility induced by iron chelation was due to twitching motility; as free iron levels decreased, twitching motility increased, resulting in the restriction of biofilm development. Subsequent research on the effects of iron using mutant strains point to the role of pyoverdine production for the successful uptake of iron, and that pyoverdine-deficient mutants exhibited markedly lower biofilm growth yields compared to the wild-type strain (Patriquin *et al.*, 2008) with a concomitant increase in levels of twitching motility.

*Pseudomonas* spp. utilize at least two extracellular signals for cell-to-cell communication. The diffusible extracellular signal *N*-(3-oxododecanoyl)-L-homoserine lactone plays an important role in the attached life of this species. Studies conducted by Davies and colleagues (1998) demonstrated that cell-to-cell signaling and QS was required for the differentiation of planktonic cells of *P. aeruginosa* into complex multicellular structures. A mutation that blocked generation of the signal molecule resulted in the formation of an abnormal biofilm that was sensitive to the detergent biocide sodium dodecyl sulfate (SDS). More recent work by Patriquin *et al.* (2008) has further highlighted the importance of QS in *P. aeruginosa* biofilm formation; *rhlI* mutants (the *rhlI* gene codes for the C4-homoserine lactone signal compound) produced biofilms with normal architecture (large mounds and channels) under iron-deficient conditions; whereas, under the same conditions, the wild-type strain formed a thin, flat biofilm. A number of Gram-negative bacteria of food origin have been demonstrated to produce acetylated homoserine lactones (Whan *et al.*, 2000; Ravn *et al.*, 2001; Gram *et al.*, 2002), and while these compounds are known to function in virulence regulation and surface colonization in some pseudomonads, the role of they play in food spoilage is not altogether clear.

### 6.3.2 Bacillus cereus

*Bacillus cereus*, a spore-forming Gram-positive motile bacterium, is an opportunistic pathogen that primarily causes food poisoning manifested by diarrheal and emetic syndromes, as well as a wide spectrum of local and chronic infections, including: septicemia, meningitis, periodontitis, endocarditis and wound infections (Drobniewski, 1993; Schoeni and Wong, 2005). The ubiquitous nature of *B. cereus*, in combination with the ability of its spores to survive pasteurization and then germinate and grow at refrigeration temperatures, make this one of the most important food spoilage organisms in milk industry (Ternström *et al.*, 1993; Crielly *et al.*, 1994; Christiansson *et al.*, 1999). The genus *Bacillus* is generally recognized to be more heterogeneous than other bacterial genera, and is comprised of six species: *B. cereus*, *B. anthracis*, *B. mycoides*, *B. pseudomycoides*, *B. thuringiensis* and *B. weihenstephanensis* (Priest *et al.*, 2004; Rasko *et al.*, 2005). The genome of *B. cereus* ATCC (American Type Culture Collection) 14579 revealed a surprising amount of genetic relatedness of this organism to the animal and human pathogen *B. anthracis*, as well as to the insect pathogen *B. thuringiensis*, suggesting a common ancestor for all three organisms (Ivanova *et al.*, 2003). The diarrheal response caused by *B. cereus* is linked to the production of heat-labile enterotoxins, and mainly linked with meat products, vegetables and milk products; whereas, the emetic (vomiting) food illness, triggered by a smaller heat-stable peptide toxin, has mainly been related to carbohydrate-rich foods like rice, noodles and pasta (Shinagawa *et al.*, 1990; Kramer and Gilbert, 1989).

*Bacillus cereus* readily forms biofilms on a variety of surfaces, including plastic, soil, glass wool, and stainless steel (Peng *et al.*, 2001; Oosthuizen *et al.*, 2002; Auger *et al.*, 2006; Vilain *et al.*, 2006) and may lead to the establishment of biocide-resistant biofilms in food processing installations (Peng *et al.*, 2002) that subsequently pose a persistent risk to public health. A recent study by Vilain and Brozel (2006), using a principal component analysis approach, compared the whole cell proteomes of 24 h-old *B. cereus* biofilms and the associated suspended population with exponential, transient, and stationary phase planktonic cultures, revealing a unique pattern of gene expression in sessile cells. The authors concluded that the biofilm-specific phenotype of *B. cereus* (and perhaps other Gram-positive bacteria) was due more to an altered profile of relative protein quantities rather than to the appearance or disappearance of unique protein spots. Interestingly, the authors failed to detect the hallmark expression of biofilm-specific genes.

### 6.3.3 Shewanella putrefaciens

*Shewanella putrefaciens* are Gram negative, 0.5–2.0 μm, rod-shaped motile organisms possessing a single, unsheathed, polar flagellum. Facultatively anaerobic, this organism uses $O_2$ as electron acceptor, whereas anaerobic growth is predominantly respiratory with the oxidation of organic carbon compounds or $H_2$ being coupled with the reduction of wide spectrum inorganic and organic electron acceptors, including $NO_3^-$, $NO_2^-$, $Fe^{3+}$, fumarate, and various sulfur compounds (Bowman, 2001). *Shewanella putrefaciens* is a ubiquitous mesophile found predominantly in marine, freshwater, sewage, soil, and natural energy reserves (oil and gas wells) (Semple *et al.*, 1989; Stenstrom and Molin, 1990; Vogel *et al.*, 1997) with growth a range between 4 and 40°C (optimum, 30–35°C). Its ubiquitous distribution, flexible metabolism and temperature tolerance make this organism a problem to the food sector (Stenstrom and Molin, 1990). *Shewanella putrefaciens* has often been linked with the spoilage of marine fish stored on ice, vacuum-packaged meats, and chicken (Debevere and Boskou, 1996; Jørgensen *et al.*, 2000). The spoilage of fish by *S. putrefaciens* is manifested by the production of various volatile sulfides: methylmercaptane ($CH_3SH$), hydrogen sulphide ($H_2S$) from cysteine, dimethylsulphide $(CH_3)_2S$ from methionine, as well as the reduction of trimethylamine oxide to the fishy-smelling compound trimethylamine (Gram and Huss, 1996; Boskou and Devebere, 1997) have all been documented. Bagge and coworkers (2001), using both batch and flow conditions, demonstrated that *S. putrefaciens* was able to attach and form biofilms on food processing surfaces. Furthermore, the numbers of adhering bacteria from a suspension of *S. putrefaciens* cells were much lower in a laminar flow system than in a static batch system, in contrast to many other organisms that require flow as an attachment stimuli. Surprisingly, Bagge *et al.* (2001) revealed that the rate of biofilm formation and the thickness of film were not dependent on the availability of carbohydrate (lactate and glucose) or on iron starvation (thereby being distinct from pseudomonads), indicating a unique form of biofilm formation and regulation by this organism.

### 6.3.4 Fungal/yeast biofilms

Fungi and yeast are eukaryotic microorganisms that occupy a wide range of environmental niches, including those with very low pH, low water activity, and extremes of temperatures. The high adaptability of fungi and yeast has obvious implications for the food industry, as these organisms spoil a wide spectrum of foods and beverages. It is known that yeast have the ability to utilize food ingredients such as lactic, citric and acetic acids, compounds generally added to foods as microbial inhibitors (Jos and Huis, 1996). For example, fruit juices, which are high in citric acid, sugar and have a low pH (~ from 2.0 and 4.5), often become spoiled due to the growth of yeast spoilage organisms (Edwards *et al.*, 1998; Tajchakavit *et al.*, 1998). Furthermore, it has been reported that pasteurized foods and beverages, even those products subjected to processes more rigorous than pasteurization (e.g., hot-fill beverages), may contain a wide range of viable molds (Mann and Beuchat, 2008). Growth of certain species of yeast and molds cannot always be inhibited by common antimycotics such as benzoate, propionate and sorbate (Miller, 1979); some *Penicillium* cheese spoilage species, for example, are capable of growth in the presence of potassium sorbate at concentrations as high as 7100 μg $ml^{-1}$ (Marth *et al.*, 1966) and 12 000 μg $ml^{-1}$ (Finol *et al.*, 1982). In addition to visible spoilage, some molds produce mycotoxins that pose a serious threat to the health of the consumer. So far, it is known that over 200 different types of molds associated with food items can generate metabolites that are orally toxic to humans (Jos and Huis, 1996). Table 6.1 provides a list of the most common mycotoxins and their association with different types of foods.

Overall, the majority of research related to mold and yeast food spoilage has focused on identification of the causative food spoilage organism and their physiological characterization. There is an urgent need for food

**Table 6.1** Association of mycotoxins and their food origin

| Food origin | Mycotoxins |
|---|---|
| Citrus fruit | Alternariols |
| | Tenuazonic acid |
| Pomaceous and stone fruit | Patulin |
| | Citrinin |
| | Alternariols |
| Potato tubers | Diacetoxyscirpenol |
| Wheat and rye grain | Deoxynivalenol |
| | Zearalenone |
| | Alternariols |
| | Ergot alkaloids |
| | Tenuazonic acid |
| | Ochratoxin A |
| | Aflatoxins |
| | Xanthomegnins |
| | Citrinin |
| Rye bread | Aflatoxins |
| | Citrinin |
| | Patulin |
| | Ergot alkaloids |
| Cheese | Sterigmatocystin |
| Fermented sausage | Citreoviridin |
| | Citrinin |
| | Cyclopiazonic acid |
| | Isofumigaclavine A |
| | Ochratoxin A |
| | Patulin |
| | Roquefortine C |
| | Rugulovasine A |

Modified from Filtenborg *et al.*, 1996.

microbiologists to carry out more detailed investigation on molds and yeast communities and their interactive behavior within bacterial biofilm consortia, with the potential benefits of developing novel preservation and control techniques.

### 6.3.5 Mixed-species biofilm communities

Although many food environments are highly selective for specific spoilage organisms due to chemical and physical factors, a community of organisms will ultimately develop given sufficient time. Relatively little is known about the ecology and development of these mixed-species biofilm communities. Under most instances, the food or food processing ecosystem involves a solid–liquid interface; thus, cells are attaching to, or becoming entrapped at, a surface (Fleet, 1999). Sensorily-speaking, food spoilage occurs following growth of microorganisms to concentrations in the range of $10^7$–$10^9$ CFU (colony forming units) per gram of food (Boddy and Wimpenny, 1992), after which different microbial species in both the planktonic and sessile state interact with each other. As in other systems, interactions between different microorganisms in food systems (with the exception of highly processed foods) will involve a range of different bacteria, yeast and fungi (Fleet, 1999), with either antagonistic and/or beneficial outcomes between the different groups (Fredrikson, 1977). Rao *et al.* (2005) revealed that AlpP, an antibacterial protein produced by the marine bacterium *Pseudoalteromonas tunicata*, played an important role during competitive biofilm formation. The wild-type *P. tunica* strain could aggressively colonize plant surfaces and thereby dominate over other AlpP-sensitive microorganisms. In contrast, a *alpP P. tunica* mutant lost the ability to successfully compete with other organisms under the same conditions. Successional transitions (or selection) of different organisms may also occur as the food environment becomes modified following the metabolic activity of individual or associations (sets) of organisms (Caldwell *et al.*, 1997).

Recently, An and colleagues (2006) found that QS-regulated functions and surface motility were important competitive factors during formation of mixed-species microbial biofilms. Christensen *et al.* (2002) demonstrated that biofilm-mediated juxtapositioning of *Pseudomonas putida* and *Acinetobacter* C6 strains facilitated the successful proliferation of both strains when strain C6 was oriented nearer the bulk liquid phase where high concentrations of benzyl alcohol existed, while *P. putida* was positioned so as to benefit from the benzoate leaking from the *Acinetobacter*. In contrast, both organisms underwent competitive growth for benzyl alcohol when cultured planktonically in a chemostat. The above examples demonstrate that the coexistence of different organisms depend on the types of organisms and the environment in which they are found. Much more work on the interactive behavior of important food spoilage organisms is necessary in order to develop sustainable and effective preservation techniques associated with different food processing environments.

## 6.4 Control of microbial biofilms

### 6.4.1 Antimicrobial agents

Antimicrobial agents are chemical or biochemical compounds that inhibit or kill viruses, bacteria, or fungi, and broadly include: i) antibiotics, which have (often specific) chemotherapeutic application for the treatment of bacterial or fungal infections in humans and animals (Russell, 2003), and ii) biocides, which include disinfectants, antiseptics, and preservatives used to disinfect inanimate objects such as food processing surfaces, for external application to the skin (antiseptics and topical antimicrobials), or for incorporation into pharmaceutical, cosmetic, or other types of products to prevent microbial growth (McDonnell and Russell, 1999; Russell, 2003). A wide range of biocidal compounds have application for the control (disinfection) of bacteria in food processing settings, and the use of these different compounds is dependent on the specific application as well as process-specific factors (e.g., food contact surfaces or non-contact surfaces, degree of soiling/fouling, target organisms). These compounds may exert momentary or residue antimicrobial effects. Different antimicrobial agents have different chemical properties, and accordingly exert different effects on bacterial cells, including: destabilization and damage to the cell envelope (cell wall and outer membrane) and cytoplasmic (inner) membrane (e.g., EDTA, gluteraldehyde, and chlorhexidine), cross-linking of macromolecules such as proteins, RNA, and DNA (e.g., formaldehyde and gluteraldehyde), interaction with thiol groups (e.g., silver compounds), DNA intercalation (e.g., acridines), breakage of DNA strands and inhibition of DNA synthesis (e.g., halogens and $H_2O_2$), and oxidation of thiol groups in enzymes and proteins (e.g., halogens and peroxygens) (*see* Table 6.2 for a list of antimicrobial agents with application to the food sector). Different bacteria have differing susceptibilities to various agents; thus, the vulnerability of a particular organism is key in choosing an antimicrobial agent. Chemical compounds such as chlorhexidine salts and quaternary ammonium compounds (QACs) have a long history of broad-spectrum inhibitory (not, however, on *Mycobacterium tuberculosis*) use as antiseptics, disinfectants, and preservatives (Houari and Di Martino, 2007), causing generalized membrane damage involving the phospholipid bilayers of the cytoplasmic membrane (McDonnell and Russell, 1999). Alternatively, gluteraldehyde and orthophthalaldehyde are used predominantly for the disinfection of endospores (Russell, 2003). There are also antimicrobial preservative compounds, some of which are considered 'natural', that may applied directly to foods, with the most typical effect being to extend the shelf-life of these products. Compounds that have

**Table 6.2** Antimicrobial agents with varying application, depending on intended use, for the control of bacteria and bacterial biofilms in the food industry

| Antimicrobial agents with residual effect | |
|---|---|
| Compound group | Compound |
| Alcohols | ethanol<br>isopropanol |
| Aldehydes | glutaraldehyde<br>formaldehyde |
| Oxidizing agents | sodium hypochlorite and chlorine-containing compounds<br>iodine compounds<br>acidic electrolyzed water<br>performic acid<br>peracetic acid<br>hydrogen peroxide<br>ozone |
| Gaseous substances | ethylene oxide<br>formaldehyde |
| **Antimicrobial agents with no residual effect** | |
| Compound group | Compound |
| Anilides | triclocarban |
| Biguanides | chlorhexidine<br>alexidine<br>polyaminopropyl biguanide<br>polymeric biguanides |
| Bisphenols | triclosan<br>hexachlorophene |
| Metals | silver compounds |
| Phenols and cresols | phenol<br>cresol<br>thymol |
| Quaternary ammonium compounds | benzalkonium chloride<br>cetylpyridinium chloride<br>cetrimide |
| **Antimicrobials compounds approved for the treatment of carcasses, meats, parts or comminuted product** | |
| Compound group | Compound |
| Polyphosphate | trisodium phosphate |
| Natural antimicrobials | organic acids (e.g., lactic, acetic, citric, and ascorbic acids), nisin, bacteriocins and other animal- or plant-derived antimicrobial peptides, lactoferrin and lactoperoxidase, herbs and spices, chitosan |
| Radiation | UV and Ionizing radiation |

seen application in North America and Europe include nisin, lactoferrin, lactoperoxidase, organic acids, etc.

### 6.4.2 Biofilm resistance to antimicrobial agents

Over the past several decades, there have been numerous reports of enhanced biofilm bacterial resistance to treatment with antimicrobial agents such as those listed above, including, for example, *Acinetobacter*, *Bacillus*, *Campylobacter*, *Delftia*, *Enterococcus*, *Escherichia*, *Legionella*, *Listeria*, *Mycobacterium*, *Salmonella*, *Pseudomonas*, *Staphylococcus*, *Streptococcus*, and *Vibrio*. This phenomenon has greatly complicated the control of biofilm-forming organisms in food processing settings (not to mention the industrial and medical sectors). *Staphylococcus aureus* biofilm cells, when grown on silicon surfaces, exhibited ten times the resistance to vancomycin than similar planktonic cells (Williams *et al.*, 1997). Ceri *et al.* (1999) showed that the minimal biofilm eradication concentrations (MBEC) for *P. aeruginosa* and *Escherichia coli* biofilms was ~1024 μg imipenem and ~512 μg of ampicillin, compared to just 1 and 2 μg, respectively, for their planktonic counterparts. Others have observed increases in resistance of biofilms and attached microorganism on food contact surfaces to various sanitizing/disinfecting agents despite their membrane-level destructive effects and broad-spectrum activity (Cargill *et al.*, 1991; Mosteller and Bishop, 1993; Yu *et al.*, 1993; Costerton *et al.*, 1995; Stewart *et al.*, 2004; Fux *et al.*, 2005).

A number of factors influence how well different antimicrobial agents function, including concentration, contact time, pH, temperature, the presence of organic matter or other interfering or enhancing materials, and the nature, numbers/density, location, age and physiological status of the microorganisms (Stewart *et al.*, 2004; Fux *et al.*, 2005; Russell, 2003). For example, Wirtanen and Mattila-Sadholm (1992) reported that the resistance of *P. fluorescens*, *L. monocytogenes*, and *B. subtilis* food borne biofilms to chlorine-based sanitizers increased with biofilm age. While much information has accrued thus far (see below and pages 184–7), a clear and universal understanding of enhanced biofilm resistance to antimicrobials agents has yet to emerge.

*Mechanisms of biofilm resistance to antimicrobial agents*

An assortment of mechanisms have been proposed over recent years to explain the resistance of biofilm bacteria to antimicrobial agents, and include, but are not limited to: (i) depletion of the antimicrobial agent from the bulk liquid phase, (ii) delayed penetration of the antimicrobial agent into the biofilm matrix, (iii) altered growth rate of biofilm organisms, and (iv) adaptive physiological changes due to the biofilm mode of growth. It is somewhat logical that for reactive-type antimicrobial agents (e.g., oxidizing agents), a relationship between degree of fouling and agent concentration will exist, typically illustrated where highly fouled surfaces require greater contact times and higher concentrations to achieve effective biofilm control. Hindered delivery of various agents is less intuitive, but in general, the high concentration/low molecular-weight agents and long contact times typically used by the food industry (e.g., benzalkonium chloride and QACs, chlorhexidine, peroxyacetogens) should not result in a situation where the biofilm cell/EPS matrix would prevent the ultimate delivery of these compounds into all regions of the biofilm (Lawrence *et al.*, 1994; Stewart *et al.*, 2004). There are examples, however, where hindered diffusion has been documented. Suci *et al.* (1994) demonstrated a considerable delay in penetration of ciprofloxacin into *P. aeruginosa* biofilms (40 s for a sterile surface versus 21 min for a biofilm-fouled surface). Aside from circumstances where hindered delivery may occur due to other bacteria-specific phenomena (chlorine reaction with biofilm organic material, catalase positive cells degrading hydrogen peroxide, β-lactamase-positive bacteria and treatment with β-lactam antibiotics, etc.) (Fux *et al.*, 2005), there is increasing evidence that many biofilm bacteria often survive despite the antimicrobial agent penetrating the matrix at the appropriate concentration. Furthermore, the microorganism may not have either the correct biochemical target for a particular antimicrobial, or the necessary transporter system required for antibiotic uptake (Volk *et al.*, 1996). It is also noteworthy that if the delivery of the antimicrobial agent is slowed, or its concentration is somehow depleted, the possibility of the microorganisms developing adaptive resistance (Mangalappalli-Illathu and Korber, 2006; Mangalappalli-Illathu *et al.*, 2008b) via various inducible stress response pathways would increase (Fux *et al.*, 2005). Thus, if biofilm bacteria are not killed outright during treatment, there is the possibility that an antimicrobial-resistant population of organisms may arise.

Biofilm bacteria live in a diffusion-dominated environment where delivery of nutrients and removal of wastes are governed by molecular diffusion (Lawrence *et al.*, 1994; Wolfaardt *et al.*, 2007); thus, pseudo-steady state chemical gradients develop. One predictable outcome of these variable microscale conditions is the development of a range of microbial responses in terms of growth rate, cell physiology and stress response. Early evidence of differences in the physiology of biofilm bacteria was provided by Caldwell and Lawrence (1989), where attached *E. coli* cells that appeared identical in terms of size, shape and phase density under light microscopy appeared markedly-different when visualized by the redox-indicator dye resazurin using confocal scanning phase microscopy. Microscale variability in the rates of growth of seemingly-identical biofilm bacteria treated with antibiotics have also been demonstrated (Sufya *et al.*, 2003), with the maximum specific growth rate in these organisms varying by more than 10 fold (from 45 minutes to 500 minutes). The variable survival of small groups of individual cells or microcolonies within the biofilm matrix may thus be due to localized variations in cell growth rate. Plate 2 (between pages 108 and 109) shows a confocal laser microscope dual-channel image obtained from the base (attachment surface) of a thick *Brochothrix thermosphacta* biofilm treated with 0.5% trisodium phosphate (TSP, a food grade antimicrobial) and stained with a fluorescent viability stain. While thin *B. thermosphacta* biofilms treated with this concentration of TSP were completely killed (red cells), the thicker biofilm contained variable numbers of surviving (green) cells embedded at the base of the biofilm.

Cellular growth rate (and active cell growth) has previously been shown to influence the resistance of bacteria to antimicrobial agents, since many antibiotics target a growth-specific, and usually essential, biochemical pathway (e.g., cell wall biosynthesis, protein synthesis, DNA replication). Eng *et al.* (1991) demonstrated that fluoroquinolone antibiotics exerted cidal effects against non-growing (stationary-phase) Gram-negative bacteria, whereas an increase in antimicrobial efficacy was seen when the growth rate of biofilm cells was increased. Korber *et al.* (1994) similarly used the fluoroquinolone, fleroxacin, which inhibits gyrase and causes growing cells to elongate, along with fluorescent probes and confocal microscopy, to show that a gradient of cell growth rates existed within thick *P. fluorescens* biofilms; cells located at the biofilm–liquid interface elongated significantly more than the slowly-growing cells located near the biofilm–substratum interface, and were accordingly more susceptible to fleroxacin.

Altered microbial physiology, mediated by adaptive and genetically programmed cell responses, may also play a role in biofilm resistance to antimicrobial agents. Different unfavorable environmental conditions trigger the expression of appropriate sets of genes and production of proteins that facilitate bacterial survival. Conserved widely throughout nature (i.e., plants, animals, insects, microorganisms), external stimuli cause the exchange of sigma subunits (there are seven such species of the $\sigma$ subunit in *E. coli*) of the RNA polymerase (RNAP) complex, with gene expression accordingly adjusted to meet emerging needs. These different sigma factors recognize various gene promoters, activating distinct gene sets, or regulons. Most housekeeping genes expressed during exponential-phase growth are transcribed by the RNA holoenzyme containing the *rpoD* gene product; whereas, the holoenzyme RpoS ($\sigma^S$), considered the general stress sigma factor, is essential for transcription of certain stationary-phase specific genes (Lange and Hennge-Aronis, 1991). RpoS becomes strongly induced in the stationary growth phase or when a cell in exponential phase is exposed to various stressors, including ultra-violet radiation, hyperosmolarity, pH downshifts, and non-optimal high or low temperature regimes (Hengge-Aronis, 2002), and is often accompanied by a cessation of growth and provides cells with the ability to survive the actual stress as well as additional stresses not yet encountered ('cross-protection'). The dual nature of the $\sigma^S$ stress response is contrary to other specific stress responses, which are triggered by a single stress signal and result in the induction of proteins that allow cells to overcome only that particular stress.

Because of its role in stress response and cell survival, RpoS has been the focus of considerable interest in the biofilm research community, and some have speculated that life in biofilms results in the constitutive expression of stress response systems like RpoS in conjunction with drug efflux pump systems. Adams and McLean (1999) observed that *E. coli* biofilms were more dense and contained a greater number of viable cells when an $RpoS^+$ strain was used. The authors hypothesized that diffusional limitation experienced within biofilms caused nutrient shortages and build-up of toxic metabolites, reducing the overall growth rate of cells and thereby activating *rpoS*. Reports of this nature have unfortunately been difficult to link with the phenomenon of enhanced biofilm resistance to antimicrobial agents for foodborne organisms. Data obtained by our group (Mangalappalli-Illathu and Korber, 2006; Mangalappalli-Illathu *et al.*, 2008b) using *Salmonella enterica* serovar Enteritidis biofilms adapted to low concentrations of benzalkonium chloride, and others (Xu *et al.*, 2001; Stewart *et al.*, 2004), do not lend much support to the above theory despite the fact that *Salmonella* biofilms were considerably more resistant to benzalkonium chloride than were otherwise equivalent planktonic cells and a range of differentially expressed (up- and down-regulated) proteins were seen in biofilm and planktonic bacteria. It is noteworthy that most examinations of gene or protein expression in biofilms have been conducted on bulk preparations of a cell population, due to the difficulties in separating otherwise identical bacteria for differential analysis; therefore, even if a significant proportion (e.g., 5–10%) of cells were expressing these genes and consequently survived a subsequent antimicrobial challenge, their presence might effectively be masked by the phenotypes of the majority of cells not expressing these genes/proteins. Thus, survival of biofilm bacteria may involve small numbers of cells with altered resistant phenotypes, a proposition somewhat consistent with the 'persister' hypothesis (Spoering and Lewis, 2001), where fewer than ~1% of any cell population (not biofilm-specific) is dormant and thus predisposed to surviving chemical or physical challenge. In either case, these resistant cells would be able to regrow after the antimicrobial challenge had been removed, contributing to a recurrent problem biofilm.

Overall, it seems logical that multiple resistance mechanisms (Mah and O'Toole, 2001), such as slow growth and/or induction of an *rpoS*-mediated stress response, along with the physical and/or chemical structure of EPS or other aspects of biofilm architecture, likely function in concert to aid in the survival of biofilm bacteria. Given the diverse range of molecular and physiological responses seen in different bacteria involved in food spoilage and disease, this conservative conclusion seems appropriate. Thus, a surface sanitation program must first dislodge cells and their EPS from the food processing surfaces, as physical removal of biofilm material is not dependent on their killing *per se*. Once cells are in a dispersed planktonic state, the efficacy of any antimicrobial treatment that is applied will increase. For more details on the resistance of biofilms to antimicrobial agents, the reader is directed to the following review article (Stewart *et al.*, 2004).

## 6.5 Biofilm problem areas in processing facilities

Biofilms that develop in food processing environments are usually mixed-microbial communities that often become thicker and more stable than monospecies biofilms (Kumar and Anand, 1998; Lebert *et al.*, 2007). In recent years, however, it has been noted that while bacteria on food-contact surfaces (which are generally cleaned and sanitized following each shift of operation) do tend to accumulate over the course of a typical shift, they don't become the foci of extensive biofilm formation. Rather, areas that are only infrequently cleaned are more likely to harbor thicker and more resilient biofilms; for example, overhead structures/shrouds with high humidity and condensation, walls, corners and drains, and moving mechanical parts with poor accessibility. For example, Gibson *et al.* (1999) reported that non-food contact surfaces that were washed infrequently (e.g., on a weekly basis) supported the extensive development of biofilms compared to food contact surfaces that were not sites of extensive growth.

Food spoilage bacteria such as *Pseudomonas* spp. (Michiels *et al.*, 1997) and *S. putrefaciens* (Bagge *et al.*, 2001; Hjelm *et al.*, 2002) colonize such sites and, via cross-contamination, contaminate food products. Once detected, the complete removal of these biofilms can be decidedly difficult (e.g., the measures spent trying to eliminate organisms such as *Listeria* from a facility may become very extreme). Dead ends in pipes, crevices, corners, gaskets, and joints, as well as pumps and storage tanks, are also particularly vulnerable to biofilm accumulation due to poor access during cleaning; such locations can then become a source of persistent potential food contamination (Mattila-Sandholm and Wirtenen, 1992; Hjelm *et al.*, 2002).

Associated issues linked with biofilm formation in fluid processing operations include mechanical blockages and reduced flow velocity, impedance in heat transfer processes, biodeterioration of metallic and polymeric components of the production facilities, and product spoilage and public health risks associated (Mittelman, 1998). In drinking water distribution systems, biofilms may lead to reduced water velocity and carrying capacity, clogging of pipes, increases in energy utilization and reduced operational efficiency (Kumar and Anand, 1998) and may also shelter primary (i.e., those that cause disease in healthy individuals, such as *Helicobacter pylori*), as well as opportunistic (*Acinetobacter* spp.) (EPA, 2002), pathogens. A similar protective relationship has been found in food processing settings between *P. fluorescens* biofilms and *L. monocytogenes* (Carpentier and Chassaing, 2004).

Dairy industry biofilms are similarly a serious problem; typical problem areas are food contact surfaces such as packaging machines and feeding units, floor drains, and the door mat (Bower *et al.*, 1996; Gunduz and Tuncel, 2006). Most of the biofilm formation ($6.5 \times 10^3$ CFU $ml^{-1}$) associated with an ice cream plant was found on the conveyor belt of the packaging machine, and typically would become noticeable 8 h following the start of production (Gunduz and Tuncel, 2006); these biofilms were composed of Gram-negative organisms of *Enterobacteriaceae* family (*Proteus*, *Enterobacter*, *Citrobacter*, *Shigella*, *Escherichia*, and *Edwardsiella* spp.), other Gram-negative microflora (*Aeromonas*, *Pleisomonas*, *Moraxella*, *Pseudomonas*, and *Alcaligenes* spp.), and various Gram-positive bacteria (*Staphylococcus*, *Bacillus*, *Listeria* spp. and lactic acid bacteria such as *Streptococcus*, *Leuconostoc* or *Pediococcus* spp.). The majority of the organisms were important food spoilage organisms and some were significant pathogens. The dairy industry is also plagued by the formation of milkstone, a thick layer of scale mainly formed by the deposition of hard water cations like calcium and magnesium and organic/protein materials from bacteria and milk on surfaces); milkstone residues are not only unclean but often entrain recalcitrant biofilm. In order to disinfect these surfaces, one must first dissolve the precipitated mineral/organic matrix using inorganic acid-containing cleaners (e.g., nitric, hydrochloric, hydrofluoric and sulphuric acids) (Sansebastiano *et al.*, 2006), and only then can the biofilm bacteria effectively be killed.

Jessen and Lammert (2003) reported that a single method or detergent/disinfectant was not sufficient to eradicate attached bacteria on meat processing surfaces. Many reports have claimed that disinfectants containing peracetic acid are most effective at removing biofilms; whereas, hypochlorous acid, which forms from addition of chlorine salts (or $ClO_2$) to water, appears more efficient in meat processing plants. Extra disinfection time in addition to regular sanitation did not improve the hygienic level, but the introduction of a scrubbing step reduced the bacterial load and is thus, as mentioned above, recommended on indirect sites (e.g., control panels and rollers along conveyor belts) and critical sites (e.g., knives and holders for keeping meat products) (Jessen and Lammert, 2003).

Biofilm formation by *Oenococcus oeni* and related lactic acid bacteria are representative of malolactic problem contamination in the wine industry. Nel *et al.* (2002) found antimicrobial peptide pediocin PD-1, which is non-toxic, colorless and odorless, was the most effective of three bacteriocins tested at killing all viable cells in established *O. oeni* biofilms that formed in acidic grape medium.

Factors like nutrient and water availability, equipment cleanability and design, and temperature control are all important in biofilm control (Chmielewski and Frank, 2003). Designing equipment with smooth, highly-polished surfaces prevents biofilm formation by making the initial adhesion step more difficult, and reduces the occurrence of protective microniches such as scratches and crevices wherein bacteria can survive disinfection treatments (Korber and Lawrence, 2004). Mean reduction values in viable counts of *L. monocytogenes* following treatment with a combination of sodium hypochlorite and heat were approximately 100 times lower in biofilms than for planktonic cells. An ongoing concern is that many of the companies that produce antimicrobial agents have no idea as to how well these agents perform against biofilm bacteria. The common practice employed, until most recently, was to examine the effectiveness of these compounds using planktonic (batch) cells as targets; the now obvious disadvantages of a batch approach is that the target cells are individually suspended rather than present in a thick matrix entrained in polymer, and that the complex cell-cell arrangements that might exist in a biofilm community cannot develop. While there is growing interest in the ecology and behaviour of biofilm communities in food systems, a universal approach to their testing against antimicrobial challenge has proven elusive. More recently, Ceri *et al.* (1999) developed a simple, 96-well microplate system for determining the MBEC (minimum biofilm inhibitory concentration) of different antimicrobial agents, with potential application for the testing of antibiotics, heavy metals and other compounds (Ceri *et al.*, 1999; Harrison *et al.*, 2005a,b).

## 6.6 Practical considerations for biofilm control

At the outset, it is necessary to clearly distinguish between cleaning (the removal of dirt from the surface), sanitation (the reduction of the contaminating microbial load to safe levels) and disinfection (the elimination of pathogenic microorganisms via chemical or physical means). It is critical to understand that you cannot disinfect a surface without first employing an effective cleaning regime. Regardless of whether biofilms are known to be present, the procedure for cleaning is the same:

i) Pre-washing or rinsing to remove gross contaminating soil or dirt.
ii) Cleaning with detergent and water; acidic detergents dissolve mineral buildup, whereas alkali-based detergents dissolve microbial polymeric substances.
iii) Rinsing with water to remove any residual detergent and soil that might interfere with subsequent sanitizing steps.
iv) Surface treatment with a sanitizing agent, which typically involves use of chlorine-based or peroxyacetic acid (POAA) treatments. The POAA treatment may be left on as peroxide and acetic acid are food grade additives, whereas chlorinated compounds must be rinsed off – this is important not only because of food flavor effects, but also

because strong alkaline compounds tend to corrode even food grade stainless steels.

v) Final water rinse (this optional step may be omitted if a food-grade sanitizer with residual action is used) (Korber and Lawrence, 2004; Sansebastiano *et al.*, 2006). Note that all the rinsing steps should use potable water.

The types of soil or dirt will dictate, to a large degree, the types of cleaning strategies and the agents necessary. For example, greasy build-up caused by passage of foodstuffs such as raw meats and fat must be emulsified using an alkaline detergent Build-up of proteinaceous materials are problematic since excessive temperatures (>60°C) will denature the proteins onto the surface; alkaline detergents with proteolytic properties, sometimes used in combination with surfactants, are well-suited for this type of job. Carbohydrate build-up may be removed via the use of hot water, perhaps along with mild detergents. Excessive temperatures during cleaning can, however, caramelize the sugars on the surface. The presence of inorganic materials such as calcium or magnesium salts can lead to the formation of precipitates or scale (*see* above reference to milkstone), necessitating the use of dissolving acid solutions. The type of detergent, temperature, contact time, and use of mechanical action all contribute to the relative efficacy of the cleaning process. It is noteworthy that there are a large number of different detergents, sanitizers (see above), and surfactants with application in the food sector. Use of compatible and appropriate agents for the various soil types is thus the foundation to a successful sanitation program. Surface type also influences bacterial adhesion and is thus an important consideration; surfaces such as glass are highly cleanable but susceptible to damage, whereas stainless steel is damage-resistant but prone to corrosion and scratching. Their respective cleanability (rubber, glass, plastics, metals, woods, etc.) is a deciding factor in their potential industrial application. The incorporation of novel antimicrobial agents, such as surface-incorporated polychlorinated compounds (e.g., triclosan) or silver compounds, provides an additional option; such treatments are intended to prevent initial colonization and growth by biofilm-forming bacteria. However, both manufacturers and opponents to this type of antimicrobial approach insist that this doesn't replace the need for proper sanitation and/or hygiene.

Overall, the above sanitation and cleaning considerations are best made during the planning stage of processing plant construction; the appropriate type of material or equipment design should be selected to prevent microbial adhesion and facilitate cleaning. Where an existing facility is of concern, a suitable starting point involves a microbiological audit in order to determine where biofilms tend to occur, followed by strict sanitation schedule in conjunction with HACCP. Such a program would typically involve the cyclic or rotational use of antimicrobial agents and sanitizers. Expert technical support may also be provided by biocide/detergent vendors, and such consultations are recommended as a logical starting point when first developing a cleaning and sanitization program. Practical coverage of cleaning and disinfection in the food industry is provided in a recent chapter by Sansebastiano *et al.* (2006).

## 6.7 Future trends

Physiological adaptation of biofilm bacteria to various environmental stress factors, such as fluid flow, mechanical disturbances (e.g., routine scrubbing and scrapping), and prolonged exposure to sub-lethal biocide concentrations could significantly influence the outcome of our efforts in biofilm control, especially when the biofilms are protected by surface scratches and imperfections (Mangalappalli-Illathu and Korber, 2006; Mangalappalli-Illathu *et al.*, 2008a,b). At present in the food industry, we see the combined use of multi-faceted approaches (e.g, a hurdle-based system) such as the mechanical disruption of biofilms and surface cleaning, followed by lethal biocide exposure for sufficient length of time. Combination of this foundational strategy with novel biofilm prevention techniques (e.g., improved design of food processing equipment to prevent biofilm formation, use of special bacteria-repelling coatings, use of super-high magnetic fields and ultrasound treatments, and use of high-pulsed electrical fields to repel bacteria) could further reduce the occurrence of problem biofilms (Kumar and Anand, 1998).

More research is needed to prevent microbial attachment to, and persistent growth on, surfaces. In order to achieve this, a much more comprehensive understanding of the ecology and diversity of mixed microbial communities that form in food industry settings, and their mechanisms of resistance to antimicrobial strategies, is needed. Research goals should also include the development of safe, efficient, and environmentally friendly biocides that are effective at low concentrations. These compounds should be non-specifically lethal to microbial cells, thereby minimizing the possibility of acquired resistance. The application of broad-spectrum, food grade bacteriocins (ribosomally synthesized, extracellular bioactive antimicrobial peptides) (Nel *et al.*, 2002), or the technological organisms that produce them (Ammor *et al.*, 2006), to food processing surfaces to prevent, or competitively exclude, both food pathogens and food spoilage organisms could function much like probiotics in the human gastrointestinal tract. The enhanced activity of these compounds at low concentrations has potential, especially in industries where conventional treatment strategies have failed or where the use of chemical disinfectants is restricted, such as in the wine industry (Bower *et al.*, 1995; Nel *et al.*, 2002). The integration of a HACCP-style system of biofilm prevention in the food processing environment is clearly needed regardless of which approach is employed.

## 6.8 Sources of further information and advice

1. World Health Organization – Food Safety Program [http://www.who.int/foodsafety/en/index.html].
2. Food and Agriculture Organization of the United Nations – Food Safety and Quality [http://www.fao.org/ag/agn/agns/].
3. The European Food Information Council: provides information on food safety and quality, and health and nutrition for a balanced diet and healthy lifestyle [http://www.eufic.org/].
4. United States Food and Drug Administration – Center for Food Safety and Applied Nutrition [http://www.cfsan.fda.gov/].
5. United States Department of Agriculture – National Agricultural Library [http://www.nal.usda.gov/].
6. Centers for Disease Control and Prevention [http://www.cdc.gov].
7. National Food Safety Database [http://www.foodsafety.org].
8. Institute of Food Technologists [http://www.ift.org].
9. Canadian Food Safety Network [http://www.foodsafetynetwork.ca/en/].
10. European Commission Research Project 'Tradisausage' that evaluates and improves the safety of traditional dry sausages while preserving their typical qualities [http://www.clermont.inra.fr/tradisausage/].

## 6.9 References

ADAMS, J.L. and R.J.C. MCLEAN. 1999. Impact of rpoS deletion on *Escherichia coli* biofilms. *Appl Environ Microbiol* 65, 4285–4287.

ALLISON, D.G., B. RUIZ, C. SANJOSE, A. JASPE, and P. GILBERT. 1998. Extracellular products as mediators of the formation and detachment of *Pseudomonas fluorescens* biofilms. *FEMS Microbiol Lett* 167, 179–184.

AMMOR, S., G. TAUVERON, E. DUFOUR, and I. CHEVALLIER. 2006. Antibacterial activity of lactic acid bacteria against spoilage and pathogenic bacteria isolated from the same meat small-scale facility. 2 – Behaviour of pathogenic and spoilage bacteria in dual species biofilms including a bacteriocin-like-producing lactic acid bacteria. *Food Control* 17, 462–468.

AN, D., T. DANHORN, C. FUQUA, and M.R. PARSEK. 2006. Quorum sensing and motility mediate interactions between *Pseudomonas aeruginosa* and *Agrobacterium tumefaciens* in biofilm cocultures. *Proc Natl Acad Sci USA* 103, 3828–3833.

AUGER, S., E. KRIN, S. AYMERICH, and M. GOHAR. 2006. Autoinducer 2 affects biofilm formation by *Bacillus cereus*. *Appl Environ Microbiol* 72, 937–941.

BAGGE, D., M. HJELM, C. JOHANSEN, I. HUBER, and L. GRAM. 2001. *Shewanella putrefaciens* adhesion and biofilm formation on food processing surface. *Appl Environ Microbiol* 67, 2319–2325.

BELOIN, C. and J.-M. GHIGO. 2004. Finding gene-expression patterns in bacterial biofilms. *Trends in Microbiology* 13, 16–19.

BÖCKELMANN, U., A. JANKE, R. KUHN, T.R. NEU, J. WECKE, J.R. LAWRENCE, and U. SZEWZYK. 2006. Bacterial extracellular DNA forming a defined network-like structure. *FEMS Microbiol Lett* 262, 31–38.

BODDY, L. and J.W.T. WIMPENNY. 1992. Ecological concepts in food microbiology. *J Appl Bacteriol* 73, 23S–38S.

BOSKOU, G. and J. DEBEVERE. 1997. Reduction of trimethylamine oxide by *Shewanella* spp. under modified atmosphere in vitro. *Food Microbiol* 14, 543–553.

BOWER, C.K., J. MCGUIRE, and M.A. DAESCHEL. 1995. Influences on the antimicrobial activity of surface-adsorbed nisin. *J Ind Microbiol* 15, 227–233.

BOWER, C.K., J. MCGUIRE, and M.A. DAESCHEL. 1996. The adhesion and detachment of bacteria and spores on food contact surfaces. *Trends Food Sci Technol* 7, 152–157.

BOWMAN, J.P. 2001. '*Genus XIII.* Shewanella'. In: *Bergey's Manual of Systematic Bacteriology* (G.M. Garrity, Ed.), Springer, volume two, 480–491.

BRYERS, J.D. 1987. Biologically active surfaces, Processes governing the formation and persistence of biofilms. *Biotechnol Prog* 3, 57–68.

BUSSCHER, H.J., BELLON-FONTAINE, J. SJOLLEMA, and H.C. VAN DER MEI. 1990. Relative importance of surface free energy as a measure of hydrophobicity in bacterial adhesion to solid surfaces. In: *Microbial Cell Surface Hydrophobicity* (R.J. Doyle and M. Rosenberg, Eds.), American Society for Microbiology Press, Washington, DC, pp. 335–359.

CALDWELL, D.E. and J.R. LAWRENCE. 1989. Microbial growth and behaviour within surface microenvironments. In: Proceedings of ISME-5 (T. Hattori, Y. Ishida, Y. Maruyama, R.Y. Morita and A. Uchida, Eds.), JSS Press, Tokyo, Japan, pp. 140–145.

CALDWELL, D.E., G.M. WOLFAARDT, D.R. KORBER, and J.R. LAWRENCE. 1997. Do microbial communities transcend Darwinism? In: *Advances in Microbial Ecology* (J.G. Jones, Ed.), Vol. 15, Plenum Press, New York, pp. 105–191.

CARGILL, K.L., B.H. PYLE, R.L. SAUCER, and G.A. MCFETERS. 1991. Effects of culture conditions and biofilm formation on the iodine susceptibility of *Legionella pneumophila*. *Can J Microbiol* 38, 423–429.

CARPENTIER, B. and D. CHASSAING. 2004. Interactions in biofilms between *Listeria monocytogenes* and resident microorganisms from food industry premises. *Int J Food Microbiol* 97, 111–122.

CERI, H., M.E. OLSON, C. STREMICK, R.R. READ, D. MORCK, and A. BURET. 1999. The Calgary Biofilm Device: new technology for rapid determination of antibiotic susceptibilities of bacterial biofilms. *J Clin Microbiol* 37, 1771–1776.

CHMIELEWSKI, R.A.N. and J.F. FRANK. 2003. Biofilm formation and control in food processing facilities. *Comp Rev Food Sci Food Safety* 2, 22–32.

CHRISTENSEN, B.B., J.A.J. HAAGENSEN, A. HEYDORN, and S. MOLIN. 2002. Metabolic commensalisms and competition in a two-species microbial consortium. *Appl Environ Microbiol* 68, 2495–2502.

CHRISTIANSSON, A., J. BERTILSSON, and B. SVENSSON. 1999. *Bacillus cereus* spores in raw milk: factors affecting the contamination of milk during the grazing period. *J Dairy Sci* 82, 305–314.

COSTERTON, J.W., Z. LEWANDOWSKI, D.E. CALDWELL, D.R. KORBER, and H.M. LAPPIN-SCOTT. 1995. Microbial biofilms. *Annu Rev Microbiol* 49, 711–745.

CRAVEN, H.M. and B.J. MACAULY. 1992. Microorganisms in pasteurised milk after refrigerated storage 1. Identification of types. *Aust J Dairy Technol* 47, 38–45.

CRIELLY, E.M., N.A. LOGAN, and A. ANDERTON. 1994. Studies on the *Bacillus* flora of milk and milk products. *J Appl Bacteriol* 77, 256–263.

DAINTY, R.H. 1996. Chemical/biochemical detection of spoilage. *Int J Food Microbiol* 33, 19–33.

DAVIES, D.G., M.R. PARSEK, J.P. PEARSON, B.H. IGLEWSKI, J.W. COSTERTON, and E.P. GREENBERG. 1998. The involvement of cell-to-cell signals in the development of a bacterial biofilm. *Science* 280, 295–297.

DEBEVERE, G. and J. BOSKOU. 1996. Effect of modified atmosphere packaging on TVB/TMA producing microflora of cod fillets. *Int J Food Microbiol* 31, 221–229.

DECHO, A. 2000. Microbial biofilms in intertidal systems: an overview. *Continental Shelf Research* 20, 1257–1273.

DOYLE, R.J. 1999. *Biofilms: Methods in Enzymology*, Volume 310, Academic Press.

DROBNIEWSKI, F.A. 1993. *Bacillus cereus* and related species. *Clin Microbiol Rev* 6, 324–338.

EDWARDS, C.G., K.M. HAAG, M.D. COLLINS, R.A. HUTSON, and Y.C. HUANG. 1998. *Lactobacillus kunkeei* sp. nov.: a spoilage organism associated with grape juice fermentations. *J Appl Microbiol* 84, 698–702.

ENG, R.H.K., F.T. PADBERG, S.M. SMITH, E.N. TAN, and C.E. CHERUBIN. 1991. Bactericidal effects of antibiotics on slowly growing and nongrowing bacteria. *Antimicrob Agents Chemother* 35, 1824–1828.

EPA. 2002. Distribution System White Paper, Health risks from microbial growth and biofilms in drinking water distribution systems, US EPA, 52 pp.

FILTENBORG, O., J.C. FRISVAD, and U. THRANE. 1996. Moulds in food spoilage. *Int J Food Microbiol* 33, 85–102.

FINOL, M.L., E.H. MARTH, and R.C. LINDSAY. 1982. Depletion of sorbate from different media during growth of *Penicillium* species. *J Food Prot* 45, 398–404.

FLEET, G.H. 1999. Microorganisms in food ecosystems. *Int J Food Microbiol* 50, 101–117.

FOSTER, T.J. and M. HÖÖK. 1999. Surface protein adhesins of *Staphylococcus aureus*. *Trends in Microbiology* 6, 484–488.

FREDRICKSON, A.G. 1977. Behavior of mixed cultures of microorganisms. *Ann Rev Microbiol* 31, 63–87.

FUX, C.A., J.W. COSTERTON, P.S. STEWART, and P. STOODLEY. 2005. Survival strategies of infectious biofilms. *Trends in Microbiology* 13, 34–40.

GHANNOUM, M. and G. O'TOOLE. 2004. *Microbial Biofilms*, Wiley-VCH.

GIBSON, H., J.H. TAYLOR, K.E. HALL, and J.T. HOLAH. 1999. Effectiveness of cleaning techniques used in the food industry in terms of the removal of bacterial biofilms. *J Appl Microbiol* 87, 41–48.

GRAM, L. and H.H. HUSS. 1996. Microbial spoilage of fish and fish products. *Int J Food Microbiol* 33, 121–137.

GRAM, L., L. RAVN, M. RASCH, J.B. BRUHN, A.B. CHRISTENSEN, and M. GIVSKO. 2002. Food spoilage – interactions between food spoilage bacteria. *Int J Food Microbiol* 78, 79–97.

GUNDUZ, G.T. and G. TUNCEL. 2006. Biofilm formation in an ice cream plant. *Antonie van Leeuwenhoek* 89, 329–336.

HARRISON, J.J., R.J. TURNER, and H. CERI. 2005a. Persister cells, the biofilm matrix and tolerance to metal cations in biofilm and planktonic *Pseudomonas aeruginosa*. *Environ Microbiol* 7, 981–994.

HARRISON, J.J., R.J. TURNER, and H. CERI. 2005b. High-throughput metal susceptibility testing of microbial biofilms. *BMC Microbiol* 5, 53–64.

HENGGE-ARONIS, R. 2002. Signal transduction and regulatory mechanisms involved in control of the σS (RpoS) subunit of RNA polymerase. *Microbiol Mol Biol Rev* 66, 373–395.

HJELM, M., L.R. HILBERT, P. MØLLER, and L. GRAM. 2002. Comparison of adhesion of the food spoilage bacterium *Shewanella putrefaciens* to stainless steel and silver surfaces. *J Appl Microbiol* 92, 903–911.

HOUARI, A. and P. DI MARTINO. 2007. Effect of chlorhexidine and benzalkonium chloride on bacterial biofilm formation. *Lett Applied Microbiol* 1–5. doi:10.1111/j.1472-765X.2007.02249.x

IVANOVA, N., A. SOROKIN, I. ANDERSON, N. GALLERON, B. CADELON, V. KAPATRAL, A. BHATTACHARYYA, G. REZNIK, N. MIKHAILOVA, A. LAPIDUS, L. CHU, M. MAZUR, E. GOLTSMAN, N. LARSEN, M. D'SOUZA, T. WALUNAS, Y. GRECHKIN, G. PUSCH, R. HASELKORN, M. FONSTEIN, S.D. EHRLICH, R. OVERBEEK, and N. KYRPIDES. 2003. Genome sequence of *Bacillus cereus* and comparative analysis with *Bacillus anthracis*. *Nature* 423, 87–91.

JESSEN, B. and L. LAMMERT. 2003. Biofilm and disinfection in meat processing plants. *Int Biodeterior Biodegrad* 51, 265–269.

JOS, H.J. and V. HUIS. 1996. Microbial and biochemical spoilage of foods. *Int J Food Microbiol* 33, 1–18.

JØRGENSEN, L.V., H.H. HUSS, and P. DALGAARD. 2000. The effect of biogenic amine production by single bacterial cultures and metabiosis on cold-smoked salmon. *J Appl Microbiol* 89, pp. 920–934.

KORBER, D.R. and J.R. LAWRENCE. 2004. Biofilm Formation. In: *Encyclopedia of Meat Sciences* (W.K. Jensen, C. Devine and M. Dikeman, Eds.), Vol. 1, Elsevier, Oxford, pp. 59–67.

KORBER, D.R., J.R. LAWRENCE, B. SUTTON, and D.E. CALDWELL. 1989. Effect of laminar flow velocity on the kinetics of surface recolonization by mot$^+$ and mot$^-$ *Pseudomonas fluorescens*. *Microb Ecol* 18, 1–19.

KORBER, D.R., G.A. JAMES, and J.W. COSTERTON. 1994. Evaluation of fleroxacin activity against established *Pseudomonas fluorescens* biofilms. *Appl Environ Microbiol* 60, 1663–1669.

KORBER, D.R., J.R. LAWRENCE, H.M. LAPPIN-SCOTT, and J.W. COSTERTON. 1995. Growth of microorganisms on surfaces. In: *Microbial Biofilms* (H.M. Lappin-Scott and J.W. Costerton, Eds.), Cambridge University Press, Cambridge, UK, pp. 15–45.

KORBER, D.R., A. CHOI, G.M. WOLFAARDT, S.C. INGHAM, and D.E. CALDWELL. 1997. Surface heterogeneities confer resistance to *Salmonella enteritidis* biofilms treated with trisodium phosphate. *Appl Environ Microbiol* 63, 3352–3358.

KORBER, D.R., G.G. GREER, G.M. WOLFAARDT, and S. KOHLMAN. 2002. Efficacy enhancement of trisodium phosphate against spoilage and pathogenic bacteria in model biofilms and on adipose tissue. *J Food Prot* 65, 627–635.

KRAMER, J.M. and R.J. GILBERT. 1989. *Bacillus cereus* and other *Bacillus* species. In: *Food borne bacterial pathogens* (M.P. Doyle, Ed.), Marcel Dekker, New York, pp. 21–70.

KUMAR, C.G. and S.K. ANAND. 1998. Significance of microbial biofilms in food industry: a review. *Int J Food Microbiol* 42, 9–27.

LANGE, R. and R. HENNGE-ARONIS. 1991. Identification of a central regulator of stationary phase gene expression in *E. coli*. *Mol Microbiol* 5, 49–59.

LANGILLE, S.E. and R.M. WEINER. 1988. Spatial and temporal deposition of *Hyphomonas* strain VP-6 capsules involved in biofilm formation. *Appl Environ Microbiol* 64, 2906–2913.

LAWRENCE, J.R., G.M. WOLFAARDT, and D.R. KORBER. 1994. Monitoring diffusion in biofilm matrices using confocal laser microscopy. *Appl Environ Microbiol* 60, 1166–1173.

LAWRENCE, J.R., D.R. KORBER, G.M. WOLFAARDT, and D.E. CALDWELL. 1996. Surface colonization strategies of biofilm-forming bacteria. In: *Advances in Microbial Ecology* (J.G. Jones, Ed.), Vol. 14, Plenum Press, New York, pp. 1–75.

LEBERT, I., S. LEROY, and R. TALON. 2007. Effect of industrial and natural biocides on spoilage, pathogenic and technological strains grown in biofilm. *Food Microbiol* 24, 281–287.

LEWIS, S.J. and A. GILMOUR. 1987. Microflora associated with the internal surfaces of rubber and stainless steel milk transfer pipeline. *J Appl Bacteriol* 62, 327–333.

MAH, T.-F. and G.A. O'TOOLE. 2001. Mechanisms of biofilm resistance to antimicrobial agents. *Trends Microbiol* 9, 34–39.

MAKIN, S.A. and T.J. BEVERIDGE. 1996. The influence of A-band and B-band lipopolysaccharide on the surface characteristics and adhesion of *Pseudomonas aeruginosa* to surfaces. *Microbiology* 142, 299–307.

MANGALAPPALLI-ILLATHU, A.K. and D.R. KORBER. 2006. Adaptive resistance and differential protein expression of *Salmonella enterica* serovar Enteritidis biofilms exposed to benzalkonium chloride. *Antimicrob Agents Chemother* 50, 3588–3596.

MANGALAPPALLI-ILLATHU, A.K., J.R. LAWRENCE, G.D.W. SWERHONE, and D.R. KORBER. 2008a. Architectural adaptation and protein expression patterns of *Salmonella enterica* serovar Enteritidis biofilms under laminar flow conditions. *Int J Food Microbiol* 123, 109–120.

MANGALAPPALLI-ILLATHU, A.K., S. VIDOVIĆ, and D.R. KORBER. 2008b. Differential Adaptive Response and Survival of *Salmonella enterica* Serovar Enteritidis Planktonic and Biofilm Cells Exposed to Benzalkonium Chloride. *Antimicrob Agents Chemother* 52, 3669–2680.

MANN, D.A. and L.R. BEUCHAT. 2008. Combination of antimycotics to inhibit the growth of molds capable of producing 1,3-pentadiene. *Food Microbiol* 25, 144–153.

MARTH, E.M., C.M. CAPP, L. HASENZAHL, H.W. JACKSON, and R.V. HUSSONG. 1966. Degradation of potassium sorbate by *Penicillium* species. *J Dairy Sci* 49, 1197–1205.

MATTILA-SANDHOLM, T. and G. WIRTANEN. 1992. Biofilm formation in the industry: a review. *Food Rev Int* 8, 573–603.

MCDONNELL, G. and A.D. RUSSELL. 1999. Antiseptics and disinfectants: activity, action, and resistance. *Clin Microbiol Rev* 12, 147–179.

MICHIELS, C.W., M. SCHELLEKENS, C.C. SOONTJENS, and K.J.A. HAUBEN. 1997. Molecular and metabolic typing of resident and transient fluorescent pseudomonad flora from a meat mincer. *J Food Prot* 60, 1515–1519.

MILLER, M.W. 1979. Yeasts in food spoilage: an update. *Food Technol.*, 33, 76–80.

MITTELMAN, M.W. 1998. Structure and functional characteristics of bacterial biofilms in fluid processing operations. *J Dairy Sci* 81, 2760–2764.

MOSTELLER, T.M. and J.R. BISHOP. 1993. Sanitizer efficacy against attached bacteria in a milk biofilm. *J Food Prot* 56, 34–41.

NEL, H.A., R. BAUER, G.M. WOLFAARDT, and L.M.T. DICKS. 2002. Effect of bacteriocins Pediocin PD-1, Plantaricin 423, and nisin on biofilms of *Oenococcus oeni* on a stainless steel surface. *Am J Enol Vitic* 53, 191–196.

OOSTHUIZEN, M.C., B. STEYN, J. THERON, P. COSETTE, D. LINDSAY, A. VON HOLY, and V.S. BROZEL. 2002. Proteomic analysis reveals differential protein expression by *Bacillus cereus* during biofilm formation. *Appl Environ Microbiol* 68, 2770–2780.

PALLERONI, N.J. 2001. '*Genus I.* Pseudomonas'. In: *Bergey's Manual of Systematic Bacteriology* (G.M, Garrity, Ed.), Vol. 2, Springer, pp. 323–379.

PATRIQUIN, G.M., E. BANIN, C. GILMOUR, R. TUCHMAN, E.P. GREENBERG, and K. POOLE. 2008. Influence of quorum sensing and iron on twitching motility and biofilm formation in *Pseudomonas aeruginosa*. *J Bacteriol* Jan, 662–671.

PENG, J.S., W.C. TSAI, and C.C. CHOU. 2001. Surface characteristics of *Bacillus cereus* and its adhesion to stainless steel. *Int J Food Microbiol* 65, 105–111.

PENG, J.S., W.C. TSAI, and C.C. CHOU. 2002. Inactivation and removal of *Bacillus cereus* by sanitizer and detergent. *Int J Food Microbiol* 77, 11–18.

PRIEST, F.G., M. BARKER, L.W. BAILLIE, E.C. HOLMES, and M.C.J. MAIDEN. 2004. Population structure and evolution of the *Bacillus cereus* group. *J Bacteriol* 186, 7959–7970.

PRINGENT-COMBARET, C., O. VIDAL, C. DOREL, and P. LEJEUNE. 1999. Abiotic surface sensing and biofilm-dependent regulation of gene expression in *Escherichia coli. J Bacteriol* 181, 5993–6002.

RAO, D., J.S. WEBB, and S. KJELLEBERG. 2005. Competitive interactions in mixed-species biofilms containing the marine bacterium *Pseudoaltermonas tunicata. Appl Environ Microbiol* 71, 1729–1736.

RASKO, D.A., M.R. ALTHERR, C.S. HAN, and J. RAVEL. 2005. Genomics of the *Bacillus cereus* group of organisms. *J FEMS Microbiol Rev* 29, 303–329.

RAVN, L., A.B. CHRISTENSEN, S. MOLIN, M. GIVSKOV, and L. GRAM. 2001. Methods for identifying and quantifying acylated homoserine lactones produced by Gram-negative bacteria and their application in studies of AHL-production kinetics. *J Microbiol Meth* 44, 239–251.

REGEZ, P., L. GALLO, R.E. SCHMITT, and W. SCHMIDT-LORENZ. 1988. Microbial spoilage of refrigerated fresh broilers. III. Effect of storage temperature on the microbial association of poultry carcasses. *Lebensm Wiss Technol* 21, 229–233.

REN, D., L.A. BEDZYK, S.M. THOMAS, R.W. YE, and T.K. WOOD. 2004. Gene expression in *Escherichia coli* biofilms. *Appl Microbiol Biotechnol* 64, 515–524.

RUSSELL, A.D. 2003. Biocide use and antibiotic resistance: the relevance of laboratory findings to clinical and environmental situations. *Lancet Infectious Diseases* 3, 794–803.

SAMPATHKUMAR, B., G.G. KHACHATOURIANS, and D.R. KORBER. 2004. Food processing biofilms and antimicrobial agents. In: *Handbook of Food Science* (Y.H. Hui Ed.), Marcel Dekker, Inc., New York, Switzerland, pp. 84-1–84-12.

SANSEBASTIANO, G., R. ZONI, and L. BIGLIARDI. 2006. Cleaning and disinfection procedures in the food industry: General aspects and practical applications. In: *Food safety: A practical and case study approach* (A. McElhatton and R.J. Marshall, Eds.), Springer, US, pp. 253–280.

SAUER, K. 2003. The genomics and proteomics of biofilm formation. *Genome Biol* 4, 219.

SAUER, K. and A.K. CAMPER. 2001. Characterization of phenotypic changes in *Pseudomonas putida* in response to surface-associated growth. *J Bacteriol* 183, 6579–6589.

SAUER, K., A.K. CAMPER, G.D. EHRLICH, J.W. COSTERTON, and D.G. DAVIES. 2002. *Pseudomonas aeruginosa* displays multiple phenotypes during development as a biofilm. *J Bacteriol* 184, 1140–1154.

SCHOENI, J.L. and A.C. WONG. 2005. *Bacillus cereus* food poisoning and its toxins. *J Food Prot* 68, 636–648.

SEMPLE, K.M., J.L. DORAN, and D.S. WESTLAKE. 1989. DNA relatedness of oil-field isolates *Shewanella putrefaciens. Can J Microbiol* 35, 925–931.

SHINAGAWA, K. 1990. Analytical methods for *Bacillus cereus* and other *Bacillus* species. *Int J Food Microbiol* 10, 125–141.

SINGH, P.K., M.R. PARSEK, E.P. GREENGERG, and M.J. WELSH. 2002. A component of innate immunity prevents bacterial biofilm development. *Nature* 417, 552–555.

SJOLLEMA, J., H.J. BUSSCHER, and A.H. WEERKAMP. 1988. Deposition of oral streptococci and polystyrene latices onto glass in a parallel plate flow cell. *Biofouling* 1, 101–112.

SJOLLEMA, J., H.C. VAN DER MEI, H.M. UYEN, and H.J. BUSSCHER. 1990. Direct observations of cooperative effects in oral streptococcal adhesion to glass by analysis of the spatial arrangement of adhering bacteria. *FEMS Microbiol Lett* 69, 263–270.

SPOERING, A.L. and K. LEWIS. 2001. Biofilms and planktonic cells of *Pseudomonas aeruginosa* have similar resistance to killing by antimicrobials. *J Bacteriol* 183, 6746–6751.

STARKEY, M., K.A. GRAY, S.I. CHANGE, and M.R. PARSEK. 2004. A sticky business: The extracellular polymeric substance matrix of bacterial biofilms. In: *Microbial biofilms* (M. Ghannoum and G.A. O'Toole, Eds.), ASM Press, pp. 174–191.

STENSTROM, M. and G. MOLIN. 1990. Classification of the spoilage flora of fish, with special references to *Shewanella putrefaciens*. *J Appl Bacterol* 68, 601–618.

STEWART, P.S., P.K. MUKHERJEE, and M.A. GHANNOUM. 2004. Biofilm antimicrobial resistance. In: *Microbial Biofilms* (M. Ghannoum and G.A. O'Toole, Eds.), ASM Press, Washington, DC, pp. 250–268.

SUCI, P.A., M.W. MITTELMAN, F.P. YU, and G.G. GEESEY. 1994. Investigation of ciprofloxacin penetration into *Pseudomonas aeruginosa* biofilms. *Antimicrob Agents Chemother* 38, 2125–2133.

SUFYA, N., D.G. ALLISON, and P. GILBERT. 2003. Clonal variation in maximum specific growth rate and susceptibility towards antimicrobials. *J Appl Microbiol* 95, 1261–1267.

TAJCHAKAVIT, S., H.S. RAMASWAMY, and P. FUSTIER. 1998. Enhanced destruction of spoilage organisms in apple juice during continuous flow microwave heating. *Food Res Int* 31, 713–722.

TERNSTRÖM, A., A.-M. LINDBERG, and G. MOLIN. 1993. Classification of the spoilage flora of raw and pasteurized bovine milk, with special reference to Pseudomonas and Bacillus. *J Appl Bacteriol* 75, 25–34.

VAN LOOSDRECHT, M.C.M., W. NORDE, and A.J.B. ZEHNDER. 1987. Influence of cell surface characteristics on bacterial adhesion to solid surfaces. *Proceeding of the 4th European Congress on Biotechnology* 4, 575–580.

VILAIN, S. and V.S. BROZEL. 2006. Multivariate approach to comparing whole-cell proteomes of *Bacillus cereus* indicates a biofilm-specific proteome. *J Proteome Res* 5, 1924–1930.

VILAIN, S., Y. LUO, M.B. HILDRETH, and V.S. BRÖZEL. 2006. Analysis of the life cycle of the soil saprophyte *Bacillus cereus* in liquid soil extract and in soil. *Appl Environ Microbiol* 72, 4970–4977.

VOGEL, B.F., K. JORGENSEN, H. CHRISTENSEN, J.E. OLSEN, and L. GRAM. 1997. Differentiation of *Shewanella putrefaciens* and *Shewanella alga* on the basis of whole-cell protein profiles, ribotyping, phenotypic characterization, and 16S rRNA gene sequence analysis. *Appl Environ Microbiol* 63, 2189–2199.

VOLK, W.A., B.M. GEBHARDT, M.L. HAMMARSKJÖLD, and R.J. KADNER. 1996. Antibiotic action and resistance: bacterial cell surfaces; and mechanisms of antibiotic action. In: *Essentials of Medical Microbiology*, 5th ed, Lippincott-Raven, Philadelphia, PA, pp. 253–265.

WHAN, L., G. DUNSTALL, and M.T. ROWE. 2000. A study of the growth kinetics of two pseudomonads from pasteurized milk and the possible roles of quorum sensing. *Milchwissenschaft* 55, 371–373.

WILLIAMS, I., W.A. VENABLES, D. LLOYD, F. PAUL, and I. CRITCHLEY. 1997. The effects of adherence to silicon surfaces on antibiotic susceptibility in *Staphylococcus aureus*. *Microbiology* 143, 2407–2413.

WILLIAMS, V. and M. FLETCHER. 1996. *Pseudomonas fluorescens* adhesion and transport through porous media are affected by lipopolysaccharide composition. *Appl Environ Microbiol* 62, 100–104.

WINGENDER, J., T.R. NEU, and H.-C. FLEMMING. 1999. Function of EPS. In: *Microbial Exopolymeric Substances* (J. Wingender, T.R. Neu, and H.-C. Flemming, Eds.), Springer-Verlag, Berlin. pp. 1–19.

WIRTANEN, G. and T. MATTILA-SANDHOLM. 1992. Effect of the growth phase of foodborne biofilms on their resistance to a chlorine sanitizer. Part II. *Lebensm Wiss U Technol* 25, 50–54.

WOLFAARDT, G.M., J.R. LAWRENCE, and D.R. KORBER. 1999. Function of EPS. In: *Microbial Exopolymeric Substances* (J. Wingender, T.R. Neu, and H.-C. Flemming, Eds.), Springer-Verlag, Berlin, pp. 171–200.

WOLFAARDT, G.M., J.R. LAWRENCE, and D.R. KORBER. 2007. Cultivation of Microbial Communities. In: *Manual of Environmental Microbiology*, (3rd Edition), ASM Press, Washington, D.C. pp. 101–111.

XU, K.D., M.D. FRANKLIN, C.-H. PARK, G.A. MCFETERS, and P.S. STEWART. 2001. Gene expression and protein levels of the stationary phase sigma factor, RpoS, in continuously fed *Pseudomonas aeruginosa* biofilms. *FEMS Microbiol Lett* 199, 67–71.

YU, F.P., B.H. PYLE, and G.A. MCFETERS. 1993. A direct viable count method for the enumeration of attached bacteria and assessment of biofilm disinfection. *J Microbiol Meth* 17, 167–180.

# 7

# Biofilm formation by *Listeria monocytogenes* and transfer to foods

**A. Rodríguez-Lozano, Campden BRI, UK and L. McLandsborough, University of Massachusetts, USA**

**Abstract**: It is generally thought that biofilms within food processing environments are a major source of *L. monocytogenes.* In processing plans, common sites of post-processing *L. monocytogenes* contamination are filling or packaging equipment, conveyors, collators used for assembling product for packaging, racks for transporting product, hand tools, gloves, and freezers. This review discusses the basic mechanisms of adhesion and biofilm formation of *L. monocytogenes* as defined from genetics and proteomic studies and the influence of lineages upon biofilm formation. In addition, the current state of knowledge of the parameters which affect bacterial transfer from food processing surfaces to foods are discussed.

**Key words**: biofilm, *Listeria*, *L. monocytogenes*, transfer.

## 7.1 Introduction

*Listeria monocytogenes* is ubiquitous in the environment and resistant to diverse environmental conditions. It is associated with plant material (alive or dead), water, and soil. Listeriosis is a rare foodborne disease, although exposure to *Listeria* is probably very common due to the wide environmental distribution on green plant material (some of which is consumed as food) (Swaminathan *et al.* 2007). It is the severity of the non-enteric disease (meningitis, septicemia and abortion) in susceptible individuals (such as immunocompromised or pregnant women) that warrants the alarm concerning this organism (Swaminathan *et al.* 2007). The very high mortality of listeriosis (20–30%) (Swaminathan *et al.* 2007) has prompted the United States to adopt a zero tolerance policy in processed foods. *L. monocytogenes* was the most frequent microbial agent associated with FDA food product recalls in the years 1994–1998 (61% of all food product recalls and 88% of all Class I), with the next most frequent recall being due to *Salmonella* (11%) (Wong *et al.* 2002). Close to 75% of the listerial recalls by the FDA occurred in dairy products, pastries, salads or sandwiches (Wong *et al.* 2002). Other food products associated with outbreaks include hot dog and ready-to-eat meats (Anonymous 2002; Anonymous 2003).

It is generally thought that biofilms within food processing environments are a major source of *L. monocytogenes* (Poulsen 1999; Tompkin *et al.* 1999; Reij *et al.* 2004). The recent FSIS rule, that required that all establishments producing ready-to-eat meat and poultry products to re-formulate products to prevent listerial growth and/or conduct environmental testing of food contact surfaces for *Listeria* sp (Anonymous 2003), acknowledges that transfer from the food processing environment is critical in the journey of this organism into processed foods after a lethal heat treatment. In processing plans, common sites of post-processing *L. monocytogenes* contamination are filling or packaging equipment, conveyors, collators used for assembling product for packaging, racks for transporting product, hand tools, gloves, and freezers (Poulsen 1999; Tompkin *et al.* 1999). One recall from a baking company involved 128 different products after it was discovered that the machine used to cut the items was contaminated with *L. monocytogenes* (Wong *et al.* 2002). Recently, there was a large outbreak in Canada with 56 confirmed cases and 20 deaths, and has been contributed to contamination of ready-to-eat deli meats by a contaminated slicer (Anonymous 2008). The presence and persistence of this organism within the food processing environment continues to be a major challenge for the food industry (Moretro *et al.* 2004).

## 7.2 Physiology of bacteria growing in biofilms

Researchers now recognize that in their natural environments, bacteria do not exist as isolated cells but grow and survive in organized communities known as biofilms (Jenkinson *et al.* 2001). Simplistically, biofilms are microorganisms growing on a solid surface. However, biofilms are generally defined as matrix-enclosed bacterial populations that adhere to a surface and/or to each other producing a dynamic environment in which the component microbial cells appear to reach homeostasis, optimally organized to make use of all available nutrients (An *et al.* 1997; Poulsen 1999; Doyle 2001; Sutherland 2001; O'Toole *et al.* 2002). Once colonizing the surface, microorganisms form a monolayer or multilayer of cells at phase interfaces. Interfaces where biofilms may grow in food processing environments include solid/liquid, gas/liquid, or in the case of solid foods gas/solid interfaces (Poulsen 1999; Jenkinson *et al.* 2001). Throughout natural ecosystems, biofilms can be found on almost any surface with a high enough level of moisture to support growth (Kim *et al.* 1995). Biofilms are formed by almost every type of microorganism under suitable conditions, including spoilage microorganisms such as *Pseudomonas* and pathogens of great concern to the food industry including the genera of *Bacillus*, *Vibrio*, *Listeria*, *Escherichia* and *Salmonella* (Marriott 1999; Poulsen 1999; Sommer *et al.* 1999).

When growing in a biofilm, bacteria are known to have a different rate of growth, cellular morphology, and physiology than their planktonic counterparts and may exhibit varied physiological responses to nutrient conditions (Hodgson *et al.* 1995; Kim *et al.* 1995; O'Toole *et al.* 1998; Kuchma *et al.* 2000; Sauer *et al.* 2002). Investigations of diffusional transport of gaseous and liquid components through the biofilm matrix have indicated that biofilm bacteria receive less oxygen and fewer nutrients than cells in suspension. Surprisingly, this leads to advantages in growth, altered physiology, and increased resistance to toxic agents compared to their planktonic forms (Frank *et al.* 1997; Vatanyoopaisarn *et al.* 2000; Sutherland 2001). The fundamental alterations of diffusional mass transport processes and biophysical interactions with components present in the neighboring aqueous phase seem to allow commensal and mutual communities of organisms to survive low nutrient and decreased temperature conditions that are often found in food processing and storage environments. The ability to resist antimicrobial agents is of particular concern to both the medical and food processing communities, since once a biofilm has been established on a surface, it becomes exceedingly difficult to completely remediate the film (Bolton *et al.* 1988; LeChevallier *et al.* 1988; Frank *et al.* 1990a; 1990b; Nickel *et al.* 1992; Reid *et al.* 1993; Bower *et al.* 1999; Donlan 2002; Donlan *et al.* 2002; Dunne 2002).

Simplistically, biofilms are composed of microbial cells that in the latter stages of the growth cycle are embedded in an exopolymer matrix (van Loosdrecht *et al.* 1995). However, it should be clearly stated that biofilms do not possess a uniform structure. For example, Wimpenny and Colasanti recognized that biofilms exist that are porous and that are non-porous (Wimpenny *et al.* 2000). The structures that are formed depend on a large variety of intrinsic and extrinsic parameters such as species, temperature, flow conditions, pH, presence of salts, etc.

## 7.3 Biofilm formation and propagation

There are several steps in the formation of bacterial biofilms: (i) transport, (ii) initial adhesion, (iii) substrate attachment, and (iv) micro-colony formation (cell-cell adhesion) leading to mature biofilms consisting of cells and a surrounding exopolymer matrix (van Loosdrecht *et al.* 1995; O'Toole *et al.* 2002; Purevdorj-Gage *et al.* 2004). The first step in biofilm formation consists of the transport of the organism to a solid surface. This can occur via motility of the organism, diffusion of the organism through the environment or natural or forced convection in the system. Biofilm-forming bacteria may use all of these mechanisms at one time or another. It is well documented that flagella mutants often have lower biofilm production under static conditions indicating that under these conditions flagella are involved in active cellular transport to surfaces and this has also been observed in *Listeria* biofilm formation (Vatanyoopaisarn *et al.* 2000; Lemon *et al.* 2007). The role of flow conditions on the attachment and growth of cells was investigated by various authors (Sasahara *et al.* 1993; Millsap *et al.* 1996; Pereira *et al.* 2002). Contrary to expectations, greater deposition of bacteria under both laminar and turbulent flow conditions has been observed when compared to static conditions (Rijnaarts *et al.* 1993). It has been speculated that turbulent flow may thrust bacterial cells onto the surface, thus enhancing probability of adhesion and biofilm formation (Donlan *et al.* 2002).

Once the organisms approach the surface, physical interaction forces are thought to influence the initial adhesion of the organisms. Typical interactions that can take place include Van der Waals interactions (>50 nm from the surface), repulsive or attractive electrostatic interactions (2–10 nm from the surface) and hydrophobic interactions (0.5–2 nm from the surface) (Fletcher 1996). Van der Waals forces are due to dipole–dipole, induced dipole–dipole and induced dipole–induced dipole interactions and are always attractive (Israelachvili 1992). Electrostatic interactions arise because the cells, the developing exopolymer matrix and the surface may carry a permanent positive or negative charge leading to the formation of a diffuse electrostatic double layer. Bacteria, as well as most natural solid surfaces, generally have an overall gross negative charge but the origin of the overall charge is due to the combination of various charges from functional groups on the membrane constituent molecules, such as amino, carboxyl, phosphate and less commonly, sulfate groups and capsular macromolecules (James 1991). Ultimately, the magnitude of the electrostatic interactions is influenced by the nature of the environment, e.g. pH, ionic strength, valency of present counterions and nature of the solvent (Israelachvili 1992).

Simplistically, the Derjaguin, Landau, Verwey and Overbeek (DLVO) theory has been suggested as a first approximation to describe the interaction of bacteria with a solid surface as a function of the separation distance between the two systems. The DLVO theory assumes that the overall interaction is a sum of the attractive Van der Waals interaction and electrostatic repulsive interactions (Olivera 1992; McClements 1999). Initially, Van der Waals attractions are thought to dominate the overall interaction leading to a reduction in the separation distance between bacteria and the substrate surface. However, as bacterial cells move closer to the substrate surface, repulsive electrostatic interactions may create an energy barrier that must be overcome before the two interacting bodies can come into close contact required for bacterial attachment (McClements 1999). Up to this point, the adhesion may be reversible, in particular if the minimum in the interaction potential is smaller than the thermal energy of the bacterial cells. The occurrence of irreversible adhesion is not taken into account in the DVLO theory, and this is thought to occur through hydrophobic interactions. Hydrophobic interactions in water are much stronger than Van der Waals attraction at small separation distances (Israelachvili 1992), and it has been suggested that hydrophobic interactions between the cell surface and the solid substrate may be responsible for overcoming the repulsive electrostatic interactions. The origin of the hydrophobic interactions is thought to be enthalpically driven, that is the exclusion of water molecules between the two surfaces and the direct interaction of the bacterial and substrate surface is believed to lower the overall free energy of the system. As a result, bacterial cells adhere irreversibly (Busscher *et al.* 1987; 1990). This strict physicochemical approach, however, should not be over interpreted. The bacterial surface is an extremely complex entity and contains a multitude of molecules that not only carry a variety of charges but are also more or less hydrophobic. In addition, the nature and composition of bacterial surfaces can vary greatly between different species. The fact that a single bacterial strain can adhere to a variety of surfaces with differing surface energies indicates that this simplified physicochemical interaction model is most likely not entirely correct. Strategies that attempted to prevent bacterial attachment by engineering the surface to be more or less hydrophobic have not led to the desired results. A large variety of bacterial cells have no difficulty attaching to both hydrophobic or hydrophilic surfaces (Fletcher 1996).

## 7.4 Biofilm formation by *Listeria monocytogenes*

Many researchers have studied the mechanisms of listerial adhesion and growth on surfaces, although some have questioned whether or not surface growth of *Listeria monocytogenes* is technically a biofilm (Hood *et al.* 1997; Kalmokoff *et al.* 2001; Donlan *et al.* 2002). This is based upon the lack of large increases in cell numbers after initial adhesion (Hood *et al.* 1997) and the heterogeneity of biofilm production, since only one strain out of thirty-six tested produced a basic biofilm under high nutrient conditions (Kalmokoff *et al.* 2001). Dolan questioned whether this organism could grow within a biofilm due to its fastidious growth requirements and inability to compete with other microflora. (Donlan *et al.* 2002), thus bringing forth the argument that this organism may adhere to other organisms in the environment rather than produce an independent biofilm (Kalmokoff *et al.* 2001). It is known that *Listeria monocytogenes* has the ability to survive within the food processing environments with many other flora over many years (Tompkin 2002).

So does *Listeria monocytogenes* produce biofilms? In pure laboratory systems, *L. monocytogenes* has been observed to produce fibrils and extracellular material production when attached to a variety of surfaces (Herald *et al.* 1988; Mafu *et al.* 1990; Kalmokoff *et al.* 2001; Marsh *et al.* 2003). The production of extracellular polymeric substances with carbohydrate components has been observed (Borucki *et al.* 2003; Hassan *et al.* 2004; Hefford *et al.* 2005; Chae *et al.* 2006; Rodriguez *et al.* 2008a). The surface growth has been shown to have a distinct three-dimensional structure in the form a honeycomb or groups of aggregates surrounded by voids (Chavant *et al.* 2002; Djordjevic *et al.* 2002; Marsh *et al.* 2003). In addition, the biofilm thickness can vary, with lower cell levels in the center of the biofilm and higher cell densities at the edges (Chae *et al.* 2000). It has been observed that the growth rate of *Listeria sp* in biofilms was slower that planktonic cells (Chae *et al.* 2000), required a greater levels of phosphate (Kim *et al.* 1995) and produce a stringent response (Taylor *et al.* 2002). In addition, cell-to-cell communication has been shown to be involved in initial adhesion of *L. monocytogenes* to surfaces using the *agr* system (Rieu *et al.* 2007). This indicates that in laboratory systems, *L. monocytogenes* has exhibited the three-benchmark biofilm characteristics: extracellular material, three-dimensional structure and altered cellular physiology.

In most foods and in food processing environments, the biofilms present will be very complex, i.e. they may be composed of multiple species that form a community of microorganisms, and that may or may not behave in a similar manner as pure laboratory systems. Similarly, studies of biofilms under laboratory conditions usually do not take into account that biofilms in food processing environments may contain proteins and fats derived from improperly cleaned or sanitized processing areas as integral part of their structure. Experimental evidence suggests that the formation of a multispecies biofilm is advantageous. For example, higher numbers of *L. monocytogenes* were counted in biofilms that were co-cultured with *Pseudomonas* (Sasahara *et al.* 1993; Hassan *et al.* 2004). Others, have observed that when *Listeria innocua* was co-cultured with *Pseudomonas aeruginosa* in a bioreactor, the levels of *L. innocua* were two log higher at day 4 than when grown in monoculture (Bourion *et al.* 1996). In other cases, the adhesion of *Listeria monocytogenes* is limited in the presence of other bacteria (Leriche *et al.* 2000; Norwood *et al.* 2001). Thus being said, the majority of basic knowledge of biofilm production by *L. monocytogenes* was obtained in the laboratory using pure culture and will be discussed in more depth below.

### 7.4.1 Attachment of *L. monocytogenes* to surfaces

Research in the attachment by pure cultures of *L. monocytogenes* has focused on the influence of environmental conditions such as growth temperature, pH and media composition (Herald *et al.* 1988; Kim *et al.* 1994; 1995; Hood *et al.* 1997), physiochemical forces of adhesion and the influence of absorbed food components (or 'conditioning layer') in listerial adhesion (Al-Makhlafi *et al.* 1994; 1995). In general, *L. monocytogenes* attach better when grown in minimal or diluted rich media (Kim *et al.* 1994; Djordjevic *et al.* 2002; Moltz *et al.* 2005). The increased production of flagella and extracellular matrix components were observed at lower incubation temperatures (20°C or 4°C), although this may be due to increased incubation times required at lower temperatures (Herald *et al.* 1988; Mafu *et al.* 1990). Neutral pH (7.0) appears to have greater attachment in comparison to basic or acidic media (Herald *et al.* 1988; Smoot *et al.* 1991).

Several different researchers have studied the involvement of physiochemical forces in adhesion. *L. monocytogenes* has been shown to attach with varying surface free energies to food processing surfaces such as stainless steel, glass, polypropylene and rubber (Herald *et al.* 1988; Mafu *et al.* 1990; Smoot *et al.* 1991). The surface charge of *L. monocytogenes* cells has been observed to be dependent upon growth temperature: as growth temperature decreased to 8°C, the electrophoretic mobility of the cells became less negative (Braindet *et al.* 1999). In addition, as the ionic strength of the suspending media decreased, the numbers of adherent *L. monocytogenes* ScottA cells increased indicating that electrical and Lewis acid base interactions are involved in adhesion to stainless steel (Braindet *et al.* 1999). Others have measured cell surface hydrophobicity and calculated interfacial free energy of adhesion to stainless steel and rubber, and found these parameters alone did not explain the attachment capabilities of *L. monocytogenes* (Mafu *et al.* 1991; Smoot *et al.* 1991), suggesting another mechanism such as production of exopolymer or protein may be important for adhesion (Mafu *et al.* 1991; Smoot *et al.* 1991). This is further supported by the observation that the presence of the proteinase trypsin could reduce listerial adhesion to rubber and stainless steel by 99.9% (Smoot *et al.* 1991). The presence of adsorbed proteins to surfaces generally reduced the adhesion of *L. monocytogenes* regardless of the surface composition or free energy (Al-Makhlafi *et al.* 1994; 1995; Barnes *et al.* 1999; Cunliffe *et al.* 1999).

Rough colony types of *L. monocytogenes* have been described from both food and clinical isolates (Gutekunst *et al.* 1992; Rowan *et al.* 2000), and it has been proposed that this phenotype is due to the lack of petidoglycan hydrolase secretion (Monk *et al.* 2004). Recently it was reported that rough colony types can be due to the presence of one of two types of filamentous cells, short chain or long chain, which had a 5- and 100-fold greater adhesion to stainless steel surfaces, respectively, than wild-type smooth colonies or revertant smooth colony types (Monk *et al.* 2004). Monk *et al.* (2004) propose that phenotypic variation may be important in the environmental survival of *L. monocytogenes.*

### 7.4.2 Genetics and proteomics of *L. monocytogenes* biofilm formation

Researchers have been investigating the genes needed for *L. monocytogenes* surface growth using both genetic and proteinomic approaches. Using a transposon mutagenesis approach, *L. monocytogenes* mutants that were deficient in attached surface growth were isolated. The genes encoding (p)ppGpp synthase (*relA*) and a gene in the purine salvage pathway (*hpt*) were identified as responsible for the reduction in surface growth and mutants were unable to accumulate (p)ppGpp and induce a stringent response (Taylor *et al.* 2002). In addition, these researchers showed that the *relA* gene was induced after bacterial adhesion suggesting that adhesive bacteria adapt physiologically to the early stages of biofilm formation (Taylor *et al.* 2002).

Like in other bacterial species, it appears that flagella play an important role in *L. monocytogenes* biofilm formation. Flagellin mutants of *L. monocytogenes* have been observed to attach to stainless steel at levels 10-fold lower than wild-type cells with short incubation times (4 h) (Vatanyoopaisarn *et al.* 2000). These authors observed that with longer incubation periods, cell coverage by flagella negative mutants reach close to the same level as flagellated cells, indicating that the flagella are important for initial attachment (Vatanyoopaisarn *et al.* 2000). Recently, Lemon *et al.* used series of *L. monocytogenes* flagellar mutants (did not produce flagella, produced flagella that were not functional or produced functional flagella that were not glycosylated) to show that the active bacterial propulsion by flagella was key for initial adhesion and subsequent biofilm formation (Lemon *et al.* 2007).

Recently, it was reported that cell-to-cell communication is involved in initial adhesion of *L. monocytogenes* to surfaces through the *agr* system (Rieu *et al.* 2007). The accessory gene regulator (*agr*) system is a series of peptide signaling pathways that was originally described in *Staphylococcus aureus* (recently reviewed by Kong *et al.* (2006)) which has orthologs in several other gram positive systems, including *L. monocytogenes*. The *agr* system is made up of four genes, *agrA*, *agrC*, *agrD and agrB. In L. monocytogenes*, mutations in the *agrD* gene (encoding the precursor to the autoinducer peptide) and *agrA* (encoding the second component of a two-component signal transduction system that is a positive response regulator) both caused a 62% decrease numbers of *L. monocytogenes* EGD-e cells attached to glass slides after 2 h of incubation (Rieu *et al.* 2007). Biofilm formation during the first 24 h of incubation was reduced in the *agrD* and *agrA* mutants; however, this reduction was not observed upon further incubation (Rieu *et al.* 2007). These results were similar to *S. aureus* in which the *agr* system has been shown to be important for the initial attachment and detachment stages of biofilm development (Kong *et al.* 2006).

Others have used a proteomic approach to compare the proteins produced within a single strain growth within broth (planktonic) and within a biofilm and identified thirty-one proteins at variable levels between the two growth conditions (Tremoulet *et al.* 2002). Twenty-two proteins were up regulated while nine others were down regulated. Of the seven proteins identified by peptide sequencing, the majority (five) had uncharacterized functions. Levels of superoxide dismutase and 30S ribosomal protein S2 were increased when grown in biofilms and the level of flagellin protein decreased when compared to protein levels isolated from planktonically grown cells (Tremoulet *et al.* 2002).

## 7.5 Influence of lineages on biofilm formation

Many researchers have studied the variation in the ability of different strains to adhere and form biofilms on processing surfaces. The rationale being that strains that have the ability to produce persistent biofilms in the food processing environment would be more likely get into foods during processing and cause foodborne disease. Moreover, the relationship between biofilms and post-processing contamination may be more complex. Researchers have suggested that different isolates of *L. monocytogenes* may have different levels of pathogenicity (Boelin *et al.* 1991; Wiedmann *et al.* 1997). Theoretically, strains with differing levels of pathogenesis may also vary in their ability to form biofilms. Therefore, virulent strains that cannot form biofilms may be less likely to cause outbreaks in processed foods since such an organism will not persist in the processing environment. On the other hand, a biofilm forming, less virulent strain of *L. monocytogenes* may also be less likely to cause foodborne outbreaks, due to differences in pathogenisity rather than ecological survival. Following the same logic, the virulent strains that can form persistent biofilms would be most the likely to cause foodborne listeriosis because of their ability to persist in the processing plant and their ability to cause human disease.

This potential relationship between pathogenesis and persistence has spurred a number of investigations into adhesion of biofilms produced from different strains of *L. monocytogenes.* Lunden *et al.* (2000) observed that persistent strains had higher levels of adhesion to stainless steel than non-persistent strains after two hours contact time, but this difference was not observed after increased incubation time (Lunden *et al.* 2000). My research group adapted a commonly used rapid microtiter plate assay that utilizes crystal violet staining as an indication of level of surface growth (Djordjevic *et al.* 2002). We obtained thirty coded strains from both clinical and food processing environments that had been characterized into three genetic lineages (Wiedmann *et al.* 1997; Norton *et al.* 2001). Coded strains were used to remove any researcher bias. After testing using the biofilm microtiter plate assay, we found that lineage 1 strains (predominately serotype 4b), that were generally associated with human disease, exhibited higher levels of biofilm formation than the other two lineages and no correlation was found between environmental persistence and biofilm staining intensity (Djordjevic *et al.* 2002). However, Borucki *et al.* used this assay to compare a larger sample size (eighty isolates) and found that Division II strains (serotypes 1/2a and 1/2c) predominantly formed greater amounts of crystal violet stained films than division I strains (serotypes 4b and 1/2b) (Borucki *et al.* 2003). They also compared a number of the same strains from the study by Djordjevic *et al.* (2002) and found similar test results, indicating that differences between the two studies were not likely due to technique, but instead be related to sample sizes and specific strains used in each study. This was confirmed by electron micrographs of highly strained biofilms dense multidimensional growth, while lesser-stained strains showed loosely distributed cells adherent on a surface, once again indicating that strains of *L. monocytogenes* can vary widely in their biofilm production under laboratory test conditions (Borucki *et al.* 2003). In addition, Borucki *et al.* observed a statistical correlation between persistent strains from a dairy processing plant and greater biofilm formation (Borucki *et al.* 2003).

A large deviation in ability of pure cultures of *L. monocytogenes* to adhere and grow on surfaces under laboratory conditions has been shown (Norwood *et al.* 1999; Lunden *et al.* 2000; Kalmokoff *et al.* 2001; Djordjevic *et al.* 2002; Borucki *et al.* 2003). Microtiter plates have become a popular method to screen and compare biofilm formation of a variety of *L. monocytogenes* strains; however, interpretation and comparison of results from different laboratories can be difficult due to varying composition of microtiter plates. For example, *L. monocytogenes* biofilm formation has been evaluated in microtiter plates made of polyvinyl choride (PVC) (Djordjevic *et al.* 2002; Call *et al.* 2003; Kim *et al.* 2007) or polystyrene (Stepanovic *et al.* 2004; Gable *et al.* 2007; Harvey *et al.* 2007; Jensen *et al.* 2007). Polystyrene plates that have been used have either been untreated (Stepanovic *et al.* 2004; Harvey *et al.* 2007) or treated for tissue culture (Jensen *et al.* 2007). Tissue culture treated polystyrene is a modification to the surface to make typically hydrophobic polystyrene more hydrophilic to enhance adherence. Without a constant testing method, it becomes very difficult in comparing results between research groups to determine the relationship between lineage and ability to form biofilms.

It is still unclear whether laboratory tests accurately evaluate the reasons behind strain persistence in the food-processing environment. Moretro and Langsrund proposed that the differences in persistence might be due to the ability of stains to adapt to the complex diversity of biofilms present in food processing environments (Moretro *et al.* 2004) and if this is the case, monospecies evaluation may not be the best approach to give the answers behind ecological persistence.

## 7.6 *L. monocytogenes* biofilms and exopolymeric substances (EPS)

Most of the biofilm research on *Listeria monocytogenes* focuses on cell growth and enumeration to describe the biofilm and study its behavior. However, there is a lack of understanding on the *L. monocytogenes* EPS composition, structure and role on biofilm formation, synthesis and composition. Recently, there has been an increase in listerial EPS interest, though the actual EPS composition and structure has not been defined yet. With regards to listerial EPS, there have been several attempts to assess and image its presence within a listerial biofilm. Imaging listerial EPS has been conducted with different macroscopic techniques such as scanning electron microscopy (SEM), fluorescence techniques, and atomic force microscopy (AFM).

SEM has traditionally been used to image listerial biofilms, after samples are fixed and prepared following a series of steps. In most of the cases, a three-dimensional network of bacterial cells associated as microcolonies is shown (Blackman *et al.* 1996; Borucki *et al.* 2003; Kalmokoff *et al.* 2001; Kim *et al.* 2001; Chavant *et al.* 2002; Tremoulet *et al.* 2002; Marsh *et al.* 2003; Hefford *et al.* 2005; Moltz *et al.* 2005; Gable *et al.* 2007). In some cases, the presence of fibrils within the listerial microcolonies can be seen, suggesting the presence of EPS (Kalmokoff *et al.* 2001; Hefford *et al.* 2005; Moltz *et al.* 2005), though it is hard to assess whether this is an artifact due to dehydration while preparing the sample, or these fimbrils are naturally present within the biofilm.

Fluorescence techniques have traditionally been used to stain listerial cells for visualization under a fluorescence microscope or confocal scanning laser microscope (CSLM) (Blackman *et al.* 1996; Folsom *et al.* 2006). Recently, more attention has been paid to the presence of carbohydrates in *Listeria* biofilms. In order to be visualized with fluorescence, several fluorescence dyes have been used such as lectins (that specifically bind to carbohydrates) (Hassan *et al.* 2004; Hefford *et al.* 2005) or ruthenium red (Borucki *et al.* 2003). As one can expect, different dyes will bind to different carbohydrates, which allows researchers to get an idea of the potential carbohydrate composition. Borucki *et al.* (2003) imaged listerial EPS using CSLM; ruthenium red was the dye used, suggesting the presence of extracellular acid polysaccharides between listerial cells. On the other hand, the use of concanavalin A as a dye (Hassan *et al.* 2004; Hefford *et al.* 2005) suggests the presence of a cloud surrounding bacterial cells composed of α-D-glucose and α-D-mannose. In a more in depth study, Chae *et al.* (2006) evaluated the difference in total extracellular glucose and total extracellular carbohydrates when comparing listerial planktonic cells and listerial adhered cells after 3 hours of incubation at 37°C. The results of this work showed a higher total extracellular carbohydrates produced by adhered cells when compared to their planktonic counter partners, though no specific composition of such carbohydrates were reported. In a recent study, (Rodriguez *et al.* 2008a) a 48 h listerial biofilm grown on stainless steel was imaged using atomic force microscopy (AFM), (Fig. 7.1). The use of AFM allows researchers to image biofilms *in situ* without the use of any specific dyes or further sample preparation (Nunez *et al.* 2005). The results of Rodriguez *et al.* strongly suggest the presence of EPS associated to the listerial cells within the microcolony (Rodriguez *et al.* 2008a). None of the tech-

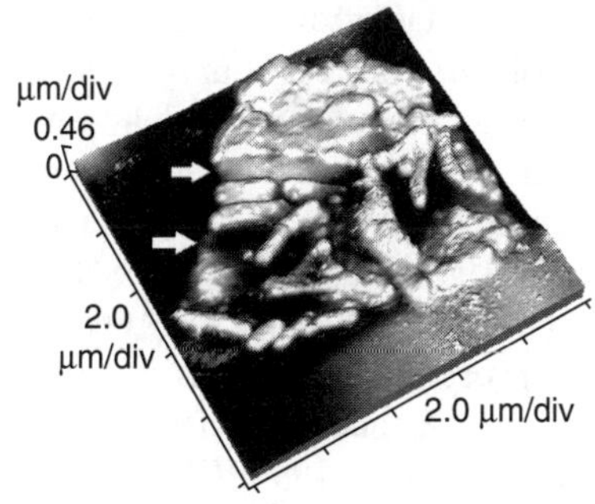


**Fig. 7.1** An atomic force microscope (3 dimensional, 10 × 10 μm) representation of a microcolony composed of *Listeria* cells and exopolymeric substances (EPS) (arrows) on a stainless steel coupon. (Originally published in (Rodriguez *et al.* 2008a). Copyright 2008 by the Intl Assn for Food Protection. Reproduced with permission of Intl. Assn. for Food Protection via Copyright Clearance Center.)

niques mentioned above used has allowed clear quantification or evaluation of the specific composition of the EPS associated to *Listeria* biofilms. In fact, the work conducted to date only suggests the presence of EPS associated to listerial biofilms. In many cases, only the presence of polysaccharides associated to bacterial cells has been reported.

Most of the work done to date has focused on the presence of carbohydrates only, but very little work has been conducted to elucidate the presence of proteins or other components in a listerial biofilm. A couple of research groups have evaluated the difference in protein expression comparing listerial biofilm vs. planktonic cells (Tremoulet *et al.* 2002; Hefford *et al.* 2005). In both cases, a higher expression of proteins was seen when bacteria were grown as biofilms. In both studies (Hassan *et al.* 2004; Hefford *et al.* 2005) the change in protein expression was assessed after separating the cells by centrifugation, further lysis and separation on a gel using a electrophoretic technique. The results of these studies only suggests the expression of stress related proteins upon biofilm formation, perhaps as actual presence of proteins within the listerial EPS. Further studies need to be completed to determine the mechanism of cell adaption or adherence to the surface tested.

### 7.6.1 Cleaning and Sanitation of *L. monocytogenes* biofilm

Means to control *L. monocytogenes* biofilms in the processing plants is a stringent cleaning and sanitation program. Physical cleaning will remove a portion of biofilms and in laboratory systems cleaning has been shown to reduce numbers by 0.9 log (Gibson *et al.* 1999). By comparing data from microscopic counts and agar plating, it was observed that cleaning products that did not increase the amount of bacteria removed from surfaces;

however, alkaline and acidic cleaning agents reduced the viability of the bacteria removed (Gibson *et al.* 1999). This is important to reduce the possibility of potential bacterial spread during cleaning procedures.

In general, biofilms are more resistant to sanitizing agent, than than planktonic cells (Bolton *et al.* 1988; LeChevallier *et al.* 1988; Frank *et al.* 1990a; 1990b; Nickel *et al.* 1992; Reid *et al.* 1993; Bower *et al.* 1999; Donlan 2002; Donlan *et al.* 2002; Dunne 2002). However, the composition of the supporting surface, the presence of competing organisms, as well as the presence of food soils can influence how effective sanitizing agents can be against *Listeria sp.* (Bourion *et al.* 1996; Chmielewski *et al.* 2004; Somers *et al.* 2004; Chmielewski *et al.* 2006). Chavant *et al.* (2004) directly compared the sensitivity of planktonicly grown cells and biofilms to a variety of sanitizers. In general, biofilms were more resistant than planktonicly grown cells; however, the results varied by sanitizer. They found quaternary ammonium cleaners to highly effective against (>90% destruction) planktonic cells and young biofilms (6 h and 1 day) but much less effective (approximately 40% destruction) against 7-day biofilms (Chavant *et al.* 2004). Alkaline cleaners were most effective (>90% destruction) on planktonic and older biofilms (>90%) but were least effective against the 6 h-old biofilms (Chavant *et al.* 2004). The effectiveness of acetic acid cleaner (pH 5), $Na_2SO_2$, and monolauren were much less effective than alkaline and quandary ammonium cleaners, with the general trend of greater resistance of the biofilms compared to the planktonicly grown cells.

The presence of food residue or soil can further reduce effectiveness of sanitizers or heat against biofilms (Aarnisalo *et al.* 2000; Chmielewski *et al.* 2004; Somers *et al.* 2004; Chmielewski *et al.* 2006; Gram *et al.* 2007). Aarnisalo *et al.* tested the effectiveness of ten commercially available disinfecting agents using a surface test method (where cells are scraped from the surface of an agar plate and added to stainless steel disks)(Aarnisalo *et al.* 2000). In most cases, the anti-biacterial activity of the surface test had a >3 log reduction of *Listeria* in 30 seconds, but antimicrobial effects were greatly reduced for almost all sanitizers tested when soil was present (in the form of a pork mixture), confirming food soil can reduce the effectiveness of sanitizers. In this study, the only sanitizer that was effective with the soiled surface was isopropanol-based (60% w/v) (Aarnisalo *et al.* 2000). Somers and Wong observed biofilms grown in the presence of food soil (frankfurter and fat residues) were more resistant to sanitizing than those grown in straight bacteriological media (Somers *et al.* 2004). Food soil also increased the resistance of *L. monocytogenes* biofilms to heat disinfection (Chmielewski *et al.* 2004; 2006).

### 7.6.2 Biological listerial control

Several research groups have investigated the possibility of using biological control to eliminate or prevent the growth of *L. monocytogenes* in the food processing environment. Competitive exclusion has been one approach to inhibit the colonization and high levels of *L. monocytogenes* in the food processing environment (Leriche *et al.* 1999; Zhao *et al.* 2004; 2006). Leriche *et al.* studied the behavior of *L. monocytogenes* in a biofilm with a nisin-producing strain of *Lactococcus lactis* (Leriche *et al.* 1999). They found that when co-inoculated the reduction of *L. monocytogenes* in the mixed biofilm was dependent upon the inoculum size of *L. monocytogenes* in comparison to *L. lactis.* However, if the nisin-producing strain *L. lactis* was allowed to establish for 24 h prior to addition of *L. monocytogenes*, the lactococcal biofilm was able to inhibit the establishment of *Listeria* to below detectable levels (Leriche *et al.* 1999). In a study by Zhao *et al.* (2004), microbial isolates from floor drains of food processing plants were screened for the ability to inhibit or destroy *L. monocytogenes*, and two isolates *Enterococcus durans* strain 152 and *Lactococcus lactis* subsp. *Lactis* strain C-1-92 were identified as highly inhibitory to listerial growth in broth and in biofilm cultures (Zhao *et al.* 2004). In phase two of the study, a mixture of the two competitive exclusion lactic acid organisms was added to an enzyme-foam-based cleaner and added to on floor drains in poultry processing plants (Zhao *et al.* 2006). The effectiveness of the competitive exclusion mixture was dependent upon the temperature of the processing environment, and may also have been dependent on floor drain material and the composition of the liquid waste flowing in the drain (Zhao *et al.* 2006). The highest exclusion of *Listeria sp.* was observed after one application to a fiberglass floor drain in a refrigerated environment (3.4°C) and levels remained very low for weeks after treatment, while in a warmer room (26°C) with a steady flow of high nutrients in the drain (blood and poultry debris), it took three applications to observe a significant reduction in *Listeria* sp. levels and after treatment ended, levels increased (Zhao *et al.* 2006). Although the success of competitive exclusion may be variable based upon the nutrients and temperature in the environment (Zhao *et al.* 2006), or possibly the level of *L. monocytogenes* present before the addition of the competitor (Leriche *et al.* 1999), this approach is a promising method to reduce and possibly inhibit listerial levels in biofilms and warrants further research.

### 7.6.3 Mechanisms and modeling of transfer of *L. monocytogenes* from surfaces to foods

When *L. monocytogenes* is present in the processing environment, it opens up the potential for environmental contamination of foods. Contamination of *L. monocytogenes* from food processing equipment and/or the processing environment is thought to be a critical mode of listerial contamination in processed foods. Its presence in ready-to-eat meals is likely caused by recontamination of the product after processing prior to packaging (Lunden *et al.* 2002). Detachment and transfer of bacteria from processing surfaces to foods and vice versa, can occur during and after processing and this represents the most common route of recontamination of processed foods. Several studies have focused on the transfer of bacteria and viruses with the majority of the work in this area focused on the transfer of *L. monocytogenes* from processing surfaces to foods (Midelet *et al.* 2002; 2006; Vorst *et al.* 2006a; 2006b; Rodriguez *et al.* 2007a; 2007b; 2007c; Keskinen *et al.* 2008). Less work has been done to study the transfer of *Listeria monocytogenes* from foods to processing materials (Vorst *et al.* 2006a; Rodriguez *et al.* 2007b).

There is an intrinsic variability when evaluating transfer of bacteria from one surface to another (Zhao *et al.* 1998; Sattar *et al.* 2001; Montville *et al.* 2003; D'Souza *et al.* 2006; Flores *et al.* 2006; Vorst *et al.* 2006a; 2006b; Rodriguez *et al.* 2007a; 2007b; 2007c; Keskinen *et al.* 2008). This variability makes comparison of transfer experiments difficult. Most of this variability comes from several factors: differences in methodology, surfaces used for transfer, inoculation method, data analysis, and hydration level prior to transfer.

It is important to notice that one of the main parameters that brings variability to the transfer research is the method used. Whereas each research group uses a different methodology, listerial transfer work done to date can easily be divided into two groups depending on the predominant force used behind the transfer designs: normal force vs. shear force. Most of the published work has been conducted using a normal force (perpendicular to the contact surface), in which a certain amount of weight has been placed on top of a food product or a food contact surface after a set amount of time (Zhao *et al.* 1998; Sattar *et al.* 2001; Montville *et al.* 2003; D'Souza *et al.* 2006; Flores *et al.* 2006; Vorst *et al.* 2006a; 2006b; Rodriguez *et al.* 2007a; 2007b; 2007c; Keskinen *et al.* 2008). The pressure applied can be easily calculated by dividing the force applied by the contact surface area. A second type of experimental design involves the use of additional forces, shear forces, where the force applied is tangential or parallel to the food; this is commonly done with the use of a cutting blade or by bringing the knife perpendicular to the food where shear forces are generated as the knife advances through the food (Vorst *et al.* 2006a; 2006b; Keskinen *et al.* 2008). This latter design makes pressure calculation more difficult, which has been underestimated when evaluating listerial transfer.

A great deal of this variability comes from the pressures and the contact times used for the transfer that vary drastically from one study to another. Vorst *et al.* (2006a) evaluated the sequential transfer of *L. monocytogenes* from inoculated product (using an inoculation level of $10^8$ CFU/cm$^2$) to sterile blade. The results of their work showed that as the pressure increased the transfer of *Listeria* to the mechanical slicing machine increased. Midelet and Carpentier (Midelet *et al.* 2002) studied the transfer of *L. monocytogenes*, *Staphylococcus sciuri*, *Pseudomonas putida* and *Comamonas sp* from three processing surfaces to beef by pressing beef onto a stainless steel slide with biofilms (25 cm$^2$) with a 500 g weight, giving a calculated pressure of approximately 1.91 kPa during the 30 s contact time (Midelet *et al.* 2002). In subsequent work, this group utilized an agar-filled syringe with a contact area of 0.12 $cm^2$ with a calculated pressure of 10 kPa (Midelet *et al.* 2006). Flores *et al.* (2006) used pressure of 0.11 kPa for transfer of *E. coli* O157:H7 from beef tissues and 0.31 kPa for transfer from HDPE boards (Flores *et al.* 2006). By comparison, Vorst *et al.* (2006a) evaluated the transfer from inoculated food to a moving cutting blade under two different static forces. Based upon the contact area of the cutting blade, the static pressure was calculated as 1.7 and 8.8 kPa. Other researchers have used a much higher pressure (45 kPa) for the transfer experiments (Rodriguez *et al.* 2007a; 2007b; 2007c). The pressure used in these studies (45 kPa) was from 3 to 363 times higher than those used in previously published work. Rodriguez *et al.* (2007c) showed the importance of contact time and pressure on listerial transfer, by showing that the transfer of listerial biofilms grown on stainless steel to bologna as a function of both contact time and pressure (Rodriguez *et al.* 2007c). As both, contact time and pressure increased, the transfer of *Listeria* from stainless steel to bologna increased. After a certain contact time (30 s) and pressure (45 kPa), transfer became constant and independent of both contact time and pressure (Rodriguez *et al.* 2007c).

*Listeria* adhesion to different materials has been shown (Smoot *et al.* 1991; Kim *et al.* 1994; 1995; Blackman *et al.* 1996; Kalmokoff *et al.* 2001; Norwood *et al.* 2001; Somers *et al.* 2004; Moltz *et al.* 2005; Rodriguez *et al.* 2008b). *Listeria* tend to adhere better to hydrophobic surfaces (ex: polyethylene) than hydrophilic (ex: stainless steel). For this reason, it is obvious to assume that bacterial transfer from and to different materials will be different depending on the material used, its characteristics and properties. Most of the transfer work on *Listeria* done to date addresses the transfer from stainless steel surfaces and plastic materials to foods and vice-versa. In most of the cases, *Listeria* tended to transfer more from and to stainless steel compared to plastic surfaces (Midelet *et al.* 2002; Rodriguez *et al.* 2007a; 2007b). A recent publication (Rodriguez *et al.* 2008b) has focused of the fundamentals of transfer at a cellular level using atomic force microscope (AFM). This was done using the colloidal probe technique where small beads (0.1 μm in diameter) made out of two different materials (polystyrene and glass) were placed on the tip of a cantilever and brought in contact against a listerial biofilm. The exact adhesive forces to each colloidal probe were recorded (Fig. 7.2). The results of this work showed that there was a greater maximum pull-off force (Fmax) to a colloidal probe made out of polystyrene (–113.38 ± 0.70 nN) which was higher than that of glass (–85.42 ± 1.05 nN) (Rodriguez *et al.* 2008b).

The inoculation method for the transfer experiments can be another source of variability. All of the transfer work done with *Listeria* uses attached cells, in which cells were forced to adhere to a processing surface or food prior to a transfer experiment for short periods of time (Midelet *et al.* 2002; 2004; 2006; Vorst *et al.* 2006a; 2006b; Rodriguez *et al.* 2007a; Keskinen *et al.*

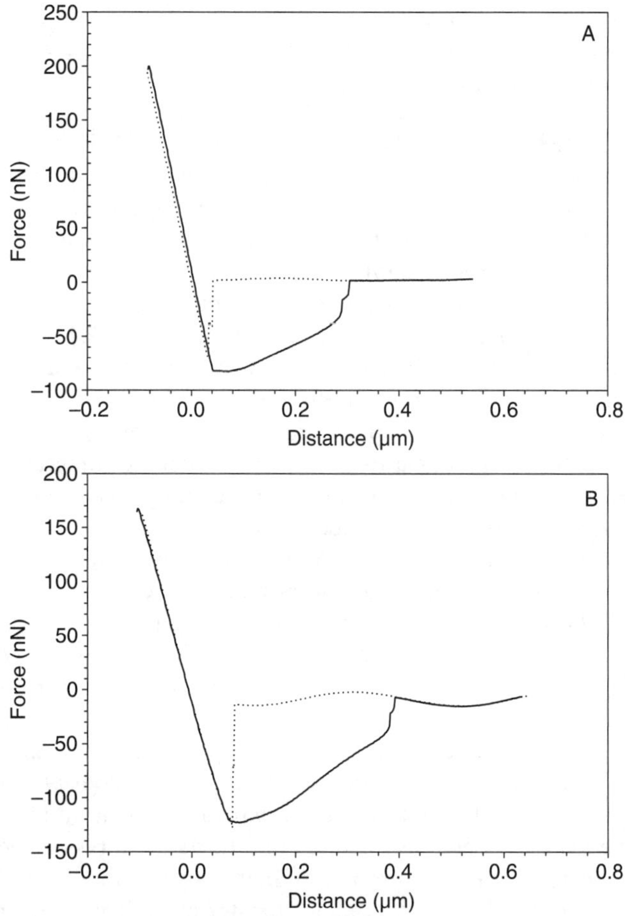


**Fig. 7.2** Typical force vs. distance curve generated when using colloidal probes composed of glass ($SiO_2$) (A) or polyethylene (B) (k = 1.75 N/m) on listerial biofilms grown on stainless steel and equilibrated at 94%RH. Dotted line represents approach and continuous line represents retraction of the cantilever from the sample. (Originally published in Rodriguez *et al.* (2008b) and reproduced with kind permission of Springer Science and Business media.)

2008; Sheen 2008). On the other hand, very little work has been conducted to assess the listerial transfer where bacteria are grown as a true biofilm (Rodriguez *et al.* 2007c; 2007b; 2008b). Recent published work has shown that biofilm grown bacteria do transfer better from inoculated surfaces to foods than adhered cells (Rodriguez *et al.* 2007c). This could be easily explained by looking at some of the forces that hold a biofilm together. When *Listeria* is grown on steel surfaces, it can exist as small microcolonies with a thickness of 2 to 3 cells (Fig. 7.1) (Rodriguez *et al.* 2008a). Therefore,

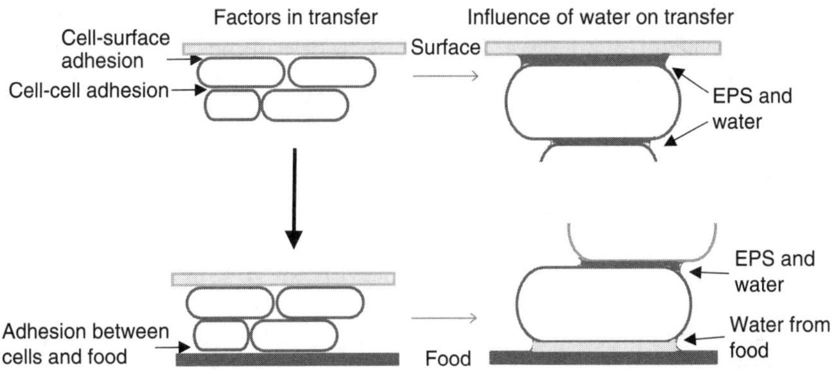


**Fig. 7.3** Working model of factors involved in transfer of bacteria from surfaces to foods. These factors include: (i) the adhesion between the cells and the initial supporting surface, (ii) cell-to-cell adhesion, and (iii) the adhesion between the cells and the food. Water appears to be an important parameter that influences the level of transfer. It is involved as a component of EPS, holding the cells to surfaces and cell-to-cells in biofilms and as a component of foods.

it is valid to assume that forces within the biofilm (cell-cell) are to be weaker that cell-surface forces (Fig. 7.3). When a biofilm is in contact with another surface, the weaker cell-cell interactions will allow bacterial cells to transfer easier when compared to only adhered cells.

The hydration level of surfaces or inoculated surfaces has been shown to be important for other bacterial species than *Listeria* when used in completely different systems (Sattar *et al.* 2001; Flores *et al.* 2006). When it comes to listerial transfer research, this phenomenon may have been overlooked, and just a fraction of the published work addresses this issue (Rodriguez *et al.* 2007a; 2007b; 2007c). The importance of the hydration level can be divided into two components: hydration of the processing surfaces and hydration of the bacterial cells prior to transfer. In a food processing scenario, the hydration state of the processing surfaces does not seem to be of importance for listerial transfer. Rodriguez *et al.* (2007b) did not find any differences when they evaluated the transfer of *Listeria* from inoculated bologna to stainless steel and high density polyethylene, either dried or pre-wetted in a buffer solution prior to transfer. The results of this work were more likely due to the constant state of hydration of the food system used (bologna) that re-wetted the processing surfaces upon contact (Rodriguez *et al.* 2007b).

Recently published work on listerial transfer shows the tremendous importance of the bacterial hydration level prior to transfer. Rodriguez *et al.* (2007c) found that biofilms grown on stainless steel that were dried prior to transfer, had higher levels of bacteria transfer to bologna; however, in this study the level of dryness was not well defined (Rodriguez *et al.* 2007c). In a subsequent study, Rodriguez *et al.* (2007a) evaluated the transfer of listerial biofilms grown on stainless steel to foods (bologna and hard salami) controlling the exact level of dryness of the biofilms by equilibrating them to known percentage relative humidities of 96, 75, 56, and 33% for a period of 24 h (Rodriguez *et al.* 2007a). The results of this work showed that as biofilms were dried, listerial transfer increased to both foods tested. More interesting was to find that the degree of water activity of the food system used did also influence the transfer. Listerial biofilms grown on stainless steel did transfer more to foods with higher water activity (bologna, $a_w = 0.96$) compared to food with lower water activity (hard salami, $a_w = 0.83$) (Rodriguez *et al.* 2007a). In a recent study, (Keskinen *et al.* 2008), dried attached cells to a relative humidities of 72% on stainless steel prior to transfer.

## 7.7 Conclusions

Based upon our research, we have developed a model for the factors that are important in bacterial transfer under normal force (Fig. 7.3). We have observed differences in transfer related to surface composition (Rodriguez *et al.* 2007c), indicating that the adhesion between the cells and the initial supporting surface will influence transfer and this was confirmed using AFM, and measuring the pull-off forces between listerial cells and materials with various compositions (Rodriguez *et al.* 2008b). In addition, we've observed differences in transfer level based upon whether the *L. monocytogenes* cells were dried onto a surface or grown as a biofilms (Rodriguez *et al.* 2007c), indicating that cell-to-cell adhesive forces, that would be encountered in biofilms, may also be a parameter of transfer. From these results we can conclude that the factors important in bacterial transfer from one surface to food are: (i) the adhesion between the cells and the initial supporting surface, (ii) the cell-to-cell adhesion, (iii) the attraction between the cells and the food (Fig. 7.3).

From our experiments there is no doubt that water is an important parameter in transfer. We have measured increasing transfer as biofilms were equilibrated to lower water activities (Rodriguez *et al.* 2007a; 2007c). This indicates that water in the biofilm can affect the transfer (Rodriguez *et al.* 2007a), either by reducing cell-to-cell adhesion and/or reducing cell-to-surface adhesion. The level of water in the food also appears to influence the transfer of *Listeria*, since we've observed higher transfer of dried biofilms to foods with higher water activities (Rodriguez *et al.* 2007a). As anyone who has ever built a sand castle can attest, water can act as a strong capillary force holding a sand structure together; however, as the structure dries, the castle shape loses cohesion and rapidly turns back into a pile of sand that can be easily dispersed. In a similar sense, we propose that water may be a critical component in holding EPS together in a listerial biofilm and/or holding the bacterial cells on the surface. Water in the food may also facilitate transfer by 'wicking' the bacteria from a drier surface to a moist food. Understanding the fundamental steps in bacterial transfer from surfaces to foods may help us design food processing facilities that may reduce the possibility of cross-contamination with this organism.

## 7.8 References


AARNISALO, K., S. SALO, H. MIETTINEN, M. L. SUIHKO, G. WIRTANEN, T. AUTIO, J. LUNDEN, H. KORKEALA and A. M. SJOBERG (2000), 'Bactericidal efficiencies of commercial disinfectants against *Listeria monocytogenes* on surfaces.' *Journal of Food Safety*, **20**(4): 237–250.

AL-MAKHLAFI, H., J. MCGUIRE and M. DAESCHEL (1994), 'Influence of preabsorbed milk proteins on adhesion of *Listeria monocytogenes* to hydrophobic and hydrophilic silica surfases' *Appl Environ Microbiol*, **65**(10): 3560–3565.

AL-MAKHLAFI, H., A. NASIR, J. MCGUIRE and M. DAESCHEL (1995), 'Adhesion of *Listeria monocytogenes* to silica surfaces after suqential and competitive adsoption of bovine serum albumin and β-lactoglobulin' *Appl Environ Microbiol*, **61**(5): 2013–2015.

AN, Y. H. and R. J. FRIEDMAN (1997), 'Laboratory methods for studies of bacterial adhesion.' *J Microbiol Meth*, **30**: 141–152.

ANONYMOUS (2002), 'Public Health Dispatch: Outbreak of Listeriosis – Northeastern United States, 2002.' *Morbidity and Mortality Weekly Report (MMWR)*, **51**(42): 950–951.

ANONYMOUS (2003), '9 CFU Part 430: Control of *Listeria monocytogenes* in ready-to-eat meat and poulty products; interim final rule.' *Federal Register*: 34207–34254.

ANONYMOUS (2008), '*Listeria monocytogenes* outbreak: Final Update http://www.phac-aspc.gc.ca/alert-alerte/listeria/listeria_2008-eng.php Public Health Agency of Canada.' Retrieved 20 February 2009.

BARNES, L.-M., M. F. LO, M. R. ADAMS and H. L. CHAMBERLAIN (1999), 'Effect of milk protein on adhesion of bacterial stainless steel surfaces.' *Appl Environ Microbiol*, **65**(10): 4543–4548.

BLACKMAN, I. C. and J. F. FRANK (1996), 'Growth of *Listeria monocytogenes* as a biofilm on various food-processing surfaces.' *J Food Prot*, **59**(8): 827–831.

BOELIN, P. and J.-C. PIFFARETTI (1991), 'Typing of human, animal, food, and environmental isolates of *Listeria monocytogenes* by multilocus enzyme electrophoresis.' *Appl Environ Microbiol*, **57**: 1624–1629.

BOLTON, K. J., C. E. R. DODD, G. C. MEAD and W. M. WAITES (1988), 'Chlorine resistance of strains of *Staphylococcus aureus* isolated from poultry processing plants.' *Lett Appl Microbiol*, **6**: 31–34.

BORUCKI, M. K., J. D. PEPPIN, D. WHITE, F. LOGE and D. R. CALL (2003), 'Variation in biofilm formation among strains of *Listeria monocytogenes*.' *Appl Environ Microbiol*, **69**(12): 7336–7342.

BOURION, F. and O. CERF (1996), 'Disinfection efficacy against pure-culture and mixed-population biofilms of *Listeria innocua* and *Pseudomonas aeruginosa* on stainless steel, Teflon® and rubber.' *Sci Des Aliments*, **16**(2): 151–166.

BOWER, C. K. and M. A. DAESCHEL (1999), 'Resistance reponses of microorganisms in food environments.' *Intl J Food Microbiol*, **50**: 33–44.

BRAINDET, R., V. LERICHE, B. CARPENTIER and M. N. BELLON-FONTAINE (1999), 'Effects of the growth procedure on the surface hydrophbicity and *Listeria monocytogenes* cells and their adhesion to stainless steel.' *J Food Prot*, **62**(9): 994–998.

BUSSCHER, H. J. and A. H. WEERKAMP (1987), 'Specific and non-specific interactions in bacterial adhesion to solid substrata.' *FEMS Microbiol Rev*, **46**: 165–173.

BUSSCHER, H. J., J. SJOLLEMA and H. C. V. D. MEI (1990), Relative importance of surface free energy as a measure of hydrophobicity in bacterial adhesion to solid surfaces. *Microbial Cell Surface Hydrophobicity*. R.J. and D. a. M. Rosenberg. Washington, DC, ASM.

CALL, D. R., M. K. BORUCKI and T. E. BESSER (2003), 'Mixed-genome microarrays reveal multiple serotype and lineage-specific differences among strains of *Listeria monocytogenes*.' *J. Clin Microbiol*, **41**(2): 632–639.

CHAE, M. S. and H. SCHRAFT (2000), 'Comparative evaluation of adhesion and biofilm formation of different *Listeria monocytogenes* strains.' *Int J Food Microbiol*, **62**(1–2): 103–111.

CHAE, M. S., H. SCHRAFT, L. T. HANSEN and R. MACKERETH (2006), 'Effects of physicochemical surface characteristics of *Listeria monocytogenes* strains on attachment to glass.' *Food Microbiol*, **23**(3): 250–259.

CHAVANT, P., B. MARTINIE, T. MEYLHEUC, M. N. BELLON-FONTAINE and M. HEBRAUD (2002), '*Listeria monocytogenes* LO28: Surface physicochemical properties and ability to form biofilms at different temperatures and growth phases.' *Appl Environ Microbiol*, **68**(2): 728–737.

CHAVANT, P., B. GAILLARD-MARTINE and M. HEBRAUD (2004), 'Antimicrobial effects of sanitizers against planktonic and sessile *Listeria monocytogenes* cells according to the growth phase.' *FEMS Microbiology Letters*, **236**(2): 241–248.

CHMIELEWSKI, R. A. N. and J. F. FRANK (2004), 'A predictive model for heat inactivation of *Listeria monocytogenes* biofilm on stainless steel.' *Journal of Food Protection*, **67**(12): 2712–2718.

CHMIELEWSKI, R. A. N. and J. F. FRANK (2006), 'A predictive model for heat inactivation of *Listeria monocytogenes* biofilm on buna-N rubber.' *LWT – Food Science and Technology*, **39**: 11–19.

CUNLIFFE, D., C. A. SMART, C. ALEXANDER and E. N. VULFSON (1999), 'Bacterial adhesion at synthetic surfaces.' *Appl Environ Microbiol*, **65**(11): 4995–5002.

D'SOUZA, D. H., A. SAIR, K. WILLIAMS, E. PAPAFRAGKOU, J. JEAN, C. MOORE and L. JAYKUS (2006), 'Persistence of caliciviruses on enviromnental surfaces and their transfer to food.' *Int J Food Microbiol*, **108**(1): 84–91.

DJORDJEVIC, D., A. WIEDMANN and L. A. MCLANDSBOROUGH (2002), 'Microtiter plate assay for assessment of *Listeria monocytogenes* biofilm formation.' *Appl Environ Microbio'*, **68**(6): 2950–2958.

DONLAN, R. M. (2002), 'Biofilms: microbial life on surfaces.' *Emerg Infect Dis'*, **8**(9): 881–890.

DONLAN, R. M. and J. W. COSTERTON (2002), 'Biofilms: survival mechanisms of clinically relevant microorganisms.' *Clin Microbiol Rev*, **15**(2): 167–193.

DOYLE, R. J. (2001), *Microbial Growth in Biofilms – Part A: Developmental and Molecular Biological Aspects.*, San Diego, CA., Academic Press.

DUNNE, W. M. (2002), 'Bacterial Adhesion: Seen any good biofilms lately?' *Clin Microbiol Rev*, **15**: 155–166.

FLETCHER, M. (1996), Bacterial attachement in aquatic environments: a diversity of surfaces and adhesion strategies. *Bacterial adhesion: Molecular and ecological diversity*. M. Fletcher. New York, NY, Wiley-Liss: 1–24.

FLORES, R. A., M. L. TAMPLIN, B. S. MARMER, J. G. PHILLIPS and P. H. COOKE (2006), 'Transfer coefficient models for *Escherichia coli* O157:H7 on contacts between beef tissue and high-density polyethylene surfaces.' *J Food Prot*, **69**(6): 1248–1255.

FOLSOM, J. P. and J. F. FRANK (2006), 'Chlorine resistance of *Listeria monocytogenes* biofilms and relationship to subtype, cell density, and planktonic cell chlorine resistance.' *J Food Prot*, **69**(6): 1292–1296.

FRANK, J. F., R. A. N. GILLETT and G. O. WARE (1990a), 'Association of *Listeria* spp. contamination in the dairy processing plant environment with the presence of staphylococci.' *J Food Prot*, **53**: 928–932.

FRANK, J. F. and R. A. KOFFI (1990b), 'Surface-adherent growth of *Listeria monocytogenes* is associated with increased resistance to surfactant sanitizers and heat.' *J Food Prot*, **53**: 550–554.

FRANK, J. F. and R. A. N. CHMIELEWSKI (1997), 'Effectiveness of sanitation with quaternary ammonium compound or chlorine on stainless steel and other domestic food-preparation surfaces.' *J Food Prot*, **60**(1): 43–47.

GABLE, R. and P. M. MURIANA (2007), 'Microplate fluorescence assay for measurement of the ability of strains of *Listeria monocytogenes* from meat and meat-processing plants to adhere to abiotic surfaces.' *Appl Environ Microbiol*, **73**: 5235–5244.

GIBSON, H., J. H. TAYLOR, K. E. HALL and J. T. HOLAH (1999), 'Effectiveness of cleaning techniques used in the food industry in terms of the removal of bacterial biofilms.' *J Appl Microbiol*, **87**(1): 41–48.

GRAM, L., D. BAGGE-RAVN, Y. Y. NG, P. GYMOESE and B. F. VOGEL (2007), 'Influence of food soiling matrix on cleaning and disinfection efficiency on surface attached *Listeria monocytogenes*.' *Food Control*, **18**(10): 1165–1171.

GUTEKUNST, K. A., L. PINE, E. WHITE, S. KATHARIOU and G. M. CARLONE (1992), 'A filamentous-like mutant of *Listeria monocytogenes* with reduced expression of a 60-kilodalton extracellular proteins invades and grows in 3T6 and Caco-2 cells.' *Can J Microbiol*, **38**(8): 843–851.

HARVEY, J., K. P. KEENAN and A. GILMOUR (2007), 'Assessing biofilm formation by *Listeria monocytogenes* strains.' *Food Microbiol*, **24**(4): 380–392.

HASSAN, A. N., D. M. BIRT and J. F. FRANK (2004), 'Behavior of *Listeria monocytogenes* in a *Pseudomona putida* biofilm on a condensate-forming surface.' *J Food Prot*, **67**(2): 322–327.

HEFFORD, M. A., S. D'AOUST, T. D. CYR, J. W. AUSTIN, G. SANDERS, E. KHERADPIR and M. L. KALMOKOFF (2005), 'Proteomic and microscopic analysis of biofilms formed by *Listeria monocytogenes* 568.' *Can J Microbiol*, **51**(3): 197–208.

HERALD, P. J. and E. A. ZOTTOLA (1988), 'Attachment of *Listeria monocytogenes* to stainless steel surfaces at various temperature and pH values.' *J Food Sci*, **53**(5): 1549–1562.

HODGSON, A. E., S. M. NELSON, M. R. W. BROWN and P. GILBERT (1995), 'A Simple in vitro Model for Growth Control of Bacterial Biofilms.' *J Appl Bacteriol*, **79**: 87–93.

HOOD, S. K. and E. A. ZOTTOLA (1997), 'Adherence to stainless steel by foodborne microorganisms during growth in model food systems.' *International Journal of Food Microbiology*, **37**(2–3): 145–153.

ISRAELACHVILI, J. N. (1992), *Intermolecular and Surface Forces*, San Diego, CA Academic Press.

JAMES, A. M. (1991), Charge properties of microbial cell surfaces. In *Microbial cell surface analysis-structural and physicochemical methods*. N. Mozes, P. S. Handley, H. J. Busscher and P. G. Rouxhet. New York, VCH Publishers, Inc.: 221–262.

JENKINSON, H. F. and H. M. LAPPIN-SCOTT (2001), 'Biofilms adhere to stay.' *Trends in Microbiology*, **9**: 9–10.

JENSEN, A., M. H. LARSEN, H. INGMER, B. F. VOGEL and L. GRAM (2007), 'Sodium chloride enhances adherence and aggregation and strain variation influences invasiveness of *Listeria monocytogenes* strains.' *J Food Prot*, **70**(3): 592–599.

KALMOKOFF, M. L., J. W. AUSTIN, X. D. WAN, G. SANDERS, S. BANERJEE and J. M. FARBER (2001), 'Adsorption, attachment and biofilm formation among isolates of *Listeria monocytogenes* using model conditions.' *J Appl Microbiol*, **91**(4): 725–734.

KESKINEN, L. A., E. C. D. TODD and E. T. RYSER (2008), 'Transfer of surface-dried *Listeria monocytogenes* from stainless steel knife blades to roast turkey breast.' *J Food Prot*, **71**(1): 176–181.

KIM, C., Y. C. HUNG, R. E. BRACKETT and J. F. FRANK (2001), 'Inactivation of *Listeria monocytogenes* biofilms by electrolyzed oxidizing water.' *J Food Proc Preserv*, **25**(2): 91–100.

KIM, K. Y. and J. F. FRANK (1994), 'Effect of growth nutrients on attachment of *Listeria monocytogenes* to stainless steel.' *J Food Prot*, **57**(8): 720–724.

KIM, K. Y. and J. F. FRANK (1995), 'Effect of nutrients on biofilm formatin by *Listeria monocytogens* on stainless steel.' *J Food Prot*, **58**(1): 24–28.

KIM, S.-H. and C. WEI (2007), 'Biofilm formation by multidrug-resistant *Salmonella enterica* serotype Typhimurium phage type DT104 and other pathogens.' *J Food Prot*, **70**: 22–29.

KONG, K. F., C. VUONG and M. OTTO (2006), 'Staphylococcus quorum sensing in biofilm formation and infection.' *Intl JMed Microbiol*, **296**(2–3): 133–139.

KUCHMA, S. L. and G. A. O'TOOLE (2000), 'Surface-induced and biofilm-induced changes in gene expression.' *Curr Opin Biotech*, **11**: 429–433.

LECHEVALLIER, M. W., C. D. CAWTHON and R. G. LEE (1988), 'Inactivation of biofilm bacteria.' *Appl Environ Microbiol*, **54**: 2492–2499.

LEMON, K. P., D. E. HIGGINS and R. KOLTER (2007), 'Flagellar motility is critical for *Listeria monocytogenes* biofilm formation.' *J Bacteriol*, **189**(12): 4418–4424.

LERICHE, V. and B. CARPENTIER (2000), 'Limitation of adhesion and growth of *Listeria monocytogenes* on stainless steel surfaces by *Staphylococcus sciuri* biofilms.' *J Appl Microbiol*, **88**(4): 594–605.

LERICHE, V., D. CHASSAING and B. CARPENTIER (1999), 'Behaviour of *L. monocytogenes* in an artificially made biofilm of a nisin-producing strain of *Lactococcus lactis*.' *Intl J Food Microbiol*, **51**(2–3): 169–182.

LUNDEN, J. M., M. K. MIETTINEN, T. J. AUTIO and H. J. KORKEALA (2000), 'Persistent *Listeria monocytogenes* strains show enhanced adherence to food contact surface after short contact times.' *J Food Prot*, **63**(9): 1204–1207.

LUNDEN, J. M., T. J. AUTIO and H. J. KORKEALA (2002), 'Transfer of persistent *Listeria monocytogenes* contamination between food-processing plants associated with a dicing machine.' *J Food Prot*, **65**(7): 1129–1133.

MAFU, A. A., D. ROY, J. GOULET and P. MAGNY (1990), 'Attachment of *Listeria monocytogenes* to stainless steel, glass, polypropylene and rubber surfaces after short contact times.' *J Food Prot*, **53**: 742–746.

MAFU, A. A., D. ROY, J. GOULET and L. SAVOIE (1991), 'Characterization of physicochemical forces involved in adhesion of *Listeria monocytogenes* to surfaces.' *Appl Environ Microbiol*, **57**(7): 1969–1973.

MARRIOTT, N. G. (1999), *Principles of Food Sanitation*, Gaithersburg, MD, Aspen Publishers.

MARSH, E. J., H. LUO and H. WANG (2003), 'A three-tiered approach to differentiate *Listeria monoctogenes* biofilm-forming abilities.' *FEMS Microbiol Lett*, **228**: 203–210.

MCCLEMENTS, D. J. (1999), *Food Emulsions: Principles, Practice and Techniques*, Boca Raton, FA, CRC Press.

MIDELET, G. and B. CARPENTIER (2002), 'Transfer of microorganisms, including Listeria monocytogenes, from various materials to beef.' *Appl Environ Microbiol*, **68**(8): 4015–4024.

MIDELET, G. and B. CARPENTIER (2004), 'Impact of cleaning and disinfection agents on biofilm structure and on microbial transfer to a solid model food.' *J Appl Microbiol*, **97**(2): 262–270.

MIDELET, G., A. KOBILINSKY and B. CARPENTIER (2006), 'Construction and analysis of fractional multifactorial designs to study attachment strength and transfer of

*Listeria monocytogenes* from pure or mixed biofilms after contact with a solid model food.' *Appl Environ Microbiol*, **72**(4): 2313–2321.

MILLSAP, K. W., G. REID, H. C. VAN DER MEI and H. J. BUSSCHER (1996), 'Adhesion of *Lactobacillus* species in urine and phosphate buffer to silicone rubber and glass under flow.' *Biomaterials*, **18**: 87–91.

MOLTZ, A. G. and S. E. MARTIN (2005), 'Formation of biofilms by *Listeria monocytogenes* under various growth conditions.' *J Food Prot*, **68**(1): 92–97.

MONK, I. R., G. M. COOK, B. C. MONK and P. J. BREMER (2004), 'Morphotypic conversion in *Listeria monocytogenes* biofilm formation: Biological significance of rough colony isolates.' *Appl Environ Microbiol*, **70**(11): 6686–6694.

MONTVILLE, R. and D. W. SCHAFFNER (2003), 'Inoculum size influences bacterial cross contamination between surfaces.' *Appl Environ Microbiol*, **69**: 7188–7193.

MORETRO, T. and S. LANGSRUD (2004), '*Listeria monocytogenes*: biofilm formation and persistence in food-processing environments.' *Biofilms*, **1**: 107–127.

NICKEL, J. C. and J. W. COSTERTON (1992), 'Bacterial biofilms and catheters: A key to understanding bacterial straegies in catheter-associated urinary tract infection.' *Can J Infect Dis*, **3**: 619–624.

NORTON, D. M., J. M. SCARLETT, K. HORTON, D. SUE, J. THIMOTHE, K. J. BOOR and M. WIEDMANN (2001), 'Characterization and pathogenic potential of *Listeria monocytogenes* isolates from the smoked fish industry.' *Appl Environ Microbiol*, **67**(2): 646–653.

NORWOOD, D. E. and A. GILMOUR (1999), 'Adherence of *Listeria monocytogenes* strains to stainless steel coupons.' *J Appl Microbiol*, **86**(4): 576–582.

NORWOOD, D. E. and A. GILMOUR (2001), 'The differential adherence capabilities of two *Listeria monocytogenes* strains in monoculture and multispecies biofilms as a function of temperature.' *Lett Appl Microbiol*, **33**(4): 320–324.

NUNEZ, M. E., M. O. MARTIN, P. H. CHAN, L. K. DUONG, A. R. SINDHURAKAR and E. M. SPAIN (2005), 'Atomic force microscopy of bacterial communities.' *Envirom Microbiol Meth Enzym*, **397**: 256–268.

O'TOOLE, G. A. and R. KOLTER (1998), 'Initiation of biofilm formation in *Pseudomonas fluorescens* WCS365 proceeds via multiple, convergent signaling pathways: a genetic analysis.' *Mol Microbiol*, **28**(3): 449–461.

O'TOOLE, G. A., H. B. KAPLAN and R. KOLTER (2002), 'Biofilm formation as microbial development.' *Annu Rev Microbiol*, **54**: 49–79.

OLIVERA, D. R. (1992), Physio-chemical aspects of adhesion. In *Biofilms – Science and Technology*. L. F. Melo, T. R. Bott, M. Fletcher and B. Capdeville. Dordecht, The Netherlands, Kluwer Academic Publishers.

PEREIRA, M. A., M. KUEHN, S. WUERTZ, T. NEU and L. F. MELO (2002), 'Effect of flow regime on the architecture of a *Pseudomonas fluorescens* biofilm.' *Biotech and Bioeng*, **78**(2): 164–171.

POULSEN, L. V. (1999), 'Microbial biofilm in food processing.' *Lebensm-Wiss Technol*, **32**(6): 321–326.

PUREVDORJ-GAGE, L. B. and P. STOODLEY (2004), Biofilm structure, behaviour, and hydrodynamics. In *Microbial biofilms*. M. Ghannoum and G. A. O'Toole. Washington, DC, ASM Press: 160–173.

REID, G., C. TIESZER, R. FOERCH, H. J. BUSSCHER, A. E. KHOURY and A. W. BRUCE (1993), 'Adsoption of ciprofloxacin to urinary catheters and effect on subsequent bacterial adheion and survival.' *Coll Surf B Biointerfaces*, **1**: 9–16.

REIJ, M. W., E. D. D. AANTREKKER and I. E. R. A. I. M. T. FORCE (2004), 'Recontamination as a source of pathogens in processed foods.' *Int J Food Microbiol*, **91**: 1–11.

RIEU, A., S. WEIDMANN, D. GARMYN, P. PIVETEAU and J. GUZZO (2007), 'Agr system of *Listeria monocytogenes* EGD-e: Role in adherence and differential expression pattern.' *Appl Environ Microbiol*, **73**(19): 6125–6133.

RIJNAARTS, H. H. M., W. NORDE, E. J. BOUWER, J. LYKLEMA and A. J. B. ZEHNDER (1993), 'Bacterial adhesion under static and dynamic conditions.' *Appl Environ Microbiol*, **59**: 3255–3265.

RODRIGUEZ, A., W. AUTIO and L. MCLANDSBOROUGH (2007a), 'Effect of biofilm dryness on the transfer of *Listeria monocytogenes* Biofilms grown on stainless steel to Bologna and hard salami.' *J Food Prot*, **70**(11): 2480–2484

RODRIGUEZ, A., W. R. AUTIO and L. A. MCLANDSBOROUGH (2007b), 'Effects of inoculation level, material hydration and stainless steel surface roughness on the transfer of *Listeria monocytogenes* from inoculated bologna to stainless steel and high density polyethylene.' *J Food Prot*, **70**(6): 1423–1428.

RODRIGUEZ, A. and L. A. MCLANDSBOROUGH (2007c), 'Evaluation of the transfer of *Listeria monocytogenes* from stainless steel and high density polyethylene to bologna and American cheese.' *J Food Prot*, **70**(3): 600–606.

RODRIGUEZ, A., W. AUTIO and L. MCLANDSBOROUGH (2008a), 'Effect of surface roughness and stainless steel finish on *Listeria monocytogenes* attachment and biofilm formation.' *J Food Prot*, **71**(1): 170–175.

RODRIGUEZ, A., W. R. AUTIO and L. A. MCLANDSBOROUGH (2008b), 'Effects of contact time, pressure, percent relative humidity (%RH), and material type on *Listeria* biofilm adhesive strength at a cellular level using atomic force microscopy (AFM).' *Food Biophys*, **3**(3): 305–311.

ROWAN, N. J., A. A. G. CANDLISH, A. BUBERT, J. G. ANDERSON, K. KRAMER and J. MCLAUCHLIN (2000), 'Virulent rough filaments of Listeria monocytogenes from clinical and food samples secreting wild-type levels of cell-free p60 protein.' *J Clin Microbiol*, **38**(7): 2643–2648.

SASAHARA, K. C. and E. A. ZOTTOLA (1993), 'Biofilm formation by *Listeria monocytogenes* utilizes a primary colonizing microorganism in flowing systems.' *J Food Prot*, **56**(12): 1022–1028.

SATTAR, S. A., S. SPRINGTHORPE, S. MANI, M. N. GALLANT, R.C., E. SCOTT and J. KAIN (2001), 'Transfer of bacteria from fabrics to hands and other fabrics: development and application of a quantitative method using *Staphylococcus aureus* as a model.' *J Appl Microbiol*, **90**(6): 962–970.

SAUER, S., A. K. CAMPER, G. D. EHRLICH, J. W. COSTERTON and D. G. DAVIES (2002), '*Pseudomonas aeruginosa* displays multiple phenotypes during development as a biofilm.' *J Bacteriol*, **184**: 1140–1154.

SHEEN, S. (2008), 'Modeling surface transfer of *Listeria monocytogenes* on salami during slicing.' *J Food Sci*, **73**: 304–311.

SMOOT, L. M. and M. D. PIERSON (1991), 'Influence of environmental stress on the kinetics and strength of attachment of *Listeria monocytogenes* Scott A to Buna-N rubber and stainless steel.' *J Food Prot*, **61**(10): 1286–1292.

SOMERS, E. B. and A. C. L. WONG (2004), 'Efficacy of two cleaning and sanitizing combinations on *Listeria monocytogenes* biofilms formed at low temperature on a variety of materials in the presence of ready-to-eat meat residue.' *J Food Prot*, **67**(10): 2218–2229.

SOMMER, P., C. MARTIN-ROUAS and E. METTLER (1999), 'Influence of the Adherent Population Level on Biofilm Population, Structure and Resistance to Chlorination.' *Food Microbiol*, **16**: 503–515.

STEPANOVIC, S., I. CIRKOVIC, L. RANIN and M. SVABIC-VLAHOVIC (2004), 'Biofilm formation by *Salmonella* spp. and *Listeria monocytogenes* on plastic surface.' *Lett Appl Microbiol*, **38**(5): 428–432.

SUTHERLAND, I. W. (2001), 'The biofilm matrix – an immobilized but dynamic microbial environment.' *Trends Microbiol*, **9**: 222–227.

SWAMINATHAN, B., D. CABANES, W. ZHANG and P. COSSART (2007), *Listeria monocytogenes*. In *Food Microbiology: Fundamentals and Frontiers 3rd Ed.* P. Doyle and L. R. Beuchat. Washington DC, ASM Press.

TAYLOR, C. A., M. BERESFORD, H. A. S. EPTON, D. C. SIGEE, G. SHAMA, P. W. ANDREW and I. S. ROBERTS (2002), '*Listeria monocytogenes* relA and hpt mutants are impaired in surface-attached growth and virulence.' *J. Bacteriol*, **184**(3): 621–628.

TOMPKIN, R. B. (2002), 'Control of *Listeria monocytogenes* in the food-processing environment.' *J Food Prot*, **65**(4): 709–725.

TOMPKIN, R. B., V. N. SCOTT, D. T. BERNARD, W. H. SVEUM and K. S. GOMBAS (1999), 'Guidelines to prevent post-processing contamination from *Listeria monocytogenes*.' *Dairy Food Environ Sanit*, **19**: 551–562.

TREMOULET, F., O. DUCHE, A. NAMANE, B. MARTINIE and J. C. LABADIE (2002), 'Comparison of protein patterns of *Listeria monocytogenes* grown in biofilm or in planktonic mode by proteomic analysis.' *FEMS Microbiol Lett*, **210**(1): 25–31.

VAN LOOSDRECHT, M. C. M., D. EIKELBOOK, A. GJALTEMA, A. MULDER, L. TIJUIS and J. J. HEIJNEN (1995), 'Biofilm Structures.' *Water Sci Technol*, **32**(8): 35–43.

VATANYOOPAISARN, S., A. NAZLI, C. E. R. DODD, C. E. D. REES and W. M. WAITES (2000), 'Effect of flagella on initial attachment of *Listeria monocytogenes* to stainless steel.' *Appl Environ Microbiol*, **66**(2): 860–863.

VORST, K. L., E. C. D. TODD and E. T. RYSER (2006a), 'Transfer of *Listeria monocytogenes* during mechanical slicing of turkey breast, bologna, and salami.' *J Food Prot*, **69**(3): 619–626.

VORST, K. L., E. C. D. TODD and E. T. RYSER (2006b), 'Transfer of *Listeria monocytogenes* during slicing of turkey breast, bologna, and salami with simulated kitchen knives.' *J Food Prot*, **69**(12): 2939–2946.

WIEDMANN, M., J. L. BRUCE, C. KEATING, A. E. JOHNSON, P. L. MCDONOUGH and C. A. BATT (1997), 'Ribotypes and virulence gene polymorphisms suggest three distinct Listeria monocytogenes lineages with differences in pathogenic potential.' *Infect Immun*, **65**(7): 2707–2716.

WIMPENNY, J., W. MANZ and U. SZEWZYK. (2000), 'Heterogeneity in Biofilms.' *FEMS Microbiol Rev*, **24**: 661–671.

WONG, S., D. STREET, S. I. DELGADO and K. C. KLONTZ (2002), 'Recalls of foods and cosmetics due to microbial contamination reported to the US Food and Drug Administration.' *J Food Prot*, **63**(8): 1113–1116.

ZHAO, P., T. ZHAO, M. P. DOYLE, J. R. RUBINO and J. MENG (1998), 'Development of a model for evaluation of microbial cross-contamination in the kitchen.' *J Appl Microbiol*, **61**(8): 960–963.

ZHAO, T., M. P. DOYLE and P. ZHAO (2004), 'Control of *Listeria monocytogenes* in a biofilm by competitive-exclusion microorganisms.' *Appl Environ Microbiol*, **70**(7): 3996–4003.

ZHAO, T., T. C. PODTBURG, P. ZHAO, B. E. SCHMIDT, D. A. BAKER, B. CORDS and M. P. DOYLE (2006), 'Control of Listeria spp. by competitive-exclusion bacteria in floor drains of a poultry processing plant.' *Applied and Environmental Microbiology*, **72**(5): 3314–3320.

# 8

# Biofilm formation by *Salmonella* in food processing environments

**C. Gamazo, University of Navarra, Spain and C. Solano and I. Lasa, Instituto de Agrobiotecnología y Recursos Naturales, Spain**

**Abstract**: *Salmonella* is a group of food-transmitted pathogens which are of considerable significance to the food processing industry, and cause highly prevalent food-borne diseases over the world. Salmonellosis can be manifested as a number of disease syndromes that include gastroenteritis and typhoid fever. Both the World Health Organization and the European Union, have laid down guidelines to eradicate this pandemic. Understanding how *Salmonella* enters and survives along the food chain is critical for the development of strategies to reduce its prevalence. This chapter is divided into two major areas. The first is focused on *Salmonella* as the causative agent of one of the most prevalent food-borne diseases. The second gets inside the survival strategies of this bacteria based on its ability to form biofilms. The study of structure, synthesis, regulation and final expression of biofilm formation and the role of cellulose, curli fimbriae and biofilm associated proteins will be discussed. The knowledge acquired throughout may help to develop better strategies for the identification and elimination of biofilms along the food chain, and thus, reduce the incidence of *Salmonella* at the farm level and throughout the rest of the food chain.

**Key words**: *Salmonella*, salmonellosis, biofilm, cellulose, Bap, fimbriae.

## 8.1 Salmonellosis, one of the most prevalent food-borne diseases

### 8.1.1 *Salmonella*: diversity and ubiquity

The *Salmonella* genus, member of the *Enterobacteriaceae* family, causes various diseases in humans and animals that range from gastroenteritis to typhoid fever. The genus is taxonomically divided into two species: *Salmonella enterica* and *Salmonella bongori*. *S. enterica* includes more than 2500 serovars that are clustered into seven subspecies designated I, II, IIIa, IIIb, IV, VI, and VII. Subspecies I contains 99% of human-pathogenic serovars, including, among others, Enteritidis and Typhimurium, which are responsible for gastroenteritis in humans, and Typhi, a human-adapted serotype that is the causal agent of typhoid fever. Owing to the clinical importance of some of these serovars, the current nomenclature accepts the omission of the species in their nomination. For example, *Salmonella enterica* subspecies *enterica* serotype Enteritidis can be referred to simply *Salmonella* Enteritidis, nomenclature that will be used throughout this chapter.

There exists a great diversity in this genus, since we are dealing with an old pathogen that probably evolved from reptiles and adapted to mammals about two hundred million years ago. In fact, currently, *Salmonella* serovars may be able to infect a wide phylogenetic range of hosts, from reptiles and birds to mammals, including wildlife animals and pets (Ashbolt and Kirk, 2006; Compton *et al.*, 2008; Handeland *et al.*, 2008; Smith *et al.*, 2002). These epidemiological facts are the main barrier to controlling the spread of this pandemic infection (2004).

Some serovars, however, are restricted to animal species where they can become systemic. For instance, Typhi and Paratyphi serovars exclusively infect human beings; *S.* Abortusequi causes abortion in horses; *S.* Abortusovis and *S.* Montevideo cause abortion in sheep; *S.* Dublin causes septicemia and abortion in cattle and abortion in sheep; and *S.* Gallinarum causes fowl typhoid (Tsolis *et al.*, 1999b). The most frequent serotypes that affect human beings are *S.* Enteritidis and *S.* Typhimurium, and domesticated fowl and their products are recognized as the most common sources of infection (Braden, 2006).

### 8.1.2 Human salmonellosis and food consumption

Salmonellosis is one of the most serious food-borne diseases affecting humans. It may be considered the most important pandemic zoonosis under natural conditions (Braden, 2006; Rodrigue *et al.*, 1990). In particular, *Salmonella* follows *Campylobacter* as a leading cause of zoonotic infections in the European Union (2004). It is estimated that the *Salmonella* worldwide incidence is higher than a thousand million cases and results in three million deaths per year. In countries, where the food supply is considered very safe and is processed under the best conditions available, i.e, Europe and USA, *S.* Enteritidis has emerged as the major etiologic agent of human salmonellosis resulting in 2–4 million cases officially reported yearly (Voetsch *et al.*, 2004), although the real number of infections is considered to be much higher (Mead *et al.*, 1999).

Salmonellosis can result from eating chicken, raw eggs, meat, dairy products, and a variety of fresh fruits and vegetables (Table 8.1). The minimal infective dose reported is highly variable: from 1 to $10^8$ CFU. Factors such as bacterial virulence (expression of mechanisms of pathogenicity), the host (including the stress, hormonal status and its immune system in general) (Bearson and Bearson, 2008; Kingsley and Baumler, 2002), and the matrix

**Table 8.1** Food and beverages as vehicles for *Salmonella* spp. infection

| Category | Food product | References |
|---|---|---|
| Vegetables | Broad-leaf: lettuce, endive, parsley, cilantro | Ercolani, 1976, Lapidot *et al.*, 2006, Duffy *et al.*, 2005, Morris *et al.*, 1998 |
| | Sprouts: alfalfa, bean | Fett, 2000, Mahon *et al.*, 1997, Barak *et al.*, 2002, Gandhi *et al.*, 2001, van Duynhoven *et al.*, 2002 |
| Fruits | Cantaloupe melon, mango, celery | Golden *et al.*, 1993, Burnett and Beuchat, 2000, Solomon, *et al.*, 2005b; Bowen *et al.*, 2006; Mohle-Boetani *et al.*, 1999 |
| | Tomato, pepper | Hedberg *et al.*, 1999, Sagoo *et al.*, 2003, Zhuang *et al.*, 1995 ; Liao and Cooke, 2001 |
| | Sesame seeds | Unicomb *et al.*, 2005 |
| | Almonds | Anon, 2004 |
| | Carrot, fennel | Dipersio *et al.*, 2007, Campbell, 2001, Brandl & Mandrell, 2002 |
| Mushroom | | Doran *et al.*, 2005 |
| Animal products | Chickens and turkey | Zhao *et al.*, 2001, Berrang *et al.*, 1998 |
| | Eating eggs and egg-related products | Braden, 2006, Schroeder *et al.*, 2005 |
| | Pork products | Davies *et al.*, 1999, Delpech *et al.*, 1998, Ojha & Kostrzynska, 2007 |
| | Beef and dairy, cattle cooked ground beef | Mead *et al.*, 1999; Foley & Lynne, 2008; Jordan *et al.*, 2006 |
| | Milk and dairy products | Hedberg *et al.*, 1992, Cahill *et al.*, 2008; Olsen *et al.*, 2004 |
| | Water effluent from farms | Thurston-Enriquez *et al.*, 2005, Soupir *et al.*, 2006, Howard & Hutcheson, 2003, Dhiaf & Bakhrouf, 2007; Hill and Sobsey, 2003 |
| Beverages | Juices | Cook *et al.*, 1998, Khan *et al.*, 2007, Burnett & Beuchat, 2000 |

where the bacteria is located during ingestion, are responsible for such variability (Humphrey, 2004). For example, a high-fat matrix found in cheese or chocolate protects *Salmonella* against the acid pH of the stomach and subsequently bacteria can cause the disease at a lower infective dose (≤100 UFC) (Blaser and Newman, 1982; D'Aoust *et al.*, 1975; Gawande and Bhagwat, 2002; Hedberg *et al.*, 1992; Humphrey *et al.*, 1993; Waterman and Small, 1998).

Contamination can occur at various stages during production, harvest, processing, and transport. The attached microorganisms, thanks to many structural factors, including the ability to form biofilms, persist even after washing or treating with antibacterial agents (see later in this review). Understanding how *Salmonella* enters and survives along the food chain is critical for the development of strategies to reduce its incidence.

The well recognized French physiologist Claude Bernard (1813–1878) emphasized the fact that: '*The cell is nothing. The environment is all important.*' This is a drastic statement, but extremely didactic. Thus, the environment also dictates the biological behaviour of *Salmonella*. Such conditions can range from the different environments found along the food chain to the following external pressures encountered along the animal gut, and the external environment where the bacteria is excreted and must survive changes in temperature, water activity, gases, radiation. The persistence of salmonellae in the environment is an important characteristic in their epidemiology, thus, *Salmonella* is highly prevalent in animals thanks to its long-term survival capacity in the environment; i.e, for up to 300 days in soil, in a dormant state, until bacteria encounter conditions suitable for growth (Hensel *et al.*, 1998; You *et al.*, 2006). Just consider, for instance, the billions of *Salmonella* that infected dairy cows may excrete and accumulate in the piles of wet manure; bacteria that if they are able to survive during storage and in the soil, will be adsorbed on growing plants (in the form of biofilms), dragged by rain water to streams and transmitted to other animals and human beings (Dhiaf and Bakhrouf, 2007; Natvig *et al.*, 2002; Sagoo *et al.*, 2003; Vanselow *et al.*, 2007; Wells and Butterfield, 1997). The asymptomatic excretion phenomenon is unquestionably involved in the spread and importance of that phenomenon (Oliveira *et al.*, 2005; van Winsen *et al.*, 2001; Wood *et al.*, 1989).

It has also been reported that *Salmonella* can persist for at least one year in empty trials or in litter and dust present in poultry unit buildings and transport containers, resulting in a source of cross-contamination for subsequent flocks (Bhatia and McNabb, 1980; Davies and Wray, 1996). Finally, foodstuff can also become contaminated with *Salmonella* during handling, washing and cutting during industrial or home processing of food (Beuchat and Ryu, 1997; Beuchat, 2002; Kaneko *et al.*, 1999; Thunberg *et al.*, 2002). At this point the bacterial adherence capacities (virulence factors) are, again, key points to consider salmonellosis as one of the most prevalent food-borne diseases.

## 8.2 *Salmonella* attachment, biofilm formation and molecular mechanisms involved

Throughout its life cycle, *Salmonella* has to survive inside the host and in the external environment, withstanding different nutrient availability, osmotic stress, pH and temperature conditions. It is now recognized that part of the ecological success of bacteria that have to face such highly variable environmental conditions lies in their ability to grow as surface-attached biofilms.

The development of a *biofilm initiates with the attachment* of the bacteria to a surface. Bacterial attachment is, therefore, a critical step for bacterial biofilm formation, and is related with the access to nutrients, and also the particular biological cycle. Adherence is not merely an electrostatic and hydrophobic mediated process, but also an active biological changing phenomenon settled by a variety of structures such as proteins (i.e; fimbriae, pili, flagella), and polysaccharides (i.e; LPS, glycocalix, colanic acid, cellulose) (Berger *et al.*, 2008; Matthysse *et al.*, 2008; Pratt and Kolter, 1999; Stanley *et al.*, 2003). Adherence is followed by the maturation of the biofilm. Within the biofilm, bacteria adhere to one another and/or to surfaces and are embedded into a self-produced sticky extracellular matrix (Costerton *et al.*, 1995) (Fig. 8.1). In general, the biofilm matrix confers increased tolerance to antimicrobials, host defences and environmental fluctuations. In the particular case of *Salmonella,* it has been shown that biofilm production supplies this enteropathogen with increased resistance to heat, desiccation, chlorination and antibiotic action (Anriany *et al.*, 2001; Dhir and Dodd, 1995; Olson *et al.*, 2002; Scher *et al.*, 2005; Solano *et al.*, 2002).

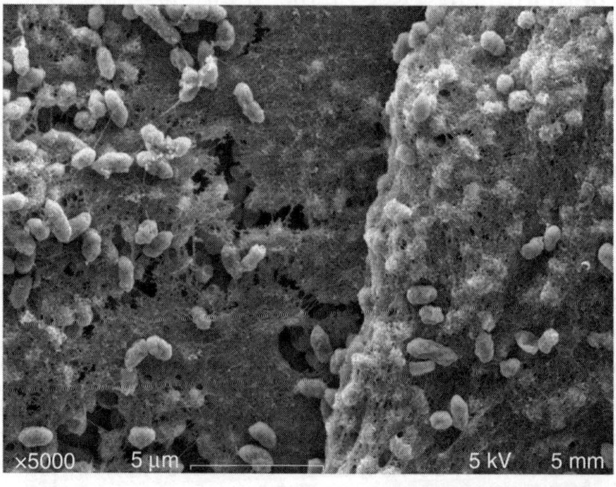


**Fig. 8.1** Scanning electron micrograph of the biofilm produced by *Salmonella* Enteritidis at the air–liquid interface in LB medium after three days of incubation at room temperature. Thick fibrous matrix layers that embed bacteria can be observed.

(a) Congo Red agar 72 hours at room temperature (b) Luria-Bertani (LB) medium 72 hours at room temperature (c) ATM medium, 4 hours at 37°C under shaking conditions (d) Calcofluor supplemented agar 48 hours at room temperature

**Fig. 8.2** Different phenotypes displayed by a biofilm positive strain of *Salmonella* Enteritidis under diverse environmental conditions. a) Rdar morphotype on Congo Red agar after 72 h of incubation at room temperature. b) Pellicle developed at the air-liquid interface of a standing culture in LB after 72 h of incubation at room temperature. c) Ring of bacterial cells adhered to the surface of a glass tube after 4 h of incubation in ATM under vigorous shaking conditions. d) Fluorescence on calcofluor agar plates as result of cellulose production after 48 h of incubation at room temperature.

Besides, biofilm formation on human gallstones facilitates gallbladder colonization and maintenance of the carrier state (Crawford *et al.*, 2008; Lai *et al.*, 1992; Prouty *et al.*, 2002; Prouty and Gunn, 2003; Raad *et al.*, 1998). However, biofilm formation seems not to be involved in virulence during acute *Salmonella* infections, though it certainly contributes to the accumulation and dispersal of a sufficient number of pathogens that may reach an infective dose.

At present, it has become clear that different types of biofilms exist, and that their characteristics might depend on the environmental conditions or on the surface available for adhesion. Accordingly, the biofilm formation process of *Salmonella* exhibits various phenotypes such as a macroscopic pellicle formed on the surface of a liquid culture (Fig. 8.2b), cell adhesion to the surface of a test tube (Fig. 8.2c), cell aggregation in liquid culture and and also the display of a certain colony morphology in the presence of Congo Red dye (Fig. 8.2a). In this medium, the biofilm positive strains produce red, dry and rough colonies known as the rdar morphotype (Romling *et al.*, 1998b). Presumably, these different multicellular behaviours are a consequence of the diverse environmental conditions that *Salmonella* encounters and senses, which trigger the production of different compounds of the extracellular matrix, such as thin aggregative fimbriae, cellulose, BapA and additional exopolysaccharides.

### 8.2.1 Role of thin aggregative fimbriae

Sequence analysis has identified multiple putative fimbrial operons in the *S.* Typhimurium genome (Chessa *et al.*, 2008; McClelland *et al.*, 2001). Under standard laboratory conditions, *Salmonella* Typhimurium expresses type 1 fimbriae (Duguid *et al.*, 1966) and thin aggregative fimbriae (Tafi) (formerly known as curli fimbriae) (Grund and Weber, 1988). Other fimbrial genes have been identified: *pef* (Friedrich *et al.*, 1993), *lpf* (Baumler and Heffron, 1995), *bcf* (Tsolis *et al.*, 1999a), *stf* (Morrow *et al.*, 1999), *saf* (Folkesson *et al.*, 1999), *stb*, *stc*, *std*, *sth*, *sti* and *stj* (McClelland *et al.*, 2001); however, the lack of *in vitro* expression has not so far allowed establishment of their role in nature.

Studies performed by the group of Ute Romling have shown that thin aggregative fimbriae are one of the compounds of the extracellular matrix of the biofilm formed by *Salmonella* when it is grown in rich media, at room temperature, under standing conditions. Thus, knocking out the operon that encodes for this type of fimbriae results in the loss of the ability to produce a biofilm at the air–medium interface and also in the production of a pink colony on Congo Red Agar (pdar phenotype; pink dry and rough) (Romling *et al.*, 1998b; Romling and Rohde, 1999). Even though, as stated above, multiple fimbrial operons are present in *S.* Typhimurium, three of which are also present in *S.* Enteritidis (Humphries *et al.*, 2001), only the presence of curli thin aggregative fimbriae and type I fimbriae, whose adhesin FimH has been linked to biofilm formation on epithelial cells (Boddicker *et al.*, 2002), are related to biofilm formation capacity. *Salmonella agf* genes (thin aggregative fimbriae) encode for thin aggregative fimbriae expression or SEF17 expression in *S.* Enteritidis. These genes are normally referred to as *csg* owing to the existing homology with *E. coli csg* (curli subunit gen) genes that encode for curli fimbriae (Collinson *et al.*, 1996; Hammar *et al.*, 1995; Romling *et al.*, 1998a). The *csg* region is organized into two operons transcribed in opposite directions, *csg*DEFG and *csg*BAC, which are connected by a 521 bp intergenic region (Hammar *et al.*, 1995; 1996; Romling *et al.*, 1998a; Sukupolvi *et al.*, 1997). The *csg*BAC operon encodes for the structural protein CsgA, the main thin aggregative subunit, and CsgB, a nucleator protein that induces the polymerization of CsgA and is also present along the filament in small amounts (Hammar *et al.*, 1996; Sukupolvi *et al.*, 1997). These proteins are secreted outside and are assembled by precipitation on the surface of bacteria, where subunits are added to the distal extreme of the fiber in formation, making up branched, thin and irregular structures (Fontanella *et al.*, 1999). *csg*BAC also encodes for the CsgC protein, whose function is not already clear, but may influence certain functional properties of thin aggregative fimbriae (Hammar *et al.*, 1995). On the other hand, *csg*DEFG encodes for CsgG, and outer membrane lipoprotein that protects CsgA and CsgB from proteolitic degradation and is needed for the assembling of the subunits (Loferer *et al.*, 1997). CsgE and CsgF are assumed to be also required for the assembling of subunits (Hammar *et al.*, 1995), and CsgD is a transcriptional LuxR like regulator that transcriptionally controls the expression of the *csg*BAC operon and other genes involved in multicellular behaviour (Hammar *et al.*, 1995; Romling *et al.*, 2000) (Fig. 8.3). Thin

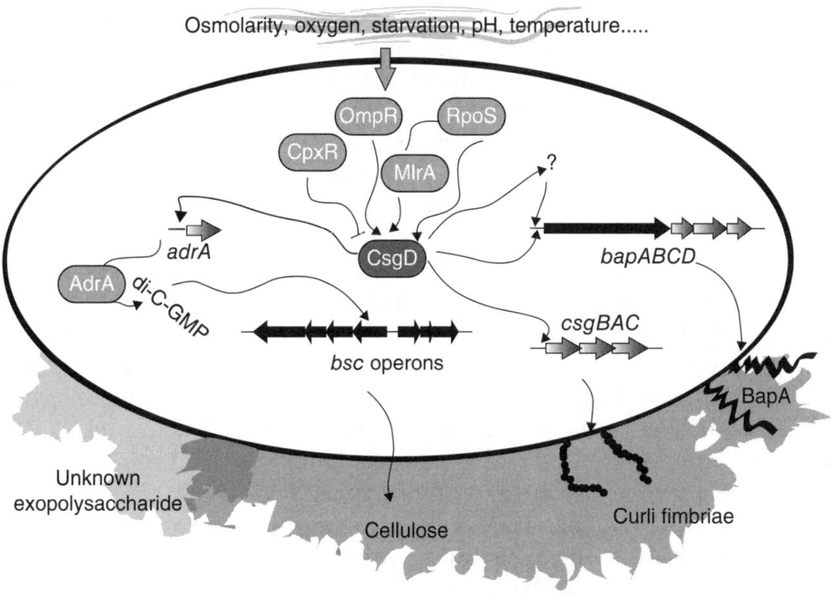


**Fig. 8.3** Genetic model for biofilm formation in *Salmonella*. CsgD, a key regulator of biofilm formation in LB, modulates and coordinates the synthesis of fimbriae, cellulose and BapA surface protein.

aggregative fimbriae establish the short range cell-to-cell interactions necessary for multicellular behaviour and their expression is required for biofilm formation on inert surfaces, pellicle formation on liquid media, cell aggregation in large clumps, and the ability to bind the hydrophobic Red Congo dye (Austin *et al.*, 1998; Collinson *et al.*, 1993; Romling *et al.*, 1998a; Romling *et al.*, 2000; Woodward *et al.*, 2000).

The vast majority of serotypes from *S.* Typhimurium and *S.* Enteritidis only express thin aggregative fimbriae at temperatures below 30°C and under low osmolarity conditions. However, it has been reported that spontaneous mutations in the promoter region of the regulator csgD can result in a constitutive or derepressed expression of thin aggregative fimbriae (Romling *et al.*, 1998a; Uhlich *et al.*, 2001), a fact that could lead to a phenotypic variation for ensuring bacterial survival under environmental changes.

### 8.2.2 Cellulose

*Salmonella* mutants deficient in thin aggregative fimbriae are still connected in an elastic fashion when they are incubated under biofilm forming conditions, evidencing the existence of a second matrix compound (Romling *et al.*, 2000). Using a screening method based on the fluorescence of colonies on calcofluor agar plates (Fig. 8.2d) two operons, *bcsABZC* and *bcsEFG* involved in the synthesis of cellulose (Solano *et al.*, 2002; Zogaj *et al.*, 2001) have been identified. Based on the homology between the *bcs* genes (previously designated as *yhj*) present in *Salmonella* and the *bcs* genes of *Acetobacter xylinum*, *bcsA* has been proposed to encode for the catalytic subunit of the cellulose synthase, which polymerizes uridine 5'-diphosphate glucose to cellulose, while *bcsB* encodes for the cellulose synthase regulatory subunit that binds to the activator cyclic-di-GMP (Ross *et al.*, 1991) (Fig. 8.3). BcsC has been shown to be required for maximal cellulose synthesis. Finally, *bcsZ* encodes for an endoglucanase that hydrolyses 1,4-beta-D-glucosidic linkages (Standal *et al.*, 1994). Similarly, a database homology search of genes in the second operon suggested that *bcsE* encodes for a putative protease, while *bcsF* and *bcsG* displayed insufficient homology at amino acidic level to be assigned a putative function (Solano *et al.*, 2002). Cellulose production confers a significant chlorine resistance in the range used on equipment surfaces in food-processing plants (Solano *et al.*, 2002) and enhances survival during desiccation in the absence of nutrients (White *et al.*, 2006). However, up to day, there is no experimental evidence showing that cellulose plays a role in virulence, as cellulose deficient mutants do not display any deficiency in their capacity to colonize human gallstones (Prouty and Gunn, 2003) or during intestinal colonization and survival assays both in BALB/c mice and in one-day old chickens (Solano *et al.*, 2002).

Solano *et al.* (1998), assuming that nutrient-stress conditions might mimic natural conditions, proposed the use of a medium (ATM) deficient in several essential elements, such as nitrogen, phosphorous, calcium, magnesium, sulphur and iron to study *Salmonella* biology. Under these conditions, aggregation of cells occurred in the context of forming thick biofilms, cellulose being the single component of the biofilm matrix (Solano *et al.*, 2002). These authors also found that some S. enterica host-adapted serovars that cause systemic diseases did not express the biofilm phenotype in ATM, supporting again the role of cellulose in long-term survival in the environment, a stage that is not required in the cell cycle of host-adapted serovars (Solano *et al.*, 2002; White *et al.*, 2006). That is to say, the expression of cellulose would facilitate the survival of *Salmonella* in non-host nutrient-poor environments enabling more cells to be available for host infection (Winfield and Groisman, 2003).

### 8.2.3 Biofilm associated proteins (BapA)

BapA is the second largest protein encoded by the *Salmonella enterica* ser. *typhimurium* (*S.* Typhimurium) genome (Latasa *et al.*, 2005). The region B of BapA (aa 159 to 3003) contains 28 tandem imperfect repeats of 86–106 amino acids. Although the precise function of the repeats remains unknown, it has been proposed that they might be needed to anchor the N-terminal region of the protein in the correct orientation at the cell surface, whereas the phase-switching differences in the number of repeats might be related to an evasion from the immune response. Deletion of *bap*A caused the loss of the capacity to form a biofilm pellicle in LB, at the air-broth interface, whereas *bapA* overexpression increased the thickness and strength of the biofilm formed (Fig. 8.3). The mechanism through which BapA promotes cell–cell interactions is not well understood, but BapA deficiency could be counteracted by overproduction of curli fimbriae but not cellulose, suggesting that BapA could play a complementary role to fimbriae in connecting cells, either by strengthening fimbriae mediated interactions or by allowing the interconnection of bacteria separated by long distances.

A murine model of infection revealed that animals inoculated orally with a lethal dose of a BapA deficient strain were able to survive longer, many of them (55%) reaching a complete clearance of the infection, in comparison to animals infected with the wild-type strain (Latasa *et al.*, 2005). Furthermore, when ileal-ligated loops of mice were simultaneously infected with the wild-type and the BapA deficient strains, the BapA mutant displayed a reduced capacity to adhere and invade the intestinal epithelium. Whether this deficiency of the BapA mutant to colonize the intestinal epithelia is due to a failing in the colonization of the apical membranes of epithelial cells (enterocytes) or of specialized M cells, remains unknown. Once the BapA mutant strain has passed through the epithelial barrier, the capacity to colonize the target organs (spleen, liver, gall bladder and caecum) is similar to that of the wild-type strain, indicating that BapA does not play a significant role at the stage of infection in the internal organs.

## 8.3 Resistance of *Salmonella* biofilms to sanitizers

As stated above, *Salmonella* strains, that are amongst the most hard-hitting of all foodborne pathogens, are capable of forming biofilms in a large variety of food processing environments (Korber *et al.*, 1997; Solano *et al.*, 2002; Stepanović *et al.*, 2003), a feature of significance from the perspective of food safety and public health that leads to important economic losses. For this reason, increasing efforts are being made to improve food safety around the world (WHO, 2007), including the review of sanitization processes. In this regard, there are concerns about the indiscriminate and inappropriate use of biocides that might contribute to the spread of bacterial resistance to this type of compound.

In contrast to most other bacteria, *Salmonella* is able to survive under diverse environmental conditions, such as variable temperature and a wide pH or water activity range, thanks to its genomic plasticity (Foster and Spector, 1995). This wide gene regulation is responsible for pleiotropic effects. For instance, chilling induces cell elongation and also heat and acid tolerance in *Salmonella* Enteritidis. In addition, it can even increase *Salmonella* virulence (Phillips *et al.*, 1998). Moreover, dehydration may lead to heat resistance (Kirby and Davies, 1990). Such an array of adaptation-resistance is further potentiated by the fact that, as stated above, *Salmonella* is able to form biofilms on materials of different nature and under different growth conditions (Korber *et al.*, 1997; Mangalappalli-Illathu and Korber, 2006; Solano *et al.*, 2002). Nutrient and water limitation, equipment design, and temperature regulation are very relevant in biofilm control (Chmielewski and Frank, 2003). However, such control techniques that imply stress bacteria may be directly or indirectly responsible of the recalcitrance of bacterial contamination against microbial control measures. Several complex factors are behind biofilm resistance to classical sanitization processes. Initially, resistance was simplistically attributed to the diffusion barrier that biofilm represents to heat, chemicals, and biocides in general. However, standing against the barrier hypothesis, the importance of circulatory channels inside hydrated biofilms has been demonstrated (Scher *et al.*, 2005). Furthermore, amongst the factors responsible for such increased resistance is the very slow growth rate displayed by cells inside a biofilm (Costerton, 1995), that may even reach the viable-nonculturable state in the nutrient-depleted interior of the biofilm (Gilbert *et al.*, 2002). Thus, under these nutrient stress-conditions *Salmonella* cells may get such new resistance properties. Accordingly, it has been reported that *Salmonella* Typhimurium reached the viable-nonculturable state after exposure to chlorine stress (Leriche and Carpentier, 1995). Trisodium phosphate, a post-chill antimicrobial agent used on raw poultry, has been found to be effective for reducing planktonic and, to a lesser extent biofilm cells of *Salmonella* Typhimurium. However, sub-lethal concentrations of this compound have been shown to induce thermotolerance, that could be due to *de novo* protein synthesis and also altered membrane fatty acid composition (Sampathkumar *et al.*, 2004).

As stated above, cellulose production confers a significant chlorine resistance to *Salmonella* in the range used on equipment surfaces in food-processing plants (Solano *et al.*, 2002) and enhances survival during desiccation in the absence of nutrients (White *et al.*, 2006). In accordance, Lang *et al.* (2004) reported that a chlorine treatment at a concentration of 200 mg/L reduced *Salmonella* contamination less than 2 log on experimentally inoculated lettuce, due to cell aggregation and the biofilm produced. Therefore, although it may be accepted that chlorine is able to depolymerize biofilm matrices (Kumar and Anand, 1998), *Salmonella* biofilms resists chlorination treatments due to cellulose production (Stevenson *et al.*, 2000; White *et al.*, 2006).

Similar observations on resistance of *Salmonella* biofilms to biocides have been reported by other groups (Korber *et al.*, 1997; Mangalappalli-Illathu *et al.*, 2008). Ukuku *et al.*, working on cantaloupes, described the inability of a variety of sanitizing rinses to control *Salmonella* owing to the formation of bacterial biofilms in experimentally contaminated fruits, and demonstrated an increased strength of attachment from day 0 to day 7 after bacterial incubation. Thus, the effect of rinsing with chlorine and hydrogen peroxide decreased significantly when the organism was allowed to reside on the melon surface for more than 24 h (Ukuku and Sapers, 2001; Ukuku and Fett, 2002, 2006).

Another interesting topic related to the inappropriate use of biocides is the emergence of cross-resistance to certain therapeutic antibiotics mediated by nonspecific alterations of the cell envelope, drug degradation, or active antibiotic efflux (Braoudaki and Hilton, 2005; Fraise, 2002; Russell, 2002). In this regard, cationic agents (i.e., quaternary ammonium compounds, chlorhexidine, and triclosan) have been involved in the selection of antibiotic resistant cells. In the particular case of *Salmonella*, it has been reported that the use of low concentrations of benzalkonium chloride (cationic surfactant widely used as a disinfectant in the food industry) may favour the development of resistance to other biocides (chlorhexidine and triclosan), and also to several antibiotics (including, amoxicillin, clavulanic acid, and tetracycline (Braoudaki and Hilton, 2005; Mangalappalli-Illathu and Korber, 2006; Szomolay *et al.*, 2005). Also, *Salmonella* cross-resistance to hypochlorous acid and $H_2O_2$ has been reported (Mokgatla *et al.*, 1998; Seymour *et al.*, 1996).

Food contact surfaces are critical to allow bacterial initial attachment. In some cases, the contact time between foods and a surface may vary between 24 and 48 hours depending on the processing conditions, and once a biofilm is formed, such a surface becomes a source of cross-contamination (Beuchat, 2002). Not only the efficiency of biofilm formation but also the resistance to biocide treatments varies depending on the surface characteristics (hydrophobicity and roughness, amongst others). In this respect, *Salmonella* biofilms formed on plastic have been shown to be more resistant to cleaning than the ones produced on a more hydrophilic surface such as stainless steel (Joseph *et al.*, 2001). Therefore, although currently there are no materials in the market inherently resistant to biofilm formation, the use of equipment with smooth surfaces may help in the prevention of biofilm formation (Giaouris and Nychas, 2006; Manijen *et al.*, 2008; Sinde and Carballo, 2000; Stepanović *et al.*, 2003).

On the other hand, a matter of significance is the cleaning procedure of the processing equipment prior to sanitation in order to ensure that cells in the nascent biofilms can be reached by sanitizers (Carpentier and Cerf, 1993; Chmielewski and Frank, 2003; Gibson *et al.*, 1999; Lomander *et al.*, 2004; Samrakandi *et al.*, 1997; Zottola and Sasahara, 1994). For this reason, new methods including high magnetic fields, ultrasound treatment, and high-pulsed electric fields have been developed (Kumar and Anand, 1998).

## 8.4 Future trends and concluding remarks

The *Salmonella* genus is highly prevalent in food products (see Table 8.1), and *Salmonella* serovars are the most frequently reported cause of food-borne gastroenteritis in many developed countries. The application of control measures to reduce the horizontal transmission and persistence of *Salmonella* in the environment requires the study of the biofilm formation process. It is accepted that biofilms are ubiquitous, in fact, most isolates of *Salmonella* spp. are able to synthesize the main components of the biofilm matrix (curli, cellulose and BapA) (Romling *et al.*, 2003; Solano *et al.*, 2002; Solomon *et al.*, 2005a). Therefore, the major concern in food-related management should be the development of methods to remove biofilms.

With regard to biofilm components, fimbriae are necessary for the initial phase of bacterial attachment to solid surfaces. After adhesion, bacteria produce several proteins and polisaccharides of different composition, which together with fimbriae contribute to the formation of a voluminous biofilm (Fig. 8.1). Various environmental factors regulate bacterial biofilms produced by *Salmonella*, which differ from each other with respect to the physicochemical structure and chemical composition of the matrix. This extracellular matrix, rich in cellulose, contributes to an increase in the bacterial resistance to different environmental stresses, including desiccation, starvation and chemical stress, such as commercial biocides.

Alternative approaches to control *Salmonella* biofilms are necessary. Some authors have achieved positive results by using salicylate-based poly anhydride esters (PAE), that mainly affect surface attachment (Rosenberg *et al.*, 2008). Blocking or modifying the matrix synthesis, as well as disrupting signalling communication, known as quorum sensing, required for normal biofilm formation (Kjelleberg and Molin, 2002; Labbate *et al.*, 2004; Parsek and Greenberg, 2005) may provide new treatment strategies to prevent biofilm prudcution (Federle and Bassler, 2003; Girennavar *et al.*, 2008; Wu *et al.*, 2004).

The ecophysiology of *Salmonella* biofilms under normal or sub-lethal concentrations of physical and chemical antimicrobial agents requires further research. New cost-effective strategies to monitor and reduce *Salmonella* biofilm contamination in the food chain will reduce the incidence of human salmonellosis, being the expression of biofilm-mediated structures a main pillar that must be taken into account.

## 8.5 Sources of further information and advice

Biofilm formation by *Salmonella* spp. may occur virtually on all surface types. This chapter has discussed how a biofilm develops and is influenced by environmental parameters. There are many concerns about how the presence of biofilms may contribute to the spread of bacterial cross-contamination in the food processing chain. However, the significance of biofilms in the industrial operations is often underestimated, especially with regard to sanitization processes. Nowadays, the availability of the genome sequences of several *Salmonella* serovars has greatly contributed to the understanding of the molecular bases and the mechanisms of expression of virulence factors related with the cellular biology of these pathogens. This knowledge may help to develop new approaches to the identification and control of hazards at all stages in the supply chain.

The following reviews describe the biofilm formation process, especially with regard to the food industry in order to control biofilm development by *Salmonella* strains and diminish the risk of diseases as a consequence of its formation.

Jordan *et al.*, 2006
The data presented in this paper will help the industry and regulatory agencies target sectors that need additional monitoring or more effective controls.

Chmielewski and Frank, 2003
Comprehensive review focused on the recognition of critical points in safety for the food industry: detection of biofilms, removal methods and strategies to prevent their formation.

Center for Disease Control and Prevention, Atlanta, GA, USA
http://www.cdc.gov/salmonella/
Foodborne outbreak response and surveillance unit. Information on surveillance and analyses of data gathered during the investigations of salmonellosis cases.

*Food safety control in the poultry industry.* Edited by G Mead. Woodhead Publishing Limited, Abington Hall, Abington, Cambridge, CB21 6AH, England. 2005
This book offers comprehensive chapters on the research focused on the identification and control of hazards at all stages in the supply chain of poultry meat and eggs. Routes of transmission of *Salmonella* into poultry flocks and eggs and sources of further information and advice are provided.

*Food consumption and disease risk: Consumer-pathogen interactions.* Edited by M Potter, US FDA, USA. Woodhead Publishing Limited, Cambridge, England. 2006
The book includes chapters on the risk of exposure to foodborne pathogens and factors which increase hazards in food and agricultural management. *Salmonella* mechanisms of colonization, attachment and invasion are also covered.

*Salmonella: Molecular Biology and Pathogenesis.* Mikael Rhen, Duncan Maskell, Pietro Mastroeni, John Threlfall. Horizon Scientific Press, 2007
In this book, cutting-edge topics in Salmonella research are reviewed. Topics include epidemiology, molecular typing, antibiotic resistance, host-interaction in the gut, adhesins and biofilms

Humphrey, 2004
The link between biocide and antibiotic resistance is of considerable importance to food safety and public health. This article is about the ability of biofilms to become adapted to sublethal concentrations of antimicrobial agents and to eventually become resistant to lethal biocide concentrations.

## 8.6 References


ANON, (2004) Outbreak of *Salmonella* serotype Enteritidis infections associated with raw almonds – United States and Canada, 2003–2004. *MMWR Morb Mortal Wkly Rep* **53**: 484–487.

ANRIANY, Y.A., WEINER, R.M., JOHNSON, J.A., DE REZENDE, C.E. and JOSEPH, S.W. (2001) Salmonella enterica serovar Typhimurium DT104 displays a rugose phenotype. *Appl Environ Microbiol* **67**: 4048–4056.

ASHBOLT, R. and KIRK, M.D. (2006) Salmonella Mississippi infections in Tasmania: the role of native Australian animals and untreated drinking water. *Epidemiol Infect* **134**: 1257–1265.

AUSTIN, J.W., SANDERS, G., KAY, W.W. and COLLINSON, S.K. (1998) Thin aggregative fimbriae enhance Salmonella enteritidis biofilm formation. *FEMS Microbiol Lett* **162**: 295–301.

BARAK, J.D., WHITEHAND, L.C. and CHARKOWSKI, A.O. (2002) Differences in attachment of Salmonella enterica serovars and Escherichia coli O157:H7 to alfalfa sprouts. *Appl Environ Microbiol* **68**: 4758–4763.

BAUMLER, A.J. and HEFFRON, F. (1995) Identification and sequence analysis of lpfABCDE, a putative fimbrial operon of Salmonella typhimurium. *J Bacteriol* **177**: 2087–2097.

BEARSON, B.L. and BEARSON, S.M. (2008) The role of the QseC quorum-sensing sensor kinase in colonization and norepinephrine-enhanced motility of Salmonella enterica serovar Typhimurium. *Microb Pathog* **44**: 271–278.

BERGER, C.N., SHAW, R.K., BROWN, D.J., MATHER, H., CLARE, S., DOUGAN, G., PALLEN, M.J. and FRANKEL, G. (2008) Interaction of Salmonella enterica with basil and other salad leaves. *ISME Journal* (2009), **3**: 26–265.

BERRANG, M.E., FRANK, J.F., BUHR, R.J., BAILEY, J.S., COX, N.A. and MAULDIN, J. (1998) Eggshell characteristics and penetration by Salmonella through the productive life of a broiler breeder flock. *Poult Sci* **77**: 1446–1450.

BEUCHAT, L.R. (2002) Ecological factors influencing survival and growth of human pathogens on raw fruits and vegetables. *Microbes Infect* **4**: 413–423.

BEUCHAT, L.R. and RYU, J.H. (1997) Produce handling and processing practices. *Emerg Infect Dis* **3**: 459–465.

BHATIA, T.R. and MCNABB, G.D. (1980) Dissemination of Salmonella in broiler-chicken operations. *Avian Dis* **24**: 616–624.

BLASER, M.J. and NEWMAN, L.S. (1982) A review of human salmonellosis: I. Infective dose. *Rev Infect Dis* **4**: 1096–1106.

BODDICKER, J.D., LEDEBOER, N.A., JAGNOW, J., JONES, B.D. and CLEGG, S. (2002) Differential binding to and biofilm formation on, HEp-2 cells by Salmonella enterica serovar Typhimurium is dependent upon allelic variation in the fimH gene of the fim gene cluster. *Mol Microbiol* **45**: 1255–1265.

BOWEN, A, FRY, A, RICHARDS, G, and BEUCHAT, L. (2006) Infections associated with cantaloupe consumption: a public health concern. *Epidemiol Infect* **134**: 675–685.

BRADEN, C.R. (2006) Salmonella enterica serotype Enteritidis and eggs: a national epidemic in the United States. *Clin Infect Dis* **43**: 512–517.

BRANDL, M.T. and MANDRELL, R.E. (2002) Fitness of Salmonella enterica serovar Thompson in the cilantro phyllosphere. *Appl Environ Microbiol* **68**: 3614–3621.

BRAOUDAKI, M. and HILTON, A.C. (2005) Mechanisms of resistance in Salmonella enterica adapted to erythromycin, benzalkonium chloride and triclosan. *Int J Antimicrob Agents* **25**: 31–37.

BURNETT, S.L. and BEUCHAT, L.R. (2000) Human pathogens associated with raw produce and unpasteurized juices, and difficulties in decontamination. *J Ind Microbiol Biotechnol* **25**: 281–287.

CAHILL, S.M., WACHSMUTH, I.K., COSTARRICA MDE, L. and BEN EMBAREK, P.K. (2008) Powdered infant formula as a source of Salmonella infection in infants. *Clin Infect Dis* **46**: 268–273.

CAMPBELL, G.R. (2001) Does drinking carrot juice affect cancer of the prostate? *Med J Aust* **175**: 51; author reply 53–54.

CARPENTIER, B. and CERF, O. (1993) Biofilms and their consequences, with particular reference to hygiene in the food industry. *J Appl Bacteriol* **75**: 499–511.

CHESSA, D., WINTER, M.G., NUCCIO, S.P., TUKEL, C. and BAUMLER, A.J. (2008) RosE represses Std fimbrial expression in Salmonella enterica serotype Typhimurium. *Mol Microbiol* **68**: 573–587.

CHMIELEWSKI, R.A.N. and FRANK, J.F. (2003) Biofilm formation and control in food processing facilities. *Comp Rev Food Sci Food Safety* **2**: 22–32.

COLLINSON, S.K., DOIG, P.C., DORAN, J.L., CLOUTHIER, S., TRUST, T.J. and KAY, W.W. (1993) Thin, aggregative fimbriae mediate binding of Salmonella enteritidis to fibronectin. *J Bacteriol* **175**: 12–18.

COLLINSON, S.K., CLOUTHIER, S.C., DORAN, J.L., BANSER, P.A. and KAY, W.W. (1996) Salmonella enteritidis agfBAC operon encoding thin, aggregative fimbriae. *J Bacteriol* **178**: 662–667.

COMPTON, J.A., BANEY, J.A., DONALDSON, S.C., HOUSER, B.A., SAN JULIAN, G.J., YAHNER, R.H., CHMIELECKI, W., REYNOLDS, S. and JAYARAO, B.M. (2008) Salmonella Infections in the common raccoon (Procyon lotor) in Western Pennsylvania. *J Clin Microbiol* **46**: 3084–3086.

COOK, K.A., DOBBS, T.E., HLADY, W.G., WELLS, J.G., BARRETT, T.J., PUHR, N.D., LANCETTE, G.A., BODAGER, D.W., TOTH, B.L., GENESE, C.A., HIGHSMITH, A.K., PILOT, K.E., FINELLI, L. and SWERDLOW, D.L. (1998) Outbreak of Salmonella serotype Hartford infections associated with unpasteurized orange juice. *Jama* **280**: 1504–1509.

COSTERTON, J.W. (1995) Overview of microbial biofilms. *J Ind Microbiol* **15**: 137–140.

COSTERTON, J.W., LEWANDOWSKI, Z., CALDWELL, D.E., KORBER, D.R. and LAPPIN-SCOTT, H.M. (1995) Microbial biofilms. *Annu Rev Microbiol* **49**: 711–745.

CRAWFORD, R.W., GIBSON, D.L., KAY, W.W. and GUNN, J.S. (2008) Identification of a bile-induced exopolysaccharide required for Salmonella biofilm formation on gallstone surfaces. *Infect Immun* **76**: 5341–5349.

D'AOUST, J.Y., ARIS, B.J., THISDELE, P., DURANTE, A., BRISSON, N., DRAGON, D., LEACHAPELLE, G., JOHNSTON, M. and LAIDLEY, R. (1975) Salmonella eastbourne outbreak associated with chocolate. *Can Inst Food Sci Technol* **8**: 181–184.

DAVIES, R.H. and WRAY, C. (1996) Persistence of Salmonella enteritidis in poultry units and poultry food. *Br Poult Sci* **37**: 589–596.

DAVIES, R.H., TEALE, C.J., WRAY, C., MCLAREN, I.M., JONES, Y.E., CHAPPELL, S. and KIDD, S. (1999) Nalidixic acid resistance in salmonellae isolated from turkeys and other livestock in Great Britain. *Vet Rec* **144**: 320–322.

DELPECH, V., MCANULTY, J. and MORGAN, K. (1998) A salmonellosis outbreak linked to internally contaminated pork meat. *Aust N Z J Public Health* **22**: 243–246.

DHIAF, A. and BAKHROUF, A. (2007) Fate of Salmonella typhimurium on rosemary and barley grown in fields treated with contaminated irrigation water. *African Journal of Microbiology Research* **1(4)**: 042–045.

DHIR, V.K. and DODD, C.E. (1995) Susceptibility of suspended and surface-attached Salmonella enteritidis to biocides and elevated temperatures. *Appl Environ Microbiol* **61**: 1731–1738.

DIPERSIO, P.A., KENDALL, P.A., YOON, Y. and SOFOS, J.N. (2007) Influence of modified blanching treatments on inactivation of Salmonella during drying and storage of carrot slices. *Food Microbiol* **24**: 500–507.

DORAN, G., SHERIDAN, F., DELAPPE, N., C, O.H., ANDERSON, W., CORBETT-FEENEY, G. and CORMICAN, M. (2005) Salmonella enterica serovar Kedougou contamination of commercially grown mushrooms. *Diagn Microbiol Infect Dis* **51**: 73–76.

DUFFY, E.A., CISNEROS-ZEVALLOS, L., CASTILLO, A., PILLAI, S.D., RICKE, S.C. and ACUFF, G.R. (2005) Survival of salmonella transformed to express green fluorescent protein on Italian parsley as affected by processing and storage. *J Food Prot* **68**: 687–695.

DUGUID, J.P., ANDERSON, E.S. and CAMPBELL, I. (1966) Fimbriae and adhesive properties in Salmonellae. *J Pathol Bacteriol* **92**: 107–138.

ERCOLANI, G.L. (1976) Bacteriological quality assessment of fresh marketed lettuce and fennel. *Appl Environ Microbiol* **31**: 847–852.

FEDERLE, M.J. and BASSLER, B.L. (2003) Interspecies communication in bacteria. *J Clin Invest* **112**: 1291–1299.

FETT, W.F. (2000) Naturally occurring biofilms on alfalfa and other types of sprouts. *J Food Prot* **63**: 625–632.

FOLEY, S.L. and LYNNE, A.M. (2008) Food animal-associated Salmonella challenges: pathogenicity and antimicrobial resistance. *J Anim Sci* **86**: E173–187.

FOLKESSON, A., ADVANI, A., SUKUPOLVI, S., PFEIFER, J.D., NORMARK, S. and LOFDAHL, S. (1999) Multiple insertions of fimbrial operons correlate with the evolution of Salmonella serovars responsible for human disease. *Mol Microbiol* **33**: 612–622.

FONTANELLA, L., POZZUOLO, S., COSTANZO, A., FAVARO, R., DEHO, G. and TORTORA, P. (1999) Photometric assay for polynucleotide phosphorylase. *Anal Biochem* **269**: 353–358.

FOSTER, J.W. and SPECTOR, M.P. (1995) How Salmonella survive against the odds. *Annu Rev Microbiol* **49**: 145–174.

FRAISE, A.P. (2002) Biocide abuse and antimicrobial resistance – a cause for concern? *J Antimicrob Chemother* **49**: 11–12.

FRIEDRICH, M.J., KINSEY, N.E., VILA, J. and KADNER, R.J. (1993) Nucleotide sequence of a 13.9 kb segment of the 90 kb virulence plasmid of Salmonella typhimurium: the presence of fimbrial biosynthetic genes. *Mol Microbiol* **8**: 543–558.

GANDHI, M., GOLDING, S., YARON, S. and MATTHEWS, K.R. (2001) Use of green fluorescent protein expressing Salmonella Stanley to investigate survival, spatial location, and control on alfalfa sprouts. *J Food Prot* **64**: 1891–1898.

GAWANDE, P.V. and BHAGWAT, A.A. (2002) Inoculation onto solid surfaces protects Salmonella spp. during acid challenge: a model study using polyethersulfone membranes. *Appl Environ Microbiol* **68**: 86–92.

GIAOURIS, E.D. and NYCHAS, G.J. (2006) The adherence of Salmonella Enteritidis PT4 to stainless steel: the importance of the air-liquid interface and nutrient availability. *Food Microbiol* **23**: 747–752.

GIBSON, H., TAYLOR, J.H., HALL, K.E. and HOLAH, J.T. (1999) Effectiveness of cleaning techniques used in the food industry in terms of the removal of bacterial biofilms. *J Appl Microbiol* **87**: 41–48.

GILBERT, P., MAIRA-LITRAN, T., MCBAIN, A.J., RICKARD, A.H. and WHYTE, F.W. (2002) The physiology and collective recalcitrance of microbial biofilm communities. *Adv Microb Physiol* **46**: 202–256.

GIRENNAVAR, B., CEPEDA, M.L., SONI, K.A., VIKRAM, A., JESUDHASAN, P., JAYAPRAKASHA, G.K., PILLAI, S.D. and PATIL, B.S. (2008) Grapefruit juice and its furocoumarins inhibits autoinducer signaling and biofilm formation in bacteria. *Int J Food Microbiol* **125**: 204–208.

GOLDEN, D.A., RHODEHAMEL, E.J. and KAUTTER, D.A. (1993) Growth of *Salmonella* spp. in cantaloupe, watermelon and honeydew melons, *J Food Prot* **56**: 194–196.

GRUND, S. and WEBER, A. (1988) A new type of fimbriae on Salmonella typhimurium. *Zentralbl Veterinarmed B* **35**: 779–782.

HAMMAR, M., ARNQVIST, A., BIAN, Z., OLSEN, A. and NORMARK, S. (1995) Expression of two csg operons is required for production of fibronectin- and congo red-binding curli polymers in Escherichia coli K-12. *Mol Microbiol* **18**: 661–670.

HAMMAR, M., BIAN, Z. and NORMARK, S. (1996) Nucleator-dependent intercellular assembly of adhesive curli organelles in Escherichia coli. *Proc Natl Acad Sci USA* **93**: 6562–6566.

HANDELAND, K., NESSE, L.L., LILLEHAUG, A., VIKOREN, T., DJONNE, B. and BERGSJO, B. (2008) Natural and experimental Salmonella Typhimurium infections in foxes (*Vulpes vulpes*). *Vet Microbiol* **132**: 129–134.

HEDBERG, C.W., KORLATH, J.A., D'AOUST, J.Y., WHITE, K.E., SCHELL, W.L., MILLER, M.R., CAMERON, D.N., MACDONALD, K.L. and OSTERHOLM, M.T. (1992) A multistate outbreak of Salmonella javiana and Salmonella oranienburg infections due to consumption of contaminated cheese. *Jama* **268**: 3203–3207.

HEDBERG, C.W., ANGULO, F.J., WHITE, K.E., LANGKOP, C.W., SCHELL, W.L., STOBIERSKI, M.G., SCHUCHAT, A., BESSER, J.M., DIETRICH, S., HELSEL, L., GRIFFIN, P.M., MCFARLAND, J.W. and OSTERHOLM, M.T. (1999) Outbreaks of salmonellosis associated with eating uncooked tomatoes: implications for public health. The Investigation Team. *Epidemiol Infect* **122**: 385–393.

HENSEL, M., SHEA, J.E., WATERMAN, S.R., MUNDY, R., NIKOLAUS, T., BANKS, G., VAZQUEZ-TORRES, A., GLEESON, C., FANG, F.C. and HOLDEN, D.W. (1998) Genes encoding putative effector proteins of the type III secretion system of Salmonella pathogenicity island 2 are required for bacterial virulence and proliferation in macrophages. *Mol Microbiol* **30**: 163–174.

HILL, V.R. and SOBSEY, M.D. (2003) Performance of swine waste lagoons for removing *Salmonella* and enteric microbial indicators. *Trans Am Soc Agric Eng* **46**: 781–788.

HOWARD, M.B. and HUTCHESON, S.W. (2003) Growth dynamics of Salmonella enterica strains on alfalfa sprouts and in waste seed irrigation water. *Appl Environ Microbiol* **69**: 548–553.

HUMPHREY, T. (2004) Salmonella, stress responses and food safety. *Nat Rev Microbiol* **2**: 504–509.

HUMPHREY, T.J., RICHARDSON, N.P., STATTON, K.M. and ROWBURY, R.J. (1993) Effects of temperature shift on acid and heat tolerance in Salmonella enteritidis phage type 4. *Appl Environ Microbiol* **59**: 3120–3122.

HUMPHRIES, A.D., TOWNSEND, S.M., KINGSLEY, R.A., NICHOLSON, T.L., TSOLIS, R.M. and BAUMLER, A.J. (2001) Role of fimbriae as antigens and intestinal colonization factors of Salmonella serovars. *FEMS Microbiol Lett* **201**: 121–125.

JORDAN, E., EGAN, J., DULLEA, C., WARD, J., MCGILLICUDDY, K., MURRAY, G., MURPHY, A., BRADSHAW, B., LEONARD, N., RAFTER, P. and MCDOWELL, S. (2006) Salmonella surveillance in raw and cooked meat and meat products in the Republic of Ireland from 2002 to 2004. *Int J Food Microbiol* **112**: 66–70.

JOSEPH, B., OTTA, S.K. and KARUNASAGAR, I. (2001) Biofilm formation by salmonella spp. on food contact surfaces and their sensitivity to sanitizers. *Int J Food Microbiol* **64**: 367–372.

KANEKO, K., HAYASHIDANI, H., TAKAHASHI, K., SHIRAKI, Y., LIMAWONGPRANEE, S. and OGAWA, M. (1999) Bacterial contamination in the environment of food factories processing ready-to-eat fresh vegetables. *J Food Prot* **62**: 800–804.

KHAN, A.A., MELVIN, C.D. and DAGDAG, E.B. (2007) Identification and molecular characterization of Salmonella spp. from unpasteurized orange juices and identification of new serotype Salmonella strain S. enterica serovar Tempe. *Food Microbiol* **24**: 539–543.

KINGSLEY, R.A. and BAUMLER, A.J. (2002) Pathogenicity islands and host adaptation of Salmonella serovars. *Curr Top Microbiol Immunol* **264**: 67–87.

KIRBY, R.M. and DAVIES, R. (1990) Survival of dehydrated cells of Salmonella typhimurium LT2 at high temperatures. *J Appl Bacteriol* **68**: 241–246.

KJELLEBERG, S. and MOLIN, S. (2002) Is there a role for quorum sensing signals in bacterial biofilms? *Curr Opin Microbiol* **5**: 254–258.

KORBER, D.R., CHOI, A., WOLFAARDT, G.M., INGHAM, S.C. and CALDWELL, D.E. (1997) Substratum topography influences susceptibility of Salmonella enteritidis biofilms to trisodium phosphate. *Appl Environ Microbiol* **63**: 3352–3358.

KUMAR, C.G. and ANAND, S.K. (1998) Significance of microbial biofilms in food industry: a review. *Int J Food Microbiol* **42**: 9–27.

LABBATE, M., QUECK, S.Y., KOH, K.S., RICE, S.A., GIVSKOV, M. and KJELLEBERG, S. (2004) Quorum sensing-controlled biofilm development in Serratia liquefaciens MG1. *J Bacteriol* **186**: 692–698.

LAI, C.W., CHAN, R.C., CHENG, A.F., SUNG, J.Y. and LEUNG, J.W. (1992) Common bile duct stones: a cause of chronic salmonellosis. *Am J Gastroenterol* **87**: 1198–1199.

LANG, M.M., HARRIS, L.J. and BEUCHAT, L.R. (2004) Survival and recovery of Escherichia coli O157:H7, Salmonella, and Listeria monocytogenes on lettuce and parsley as affected by method of inoculation, time between inoculation and analysis, and treatment with chlorinated water. *J Food Prot* **67**: 1092–1103.

LAPIDOT, A., ROMLING, U. and YARON, S. (2006) Biofilm formation and the survival of Salmonella Typhimurium on parsley. *Int J Food Microbiol* **109**: 229–233.

LATASA, C., ROUX, A., TOLEDO-ARANA, A., GHIGO, J.M., GAMAZO, C., PENADES, J.R. and LASA, I. (2005) BapA, a large secreted protein required for biofilm formation and host colonization of *Salmonella enterica* serovar Enteritidis. *Mol Microbiol* **58**: 1322–1339.

LERICHE, V. and CARPENTIER, B. (1995) Viable but nonculturable *Salmonella typhimurium* in single- and binary-species biofilms in response to chlorine treatment. *J Food Prot* **58**: 1186–1191.

LIAO, C.H. and COOKE, P.H. (2001) Response to trisodium phosphate treatment of *Salmonella Chester* attached to fresh-cut green pepper slices, *Canadian Journal of Microbiology* **47**: 25–32.

LOFERER, H., HAMMAR, M. and NORMARK, S. (1997) Availability of the fibre subunit CsgA and the nucleator protein CsgB during assembly of fibronectin-binding curli is limited by the intracellular concentration of the novel lipoprotein CsgG. *Mol Microbiol* **26**: 11–23.

LOMANDER, A., SCHREUDERS, P., RUSSEK-COHEN, E. and ALI, L. (2004) Evaluation of chlorines' impact on biofilms on scratched stainless steel surfaces. *Bioresour Technol* **94**: 275–283.

MAHON, B.E., PONKA, A., HALL, W.N., KOMATSU, K., DIETRICH, S.E., SIITONEN, A., CAGE, G., HAYES, P.S., LAMBERT-FAIR, M.A., BEAN, N.H., GRIFFIN, P.M. and SLUTSKER, L. (1997) An international outbreak of Salmonella infections caused by alfalfa sprouts grown from contaminated seeds. *J Infect Dis* **175**: 876–882.

MANGALAPPALLI-ILLATHU, A.K. and KORBER, D.R. (2006) Adaptive resistance and differential protein expression of Salmonella enterica serovar Enteritidis biofilms exposed to benzalkonium chloride. *Antimicrob Agents Chemother* **50**: 3588–3596.

MANGALAPPALLI-ILLATHU, A.K., VIDOVIC, S. and KORBER, D.R. (2008) Differential adaptive response and survival of Salmonella enterica serovar enteritidis planktonic and biofilm cells exposed to benzalkonium chloride. *Antimicrob Agents Chemother* **52**: 3669–3680.

MANIJEN, M., MOHAMMAD, J. and KERMANSHAHI ROHA, K. (2008) Biofilm Formation by Salmonella enteritidis on Food Contact Surfaces. *Journal of Biological Sciences* **8**: 502–505.

MATTHYSSE, A.G., DEORA, R., MISHRA, M. and TORRES, A.G. (2008) Polysaccharides cellulose, poly-beta-1,6-n-acetyl-D-glucosamine, and colanic acid are required for optimal binding of Escherichia coli O157:H7 strains to alfalfa sprouts and K-12 strains to plastic but not for binding to epithelial cells. *Appl Environ Microbiol* **74**: 2384–2390.

MCCLELLAND, M., SANDERSON, K.E., SPIETH, J., CLIFTON, S.W., LATREILLE, P., COURTNEY, L., PORWOLLIK, S., ALI, J., DANTE, M., DU, F., HOU, S., LAYMAN, D., LEONARD, S., NGUYEN, C., SCOTT, K., HOLMES, A., GREWAL, N., MULVANEY, E., RYAN, E., SUN, H., FLOREA, L., MILLER, W., STONEKING, T., NHAN, M., WATERSTON, R. and WILSON, R.K. (2001) Complete genome sequence of *Salmonella enterica* serovar Typhimurium LT2. *Nature* **413**: 852–856.

MEAD, P.S., SLUTSKER, L., DIETZ, V., MCCAIG, L.F., BRESEE, J.S., SHAPIRO, C., GRIFFIN, P.M. and TAUXE, R.V. (1999) Food-related illness and death in the United States. *Emerg Infect Dis* **5**: 607–625.

MOHLE-BOETANI, J.C., REPORTER, R., WERNER, S.B., ABBOTT, S., FARRAR, J., WATERMAN, S.H. and VUGIA, D.J. (1999) An outbreak of *Salmonella* serogroup Saphra due to cantaloupes from Mexico, *J Infect Dis* **180**: 1361–1364.

MOKGATLA, R.M., BROZEL, V.S. and GOUWS, P.A. (1998) Isolation of Salmonella resistant to hypochlorous acid from a poultry abattoir. *Lett Appl Microbiol* **27**: 379–382.

MORRIS, C.E., MONIER, J.M. and JACQUES, M.A. (1998) A technique to quantify the population size and composition of the biofilm component in communities of bacteria in the phyllosphere. *Appl Environ Microbiol* **64**: 4789–4795.

MORROW, B.J., GRAHAM, J.E. and CURTISS, R., 3RD (1999) Genomic subtractive hybridization and selective capture of transcribed sequences identify a novel Salmonella typhimurium fimbrial operon and putative transcriptional regulator that are absent from the Salmonella typhi genome. *Infect Immun* **67**: 5106–5116.

NATVIG, E.E., INGHAM, S.C., INGHAM, B.H., COOPERBAND, L.R. and ROPER, T.R. (2002) Salmonella enterica serovar Typhimurium and Escherichia coli contamination of root and leaf vegetables grown in soils with incorporated bovine manure. *Appl Environ Microbiol* **68**: 2737–2744.

OJHA, S. and KOSTRZYNSKA, M. (2007) Approaches for reducing Salmonella in pork production. *J Food Prot* **70**: 2676–2694.

OLIVEIRA, C.J., CARVALHO, L.F., FERNANDES, S.A., TAVECHIO, A.T. and DOMINGUES, F.J., JR. (2005) Prevalence of pigs infected by Salmonella Typhimurium at slaughter after an enterocolitis outbreak. *Int J Food Microbiol* **105**: 267–271.

OLSEN, S.J., YING, M., DAVIS, M.F., DEASY, M., HOLLAND, B., IAMPIETRO, L., BAYSINGER, C.M., SASSANO, F., POLK, L.D., GORMLEY, B., HUNG, M.J., PILOT, K., ORSINI, M., VAN DUYNE, S., RANKIN, S., GENESE, C., BRESNITZ, E.A., SMUCKER, J., MOLL, M. and SOBEL, J. (2004)

Multidrug-resistant Salmonella Typhimurium infection from milk contaminated after pasteurization. *Emerg Infect Dis* **10**: 932–935.

OLSON, M.E., CERI, H., MORCK, D.W., BURET, A.G. and READ, R.R. (2002) Biofilm bacteria: formation and comparative susceptibility to antibiotics. *Can J Vet Res* **66**: 86–92.

PARSEK, M.R. and GREENBERG, E.P. (2005) Sociomicrobiology: the connections between quorum sensing and biofilms. *Trends Microbiol* **13**: 27–33.

PHILLIPS, L.E., HUMPHREY, T.J. and LAPPIN-SCOTT, H.M. (1998) Chilling invokes different morphologies in two Salmonella enteritidis PT4 strains. *J Appl Microbiol* **84**: 820–826.

PRATT, L.A. and KOLTER, R. (1999) Genetic analyses of bacterial biofilm formation. *Curr Opin Microbiol* **2**: 598–603.

PROUTY, A.M. and GUNN, J.S. (2003) Comparative analysis of Salmonella enterica serovar Typhimurium biofilm formation on gallstones and on glass. *Infect Immun* **71**: 7154–7158.

PROUTY, A.M., SCHWESINGER, W.H. and GUNN, J.S. (2002) Biofilm formation and interaction with the surfaces of gallstones by Salmonella spp. *Infect Immun* **70**: 2640–2649.

RAAD, II, DAROUICHE, R.O., HACHEM, R., ABI-SAID, D., SAFAR, H., DARNULE, T., MANSOURI, M. and MORCK, D. (1998) Antimicrobial durability and rare ultrastructural colonization of indwelling central catheters coated with minocycline and rifampin [see comments]. *Crit Care Med* **26**: 219–224.

RODRIGUE, D.C., TAUXE, R.V. and ROWE, B. (1990) International increase in Salmonella enteritidis: a new pandemic? *Epidemiol Infect* **105**: 21–27.

ROMLING, U. and ROHDE, M. (1999) Flagella modulate the multicellular behavior of *Salmonella typhimurium* on the community level. *FEMS Microbiol Lett* **180**: 91–102.

ROMLING, U., BIAN, Z., HAMMAR, M., SIERRALTA, W.D. and NORMARK, S. (1998a) Curli fibers are highly conserved between *Salmonella typhimurium* and *Escherichia coli* with respect to operon structure and regulation. *J Bacteriol* **180**: 722–731.

ROMLING, U., SIERRALTA, W.D., ERIKSSON, K. and NORMARK, S. (1998b) Multicellular and aggregative behaviour of Salmonella typhimurium strains is controlled by mutations in the *agf*D promoter. *Mol Microbiol* **28**: 249–264.

ROMLING, U., ROHDE, M., OLSEN, A., NORMARK, S. and REINKOSTER, J. (2000) AgfD, the checkpoint of multicellular and aggregative behaviour in *Salmonella typhimurium* regulates at least two independent pathways. *Mol Microbiol* **36**: 10–23.

ROMLING, U., BOKRANZ, W., RABSCH, W., ZOGAJ, X., NIMTZ, M. and TSCHAPE, H. (2003) Occurrence and regulation of the multicellular morphotype in Salmonella serovars important in human disease. *Int J Med Microbiol* **293**: 273–285.

ROSENBERG, L.E., CARBONE, A.L., ROMLING, U., UHRICH, K.E. and CHIKINDAS, M.L. (2008) Salicylic acid-based poly(anhydride esters) for control of biofilm formation in Salmonella enterica serovar Typhimurium. *Lett Appl Microbiol* **46**: 593–599.

ROSS, P., MAYER, R. and BENZIMAN, M. (1991) Cellulose biosynthesis and function in bacteria. *Microbiol Rev* **55**: 35–58.

RUSSELL, A.D. (2002) Introduction of biocides into clinical practice and the impact on antibiotic-resistant bacteria. *J Appl Microbiol* **92 Suppl**: 121S-135S.

SAGOO, S.K., LITTLE, C.L., WARD, L., GILLESPIE, I.A. and MITCHELL, R.T. (2003) Microbiological study of ready-to-eat salad vegetables from retail establishments uncovers a national outbreak of salmonellosis. *J Food Prot* **66**: 403–409.

SAMPATHKUMAR, B., KHACHATOURIANS, G.G. and KORBER, D.R. (2004) Treatment of Salmonella enterica serovar Enteritidis with a sublethal concentration of trisodium phosphate or alkaline pH induces thermotolerance. *Appl Environ Microbiol* **70**: 4613–4620.

SAMRAKANDI, M.M., ROQUES, C. and MICHEL, G. (1997) Influence of trophic conditions on exopolysaccharide production: bacterial biofilm susceptibility to chlorine and monochloramine. *Can J Microbiol* **43**: 751–758.

SCHER, K., ROMLING, U. and YARON, S. (2005) Effect of heat, acidification, and chlorination on Salmonella enterica serovar typhimurium cells in a biofilm formed at the air-liquid interface. *Appl Environ Microbiol* **71**: 1163–1168.

SCHROEDER, C.M., NAUGLE, A.L., SCHLOSSER, W.D., HOGUE, A.T., ANGULO, F.J., ROSE, J.S., EBEL, E.D., DISNEY, W.T., HOLT, K.G. and GOLDMAN, D.P. (2005) Estimate of illnesses from Salmonella enteritidis in eggs, United States, 2000. *Emerg Infect Dis* **11**: 113–115.

SEYMOUR, R.L., MISHRA, P.V., KHAN, M.A. and SPECTOR, M.P. (1996) Essential roles of core starvation-stress response loci in carbon-starvation-inducible cross-resistance and hydrogen peroxide-inducible adaptive resistance to oxidative challenge in Salmonella typhimurium. *Mol Microbiol* **20**: 497–505.

SINDE, E. and CARBALLO, J. (2000) Attachment of *Salmonella* spp. and *Listeria monocytogenes* to stainless steel, rubber and polytetrafuorethylene. *J Food Microbiol* **17**(4): 439–447.

SMITH, W.A., MAZET, J.A. and HIRSH, D.C. (2002) Salmonella in California wildlife species: prevalence in rehabilitation centers and characterization of isolates. *J Zoo Wildl Med* **33**: 228–235.

SOLANO, C., SESMA, B., ALVAREZ, M., HUMPHREY, T.J., THORNS, C.J. and GAMAZO, C. (1998) Discrimination of strains of *Salmonella enteritidis* with differing levels of virulence by an in vitro glass adherence test. *J Clin Microbiol* **36**: 674–678.

SOLANO, C., GARCIA, B., VALLE, J., BERASAIN, C., GHIGO, J.M., GAMAZO, C. and LASA, I. (2002) Genetic analysis of *Salmonella enteritidis* biofilm formation: critical role of cellulose. *Mol Microbiol* **43**: 793–808.

SOLOMON, E.B., NIEMIRA, B.A., SAPERS, G.M. and ANNOUS, B.A. (2005a) Biofilm formation, cellulose production, and curli biosynthesis by Salmonella originating from produce, animal, and clinical sources. *J Food Prot* **68**: 906–912.

SOLOMON, E.B., COOKE, P.H., BURKE, A.M., UKUKU, D.O. and ANNOUS, B.A. (2005b) Biofilm formation by salmonella spp. on cantaloupe surfaces. *Journal of Food Safety* **25**: 276–287.

SOUPIR, M.L., MOSTAGHIMI, S. and YAGOW, E.R. (2006) Nutrient transport from livestock manure applied to pastureland using phosphorus-based management strategies. *J Environ Qual* **35**: 1269–1278.

STANDAL, R., IVERSEN, T.G., COUCHERON, D.H., FJAERVIK, E., BLATNY, J.M. and VALLA, S. (1994) A new gene required for cellulose production and a gene encoding cellulolytic activity in Acetobacter xylinum are colocalized with the bcs operon. *J Bacteriol* **176**: 665–672.

STANLEY, N.R., BRITTON, R.A., GROSSMAN, A.D. and LAZAZZERA, B.A. (2003) Identification of catabolite repression as a physiological regulator of biofilm formation by Bacillus subtilis by use of DNA microarrays. *J Bacteriol* **185**: 1951–1957.

STEPANOVIĆ, S., ĆIRCOVIĆ, I., MIJAČ, V. and ŠVABIC-VLAHOVIC, M. (2003) Influence of incubation temperature, atmosphere and dynamic conditions on biofilm formation by Salmonella spp. *Food Microbiol* **20**: 339–343.

STEVENSON, G., LAN, R. and REEVES, P.R. (2000) The colanic acid gene cluster of Salmonella enterica has a complex history. *FEMS Microbiol Lett* **191**: 11–16.

SUKUPOLVI, S., LORENZ, R.G., GORDON, J.I., BIAN, Z., PFEIFER, J.D., NORMARK, S.J. and RHEN, M. (1997) Expression of thin aggregative fimbriae promotes interaction of Salmonella typhimurium SR-11 with mouse small intestinal epithelial cells. *Infect Immun* **65**: 5320–5325.

SZOMOLAY, B., KLAPPER, I., DOCKERY, J. and STEWART, P.S. (2005) Adaptive responses to antimicrobial agents in biofilms. *Environ Microbiol* **7**: 1186–1191.

THUNBERG, R.L., TRAN, T.T., BENNETT, R.W., MATTHEWS, R.N. and BELAY, N. (2002) Microbial evaluation of selected fresh produce obtained at retail markets. *J Food Prot* **65**: 677–682.

THURSTON-ENRIQUEZ, J.A., GILLEY, J.E. and EGHBALL, B. (2005) Microbial quality of runoff following land application of cattle manure and swine slurry. *J Water Health* **3**: 157–171.

TSOLIS, R.M., ADAMS, L.G., FICHT, T.A. and BAUMLER, A.J. (1999a) Contribution of Salmonella typhimurium virulence factors to diarrheal disease in calves. *Infect Immun* **67**: 4879–4885.

TSOLIS, R.M., KINGSLEY, R.A., TOWNSEND, S.M., FICHT, T.A., ADAMS, L.G. and BAUMLER, A.J. (1999b) Of mice, calves, and men. Comparison of the mouse typhoid model with other Salmonella infections. *Adv Exp Med Biol* **473**: 261–274.

UHLICH, G.A., KEEN, J.E. and ELDER, R.O. (2001) Mutations in the csgD promoter associated with variations in curli expression in certain strains of Escherichia coli O157:H7. *Appl Environ Microbiol* **67**: 2367–2370.

UKUKU, D.O. and FETT, W.F. (2002) Relationship of cell surface charge and hydrophobicity to strength of attachment of bacteria to cantaloupe rind. *J Food Prot* **65**: 1093–1099.

UKUKU, D.O. and FETT, W.F. (2006) Effects of cell surface charge and hydrophobicity on attachment of 16 Salmonella serovars to cantaloupe rind and decontamination with sanitizers. *J Food Prot* **69**: 1835–1843.

UKUKU, D.O. and SAPERS, G.M. (2001) Effect of sanitizer treatments on Salmonella Stanley attached to the surface of cantaloupe and cell transfer to fresh-cut tissues during cutting practices. *J Food Prot* **64**: 1286–1291.

UNICOMB, L.E., SIMMONS, G., MERRITT, T., GREGORY, J., NICOL, C., JELFS, P., KIRK, M., TAN, A., THOMSON, R., ADAMOPOULOS, J., LITTLE, C.L., CURRIE, A. and DALTON, C.B. (2005) Sesame seed products contaminated with Salmonella: three outbreaks associated with tahini. *Epidemiol Infect* **133**: 1065–1072.

VAN DUYNHOVEN, Y.T., WIDDOWSON, M.A., DE JAGER, C.M., FERNANDES, T., NEPPELENBROEK, S., VAN DEN BRANDHOF, W., WANNET, W.J., VAN KOOIJ, J.A., RIETVELD, H.J. and VAN PELT, W. (2002) Salmonella enterica serotype Enteritidis phage type 4b outbreak associated with bean sprouts. *Emerg Infect Dis* **8**: 440–443.

VAN WINSEN, R.L., VAN NES, A., KEUZENKAMP, D., URLINGS, H.A., LIPMAN, L.J., BIESTERVELD, S., SNIJDERS, J.M., VERHEIJDEN, J.H. and VAN KNAPEN, F. (2001) Monitoring of transmission of Salmonella enterica serovars in pigs using bacteriological and serological detection methods. *Vet Microbiol* **80**: 267–274.

VANSELOW, B.A., HUM, S., HORNITZKY, M.A., EAMENS, G.J. and QUINN, K. (2007) Salmonella Typhimurium persistence in a Hunter Valley dairy herd. *Aust Vet J* **85**: 446–450.

VOETSCH, A.C., VAN GILDER, T.J., ANGULO, F.J., FARLEY, M.M., SHALLOW, S., MARCUS, R., CIESLAK, P.R., DENEEN, V.C. and TAUXE, R.V. (2004) FoodNet estimate of the burden of illness caused by nontyphoidal Salmonella infections in the United States. *Clin Infect Dis* **38 Suppl 3**: S127–134.

WATERMAN, S.R. and SMALL, P.L. (1998) Acid-sensitive enteric pathogens are protected from killing under extremely acidic conditions of pH 2.5 when they are inoculated onto certain solid food sources. *Appl Environ Microbiol* **64**: 3882–3886.

WELLS and BUTTERFIELD, J.E. (1997) Salmonella contamination associated with bacterial soft rot of fresh fruits and vegetables in the marketplace. *Plant Dis* **81**: 867–872.

WHITE, A.P., GIBSON, D.L., KIM, W., KAY, W.W. and SURETTE, M.G. (2006) Thin aggregative fimbriae and cellulose enhance long-term survival and persistence of *Salmonella*. *J Bacteriol* **188**: 3219–3227.

WINFIELD, M.D. and GROISMAN, E.A. (2003) Role of nonhost environments in the lifestyles of Salmonella and Escherichia coli. *Appl Environ Microbiol* **69**: 3687–3694.

WOOD, R.L., POSPISCHIL, A. and ROSE, R. (1989) Distribution of persistent Salmonella typhimurium infection in internal organs of swine. *Am J Vet Res* **50**: 1015–1021.

WOODWARD, M.J., SOJKA, M., SPRIGINGS, K.A. and HUMPHREY, T.J. (2000) The role of SEF14 and SEF17 fimbriae in the adherence of Salmonella enterica serotype Enteritidis to inanimate surfaces. *J Med Microbiol* **49**: 481–487.

WORLD HEALTH ORGANIZATION (2007) Food Safety and Foodborne Illness. *WHO* Sheet No. 237.

WU, H., SONG, Z., HENTZER, M., ANDERSEN, J.B., MOLIN, S., GIVSKOV, M. and HOIBY, N. (2004) Synthetic furanones inhibit quorum-sensing and enhance bacterial clearance in Pseudomonas aeruginosa lung infection in mice. *J Antimicrob Chemother* **53**: 1054–1061.

YOU, Y., RANKIN, S.C., ACETO, H.W., BENSON, C.E., TOTH, J.D. and DOU, Z. (2006) Survival of Salmonella enterica serovar Newport in manure and manure-amended soils. *Appl Environ Microbiol* **72**: 5777–5783.

ZHAO, C., GE, B., DE VILLENA, J., SUDLER, R., YEH, E., ZHAO, S., WHITE, D.G., WAGNER, D. and MENG, J. (2001) Prevalence of Campylobacter spp., Escherichia coli, and Salmonella serovars in retail chicken, turkey, pork, and beef from the Greater Washington, D.C., area. *Appl Environ Microbiol* **67**: 5431–5436.

ZHUANG, R.Y., BEUCHAT, L.R. and ANGULO, F.J. (1995) Fate of Salmonella montevideo on and in raw tomatoes as affected by temperature and treatment with chlorine. *Appl Environ Microbiol* **61**: 2127–2131.

ZOGAJ, X., NIMTZ, M., ROHDE, M., BOKRANZ, W. and ROMLING, U. (2001) The multicellular morphotypes of *Salmonella typhimurium* and *Escherichia coli* produce cellulose as the second component of the extracellular matrix. *Mol Microbiol* **39**: 1452–1463.

ZOTTOLA, E.A. and SASAHARA, K.C. (1994) Microbial biofilms in the food processing industry – should they be a concern? *Int J Food Microbiol* **23**: 125–148.

# 9

# Biofilm formation by Gram-positive bacteria including *Staphylococcus aureus*, *Mycobacterium avium* and *Enterococcus* spp. in food processing environments

**S. Langsrud, Nofima Mat, Norwegian Food Research Institute, Norway**

**Abstract**: Biofilm formation of Gram-positive food pathogens and spoilage bacteria is important for persistence in food production environments and transmission to food. *Staphylococcus aureus* and other staphylococci are frequently isolated from surfaces in food production environments. In most cases, survival on surfaces has been explained by high ability to resist drying. However, survival of *S. aureus* after practical disinfection has also been associated with biofilm formation on the food processing equipment. Several mechanisms are involved in biofilm formation and vary much between strains and growth conditions. The role of biofilm formation for survival and spread of two emerging pathogens, the *Mycobacterium avium* group and enterococci, is briefly discussed.

**Key words**: *Staphylococcus aureus*, biofilm, food, *Mycobacterium*, *Enterococcus*, MRSA, disinfection, food poisoning.

## 9.1 Introduction

Biofilm formation is important for persistence in food production environments and transmission to food for a range of Gram-positive food pathogens and spoilage bacteria. The main part of this chapter will focus on *Staphylococcus aureus*. An overview of the main transmission routes of *S. aureus* to food will be given. The incidence of *S. aureus* in food production environments and association with contamination of food and food poisoning will be discussed. *Staphylococcus aureus* is not only a food poisoning bacteria, but can cause a range of infectious diseases. Therefore, the majority of publications on biofilm formation of *S. aureus* focus on clinical strains and infection-related conditions. The chapter describing conditions promoting biofilm formation and mechanisms involved, will include results from both clinical and food-related studies but focusing on relevance for food production conditions. The role of biofilm formation for survival after cleaning and disinfection will be discussed in relation to other described resistance mechanisms. *S. aureus* is not the most commonly isolated species of *Staphylococcus* from food production environments, and biofilm formation in coagulase-negative staphylococci will be briefly presented. In addition, biofilm formation for two emerging groups of pathogens, the *Mycobacterium avium* group and enterococci will be briefly discussed. Biofilm formation of *Listeria monocytogenes, Bacillus cereus, Clostridium perfringens, Clostridium botulinum* and lactic acid bacteria will be described elsewhere in this book.

## 9.2 *Staphylococcus aureus*

### 9.2.1 Characteristics

*Staphylococcus aureus* is a non-motile, Gram-positive, catalase-positive and facultative anaerobic coccus. The bacterium occurs widely, in low numbers, on the skin and mucous membranes of warm-blooded animals. Normally, the bacterium is transient and not a part of the resident flora on skin. If the normal balance of the skin flora is disturbed (frequent washing, antibiotic treatment), *S. aureus* may become established as a part of the resident flora.

Staphylococci survive well under dry conditions and have been isolated from surfaces in a range of environments, such as the domestic environment (Kennedy *et al.*, 2005), healthcare facilities (Sneed *et al.*, 2004, Boyce, 2007), and food preparation facilities (Jorgensen *et al.*, 2005, Sharma and Anand, 2002).

*Staphylococcus aureus* can cause a wide range of diseases, including skin infections, infections in internal organs, and intoxications such as toxic shock syndrome and food poisoning (Baird-Parker, 2000). Antibiotic resistance has mainly been associated with nosocomial isolates, but in recent years, there has been an increased frequency of community-acquired methicillin-resistant *S. aureus* (MRSA) (Zetola *et al.*, 2005). Some of the community-acquired clones differ from nosocomial clones and the source and transmission of these new clones is often not known. In the Netherlands, a new MRSA-clone found frequently in pigs and cattle could also be isolated from meat, indicating food as one possible vector for the spread of MRSA (Kluytmans *et al.*, 1995). Food-borne outbreaks caused by community-acquired and nosocomial MRSA have been reported (Jones *et al.*, 2002, Kluytmans *et al.*, 1995), and there is therefore not a clear separation between food and clinical *S. aureus* strains.

Symptoms of food poisoning with *S. aureus* enterotoxins include nausea, vomiting, abdominal cramps and diarrhoea occurring 1–8 hours after consumption of contaminated food. The recovery from the illness is relatively fast (1–2 days), and the lethality low.

Although *Staphylococcus aureus* is one of the most frequent causes of foodborne disease (Loir *et al.*, 2003, Mead *et al.*, 1999), it is rarely mentioned among the most important food pathogens, probably because of the limited severity of the illness. A range of foods have been involved in outbreaks, including cooked meat, milk-products, poultry, seafood and baked foods (Baird-Parker, 2000). Large outbreaks have been reported, such as the outbreak in 2000 when contamination of milk from a dairy in Hokkaido (Japan) resulted in more than 13 000 cases of staphylococcal food poisoning (Asao *et al.*, 2003). Food poisoning outbreaks with *S. aureus* are associated with contaminated food stored under conditions allowing growth and toxin production. Direct contamination by food handlers and the use of low quality raw materials (for example, milk from cows with mastitis) are often regarded as the obvious sources in *S. aureus* food poisoning. However, a large retrospective study from the US indicated that contaminated equipment contributes to a significant proportion of the outbreaks and the role of indirect contamination routes should not be neglected (Holmberg and Blake, 1984). The bacterium grows at temperatures between 7 and 48°C, pH 4–10, and at water activities down to 0.86. Conditions allowing toxin production are more narrow. The staphylococcal enterotoxins are more heat stable than the bacterium, and the large outbreak in Japan is a good example of food poisoning by heat-treated products with no viable bacteria.

### 9.2.2 Association with food production environments and foodborne disease

Direct and indirect transmission of *S. aureus* from hands or raw materials to ready-to-eat food can be regarded as the classical contamination routes behind outbreaks with *S. aureus*. The incidence of enterotoxiogenic *S. aureus* on hands, throat, and mucous membranes of nasopharynx in humans has been reported to be between 10 and 40% in healthy individuals (Soriano *et al.*, 2002). One example of the contamination of food by *S. aureus* was published by Kishimoto and co-workers who studied transmission of *S. aureus* in a food preparation facility (Kishimoto *et al.*, 2004). Direct and indirect (via cooking utensils) transmission of *S. aureus* from the hands of a worker with atopic dermatitis with skin inflammation to the food was identified as the main contamination route. Transmission via shared equipment to a third person and then to food was also observed.

The frequency of *S. aureus* on surfaces in food production environments varies significantly. Normanno *et al.* (2005) investigated the presence of *S. aureus* on food contact surfaces in Italy (the type of surfaces was not provided). Out of a total of more than 1500 swab-samples, only six strains of coagulase positive staphylococci were isolated and out of these, only three produced enterotoxin (SEC). The incidence of *S. aureus* was much higher on foods, but the proportion of enterotoxiogenic strains was similar (about 50%). Other studies have shown higher incidence of *S. aureus.* A survey on milk pasteurisation lines in India showed that *S. aureus* represented about 16% of the isolated bacteria (Sharma and Anand, 2002). About 20% of the samples from equipment at a small-scale cheese producer in Norway were positive for the bacterium, but none of the isolates were toxin-producing (Jorgensen *et al.*, 2005). About 15% of samples from production equipment were positive for *S. aureus* after 4 hours of production in an Iberian pork cutting room, and out of these 50% were positive for enterotoxin production. In a study on food refrigeration equipment in the UK, *S. aureus* was isolated from 8 out of 15 plants (Evans *et al.*, 2004). Toxin production was not tested. In domestic refrigerators, a high incidence of *S. aureus* (40%) has been reported (Kennedy *et al.*, 2005). In a recent survey on the microbial ecology from small-scale producers of sausages, *S. aureus* was detected from 12% of the food contact surfaces after cleaning (Gounadaki *et al.*, 2008). The environmental samples with *S. aureus* also showed high numbers of spoilage bacteria and the composition of the bacterial flora on the equipment was similar to what was found in the batter. The authors therefore suggested that the cleaning process was insufficient to eliminate bacteria originating from raw materials processed during the day.

The investigations described above do not demonstrate the association of staphylococci with niches commonly known to support biofilm growth or the survival after processes that would kill free-living bacteria but not biofilm bacteria (for example, disinfection). The relatively frequent findings of *S. aureus* in these surveys most probably reflect the ability of survival on surfaces and do not necessarily indicate biofilm formation. There are only few studies describing biofilm-associated *S. aureus* in food processing environments. In the late 70s and the 80s, several investigators reported on the presence of persistent strains of *S. aureus* in poultry processing lines in the UK. The endemic strains could be detected in the same plant for years and were characterised by a tendency for clumping in growth media (Dodd *et al.*, 1988, Mead *et al.*, 1989). The main contamination source found was defeathering equipment. The defeathering process increased the level of *S. aureus* on poultry 100–1000 fold (Notermans *et al.*, 1982, Dodd *et al.*, 1988). Gibbs *et al.* (1978) reported a level of $10^5/cm^2$ *S. aureus* on rubber fingers and ineffective disinfection (Gibbs *et al.*, 1978). Further investigations showed that persistent strains were eight times more resistant to chlorine than poultry skin isolates and that this could be explained by production of extracellular slime and growth in macroclumps. Another example of biofilm-associated *S. aureus* from the food industry is an investigation of disinfection footbaths in Norwegian dairies, where staphylococci were isolated from swab samples from two out of 12 footbaths tested (Langsrud *et al.*, 2006). One of the strains was identified as *S. aureus*. The staphylococci were not found in the used disinfectant and were susceptible to hypochlorite in suspension. It was therefore hypothesized that the bacteria survived due to biofilm formation on the footbath surface. A large study of the surface microbiota in a range of food factories in the UK showed that *Staphylococcus* spp. was the second most frequent genus isolated from biofilms (8.6% of the isolates) (Gibson *et al.*, 1999). Strains of *Pseudomonas* spp. were the most common. However, none of the staphylococcal isolates were identified as *S. aureus*. A characterisation of the staphylococcal community in a small factory producing dry fermented sausages in France, showed establishment of persistent house-strains of *Staphylococcus equorum* and *S. succinus* (Morot-Bizot *et al.*, 2006). The authors stated that these species are rarely found in meat and the environment and showed a high capacity to colonize surfaces. No *S. aureus* was detected in the investigation.

In summary, several studies have shown a relatively frequent occurrence of *S. aureus* on food contact surfaces in food processing facilities. There are few studies on biofilm-formation of *S. aureus* on food processing equipment; however, species belonging to the genus *Staphylococcus* have been demonstrated to form biofilms. While infections caused by biofilm formation of *S. aureus* and *S. epidermidis* on indwelling devices are undebatable, the role of biofilm formation in food poisoning outbreaks is uncertain.

### 9.2.3 Biofilm formation

The isolation of staphylococci, which do not grow well at low temperatures, from food processing surfaces raises the question of whether they are better biofilm formers than many other food associated bacteria or if the findings are only a result of their resistance to dry conditions. *S. aureus* may survive for days under dry conditions and are more resistant to drying than for example *Enterobacter cloacae* and *Pseudomonas aeruginosa* (Fuster-Valls *et al.*, 2008). At saturated relative humidity they also grow at stainless steel at 22°C. Unfortunately, there is little information in the literature about specific niches in food production environments supporting biofilm formation of *S. aureus*. The present knowledge about influencing environmental factors and mechanisms behind attachment and biofilm formation is therefore mainly based on studies with clinical strains and in conditions simulating infections or growth on indwelling devices.

Attachment and biofilm formation is influenced by several external factors, such as the surface material, salt, glucose, ethanol, limited availability of ferric iron, anaerobic conditions and temperature (Kennedy, 2004, O'Neill, 2007, Johnson *et al.*, 2005, Valle *et al.*, 2003, Fitzpatrick *et al.*, 2006, Rode *et al.*, 2007, Cramton, 2001). The most commonly used material in laboratory biofilm models is polystyrene. Kennedy (2004) showed that *S. aureus* formed biofilm on tissue-culture-treated microtiter plates (Nunc, hydrophilic, negatively charged surface), but not on untreated tissue plates

(Sterilin, more hydrophobic, uncharged) under the same growth conditions. Other studies have shown that *S. aureus* is capable of growing as biofilms on a range of abiotic materials, such as glass, stainless steel, polystyrene, polypropylene and polyvinyl chloride (Cucarella *et al.*, 2001, Gross, 2001, Yarwood *et al.*, 2007, Herrera *et al.*, 2007, Oulahal *et al.*, 2008). Therefore, biofilm formation on materials used in the food industry can be expected.

Glucose-concentration affects biofilm production and glucose is therefore included in the growth medium in most laboratory investigations. The effect of glucose varies between strains and with temperature (Rode *et al.*, 2007).

Induction of biofilm by sodium chloride has been demonstrated in several studies, although the effect varies between strains and temperature (Fitzpatrick *et al.*, 2005, Rode *et al.*, 2007, Valle *et al.*, 2003). *Staphylococcus* spp. has high salt tolerance and is therefore commonly found in production environments for salted food. Induction of biofilm by salt is therefore possibly of importance for biofilm formation in the food industry. A study of biofilm formation on polypropylene at 22°C showed a gradual increase in biofilm growth during the first 10 hours followed by detachment from the surface (Herrera *et al.*, 2007). Using a conditioning film of dried mussel cooking juice resulted in a gradual increase in biofilm development for 25 hours. The biofilm appeared more stable (less detachment) than biofilms produced in culture medium or in PBS. This was attributed to the higher salt content of mussel juice compared to medium and PBS (Herrera *et al.*, 2007). With a few exceptions, most investigations of biofilm formation in *S. aureus* have been performed at 37°C, a temperature relevant in infectious disease, but not as relevant for survival in food production environments. It has been shown that the initial adherence of *S. aureus* to polypropylene is higher at 12°C than 30°C (Pompermayer and Gaylarde, 2000). It has also been shown that biofilm formation on polypropylene and stainless steel is slower at 12°C than 20°C, but the same amount of biofilm is obtained (Oulahal *et al.*, 2008). Consistent with these findings, *S. aureus* from infections and food poisoning cases showed higher biofilm formation on polystyrene at 25 and 30°C than at 37°C (Rode *et al.*, 2007). At a temperature of 20°C, which is not uncommon in some food production environments, the level of biofilm formation was comparable with what was found at 37°C, and was similar for strains from food poisoning cases and infections not associated with food (Rode *et al.*, 2007). In summary, reducing temperature to 12°C in food production environments is not sufficient to prevent biofilm formation in *S. aureus*.

Production of biofilm on surfaces requires both the ability to attach to the surface and to produce a multilayer matrix. The ability of initial attachment to abiotic surfaces has been linked to surface proteins such as Bap (biofilm associated protein) (Cucarella *et al.*, 2001) and teichoic acid. **Bap** in *S. aureus* is a member of a family of surface proteins associated with biofilm formation in several bacterial species (Lasa and Penades,

2006). It has so far only been found in *S. aureus* mastitis isolates and not in isolates from human infections (Cucarella *et al.*, 2001, Cucarella, 2004). Bap is present in 5% of mastitis isolates, and Bap-positive strains have been shown to be more able to colonise and persist in the bovine mammary gland (Cucarella, 2004). They also have an increased ability to cause infection and persist in a mouse infection model. Bap is involved in both the initial attachment and cell-to-cell binding. Its presence can compensate for lack of other biofilm production mechanisms, such as expression of the *ica* (intercellular adhesin) operon, Since Bap-positive isolates may be present in raw milk, one might speculate that Bap could have a role in biofilm formation in dairy production environments, but this have to be confirmed.

**Teichoic acid** is important for the initial attachment to surfaces and cell-to-cell interactions. Mutation in *tagO*, encoding the first enzyme in the pathway for production of teichoic acid, results in cells that aggregate in growth medium but that do not attach to glass or polystyrene (Vergara-Irigaray *et al.*, 2008). The operon *dltABCD* is necessary for D-alanylation of teichoic acids. Mutation in *dltA* leads to reduction in initial attachment, and this has been explained by a more hydrophobic cell surface (Gross, 2001). Mutations in a *dlt*-repressor (ArlRS) lead to increased initial attachment (Toledo-Arana *et al.*, 2005).

**Delta-toxin** production has a negative impact on initial attachment. It has been proposed that delta-toxin acts like a surfactant and prevents the interaction between the cells and abiotic surfaces such as polystyrene. Delta-toxin production is regulated by the *agr* quorum sensing system and lack of, or down-regulation of *agr* is associated with attachment and decreased production of delta-toxin (Vuong *et al.*, 2000).

Accumulation of cells in biofilms is linked to the production of polysaccharide intercellular adhesin (PIA), surface proteins, teichoic acid, and extracellular DNA (ecDNA).

**PIA** is the most studied biofilm production mechanism in staphylococci (O'Gara, 2007). PIA consists mainly of polymeric N-acetyl glucosamine (PNAG) and its production is encoded by genes in the *icaADBC* operon. A transcriptional repressor, *icaR*, is situated upstream the *icaADBC* operon. Although all *S. aureus* strains investigated have the *ica* operon, its importance for biofilm formation varies. The expression of *ica* is strictly regulated by environmental conditions and differs between strains. Deletion of the *ica* operon results in lack of biofilm formation in a microtiter plate model with *S. aureus* ATCC35556 (NCTC8325 derivate) (Cramton *et al.*, 1999). Other NCTC8325-derived laboratory strains also show *ica*-dependent biofilm formation in microtiter plate assays (Fitzpatrick *et al.*, 2005). However, Fitzpatrick (2005) found no association between expression of *ica* and biofilm formation for MRSA from intensive care patients. Several studies have indicated that for infection, *ica*-independent mechanisms are involved and may be more important than *ica*. It has been shown for *S. aureus* ATCC35556 that deletion in *dlt* (D-alanylation of teichoic acid) but not *ica* is important for virulence in mouse models (Kristian, 2004). *S. aureus* biofilms from prosthetic hip and knee joint infections contain PIA, but proteins are also important for biofilm formation as the biofilms ca be dispersed by proteinases (Rohde, 2007). Mastitis isolates show *ica*-independent mechanisms for biofilm formation as deletion of *ica* has no effect on biofilm formation in vitro or in vivo. Salt-induced PIA-production appears to be a more important mechanism in methicillin sensitive *S. aureus* than in MRSA (O'Neill, 2007). In MRSA, biofilm formation is more often *ica*-independent and involves protein adhesins regulated by *SarA* and *agr*. A recent report showed that biofilms of *S. epidermidis* could be resolved by a PNAG-degrading enzyme, but not biofilms of *S. aureus* (Izano *et al.*, 2008). Boles and Horswill (2008) showed that the dispersal process in *S. aureus* biofilms required extracellular proteinases and suggested that the target proteins were Atl, Bap and SasG. In summary, for clinical strains and in infection, the importance of the *ica* operon for biofilm formation is more uncertain for *S. aureus* than for *S. epidermidis*. The role of *ica* for biofilm formation in *S. aureus* associated with food poisoning has been less studied. In a *S. aureus* isolate from a foodborne outbreak associated with sour cream (SA50076), *ica*-expression was induced by NaCl and this was followed by higher biofilm formation (Rode *et al.*, 2007). The same tendency was found for another strain (SA50583) from a pizza-outbreak. The importance of *ica* for biofilm formation on surfaces in food production can therefore not be ruled out and more investigation is needed.

***rbf*** encodes a 80-kDa protein that has been shown to be an important regulator for induction of biofilm formation by glucose and salt without parallel changes in PIA production (Lim, 2004). Interestingly, salt- and glucose-induced biofilm formation was associated with expression of *rbf* in the food poisoning strain SA50076 (Rode *et al.*, 2007). The genes regulated by *rbf* are still not known.

**ecDNA** has recently been identified as an important component in *S. aureus* biofilms (Rice 2007, Izano *et al.* 2008). DNAase I inhibits *S. aureus* biofilm formation, resolves biofilms, and sensitises the biofilm to killing with a cationic detergent (Izano *et al.*, 2008). Izano *et al.* (2008) suggested that ecDNA-fragments of >11 kb can function as intercellular adhesins.

Many regulators have been reported to be involved in biofilm formation, including *sigmaB*, *sarA*, *agr*, *arlRS* and *rbf*. Some of the published reports are conflicting, probably because of differences between strains and experimental design. As an example of the complexity, Cafiso and co-workers showed that *S. aureus* could be separated into three groups according to *agr* genotypes (Cafiso *et al.*, 2007). The genotypes showed different regulation profiles and levels of biofilm formation. For food poisoning strains, the situation is probably equally complex, as illustrated by the highly diverse responses to external factors such as salt, glucose, temperature, and ethanol and variable gene responses between different strains (Rode *et al.*, 2007).

### 9.2.4 Decontamination issues

Sanitation programmes in the food industry usually consist of a mechanical/chemical cleaning step to remove soil and biofilm. This is often followed by a disinfection step to eliminate residual microorganisms. Disinfection is rarely performed without a prior cleaning step. Although cleaning is an crucial measure for elimination of bacteria in the food industry, there are few reports on the efficacy of cleaning agents and technologies in the scientific literature compared to investigations on disinfection effects.

Compared with planktonic cells, bacteria within biofilms are resistant to both the mechanical/chemical cleaning step and to disinfection. There are few studies on the attachment strength of *S. aureus* to surfaces compared to other bacteria. *S. aureus* monolayer biofilms on stainless steel are easier to detach by chemical cleaning agents than *P. aeruginosa* (Gibson *et al.*, 1999). Laboratory studies indicate that acidic cleaning agents are more efficient than alkaline or neutral cleaners, and a pressure of at least 20 bar is needed for effective removal of biofilm. The need for mechanical energy in addition to chemicals for the removal of biofilm has been confirmed by cleaning trials on industrial biofilms. A new technique, intended for removal of biofilm in pipes in connection with hygienic monitoring has shown promising results. This technique is based on a combination of ultrasound, EDTA, and enzymes (Oulahal *et al.*, 2007). This approach could potentially be further developed as a mechanical treatment for cleaning.

The most commonly used disinfectants in the food industry can be separated into tenside-based (e.g., quaternary ammonium compounds (QAC), amphoterics), oxidative (e.g., hypochlorite, peracetic acid) and alcohol-based disinfectants. Most investigations on disinfectant effects are based on bactericidal suspension tests. In these tests, *S. aureus* is equally or less susceptible than other pathogens to oxidative disinfectants (Sagripanti *et al.*, 1997, Rodgers *et al.*, 2001, Langsrud and Sundheim, 1998, Orth and Mrozeck, 1989). On the contrary, *S. aureus* (and other Gram-positive bacteria) are more susceptible than Gram-negative bacteria to tenside-based disinfectants (Langsrud *et al.*, 2003, Jono *et al.*, 1986, Gelinas *et al.*, 1984). It should be noted that the reported large differences in susceptibility between *S. aureus* and Gram-negative bacteria are often based on measuring the minimum inhibitory concentration (MIC) of disinfectants, and the differences are smaller in biocidal tests (Langsrud *et al.*, 2003). Also, the MIC is only relevant for a few applications in food production (for example, disinfecting baths for small equipment), and for most applications biocidal tests give more relevant information.

*S. aureus* survive much higher concentrations of disinfectants in biofilm than in a suspension (Kolawole, 1984, Luppens *et al.*, 2002, Oie, 1996). It is important to be aware that the differences in strain susceptibility or disinfectant efficacy found in suspension tests cannot necessarily be extrapolated to bacteria on surfaces. Although *S. aureus* do not appear resistant compared to other bacteria in suspension tests, it may be among the most resistant bacteria when attached to a surface (Langsrud *et al.*, 2006). Also, the increase in concentration needed for eliminating *S. aureus* in a biofilm compared to a suspension varies between disinfectants (Luppens *et al.*, 2002). For example, the difference in activity toward suspended and sessile cells may be larger for a reactive disinfectant that is neutralised by extracellular material. It appears that the susceptibility to hypochlorite is similar for *S. aureus* and other bacteria attached to surfaces (Rossoni and Gaylarde, 2000, Kim *et al.*, 2007). As also found from suspension tests, attached *S. aureus* are relatively resistant to peracetic acid compared to other bacteria and relatively susceptible to quaternary ammonium compounds (Rossoni and Gaylarde, 2000, Oie, 1996). Results from comparisons on the bactericidal effect of various types of disinfectants are conflicting (Rodgers *et al.*, 2001, Oie, 1996). The effect of disinfectants is in variable degree dependent on the concentration, temperature and organic matter, and commercial disinfectants are often composed of several active components. For this reason, it is difficult to draw any conclusions about which disinfectants should be used to eliminate *S. aureus* (or any other bacteria).

Resistance of *S. aureus* and other staphylococci to quaternary ammonium compounds (QAC) has been extensively studied because some strains harbour plasmid-borne genes for QAC efflux pumps. The presence of these genes resulted in a two-fold increase in the minimum biocidal concentration (MBC) of benzalkonium chloride (BC) in suspension (Heir *et al.*, 1999b). The importance of these genes in *S. aureus* grown in biofilms is not known, but for other staphylococcal species, biofilm-grown strains with and without efflux pumps shows similar resistance to BC (not published). There are several examples of plasmids with QAC-efflux genes that also harbour antibiotic resistance genes, both for strains from food and clinical environments (Sidhu *et al.*, 2001, 2002). Co-selection of strains with QAC and antibiotic resistance is therefore a possibility if the disinfection is inefficient. Resistance to oxidative agents is associated with clumping and biofilm formation (Bolton *et al.*, 1988, Langsrud *et al.*, 2006).

Proposed new disinfection strategies, some of them directed toward use in the clinical area, include fogging with oxidative disinfectants (for example, Virkon S, super-oxidized water, hydrogen peroxide) and ozonation (Otter *et al.*, 2007, Deza *et al.*, 2005, Dunowska *et al.*, 2005, Guzel-Seydim *et al.*, 2004). The main advantage with these technologies is that they are automated: There may be a reduction in costs because of less need for labour, and the disinfection process is more standardised. It may also be possible to use disinfectants or disinfectant concentrations that cannot be used in manual disinfection for safety reasons. The efficacy of these methods on bacterial staphylococcal biofilms is not known.

In conclusion, attachment of *S. aureus* and biofilm formation appears to increase resistance to cleaning and disinfection significantly and may contribute to survival in food production cnvironmcnt.

## 9.3 Coagulase-negative staphylococci

*Staphylococcus aureus* is rarely isolated from food production environments compared to other species of *Staphylococcus* (Gibson *et al.*, 1995, Gounadaki *et al.*, 2008, Kleiss *et al.*, 1994, Morot-Bizot *et al.*, 2006, Heir *et al.*, 1999a). The ability to form biofilm varies between strains of coagulase-negative staphylococci (CNS) from the food industry (Moretro *et al.*, 2003), and biofilm formation of strong biofilm formers is associated with high expression of the *ica* genes. The most studied CNS with regard to biofilm formation is *S. epidermidis* because of its importance in infections associated with colonisation of medical devices. This emerging pathogen has also been isolated from food environments (Shale *et al.*, 2005, Moretro *et al.*, 2003). Initial attachment of *S. epidermidis* to abiotic surfaces involves adhesive proteins (such as adhesion Aae and autolysin AtlE) (Mack *et al.*, 2004). It has recently been suggested that the main function of autolysin AtlE is lysis of a subpopulation of bacteria and that the extracellular DNA promotes biofilm formation (Qin *et al.*, 2007). The second phase of biofilm formation involves production of PIA or alternatively the accumulation associated protein Aap. Recent studies have indicated that production of PIA may be a more important mechanism for biofilm production in *S. epidermidis* than *S. aureus* (Izano *et al.*, 2008).

Biofilm formation could contribute to survival and persistence of CNS in food production facilities. However, the staphylococcal microbiota in the food industry is not dominated by strains appearing as strong biofilm formers in laboratory studies and further research is necessary to determine the role of biofilm formation for persistence. Another consequence of CNS in food industry could possibly be protection of other pathogens, such as *L. monocytogenes*. Co-aggregation of some strains of CNS and *L. monocytogenes* resulting in increased biofilm formation of the latter has been reported (Carpentier and Chassaing, 2004). In the clinical setting, biofilm formation of CNS is a major pathogenesis factor. There is no evidence that clinical infections are linked to food associated CNS.

## 9.4 *Mycobacterium avium*

Mycobacteria are Gram-positive, slow-growing aerobic and nonmotile bacteria with a characteristic thick hydrophobic cell wall rich in mycolic acid. *Mycobacterium avium* and other members of the '*M. avium* complex' are mycobacteria that are potentially pathogenic to humans and animals. These bacteria are found in dust, soil and water and are transmitted by ingestion, inhalation or inoculation from environmental sources, contrary to *Mycobacterium tuberculosis,* which is mainly transmitted person-to-person. They can cause a broad range of infections, mainly in immune-compromised persons and less commonly in healthy individuals. Environmental mycobacteria have been frequently isolated from drinking water (Vaerevijck *et al.*, 2005). Biofilm formation in piping systems has been identified as an important factor contributing to survival, colonisation, and persistence of environmental mycobacteria in water distribution systems. It has been suggested that drinking water and food can be sources of intake of environmental mycobacteria and this can have serious implications for AIDS patients. The potential health risk for healthy individuals is, however, unsure.

*Mycobacterium avium spp. paratuberculosis* is widespread in both healthy and clinically infected ruminants and is likely to be present in bulk milk (Skovgaard, 2007). It has been suggested that there is a link between *M. avium spp. paratuberculosis* and Crohn's disease, a chronic incurable inflammation of the terminal ileum. Unpasteurised milk products can be regarded as the most likely transmission route, but the definite association with human disease is not proven (Skovgaard, 2007).

*Mycobacterium avium* grows at temperatures between 14 and 37°C and is relatively resistant to desiccation on surfaces (Archuleta *et al.*, 2002). Biofilm growth in water distribution pipes is well documented and contributes to disinfection resistance (Steed and Falkinham, 2006). Persistence in the hospital environment may have medical importance (Rowe and Grant, 2006). The *M. avium* group of organisms are emerging pathogens, but biofilm formation in the food industry has not been investigated, and the role of food and food contact surfaces for transmission and subsequent infection is not known.

## 9.5 *Enterococcus* spp.

Enterococci are Gram-positive, facultative anaerobic cocci that are ubiquitous but have a predominant habitat in the gastrointestinal tract of humans and animals. Enterococci have traditionally been regarded as food-grade microorganisms contributing to the ripening of cheese and are used as probiotics and starter-cultures (Moreno *et al.*, 2006). Contamination of meat products, however, contributes to spoilage. Because of their relatively high heat tolerance, enterococci can survive heating processes and grow in cooked food in the absence of competing flora (Moreno *et al.*, 2006). Enterococci can cause food intoxication through the production of biogenic amines, but the main concern over the widespread presence of enterococci in the food chain is their role as a reservoir for opportunistic infections and antibiotic resistance genes (Giraffa, 2002). Vancomycin-resistant enterococci is an increasing problem, and enterococci is now regarded as one of the leading causes of nosocomial infections (Top *et al.*, 2008). Attachment and biofilm formation has been identified as an important factor in the pathogenesis of enterococcal infection. As for staphylococci, a number of surface proteins, teichoic acid and polysaccharides are involved in biofilm formation (Mohamed and Huang, 2007). Utilisation of the biofilm-producing ability of enterococci to control *L. monocytogenes* in food production environments has been suggested (Zhao *et al.*, 2004, Minei *et al.*, 2008). Zhao and co-workers screened for antagonistic effects of microorganisms from drains in food processing facilities and isolated an *Enterococcus durans* capable of inhibiting growth of *L. monocytogenes* biofilms. In another study, *Enterococcus faecium* reduced attachment and biofilm growth of *L. monocytogenes*. The role as a spoilage organism and potential opportunistic pathogen, however, may limit the use of enterococci for combating *Listeria monocytogenes* in food production.

## 9.6 Future trends

Most studies are strictly focusing on *S. aureus* either in a clinical context or a food context. In some sense this is necessary because of the differences in factors leading to disease or intoxication and the nature of the infectious diseases compared to food poisoning. However, recent trends indicate future needs for a more trans-disciplinary approach. It is important to bear in mind that the same clone of *S. aureus* can be involved in both infection and food poisoning. The increasing problem of community acquired MRSA makes it necessary to study *S. aureus* in a wider perspective:

- In the past decade new strains of MRSA have emerged in the community, causing aggressive infections in young otherwise healthy people. Some of the clones originate from hospitalisation whereas others are genotypically distinct from nosocomial isolates.
- One of the possible sources and a transmission pathway of MRSA in the community is food. A MRSA clone isolated from meat in the Netherlands was shown to originate from pork and cattle farming.
- *S. aureus* is commonly found in kitchens indicating fitness for surviving and persisting in consumer environments. This fitness is possibly linked to the ability to attach to surfaces, to form biofilms in humid niches, and to resist dry conditions.
- Food poisoning outbreaks associated with community-acquired or nosocomial MRSA has been reported.

The contribution of food, both as a source of and a vector for spread of MRSA should be further investigated. The ability of *S. aureus* to form biofilms is considered important both for persistence in infections and on abiotic surfaces, for example in the food production and clinical environments. Some of the same mechanisms for attachment and biofilm formation are involved. More research on the level and mechanisms of biofilm formation on inorganic surfaces in conditions simulating food production environments are needed. Similarities and differences between CNS and enterococci from food and clinical infections should also be investigated further.

It should be emphasised that *S. aureus* is resistant to drying, and the resistance to disinfection increases significantly from the attachment phase. Thus, formation of thick multilayer biofilms is not necessary for survival. There are few reports indicating this phenomenon in the food industry. Disinfection does not have a sufficient effect on attached bacteria or bacteria in biofilms. Combating *S. aureus* in food processing environments should therefore focus on more effective cleaning through the use of improved cleaning technologies, easy-to-clean surface materials, and hygienic design. There is an increasing interest for activating quorum sensing systems in staphylococci for dispersal of biofilms to enhance the effect of antibacterial treatments. This concept may have a potential application in the clinical area. For the food industry, the diversity of microbial contaminants and dominance of mixed-species biofilm demands less species-specific elimination methods.

## 9.7 Sources of further information and advice

- **Biofilms**: Additional and continuously updated information about staphylococcal biofilms can be found at 'Biofilms online': (http://www.biofilmsonline.com). The web site is published by Center for Biofilm Engineering at Montana State University and is an electronic serial publication for news, trade information, and applied research on microbial biofilms.
- ***S. aureus*** **and other pathogenic bacteria**: More information about *S. aureus* can be found on web sites published by the US Food and Drug administration (BadBugBook, http://vm.cfsan.fda.gov/~mow/intro.html), U.K. Health Protection agency (http://www.camr.org.uk/webw/HPAweb&Page&HPAwebAutoListName/Page/1191942169197), and K Todar at the University of Wisconsin-Madison Department of Bacteriology (http://www.textbookofbacteriology.net/). *S. aureus* toxin-mediated diseases has been reviewed by R.J. Murray (Murray, 2005), and an overview of toxin production in restaurant food has been published by J.M. Soriano and coworkers (Soriano *et al.*, 2002).
- ***S. aureus*** **biofilm formation**: Mechanisms behind biofilm formation in *S. aureus* and *S. epidermidis* have been reviewed by O'Gara (O'Gara, 2007). The review of K. F. Kong and co-workers focuses on quorum sensing in biofilm formation (Kong *et al.*, 2006).
- **Biofilm formation by other Gram-positive bacteria**: D. Mack and co-workers have published a review describing microbial interactions in

*S. epidermidis* biofilms (Mack *et al.*, 2007). Further information about biofilm formation in enterococci can be found in the review of J. A. Mohamed and D. B. Hung (2007). M.R. Moreno and co-workers discuss the role of enterococci in food and health in their review from 2006 (Moreno *et al.*, 2006). The significance of mycobacteria in drinking water systems has been reviewed by M. J. M. Vaerewijck and co-workers (2005).

## 9.8 References

ARCHULETA, R. J., MULLENS, P. & PRIMM, T. P. (2002) The relationship of temperature to desiccation and starvation tolerance of the *Mycobacterium avium* complex. *Archives of Microbiology*, 178, 311–314.

ASAO, T., KUMEDA, Y., KAWAI, T., SHIBATA, T., ODA, H., HARUKI, K., NAKAZAWA, H. & KOZAKI, S. (2003) An extensive outbreak of staphylococcal food poisoning due to low-fat milk in Japan: estimation of enterotoxin A in the indiscriminated milk and powdered skim milk. *Epidemiology and Infection*, 130, 33–40.

BAIRD-PARKER, T. C. (2000) *Staphylococcus aureus*. In Lund, B. M., Baird-Parker, T. C. & Gould, G. W. (Eds.) *The microbiological safety and quality of food.* Gaithersburg, Aspen Publishers.

BOLES, B. R., & HORSWILL, A. R. (2008) *agr*-Mediated Dispersal of *Staphylococcus aureus* Biofilms: PLoS Pathogens, v. 4, p. e1000052.

BOLTON, K. J., DODD, C. E. R., MEAD, G. C. & WAITES, W. M. (1988) Chlorine resistance of strains of *Staphylococcus aureus* isolated from poultry processing plants. *Letters in Applied Microbiology*, 6, 31–34.

BOYCE, J. M. (2007) Environmental contamination makes an important contribution to hospital infection. *Journal of Hospital Infection*, 65, 50–54.

CAFISO, V., BERTUCCIO, T., SANTAGATI, M., DEMELIO, V., SPINA, D., NICOLETTI, G. & STEFANI, S. (2007) agr-Genotyping and transcriptional analysis of biofilm-producing *Staphylococcus aureus*. *FEMS Immunology and Medical Microbiology*, 51, 220–227.

CARPENTIER, B. & CHASSAING, D. (2004) Interactions in biofilms between *Listeria monocytogenes* and resident microorganisms from food industry premises. *International Journal of Food Microbiology*, 97, 111–122.

CRAMTON, S. E. (2001) Anaerobic conditions induce expression of polysaccharide intercellular adhesin in *Staphylococcus aureus* and *Staphylococcus epidermidis*. *Infection and Immunity*, 69, 4079.

CRAMTON, S. E., GERKE, C., SCHNELL, N. F., NICHOLS, W. W. & GOTZ, F. (1999) The intercellular adhesion (ica) locus is present in *Staphylococcus aureus* and is required for biofilm formation. *Infection and Immunity*, 67, 5427–5433.

CUCARELLA, C. (2004) Role of biofilm-associated protein bap in the pathogenesis of bovine *Staphylococcus aureus*. *Infection and Immunity*, 72, 2177.

CUCARELLA, C., SOLANO, C., VALLE, J., AMORENA, B., LASA, I. & PENADES, J. R. (2001) Bap, a *Staphylococcus aureus* surface protein involved in biofilm formation. *Journal of Bacteriology*, 183, 2888–2896.

DEZA, M. A., ARAUJO, M. & GARRIDO, M. J. (2005) Inactivation of *Escherichia coli, Listeria monocytogenes, Pseudomonas aeruginosa* and *Staphylococcus aureus* on stainless steel and glass surfaces by neutral electrolysed water. *Letters in Applied Microbiology*, 40, 341–346.

DODD, C. E. R., MEAD, G. C. & WAITES, W. M. (1988) Detection of the site of contamination by *Staphylococcus aureus* within the defeathering machinery of a poultry-processing plant. *Letters in Applied Microbiology*, 7, 63–66.

DUNOWSKA, M., MORLEY, P. S. & HYATT, D. R. (2005) The effect of Virkon®S fogging on survival of *Salmonella enterica* and *Staphylococcus aureus* on surfaces in a veterinary teaching hospital. *Veterinary Microbiology*, 105, 281–289.

EVANS, J. A., RUSSELL, S. L., JAMES, C. & CORRY, J. E. L. (2004) Microbial contamination of food refrigeration equipment. *Journal of Food Engineering*, 62, 225–232.

FITZPATRICK, F., HUMPHREYS, H. & O'GARA, J. P. (2005) Evidence for *ica*ADBC-independent biofilm development mechanism in methicillin-resistant *Staphylococcus aureus* clinical isolates. *Journal of Clinical Microbiology*, 43, 1973–1976.

FITZPATRICK, F., HUMPHREYS, H. & O'GARA, J. P. (2006) Environmental regulation of biofilm development in methicillin-resistant and methicillin-susceptible *Staphylococcus aureus* clinical isolates. *Journal of Hospital Infection*, 62, 120–122.

FUSTER-VALLS, N., HERNADEZ-HERRERO, M., MARIN-DE-MATEO, M. & RODRIGUEZ-JEREZ, J. J. (2008) Effect of different environmental conditions on the bacteria survival on stainless steel surfaces. *Food Control*, 19, 308–314.

GELINAS, P., GOULET, J., TASTAYRE, G. M. & PICARD, G. A. (1984) Effect of temperature and contact time on the activity of eight disinfectants – a classification. *Journal of Food Protection*, 47, 841–847.

GIBBS, P. A., PATTERSON, J. T. & THOMPSON, J. K. (1978) Distribution of *Staphylococcus aureus* in a poultry-processing plant. *Journal of Applied Bacteriology*, 44, 401–410.

GIBSON, H., TAYLOR, J. H., HALL, K. H. & HOLAH, J. T. (1995) *Biofilms and their detection in the food industry*, Campden and Chorleywood Food Research Association (CCFRA), Gloucesterchire, UK.

GIBSON, H., TAYLOR, J. H., HALL, K. E. & HOLAH, J. T. (1999) Effectiveness of cleaning techniques used in the food industry in terms of removal of bacterial biofilms. *Journal of Applied Microbiology*, 87, 41–48.

GIRAFFA, G. (2002) Enterococci from foods. *FEMS Microbiology Reviews*, 26, 163–171.

GOUNADAKI, A. S., SKANDAMIS, P. N., DROSINOS, E. H. & NYCHAS, G. J. E. (2008) Microbial ecology of food contact surfaces and products of small-scale facilities producing traditional sausages. *Food Microbiology*, 25, 313–323.

GROSS, M. (2001) Key role of teichoic acid net charge in *Staphylococcus aureus* colonization of artificial surfaces. *Infection and Immunity*, 69, 3423.

GUZEL-SEYDIM, Z., BEVER, P. I. & GREENE, A. K. (2004) Efficacy of ozone to reduce bacterial populations in the presence of food components. *Food Microbiology*, 21, 475–479.

HEIR, E., SUNDHEIM, G. & HOLCK, A. L. (1999a) Identification and characterization of quaternary ammonium compound resistant staphylococci from the food industry. *International Journal of Food Microbiology*, 48, 211–219.

HEIR, E., SUNDHEIM, G. & HOLCK, A. L. (1999b) The qacG gene on plasmid pST94 confers resistance to quaternary ammonium compounds in staphylococci isolated from the food industry. *Journal of Applied Microbiology*, 86, 378–388.

HERRERA, J. J. R., CABO, M. L., GONZALEZ, A., PAZOS, I. & PASTORIZA, L. (2007) Adhesion and detachment kinetics of several strains of *Staphylococcus aureus* subsp *aureus* under three different experimental conditions. *Food Microbiology*, 24, 585–591.

HOLMBERG, S. D. & BLAKE, P. A. (1984) Staphylococcal food poisoning in the United States – new facts and old misconceptions. *Jama – Journal of the American Medical Association*, 251, 487–489.

IZANO, E. A., AMARANTE, M. A., KHER, W. B. & KAPLAN, J. B. (2008) Differential roles of Poly-N-acetylglucosamin surface polysaccharide and extracellular DNA in *Staphylococcus aureus* and *Staphylococcus epidermidis* biofilms. *Applied and Environmental Microbiology*, 74, 470–476.

JOHNSON, M., COCKAYNE, A., WILLIAMS, P. H. & MORRISSEY, J. A. (2005) Iron-responsive regulation of biofilm formation in *Staphylococcus aureus* involves Fur-dependent and Fur-independent mechanisms. *Journal of Bacteriology*, 187, 8211–8215.

JONES, T. F., KELLUM, M. E., PORTER, S. S., BELL, M. & SCHAFFNER, W. (2002) An outbreak of community-acquired foodborne illness caused by methicillin-resistant *Staphylococcus aureus. Emerging Infectious Diseases*, 8, 82–84.

JONO, K., TAKAYAMA, T., KUNO, M. & HIGASHIDE, E. (1986) Effect of alkyl chain length of benzalkonium chloride on the bactericidal activity and binding to organic materials. *Chemical and Pharmaceutical Bulletin*, 34, 4215–4224.

JORGENSEN, H. J., MORK, T. & RORVIK, L. M. (2005) The occurrence of *Staphylococcus aureus* on a farm with small-scale production of raw milk cheese. *Journal of Dairy Science*, 88, 3810–3817.

KENNEDY, C. A. (2004) Contribution of culture media and chemical properties of polystyrene tissue culture plates to biofilm development by *Staphylococcus aureus*. *Journal of Medical Microbiology*, 53, 1171.

KENNEDY, J., JACKSON, V., BLAIR, I. S., MCDOWELL, D. A., COWAN, C. & BOLTON, D. J. (2005) Food safety knowledge of consumers and the microbiological and temperature status of their refrigerators. *Journal of Food Protection*, 68, 1421–1430.

KIM, J. H., JO, C., RHO, Y. T., LEE, C. B. & BYUN, M. W. (2007) Comparison of gamma irradiation and sodium hypochlorite treatments to inactivate *Staphylococcus aureus* and *Pseudomonas aeruginosa* biofilms on stainless steel surfaces. *Food Science and Biotechnology*, 16, 315–319.

KISHIMOTO, M., HIOKI, Y., OKANO, T., KONUMA, H., TAKAMIZAWA, K., KASHIO, H. & KASUGA, F. (2004) Ribotyping and a study of transmission of *Staphylococcus aureus* collected from food preparation facilities. *Journal of Food Protection*, 67, 1116–1122.

KLEISS, T., VAN SCHOTHORST, M., CORDIER, J.-L. & UNTERMANN, F. (1994) Staphylococci in a whey powder plant environment: an ecological survey as a contribution to HACCP studies. *Food Control*, 5, 196–199.

KLUYTMANS, J., VANLEEUWEN, W., GOESSENS, W., HOLLIS, R., MESSER, S., HERWALDT, L., BRUINING, H., HECK, M., ROST, J., VANLEEUWEN, N., VANBELKUM, A. & VERBRUGH, H. (1995) Food-initiated outbreak of methicillin-resistant *Staphylococcus aureus* analyzed by phenotyping and genotyping. *Journal of Clinical Microbiology*, 33, 1121–1128.

KOLAWOLE, D. O. (1984) Resistance mechanisms of mucoid-grown *Staphylococcus aureus* to the antibacterial action of some disinfectants and antiseptics. *FEMS Microbiology Letters*, 25, 205–209.

KONG, K. F., VUONG, C. & OTTO, M. (2006) Staphylococcus quorum sensing in biofilm formation and infection. *International Journal of Medical Microbiology*, 296, 133–139.

KRISTIAN, S. A. (2004) The ability of biofilm formation does not influence virulence of *Staphylococcus aureus* and host response in a mouse tissue cage infection model. *Microbial Pathogenesis*, 36, 237.

LANGSRUD, S. & SUNDHEIM, G. (1998) Factors contributing to a suspension test method for antimicrobial activity of disinfectants. *Journal of Applied Microbiology*, 85, 1006–1012.

LANGSRUD, S., SIDHU, M. S., HEIR, E. & HOLCK, A. L. (2003) Bacterial disinfectant resistance – a challenge for the food industry. *International Biodeterioration & Biodegradation*, 51, 283–290.

LANGSRUD, S., SEIFERT, L. & MØRETRØ, T. (2006) Characterization of the microbial flora in disinfecting footbaths with hypochlorite. *Journal of Food Protection*, 69, 2193–2198.

LASA, I. & PENADES, J. R. (2006) Bap: A family of surface proteins involved in biofilm formation. *Research in Microbiology*, 157, 99–107.

LIM, Y. (2004) Control of glucose- and NaCI-induced biofilm formation by rbf in *Staphylococcus aureus*. *Journal of Bacteriology*, 186, 722.

LOIR, Y., BARON, F. & GAUTIER, M. (2003) *Staphylococcus aureus* and food poisoning. *Genetics and Molecular Research*, 2, 63–76.

LUPPENS, S. B. I., REIJ, M. W., VAN DER HEIJDEN, R. W. L., ROMBOUTS, F. M. & ABEE, T. (2002) Development of a standard test to assess the resistance of *Staphylococcus aureus* biofilm cells to disinfectants. *Applied and Environmental Microbiology*, 68, 4194–4200.

MACK, D., BECKER, P., CHATTERJEE, I., DOBINSKY, S., KNOBLOCH, J. K. M., PETERS, G., ROHDE, H. & HERRMANN, M. (2004) Mechanisms of biofilm formation in *Staphylococcus epidermidis* and *Staphylococcus aureus*: functional molecules, regulatory circuits, and adaptive responses. *International Journal of Medical Microbiology*, 294, 203–212.

MACK, D., DAVIES, A. P., HARRIS, L. G., ROHDE, H., HORSTKOTTE, M. A. & KNOBLOCH, J. K. M. (2007) Microbial interactions in *Staphylococcus epidermidis* biofilms. *Analytical and Bioanalytical Chemistry*, 387, 399–408.

MEAD, G. C., NORRIS, A. P. & BRATCHELL, N. (1989) Differentiation of *Staphylococcus aureus* from freshly slaughtered poultry and strains endemic to processing plants by biochemical and physiological tests. *Journal of Applied Bacteriology*, 66, 153–159.

MEAD, P. S., SLUTSKER, L., DIETZ, V., MCCAIG, L. F., BRESEE, J. S., SHAPIRO, C., GRIFFIN, P. M. & TAUXE, R. V. (1999) Food-related illness and death in the United States. *Emerging Infectious Diseases*, 5, 607–625.

MINEI, C. C., GOMES, B. C., RATTI, R. P., D'ANGELIS, C. E. M. & DE MARTINIS, E. C. P. (2008) Influence of peroxyacetic acid and nisin and coculture with *Enterococcus faecium* on *Listeria monocytogenes* biofilm formation. *Journal of Food Protection*, 71, 634–638.

MOHAMED, J. A. & HUANG, D. B. (2007) Biofilm formation by enterococci. *Journal of Medical Microbiology*, 56, 1581–1588.

MORENO, M. R. F., SARANTINOPOULOS, P., TSAKALIDOU, E. & DE VUYST, L. (2006) The role and application of enterococci in food and health. *International Journal of Food Microbiology*, 106, 1–24.

MORETRO, T., HERMANSEN, L., HOLCK, A. L., SIDHU, M. S., RUDI, K. & LANGSRUD, S. (2003) Biofilm formation and the presence of the intercellular adhesion locus ica, among staphylococci from food and food processing environments. *Applied and Environmental Microbiology*, 69, 5648–5655.

MOROT-BIZOT, S. C., LEROY, S. & TALON, R. (2006) Staphylococcal community of a small unit manufacturing traditional dry fermented sausages. *International Journal of Food Microbiology*, 108, 210–217.

MURRAY, R. J. (2005) Recognition and management of *Staphylococcus aureus* toxin-mediated disease. *International Medicine journal*, 35, S106–S119.

NORMANNO, G., FIRINU, A., VIRGILIO, S., MULA, G., DAMBROSIO, A., POGGIU, A., DECASTELLI, L., MIONI, R., SCUOTA, S., BOLZONI, G., DI GIANNATALE, E., SALINETTI, A. P., LA SALANDRA, G., BARTOLI, M., ZUCCON, F., PIRINO, T., SIAS, S., PARISI, A., QUAGLIA, N. C. & CELANO, G. V. (2005) Coagulase-positive Staphylococci and *Staphylococcus aureus* in food products marketed in Italy. *International Journal of Food Microbiology*, 98, 73–79.

NOTERMANS, S., DUFRENNE, J. & VANLEEUWEN, W. J. (1982) Contamination of broiler-chickens by *Staphylococcus aureus* during processing – incidence and origin. *Journal of Applied Bacteriology*, 52, 275–280.

O'GARA, J. P. (2007) ica and beyond: biofilm mechanisms and regulation in *Staphylococcus epidermidis* and *Staphylococcus aureus*. *FEMS Microbiology Letters*, 270, 179–188.

OIE, S. (1996) Efficacy of disinfectants against biofilm cells of methicillin-resistant *Staphylococcus aureus*. *Microbios*, 85, 223–230.

O'NEILL, E. (2007) Association between methicillin susceptibility and biofilm regulation in *Staphylococcus aureus* isolates from device-related infections. *Journal of Clinical Microbiology*, 45, 1379.

ORTH, R. & MROZECK, H. (1989) Is the control of *Listeria, Campylobacter* and *Yersinia* a disinfection problem? *Fleischwirtsch*, 69, 1575–1576.

OTTER, J. A., CUMMINS, M., AHMAD, F., VAN TONDER, C. & DRABU, Y. J. (2007) Assessing the biological efficacy and rate of recontamination following hydrogen peroxide vapour decontamination. *Journal of Hospital Infection*, 67, 182–188.

OULAHAL, N., MARTIAL-GROS, A., BONNEAU, M. & BLUM, L. J. (2007) Removal of meat biofilms from surfaces by ultrasounds combined with enzymes and/or a chelating agent. *Innovative Food Science & Emerging Technologies*, 8, 192–196.

OULAHAL, N., BRICE, W., MARTIAL, A. & DEGRAEVE, P. (2008) Quantitative analysis of survival of *Staphylococcus aureus* or *Listeria innocua* on two types of surfaces: Polypropylene and stainless steel in contact with three different dairy products. *Food Control*, 19, 178–185.

POMPERMAYER, D. M. C. & GAYLARDE, C. C. (2000) The influence of temperature on the adhesion of mixed cultures of *Staphylococcus aureus* and *Escherichia coli* to polypropylene. *Food Microbiology*, 17, 361–365.

QIN, Z. Q., OU, Y. Z., YANG, L. A., ZHU, Y. L., TOLKER-NIELSEN, T., MOLIN, S. & QU, D. (2007) Role of autolysin-mediated DNA release in biofilm formation of *Staphylococcus epidermidis*. *Microbiology-Sgm*, 153, 2083–2092.

RICE, K. C., MANN, E. E., ENDRES, J. L., WEISS, E. C., CASSAT, J. E., SMELTZER, M. S. & BAYLES, K. W. (2007) The cidA murein hydrolase regulator contributes to DNA release and biofilm development in *Staphylococcus aureus*. *Proceedings of the National Academy of Sciences of the United States of America*, 104, 8113–8118.

RODE, T. M., LANGSRUD, S., HOLCK, A. & MØRETRØ, T. (2007) Different patterns of biofilm formation in *Staphylococcus aureus* under food-related stress conditions. *International Journal of Food Microbiology*, 116, 372–383.

RODGERS, J. D., MCCULLAGH, J. J., MCNAMEE, P. T., SMYTH, J. A. & BALL, H. J. (2001) An investigation into the efficacy of hatchery disinfectants against strains of *Staphylococcus aureus* associated with the poultry industry. *Veterinary Microbiology*, 82, 131–140.

ROHDE, H. (2007) Polysaccharide intercellular adhesin or protein factors in biofilm accumulation of *Staphylococcus epidermidis* and *Staphylococcus aureus* isolated from prosthetic hip and knee joint infections. *Biomaterials*, 28, 1711.

ROSSONI, E. M. M. & GAYLARDE, C. C. (2000) Comparison of sodium hypochlorite and peracetic acid as sanitising agents for stainless steel food processing surfaces using epifluorescence microscopy. *International Journal of Food Microbiology*, 61, 81–85.

ROWE, M. T. & GRANT, I. R. (2006) *Mycobacterium avium* spp. *paratuberculosis* and its potential survival tactics. *Letters in Applied Microbiology*, 42, 305–311.

SAGRIPANTI, J. L., EKLUND, C. A., TROST, P. A., JINNEMAN, K. C., ABEYTA, C., KAYSNER, C. A. & HILL, W. E. (1997) Comparative sensitivity of 13 species of pathogenic bacteria to seven chemical germicides. *American Journal of Infection Control*, 25, 335–339.

SHALE, K., LUES, J. F. R., VENTER, P. & BUYS, E. M. (2005) The distribution of *Staphylococcus* sp on bovine meat from abattoir deboning rooms. *Food Microbiology*, 22, 433–438.

SHARMA, M. & ANAND, S. K. (2002) Characterization of constitutive microflora of biofilms in dairy processing lines. *Food Microbiology*, 19, 627–636.

SIDHU, M., HEIR, E., K, L., WIGER, K. & HOLCK, A. (2002) Frequency of disinfectant resistance genes and genetic linkage with beta-lactamase transposon Tn552 among clinical staphylococci. *Antimicrobial Agents and Chemotherapy*, 46, 2797–2803.

SIDHU, M. S., HEIR, E., SORUM, H. & HOLCK, A. (2001) Genetic linkage between resistance to quaternary ammonium compounds and beta-lactam antibiotics in food-related *Staphylococcus* spp. *Microbial Drug Resistance*, 7, 363–371.

SKOVGAARD, N. (2007) New trends in emerging pathogens. *International Journal of Food Microbiology*, 120, 217–224.

SNEED, J., STROHBEHN, C., GILMORE, S. A. & MENDONCA, A. (2004) Microbiological evaluation of foodservice contact surfaces in Iowa assisted-living facilities. *Journal of the American Dietetic Association*, 104, 1722–1724.

SORIANO, J. M., FONT, G., MOLTO, J. G. & MANES, J. (2002) Enterotoxigenic staphylococci and their toxins in restaurant foods. *Trends in Food Science & Technology*, 13, 60–67.

STEED, K. A. & FALKINHAM, J. O. (2006) Effect of growth in biofilms on chlorine susceptibility of *Mycobacterium avium* and *Mycobacterium intracellulare*. *Applied and Environmental Microbiology*, 72, 4007–4011.

TOLEDO-ARANA, A., MERINO, N., VERGARA-IRIGARAY, M., DEARBOUILLE, M., PENADES, J. R. & LASA, H. (2005) *Staphylococcus aureus* develops an alternative, ica-independent biofilm in the absence of the arlRS two-component system. *Journal of Bacteriology*, 187, 5318–5329.

TOP, J., WILLEMS, R. & BONTEN, M. (2008) Emergence of CC17 *Enterococcus faecium*: from commensal to hospital-adapted pathogen. *FEMS Immunology and Medical Microbiology*, 52, 297–308.

VAEREVIJCK, M. J. M., HUYS, G., PALOMINO, J. C., SWINGS, J. & PORTAELS, F. (2005) Mycobacteria in drinking water systems: ecology and significance for human health. *FEMS Microbiology Reviews*, 29, 911–934.

VALLE, J., TOLEDO-ARANA, A., BERASAIN, C., GHIGO, J. M., AMORENA, B., PENADES, J. R. & LASA, I. (2003) SarA and not sigma(B) is essential for biofilm development by *Staphylococcus aureus*. *Molecular Microbiology*, 48, 1075–1087.

VERGARA-IRIGARAY, M., MAIRA-LITRAN, T., MERINO, N., PIER, G. B., PENADES, J. R. & LASA, I. (2008) Wall teichoic acids are dispensable for anchoring the PNAG exopolysaccharide to the *Staphylococcus aureus* cell surface. *Microbiology-Sgm*, 154, 865–877.

VUONG, C., SAENZ, H. L., GOTZ, F. & OTTO, M. (2000) Impact of the agr quorum-sensing system on adherence to polystyrene in *Staphylococcus aureus*. *Journal of Infectious Diseases*, 182, 1688–1693.

YARWOOD, J. M., PAQUETTE, K. M., TIKH, I. B., VOLPER, E. M. & GREENBERG, E. P. (2007) Generation of virulence factor variants in *Staphylococcus aureus* biofilms. *Journal of Bacteriology*, 189, 7961–7967.

ZETOLA, N., FRANCIS, J. S., NUERMBERGER, E. L. & BISHAI, W. R. (2005) Community-acquired meticillin-resistant *Staphylococcus aureus*: an emerging threat. *Lancet Infectious Diseases*, 5, 275–286.

ZHAO, T., DOYLE, M. P. & ZHAO, P. (2004) Control of *Listeria monocytogenes* in a biofilm by competitive-exclusion microorganisms. *Applied and Environmental Microbiology*, 70, 3996–4003.

# 10

# Biofilm formation by spore-forming bacteria in food processing environments

**D. Lindsay, Fonterra Co-operative Group Limited, New Zealand and S. Flint, Massey University, New Zealand**

**Abstract**: Endospore-forming bacteria present a unique challenge in biofilm studies, as well as to food and beverage manufacturers within processing stages. This chapter discusses the mesophilic endospore formers, such as *Bacillus* species, with respect to attachment of spores to surfaces and subsequent biofilm development. In addition, many manufacturing processes utilize processing temperatures which are suitable for the growth of thermophilic endospore-forming bacteria, such as *Geobacillus* and *Anoxybacillus* strains, increasing concerns of contamination of product. This chapter also highlights the biofilm-forming ability coupled with spore development in these strains. From results obtained to date, biofilm-spore development in themophilic strains seems to be very different from that of mesophilic *Bacillus* isolates. Future studies will further elucidate the spore-biofilm interaction and allow for better control of such biofilms within industrial settings.

**Key words**: biofilms, endospores, *Bacillus*, *Geobacillus*, *Anoxybacillus*.

## 10.1 Introduction

Endospores are highly resistant cell types produced by mostly Gram-positive bacteria, with a few exceptions (Biebl *et al.* 2000), to withstand extreme environmental pressure, such as nutrient starvation. It has been suggested that endospores offer bacteria producing these structures life-style advantages including survival for long periods of time in unfavourable conditions, and a means for dissemination in the environment, as spores are easily carried on air currents and in water (Marahiel and Zuber, 1999). As a result, endospore producers occupy a variety of habitats and range from aerobic heterotrophs and anaerobes important in food processing (e.g., *Bacillus* and *Clostridium*) to more exotic species of importance in environmental microbiology (e.g., halophiles and sulphate reducers) (Nicholson *et al.* 2000). Although it has been noted that the ultrastructure and surface topography of *Bacillus* spores may be species-specific, in general, the structure of the *Bacillus* spore, as an example, can be divided into several layers including: an inner core containing DNA and dipicolinic acid surrounded by a cytomembrane, the cortex made up of a thick layer of peptidoglycan, the outer membrane and the spore coat comprising structural proteins and carbohydrates (Plomp *et al.* 2005). In the case of some *Bacillus* species, the spore coat is also surrounded by a loose exosporium which may exhibit appendages (Plomp *et al.* 2005). During this dormant state, bacterial endospores are highly resistant to a myriad of treatments which would destroy vegetative bacterial cells, including UV treatment, desiccation, heat treatment and various chemical bombardments (Brown, 1994, Atrih *et al.* 1996, Nicholson *et al.* 2000). Several factors have been identified as determinants for spore resistance which include: the genetics of the spore-former, temperature of sporulation, general robustness and water and/or mineral content of the spore coat and core, the presence of small acid soluble proteins in the spore DNA and repair of damage to macromolecules during spore germination and outgrowth. For a more in-depth look at these factors, refer to the review by Nicholson *et al.* (2000). As a result, these structures present problems in food processing environments due to this intrinsic resistance.

Spore formation by *Bacillus* and *Clostridium* species, in particular, have long been studied to aid with understanding cell differentiation processes (Piggot, 1996, Stragier and Losick, 1996, Levin and Grossman, 1998, Durre and Hollergschwandner, 2004). Studying the biofilm formation process in endospore-forming bacteria is a unique challenge. Vegetative cells and spores represent two distinct morphotypes, each with different attachment abilities, and/or mechanisms. In most cases in these biofilms, it is difficult to separate these two distinct forms, and as we will see in the discussion below, endospore formation and biofilm development seem to be intrinsically linked in these bacterial species. From a food processing perspective, some of the most important endospore-forming bacteria will be described and their biofilm-forming ability discussed including mesophilic strains of *Bacillus subtilis*, *Bacillus cereus*, and thermophilic strains of *Anoxybacillus flavithermus and Geobacillus.*

## 10.2 Mesophilic endospore formers, such as *Bacillus (B.)* species

### 10.2.1 Characteristics and food-borne disease/spoilage potential

It is said that *Bacillus (B.) subtilis* was one of the first bacteria to be described by Ehrenberg in 1835 (Slepecky and Hemphill, 1992). Since then, it is arguably one of the most popular bacteria used for microbiological studies besides *Escherichia coli*. *Bacillus* species are typical Gram-positive rods comprising a range of DNA base compositions (32–69 mol% G + C) (Claus and Fritze, 1989). In general, *Bacillus* are ubiquitously distributed in nature, and *B. anthracis* is the most important member of the group due to its disease-causing potential. However, within food processing, several other members including *B. subtilis*, *B. licheniformis* and *B. cereus*, are potential foodborne pathogens and have economic implications due to spoilage.

*Bacillus cereus*

*Bacillus cereus* is the best known *Bacillus* species associated with foodborne illness. Traditionally, strains of toxin producing *B. cereus* are most often associated with starchy foods, such as pasta, rice or maize dishes (Mossel *et al.* 1995, Jay, 2000). In truth, a wide variety of foods can act as vehicles for *B. cereus* toxins including vegetable dishes, cream-based pastries, meat dishes and dairy products (Jay, 2000). It has further been suggested that, based on flagellar antigens, 26 different serotypes of *B. cereus* may be able to cause food poisoning (Mossel *et al.* 1995). Most strains of *B. cereus* can be divided into two groups, the mesophiles and the psychrophiles, the latter being members of the newly suggested *B. weihenstephanensis* species within the *B. cereus* group (Lechner *et al.* 1998, Borge *et al.* 2001). Two types of gastroenteritis syndromes are produced by these *B. cereus* strains: a diarrhoeal syndrome due to production of heat labile hemolysin HBL, and an emetic syndrome due to the production of heat stable cereulide (Granum, 1994, Jay, 2000). Members of the *B. cereus* group are also often common contaminants in dairy processing environments (Batoszewicz *et al.* 2008). Of further concern is the apparent ability of these strains to produce toxins in various dairy products at different storage temperatures. For example, cytotoxicity was expressed by some strains in whipped cream but not in milk at 8°C (Christiansson *et al.* 1989), diarrhoeal cytotoxin was produced by strains of *B. cereus* and *B. weihenstephanensis* isolated from whipping cream at 25 and 32°C (Arnesen *et al.* 2007), while Svensson *et al.* (2006) showed the presence of emetic toxin producing *B. cereus* strains in a milk silo tank.

Spoilage of milk by *Bacillus* is primarily the result of production of protease and lipase enzymes (Frank, 1997). In particular, *B. cereus* is associated with spoilage phenomena of milk including sweet curdling and bitty cream due to the action of proteolytic, lipolytic and phospholipase enzymes (Meer *et al.* 1991). The resulting products also often develop bitter flavours due to proteolysis (Meer *et al.* 1991). Lipases, such as phospholipase-C, degrades fat globule membranes, resulting in fat aggregation in cream (Frank, 1997). Of further interest is the reported ability of proteolytic and lipolytic *B. cereus* strains to produce these enzymes in both the free-living (planktonic) and biofilm state and even under differing pH conditions (Lindsay *et al.* 2000a)

*B. subtilis* and *B. licheniformis*
*Bacillus subtilis* and *B. licheniformis* are Gram-positive rods which form endospores, and vegetative cells are generally mesophilic. By contrast, endospores formed by these two species may be highly heat resistant, which is particularly true in bakery environments (Norris *et al.* 1981, Kirschner and von Holy, 1989). *B. subtilis* tends to aerobic growth, with negative results in anaerobic agar, while *B. licheniformis* tends to grow facultatively anaerobically, with positive results in anaerobic agar (Slepecky and Hemphill, 1992). Within the context of bread making, *B. subtilis* and *B. licheniformis* are the most commonly isolated strains resulting in spoilage of baked goods, especially wheat breads. These strains growing within freshly baked loaves produce amylases and proteases which degrade the dough structure, and produce a capsule layer resulting in a sticky crumb (Kirschner and von Holy, 1989, Kirschner, 1993, ICMSF, 2000).

Most *Bacillus* strains, including *B. subtilis* and *B. licheniformis*, are considered safe and non-pathogenic to humans (Pederson *et al.* 2002). However, there have been a few instances where *B. subtilis* and *B. lichniformis* have been linked to cases of food poisoning, and it is known that several species of *Bacillus* outside the *B. cereus* group, do produce cytotoxic compounds (Phelps and McKillip, 2002). For example, consumption of starchy foods containing in excess of $>10^5$ $g^{-1}$ vegetative *B. subtilis* cells may result in enteritis and vomiting (Mossel *et al.* 1995). In a further example, an interesting case report of emesis relating to *B. subtilis* and *B. cereus* in infant cereal has been reported (Duc *et al.* 2005). The emesis syndrome was thought to be as a result of contamination with *B. cereus*. However, no emetic toxin could be detected in the affected food. Thus the authors speculated that *B. subtilis* may also have played a role in this case of food poisoning. Further to this, Beattie and Williams (1999) and Lindsay *et al.* (2000b) were also able to show the cytotoxin producing ability of various strains of *Bacillus*, including *B. subtilis* and *B. licheniformis*, based on a tetrazolium salt MTT assay. Indeed, a study by Østensvik *et al.* (2004) also showed that several strains of *B. subtilis* exhibited moderate cytotoxicity (20–80% protein inhibition) against Vero cell monolayers.

In addition, *in vitro* studies have demonstrated the production of lichenysin A, a surfactin-like molecule, by *B. licheniformis*. This compound was isolated from *B. licheniformis* strains associated with a fatal case of food poisoning in an infant (Salkinoja-Salonen *et al.* 1999) and was reportedly cytotoxic to sperm cells (Salkinoja-Salonen *et al.* 1999, Mikkola *et al.* 2000). However, its toxicity seemed to be approximately 100-fold less than cereulide, produced by *B. cereus* (Salkinoja-Salonen *et al.* 1999). In addition, a further study evaluated the occurrence of heat-stable toxin-producing strains among mastitic *Bacillus* isolates in milk samples in Finland. This study found two strains of *Bacillus licheniformis* which again produced the lipopeptide lichenysin A and were cytotoxic based on a sperm cell motility inhibition assay (Nieminen *et al.* 2007). A recent study on cytotoxic *Bacillus* strains associated with *okpehe*, a traditional fermented soup condiment produced from *Prosopis africana* seeds in West Africa, indicated the presence of DNA sequences encoding the haemolysin BL (hblD) enterotoxin complex in *B. licheniformis* by PCR amplification of virulence genes (Oguntoyinbo and Sanni, 2007). Thus, it seems that toxin production potential by strains of *B. subtilis* and *B. licheniformis* may be an underestimated phenomenon in food processing environments.

### 10.2.2 Association with food processing environments

*Bacillus* are widespread in the environment, and easily spread due to their resistant endospores. Thus, *Bacillus* spp. may be found in food as a result of contamination from a large variety of sources including soil, water, air, dust, plants and their products, food utensils and equipment surfaces, animal hides, animal feeds and food handlers (Jay, 2000). As a result, several food processing industries may be affected by the presence of *Bacillus* spp. at some point during processing. For example, processors of ready-to-eat foods, such as salads, have found *B. cereus* in the finished product possibly due to raw vegetables as the primary source of contamination within the production line (Valero *et al.* 2007). Other vegetable product manufacturers have also found *Bacillus* strains to be contaminants. For example, *B. cereus* strains were shown to be the major contaminants in a food processing plant producing pasteurized zucchini purées (Guinebretiere *et al.* 2003). *Bacillus* spp. found in liquid egg products were shown to be good biofilm producers and suggested as possible common contaminants in egg breaking factories in the form of biofilms on stainless steel processing surfaces (Baron *et al.* 2007). Studies have shown that endospore-forming bacteria, belonging to the genus *Bacillus* or related genera, are important contaminants of gelatine products, as they survive the production process (De Clerck and De Vos 2002, De Clerck *et al.* 2004). Rice commodities may become contaminated during growth, harvesting, processing and handling with strains of *B. cereus* in the form of endospores with an estimated prevalence of 46 to 100% (Haque and Russell, 2005). *Bacillus* strains have also been found as transient members within mixed population biofilms in the balance tank and feeding unit within an ice-cream factory (Gunduz and Tuncel, 2006). However, two of the most important industries where *Bacillus* strains are endemic sources of product contamination include dairy processing and the bread baking industry.

The term 'rope' is used to describe the characteristic phenomenon of a rotten pineapple-like odour and a distinct 'stickiness' associated with the spoiled crumb of brown bread (Kirschner and von Holy, 1989, Thompson *et al.* 1993). Stringent process and sanitation control, cooler climates and use of fermented, sourdough breads makes rope spoilage an uncommon problem in most developed nations. By contrast, rope spoilage of bread leads to large-scale economic losses, particularly in developing countries

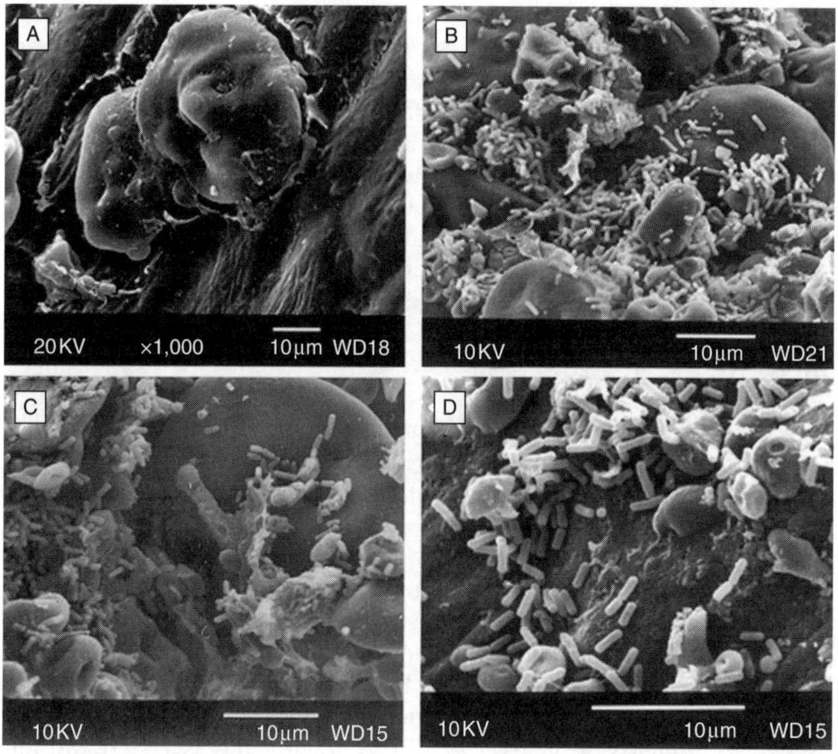


**Fig. 10.1** Scanning electron microscope images of ropey bread. The un-colonized breadcrumb (A), and *Bacillus* strains attached to the breadcrumb after development of rope symptoms (B–D). Adapted from Kirschner (1993).

(Kirschner and von Holy, 1989) (Fig. 10.1). Contamination of bread with rope-inducing *Bacillus* strains is thought to result from spore contamination of the flour (originating from ears of wheat) (Kirschner and von Holy, 1989, Sorokulova *et al.* 2003) and spore attachment to equipment surfaces (Collins *et al.* 1991, Bailey and von Holy, 1993, ICMSF, 2000). Surfaces of bakery equipment, particularly those constructed of non-metallic materials (e.g., polyurethane), have been implicated. In particular, rope-inducing *B. subtilis* strains have been isolated predominantly from non-metallic conveyor surfaces associated with the divider, rounder, moulder and the depanning and packaging processes (Collins *et al.* 1991, Bailey and von Holy, 1993). The endospores produced by rope-inducing *Bacillus* strains are highly heat-resistant and not only survive baking temperatures, but are activated during the baking process (Kirschner and von Holy, 1989, Collins *et al.* 1991). Activated spores in turn germinate in the freshly baked and warm bread within 36 to 48 h, producing enzymes which result in the rope-spoilage phenomenon (Kirschner and von Holy, 1989, Kirschner, 1993, ICMSF, 2000) (Fig. 10.1).

A variety of *Bacillus* species have been isolated from milk and milk products including *B. brevis*, *B. circulans*, *B. lentus*, *B. licheniformis*, *B. mycoides*, *B. polymyxa*, *B. pumilus*, *B. subtilis*, *B. thuringiensis* and *B. cereus* (García-Armesto and Sutherland, 1997). It has been suggested that the two main ways in which *Bacillus* species survive modern milk pasteurization processes are due to the production of heat-resistant endospores, and as attached populations to surfaces of dairy processing equipment (Frank, 1997, García-Armesto and Sutherland, 1997, Larsen and Jørgensen, 1997, Austin and Bergeron, 1995, Lindsay and von Holy, 2006).

## 10.3 Biofilm formation

It is generally accepted that attachment of bacteria to surfaces is their predominant form of survival in nature and man-made ecosystems (Costerton *et al.* 1987, Marshall, 1992, Gilbert *et al.* 1993, Bradshaw *et al.* 1996, Lindsay and von Holy, 2006). Such ecosystems include food processing factories. Several examples can be found where *in situ* endemic biofilms containing *Bacillus* strains have formed in food processing plants, such as on dairy processing equipment (Austin and Bergeron, 1995), in a milk powder manufacturing plant (Stadhouders *et al.* 1982) and in a whey processing plant (Flint *et al.* 1997). Contamination of food products in contact with these biofilms may in turn lead to foodborne illness or shorter product shelf-life (Notermans *et al.* 1991, Gilmour and Fraser, 1991, Mosteller and Bishop, 1993, Hood and Zottola, 1995). Both Gram-positive and Gram-negative bacteria are known to form biofilms. All species of *Bacillus* are also known to form biofilm, both *in vitro* and *in situ* (Flint *et al.* 1997, Lindsay and von Holy, 1997, 1998, 1999, Oosthuizen *et al.* 2001, Lindsay *et al.* 2002) (Fig. 10.2). When studying endospore-forming bacteria, such as *Bacillus*, another dimension is added to their biofilm-forming capability, that of endospores. Thus, this section will look at both the ability of endospores and vegetative *Bacillus* cells to attach to surfaces and subsequently form biofilms on these surfaces.

### 10.3.1 Attachment to and outgrowth of spores on surfaces

*Adhesion of spores to surfaces*

It is well known that bacterial spores attach to a variety of surfaces (Rönner *et al.* 1990, Anderssen *et al.* 1995, Faille *et al.* 2001, 2002, Lindsay *et al.* 2004, Faille *et al.* 2007). Two suggested mechanisms accounting for spore adhesion are presented below.

Hydrophobicity

In chemical terms, hydrophobic interactions are considered strong bonds which act over long distances, and it has long been held that hyrophobicity

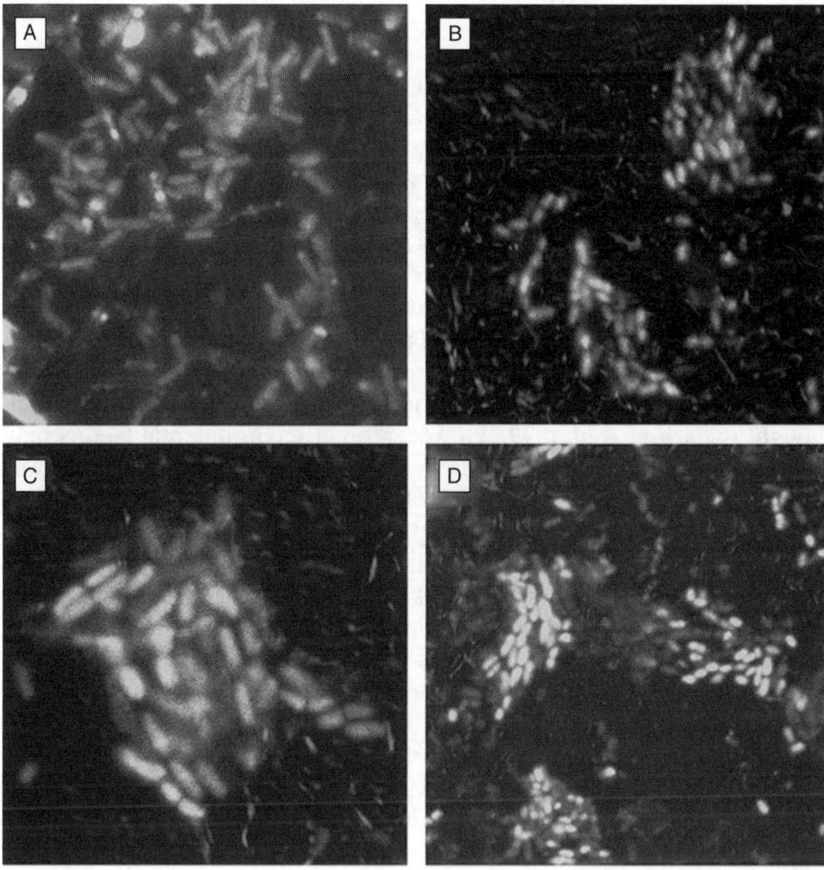


**Fig. 10.2** Confocal scanning laser microscope images (green channel) of *B. cereus* DL5 biofilms attached to stainless steel surfaces in a flow system after three days in single culture (A), and after 3 (B, C) and 5 (D) days in mixed culture with *Pseudomonas fluorescens* M2. Biofilms were stained with the *Bac*light Bacterial Viability Kit™ (Molecular Probes, Leiden, The Netherlands). For a full study, refer to Lindsay *et al.* 2002.

is a determining factor when dealing with bacterial adhesion to surfaces (van Loosdrecht *et al.* 1990). *B. cereus* spores are highly hydrophobic, mainly due to the nature of the exosporium which is reportedly composed of 53% protein, 20% amino and neutral polysaccharide, 18% lipids, and approximately 4% ash (Matz *et al.* 1970, Koshikawa *et al.* 1989). Several proteins have been identified in the exosporium layer including a collagen-like glycoprotein and two other glycosolated proteins which are important for structure (Steichen *et al.* 2003). Studies have found that the hydrophobicity of *B. cereus* spores influences adherence to surfaces and subsequent resistance to cleaning (Husmark and Rönner, 1990, Faille *et al.* 2001, 2002). Overall, it has been shown that the most hydrophobic spores will adhere to hydrophobic and hydrophilic surfaces to a greater extent than corresponding vegetative cells (Rönner *et al.* 1990). Furthermore, even within strains of *B. cereus*, spore hydrophobicity, and hence surface adhesion, may differ, where more hydrophobic strains exhibit better attachment (Andersson and Rönner, 1998). A recent study has further confirmed the role the exosporium plays in attachment of *B cereus* spores to surfaces. Spores exhibiting damaged spore exosporia, or spores without exosporia, were shown to adhere to stainless steel surfaces in significantly fewer numbers than spores with intact exosporia (Faille *et al.* 2007). Observations have also found that several proteins are present in the spore surface layer which may indicate the exosporium as a more functional layer than previously thought (Charlton *et al.* 1999). However, other environmental factors may also influence *Bacillus* spore adhesion to surfaces. For example, pH was shown to influence the attachment of alkaline tolerant *B. cereus* DL5 to polyurethane and glass surfaces, compared with the attachment of *B. cereus* ATCC 14579 spores under similar conditions. Acidic pH 4 and alkaline pH 10 enhanced spore adhesion of the alkaline tolerant strain, while pH apparently had no effect on adhesion of the ATCC strain (Lindsay *et al.* 2004). Ionic strength and different polarities of the growth environment have also been shown to similarly affect *Bacillus* spore attachment to surfaces (Husmark and Rönner, 1990).

In the case of *B. subtilis*, spores do not possess a distinct exosporium, but are generally still more hydrophobic than corresponding vegetative cells (Wiencek *et al.* 1990, Ahimou *et al.* 2001, Bailey-Smith *et al.* 2005). Early studies have shown that the spore coat of *B. subtilis* contains several proteins which are rich in hydrophobic amino acids (Munoz *et al.* 1978, Yasuda-Yasaki *et al.* 1978, Pandey *et al.* 1980), thus accounting for the overall hydrophobicity of these spores (Doyle *et al.* 1984, Wiencek *et al.* 1990). Environmental treatments have also been shown to influence the hydrophobicity, and hence adhesion, of *B. subtilis* spores. For example, increased spore hydrophobicity has been reported after sublethal thermal treatments, which has important implications for *Bacillus* spore adhesion in food processing environments (Wiencek *et al.* 1990).

#### Appendages

Studies on the ultrastructure of *Bacillus* spores using modern techniques, such as atomic force microscopy, have revealed that *B. cereus* and *B. licheniformis* strains exhibit appendages on the exosporium surface (Plomp *et al.* 2005). These appendages have been described as having a distinct tubular morphology, and being mostly composed of proteins, which may be quite complex and differ between different strains of the same species (Stalheim and Granum, 2001, Plomp *et al.* 2005). Conflicting evidence exists for the role that the presence of these structures may play in spore adhesion to surfaces. It was initially suggested that they may act in a similar manner as pili in attachment to surfaces by penetrating potential barriers and by enhancing the overall hydrophobicity of the spore (Stalheim and Granum, 2001). Indeed, Tauveron *et al.* (2006) showed that adhesion of *B. cereus* spores to stainless steel surfaces was enhanced when spores exhibited longer appendages. Other studies have also indicated that removal of *B. cereus* spore appendages significantly reduces spore adhesion to various surfaces, such as stainless steel and polypropylene (Hüsmark and Ronner, 1992, Klavenes *et al.* 2002). However, this phenomenon may be strain specific as some appendage-free strains generally adhere poorly (Klavenes *et al.* 2002). By contrast, it has been suggested that spore appendages may not affect adhesion to surfaces after all. This was shown in a study by Stalheim and Granum (2001) in which removal of appendages of *B. cereus* spores by sonication did not affect adhesion to stainless steel surfaces.

*Outgrowth of spores attached to surfaces*

One of the main concerns of *Bacillus* spores attached to surfaces in food processing factories is the suggested notion that once *Bacillus* spores have attached to surfaces, they may germinate under favourable conditions to vegetative cells, which may in turn result in biofilms (Lindsay *et al.* 2006). Indeed, a recent study showed that attached spores of both *B. cereus* DL5 and *B. subtilis* 168 on glass wool germinated under nutrient-rich, and even more surprisingly, nutrient-limited conditions. The resultant attached biofilm population contained both spores and vegetative cells (Lindsay *et al.* 2006). The authors speculated that the slow germination response of the *B. cereus* strain under nutrient limited conditions may contribute to the renowned resistance of such dairy *Bacillus* biofilms to cleaning and sanitation regimes (i.e., spores in attached form that are slow to form vegetative cells). Another study by Hornstra *et al.* (2007) further confirmed the ability of attached *B. cereus* spores to germinate. Spores of *B. cereus* ATCC 14579 and the environmental isolate *B. cereus* CMCC 3328 were attached to stainless steel surfaces and various germination conditions tested. Spores germinated under favourable conditions in the presence of known germinants (l-alanine and inosine), and also under a number of suboptimal conditions, such as a lower germination temperature (10°C), a short incubation time of 5 minutes and a low induction temperature (30°C). The authors concluded their study with a novel suggestion that a germination step before a cleaning-in-place regime in a food processing line may help to reduce the numbers of adherent *Bacillus* spores (Hornstra *et al.* 2007).

### 10.3.2 Biofilm development by vegetative cells

Based on numerous studies in the literature, it is generally accepted that biofilm development occurs in several steps (Watnick and Kolter, 2000).

*Surface pre-conditioning*
During turbulant flow of liquids or movement of food product over contact surfaces, carbohydrates, proteins and fats from the food matrix are deposited on equipment surfaces. This is considered to be the pre-conditioning phase of biofilm development, providing attaching bacteria with nutrients and influencing which microorganism will attach (Holah and Gibson, 2000, Donlan, 2002). Biofilm development has been shown for most, if not all, *Bacillus* spp. where attachment to a variety of surfaces occurs readily both with and without pre-conditioning (Lindsay and von Holy, 1997–1999, Oosthuizen *et al.* 2001, 2002, Lindsay *et al.* 2002, Peta *et al.* 2003).

*Biofilm initiation and reversible attachment*
Several environmental factors are thought to trigger the initial phases of *Bacillus* biofilm development. For example, high levels of rapidly metabolized carbon sources, such as glucose, inhibit *Bacillus* biofilm initiation. This finding led to the idea that *Bacillus* biofilm formation might be subject to catabolite repression. Stanley *et al.* (2003) found that the catabolite control protein, CcpA, of *B. subtilis*, which is a transcriptional regulator mediating catabolite repression in response to glucose and fructose, is involved in *Bacillus* biofilm formation. Another important signal for biofilm formation is postulated to be oxygen depletion. Indeed, a study by Oosthuizen *et al.* (2002) found two proteins linked with oxygen depletion and anaerobic growth (catabolic ornithine carbamoyltransferase and L-lactate dehydrogenase) which were expressed in biofilms of *B. cereus*, but not in corresponding planktonic cells. A further protein, YhbH, was also found. This protein, which is a member of the sigma 54 modulation protein family, is strongly induced in response to environmental stresses and energy depletion (Oosthuizen *et al.* 2002). Other enzymes purported to play a role in *B. subtilis* biofilm initiation and development include glutamate synthase, GltAB, and aminopeptidase, AmpS (Chagneau and Saier, 2004). Glutamate synthase functions in *B. subtilis* cells for glutamate synthesis, the assimilation of ammonium ions, and as an osmoprotector in *B. subtilis* spores (Bohannon *et al.* 1985, Ruzal and Sanchez-Rivas, 2003), while AmpS is considered an exopeptidase. The loss of these enzymes was shown to negatively affect biofilm formation (Chagneau and Saier, 2004). Similarly, glutamate synthase genes have also been shown to be differentially expressed in biofilms of other bacteria (Beenken *et al.* 2004, Pysz, *et al.* 2004).

Reversible attachment is considered the weakest link in the steps connecting biofilm cells to a conditioned surface and often results in an equilibrium distribution between attached and planktonic cells (van Loosdrecht *et al.* 1990, Busscher *et al.* 1995). Bacterial cells may be drawn to a surface by sedimentation and Brownian motion; convection currents; or electrostatic and physical interactions including cell surface hydrophobicity; or active movement by motile bacteria (van Loosdrecht *et al.* 1990, Flemming *et al.* 1998). Indeed, in various bacteria, motility has been shown to play an integral role in initial attachment (Morisaki *et al.* 1999, Vatanyoopaisarn *et al.* 2000, Kalmokoff *et al.* 2006). A similar finding has been made for *B. subtilis* biofilms, where motility was found to be important for successful biofilm development (Chagneau and Saier, 2004). Flagellar genes in *B. subtilis* were found to be regulated by *degU*, a gene which encodes a response regulator (Kobayashi, 2007). It has been suggested that DegU and phosphorolated-DegU levels are low in planktonic *B. subtilis* cells, but increase in response to quorum-signalling to repress flagella production and help stimulate cells for the biofilm form (Kobayashi, 2007). It has even been suggested that DegU may aid in co-ordinating the multicellular behaviours seen in *B. subtilis* swarming and biofilm development (Verhamme *et al.* 2007). In addition to this, cell surface hydrophobicity may also play a role. Generally, vegetative cells of *Bacillus* are more hydrophilic when compared to their spore forms (Garry *et al.* 1998). In order to enhance cell surface hydrophobicity, it has been suggested that *Bacillus* produce surface active compounds, such as surfactin and iturin A, which may further aid in this initial attachment stage (Garry *et al.* 1998, Ahimou *et al.* 2000).

Bacterial cell-to-cell communication plays a major role during both the reversible and irreversible phases of biofilm development (Winans, 2002, Hardie and Heurlier, 2008). Gram-positive bacteria use autoinducing peptides (or peptide pheromones) as cell-to-cell signalling molecules. This peptide-pheromone interacts with sensor proteins on the outer surface of the Gram-positive cell cytoplasmic membrane, e.g. the two-component sensor histidine kinases in *B. subtilis* (Okada *et al.* 2005, Bassler and Losick, 2006, Gobetti *et al.* 2007). ComX is a typical population density quorum-signalling molecule produced by *B. subtilis* which is associated with swarming behaviour and cell competence (Magnuson *et al.* 1994). It comprises a 6 amino acid peptide where tryptophan is modified by a geranyl group in the *B. subtilis* strain ROE2, or a farnesyl group in *B. subtilis* strains 168 and ROC2 (Okada *et al.* 2005, Bassler and Losick, 2006). ComX has also been shown to activate surfactin production (Cosby *et al.* 1998), which, as indicated previously, may influence cell surface hydrophobicity, and hence *B. subtilis* attachment to surfaces. In *B. cereus*, PapR (a small diffusible peptide) acts as a quorum-sensing molecule that activates PlcR, a pleiotropic regulator (Slamti and Lereclus, 2002, Hsueh *et al.* 2006). In turn, PlcR activates the expression of various genes in *B. cereus* coding for virulence factors, and also directly or indirectly represses biosurfactant production in *B. cereus*. Reportedly, biosurfactants are required for proper *B. cereus* biofilm formation (Hsueh *et al.* 2006).

*Irreversible attachment, colonization and biofilm maturity*

The final stage in biofilm development is colonization, where attached bacteria grow and divide, forming microcolonies, and produce extracellular polymeric substances (EPS) which allow for cell-to-cell binding and cement cells to the surface (Costerton *et al.* 1994). In *B. subtilis* biofilms, the *eps* operon is responsible for the biosynthesis of EPS, and is negatively regulated by SinR, which binds to its promotor region (Ren *et al.* 2004, Kearns *et al.* 2005, Chai *et al.* 2008). The EPS is reportedly composed mainly of a protein component (TasA, encoded by the *yqxM-sipW-tasA* operon) and an exopolysaccharide component (encoded by the *epsA–O* operon) (Branda *et al.* 2006, Chu *et al.* 2008). Both these components are necessary for successful formation of complex *Bacillus* biofilms.

*Bacillus* biofilm cells are genetically and physiologically distinct from their corresponding planktonic counterparts, exhibiting a unique pattern of gene expression (Vilain and Brözel, 2006). For example, studies on *B. cereus* biofilms have indicated that several unique proteins are produced in biofilm cells, which do not appear in the corresponding planktonic populations (Oosthuizen *et al.* 2002). In addition, individual members of *Bacillus* biofilms seem to play their specific roles within the biofilm community (Kearns, 2008). For example, genes encoding the components of the EPS in *B. subtilis* biofilms are expressed in only a sub-population of attached cells (Chai *et al.* 2008). This highlights the interactive consortium nature of *Bacillus* biofilms. Several global regulatory proteins have also been suggested in the control of *B. subtilis* biofilm development. Sigma-H and Spo0A, seem to be involved in early stage initiation of *B. subtilis* biofilm development and the transition to a normal 3-dimensional structure (Hamon and Lazazzera 2001, Hamon *et al.* 2004). Interestingly, Spo0A, as the major early sporulation transcriptional factor, is required by *B. subtilis* cells during sporulation, as it activates stage II sporulation genes. Related to this, the phosphorelay proteins, Spo0F and Spo0B, which activate Spo0A are also required for biofilm development. However, the major kinases (*kinA* and *kinB*) which are required for sporulation, are not required for *B. subtilis* biofilm development. Thus, sporulation and biofilm development have been shown to be intrinsically linked in *Bacillus* cells. In fact, it has even been suggested that *Bacillus* cells choose either to follow the path of spore formation, or that of biofilm development (Hamon and Lazazzera, 2001, Lindsay *et al.* 2005). Further evidence for this is found in the operon, *sdpA*, which is known to delay sporulation. In *Bacillus* biofilms, this operon is differentially expressed, thus aiding in biofilm development instead of sporulation (Gonzalez-Pastor *et al.* 2003, Hamon *et al.* 2004).

Biofilms are structurally organized formations with a complex architecture. Several genes are expressed in *Bacillus* biofilms which are responsible for maintenance of the 3-dimensional configuration. AbrB, also suggested as a third global regulatory protein, is a negative regulator of *Bacillus* biofilm formation. This transcriptional factor seems to regulate a secreted protein, YoaW, which is important for depth and hence the 3-dimensional structure of late stage *B. subtilis* biofilms (Hamon *et al.* 2004). In addition, it also regulates the signal transpeptidase, SipW, which is needed for mature structure development in *B. subtilis* biofilms (Hamon *et al.* 2004). It has also been suggested that the operon, *ecsABC*, may indirectly affect *B. subtilis* biofilm structure by promoting maturation of the biofilm and secretion of protein components of the EPS (Branda *et al.* 2004). Further to this, the *yve* operon is thought to be involved in polysaccharide production and hence biofilm maintenance (Ren *et al.* 2004).

### 10.3.3 Development of spores within a biofilm

Within the *Bacillus* group, evidence suggests that actual sporulation is not required for stable biofilm formation (Hamon and Lazazzera, 2001). In fact, results from several studies indicate that spore formation within established vegetative *Bacillus* biofilms may not be favoured compared to corresponding planktonic cells. For example, the sporulation rate within *B. subtilis* biofilms has been estimated at 1% of the attached population, compared with 48% for planktonic cells (Hamon and Lazazzera, 2001, Lindsay *et al.* 2005, 2006). Spore formation within *B. cereus* biofilms seems to be strain and growth condition specific. A recent study by Wijman *et al.* (2007) indicated a wide range of biofilm and spore-forming ability among several strains of *B. cereus*, ranging from one strain which did not form biofilms at all, to strains which constituted 90% spores within existing biofilms. By contrast, other studies have shown that spores within *B. cereus* biofilms may be as low as 0.01–10% of the attached population (Ren *et al.* 2004, Lindsay *et al.* 2005, Ryu and Beuchat, 2005, Lindsay *et al.* 2006).

### 10.3.4 Control of biofilms of mesophilic endospore-forming bacteria

It is well understood that bacterial biofilms are generally more resistant to antimicrobial treatments than their corresponding planktonic counterparts. In the case of spore-formers, endospores further increase this tendency as these structures are themselves highly resistant (Nicholson *et al.* 2000). Several studies have evaluated antimicrobial efficacies of commonly used food-grade sanitizers against biofilms of *Bacillus* (Lindsay and von Holy, 1999, Lindsay *et al.* 2002, Peta *et al.* 2003, Beuchat *et al.* 2004, Ryu and Beuchat, 2005, Kreske *et al.* 2006). Results from these studies have highlighted three important phenomena: i) biofilm cells of *Bacillus* strains are more resistant to treatment with various sanitizers than planktonic cells, ii) spores within such biofilms may exacerbate resistance and iii) co-cultures of *Bacillus* with other spoilage bacteria in biofilms may also alter resistance profiles (Lindsay and von Holy, 1999, Lindsay *et al.* 2002, Peta *et al.* 2003, Beuchat *et al.* 2004, Ryu and Beuchat, 2005, Kreske *et al.* 2006). Effective cleaning of bacterial biofilms involves a combination of detergents, mechanical action and sanitizers (Hood and Zottola, 1995). This regime still seems to be the most effective in combatting *Bacillus* biofilms where cleaning-in-pace procedures at elevated temperatures using detergents and followed by sanitizer application can effectively inactivate and remove such biofilms from soiled food processing surfaces (Peng *et al.* 2002).

## 10.4 Thermophilic endospore formers

Thermophilic endospore-forming bacteria are potential contaminants of food manufacturing plants involving hot processes. The formation of heat-resistant spores makes them difficult to control and growth of these organisms has potential to spoil food.

### 10.4.1 Characteristics and food spoilage potential

Thermophilic endospore-forming bacteria can be found in several foods including canned food (Denny *et al.* 1961, Weddig *et al.* 2007), juice (Yokota *et al.* 2007) and milk products (Basappa, 1974, Flint *et al.* 2001a, Hill, 2004, Scott *et al.* 2007). Thermophilic spore-forming bacilli associated with heat treated foods are defined as anaerobic or facultatively anaerobic endospore-forming bacteria with optimum growth temperatures of between 45 and 65°C, although specific definitions do vary in the food industry (Ito, 1981). Typical bacterial types include *Geobacillus* species, *Anoxybacillus flavithermus*, and several *Bacillus* species, *Alicyclobacillus* and *Clostridium thermosaccarolyticum*, with most biofilm studies focussing on *Geobacillus* species and *A. flavithermus* associated with the dairy industry, and in particular, milk powder manufacture.

The dominant thermophile in milk powder manufacture internationally is *A. flavithermus* although *Bacillus licheniformis*, *Geobacillus stearothermophilus* and *Bacillus subtilis* are also frequently isolated (Crielly *et al.* 1994, Ronimus *et al.* 2003, Rueckert *et al.* 2004, Seale *et al.* 2008, Seale 2009). It is likely that many thermophilic bacteria formerly referred to as *Bacillus stearothermophilus* in early literature, were in fact *A. flavithermus* as the typing of thermophilic isolates was imprecise (Flint *et al.* 2001b). The growth characteristics of many of these bacteria in planktonic culture are interesting in that there is often no lag phase and maximum cell numbers are often obtained within 5–6 h and this is followed by a very rapid death phase (Basappa, 1974). The reason for these peculiar growth patterns is unknown. These organisms were originally isolated from environmental sources such as hot springs in the case of *A. flavithermus* (Pikuta *et al.* 2000). The association of these microorganisms with food processing environments has occurred as the bacteria have become adapted to this alternative niche. These thermophiles are non-pathogenic although their existence in high numbers in foods is regarded by many as unacceptable (Basappa, 1974). There is potential for enzyme activity and by products from these organisms that may cause off flavours in food (Chopra and Mathur, 1984, Chen *et al.* 2003, 2004, Yokota *et al.* 2007). The formation of spores is a particular concern as they are generally more heat resistant than spores from mesophilic organisms. *Geobacillus* spores can survive canning and UHT processing although remain dormant once the product is cooled and stored at room temperature (Denny *et al.* 1961, Hill, 2004).

The growth of these organisms in food stored at ambient temperatures or above is unlikely unless the food is stored hot or in a tropical climate. The growth of these organisms is normally associated with the food processing/manufacturing plant.

### 10.4.2 Association with food processing

In any manufacturing process where temperatures are suitable for the growth of thermophilic bacteria, there are concerns with the contamination of product. Typical examples are canning, juice pasteurization and dairy processing industries, although specific reports of biofilm growth are limited to the dairy industry.

In the canning industry, thermophilic bacteria have been reported in sugar from the sugar refining process, and have flow on effects in canning of foods (Ito, 1981). Similarly, starch containing thermophilic organisms can also cause problems with thermophilic spoilage of canned foods. *Geobacillus stearothermophilus* can grow and spoil many low-acid canned foods such as vegetables. Soil and food ingredients are identified as the source, presumably from compost where conditions are warm enough for the growth of these organisms. Once in the canning plant, they will grow in areas such as holding tanks and blanchers where temperatures are suitable for thermophile growth (Ito, 1981). Growth on the surfaces of a canning processing plant is acknowledged as a likely source of spores, controlled through sanitiation; however, there is no specific information on the growth of biofilms in this environment (Weddig *et al.* 2007). Holding product at warm temperatures is also known to be a source of thermophile contamination in canned foods (Denny, 1981).

A thermophilic anaerobe, *Clostridium thermosaccarolyticum*, is responsible for hard swell spoilage of canned foods. These organisms originate from the general environment (soil) and typically contaminate vegetable products. They do not grow on equipment surfaces unless there is an anaerobic environment containing nutrients at elevated temperatures. Most growth is believed to occur in product during storage stages during processing (Ashton, 1981). Similarly, *Clostridium perfringens* contamination of broiler chicken carcass processing plants may arise from the presence of heat-resistant spores in water used during processing, such as in the scald and chill tanks (Craven, 2001). From a biofilm-formation perspective, it has been speculated that biofilms of *Clostridium* strains may be possible in meat processing environments due to the fact that Clostridia are unable to synthesise 13 different amino acids, which such a processing environment would provide (Andersson *et al.* 1995). In addition, it has been shown that spores and vegetative cells of *Clostridium difficile*, a *Clostridium* species of medical importance, are able to attach to and survive on stainless steel and, to a lesser extent, copper surfaces (Weaver *et al.* 2008). Various species of Clostridia are also known to form biofilms in industrial settings, e.g. for the generation of biohydrogen (Zhang *et al.* 2006, Toluanen *et al.* 2008), and in wastewater treatment (Park *et al.* 2007); however, from a practical food perspective, such biofilms of anaerobic *Clostridia* on food processing equipment surfaces are unreported in the literature to date.

*Alicyclobacillus* species are a problem in the juice industry during the pasteurization and storage of product (Yokota *et al.* 2007). Growth during the processing of juice is likely to be due to biofilm formation, although biofilm growth of these organisms does not appear to have been studied.

Most reports of biofilm formation of thermophilic bacteria in food manufacture concern the dairy industry. In the dairy industry, thermophilic endospore-forming bacteria survive heat treatments, such as pasteurization, and may grow in regeneration sections of heat exchangers and evaporators where temperatures are suitable for their growth (Scott *et al.* 2007). Milk powders are used by customers to manufacture a range of consumer products and the bacteriological quality of these depends, in part, on the quality of the milk powder. The origin of thermophlic bacteria in milk powder is often attributed to persistent colonization of the manufacturing plant that has not been cleaned from the previous manufacturing run (Austin and Bergeron, 1995), but the ultimate origin is low levels in the raw milk that are believed to originate from the farm environment, in particular silage (Deak and Timar, 1988, te Giffel *et al.* 2002). However, thermophile contamination of milk powder cannot be related to raw milk quality (Muir *et al.* 1986).

Thermophilic bacteria can reach high levels ($>10^5$/g) during the manufacture of milk powder (Stadhouders *et al.* 1982, Murphy *et al.* 1999, Scott *et al.* 2007). The residence time of milk in the manufacturing plant is less than 30 min, therefore there is limited opportunity for thermophilic bacteria, with a generation time of 15–20 min, to grow to such high levels, except as biofilms on the surfaces of the manufacturing plant. (Scott *et al.* 2007).

Dairy manufacture involves several heat treatment steps such as pasteurization and evaporation. The spores of thermophilic bacteria, found in very low levels in raw milk can survive these heat treatment steps (Janstova and Lukasova, 2001), germinate and grow in the manufacturing plant where conditions are optimum (Hinton *et al.* 2003). The conditions during the manufacture of milk powder are ideal for the growth of thermophilic bacteria (Stadhouders *et al.* 1982, Kwee *et al.* 1986, Murphy *et al.* 1999). Release of these bacteria from these biofilms results in contamination of the final product. In addition there is potential for inefficient heat transfer (Russell, 1993) and flow restriction (Criado *et al.* 1994) resulting from biofilm colonization.

Thermophilic bacterial colonization of dairy plant surfaces appears to occur with and without the presence of milk foulant (Wirtanen *et al.* 1996, Flint *et al.* 2001b, Parkar *et al.* 2001, Scott *et al.* 2007). Biofilms in dairy systems occur on any surface with those occurring on gasket material being particularly difficult to remove (Austin and Bergeron, 1995).

## 10.5 Biofilm formation

The only food industry in which biofilm growth of thermophilic spore-forming bacteria has been studied is the dairy industry (Langeveld *et al.* 1995, Murphy *et al.* 1999, Scott *et al.* 2007). Thermophile growth associated with milk fouling in a milk powder manufacturing plant may be greater than growth on an un-fouled surface (true biofilm). Thermophilie numbers appeared to be two orders of magnitude greater on fouled surfaces compared with un-fouled surfaces in a pilot plant study (Hinton *et al.* 2003). In this trial, thermophiles were more difficult to remove from the fouled surface than the un-fouled surface. However, in another trial, using a different experimental design, thermophilic bacteria were shown to adhere more firmly to an un-fouled stainless steel surface compared with a fouled surface (Wirtanen *et al.* 1996). This difference between the two trials may depend on whether the bacteria are embedded in the fouling (more difficult to remove) or growing on the surface of fouling (easier to remove).

### 10.5.1 Attachment of spores to surfaces

The attachment of spores to the surface of food manufacturing plant is believed to be one source of biofilms of thermophilic bacilli, therefore this is an important part of biofilm development. Spores of thermophilic bacteria attach to surfaces more readily (in greater numbers) than vegetative cells (Flint *et al.* 2001a). The mechanism of the initial attachment process is believed to be similar to that of vegetative cells as described by the DELVO theory (van Loosdrecht *et al.* 1989), involving Lifshitz-van der Waal's forces, acid/base interactions and electrostatic interactions. Proteins are believed to play a major role in the initial attachment process (Parkar *et al.* 2001). The attachment of spores of dairy isolates of *Geobacillus* species under flow conditions can vary between isolates with one isolate attaching in higher numbers to stainless steel than glass (Seale *et al.* 2008, Seale 2009). An investigation of the surface chemistry and structure of spores has shown differences in the amounts of carboxylated glocoproteins and polysaccharides although there is no clear link between these results and attachment to a substrate (Seale *et al.* 2008, Seale 2009).

### 10.5.2 Biofilm development of vegetative cells

While the development of a biofilm of thermophilic bacilli in a food system is likely to originate from spores, vegetative cells can also attach readily to stainless steel surfaces, forming mature biofilms within six hours of incubation in the presence of pasteurized milk (Stadhouders *et al.* 1982, Parkar *et al.* 2001, Burgess *et al.* 2009). Cell surface proteins appear to play a key role in the attachment of *A. flavithermus* to stainless steel with pyruvate formate lyase expressed on the surface of cells influencing the number of cells attaching to stainless steel (Palmer *et al.* 2007). In a laboratory biofilm reactor, biofilms of *A. flavithermus* grown on stainless steel coupons in the presence of pasteurized skim milk, showed a cyclical trend in the number of cells colonizing the coupons with numbers ranging from $10^2$–$10^6$ cells/$cm^2$ – clearly different from the rapid growth and decline in cell numbers observed in a static liquid culture (Parkar *et al.* 2003). In an earlier laboratory experiment over a shorter period of time (6 h), cell numbers of a *Geobacillus* species also reached a maximum of $10^6$ /$cm^2$ (Flint *et al.* 2001a).

In laboratory experiments, with *Geobacillus* species grown at 55°C in a continuously flowing reactor, the biofilm cells were predominantly vegetative (Flint *et al.* 2001a). Spores can be formed with difficulty in planktonic culture in the laboratory (Seale *et al.* 2008, Seale 2009). Further work is needed to determine the conditions under which biofilms of *Geobacillus* species from dairy manufacturing plant form spores.

### 10.5.3 Development of spores within a biofilm

The rate and extent of spore development in biofilms of *A. flavithermus* was recently reported (Burgess *et al.* 2009). This was done under controlled conditions in a laboratory reactor system and explored the effect of temperature on spore formation. Sporulation occurred rapidly, within 6–8 h at 55°C in conjunction with vegetative growth. At 8 h, up to 50% of the biofilm was composed of spores. In a dairy manufacturing plant, the formation of spores from thermophilic bacteria was reported to occur within 11 h (Scott *et al.* 2007). The difference between the results from the manufacturing plant and the laboratory experiments may be due to the origin of the biofilm. In the laboratory trials, vegetative cells were introduced to the model system while in a dairy manufacturing plant, spores surviving heat treatment of the milk or cleaning chemicals would provide the inoculum.

The types of thermophilic bacteria forming spores in a milk powder plant may vary at different stages in the manufacturing process (Scott *et al.* 2007). In this trial, *A. flavithermus* spores predominated in the pre-heating section of the evaporator in a whole milk powder manufacturing plant. The *Geobacillus* species did not appear until the second pass of the evaporator. This presumably reflects the conditions in the different sections of the manufacturing plant, favouring the growth of specific bacteria.

Temperature appears to be important in determining spore formation with spores of *A. flavithermus* formed at 55°C but not at 48°C or 60°C

(Burgess *et al.* 2009). This could be important to manufacturers of milk powder to avoid 55°C as much as possible in the manufacturing process to prevent spore formation. This observation is interesting as it suggests that spore formation is not produced under stress but under ideal growing conditions. This observation differs from the formation of spores seen in biofilms of *Bacillus subtilis* where spore formation occurred later in the growth of the biofilm (Lindsay *et al.* 2005). Similar studies are needed for the other thermophlic spore formers in the dairy industry, in particular the *Geobacillus* species.

### 10.5.4 Control of biofilms of thermophilic endospore forming bacteria

There are several potential options for controlling biofilms of thermophilic spore-forming bacilli and these can be categorised as temperature manipulation to limit growth, improved cleaning/sanititation and treating surfaces to prevent attachment. Most of the published research focuses on the effect of temperature and cleaning as control measures.

In milk powder manufacture, most of the growth of thermophilic endospore forming bacteria is in the pre-heat section of evaporators. By-passing plate heater pre-heaters through the use of direct steam injection was found to limit spore formation and vegetative cells that were present were inactivated during spray drying (Murphy *et al.* 1999). The success is most likely due to limiting the surface area of the manufacturing plant where conditions are suitable for the growth of thermophilic bacteria. However, thermophilic spores have also been found in foulant in a direct steam injection unit (Scott *et al.* 2007).

The use of heat to control thermophilic spore-forming bacteria in a dairy manufacturing plant is difficult. The spores survive UHT treatment (Mostert *et al.* 1979) and are highly heat resistant, particularly when adhered to stainless steel. For example, a 400% increase in the $D_{100}$ – value for *G. stearothermophilus* was recorded when the spores were attached to stainless steel compared with unattached spores (Simmonds *et al.* 2003).

Including a mini-wash with 0.2% hydrogen peroxide for 45 min, helped reduce thermophile contamination in the milk powder evaporator (Murphy *et al.* 1999). This suggests that the use of a sanitizer may help in controlling biofims of thermophiles in milk powder manufacturing plants. Sanitizers are often not used in milk powder dairy manufacturing plant.

Cleaning with caustic to remove biofilms of thermophilic endospore-forming bacilli is not always successful (Parkar *et al.* 2004) and when it is, it appears from confocal laser scanning microscopy, that the intact biofilm structure may remain, suggesting that cells have been killed but not removed from the surface (Hinton *et al.* 2003). Where cells are attached to milk foulant, removing that foulant will remove the biofilm (Wirtanen *et al.* 1996).

## 10.6 Future trends

### 10.6.1 Fundamental aspects of study

A better understanding of the spore formation process within endospore-forming bacterial biofilms is necessary in order to evaluate better means of control within industry. This is true especially for *B. cereus* biofilms, where a wide diversity of spore formation and biofilm development is apparent. Of further importance is spore development within themophilic strains, such as *Geobacillus*, both from an *in vitro* and *in situ* industry perspective. Furthermore, studies on other, 'newer' spore formers of importance in food processing should be evaluated, for example *Alicyclobacillus* in juice processing.

### 10.6.2 Practical aspects

Future innovations to control the growth of biofilms of endospore-forming bacteria on industrial plant equipment surfaces are likely to be developments that reduce attachment and aim to prevent biofilm-formation in the first place, as opposed post-biofilm treatments. These would include evaluating anti-adhesive surfaces, such as those containing inherent or coated antimicrobials (silver or bacteriocins), as well as longer-term views on more hygienic plant layout and design.

## 10.7 Sources of further information and advice

For excellent reviews on the process of sporulation in *Bacillus* refer to Stragier and Losick (1996), and on spore resistance refer to Nicholson *et al.* (2000).

## 10.8 References

AHIMOU F, JACQUES P, and DELEUA M, (2000), 'Surfactin and iturin A effects on *Bacillus subtilis* surface hydrophobicity', *Enzyme Microbial Technol*, 27, 749–754.

AHIMOU F, PAQUOT M, JACQUES P, THONART P, and ROUXHET P G, (2001), 'Influence of electrical properties on the evaluation of the surface hydrophobicity of *Bacillus subtilis*', *J Microbiol Methods*, 45, 119–126.

ANDERSSON A, and RÖNNER U, (1998), 'Adhesion and removal of dormant, heat-activated, and germinated spores of three strains of *Bacillus cereus*,' *Biofouling*, 13, 51–67.

ANDERSSON A, RÖNNER U, and GRANUM P E, (1995), 'What problems does the food industry have with the spore-forming pathogens *Bacillus cereus* and *Clostridium perfringens*', *Int J Food Microbiol*, 28, 145–155.

ARNESEN L P S, O'SULLIVAN K, and GRANUM P E, (2007), 'Food poisoning potential of *Bacillus cereus* strains from Norwegian dairies' *Int J Food Microbiol*, 116, 292–296.

ASHTON D H, (1981), 'Thermophilic organisms involved in food spoilage: thermophilic anaerobes not producing hydrogen sulfide', *J Food Protect*, 44, 146–148.

ATRIH A, ZÖLLNER P, ALLMAIER G, and FOSTER S J, (1996), 'Structural analysis of *Bacillus subtilis* 168 endospore peptidoglycan and its role during differentiation', *J Bacteriol*, 178, 6173–6183.

AUSTIN J W, and BERGERON G, (1995), 'Development of bacterial biofilms in dairy processing lines', *J Dairy Res*, 62, 509–519.

BAILEY C P, VON HOLY A, (1993), '*Bacillus* spore contamination associated with commercial bread manufacture', *Food Microbiol*, 10, 287–294.

BAILEY-SMITH K, TODD S J, SOUTHWORTH T W, PROCTOR J, and MOIR A, (2005), 'The ExsA protein of *Bacillus cereus* is required for assembly of coat and exosporium onto the spore surface', *J Bacteriol*, 187, 3800–3806.

BARON F, COCHET M F, GROSSET N, MADEC M N, BRIANDET R, DESSAIGNE S, CHEVALIER S, GAUTIER M, and JAN S, (2007), 'Isolation and characterization of a psychrotolerant toxin producer, *Bacillus weihenstephanensis*, in liquid egg products', *J Food Protect*, 70, 2782–2791.

BASAPPA P, (1974), 'Growth characteristics of thermophilic micro-organisms in milk', *Indian J Dairy Science*, 27, 103–108.

BASSLER B L, and LOSICK R, (2006), 'Bacterially speaking', *Cell*, 125, 237–246.

BATOSZEWICZ M, HANSEN B M, and SWIECICKA I, (2008), 'The members of the *Bacillus cereus* group are commonly present contaminants of fresh and heat-treated milk', *Food Microbiol*, 25, 588–596.

BEATTIE S H, and WILLIAMS A G, (1999), 'Detection of toxigenic strains of *Bacillus cereus* and other *Bacillus* spp. with an improved cytotoxicity assay', *Lett Appl Microbiol*, 28, 221–225.

BEENKEN K E, DUNMAN P M, MCALEESE F, MACAPAGAL D, MURPHY E, PROJAN S J, BLEVINS J S, and SMELTZER M S, (2004), 'Global gene expression in *Staphylococcus aureus* biofilms', *J Bacteriol*, 186, 4665–4684.

BEUCHAT, L R., PETTIGREW C A, TREMBLAY M E, ROSELLE B J and SCOUTEN A J, (2004), 'Lethality of chlorine, chlorine dioxide and a commercial fruit and vegetable sanitizer to vegetative cells and spores of *Bacillus cereus* and spores of *Bacillus thuringiensis*', *J Food Protect*, 67, 1702–1708.

BIEBL H, SCHWAB-HANISCH H, SPRÖER C, and LÜNSDORF H, (2000), '*Propionispora vibrioides*, nov. gen., nov. sp., a new gram-negative, spore-forming anaerobe that ferments sugar alcohols', *Arch Microbiol*, 174, 239–247.

BOHANNON D E, ROSENKRANTZ, M S, and SONENSHEIN, A L, (1985), 'Regulation of *Bacillus subtilis* glutamate synthase genes by the nitrogen source', *J Bacteriol*, 163, 957–964.

BORGE G I A, SKEIE M, SØRHAUG T, LANGSRUD T, and GRANUM P E, (2001), 'Growth and toxin profiles of *Bacillus cereus* isolated from different food sources', *Int J Food Microbiol*, 69, 237–246.

BRADSHAW D J, MARSH P D, ALLISON C, and SCHILLING K M, (1996), 'Effect of oxygen, inoculum composition and flow rate on development of mixed-culture oral biofilms', *Microbiol*, 142, 623–629.

BRANDA S S, GONZÁLEZ-PASTOR J E, DERVYN E, EHRLICH S D, LOSICK R, and KOLTER R, (2004), 'Genes Involved in formation of structured multicellular communities by *Bacillus subtilis*', *J Bacteriol*, 186, 3970–3979.

BRANDA S S, CHU F, KEARNS D B, LOSICK R, and KOLTER R, (2006), 'A major protein component of the *Bacillus subtilis* biofilm matrix', *Mol Microbiol*, 59, 1229–1238.

BROWN K L, (1994), 'Spore resistance and ultra heat treatment processes', *J Appl Bacteriol*, 76, 67S–80S.

BURGESS S A, BROOKS J D, RAKONJAC J, WALKER K M, and FLINT S H, (2009), 'The formation of spores in biofilms of *Anoxybacillus flavithermus*,' *J Appl Microbiol*, (published online: 25 Mar 2009; DOI: 10 1111/j. 1365-2672 2009.04282.x).

BUSSCHER H J, BOS R, and VAN DER MEI H C, (1995), 'Microbial adhesion is a determinant for the strength of biofilm adhesion', *FEMS Microbiol Lett*, 128, 229–234.

CHAGNEAU C, and SAIER M H JR, (2004), 'Biofilm-defective mutants of *Bacillus subtilis*', *J Mol Microbiol Biotechnol*, 8, 177–188.

CHAI Y, CHU F, KOLTER R, and LOSICK R, (2008), 'Bistability and biofilm formation in *Bacillus subtilis*', *Molecular Microbiol*, 67, 254–263.

CHARLTON S, MOIR A J G, BAILLIE L, and MOIR A, (1999), 'Characterization of the exosporium of *Bacillus cereus*', *J Appl Microbiol*, 87, 241–245.

CHEN L, DANIEL R M, and COOLBEAR T, (2003), 'Detection and impact of protease and lipase activities in milk and milk powders', *Int Dairy J*, 13, 255–275.

CHEN L, COOLBEAR T, and DANIEL R M, (2004), 'Characteristics of proteinases and lipases produced by seven *Bacillus* sp. isolated from milk powder production lines', *Int Dairy J*, 14, 495–504.

CHOPRA A K, and MATHUR D K, (1984), 'Isolation, screening and characterization of thermophilic *Bacillus* species isolated from dairy products', *J Appl Bacteriol*, 57, 263–271.

CHRISTIANSSON A, NAIDU S A, NILSSON I, WADSTRÖM T, and PETTERSON H-E, (1989), 'Toxin production by *Bacillus cereus* dairy isolates in milk at low temperatures', *Appl Environ Microbiol*, 55, 2595–2600.

CHU F, KEARNS D B, MCLOON A, CHAI Y, KOLTER R, and LOSICK L, (2008), 'A novel regulatory protein governing biofilm formation in *Bacillus subtilis*', *Molecular Microbiol*, 68, 1117–1127.

CLAUS D, and FRITZE D, (1989), 'Taxonomy of *Bacillus*', in Harwood C R (ed), *Biotechnology Handbooks 2*: *Bacillus*, Plenum Press, New York.

COLLINS N E, KIRSCHNER L A M, and VON HOLY A, (1991), 'Characterization of *Bacillus* isolates from ropey bread, bakery equipment and raw materials', *S Afr J Sci*, 87, 62–66.

COSBY W M, VOLLENBROICH D, LEE O H, and ZUBER P, (1998), 'Altered srf expression in *Bacillus subtilis* resulting from changes in culture pH is dependent on the Spo0K oligopeptide permease and the ComQX system of extracellular control', *J Bacteriol*, 180, 1438–1445.

COSTERTON J W, CHENG K-J, GEESEY G G, LADD T I, NICKEL J C, DASGUPTA M, and MARRIE T J, (1987), 'Bacterial biofilms in nature and disease', *Ann Rev Microbiol*, 41, 435–464.

COSTERTON J W, LEWANDOWSKI Z, DE BEER D, CALDWELL D, KORBER D, and JAMES G, (1994), 'Biofilms, the customised microniche', *J Bacteriol* 176, 2137–2142.

CRAVEN S E, (2001), 'Occurrence of *Clostridium perfringens* in the broiler chicken processing plant as determined by recovery in iron milk medium', *J Food Protect*, 64, 1956–1960.

CRIADO M T, SUAREZ B, and FERREIROS C M, (1994), 'The importance of bacterial adhesion in the dairy industry', *Food Techno*, 48, 123–126.

CRIELLY E M, LOGAN N A, and ANDERTON A, (1994), 'Studies on the *Bacillus* flora of milk and milk products', *J Appl Bacteriol*, 77, 256–263.

DE CLERCK E, and DE VOS P, (2002), 'Study of the bacterial load in a gelatine production process focussed on *Bacillus* and related endosporeforming genera', *Syst Appl Microbiol*, 25, 611–617.

DE CLERCK E, GEVERS D, DE RIDDER K, and DE VOS P, (2004), 'Screening of bacterial contamination during gelatine production by means of denaturing gradient gel electrophoresis, focussed on Bacillus and related endospore-forming genera,' *J Appl Microbiol*, 96, 1333–1341.

DEAK T, and TIMAR E, (1988), 'Simplified identification of aerobic spore-formers in the investigation of foods', *Int J Food Microbiol*, 6, 115–125.

DENNY C B, (1981), 'Thermophilic organisms involved in food spoilage: an introduction', *J Food Protect*, 44, 144–145.

DENNY C B, BOHRER C W, and SHARPE L E, (1961), 'Effects of tylosin and nisin on canned food spoilage bacteria', *Appl Microbiol*, 9, 108–110.

DONLAN R.M, (2002), 'Biofilms: microbial life on surfaces', *EID*, 8, 881–890.

DOYLE R J, NEDJAT-HAIEM F, and SINGH J S, (1984), 'Hydrophobic characteristics of *Bacillus* spores', *Curr Microbiol*, 10, 329–332.

DUC L H, DONG T C, LOGAN N A, SUTHERLAND A D, TAYLOR J, and CUTTING S M, (2005), 'Cases of emesis with bacterial contamination of an infant breakfast cereal product', *Int J Food Microbiol*, 102, 245–251.

DURRE, P, and HOLLERGSCHWANDNER, C, (2004), 'Initiation of endospore formation in *Clostridium acetobutylicum*', *Anaerobe*, 10, 69–74. ed. Springer Verlag, New York, 1663–1696.

FAILLE C, FONTAINE F, and BÉNÉZECH T, (2001), 'Potential occurrence of adhering living *Bacillus* spores in milk product processing lines', *J Appl Microbiol*, 90, 892–900.

FAILLE C, JULLIEN C, FONTAINE F, BELLON-FONTAINE M-N, SLOMIANNY C, and BENEZECH T, (2002), 'Adhesion of *Bacillus* spores and *Escherichia coli* cells to inert surfaces: role of surface hydrophobicity', *Can J Microbiol*, 48, 728–738.

FAILLE C, TAUVERON G, LE GENTIL-LELIÈVRE C, SLOMIANNY C, (2007), 'Occurrence of *Bacillus cereus* spores with a damaged exosporium: consequences on the spore adhesion on surfaces of food processing lines', *J Food Protect*, 70, 2346–2353.

FLEMMING C A, PALMER R J JR, ARRAGE A A, VAN DER MEI H C, and WHITE D C, (1998), 'Cell surface physiochemistry alters biofilm development of *Pseudomonas aeruginosa* lipopolysaccharide mutants', *Biofouling*, 13, 213–231.

FLINT S H, BREMER P J, BROOKS J D, (1997), 'Biofilms in dairy manufacturing plant description, current concerns and methods of control', *Biofouling*, 11, 81–97.

FLINT S, PALMER J, BLOEMEN K, BROOKS J, and CRAWFORD R, (2001a), 'The growth of *Bacillus stearothermophilus* on stainless steel', *J Appl Microbiol*, 90, 151–157.

FLINT S H, WARD L J H, and WALKER K M R, (2001b), 'Functional grouping of thermophilic *Bacillus* strains using amplification profiles of the 16S–23S internal spacer region', *Systematic Appl Microbiol*, 24, 539–548.

FRANK J F, (1997), 'Milk and dairy products', in Doyle M P, Beuchat L R, Montville T J (eds.), *Food microbiology: fundamentals and frontiers*, ASM Press, Washington, DC, 101–116.

GARCÍA-ARMESTO M R, and SUTHERLAND A D, (1997), 'Temperature characterization of psychrotrophic and mesophilic *Bacillus* species from milk', *J Dairy Res*, 64, 261–270.

GARRY P, VENDEUVRE J L, BELLON-FONTAINE M N, (1998), 'Surface properties and adhesion of *Bacillus cereus* and *Bacillus subtilis* to polyurethane – influence of growth temperature', *J Dispersion Sci Technol*, 19, 1175–1197.

GILBERT P, EVENS D J, and BROWN M R W, (1993), 'Formation and dispersal of bacterial biofilms *in vivo* and *in situ*', *J Appl Bacteriol Symp Suppl*, 74, 67S–78S.

GILMOUR A, and FRASER T W, (1991), 'The study of the interaction of microorganisms with dairy equipment surfaces using scanning electron microscopy', *Biofouling*, 5, 133–141.

GOBBETTI M, DE ANGELIS M, DI CAGNO R, MINERVINI F, and LIMITONE A, (2007), 'Cell–cell communication in food related bacteria', *Int J Food Microbiol*, 120, 34–45.

GONZALEZ-PASTOR J E, HOBBS EC, and LOSICK R, (2003), 'Cannibalism by sporulating bacteria', *Science* 301, 510–513.

GRANUM P E, (1994), '*Bacillus cereus* and its toxins', *J Appl Bacteriol*, 76, 61–66.

GUINEBRETIERE M H, GIRARDIN H, DARGAIGNARATZ C, CARLIN F, NGUYEN-THE C, (2003), 'Contamination flows of *Bacillus cereus* and spore-forming aerobic bacteria in a cooked, pasteurized and chilled zucchini purée processing line,' *Int J Food Microbiol*, 82, 223–232.

GUNDUZ G T, and TUNCEL G, (2006), 'Biofilm formation in an ice cream plant', *Antonie Van Leeuwenhoek*, 89, 329–336.

HAMON M A, and LAZAZZERA B A, (2001), 'The sporulation transcription factor Spo0A is required for biofilm development in *Bacillus subtilis*', *Molecular Microbiol*, 42, 1199–1209.

HAMON M A, STANLEY N R, BRITTON R A, GROSSMAN A D, and LAZAZZERA B A, (2004), 'Identification of ArB-regulated genes involved in biofilm formation by *Bacillus subtilis*', *Molecular Microbiol*, 52, 847–860.

HAQUE A, and RUSSELL N J, (2005), 'Phenotypic and genotypic characterisation of *Bacillus cereus* isolates from Bangladeshi rice', *Int J Food Microbiol*, 98, 23–34.

HARDIE K R, and HEURLIER K, (2008), 'Establishing bacterial communities by "word of mouth": LuxS and autoinducer 2 in biofilm development', *Nat Rev Microbiol*, 6, 635–43.

HILL B M, (2004), 'Thermophilic spores: UHT and retort processing of recombined milks', Proceedings of the 4th international symposium on recombined milk and milk products. Cancun, Mexico, US Dairy Export Council.

HINTON A R, TRINH K T, BROOKS J D, and MANDERSON G J, (2003), 'Thermophile survival in milk fouling and on stainless steel during cleaning', *Transactions of the Institution of Chemical Engineers*, 80 (C), 299–304.

HOLAH J, and GIBSON H, (2000), 'Food industry biofilms' in Evans L V, (ed), *Biofilms: Recent Advances in Their Study and Control*, CRC Press, 211–236.

HOOD S K, and ZOTTOLA E A, (1995), 'Biofilms in food processing', *Food Control*, 1, 9–18.

HORNSTRA L M, DE LEEUW P L A, MOEZELAAR R, WOLBERT E J, DE VRIES Y P, DE VOS W M, and ABEE T, (2007), 'Germination of *Bacillus cereus* spores adhered to stainless steel', *Int J Food Microbiol*, 116, 367–371.

HSUEH Y-H, SOMERS E B, LERECLUS D, and WONG A C L, (2006), 'Biofilm formation by *Bacillus cereus* is influenced by PlcR, a pleiotropic regulator', *Appl Environ Microbiol*, 72, 5089–5092.

HÜSMARK U, and RÖNNER U, (1990), 'Forces involved in adhesion of *Bacillus cereus* spores to solid surfaces under different environmental conditional', *J Appl Microbiol*, 69, S57–S62.

HÜSMARK U, and RÖNNER U, (1992), 'The influence of hydrophobic, electrostatic and morphologic properties on the adhesion of *Bacillus* spores', *Biofouling*, 5, 335–344.

ICMSF, (2000), 'Cereals and cereal products' in *Microorganisms in foods 6: microbial ecology of food commodities*, Aspen Publishers Inc. Maryland, USA.

ITO K A, (1981), 'Thermophlic organisms in food spoilage: Flat – sour aerobes', *J Food Protect*, 44, 157–163.

JANSTOVA B, and LUKASOVA J, (2001), 'Heat resistance of *Bacillus* spp spores isolated from cow's milk and farm environment', *Acta Veterinaria Brno*, 70, 179–184.

JAY J J, (2000), '*Modern food microbiology*', 6th edition, Aspen Publishers, New York.

KALMOKOFF M, LANTHIER P, TREMBLAY T L, FOSS M, LAU P C, SANDERS G, AUSTIN J, KELLY J, SZYMANSKI C M, (2006), 'Proteomic analysis of *Campylobacter jejuni* 11168 biofilms reveals a role for the motility complex in biofilm formation', *J Bacteriol*, 188, 4312–4320.

KEARNS D B, (2008), Division of labour during *Bacillus subtilis* biofilm formation', *Molecular Microbiol*, 67, 229–231.

KEARNS D B, CHU F, BRANDA S S, KOLTER R, and LOSICK R, (2005), 'A master regulator for biofilm formation by *Bacillus subtilis*', *Molecular Microbiol*, 55, 739–749.

KIRSCHNER L M, (1993), 'Microbial ecology of rope spoilage of bread', MSc dissertation, University of the Witwatersrand, Johannesburg, South Africa.

KIRSCHNER L M, and VON HOLY, A, (1989), 'Rope spoilage of bread', *S Afr J Sci*, 85, 425–427.

KLAVENES A, STALHEIM T, SJØVOLD O, JOSEFSEN K, GRANUM P E, (2002), 'Attachment of *Bacillus cereus* spores with and without appendages to stainless steel surfaces', *Food and Bioproducts Processing*, 80, 312–318.

KOBAYASHI K, (2007), 'Gradual activation of the response regulator DegU controls serial expression of genes for flagellum formation and biofilm formation in *Bacillus subtilis'*, *Molecular Microbiol*, 66, 395–409.

KOSHIKAWA T, YAMAZAKI M, YOSHIMI M, OGAWA S, YAMADA A, WATABE K, and TORII M, (1989), 'Surface hydrophobicity of spores of *Bacillus* species', *J Gen Microbiol*, 135, 2717–2722.

KRESKE A C, RYU J-H, and BEUCHAT L R, (2006), 'Evaluation of chlorine, chlorine dioxide, and a peroxyacetic acid-based sanitizer for effectiveness in killing *Bacillus cereus* and *Bacillus thuringiensis* spores in suspensions, on the surface of stainless steel, and on Apples', *J Food Protect*, 69, 1892–1903.

KWEE W S, DOMMETT T W, GILES J E, SMITH R A D, and ROBERTS R, (1986), 'Microbiological parameters during powdered milk manufacture 2: Relationships and predictability among counts', *Aus J Dairy Technol*, 41, 6–8.

LANGEVELD L P M, VANMONTFORTQUASIG R, WEERKAMP A H, WAALEWIJN R, and WEVER J S, (1995), 'Adherence, growth and release of bacteria in a tube heat exchanger for milk', *Neth Milk Dairy J*, 49, 207–220.

LARSEN H D, and JØRGENSEN K, (1997), 'The occurrence of *B. cereus* in Danish pasteurized milk', *Int J Food Microbiol*, 34, 179–186.

LECHNER S, MAYR R, FRANCIS K P, PRUSS B M, KAPLAN T, WIESSNER-GUNKEL E, STEWART G S, and SCHERER S, (1998), '*Bacillus weihenstephanensis* sp. nov. is a new psychrotolerant species of the *Bacillus cereus* group', *Int J Syst Bacteriol*, 48, 1373–1382.

LEVIN P A, and GROSSMAN A D, (1998), 'Cell cycle: The bacterial approach to coordination', *Curr Biol*, 8, R28–R31.

LINDSAY D, and VON HOLY A, (1997), 'Evaluation of dislodging methods for laboratory-grown bacterial biofilms', *Food Microbiol*, 14, 383–390.

LINDSAY D, and VON HOLY A, (1998), Nutrient limitation affects growth and attachment of two food spoilage bacteria, *Bacillus subtilis* and *Pseudomonas fluorescens*. *S Afri J Sci* 94, 393–396.

LINDSAY D, and VON HOLY A, (1999), 'Different responses of planktonic and attached *Bacillus subtilis* and *Pseudomonas fluorescens* to sanitizer treatment' *J Food Protect*, 62, 368–379.

LINDSAY D, and VON HOLY A, (2006), 'What food safety professionals should know about bacterial biofilms', *British Food J*, 108, 27–37.

LINDSAY D, BRÖZEL V S, MOSTERT J F, and VON HOLY A, (2000a), 'Physiology of dairy-associated *Bacillus* spp. over a wide pH range', *Int J Food Microbiol*, 54, 49–62.

LINDSAY D, MOSUPYE, F M, BRÖZEL V S, and VON HOLY A, (2000b), 'Cytotoxicity of alkaline-tolerant dairy-associated *Bacillus* spp.', *Lett Appl Microbiol*, 30, 364–369.

LINDSAY D, BRÖZEL V S, MOSTERT J F, and VON HOLY A, (2002), 'Differential efficacy of a chlorine dioxide-containing sanitizer against single species and binary biofilms of a dairy-associated *Bacillus cereus* and a *Pseudomonas fluorescens* isolate', *J Appl Microbiol*, 92, 352–361.

LINDSAY D, BRÖZEL V S, and VON HOLY A, (2004), 'Acid and alkaline pH enhance adhesion of spores of alkaline tolerant *Bacillus cereus* to different surface types', *Food Protect Trends*, 24, 487–491.

LINDSAY D, BRÖZEL V S, and VON HOLY A, (2005), 'Spore formation in *Bacillus subtilis* biofilms', *J Food Protect*, 68, 860–865.

LINDSAY D, BRÖZEL V S, and VON HOLY A, (2006), 'Biofilm/spore response in *Bacillus cereus* and *Bacillus subtilis* during nutrient limitation', *J Food Protect*, 69, 1168–1172.

MAGNUSON R, SOLOMON T, and GROSSMAN A, (1994), Biochemical and genetic characterization of a competence pheromone from *B. subtilis*, *Cell*, 77, 207–216.

MARAHIEL M A, and ZUBER P, (1999), 'Sporulation and cell differentiation', in Lengeler, J W, G Drews and H G Schlegel (eds), *Biology of the Prokaryotes*, Blackwell Science and Georg Thieme Verlag, Germany, 586–601.

MARSHALL K C, (1992), 'Biofilms: an overview of bacterial adhesion, activity and control at surfaces', *ASM News*, 58, 202–207.

MATZ L L, BEAMAN T C, and GERHARDT P, (1970), 'Chemical composition of exosporium from spores of *Bacillus cereus*', *J Bacteriol*, 101, 196–201.

MEER R R, BAKER J, BODYFELT F W, and GRIFFITHS M W, (1991), 'Psychrotrophic *Bacillus* spp. in fluid milk products: A review', *Journal Food Protect*, 54, 968–979.

MIKKOLA R, KOLARI M, ANDERSSON M A, HELIN J, and SALKINOJA-SALONEN M S, (2000), 'Toxic lactonic lipopeptide from food poisoning isolates of *Bacillus licheniformis*', *European J Biochem*, 267, 4068–4074.

MORISAKI H, NAGAI S, OHSHIMA H, IKEMOTO E, and KOGURE K, (1999), 'The effect of motility and cell-surface polymers on bacterial attachment', *Microbiol*, 145, 2797–2802.

MOSSEL D A A, CORRY J E L, STRUIJK C B, and BAIRD R M, (1995), *'Essentials of the microbiology of foods: a textbook for advanced studies'*, John Wiley, United Kingdom.

MOSTELLER T M, and BISHOP J R, (1993), 'Sanitizer efficacy against attached bacteria in a milk biofilm', *J Food Protect*, 56, 34–41.

MOSTERT J F, LUECK H, and HUSMANN R A, (1979), 'Isolation, identification and practical properties of *Bacillus* species from UHT and sterilized milk', *S Afri J Dairy Technol*, 11, 125–132.

MUIR D D, GRIFFITHS M W, PHILLIPS J D, SWEETSUR A W M, ANDWEST I G, (1986), 'Effect of the bacterial quality of raw milk on the bacterial quality and some other properties of low-heat and high-heat dried milk', *J Soc Dairy Technol*, 39, 115–118.

MUNOZ L, SADAIE Y, and DOI R H, (1978), 'Spore coat protein of *Bacillus subtilis.* Structure and precursor synthesis,' *J Biol Chem*, 253, 6694–6701.

MURPHY P M, LYNCH D, and KELLY P M, (1999), 'Growth of thermophilic spore forming bacilli in milk during the manufacture of low heat powders', *Int J Dairy Technol*, 52, 45–50.

NICHOLSON W L, MUNAKATA N, HORNECK G, MELOSH H J, and SETLOW P, (2000), 'Resistance of *Bacillus* endospores to extreme terrestrial and extraterrestrial environments', *Microbiol Mol Biol Rev*, 64, 548–572.

NIEMINEN T, RINTALUOMA N, ANDERSSON M, TAIMISTO A-M, ALI-VEHMAS T, SEPPÄLÄ A, PRIHA O, and SALKINOJA-SALONEN, M, (2007), 'Toxinogenic *Bacillus pumilus* and *Bacillus licheniformis* from mastitic milk', *Vet Microbiol*, 124, 329–339.

NORRIS J R, BERKELEY R W C, LOGAN N A, and O'DONNELL A J, (1981), 'The genus *Bacillus* and *Sporolactobacillus*', in Starr M P, Stolp H, Trüper H G, Balows A, Schlegel H G, (eds), *The Prokaryotes*, Springer-Verlag, New York, 1711–1742.

NOTERMANS S, DORMANS J A M A, and MEAD G C, (1991), 'Contribution of surface attachment to the establishment of microorganisms in food processing plants: a review', *Biofouling*, 5, 21–36.

OGUNTOYINBO F A, and SANNI A I, (2007), 'Determination of toxigenic potentials of *Bacillus* strains isolated from *Okpehe*, a Nigerian fermented condiment', *World J Microbiol Biotechnol*, 23, 65–70.

OKADA M, SATO I, CHO S J, IWATA H, NISHIO T, DUBNAU D, and SAKAGAMI Y, (2005), 'Structure of the *Bacillus subtilis* quorum-sensing peptide phereomone ComX', *Nat Chem Biol*, 1, 23–24.

OOSTHUIZEN M C, STEYN B, LINDSAY D, BRÖZEL V S, and VON HOLY A, (2001), 'Novel method for the proteomic investigation of a dairy-associated *Bacillus cereus* biofilm', *FEMS Microbiol Lett*, 194, 47–51.

OOSTHUIZEN M C, STEYN B, THERON J, COSETTE P, LINDSAY D, VON HOLY A, and BRÖZEL V S, (2002), 'Proteomic analysis reveals differential protein expression by *Bacillus cereus* during biofilm formation', *Appl Environ Microbiol*, 68, 2770–2780.

ØSTENSVIK Ø, FROM C, HEIDENREICH B, O'SULLIVAN K, and GRANUM P E, (2004), 'Cytotoxic *Bacillus* belonging to the *B. cereus* and *B. subtilis* groups in Norwegian surface waters', *J Appl Microbiol*, 96, 987–993.

PALMER J, BROOKS I J, and FLINT S H, (2007), 'Surface proteins of an adhesive thermophilic spore-forming *Bacillus* from milk powder' *Proceedings of the Fourth ASM Conference on Biofilms*, March 25–29, Quebec City, Quebec, Canada, p53.

PANDEY, N K, (1980), 'Spore coat protein of *Bacillus subtilis* – A summary of recent findings', *Int J Biochem*, 12, 553–558.

PARK S J, YOON J C, SHIN K-S, KIM E H, YIM S, CHO Y-J, SUNG G M, LEE D-G, KIM S B, LEE D-U, WOO S-H, and KOOPMAN B, (2007), 'Dominance of endospore-forming bacteria on a rotating activated *Bacillus* contactor biofilm for advanced wastewater treatment', *J Microbiol*, 45, 113–121.

PARKAR S G, FLINT S H, PALMER J S, and BROOKS J D, (2001), 'Factors influencing attachment of thermophilic bacilli to stainless steel', *J Appl Microbiol*, 90, 901–908.

PARKAR S G, FLINT S H, and BROOKS J D, (2003), 'Physiology of biofilms of thermophilic bacilli – potential consequences for cleaning', *J Indust Microbiol Biotechnol*, 30, 553–560.

PARKAR S G, FLINT S H, and BROOKS J D, (2004), 'Evaluation of the effect of cleaning regimes on biofilms of thermophilic bacilli on stainless steel', *J Appl Microbiol*, 96, 110–116.

PEDERSEN P B, BJØRNVAD M E, RASMUSSEN M D, and PETERSEN J N, (2002), 'Cytotoxic potential of industrial strains of *Bacillus* sp.', *Regul Toxicol Pharmacol*, 36, 155–161.

PENG J-S, TSAI W-C, and CHOU C-C, (2002), 'Inactivation and removal of *Bacillus cereus* by sanitizer and detergent', *Int J Food Microbiol*, 77, 11–18.

PETA M E, LINDSAY D, BRÖZEL V S, and VON HOLY A, (2003), 'Susceptibility of food spoilage *Bacillus* spp. to chlorine dioxide and other sanitizers', *S Afri J Sci*, 99, 375–380.

PHELPS R J, and MCKILLIP J L (2002), 'Enterotoxin production in natural isolates of Bacillaceae outside the *Bacillus cereus* group', *Appl Environ Microbiol*, 68, 3147–3151.

PIKUTA E, LYSENKO A M, CHUVILSKAYA N, MENDROCK U, HIPPE H, SUZINA N, NIKITIN D, OSIPOV G A, and LAURINAVICHIUS K, (2000), '*Anoxybacillus pushcinensis* gen. nov., sp nov., a novel anaerobic, alkaliphilic, moderately thermophilic bacterium from manure, and description of *Anoxybacillus flavothermus* comb. nov.', *Int J Syst Evol Microbiol*, 50, 2109–2117.

PIGGOT P J, (1996), 'Spore development in *Bacillus subtilis*', *Curr Opin Genetics Develop*, 6, 531–537.

PLOMP M, LEIGHTON T J, WHEELER K E, and MALKIN A J, (2005), 'Architecture and high-resolution structure of *Bacillus thuringiensis* and *Bacillus cereus* spore coat surfaces', *Langmuir*, 21, 7892–7898.

PYSZ M A, CONNERS S B, MONTERO C I, SHOCKLEY K R, JOHNSON M R, WARD D E, and KELLY R M, (2004), 'Transcriptional analysis of biofilm formation processes in the anaerobic, hyperthermophilic bacterium *Thermotoga maritima*', *Appl Environ Microbiol*, 70, 6098–6112.

REN D, BEDZYK L A, STELOW P, THOMAS S M, YE R W, and WOOD T K, (2004),' Gene expression in *Bacillus subtilis* surface biofilms with and without sporulation and the importance of *yve*R for biofilm maintenance', *Biotechnol Bioeng*, 86, 344–364.

RONIMUS R S, PARKER L E, TURNER N, POUDEL S, RUCKERT A, and MORGAN H W, (2003), 'A RAPD-based comparison of thermophilic bacilli from milk powders', *Int J Food Microbiol*, 85, 45–61.

RÖNNER U, HUSMARK U, and HENRIKSSON A, (1990), 'Adhesion of *Bacillus* spores in relation to hydrophobicity', *J Appl Bacteriol*, 69, 550–556.

RUECKERT A, RONIMUS R S, and MORGAN H W, (2004), 'A RAPD-based survey of thermophilic bacilli in milk powders from different countries', *Int J Food Microbiol*, 96, 263–272.

RUSSELL P, (1993), 'The formation of biofilms', *Milk Industry (London)*, 95, 10–11.

RUZAL S M, and SANCHEZ-RIVAS C, (2003), 'Effect of glutamate synthase (GOGAT) activity on *Bacillus subtilis* spore properties', *Curr Microbiol*, 47, 208–213.

RYU J-H, and BEUCHAT L R, (2005), 'Biofilm formation and sporulation by *Bacillus cereus* on a stainless steel surface and subsequent resistance of vegetative cells and spores to chlorine, chlorine dioxide, and a peroxyacetic acid-based sanitizer', *J Food Prot*, 68, 2614–2622.

SALKINOJA-SALONEN M S, VUORIO R, ANDERSSON M A, KÄMPFER P, ANDERSSON M C, HONKANEN-BUZALSKI T, and SCOGING A C, (1999), 'Toxigenic strains of *Bacillus licheniformis* related to food poisoning', *Appl Environl Microbiol*, 65, 4637–4645.

SCOTT S A, BROOKS J D, RAKONJAC J, WALKER K M R, and FLINT S H, (2007), 'The formation of thermophilic spores during the manufacture of whole milk powder', *Int J Dairy Technol*, 60, 109–117.

SEALE R B, (2009), The surface characteristics of spores from thermophilic bacilli isolated from a milk powder production line and their influence on adhesion to surfaces, PhD Thesis, University of Otago, Dunedin, New Zealand.

SEALE R B, FLINT S H, MCQUILLAN A J, and BREMER P J, (2008), 'Recovery of spores from thermophilic dairy bacilli and effects of their surface characteristics on attachment to different surfaces', *Appl Environ Microbiol*, 74, 731–737.

SIMMONDS P, MOSSEL B L, INTARAPHAN T, and DEETH H C, (2003), 'Heat resistance of *Bacillus* spores when adhered to stainless steel and its relationship to spore hydrophobicity', *J Food Protect*, 66, 2070–2075.

SLAMTI L, and LERECLUS D, (2002), 'A cell–cell signaling peptide activates the PlcR virulence regulon in bacteria of the *Bacillus cereus* group', *EMBO J.*, 21, 4550–4559.

SLEPECKY R A, HEMPHILL H E, (1992), 'The genus *Bacillus* – nonmedical', in Balows A, Trüper H G, Dworkin M, Harder W, Schleifer K H (eds), *The Prokaryotes, a handbook on the biology of bacteria: ecophysiology, isolation, identification, applications*, 2nd edition, Springer-Verlag, New York, 1663–1697.

SOROKULOVA I B, REVA O N, SMIRNOV V V, PINCHUL I V, LAPA S V, and URDACI M C, (2003), 'Genetic diversity and involvement in bread spoilage of *Bacillus* strains isolated from flour and ropey bread', *Lett Appl Microbiol*, 37, 169–173.

STADHOUDERS J, HUP G, and HASSING F, (1982), 'The conceptions index and indicator organisms discussed on the basis of the bacteriology of spray-dried milk powder', *Neth Milk Dairy J*, 36, 231–260.

STALHEIM T, and GRANUM P E, (2001), Characterization of spore appendages from *Bacillus cereus* strains', *J Appl Microbiol*, 91, 839–845.

STANLEY N R, BRITTON R A, GROSSMAN A D, LAZAZZERA B A, (2003), 'Identification of catabolite repression as a physiological regulator of biofilm formation by *Bacillus subtilis* by use of DNA microarrays', *J Bacteriol*, 185, 1951–1957.

STEICHEN C, CHEN P, KEARNEY J F, TURNBOUGH C L JR, (2003), 'Role of sialic acid and complex carbohydrate biosynthesis in biofilm formation by nontypeable *Haemophilus influenzae* in the chinchilla middle ear', *J Bacteriol*, 185, 1903–1910.

STRAGIER P, and LOSICK R, (1996), 'Molecular genetics of sporulation in *Bacillus subtilis*', *Ann Rev Genetics*, 30, 297–341.

SVENSSON B, MONTHÁN M, SHAHEEN R, ANDERSSON M A, SALKONIJA-SALONEN M, and CHRISTIANSSON A, (2006), 'Occurrance of emetic producing *Bacillus cereus* in the dairy production chain', *Int Dairy J*, 16, 740–749.

TAUVERON G, SLOMIANNY C, HENRY C, and FAILLE C, (2006), 'Variability among *Bacillus cereus* strains in spore surface properties and influence on their ability to contaminate food surface equipment', *Int J Food Microbiol*, 110, 254–262.

TE GIFFEL M C, WAGENDORP A, HERREWEGH A, and DRIEHUIS F, (2002), 'Bacterial spores in silage and raw milk', *Antonie van Leeuwenhoek*, 81, 625–630.

THOMPSON J M, DODD C E R, and WAITES W M, (1993), 'Spoilage of bread by *Bacillus*', *Int Biodeterior Biodeg*, 32, 55–66.

TOLUANEN K E S, KOSKINEN P E P, YLIKOSKI A I, OLLIKKA P K, HEMMILÄ I A, PUHAKKA J A, KARP M T, (2008), 'Quantitative monitoring of a hydrogen-producing *Clostridium butyricum* strain from a continuous-flow, mixed culture bioreactor employing real-time PCR', *Int J Hydrogen Energy*, 33, 542–549.

VALERO M, HERNÁNDEZ-HERRERO L A, and GINER M J, (2007), 'Survival, isolation and characterization of a psychrotrophic *Bacillus cereus* strain from a mayonnaise-based ready-to-eat vegetable salad', *Food Microbiol*, 24, 671–677.

VAN LOOSDRECHT M C M, LYKLEMA J, NORDE W, and ZEHNDER A J B, (1989), 'Bacterial adhesion: a physiochemical approach', *Microbial Ecol*, 17, 1–5.

VAN LOOSDRECHT M C M, LYKLEMA J, NORDE W, and ZEHNDER A J B, (1990), 'Influence of interfaces on microbial activity', *Microbiol Rev*, 54, 75–87.

VATANYOOPAISARN S, NAZLI A, DODD C E R, REES C E D, and WAITES W M, (2000), 'Effect of flagella on initial attachment of *Listeria monocytogenes* to stainless steel', *Appl Environ Microbiol*, 66, 860–863.

VERHAMME D T, KILEY T B, and STANLEY-WALL N R, (2007), 'DegU co-ordinates multicellular behaviour exhibited by *Bacillus subtilis*', *Molecular Microbiol*, 65, 554–568.

VILAIN S, and BRÖZEL V S, (2006), 'Multivariate approach to comparing whole-cell proteomes of *Bacillus cereus* indicates biofilm-specific proteome', *J Proteome*, 5, 1924–1930.

WATNICK P, and KOLTER R, (2000), 'Biofilm, city of microbes', *J Bacteriol*, 182, 2675–2679.

WEAVER L, MICHELS H T, KEEVIL C W, (2008), 'Survival of *Clostridium difficile* on copper and steel: futuristic options for hospital hygiene', *J Hospital Infect*, 68, 145–151.

WEDDIG L M, CBALESTRINI C G, and SHAFER B D, (EDS.) (2007), *'Principles of thermal process control, acidification and container closure evaluation'*, Washington DC, GMA Science and Education Foundation.

WIENCEK K M, KLAPES N A, and FOEGEDING P M, (1990), 'Hydrophobicity of *Bacillus* and *Clostridium* spores', *Appl Environ Microbiol*, 56, 2600–2605.

WIJMAN J G E, DE LEEUW P P L A, MOEZELAAR R, ZWIETERING M H, and ABEE T, (2007), 'Air-liquid interface biofilms of *Bacillus cereus*: formation, sporulation, and dispersion', *Appl Environ Microbiol*, 73, 1481–1488.

WINANS S C, (2002), 'Bacterial Esperanto', *Nat Structural Biol*, 9, 83–84.

WIRTANEN G, HUSMARK U, and MATTILA-SANDHOLM T, (1996), 'Microbial evaluation of the biotransfer potential from surfaces with *Bacillus* biofilms after rinsing and cleaning procedures in closed food-processing systems', *J Food Protect*, 59, 727–733.

YASUDA-YASAKI Y, NAMIKI-KANIE S, and HACHISUKA Y, (1978), 'Inhibition of *Bacillus subtilis* spore germination by various hydrophobic compounds: demonstration of hydrophobic character of the L-alanine receptor site', *J Bacteriol*, 136, 484–490.

YOKOTA A, FUJII T, and GOTO K, (EDS.) (2007), '*Alicyclobacillus – Thermophilic Acidophilic Bacilli*', Tokyo, Springer.

ZHANG H, BRUNS M A, and LOGAN B E, (2006), 'Biological hydrogen production by *Clostridium acetobutylicum* in an unsaturated flow reactor', *Water Res*, 40, 728–734.

# Part III

# Biofilm prevention, inactivation and removal

# 11

# Food contact surfaces, surface soiling and biofilm formation

**C. Faille, INRA-UR638, France and B. Carpentier, Agence Française de Sécurité Sanitaire des Aliments, France**

**Abstract**: The ability of microorganisms to adhere to inert surfaces, to form biofilms and to resist further cleaning and disinfection procedures varies greatly according to the substratum composition. This chapter first presents materials (stainless steels, elastomers or plastics) used in food industry as well as regulation and requirements regarding food contact surfaces. Materials in contact with food are subjected to conditioning and wear during use and the possible consequences on bacterial contamination are described. The chapter then discusses properties (topography, physico-chemistry) of new and conditioned materials affecting bacterial contamination and persistence in the food environment. It also includes new trends in hygienic materials for food purposes.

**Key words**: food contact materials, material topography, physico-chemistry and chemical composition, bacterial adhesion and biofilms, cleaning and disinfection, material conditioning and ageing.

## 11.1 Introduction

Surfaces of equipment used in food and beverage processing and handling are commonly contaminated by microorganisms (possibly including pathogenic bacteria), even following cleaning and disinfection procedures (Bagge-Ravn *et al.*, 2003, Gibson *et al.*, 1999, Marouani-Gadri *et al.*, 2008). These contaminating microorganisms appear as adherent microorganisms or as more complex structures, called biofilms. Adherent spores and bacteria, as well as biofilms, can be observed on every surface of food-industry plants such as stainless steel surfaces, floor, belts or rubber seals (Costerton *et al.*, 1995, Kumar and Anand, 1998).

Among those microorganisms contaminating food processing equipment, some are only transiently attached, i.e. do not survive, or are easily detached, by cleaning and disinfection procedures (Dunsmore *et al.*, 1981). This transient contamination does not represent a major concern in the food and beverage industry. Conversely, other contaminants such as bacterial spores (Jullien *et al.*, 2008) and bacteria embedded in biofilms (Chmielewski and Frank, 2003) are able to persist on surfaces despite the many drastic environmental conditions encountered during food processing and/or hygienic procedures. Indeed, both persistent surface contamination, and to a greater extent biofilms where cell multiplication occurs, therefore represent a potential source of cross-contamination for any food in contact, entailing both economic and health consequences. It has also to be remembered that non-food contact surfaces can be a source of food contamination following microorganisms'transfer to food by liquids or aerosols. According to Haeghebaert *et al.* (2001) equipment contamination would have been behind around 40% of foodborne diseases caused by bacteria in France between 1996 and 1998. As an example, a notorious food poisoning episode involving 14 000 people, which occurred in Japan in 2000 (Watts, 2000) was caused by a contaminated valve connecting a supply pipe to a tank used to recycle left-over milk. Other equipment such as slicing and cutting equipment, conveyors, filling and packing machines is also often incriminated in food contamination. Similar observations have been made in homes, where kitchen surfaces are frequently contaminated (Kusumaningrum *et al.*, 2003). For instance, cross-contamination involving food contact with contaminated surfaces has been identified as a contributory factor in outbreaks of *Campylobacter* and *Escherichia coli* O157 infection (Humphrey, 2001).

In order to limit such events in the food and beverage industries, materials in contact with food and beverages must be food grade (US Food and Drug Administration (FDA, 2007) or European Food Safety Authority (EFSA, 2008) approved or national equivalent) and fulfil the following requirements: they have to be inert in the foreseeable conditions of use, during food processing, cleaning and disinfection procedures, and packaging. This means that such materials must be corrosion-resistant, essentially during cleaning and disinfection processes, mechanically stable, and capable of preventing the transfer of any constituents in quantities which could endanger human health or alter the organoleptic properties of the food products. Lastly, such materials are expected to be hygienic, in the sense that they limit food and microorganism accumulation and/or facilitate their removal during cleaning, in order to prevent the microbial contamination of food during processing or packaging.

As described below, materials used in the food industry have greatly varying surface properties (Section 11.2). Some of these properties, such as physico-chemical and topographical properties, are assumed to be directly related to hygiene criteria (microbial adhesion, biofilm formation and, most important, the ability to be cleaned) (Section 11.3). However, in 'real life', these surface properties may change during use, depending on the environmental conditions to which these materials are subjected (Section 11.4). Therefore, the hygienic status of a specific material may change over a period of time. These following topics will give information on the role of substratum properties on bacterial attachment (Section 11.5) and the consequence of substratum conditioning and ageing (Section 11.6), as well as the role of substratum properties on bacterial persistence (Section 11.7) and the consequence of substratum conditioning and ageing (Section 11.8). Many research paths are currently being explored in order to improve the hygienic status of materials used in the food and beverage industries. Some of the main interesting research areas will be described at the end of this chapter (Section 11.9).

## 11.2 Range of surfaces encountered in food processing

A variety of materials are used in the construction of food equipment. These materials vary in workability, compatibility, and sanitary design features. Depending upon the application, various metals as well as non-metals (e.g., plastics, rubber) are used. Of the materials widely used in food processing lines (open or closed surfaces), stainless steels remain the material of choice, as they are not only resistant to corrosion in alkaline or acidic solutions, but are also extremely hygienic (Boulangé-Petermann *et al.*, 1997). Stainless steels are a family of ferrous alloys containing less than 1.2% carbon and over 10.5% chromium and are protected by a passive surface layer of chromium and iron oxides and hydroxides that protects them efficiently from corrosion. It is therefore recommended that stainless steel food contact surfaces be passivated (using nitric acid or other strong oxidizing agents) to maintain the passive film on the surface and to assure their efficient resistance to corrosion.

Stainless steel is available in various alloys, and the choice is dictated by the expected use (tubing, complex equipment, welding) and the process (temperature, corrosive products). Depending on the chromium content and the addition of other elements such as molybdenum or nickel, alloys will exhibit specific characteristics, such as a enhanced resistance to corrosion, or will be easy to form or weld.

Austenitic stainless steels (containing nickel) are the most commonly used alloys in the food and beverage industries, in open or in closed systems, because of their excellent corrosion resistance, formability, weldability and hygiene characteristics. The 300 series (iron-chromium-nickel) is recommended by the American Iron and Steel Institute (AISI) for food contact surfaces. The most common alloy AISI 304 (but mainly AISI 304L, modified for welding), contains chromium (16–18%) and nickel (8–13%), and is durable in typical food, dairy and beverage processing and storage environments (piping, vessels, containers . . .) and in domestic kitchens. The inclusion of molybdenum will give greater resistance to surface pitting in aggressive environments. With around 2% molybdenum, AISI 316 (and AISI 316L) is the second most common grade in the food industry, being more resistant to chlorides and sulphur dioxide and usable in moderately corrosive media such as wines or in the presence of high concentrations of salt. Otherwise, 3A Sanitary Standards require 316 grade stainless steel for most surfaces in the food industry, and allow the use of 304 grade for utility usage (e.g., pipes).

The different grades can be found in different surface finishes according to the treatment to which stainless steels have been subjected, either during or after the production process. In Europe, two mill finishes (unpolished), 2B pickling finish and 2R bright annealed (BA) finish, are commonly used in the food and beverage industries. However, for the same finishing operation, surface topography depends on the grade and thickness of the stainless steel. When stainless steel surfaces do not comply with required roughness values, a subsequent abrasive treatment on one or both sides may be applied. Abrasion (polishing) is generally performed with abrasive-coated belts defined by their grits. The #4 finish, which is obtained with a 120–150 grit size abrasive, is recommended in US (US 3-A-Sanitary Standards) for stainless steel that comes directly into contact with food. For an electro-polished finish, seldom used in the food industry, the surface material is electro-chemically dissolved, giving a mirror-like appearance.

The process of welding two stainless steel components together, e.g., by gas tungsten arc welding, the conventional welding technique used in the food industry, will affect the mechanical properties and the corrosion resistance of both the joint itself and the parent metal immediately adjacent to the weld (heated to a temperature approaching its melting point). For instance, during welding, the formation of chromium-rich carbides on the grain boundaries may compromise the resistance to corrosion. These welds can also produce an irregular topography and need further grinding and polishing to achieve compliance with requirements.

In addition, a variety of non-metal materials such as elastomers (also referred to as rubbers) and plastics are increasingly used in the food and beverage industry for specific applications, such as conveyor belts, containers, seals, gaskets or cutting boards. Rubbers, such as ethylene propylene diene monomer rubber (EPDM), nitrile butyl rubber (NBR, known as Buna-n®), silicon rubber, or fluoroelastomer (Viton) are used in both closed equipment (seal gaskets, membranes, fittings, containers) and in open equipment such as conveyor belts. Among these materials, the most frequently used gasket materials in milk processing equipment are EPDM and NBR.

Some plastics are also used for various purposes in the food industry, such as processing tanks, conveyor belts or hoses. A wide range of plastics is available, but only a few of them are food approved, such as polypropylene (PP), polycarbonate (PC), high density polyethylene

(HPDE), unplasticised polyvinyl chloride (PVC) and fluoropolymers, such as polytetrafluoroethylene (PTFE, Teflon®). The latter, used for gaskets in the food industry, is porous and lacks resilience and must be used with care.

## 11.3 Regulation, standards and other requirements regarding food contact surfaces for food industries

Materials and articles intended for food and beverage contact must be capable of preventing the transfer of any constituents in quantities which could endanger human health or alter the organoleptic properties of the food products as required by the USA (FDA, 2007) or the European Regulation (EFSA, 2008) or national equivalent. Consequently, materials used in the food and beverage industries must fulfil certain specific topography requirements to ensure cleanability. The surface topography of materials in contact with food has been taken into account by regulatory authorities: in the food and beverage industries, product contact surfaces should be smooth and resistant, and be free of crevices, pitting and cracking. Maximal surface roughness of Ra (average peak-to-valley height value, ISO 4287:1997 standard (ISO, 1997)) is specified in the appropriate food standards, such as ISO 14159 standard (ISO, 2002), which specifies that Ra values must not exceed 0.8 μm for food-contact surfaces, and EN 12462 (CEN, 1996) or ISO 4287 (ISO, 1997) standards, which authorize Ra values up to 1.6 μm for internal products contact areas. In the US, the finish of #4 grade recommended for food contact surfaces has a Ra value of 1 μm or less. Lastly, a Ra of up to 3.2 μm would be acceptable for specific use 'provided the flow is sufficient enough to remove the soil from the surfaces', according to a European Hygienic Engineering and Design Group (EHEDG) guideline (EHEDG, Doc 17, 2004). Furthermore, all joints should be smooth, continuous and free of gaps and crevices. Equipment standards generally require that welded joints on stainless steel surfaces be continuous, and ground to at least a #4 finish smoothness.

Conversely, no particular specifications could be found among the numerous standards and recommendations about surface free energies of food-contact surfaces. Among these materials, stainless steels commonly used in the food sector have surface energies in the range of about 40 to 200 mJ $m^{-2}$ and polymers and organic materials are characterized by surface free energies lower than 100 mJ $m^{-2}$.

However, this classification is more complex than it would seem, since materials often display a great surface heterogeneity, which has a possible effect on their hygienic status. For example, the stainless steel 2B finish is composed of grains and grain boundaries, making it non-homogeneous both in terms of topography and of surface free energy (Geesey *et al.*, 1996).

## 11.4 Material conditioning and ageing

### 11.4.1 Surface conditioning

Obviously, the surface properties of materials in contact with food and beverages change with use, giving rise to a greatly modified surface status.

The very first event to occur when materials are placed in the food environment is the appearance of a so-called conditioning film. Indeed, organic and inorganic compounds in contact with materials can accumulate very quickly at the solid–liquid interface and give rise to a conditioning film. These adsorbent compounds originate in the food products themselves and also in the detergents, surfactants, and sanitizers used during cleaning and disinfection procedures.

The quantity and configuration of the adsorbed molecules strongly depend on the substratum properties (Redsven *et al.*, 2007) and the molecular species present in the aqueous medium (Storgards *et al.*, 1999b). Interestingly, most molecules are able to adsorb spontaneously onto surfaces, regardless of their nature or surface charges. The level of adsorbed materials is affected by the nature and/or the hydrophobic property of the material (Boulangé-Petermann *et al.*, 2006) and also by the grade of stainless steel surface (Boyd *et al.*, 2001). Furthermore, the persistence level of the adsorbed material is facilitated by the presence of surface irregularities such as scratches or grain boundaries as shown in Fig. 11.1, and may even mask the substratum topography (Boyd *et al.*, 2001, Verran *et al.*, 2001a). In food processing lines, standard cleaning procedures are not fully effective in eliminating all adsorbed material, and conversely products involved in these procedures may further participate in the surface conditioning. Consequently, successive runs of food processing and cleaning procedures will result in the continuous accumulation of molecular species on the equipment surfaces, varying with the food and the detergents used, as shown by Jullien *et al.* (2008).

### 11.4.2 Material ageing

Besides surface contamination by organic and/or mineral compounds, material ageing under conditions encountered during food processes and/or cleaning procedures provokes deeper modifications of the substrata.

Considering stainless steels surfaces, modifications of the passive film can be induced by any aggressive environmental conditions (Munoz and Mischler, 2007), leading to an increase in the carbon/iron ratio. This would be the case following repeated cleaning procedures with nitric acid and sodium hydroxide (Leclercq-Perlat and Lalande, 1994) or during conditioning procedures involving food and detergent (Jullien *et al.*, 2008). This modification was first attributed to the carbonization of stainless steel following repeated uses of soda (Leclercq-Perlat and Lalande, 1994). More recently, the passive layer has been demonstrated to be deeply modified, as shown by chromium enrichment and a decrease in the film thickness

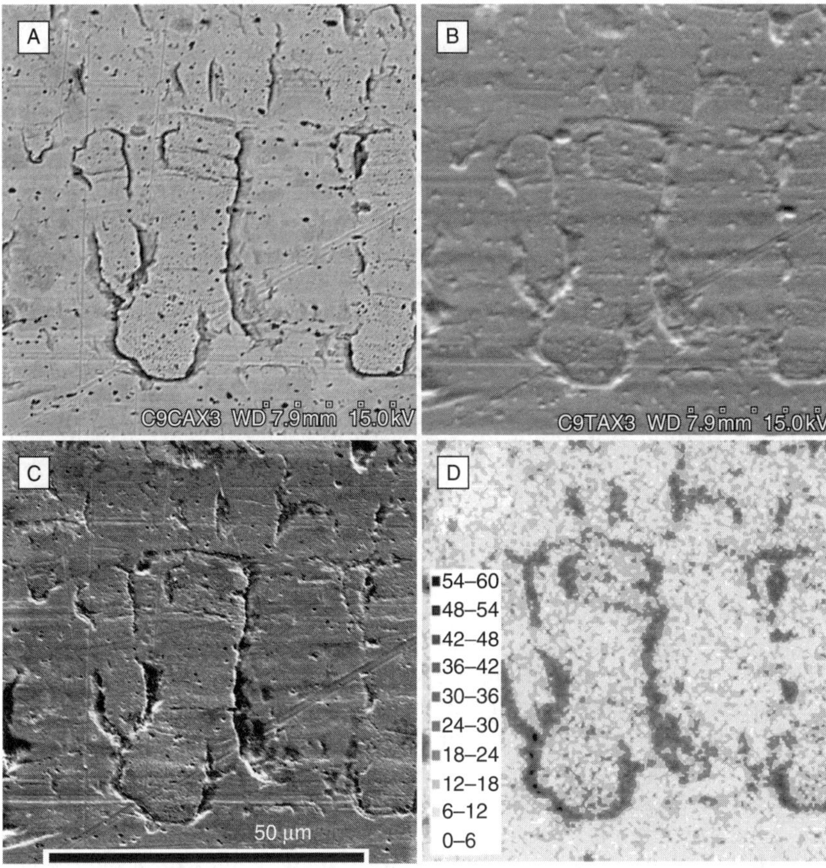


**Fig. 11.1** Surface topography and localization of carbon surface contamination on a 430 stainless steel (2B finish) coupon after cleaning with an alkaline detergent by scanning electron microscopy with X-ray microanalysis. A–B: detection of back scattered electron (composition [A] and topography [B] modes). C: detection of secondary electron. D: energy dispersive X-ray spectroscopy (EDS) showing the presence of carbon.

(Jullien *et al.*, 2008). In the opinion of these authors, the cleaning agent would induce surface peeling and dissolve the less passivable element present in the passive layer, leading to a decrease in the surface iron content. Despite its strong resistance to corrosion, austenitic stainless steel may also suffer from localized corrosion, especially in the presence of chlorides (in detergents, food products . . .). In food and beverage industries, the most common forms of localized corrosion are pitting (corresponding to localized damage to the passive film), crevice corrosion (relatively frequent in areas devoid of oxygen, such as those associated with gaskets and seals) and stress corrosion cracking (in specific environments such as in the presence of chlorides and high temperatures).

Otherwise, rubbers and plastics can suffer extensive physical degradation with time, mainly when subjected to drastic conditions (high temperatures during pasteurization, rapid changes in temperature and pressure during cleaning procedures, mechanical constraints, chemicals . . .). For example, among the various conditions encountered during cleaning procedures, gasket materials are prone to degradation by oxidizing agents such as chlorine, peracetic acid or nitric acid, leading to crack formation and even further dissolution and softening. Even if resistance to wear differs depending on the material, aged surfaces often become rougher with visible cracks, flanging, abrasion or holes (Austin and Bergeron, 1995, Czechowski, 1990). Furthermore, these surface modifications are observable not only on the surface in contact with fluids, but also on surfaces in contact with stainless steel (Mettler and Carpentier, 1997). Indeed, gasket ageing due to temperature and deformation often provokes a loss in resilience (vulcanization process), which permits not only the further accumulation of bacteria, but also of organic or mineral material on the surfaces in contact with metal (Austin and Bergeron, 1995, Mettler and Carpentier, 1997).

### 11.4.3 Impact of surface conditioning on substratum physico-chemical properties

These substratum modifications induced by conditioning and/or ageing are accompanied by changes in the physico-chemical surface properties. The conditioning film may play a major role in these modifications by masking the intrinsic properties of the underlying substratum, thereby uniformizing the physico-chemical properties of different substrata. This trend was observed for different stainless steel surfaces (Boulangé-Petermann, *et al.*, 1993, Jullien *et al.*, 2008), as well as for substrata with quite different surface properties, such as stainless steel and polymers (Sinde and Carballo, 2000) or flooring materials (Mettler and Carpentier, 1999). Consequently, the physico-chemical properties of conditioned materials will rely on the conditioning molecules rather than on the underlying substrata and will therefore strongly depend on the food with which they are in contact. Finally, it must be pointed out that the conditioning molecules are nutrients allowing microbial growth and biofilm formation even on cleaned surfaces. Lastly, Storgards *et al.* (1999b) observed that the hydrophobic character of some materials changed with ageing, depending on the material.

In conclusion, special consideration should be paid to the use of 'aged' surfaces in comparing the hygienic properties of various substrata.

## 11.5 Surface properties and microbial attachment

Microbial adhesion and colonization of substrata occur at any place along the processing lines, on all materials whatever their surface properties, both during the food process as well as during cleaning and disinfection procedures (Lelièvre *et al.*, 2002). Nonetheless, the ability of microorganisms to adhere to inert surfaces and to colonize them varies greatly according to the substratum composition. The influence of material on initial microbial adhesion has often been displayed. Differences in attached counts were observed among food equipment materials such as stainless steel, PTFE, PEHD, PVC (Faille *et al.*, 2002a) or materials used in kitchens such as granite, polypropylene or polyethylene (Silva *et al.*, 2008). Among a variety of food equipment materials, stainless steel has been found to be one of the less contaminated materials, e.g., when challenged for *Bacillus* spores adhesion (Faille *et al.*, 2002a). Conversely, in spite of the general assumption that welds would support bacterial adhesion (Sreekumari *et al.*, 2005), no significant difference in the number of adherent bacteria on various welds and the parent metal surfaces could be evidenced (Tide *et al.*, 1999), except in the case of oxidized welds (Mai *et al.*, 2006). Besides the adhesion level, adhesion kinetics (Smoot and Pierson, 1998) and distribution of adherent cells (Senechal *et al.*, 2004) are also clearly affected by the material. Other results have also shown the influence of material on biofilm formation (Storgards *et al.*, 1999a, Somers and Wong, 2004). However, it seems hard to draw any general conclusion as to which materials are the least prone to bacterial colonization, as the classification of materials is largely influenced by environmental conditions and bacterial species or even strains. For example, Marouani-Gadri *et al.* (2008) found that stainless steel and a polyurethane conveyor belt material were not significantly different with respect to *E. coli* O157:H7 biofilm formation in 6 of 8 environmental conditions tested. When differences were observed between materials, the greatest biofilms were observed either on stainless steel or on polyurethane, depending to the bacterial species in use (Marouani-Gadri *et al.*, 2008, Midelet and Carpentier, 2002).

Finally, stainless steel grade does not seem to have a clear effect on hygienic status. For instance, it is difficult to draw conclusions as to the effect of the addition of molybdenum to steels on bacterial adhesion and biofilm formation: 316 alloy was sometimes found to be less hygienic than 304 alloy (Flint *et al.*, 2000) but other results were less conclusive, or even opposite (Kielemoes *et al.*, 2000, Percival *et al.*, 1998), which suggest that stainless steel behaviour is greatly affected by environmental conditions and/or the microorganism involved.

It may be hypothesized that differences between materials are at least partly due to substratum components, which could entail physiological changes in adherent microorganisms, such as stainless steel alloys containing nickel to increase their corrosion resistance (Lopes *et al.*, 2005) or NBR, an elastomer which as been proven to be bacteriostatic towards different genera (Rönner and Wong, 1993, Storgards *et al.*, 1999a).

Further reported works have focused either on the investigation of the role of physico-chemical interactions between microorganisms and substrata or on that of substratum topography.

### 11.5.1 Physico-chemical properties

To begin, much has been written about the relationship between bacterial adhesion and the physico-chemical properties of the inert surfaces, namely acid-base interactions, electrostatic interactions and surface free energy. Two major theories involving physico-chemical interactions have been proposed in predicting the level of bacterial attachment to inert surfaces. One of the most commonly used theories is the Derjaguin-Landau-Verwey-Overbeek (DLVO) theory, which describes the total Gibbs energy of interaction between a microorganism and a surface as a function of the separation distance. The Gibbs interaction is considered to be the result of electrostatic and Lishitz-Van der Waals interactions and, in the extended DLVO theory, the acid-base interaction (for a review of the DLVO theory applied to bacterial adhesion see Hermansson (1999)). This theory, although interesting when exploring the phenomena involved in adhesion, is difficult to be apply when investigating microorganisms, because of the complexity and the wide variability of microbial cell surface composition. The other widely used approach, which can be experimentally investigated, is the thermodynamic one. This is based on the comparison of the interfacial surface free energies for bacteria in suspension and bacteria in an adhered state, thereby taking into account the Lishitz-Van der Waals and acid-base interactions (van Oss, 2006). However, theoretical predictions do not always fit observations, and existing literature is still contradictory when it comes to ascertaining which substratum properties affect adhesion. Along with the interactions taken into account in the two major theories, hydrophobic interactions are also supposed to play a major role, due to the unusually strong attraction of hydrophobic surfaces and groups in water (Meyer *et al.*, 2006).

First of all, the substratum surface charge is expected to contribute to the adhesion process. Indeed, in real conditions, microorganisms and materials are generally both negatively charged, generating an electrostatic repulsion. Microbial adhesion has actually been shown to be impaired (occurring less rapidly or to a lower extent) on negatively charged substrata in comparison to positively charged ones (Gottenbos *et al.*, 2001, Mercier-Bonin *et al.*, 2004). Furthermore, the level of this repulsion may affect both the adhesion and subsequent biofilm formation (Gottenbos *et al.*, 2001). However, in the food industry, electrostatic repulsion is frequently dramatically reduced, due to the high electrolyte concentration, which means that its role in bacterial adhesion may become negligible.

Much has been written on the role of substratum hydrophobicity in the attachment of bacteria, but little hard evidence has been published stating that substratum surface energy is correlated to bacterial adhesion. Indeed, depending on the substrata and bacterial strain, bacterial adhesion and biofilm formation have been reported to be more pronounced either on hydrophilic surfaces (Flint *et al.*, 2000) or, most often, on hydrophobic surfaces (Hüsmark and Rönner, 1993, Ong *et al.*, 1999, Vadillo-Rodriguez *et al.*, 2005). Other works suggest that intermediate values of the surface free energy, in the range of 20–30 mJ $m^{-2}$ would inhibit bacterial adhesion (Zhao *et al.*, 2007) or that adhesion is more likely to be driven by the balance between Lifshitz-van der Waals and polar interactions (Boulangé-Petermann *et al.*, 1993).

To summarize, none of the substratum properties can be clearly linked either to a facilitated microbial attachment and/or to biofilm formation of the numerous strains in the various environments encountered in the food environment. The discrepancies observed must in part be attributed to substratum heterogeneities that provide localized sites favourable for bacterial attachment. For example, the stainless steel 2B finish is composed of grains and grain boundaries, making it non-homogeneous both in terms of topography and of surface free energy (Geesey *et al.*, 1996). One can also hypothesize that bacterial properties such as mechanical ones (Mendez-Vilas *et al.*, 2006) or surface polymer properties (Abu-Lail and Camesano, 2006, Bell *et al.*, 2005) play more predominant roles in bacterial adhesion. Similarly, one can hypothesize that ability to produce slime (Méndez-Vilas *et al.*, 2004) or the nature of excreted polymers (Lequette, personal communication) play more predominant roles in biofilm formation. This latter hypothesis is emphasized by the fact that no relevant relationship has been found between the amount of biofilm formed and the initial adhesion extent (Cerca *et al.*, 2005).

### 11.5.2 Topography

Although it has been suggested that adhesion could be enhanced with increased roughness and although bacteria are sometimes found in surface flaws such as depressions, cracks or crevices, or grain boundaries (Sreekumari *et al.*, 2005), most published studies have failed to demonstrate a major influence of surface roughness or defects on bacterial adhesion (Barnes *et al.*, 1999, Flint *et al.*, 2000, Gubner and Beech, 2000, Guobjornsdottir *et al.*, 2005, Oliveira *et al.*, 2006, Silva *et al.*, 2008). This can be illustrated by Fig. 11.2 where materials were either vertically or horizontally immersed in a bacterial suspension. Many authors have tried to relate the bacterial load to a roughness parameter such as those described in the International Standard ISO 4287:1997 (ISO, 1997) or derived from the Abbott–Firestone curve (Standard ISO 13565–2 (ISO, 1998)). The average peak-to-valley height value (Ra), the most-used roughness parameter, is also the most investigated, although it is not sensitive to the presence of single peaks or valleys (Faille *et al.*, 2000). In fact, little evidence supports the assumption that minimal Ra reflects a hygienic character. Although increased microbial adhesion and biofilm formation (Wirtanen *et al.*, 1996) have sometimes been reported with increasing Ra, many results have gone against this assumption (Faille *et al.*, 2000, Guobjornsdottir *et al.*, 2005, Rodriguez *et al.*, 2008). The relevance of other roughness parameters, supposed to accurately describe surface topography (shape and depth of peaks and valleys), has

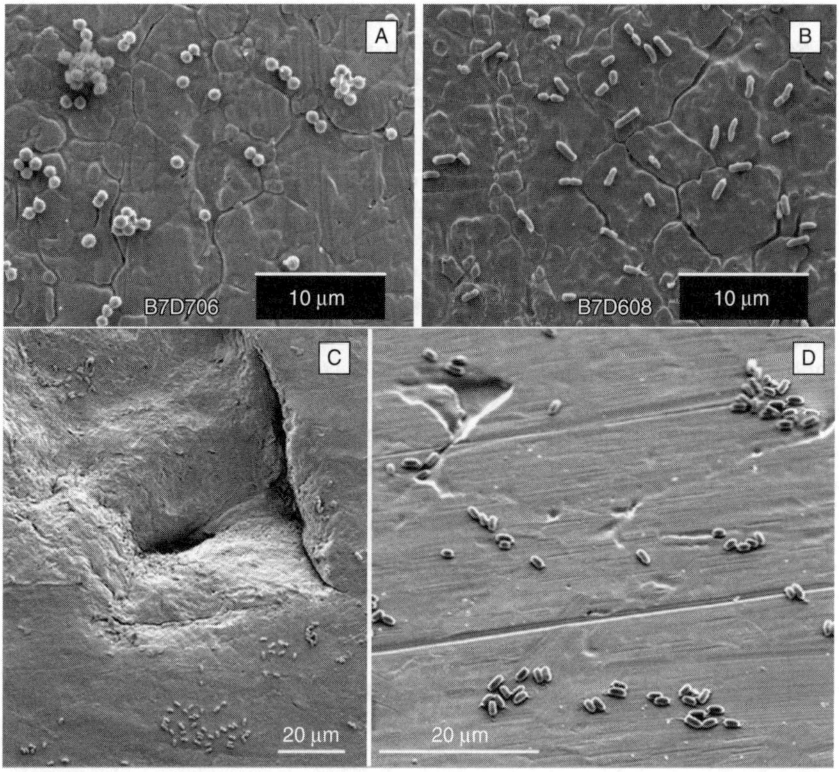


**Fig. 11.2** Adhesion of various microorganisms on stainless steel surfaces. A–B: adhesion of *Staphylococcus caprea* and *Pseudomonas fluorescens* under static condition on stainless steel with a 2B finish (horizontally immersed). C–D: adhesion of *B. cereus* spores on substrata with irregular topography, vertically immersed in a spore suspension (presence of scratches, flaws . . .).

been poorly investigated. According to Dürr (2007), Rz, the sum of the height of the largest profile peak height and the largest profile valley depth describes surface topography more adequately than Ra and is more relevant to explaining bacterial adhesion. Conversely, none of the parameters derived from the Abbott–Firestone curve have been investigated for their relevance to this question.

It is also difficult to draw conclusions concerning the influence of stainless steel finish on bacterial adhesion. For example, bacterial adhesion has been found to be reduced (Arnold *et al.*, 2004) or not affected (Rodriguez *et al.*, 2008) on electro-polished stainless steel surfaces as compared to 2B or 2R finishes.

The apparent conflict in these numerous observations is also probably strongly related to the different experimental protocols that affect the transport of microorganisms from bulk liquid to solid surfaces as well as the release of adherent microorganisms. For example, when microbial adhesion is investigated under dynamic conditions, one would expect some effect of surface features on microbial adhesion, either directly by affecting the transport of microorganisms (Rijnaarts *et al.*, 1993, Faille *et al.*, 1999), or indirectly by providing protection from flow to adhering bacteria (Whitehead and Verran, 2007). Otherwise, under static conditions, bacterial adhesion on horizontally immersed substrata may occur to a greater extent due to sedimentation, than would be the case for vertically immersed bacteria.

To summarize, none of the various properties of materials in contact with food and beverages can be clearly related to their ability to be contaminated by microorganisms. Observed discrepancies in the substratum hygienic status may be largely expressed not only by the great diversity of microorganisms and environments found in the food and beverage industries, but also by conditioning and ageing of substrata.

## 11.6 Impact of material conditioning and ageing on microbial attachment to surfaces

While it is now well recognized that surface conditioning may deeply affect substratum surface properties, most of the studies reported above (Faille *et al.*, 2002a, Oliveira *et al.*, 2006, Silva *et al.*, 2008) have been performed on unconditioned materials and the conclusions drawn from these studies may not be extrapolated to food environments.

Indeed, substratum conditioning can promote or inhibit bacterial adhesion, depending on the process conditions (Storgards *et al.*, 1999a) and the food involved (Barnes *et al.*, 1999, Flint *et al.*, 2001, Herrera *et al.*, 2007, Parkar *et al.*, 2001). Along with organic material, any other molecules to which substrata are exposed in food and beverage processing lines, such as detergents (Boulangé-Petermann, 1996), surfactants (Meylheuc *et al.*, 2006, Nitschke and Costa, 2007) or even disinfectants (Sinde and Carballo, 2000) also affect the microbial attachment to substrata. The changes observed following conditioning may also rely upon the intrinsic substratum properties. Data produced by Hüsmark and Rönner (1993) support this hypothesis (under the experimental conditions in use), i.e. that conditioning hydrophobic substrata would impair bacterial adhesion, while the conditioning of hydrophilic substrata would promote bacterial spore adhesion.

Susceptibility to bacterial contamination is also affected by the physical deterioration of polymeric materials (Storgards *et al.*, 1999b). On gaskets situated in food processing lines, many of the attached bacteria are found in crevices created by the insertion of the gasket into the pipeline (Austin and Bergeron, 1995), but also outside the cracks and holes (Czechowski, 1990). Areas in contact with pipes are also highly contaminated (Austin and Bergeron, 1995, Mettler and Carpentier, 1997) due to inadequate pressure on the metal, thereby inducing defects in impermeability to microorganisms (bacteria-tightness) and consequently allowing bacterial migration through cracks, gauges and holes.

## 11.7 Surface properties and microbial persistence

Apart from the adhesion level, hygienic status strongly relies on the ability of adherent microorganisms to persist on contaminated surfaces: in other words, on the ability of materials to be cleaned and/or disinfected.

### 11.7.1 Resistance to cleaning

As was the case for bacterial adhesion, materials clearly affect the ease of removal of adherent bacteria and biofilms. According to most of the reported studies, stainless steel promotes interactions between adherent microorganisms and substrata, similar to or weaker than other materials (PEHD, Teflon®, PVC, NBR, EPDM, Viton . . .) (Midelet and Carpentier, 2002, Somers and Wong, 2004) and, as for microbial adhesion, no noticeable difference was observed between the most commonly used 304 and 316 grades (Jullien *et al.*, 2003, Whitehead and Verran, 2007). Conversely, the ease of bacterial removal was quite different between polymers (Senechal *et al.*, 2004), and flooring and wall materials (Mettler and Carpentier, 1999, Taylor and Holah, 1996). Surface properties of materials have a predominant role among the numerous factors governing the adhesion strength between microorganisms and inert surfaces and/or the cleanability of soiled surfaces.

The role of physico-chemical properties of substrata has been extensively investigated. The ease of removal of adherent microorganisms has generally been reported to be impaired on hydrophobic substrata, as observed by cell counting after cleaning (Boulangé-Petermann *et al.*, 2006, Senechal *et al.*, 2004), or by estimation of interaction strength by AFM (Bowen *et al.*, 2002, Xiaoxia *et al.*, 2006). The weak correlation between hydrophobicity and the adhesion strength sometimes reported, as well as the few contradictory results (Bos *et al.*, 2000) may reflect non-homogeneous surface free energies, for example on stainless steel grain boundaries (Geesey *et al.*, 1996) or the involvement of other physico-chemical interactions (Sheng *et al.*, 2007).

Besides physico-chemical properties, surface topography is also expected to play a major role in bacterial retention, mainly because most cleaning procedures involve shear stresses. Intuitively, one would then expect microbial removal to be hindered on irregular surfaces, at least due to the protection from shear stresses potentially afforded by surface features under flow conditions. While correlation of surface cleanability and Ra is often unsuccessful (Frank and Chmielewski, 2001), the ease of removal of adherent bacteria is clearly impaired by the presence of surface features such as pits, holes, crevices or grooves (Holah and Thorpe, 1990, Verran *et al.*, 2001b), and grain boundaries (Boyd *et al.*, 2002). Indeed, as shown in Fig. 11.3, Ra values are not necessarily correlated to the presence of surface defects affecting substratum hygienic status. Depending on their topography, the presence of welds or gaskets also induces modified local hydrodynamics

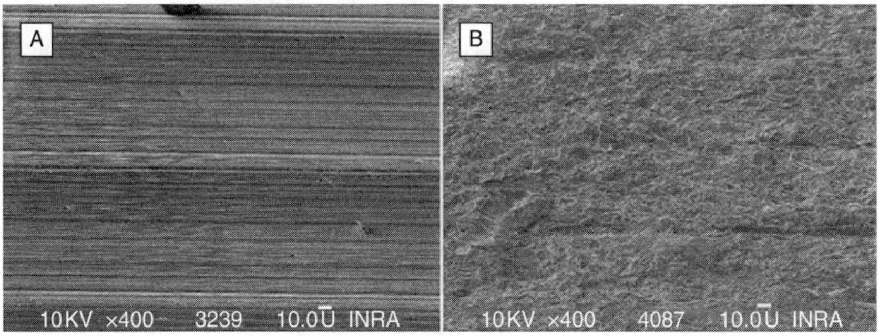


**Fig. 11.3** SEM observation of stainless steels characterized by similar Ra values (A: Ra = 1.17 μm, B: Ra = 1.12 μm) but quite different topography.

and consequently it becomes more difficult to remove bacteria by a cleaning procedure (Blel *et al.*, 2008). Conversely, the presence of a regular large waviness does not affect stainless steel cleanability (Faille *et al.*, 2000, Mettler and Carpentier, 1999). Some parameters derived from the Abbott-Firestone curve (ISO 13565 (ISO, 1998)), known to better describe the presence of surface features when peaks and valleys are not symmetrically split on both sides of the roughness profile, have been found to correlate to the ease of removal of adherent bacteria. The reduced valley depth (Rvk), a parameter used to assess the potential for lubricant retention, has been reported to influence the cleanability of floor materials (Mettler and Carpentier, 1999) and of stainless steels commonly used in pump manufacturing (Faille *et al.*, 2000), and the reduced peak height (Rpk) could discriminate between more or less cleanable stainless steel surfaces (Jullien *et al.*, 2003).

The size of surface features is also of importance as noted by Boyd *et al.* (2002), who determined the lateral force (mimicking shear stresses) required to displace *S. aureus* cells adhering to stainless steel surfaces with different topographies. When the size of the surface features was similar to the size of the cells, bacteria were hardly displaced at all, probably due to a maximal area of contact between cells and substratum or bacterial entrapment. Similar results have been reported by many authors, e.g. Whitehead *et al.* in 2005. Conversely, in the presence of very small features, the bacteria may not be able to conform to the substrate topography. The resulting discontinuous contact between the bacterium and substrate would lead to a decrease in the interaction strength (Emerson *et al.*, 2006), except if such features are filled in with organic fouling, which would clearly be the case in the food industry.

As suggested by Whitehead and Verran (2006), further works on the effect of surface roughness on microbial retention will require surfaces designed with features of specific dimensions present in a regular pattern, on both micrometric and nanometric scales. The widespread use of atomic force microscopy (AFM) and of flow cells under the microscope, which permits the observation of adherent bacteria under flow and the estimation of forces required to displace or detached single cells, will also undoubtedly allow a better understanding of the correlation between substratum topography and ease of adherent bacteria removal.

### 11.7.2 Resistance to disinfection

Besides cleanability, different materials also exhibit inherently different abilities to be sanitized. Once more, stainless steel is one of the more hygienic materials, the efficacy of disinfectants such as hypochlorite or peracetic acid being lowered on polymers and plastics including PTFE, NBR, Buna-N, polyester or polyester/polyurethane (Krysinski *et al.*, 1992, Rönner and Wong, 1993). Among the various hypotheses that have been put forward to explain these differences between materials, many rely on the phenotypic characteristics of adherent bacteria and bacteria in biofilms. One could hypothesize that the biofilm structure, affected by the substratum, would offer more or less protection to anti-microbial products. This hypothesis agrees with more recent findings that differences in resistance to sanitizers could be related to differences in the EPS composition and/or to biofilm thickness on different materials (DeQueiroz and Day, 2007). The physiological status of the adherent bacteria may also be directly affected by the substratum, which could induce further differences in the disinfectant efficacy. For example, as suggested by Rönner and Wong (1993), the increased resistance observed on Buna-n would rely on the bacteriostatic effect of this substratum, and the ensuing slow bacterial growth. The presence of surface features (Szabo *et al.*, 2007), as well as the presence of corroded surfaces (Korber *et al.*, 1997), would also afford protection to anti-microbial products.

In short, in contrast to observations of microbial adhesion, the ease of removal of adherent microorganisms is significantly affected by some of the substratum surface properties, i.e. an irregular topography or probably a hydrophobic character, at least for new materials.

## 11.8 Impact of material conditioning and ageing on microbial persistence; implications for food safety

As for the ability of substratum to be colonized by microorganisms, the ease of cleaning of materials in contact with food can be strongly compromised by the presence of a conditioning film, but also depending on material wear during use, thus impairing their hygienic status. First of all, the interaction force between microorganisms and substrata can be strongly affected by surface conditioning (van Loosdrecht *et al.*, 1990). For example, the resistance of *Staphylococcus aureus* to a cleaning procedure has been shown to be increased by a previous conditioning procedure with mussel cooking juices (Herrera *et al.*, 2007). On the other hand, the adsorption of detergent onto surfaces also impairs their hygienic properties (Boulangé-Peterman, 1996, Cloete and Jacobs, 2001). In order to investigate the influence of real conditions encountered in the food and beverage industries, several teams have performed surface conditioning consisting in successive steps of soiling with food and cleaning with detergent. They reported that the ease of cleaning of the conditioned stainless steel surfaces is compromised by the presence of the conditioning film (Herrera *et al.*, 2007, Verran and Whitehead, 2006) and that the decrease in the hygienic status depends on the amount of surface soil (Dunsmore *et al.*, 1981) and the food in use (Jullien *et al.*, 2008).

Despite the presence of a conditioning layer, the number of residual adhering bacteria after a cleaning in place procedure is still affected by the properties of the underlying substratum. As shown by Jullien *et al.* (2008), even if the number of adherent *B. cereus* spores on different stainless steel surfaces was less scattered when substrata were conditioned with food and detergent, differences in the hygienic status between substrata were still observed following conditioning. Furthermore, as suggested above, the presence of organic material, entrapped in surface features promotes the persistence of adherent bacteria, even in large otherwise easily cleanable features (Verran and Whitehead, 2006).

Material ageing is also expected to alter cleanability properties, due to the appearance of new surface features providing enhanced protection from cleaning shear forces. Along with the physical deterioration, Storgards *et al.* (1999b) have shown that cleanability is reduced on NBR, EPDM and Viton with ageing. Holah and Thorpe (1990) also found increases in bacterial retention on artificially aged materials (stainless steel, mineral resins and polycarbonate) related to the presence of pits and crevices, but, among these abraded materials, stainless steel was the most cleanable due to its resistance to abrasion. Similarly, on gaskets situated in food processing lines, attached bacteria are found on the whole gasket surfaces (Czechowski, 1990), but most of them are found in crevices created by the insertion of the gaskets into the pipeline (Austin and Bergeron, 1995).

Looking at these results, it emerges that data obtained with unused materials must be considered carefully. Consequently, in order to obtain relevant data to estimate the hygienic status of various materials and the efficacy of cleaning conditions, it is heartily recommended to perform experiments in conditions as close as possible to those encountered in the expected environment (food, cleaning procedure, flow rate, temperature . . .).

## 11.9 Future trends

### 11.9.1 Novel contact surfaces and surface treatments for the food industry to reduce biofilm formation

Many research paths are currently being explored in order to limit the colonization of materials by microorganisms. Two major strategies have been developed, the limitation of bacterial contamination and the inactivation of adherent microorganisms. Most of the available anti-microbial surfaces have been designed for medical or environmental purposes, but some antimicrobial surfaces have also been developed successfully for the food and beverage industries and are already available or under development. The possible use of such materials in food environments requires a consideration of their durability, their resistance to mechanical and chemical wear, and their harmlessness. Lastly, novel contact surfaces that may come in contact with food or beverages will have to comply with food regulations.

### 11.9.2 Anti-adhesive materials

The first strategy is to produce materials which would be resistant to microbial adhesion. Manipulation of surface energy or topography has been investigated in order to meet this requirement. If the use of smooth materials has unanimous support, even if some bacteria can still attach even to the smoothest substrata, the approach consisting in modifying substratum surface free energy is still controversial. Indeed, as expected from the results reported above, the use of hydrophobic or hydrophilic coatings has produced contradictory results. For example, the use of hydrophobic coatings (low surface energy) on stainless steels (Storgards *et al.*, 2007) and glass (Tsibouklis *et al.*, 1999) or of coatings with intermediate surface energies of around 24 mN/m (Zhao, 2004) has been shown to reduce adhesion of some bacteria, but not of the highly hydrophobic *B. cereus* spores (Rosmaninho *et al.*, 2007). Anyway, due to thermo-dynamical reasons, and to the fact that surfaces are quickly conditioned in real environments, it would be a mistake to believe that one could develop a surface which would exhibit absolute anti-adhesive properties.

### 11.9.3 Antimicrobial materials

Another attractive strategy that has emerged as a potential means of limiting bacterial contamination involves the use of organic or inorganic antimicrobial agents (silver ions, photocatalytic titanium dioxide, triclosan . . .). The antimicrobial compounds should be selected to provide a broad activity against the microorganisms commonly found in the target application and that ensure that harmful microorganisms are not selected. Materials should also retain their antimicrobial activity for long periods and, depending on the application, resist the expected drastic environmental conditions such as those of hygiene procedures.

Among the inorganic agents, the most common antimicrobial materials to be incorporated into solid materials is silver, which is safe for humans, but has strong antimicrobial properties, as it interferes with electron transport and it leads to protein and DNA damage. Silver-loaded antimicrobial materials such as zeolites (ceramic alumino-silicate matrix) have been extensively used as polymer additives. For instance, Zeomic (SINANEN ZEOMIC Co) has been commercially available since 1983, and other products, also containing zinc which provides additional antimicrobial activity, have also been developed, such as Milliken's AlphaSan® Antimicrobial Agent. These silver-based products have been approved by several bodies such as the FDA for food contact applications. They have been successfully incorporated into food contact surfaces such as fluoroelastomers, nitrile, ethylene propylene . . . (e.g., Elastoguard from Milliken Chemical) and used for gaskets or conveyor belts or coated onto stainless steel surfaces (e.g., AgION™ antimicrobial coated steel from AK coatings, Inc., a subsidiary of AK Steel Corporation) for food processing facilities. The latter surfaces are also recommended for use in air systems, including air conditioners, and ventilation ducts to inhibit *Legionella pneumophila* (Rusin *et al.*, 2003).

Copper-alloyed stainless steels with antimicrobial properties have also been developed (Nisshin Steel, Japan). Despite the lower efficacy of copper than silver-alloyed stainless steel in inhibiting bacteria (Sreekumari *et al.*, 2005), interesting laboratory results have been reported on bacterial survival on copper-alloyed stainless steels (Wilks *et al.*, 2006). Conversely, in production environments, efficacy depends greatly on the bacterial group and remains lower than that of silver-alloyed stainless steel (Storgards *et al.*, 2007). However, antimicrobial activity would be enhanced by the addition of niobium, which stimulates the formation of precipitated copper particles (Baena *et al.*, 2006). Furthermore, the addition of nickel and zinc to copper alloys would increase their resistance to corrosion, making them suitable for general use in industrial environments (Wilks *et al.*, 2005). Works are still in progress in estimating the antibacterial efficacy of different stainless steel alloys and further works will be required to estimate if these new alloys present acceptable durability, long-term bactericidal effect, and resistance to corrosion. Lastly, photocatalytic titanium dioxide ($TiO_2$) coatings that generate reactive oxygen species and render the surfaces super-hydrophilic upon illumination with UV (below 388 nm), have been successfully used to inhibit the bacterial colonisation of ceramics (Raulio *et al.*, 2006) and stainless steels (Storgards *et al.*, 2007). Of course, photocatalytic materials can only be used for open surfaces and the main obstacle to their use in the food environment remains the need for effective UV lamps. As suggested by Storgards *et al.* (2007), combining antimicrobial alloys with photocatalysis would permit the use of weak UV light or even strip lights in the food processing lines. However, recent findings have shown that the anti-microbial effect of photocatalytic $TiO_2$ coated steel on a brewery filler did not last for 18 months. Material characterizations revealed that surfaces were covered by Si-O containing precipitates, which probably decreased photocatalytic activity. It was also seen that silver ions had dissolved from the surfaces, thus suggesting improvements are needed to overcome these problems (Storgards *et al.*, 2008).

A main advantage of the use of inorganic compounds to limit the microbial contamination of surfaces is that microorganism-reaction mechanisms are highly non-specific, making it rare for microorganisms to develop resistance. Nevertheless, these materials must meet the regulations concerning maximal permitted release or intake. For example, in US, silver concentration in water should not exceed 0.1 mg/l (US EPA, 2008), whilst the joint FAO/WHO Expert Committee on Food additives (JECFA) suggests a daily intake value for copper of 0.05 mg/kg (WHO, 1967). Several attempts have also been made to avoid bacterial contamination by the incorporation of, or coating with biocides. However, despite the numerous studies performed at laboratory level on many anti-microbial compounds (nisin, lysozyme, quaternary ammonium groups . . .) coated onto food-contact surfaces, very few anti-microbial materials have come onto the market. Several reasons account for this lack of development, such as the risk of bacterial resistance development or the leaching of the anti-microbial compounds into the surrounding environment, which would make materials inefficient as soon as the agent was exhausted. Elsewhere, the anti-microbial effect could be masked by the formation of a conditioning film, as reported by Flemming (2002) for quaternary ammonium groups.

For example, the anti-microbial compound Triclosan (CAS n° 3380-34-5, $C_{12}H_7Cl_3O_2$, Ciba Specialty Chemicals, High Point, NC, USA), which has been used as a disinfectant in hospitals for over 25 years, is currently widely integrated into plastics (Microban®, Microban Products Co., UK). The main food application concerns food-contact surfaces such as conveyor belts, cutting areas, and food storage containers. However, the use of Triclosan is still highly controversial, mainly due to the possible ability of bacteria to acquire resistance and the detrimental effects on human health and aquatic ecosystems. Despite the fact that Triclosan has been approved by the European Union (Anon, 1990) for use in food contact applications, a number of environmental and health organizations are now calling for legislation to limit the use of organic anti-microbial compounds such as Triclosan.

In summary, different strategies are currently being successfully developed to reduce surface microbial colonization. However, in food and beverage industries, only a few functional materials have been developed, due to the numerous requirements that these antimicrobial materials must fulfil. In addition to the general requirements, such as a lack of induced bacterial resistance, the presence of food requires an absence of any compounds release, which could potentially constitute a risk for consumers. Furthermore, owing to the drastic conditions encountered in the food and beverage industries, these materials must resist mechanical and chemical wear and tear, since the coatings are often found to peel off during fouling tests.

Lastly, it is noteworthy that this approach carries with it the danger that users may become less vigilant. If it is true that easily cleanable materials and equipment will allow less aggressive cleaning procedures to be set up, these functional surfaces are not intended as substitutes for good hygienic practices and must still be cleaned to ensure the surfaces will be free of unwanted microorganisms.

## 11.10 Sources of further information and advice

The main European regulations and directives governing the food industry (http://europa.eu.int/comm/food/food/chemicalsafety/foodcontact) are the following:

- General regulation for food hygiene 852/2004 which replaces directive 93/43/EC. Annex II, chapter 2.
- Regulation (EC) N° 1935/2004 of the European Parliament and of the Council of 27 October 2004 on materials and articles intended to come into contact with food and repealing Directives 80/590/EEC and 89/109/EEC.
- EU directive 2006-42-CE of the European Parliament and of the Council of 17 May 2006 on machinery, amending Directive 95/16/EC. Annex 1, part 2.1.

International standards organizations (e.g., Codex Alimentarius and the International Standardization Organization (ISO)) as well as International trade associations (e.g., International Dairy Federation (IDF)) are involved in equipment hygiene standards. The most significant standard regarding the food industry is the machine standard ISO 14159 and the similar standards EN 1672–2:1997 in Europe and NSF/ANSI/3A 14159–1-2002 in US.

For further information on standards or guidelines, the following web sites can be consulted: ISO (www.iso.org), ASTM (www.astm.org), the Japanese standard association (www.jsa.or.jp), the European Committee for standardization (www.cen.eu), the European Hygienic Engineering and Design Group (www.ehedg.org/guidelines.htm), the 3-A Sanitary Standards (www.3-A.org), the National Sanitation Foundation (www.nsf.org).

As resistance of materials to the constraints due to the food process and to the cleaning and disinfection is of prime importance, readers are also strongly advised to refer to the EHEDG guideline number 32: 'Construction material' which is available through the CCFRA web site:

www.campden.co.uk/publ/pubfiles/ehedg.htm (accessed on the 16 November 2008).

## 11.11 Acknowledgements

The author would like to thank Yannick Lequette, Muriel Mercier-Bonin, Thierry Bénézech and Jean-Pierre Tissier for their help in preparing this manuscript and Helen Selliez for English proofreading.

## 11.12 References

ABU-LAIL NI and CAMESANO TA (2006), The effect of solvent polarity on the molecular surface properties and adhesion of *Escherichia coli*, *Colloids Surf B Biointerfaces*, 51, 62–70.

ANON (1990), EU 90/128/EEC Commission directive relating to plastics materials and articles intended to come into contact with foodstuffs, *Off. J. Eur. Comm.*, L75, 19–40.

ARNOLD JW, BOOTHE DH, SUZUKI O and BAILEY GW (2004), Multiple imaging techniques demonstrate the manipulation of surfaces to reduce bacterial contamination and corrosion, *J. Microscopy-Oxford*, 216, 215–221.

AUSTIN JW and BERGERON G (1995), Development of bacterial biofilms in dairy lines, *J. Dairy Res.*, 62, 509–519.

BAENA MI, MARQUEZ MC, MATRES V, BOTELLA J and VENTOSA A (2006), Bactericidal activity of copper and niobium-alloyed austenitic stainless steel, *Curr. Microbiol.*, 53, 491–495.

BAGGE-RAVN D, NG Y, HJELM M, CHRISTIANSEN JN, JOHANSEN C and GRAM L (2003), The microbial ecology of processing equipment in different fish industries – analysis of the microflora during processing and following cleaning and disinfection, *Int. J. Food Microbiol.*, 87, 239–250.

BARNES LM, LO MF, ADAMS MR and CHAMBERLAIN AHL (1999), Effect of milk proteins on adhesion of bacteria to stainless steel surfaces, *Appl. Environ. Microbiol.*, 65, 4543–4548.

BELL CH, ARORA BS and CAMESANO TA (2005), Adhesion of *Pseudomonas putida* KT2442 is mediated by surface polymers at the nano- and microscale, *Environ. Eng. Sci.*, 22, 629–641.

BLEL W, BÉNÉZECH T, LEGENTILHOMME P and LE GENTIL C (2008), Hygienic design: effect of hydrodynamics on the cleanability of a food processing line, *AICHE J.* 54, 2553–2566.

BOS R, VAN DER MEI HC, GOLD J and BUSSCHER HJ (2000), Retention of bacteria on a substratum surface with micro-patterned hydrophobicity, *FEMS Microbiol. Lett.*, 189, 311–315.

BOULANGÉ-PETERMANN L (1996), Processes of bioadhesion on stainless steel surfaces and cleanability: a review with special reference to the food industry, *Biofouling*, 10, 275–300.

BOULANGÉ-PETERMANN L, BAROUX B and BELLON-FONTAINE MN (1993), The influence of metallic surface wettability on bacterial adhesion, *J. Adhesion Sci. Technol.*, 7, 221–230.

BOULANGÉ-PETERMANN L, RAULT J and BELLON-FONTAINE MN (1997), Adhesion of *Streptococcus thermophilus* to stainless steel with different surface topography and roughness, *Biofouling*, 11, 201–216.

BOULANGÉ-PETERMANN L, GABET C and BAROUX B (2006), On the respective effect of the surface energy and micro-geometry in the cleaning ability of bare and coated steels, *Colloid Surface A*, 272, 56–62.

BOWEN WR, FENTON AS, LOVITT RW and WRIGHT CJ (2002), The measurement of *Bacillus mycoides* spore adhesion using atomic force microscopy, simple counting methods, and a spinning disk technique, *Biotech. Bioeng.*, 79, 170–179.

BOYD RD, COLE D, ROWE D, VERRAN J, PAUL AJ and WEST RH (2001), Cleanability of soiled stainless steel as studied by atomic force microscopy and time of flight secondary ion mass spectrometry, *J. Food Prot.*, 64, 87–93.

BOYD RD, VERRAN J, JONES MV and BHAKOO M (2002), Use of the atomic force microscope to determine the effect of substratum surface topography on bacterial adhesion, *Langmuir*, 18, 2343–2346.

CEN (1996), Biotechnology. Performance criteria for pumps – EN 12462. European committee for standarization, Brussels (Belgium).

CERCA N, PIER GB, VILANOVA M, OLIVEIRA R and AZEREDO J (2005), Quantitative analysis of adhesion and biofilm formation on hydrophilic and hydrophobic surfaces of clinical isolates of *Staphylococcus epidermidis*, *Res. Microbiol.*, 156, 506–14.

CHMIELEWSKI RAN and FRANK JF (2003), Biofilm formation and control in food processing facilities, *Comprehensive Rev. Food Sci. Food Safety* 2, 22–32.

CLOETE TE and JACOBS L (2001), Surfactants and the attachment of *Pseudomonas aeruginosa* to 3CR12 stainless steel and glass, *Water SA*, 27, 21–26.

COSTERTON JW, LEWANDOWSKI Z, CALDWELL DE, KORBER DR and LAPPIN-SCOTT H (1995), Microbial biofilms, *Annu. Rev. Microbiol.*, 49, 711–745.

CZECHOWSKI MH (1990), Bacterial attachment to Buna-n gaskets in milk processing equipment, *Australian J. Dairy Technol.*, 113–114.

DEQUEIROZ GA and DAY DF (2007), Antimicrobial activity and effectiveness of a combination of sodium hypochlorite and hydrogen peroxide in killing and removing *Pseudomonas aeruginosa* biofilms from surfaces, *J. Appl. Microbiol.*, 103, 794–802.

DUNSMORE DG, TWOMEY A, WHITTLESTONE WG and MORGAN HW (1981), Design and performance of systems for cleaning product-contact surfaces of food equipment: A review, *J. Food Prot.*, 44, 220–240.

DÜRR H (2007), Influence of surface roughness and wettability of stainless steel on soil adhesion, cleanability and microbial inactivation, *Trans IChemE, Part C*, 85, 49–56.

EFSA (2008), Guidance document on the submission of a dossier on a substance to be used in Food Contact Materials for evaluation by EFSA by the Panel on additives, flavourings, processing aids and materials in contact with food (AFC). European Food Safety Authority.

EHEDG DOC 17 (2004), Hygienic design of pumps, homogenizers and dampening devices.

EMERSON RJ, BERGSTROM TS, LIU Y, SOTO ER, BROWN CA, MCGIMPSEY WG and CAMESANO TA (2006), Microscale correlation between surface chemistry, texture, and the adhesive strength of *Staphylococcus epidermidis*, *Langmuir,* 22, 11311–11321.

FAILLE C, DENNIN L, BELLON-FONTAINE MN and BÉNÉZECH T (1999), Cleanability of stainless steel surfaces soiled by *Bacillus thuringiensis* spores under various flow conditions, *Biofouling*, 14, 143–151.

FAILLE C, MEMBRÉ JM, TISSIER JP, BELLON-FONTAINE MN, CARPENTIER B, LAROCHE MA and BÉNÉZECH T (2000), Influence of physicochemical properties on the hygienic status of stainless steel with various finishes, *Biofouling*, 15, 261–274.

FAILLE C, JULLIEN C, FONTAINE F, BELLON-FONTAINE MN, SLOMIANNY C and BÉNÉZECH T (2002a), Adhesion of *Bacillus* spores and *Escherichia coli* cells to inert surfaces: role of surface hydrophobicity, *Can. J. Microbiol.*, 48, 728–738.

FAILLE C, MEMBRÉ JM, KUBACZKA M and GAVINI F (2002b), Altered ability of *Bacillus cereus* spores to grow under unfavorable conditions (nisin, temperature, pH, NaCl) following a heat-treatment during sporulation, *J. Food Prot.*, 65, 1930–1936.

FDA (2007), Determining the Regulatory Status of Components of a Food Contact Material, US Food and Drug Administration.

FLEMMING HC (2002), Biofouling in water systems cases, causes and countermeasures, *Appl. Microbiol. Biotechnol.*, 59, 629–640.

FLINT SH, BROOKS JD and BREMER PJ (2000), Properties of the stainless steel subtrate, influencing the adhesion of thermo-resistant streptococci, *J. Food Eng.*, 43, 235–242.

FLINT S, PALMER J, BLOEMEN K, BROOKS J and CRAWFORD R (2001), The growth of *Bacillus stearothermophilus* on stainless steel, *J. Appl. Microbiol.*, 90, 151–157.

FRANK JK and CHMIELEWSKI R (2001), Influence of surface finish on the cleanability of stainless steel, *J. Food Prot.*, 64, 1178–1182.

GEESEY GG, GILLIS RJ, AVCI R, DALY D, HAMILTON M, SHOPE P and HARKIN G (1996), The influence of surface features on bacterial colonization and subsequent substratum chemical changes of 316L stainless steel, *Corros. Sci.*, 38, 73–95.

GIBSON H, TAYLOR JH, HALL KE and HOLAH JT (1999), Effectiveness of cleaning techniques used in the food industry in terms of the removal of bacterial biofilms, *J. Appl. Microbiol.*, 87, 41–48.

GOTTENBOS B, GRIJPMA DW, VAN DER MEI HC, FEIJEN J and BUSSCHER HJ (2001), Antimicrobial effects of positively charged surfaces on adhering Gram-positive and Gram-negative bacteria, *J. Antimicrob. Chemother.*, 48, 7–13.

GUBNER R and BEECH IB (2000), The effect of extracellular polymeric substances on the attachment of *Pseudomonas* NCIMB 2021 to AISI 304 and 316 stainless steel, *Biofouling,* 15, 25–36.

GUOBJORNSDOTTIR B, EINARSSON H and THORKELSSON G (2005), Microbial adhesion to processing lines for fish fillets and cooked shrimp: Influence of stainless steel surface finish and presence of gram-negative bacteria on the attachment of *Listeria monocytogenes*, *Food Technol. Biotech.*, 43, 55–61.

HAEGHEBAERT S, LE QUERREC F, VAILLANT V, DELAROCQUE ASTAGNEAU E and BOUVET P (2001), Les toxi-infections alimentaires collectives en France en 1998, *BEH* 15, 1–12.

HERMANSSON M (1999), The DLVO theory in microbial adhesion, *Colloid Surf. B Biointerfaces*, 14, 105–119.

HERRERA JJR, CABO ML, GONZALEZ A, PAZOS I and PASTORIZA L (2007), Adhesion and detachment kinetics of several strains of *Staphylococcus aureus* subsp aureus under three different experimental conditions, *Food Microbiol.,* 24, 585–591.

HOLAH JT and THORPE RH (1990), Cleanability in relation to bacterial retention on unused and abraded domestic sink materials, *J. Appl. Bacteriol.*, 69, 599–608.

HUMPHREY T (2001), The spread and persistence of *Campylobacter* and *Salmonella* in the domestic kitchen, *J. Infection*, 43, 50–53.

HÜSMARK U and RÖNNER U (1993), Adhesion of *Bacillus cereus* spores to different solid surfaces: cleaned or conditioned with various food agents, *Biofouling*, 7, 57–65.

ISO 14159 (2002) Safety of machinery. Hygiene requirements for the design of machinery. ISO 14159. International Standardization Organization, Geneva (Switzerland).

ISO 4287 (1997), Geometrical product specifications. Surface texture. Profile method: terms, definitions and surface texture parameters – ISO 4287. International Standardization Organization, Geneva (Switzerland).

ISO 13565 (1998) Geometrical Product Specifications (GPS). Surface texture: Profile method; surfaces having stratified functional properties. Part 2 : Height characterization using the linear material ratio curve. ISO 13565. International Standardization Organization, Geneva (Switzerland).

JULLIEN C, BÉNÉZECH T, CARPENTIER B, LEBRET V and FAILLE C (2003), Identification of surface characteristics relevant to the hygienic status of stainless steel for the food industry, *J. Food Eng.*, 56, 77–87.

JULLIEN C, BÉNÉZECH T, LE GENTIL C, BOULANGÉ-PETERMANN L, DUBOIS P, TISSIER J, TRAISNEL M and FAILLE C (2008), Physico-chemical and hygienic properties modifications of stainless steel surfaces induced by conditioning with food and detergent, *Biofouling*, 24, 163–172.

KIELEMOES J, HAMMES F and VERSTRAETE W (2000), Measurement of microbial colonisation of two types of stainless steel, *Environ. Technol.*, 21, 831–843.

KORBER DR, CHOI A, WOLFAARDT GM, INGHAM SC and CALDWELL DE (1997), Substratum topography influences susceptibility of *Salmonella enteritidis* biofilms to trisodium phosphate, *Appl. Environ. Microbiol.*, 63, 3352–3358.

KRYSINSKI EP, BROWN LJ and MARCHISELLO TJ (1992), Effect of cleaners and sanitizers on *Listeria monocytogenes* attached to product contact surfaces, *J. Food Prot.*, 55, 246–251.

KUMAR CG and ANAND SK (1998), Significance of microbial biofilms in food industry: a review, *Int. J. Food Microbiol.* 42, 9–27.

KUSUMANINGRUM HD, RIBOLDI G, HAZELEGER WC and BEUMER RR (2003), Survival of foodborne pathogens on stainless steel surfaces and cross-contamination to foods, *Int. J. Food Microbiol.*, 85, 227–236.

LECLERCQ-PERLAT MN and LALANDE M (1994), Cleanability in relation to surface chemical composition and surface finishing of some materials commonly used in food industries, *J. Food Eng.*, 23, 501–517.

LELIÈVRE C, LEGENTILHOMME P, GAUCHER C, LEGRAND J, FAILLE C and BÉNÉZECH T (2002), Cleaning in place: effect of local wall shear stress variation on bacterial removal from stainless steel equipment, *Chem. Eng. Sci.*, 57, 1287–1297.

LOPES FA, MORIN P, OLIVEIRA R and MELO LF (2005), The influence of nickel on the adhesion ability of Desulfovibrio desulfuricans, *Colloids Surf. B Biointerfaces*, 46, 127–133.

MAI TL, SOFYAN NI, FERGUS JW, GALE WF and CONNER DE (2006), Attachment of *Listeria monocytogenes* to an austenitic stainless steel after welding and accelerated corrosion treatments, *J. Food Prot.*, 69, 1527–1532.

MAROUANI-GADRI N, CHASSAING D and CARPENTIER B (2009), Comparison of biofilm formation and susceptibility to a chemical treatment of pathogenic and non-pathogenic *Escherichia coli* O157:H7 strains, *J. Food Prot*, 72(1), 157–164.

MÉNDEZ-VILAS A, GALLARDO-MORENO AM, GONZÁLEZ-MARTÍN ML, CALZADO-MONTERO R, NUEVO MJ, BRUQUE JM and PÉREZ-GIRALDO C (2004), Surface characterisation of two strains of *Staphylococcus epidermidis* with different slime-production by AFM, *Appl. Surf. Sci.*, 238, 18–23.

MENDEZ-VILAS A, GALLARDO-MORENO AM and GONZALEZ-MARTIN ML (2006), Nano-mechanical exploration of the surface and sub-surface of hydrated cells of *Staphylococcus epidermidis*, *Antonie Van Leeuwenhoek*, 89, 373–386.

MERCIER-BONIN M, OUAZZANI K, SCHMITZ P and LORTHOIS S (2004), Study of bioadhesion on a flat plate with a yeast/glass model system, *J. Colloid Interf. Sci.*, 271, 342–350.

METTLER E and CARPENTIER B (1997), Location, enumeration and identification of the microbial contamination after cleaning of EPDM gaskets introduced into a milk pasteurization line, *Lait*, 77, 489–503.

METTLER E and CARPENTIER B (1999), Hygienic quality of floors in relation to surface texture, *Trans. Inst. Chem. Eng. Part C:Food Bioprod. Process.*, 77, 90–96.

MEYER EE, ROSENBERG KJ and ISRAELACHVILI J (2006), Recent progress in understanding hydrophobic interactions, *PNAS*, 103, 15739–15746.

MEYLHEUC T, METHIVIER C, RENAULT M, HERRY JM, PRADIER CM and BELLON-FONTAINE MN (2006), Adsorption on stainless steel surfaces of biosurfactants produced by gram-negative and gram-positive bacteria: Consequence on the bioadhesive behavior of *Listeria monocytogenes*, *Colloids Surf. B Biointerfaces*, 52, 128–137.

MIDELET G and CARPENTIER B (2002), Transfer of microorganisms, including *Listeria monocytogenes*, from various materials to beef, *Appl. Environ. Microbiol.*, 68, 4015–4024.

MUNOZ AI and MISCHLER S (2007), Interactive effects of albumin and phosphate ions on the corrosion of CoCrMo implant alloy, *Electrochem. Soc.*, 154, C562–C570.

NITSCHKE M and COSTA SG (2007), Biosurfactants in food industry, *Trends Food Sci Tech*, 18, 252–259.

OLIVEIRA K, OLIVEIRA T, TEIXEIRA P, AZEREDO J, HENRIQUES M and OLIVEIRA R (2006), Comparison of the adhesion ability of different *Salmonella enteritidis* serotypes to materials used in kitchens, *J Food Prot*, 69, 2352–2356.

ONG YL, RAZATOS A, GEORGIOU G and SHARMA MM (1999), Adhesion forces between *E. coli* bacteria and biomaterial surfaces, *Langmuir*, 15, 2719–2725.

PARKAR SG, FLINT SH, PALMER JS and BROOKS JD (2001), Factors influencing attachment of thermophilic bacilli to stainless steel, *J. Appl. Microbiol.*, 90, 901–908.

PERCIVAL SL, KNAPP JS, EDYVEAN R and WALES DS (1998), Biofilm development on stainless steel in mains water, *Wat. Res.*, 32, 243–253.

RAULIO M, PORE V, AREVA S, RITALA M, LESKELA M, LINDEN M, ROSENHOLM JB, LOUNATMAA K and SALKINOJA-SALONEN M (2006), Destruction of *Deinococcus geothermalis* biofilm by photocatalytic ALD and sol-gel TiO2 surfaces, *J. Ind. Microbiol. Biotechnol.*, 33, 261–268.

REDSVEN I, KYMALAINEN HR, PESONEN-LEINONEN E, KUISMA R, OJALA-PALOPOSKI T, HAUTALA M and SJOBERG AM (2007), Evaluation of a bioluminescence method, contact angle measurements and topography for testing the cleanability of plastic surfaces under laboratory conditions, *App. Surf. Sci.*, 253, 5536–5543.

RIJNAARTS HHM, NORDE W, BOUWER EJ, LYKLEMA J and ZEHNDER AJB (1993), Bacterial adhesion under static and dynamic conditions, *Appl. Environ. Microbiol.*, 59, 3255–3265.

RODRIGUEZ A, AUTIO WR and MCLANDSBOROUGH LA (2008), Effect of surface roughness and stainless steel finish on *Listeria monocytogenes* attachment and biofilm formation, *J Food Prot*, 71, 170–175.

RÖNNER AB and WONG ACL (1993), Biofilms development and sanitizer inactivation of *Listeria monocytogenes* and *Salmonella typhimurium* on stainless steel and Buna-N rubber, *J. Food Prot.*, 56, 750–758.

ROSMANINHO R, SANTOS O, NYLANDER T, PAULSSON M, BEUF M, BÉNÉZECH T, YIANTSIOS S, ANDRITSOS N, KARABELAS A, RIZZO G, MÜLLER-STEINHAGEN H and MELO LF (2007), Modified stainless steel surfaces targeted to reduce fouling : Evaluation of fouling by milk components, *J. Food Eng.*, 80, 1176–1187.

RUSIN P, BRIGHT K and GERBA C (2003), Rapid reduction of *Legionella pneumophila* on stainless steel with zeolite coatings containing silver and zinc ions, *Lett. Appl. Microbiol.*, 36, 69–72.

SENECHAL A, CARRIGAN SD and TABRIZIAN M (2004), Probing surface adhesion forces of *Enterococcus faecalis* to medical-grade polymers using atomic force microscopy, *Langmuir*, 20, 4172–4177.

SHENG XX, TING YP and PEHKONEN SO (2007), Force measurements of bacterial adhesion on metals using a cell probe atomic force microscope, *J. Coll. Interf. Sci.*, 310, 661–669.

SILVA SP, TEIXEIRA P, OLIVEIRA R and AZEREDO J (2008), Adhesion to and viability of *Listeria monocytogenes* on food contact surfaces, *J. Food Prot.*, 71, 1379–1385.

SINDE E and CARBALLO J (2000), Attachment of *Salmonella* spp. and *Listeria monocytogenes* to stainless steel, rubber and polytetrafluorethylene: the influence of free energy and the effect of commercial sanitizers, *Food Microbiol.*, 17, 439–447.

SMOOT LM and PIERSON MD (1998), Influence of environmental stress on the kinetics and strength of attachment of *Listeria monocytogenes* ScottA to Buna-N rubber and stainless steel, *J. Food Prot.*, 61, 1286–1292.

SOMERS EB and WONG ACL (2004), Efficacy of two cleaning and sanitizing combinations on *Listeria monocytogenes* biofilms formed at low temperature on a variety of materials in the presence of ready-to-eat meat residue, *J. Food Prot.*, 67, 2218–2229.

SREEKUMARI KR, SATO Y and KIKUCHI Y (2005), Antibacterial metals – A viable solution for bacterial attachment and microbiologically influenced corrosion, *Mater. Trans.*, 46, 1636–1645.

STORGARDS E, SIMOLA H, SJOBERG AM and WIRTANEN G (1999a), Hygiene of gasket materials used in food processing equipment. Part 1: New materials, *Trans. IChemE*, 77, 137–145.

STORGARDS E, SIMOLA H, SJOBERG AM and WIRTANEN G (1999b), Hygiene of gasket materials used in food processing equipment. Part 2: Aged materials, *Trans IChemE*, 77, 146–155.

STORGARDS E, JUVONEN R and MANTYLA T (2007), Functional materials. A possibility to improve brewery hygiene?, *Functional materials – a possibility to improve brewery hygiene?. Proceedings of the 31st European Brewery Convention Congress. Venice, Italy, 6–10 May 2007. Fachverlag Hans Carl,* 158, 1385–1392.

STORGARDS E, PRIHA O, LAAKSO J and MÄNTYLÄ T (2008), The effect of functional materials on microbial attachment in breweries, Society for General Microbiology – Autumn Meeting – Food & Beverages Group Session: Microbial adhesion to food matrices, Trinity College, Dublin.

SZABO JG, RICE EW and BISHOP PL (2007), Persistence and decontamination of *Bacillus atrophaeus* subsp globigii spores on corroded iron in a model drinking water system, *Appl. Environ. Microbiol.*, 73, 2451–2457.

TAYLOR JH and HOLAH JT (1996), A comparative evaluation with respect to the bacterial cleanability of a range of wall and floor surface materials used in the food industry, *J. Appl. Bacteriol.*, 81, 262–266.

TIDE C, HARKIN SR, GEESEY GG, BREMER PJ and SCHOLTZ W (1999), The influence of welding procedures on bacterial colonization of stainless steel weldments, *J. Food Eng.*, 42, 85–96.

TSIBOUKLIS J, STONE M, THORPE AA, GRAHAM P, PETERS V, HEERLIEN R, SMITH JR, GREEN KL and NEVELL TG (1999), Preventing bacterial adhesion onto surfaces: the low-surface-energy approach, *Biomaterials*, 20, 1229–1235.

US 3-A-SANITARY STANDARDS, 3-A SANITARY STANDARDS INC. Mclean, VA (www.3-a.org).

US ENVIRONMENTAL PROTECTION AGENCY (2008) National Secondary Drinking Water Regulations, www.epa.gov/safewater/contaminants/index.html

VADILLO-RODRIGUEZ V, BUSSCHER HJ, VAN DER MEI HC, DE VRIES J and NORDE W (2005), Role of *Lactobacillus* cell surface hydrophobicity as probed by AFM in adhesion to surfaces at low and high ionic strength, *Colloids Surf. B Biointerfaces*, 41, 33–41.

VAN LOOSDRECHT MCM, LYKLEMA J, NORDE W and ZEHNDER AJB (1990), Influence of interfaces on microbial activity, *Microbiol. Rev.,* 54, 75–87.

VAN OSS CJ (2006), *Interfacial Forces in Aqueous Media*. CRC Press Taylor and Francis Group, Boca Raton, London, New York, Washington.

VERRAN J and WHITEHEAD KA (2006), Assessment of organic materials and microbial components on hygienic surfaces, *Food Bioprod. Process.*, 84, 260–264.

VERRAN J, BOYD RD, HALL K and WEST RH (2001a), Microbiological and chemical analyses of stainless steel and ceramics subjected to repeated soiling and cleaning treatments, *J. Food Prot.*, 64, 1377–1387.

VERRAN J, ROWE DL and BOYD RD (2001b), The effect of nanometer dimension topographical features on the hygienic status of stainless steel, *J. Food Prot.*, 64, 1183–1187.

WATTS J (2000), More than 13 000 reported ill in Japan after drinking milk products, *Bull. World Health Organ.*, 78, 1063.

WHITEHEAD KA and VERRAN J (2006), The effect of surface topography on the retention of microorganisms, *Food Bioprod. Process.*, 84, 253–259.

WHITEHEAD KA and VERRAN J (2007), The effect of surface properties and application method on the retention of *Pseudomonas aeruginosa* on uncoated and titanium-coated stainless steel, *Int. Biodeter. Biodegr.*, 60, 74–80.

WHITEHEAD KA, COLLIGON J and VERRAN J (2005), Retention of microbial cells in substratum surface features of micrometer and sub-micrometer dimensions, *Colloids Surf. B Biointerfaces*, 41, 129–138.

WHO (WORLD HEALTH ORGANIZATION) (1967), Specifications for the Identity and Purity of Food Additives and Their Toxicological Evaluation: Some Emulsifiers and Stabilizers and Certain Other Substances. *Tenth report of the Joint FAO/WHO Expert Committee on Food Additives. FAO Nutrition Meetings Series*, No. 43, WHO Technical Report Series No. 373. Geneva, Switzerland: World Health Organization.

WILKS SA, MICHELS H and KEEVIL CW (2006), The survival of *Escherichia coli* O157 on a range of metal surfaces. *Int J Food Microbial*, 105(3), 445–454.

WILKS SA, MICHELS HT and KEEVIL CW (2006), Survival of *Listeria monocytogenes* Scott a on metal surfaces: Implications for cross-contamination, *International Journal of Food Microbiology*, 111, 93–98.

WIRTANEN G, HÜSMARK U and MATTILA-SANDHOLM T (1996), Microbial evaluation of the biotransfer potential from surfaces with *Bacillus* biofilms after rinsing and cleaning procedures in closed food-processing systems, *J. Food Prot.*, 59, 727–733.

XIAOXIA S, PENG TY and OLAVI PS (2006), Direct force measurement of bacteria adhesion on metal in aqueous media, *Water Sci. Technol.*, 54, 17–25.

ZHAO Q (2004), Effect of surface free energy of graded NI-P-PTFE coatings on bacterial adhesion, *Surface & Coatings Technology*, 185, 199–204.

ZHAO Q, WANG C, LIU Y and WANG S (2007), Bacterial adhesion on the metal-polymer composite coatings, *International Journal of Adhesion and Adhesives*, 27, 85–91.

# 12

# Cleaning and sanitation in food processing environments for the prevention of biofilm formation, and biofilm removal

**D. Grinstead, JohnsonDiversey, Inc, USA**

**Abstract**: Biofilms can cause problems for food processors. They can be a source of microorganisms that contaminate food, and because pathogenic microorganisms can form biofilms they are a food safety risk. Microorganisms from biofilms may also reduce the quality of food even if they do not represent a food safety risk. The best way to eliminate biofilms and to prevent them from forming is not the biocides, disinfectants, and sanitizers that are often used. The best way to control biofilm is effective cleaning. If the cleaning practices that are used by a food processor are optimized to remove the most soil, those same practices will also be able to remove the most biofilm. Biocides alone cannot control biofilms; however, they do provide an addition level of protection beyond that coming from just cleaning. Effective cleaning practices will also help assure that biocides such as sanitizers and disinfectant can be most effective. Removing as much soil possible will also help prevent biofilm regrowth by limiting nutrients. The removal of biofilm will help slow regrowth of biofilms by limiting attachment of new potential biofilm-producing microorganisms to surfaces.

**Key words**: biofilm, cleaning, sanitation, biocide, TACT.

## 12.1 Introduction

Biofilms are a serious concern for almost all food processors yet most food processing environments are, ironically, very conducive to biofilm growth. Large amounts of food soil and water combined with equipment that is complex and difficult to clean makes food processing environments very hospitable for biofilms. Once a biofilm has formed, it can have numerous negative impacts on the food that is being processed in that facility.

The most notable issue with biofilms is that they pose a food safety threat because they can be formed by many food pathogens including *Listeria*, *E. coli*, and *Salmonella*. *Listeria*, in particular, is a concern because there is mounting evidence that many cases of *Listeria* contamination occur during food processing, and the organisms that are contaminating the food are more likely to come from the food handling environment than from the incoming raw material. A study by Lappi *et al.* published in 2004 examining *Listeria* contamination in smoked fish processors found that *Listeria* could persist in a food processing environment for weeks, months or even years, and that the *Listeria* in the food handling environment could then contaminate food processed in that facility. *Listeria* is a well known biofilm producer and its survival in such a community could easily account for its persistence over long periods of time in food processing environments.

Of course non-pathogenic organisms can readily form biofilms as well. Certain members of the genus *Pseudomonas* are just such an example. Even though *Pseudomonads* are not typically food pathogens, the presence of such organisms and the ability to form biofilms in a food processing environment can still have serious implications. Microorganisms from biofilms can contaminate food, leading to reduced product quality, limited shelf life of the food, damaged company reputations, and loss of sales.

Francis Bacon once said that, 'A prudent question is one-half of wisdom.' The 'prudent question' then is: Since biofilms in a food processing facility can serve as a source of contamination that may reduce food product quality or create a health hazard, what is the best way to address that concern? The answer is clear, biofilms must be eliminated. That answer, as many answers seem to do, raises another question: How can biofilms be eliminated? Simply put, biofilms can largely be eliminated with effective and optimized cleaning practices. At this point, the questions can come faster than the answers. How can cleaning be optimized? How does one know that the cleaning is optimized? How does cleaning affect biofilms? What role do sanitizers and disinfectants play? Is there an easy way to eliminate biofilms, a 'silver bullet' that will kill them? Who can I turn to for help with this problem? Answering these questions and others is the intent of this chapter. Those answers should provide an understanding of the rather simple, but not always easy, ways that biofilms can be controlled in a food processing setting.

## 12.2 The cleaning process

Cleaning is an often undervalued step in the environmental sanitation process. Mitigation of microorganisms on surfaces, including those in biofilms, is strongly impacted by cleaning. And in the case of biofilms, cleaning may actually play a larger role in controlling or eliminating them than antimicrobial sanitizers and disinfectants. It is important, therefore, to understand the fundamentals of cleaning, what factors can impact cleaning effectiveness, and how the cleaning process can be optimized.

While there are many factors that influence cleaning performance, such as soil type, the surface to be cleaned, water quality and environmental conditions, there are only four variables that can be controlled during cleaning itself that directly impact the result. These are time, action, chemical, and temperature. Cleaning performance can be improved by cleaning a surface for a longer time. Another way to improve cleaning is to apply additional mechanical action. The choice of cleaning chemical and its concentration can strongly impact cleaning performance, as can the temperature at which cleaning is conducted. The cleaning performance that will be delivered by a cleaning practice could be loosely described as an equation: Cleaning performance = Time + mechanical Action + Chemical + Temperature. A useful acronym to help remember these variables is TACT. Effective cleaning is a balancing act in which these four variables are carefully tuned for optimum results. Each of these variables is examined below to see how they impact cleaning.

### 12.2.1 Impact of time on cleaning

In general, the more time that is spent cleaning a surface, the cleaner a surface will become. However, not all of that time is necessarily spent actually applying detergent or scrubbing a surface. One way of increasing time spent cleaning is to simply increase the amount of time that a cleaner is in contact with a soiled surface. This can be accomplished by circulating a cleaning solution through a closed system such as a clean-in-place (CIP) process, or soaking soiled items in a detergent solution as in a clean-out-of-place tank or a hand dish sink. The nature of the cleaning chemical can also influence the amount of time the detergent is in contact with soiled surfaces. Foam cleaners are commonly used for general environmental cleaning in food processing facilities, and contact time between detergents and soiled surfaces varies greatly across chemical products. Some foam cleaners have higher viscosity and will therefore cling to surfaces longer. Figure 12.1 shows two different foam cleaners 10 minutes after they were applied to stainless steel panels. The cleaner applied to the left side of the panel is designed to maintain contact on vertical surfaces longer than the one on the right.

### 12.2.2 Impact of mechanical action on cleaning

Another of the variables that can be controlled to influence cleaning is mechanical action. The more energy input into cleaning in the form of manual scrubbing, automated impingement, or flow, the cleaner a surface will become. High pressure spray cleaning, turbulent flow in CIP systems,

**Fig. 12.1** Photograph of a stainless steel panel ten minutes after two different cleaners were applied to it. The cleaner on the right is a standard chlorinated alkaline foam cleaner. The cleaner applied to the left is a chlorinated alkaline cleaner that is formulated to cling longer to vertical surfaces.

and pumping cleaning solutions though properly positioned spray balls are all ways to increase the mechanical action when cleaning.

The formulation of a cleaner can also affect the mechanical action of the cleaning process. Inclusion of particulates in the cleaner can increase the 'scouring' effect in hand cleaning applications. Similarly, the use of water ice or dry ice can provide mechanical action to help remove soils. Effervescent action of certain cleaning solutions also provides mechanical action.

Care must be taken when increasing the mechanical action of cleaning processes as the aforementioned methods can damage surfaces if they are used improperly. Not only can this make equipment unsightly or even affect operation and jeopardize investments, but a damaged surface provides scratches, dents, and other abrasions which can lead to corrosion and can harbor soil and biofilms.

### 12.2.3 Impact of chemicals on cleaning

The type of chemical used to clean a surface will have a tremendous impact on cleaning effectiveness. Cleaners are formulated to clean particular soils and surfaces under certain use conditions. For example, surfactants can be useful for cleaning soils high in fat. Surfactant molecules can dissolve in both fat and water at the same time, pulling the soil into the water which helps water emulsify and rinse away fatty soils. Emulsification of soils by surfactants also helps prevent most soils from redepositing onto surfaces once the soil has been removed. Many cleaners intended for use on high fat soils also contain alkali. The alkali helps solubilize soil and prevents redeposition and is effective on crosslinked fats or caramelized soils. Where the primary soil concern is mineral scale, acids are the cleaner of choice.

Acids dissolve the relatively insoluble mineral salts that comprise most scale build up. Very strong inorganic acids are appropriate in some environments, while weaker organic acids are preferred in others.

Other important ingredients in cleaner formulations include materials such as chelating agents (like EDTA), solvents, builders (like phosphate), corrosion inhibitors, oxidizers (such as chlorine), enzymes, and others (Rouillard, 1994). Chelants help to remove mineral scale, loosen some types of soil, and condition hard water that would otherwise cause cleaners to be less effective. Solvents such as alcohol and glycol ether can help remove grease and oils. Chlorine is used in some cleaners to help remove protein films. Chlorine and other oxidizers, such as hydrogen peroxide, can be used in cleaners to bleach stains. Enzymes can improve cleaning performance on specific soils they target, including fats, proteins, and carbohydrates.

The concentration of the cleaner also has a large impact on cleaning performance. Typically, a higher concentration of cleaner will remove more soil than a lower concentration, but only to a point. Most cleaners have an optimum concentration. Additional cleaner beyond this threshold will not remove additional soil and in fact can lead to redeposition of soil on surfaces and reduced cleaning performance. Figure 12.2 shows the cleaning performance of a heat exchanger in a dairy as a function of time and cleaner concentration, with the concentration centered around the optimum. Clearly too much or too little chemical reduces the efficacy of the cleaning process.

### 12.2.4 Impact of temperature on cleaning

The final TACT variable impacting cleaning performance is temperature. Higher temperatures usually result in better cleaning. For example, a rough rule of thumb when cleaning milk soils is that you can double the cleaning result for every 10°C increase in temperature. However, as with chemical

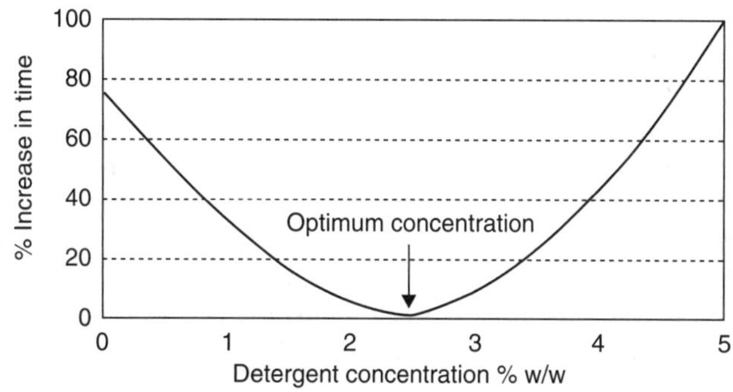


**Fig. 12.2** Effect of detergent concentration on time required to clean a dairy heat exchanger (Rouillard, 1994).

concentration, there is an optimum temperature. Proteins, in particular, can be more difficult to remove at very high temperatures. If the cleaning temperature is high enough to denature proteins, it can make them more difficult to remove and very high temperatures can bake soils onto surfaces and make them extremely difficult to remove. The optimum cleaning temperature is also influenced by the chemistry of the cleaner being used and the surface being cleaned. Volatile components of cleaners such as chlorine or solvents can be driven out of solution by temperatures that are too high, and enzymes in cleaners can be inactivated by excessive temperatures. Corrosion of metals and deterioration of plastic and rubber components are also influenced by temperature.

Collectively, time, mechanical action, chemical selection, and temperature will determine the total performance that any cleaning process will deliver. If one variable in the equation is not optimized, then another may need to be changed to give the best cleaning performance. For example, it may not be possible to apply as much mechanical action as is needed to clean a surface, but the total cleaning performance may be maintained if a stronger cleaner, longer contact time, or higher temperature is used to clean. Safety, equipment compatibility, equipment manufacturer recommendations, cleaning frequency, production schedules, regulations and environmental restrictions are also important considerations that affect or even dictate the choice of cleaner and procedure. In addition, utility availability, resource consumption (water and energy), and adjacent production area cleaning conflicts (dry production, storage or packaging areas) are just a few of the constraints that must be taken into account when selecting a cleaner or procedure. It is necessary to manipulate all four TACT variables to develop a cleaning protocol that provides the optimum soil removal given the constraints of a particular cleaning system. Therefore, optimization of the TACT variables often has to be done empirically. Of course, if the cleaning efficacy is going to be optimized empirically, then it is important that there be a way to measure the performance of a cleaning system.

### 12.2.5 Measuring the cleanliness of surfaces

In order to quantify the performance of a cleaning system it is necessary to be able to measure the cleanliness of a surface. There are many ways to accomplish this, some of which are highly subjective and others are more objective and quantitative. One way to measure cleaning is to make a strictly subjective assessment. This is probably the most common way to measure the cleanliness of a surface and is how the 2005 FDA Model Food Code recommends the cleanliness of a surface be determined. In section 4-601.11(A) the Food Code says that 'Equipment, food contact surfaces, and utensils shall be clean to sight and touch.' Although this is a very subjective measure of cleaning, it does have some advantages. It is an intuitive way to evaluate a surface and it takes no special equipment. There are other more objective ways to evaluate soil levels, however. Some of these methods include ATP measurement, protein swabs, carbohydrate dyes or stains, and microbial counts. All of these methods to measure cleaning performance have advantages and disadvantages. ATP tests are fairly sensitive, easy to do, and fast. However, they are relatively expensive and will not detect all soil. Protein swabs are easy and inexpensive but they are very specific and will detect only protein soils. Likewise, carbohydrate tests only detect carbohydrates. Microbiological tests often require significant time to get results and can be more complex to do than other tests. Also microbiological tests are only indirect measures of soil levels. It is entirely possible to have a lot of soil on a surface yet have very low microorganism counts or vice versa.

A weakness that all environmental sampling methods for soil, microorganisms, or specific soil components such as ATP share is sampling error. If samples are not taken from a location where the soil is present, it simply will not be detected. This may seem like belaboring the obvious; however, sampling error is one of the biggest shortcomings of all environmental testing programs. Surfaces that are easy to clean and easy to sample are what often get cleaned and sampled in a food processing facility. Not surprisingly then, those surfaces are rarely where biofilms can become a problem. Surfaces that are difficult to clean, hard to reach, and almost impossible to sample on a regular basis are the ones that tend to become contaminated. Therefore, any sampling program must take that into account. If a sampling program is going to provide a useful assessment of cleaning measures then it needs to be sure to include all the hidden and hard to reach places where soil or biofilms can avoid cleaning. If those hidden locations are identified and included in a sampling program, it may also help assure that the same locations are included in cleaning process. Knowledge on the part of a cleaning and sanitation crew that they will be accountable for effective sanitation in the hard to reach places may do more to assure that biofilms in those places are controlled than that actual testing itself.

The lack of a single, simple method that can rapidly provide a quantitative assay of the amount of any type of soil on any surfaces often forces the evaluation of a cleaning systems performance to be subjective or 'clean to sight and touch' as is specified in the model Food Code. While this can make it difficult to truly optimize all of the TACT variables, it is often possible to achieve a nearly optimized process, even when using strictly subjective measurements of cleaning.

## 12.3 The impact of cleaning on biofilms

The preceding discussion of cleaning and how to optimize cleaning efficacy may seem out of place in a discussion of biofilms and ways to control them. However, that is not actually the case. It is worth considering the

**Table 12.1** Soil constituents and recommended cleaning chemistry for removal (Rouillard, 1994)

| Type of soil | Example | Optimum cleaner chemistry |
|---|---|---|
| Organic soils | | |
| Carbohydrate | Sugar, starch, caramel | Alkaline, enzyme |
| Protein | Casein | Chlorinated alkaline or very strong alkaline, enzyme |
| Fat | Talo, lard, seed oils | Alkaline, enzyme |
| Petroleum | Greases, oils, lubricants | Solvent |
| Inorganic soils | | |
| Monovalent salts | Sodium chloride | Acid |
| Water stone | Calcium nitrate | Acid |
| Food stone | Calcium oxalate (beer and celery stone), tricalcium phosphate (milk stone) | Acid or high levels of chelant |
| Metallic deposits | Rust, aluminum oxide | Acid |

constituents of soil in a typical food processing facility. Table 12.1 lists the components of soils that are found in food processors and the best cleaners to remove them. Clearly there are a variety of soil types that can be encountered in a food processor and most often the soil will consist of a mixture of types listed in Table 12.1.

An obvious next question is: What are the chemical constituents of a biofilm? The extra cellular matrix that holds a biofilm together is composed primarily of carbohydrate but it will also contain protein, lipid, and nucleic acids. A biofilm will also contain microorganisms, which are themselves composed primarily of carbohydrates, protein, fat, and salts. In short the chemical constituents of a biofilm are identical to the soils that are typically found in a food processor. Biofilms can be thought of as just another type of soil that needs to be cleaned in a food plant. If indeed, biofilms are just another kind of soil it seems logical to wonder if cleaners can remove biofilms in the same way that they can remove other soils.

### 12.3.1 Effectiveness of cleaners at removing biofilms

The effect of cleaning chemicals on biofilms has been examined by several researchers. Frank *et al.* (2003) found that alkaline cleaners were more effective for removal of *Listeria monocytogenies* (LM) biofilms than neutral cleaners. A neutral cleaner produced a 5.91 $\log_{10}$ reduction of LM in 10 minutes static cleaning of biofilm that had been coated with protein. An alkaline cleaner was able to reduce the microbial levels in the same type of biofilm by 7.50 $\log_{10}$. Water yielded less than a two $\log_{10}$ reduction in this study. A study conducted by Somers and Wong (2004) also examined the reduction of LM biofilms by cleaners. This study looked at LM biofilms on various substrates including different grades of stainless steel, silicon, rubber, and polyester. Two- and five-day biofilms in the presence of various levels of ready to eat meat soil were tested. The cleaners tested were a chlorinated alkaline detergent and a solvated-alkaline cleaner. The Somers and Wong study indicated that cleaning with either product produced a significant reduction of both two- and five-day biofilms on all surfaces even in the presence of the highest level of soil tested (5.0% meat residue). The actual reduction varied depending on the substrate. For example, biofilms on stainless steel were reduced by 2.5–3 $\log_{10}$, and in some cases were reduced to non-detectable levels. Biofilms on TURE-2 (a material used for conveyer belts consisting of thermoplastic polyurethane with polyester fabric reinforcement and a polyurethane surface) were more difficult to remove with chemical cleaners, usually only achieving a one-log reduction or less. All the biofilms in this study were cleaned by either immersion in a static solution of the cleaner or sprayed with the solvated-alkaline detergent for three seconds. The differing biofilm reduction on different surfaces seen in this study is an indication of how important it is to optimize the TACT variables for each cleaning process. In the case of the TURE-2 surfaces, it may be necessary to supply additional mechanical action, change the concentration of cleaning chemical, clean for a longer time or change the temperature of the cleaning process to achieve the same reduction observed on stainless steel.

A study conducted by Antoniou and Frank (2005) looked at removal of *Pseudomonas pudita* biofilm from stainless steel surfaces. Rather than evaluating removal of biofilm by enumerating microorganisms, this study stained the microorganisms in the biofilm or stained the extra cellular polymeric substances (EPS) and then examined the surfaces for the stained material. The percent of the surfaces covered with cells and EPS could then be directly evaluated after various cleaning regimens. Several different cleaning chemicals were used in this study, including a simple alkali (NaOH) wash, an alkaline cleaning agent, or neutral cleaning detergent. When NaOH was used to remove biofilm, the temperature was critical for removal of the EPS. However when a formulated alkali cleaner that contained ingredients such as surfactants and solvents in addition to the NaOH was used, 98% or more of the EPS was removed in one minute, even at 4°C with no agitation. By comparison, a solution of 2.0% NaOH even with agitation was only able to remove about 70% of the EPS at 66°C. Increasing the temperature to 70°C resulted in over 98% removal of the EPS.

The ability of cleaners to remove the EPS produced by biofilm may be more important than the removal of microorganism themselves. The EPS plays a role in biofilms that far exceeds mere glue that holds the biofilm together. Beyond just being a structural matrix for a biofilm, the EPS may be involved with cellular transport, ion exchange, and may be a source of nutrients. An analogy proposed by Flemming *et al.* (2007) is that if biofilms are metaphorical cities of bacteria, then the EPS is the house in which the cells live. The ability of cleaning to remove this critical component of a biofilm may be the primary way in which a cleaning process can eliminate a biofilm.

In addition to the ability of cleaning to remove EPS, some cleaning chemicals also have their own strong antimicrobial activity. This can be particularly true of highly alkaline and chlorinated alkaline cleaners. Although such cleaners are not formulated to be biocides, the fact is that the chemicals that are important for good cleaning performance can also have antimicrobial activity. In a 2002 study reported by Taormina and Beuchat, incubation of LM strains in several cleaners including a chlorinate alkaline cleaner and a highly alkaline cleaner with a pH greater than 12 at use concentration could reduce microorganism levels by up to five-six $\log_{10}$ in 30 minutes at 4°C.

### 12.3.2 Response of biofilms to changes in the TACT variables

Obviously cleaners can remove biofilms and EPS from surfaces. If biofilms can be considered another type of soil, then a biofilm should respond to changes in the TACT variables in an analogous manner to soils. In fact, they do just that. In an unpublished study conducted at JohnsonDiversey a *Pseudomonas* biofilm was generated on small stainless steel plugs that could be placed into a model clean-in-place (CIP) system. The cleaning conditions of the CIP process could be varied in order to determine how changing the TACT variables impacted the removal of biofilms on the plugs. When each of the TACT variables was changed the biofilm responded exactly as a soil would. Figure 12.3 shows how the first variable, Time, impacts biofilm removal. In this experiment a chlorinated alkaline detergent was circulated

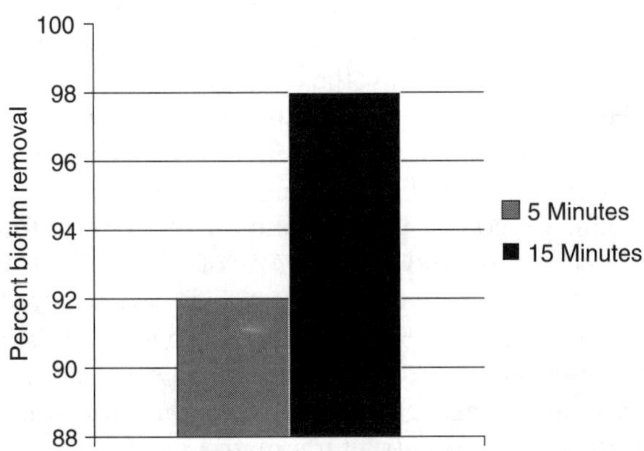


**Fig. 12.3** Effect of increased cleaning time on percent removal of a *Pseudomonas* biofilm by circulation of a chlorinated alkaline detergent at constant concentration, temperature, and velocity.

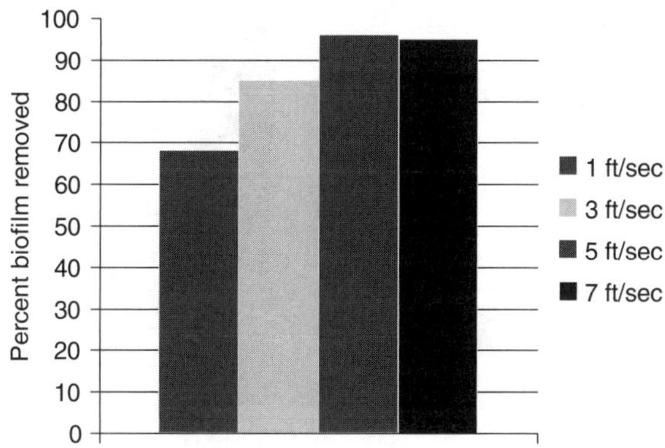


**Fig. 12.4** Effect of increased solution velocity on percent removal of a *Pseudomonas* biofilm by circulation of a chlorinated alkaline detergent at constant concentration, temperature, and time.

though the CIP system at a constant flow rate (five ft/second) and constant temperature (62.7°C). A longer time clearly resulted in greater biofilm removal.

Biofilms also respond like other soils to increased mechanical action. In the case of a CIP system such as was used in the JohnsonDiversey study, mechanical action can be increased by increasing the velocity at which the cleaning solution is moving though the system. In this experiment a 0.4% solution of a chlorinated alkaline detergent was circulated for five minutes at various flow rates. Figure 12.4 summarizes the results of that experiment.

Increased velocity clearly increases the quantity of biofilm removed. However, as is seen when cleaning soils, increasing the mechanical action does have a point of diminishing returns. In the example in Fig. 12.4, there is little difference between a flow rate of five feet/second and seven feet/second. This observation may result from the fact that the CIP system had achieved turbulent flow at the lower velocity and additional flow did not increase the mechanical action. Excessive mechanical action when removing biofilms also carries some risks just as it does when cleaning soil. In addition to the potential to damage surfaces mentioned earlier, too much mechanical action on a surface with a biofilm may inadvertently spread the biofilm by creating aerosols of organisms possibly resulting in cross contamination to other surfaces.

The next TACT variable is the type of cleaning chemical and its concentration. Once again the biofilm in this study responded exactly like a soil. Chlorinated alkaline cleaner at various concentrations was circulated through the CIP rig at a constant flow rates, temperature, and time. Figure 12.5 shows that both the presence or absence of the cleaner and its concentration had a strong impact on biofilm removal.

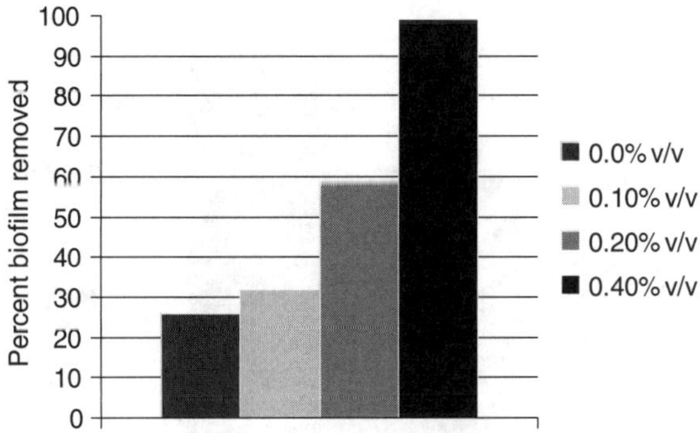


**Fig. 12.5** Effect of increased concentration of chlorinated alkaline detergent on percent removal of a *Pseudomonas* biofilm by circulation of the cleaner at constant temperature, time, and velocity.

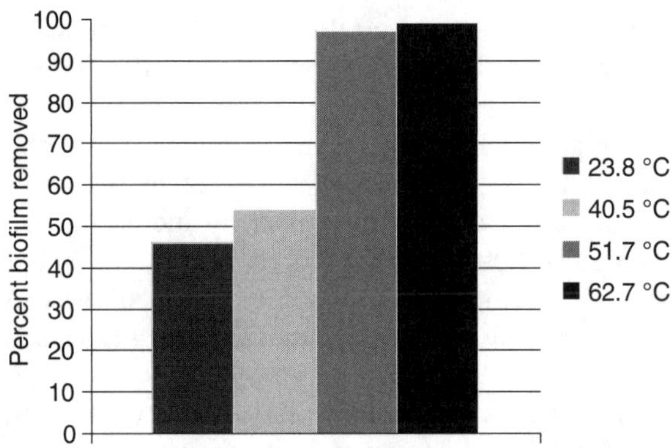


**Fig. 12.6** Effect of increasing temperature on percent removal of a *Pseudomonas* biofilm by circulation of chlorinated alkaline detergent cleaner at constant concentration, time, and velocity.

The final TACT variable is temperature and once again a biofilm responds just as a soil would to changes in that variable. Figure 12.6 shows that when a 0.4% solution of chlorinated alkaline detergent is circulated through a CIP system at a constant flow rate for five minutes, the temperature has a large effect on biofilm removal.

In the JohnsonDiversey study, the biofilm responded to every one of the TACT variables exactly like a soil would. Other studies have noted similar results. Wong (1998) reported that changing the temperature of a CIP system from 48.9°C to 60°C and the mechanical shear force at the surface of the CIP system from 84 to 114 N/m$^2$ improved the cleaning results on milk soil containing microorganisms. Parkar *et al.* (2004) found results similar to those observed in the JohnsonDiverey study. Removal of a *Bacillus* biofilm in a CIP system was influenced by the cleaning chemical chosen and the process parameters such as concentration, time, temperature, and flow rate were also important for satisfactory removal of the biofilm.

In all of these studies, biofilms respond to cleaners just as soil does. Therefore, optimizing the TACT variables in the cleaning process to provide an optimum soil removal will also go a long way toward optimizing the process for removing biofilm. Cleaners will remove biofilm from surfaces but is that all cleaners do? Are there ways that they can help control biofilm other than by directly removing them?

### 12.3.3 Alternative ways in which cleaners contribute to biofilm control

Clearly cleaners are an effective means to remove biofilms from surfaces and in some cases, cleaners are even biocidal on their own. There are also less direct ways that cleaning can help eliminate biofilms. Sanitizers are most effective on previously cleaned surfaces. In addition to removing biofilms, cleaners also remove other soils that would reduce the effectiveness of sanitizers. A study by Best *et al.* (1990) examined the effectiveness of a variety of disinfectants against *Listeria*. When the *Listeria* was suspended in growth medium and dried onto test surfaces, nine of the fourteen disinfectants tested were effective. However, when the organisms were suspended in human serum or 2% milk and then dried onto a surface, only three or one, respectively, of the disinfectants were effective. This demonstrates that the presence of soil can interfere with chemical disinfectant and sanitizer efficacy. The interference by soil with the performance of antimicrobial agents has been understood for some time. Effective cleaning will remove soil and so can improve the biocidal performance of sanitizers and disinfectants.

Yet another indirect way that cleaners may help control biofilms is by making the microorganisms in them more sensitive to sanitizers and disinfectants. Taormina and Beuchat (2002) examined the effect of preincubation in various cleaners on biocidal performance of sanitizers against *L. monocytogenes.* Preincubation in any of the cleaners tested resulted in significantly greater sensitivity of the *Listeria monocytogenes* cells to chlorine and quaternary ammonium chloride (QAC). In the case of the QAC, the reduction of cells incubated in cleaners before exposure to the biocide was more than four $\log_{10}$ greater than the reduction of cells that were preincubated with water. Although this study was conducted with organisms in suspension rather than on actual biofilms, it suggests another way in which cleaning may play a critical role in biofilm control.

It is possible that, for some microorganisms, effective cleaning will help prevent biofilm formation as well as remove biofilms that have already formed; however, that is not always the case. While a clean environment does provide limited nutrients for microorganisms, that does not necessarily mean less microbial attachment to surfaces and a reduced ability to form biofilm. In fact some microorganisms will attach to surfaces with very low nutrient levels if the conditions are right. A study by Sanders *et al.* (2008) explored the effect of nutrients and other variables on attachment of *Campylobacter jejuni* to stainless steel surfaces. *C. jejuni* was able to attach to surfaces with no preexisting biofilm at very low nutrient levels. However, if there were a pre-existing biofilm, attachment of *C. jejuni* was reduced under nutrient limited conditions. The response of microorganisms to limited nutrients is too variable to depend on cleaning to prevent microbial attachment to surfaces. Cleaning may also be too infrequent to be 100% effective for preventing biofilm formation. Many food processors manufacture for 16 hours with a single 8-hour clean up. During that 16-hour production it is possible for microorganisms to attach to surfaces and start to form biofilms. Although cleaning may not always prevent microbial attachment and biofilm formation, it can prevent biofilms from becoming problematic.

Effective cleaning has a major impact on biofilms. It removes the cells in a biofilm, helps clean away the EPS produced by the biofilm, removes soils that can inhibit sanitizer and disinfectant activity, and may even make microorganisms more susceptible to biocides. In some cases the cleaners may be antimicrobial in their own right. In the long run, attempting to control biofilms in a food processing facility only through sanitizers or disinfectants rather than cleaning will result in lower quality, less safe food. Biofilms can be considered just another soil that needs to be removed in the food processing environment. The most effective control of biofilms can be achieved by optimizing the cleaning process through the modification of the TACT variables.

### 12.3.4 Measuring biofilm removal

Although there are many ways to measure biofilm removal in a laboratory setting, measurement of biofilm removal in a food processing setting is far more problematic. Biofilms are not necessarily visible to the unaided eye. It may not be possible to tell if a surface is contaminated by a biofilm just by looking at it or touching it. An obvious way to evaluate a surface for biofilm is by microbiological testing of the surface. Unfortunately many of the same limitations to microbiological testing that apply when using it as a way to evaluate cleaning performance also apply to using it to evaluate biofilm removal. Environmental testing for microorganism can be difficult, time consuming and requires some expertise. Most of the methods that are available are relatively slow and require at least hours and sometimes days to produce results. If the method is more rapid, it may not be very sensitive and may need thousands of organisms in the sample in order to detect them. This is particularly true if the sampling includes testing for specific microorganisms.

Because biofilm has most of the same constituents as other food soils, it is also possible to use many of the soil test methods discussed earlier. ATP tests, protein assays, and carbohydrate swabs can all be used to look for biofilm. Of course because these are non-specific tests, they will not indicate the actual source of the ATP, protein, or carbohydrate. It could be from a biofilm or it could be from residual soil. From a practical standpoint, however, that may not be a particularly important distinction. The source of the ATP or whatever else may be detected really does not matter. If it is detected, the surface is clearly not clean and needs to be cleaned again. In this way the similarity between biofilm and soil, and the similar responses that they have to cleaning can work to our advantage.

Sampling error, as was discussed earlier in the context of evaluating soil removal, is also an issue when sampling for biofilms. The areas that are hard to sample and hard to clean are usually one and the same and so tend to be the areas most likely to be contaminated with biofilm. Any biofilm sampling program needs to take that into account.

## 12.4 Use of biocides

When microorganisms become a problem, in a food processing facility, kitchen, hospital, or almost any other setting, it seems like the first response is to apply more biocide around the environment. After 150 years of continual use of antimicrobial chemicals, it seems almost hardwired into people that only those chemicals are effective to control microorganisms. As was discussed above, however, the best solution to control biofilms is to focus on cleaning effectively rather than attempting to just sanitize or disinfect them away. With that said, the use of antimicrobial agents still plays an important part in control of biofilms in food processing settings. It is important to understand the strengths and limitations of sanitizers and disinfectants in order to understand how they can best be used to help control biofilms.

### 12.4.1 Biocides commonly used by food processors

There are many different biocides that are used in food processing settings. There is insufficient room to discuss them all in detail here. However, there are several classes of biocides and some specific antimicrobial agents that are used more commonly than others, and it is worthwhile to have a short discussion of some of those common chemicals. Table 12.2 summarizes the characteristics of several biocides that are commonly used in food processing environments.

One of the most common classes of biocides used are oxidizing biocides. This includes halogens such as chlorine, chlorine dioxide, and iodine. Other oxidizing biocides include peroxides such as hydrogen peroxide and

**Table 12.2** Characteristics of several biocides commonly used by food processors

| Biocide class | Active agent | Characteristics |
|---|---|---|
| Oxidizers | Hypochlorous acid | Strong broad spectrum biocide. Unstable and can be inactivated organic matter. |
| | Chlorine dioxide | Strong broad spectrum biocide. A dissolved gas that is volatile and must be generated at point of use. |
| | Iodine | Broad spectrum activity and relatively stable. Can discolor surfaces. |
| | Ozone | Strong activity; however, very unstable and reacts quickly with organic matter. Must be generated at point of use. |
| | Hydrogen peroxide | Low toxicity biocide. Must be used at high concentrations. Combination of low toxicity and efficacy at high concentrations makes this biocide useful for some specialized applications. |
| | Peroxyacetic acid | Low toxicity biocide. Not as easily inactivated as hydrogen peroxide so is effective at lower concentrations. |
| Surfactants | Quaternary ammonium chloride | Effective biocide against a wide range of vegetative organisms; however, not strongly active against spores or non-enveloped viruses. Less sensitive to inactivation by soil than oxidizing biocides. Hard water can limit efficacy. |
| | Acid anionic | Effective against a wide range of vegetative organisms. Can also function as good cleaners. Used at low pH, often 3.0 or less, which can limit applications. |

peroxyacetic acid. Chlorine, in one form or another is an oxidizing biocide that has been used since the 19th century (Dychdala, 2001). Today it is most often used as hypochlorite or hypochlorous acid. Hypochlorite is the basic form of the weak acid, hypochlorous acid. The pKa of hypochlorous acid is 7.4, therefore above that pH the unprotinated more stable hypochlorite anion predominates. The biocidally active species is the protonated hypochlorous acid. Although hypochlorous acid has strong biocidal activity it does have one significant weakness – it is unstable. Because it is unstable, it must be produced at the point of use. Hypochlorous acid can be inactivated by metals such as copper or nickel, ultraviolet light, and it can react with organic matter.

Iodine has as long a history of use as chlorine (Gottardi, 2001). Iodine is used in a variety of forms. Aqueous solutions of iodine, solutions of iodine in alcohol, and iodophors (organic polymers that are complexed with iodine) are all used as sanitizers or disinfectants. Although iodine is an effective biocide and is used in surface sanitizers and disinfectants, its tendency to discolor surfaces does limit its use as an environmental disinfectant in food processors. The most common use of iodine is in human and veterinary medicine as a skin or wound antiseptic.

Chlorine dioxide is a gas that has strong biocidal properties (Knapp and Battisti, 2001). Because it is a gas, it is usually generated at the point of use, much as hypochlorous acid must be. Chlorine dioxide is commonly used as a water treatment for drinking water. It is also used in the food industry as a treatment for process water and is used occasionally as a surface sanitizer or disinfectant. The fact that chlorine dioxide is a dissolved gas does limit its usefulness somewhat as a surface biocide. Its very strong antimicrobial, including sporicidal, activity does still make it a good choice for some applications in a food processor.

Peroxygen compounds have also been used as a biocide for more than one hundred years. Hydrogen peroxide has a long history of use (Block, 2001). Although hydrogen peroxide was originally very popular as a disinfectant, its use has waned considerably through much of the 20th century as biocides that were effective at lower concentrations became popular. However, the recent decades have seen the rise of some special applications for which hydrogen peroxide is particularly well suited as a disinfectant. A nearly unique property of hydrogen peroxide is its extraordinarily low toxicity. Hydrogen peroxide is also sporicidal at high concentrations. This has led to its use to sterilize packaging material for aseptically processed foods. One of the disadvantages of hydrogen peroxide is that $H_2O_2$ is a natural byproduct of aerobic metabolism, therefore most organisms that respire have enzymes (such as catalase and superoxide dismutase) that can detoxify hydrogen peroxide. This means that hydrogen peroxide needs to be used at relatively high levels to be effective against many aerobic organisms.

A peroxygen compound that is commonly used as a biocide in food processors that has many of the advantages of hydrogen peroxide yet avoids the disadvantage of enzymatic inactivation is peroxyacetic acid (PAA) (Block, 2001). Because it is not inactivated as easily, PAA is biocidal at lower concentrations than hydrogen peroxide. Like hydrogen peroxide, PAA is sporicidal at high concentration, and its equally favorable toxicology profile means that PAA is also used in aseptic processing of foods. The one way in which PAA is not preferable to hydrogen peroxide is stability. PAA has limited stability, and a solution of 40% PAA will lose 1–2% of its activity per month. In order to overcome that limitation, PAA is often formulated with excess acetic acid and hydrogen peroxide. Sequestering agents can also help stabilize a PAA solution by reduction decomposition due to trace metals.

Ozone ($O_3$) is an unstable three-atom allotrope of oxygen (Weavers and Wickramanayake, 2001). Ozone has many traits in common with chlorine dioxide. Ozone is a gas that can be dissolved in solution much like chlorine dioxide. Although ozone is highly biocidal it is also unstable with a half life in water that is measured in seconds to minutes depending on the quality of the water. Because ozone is unstable, it is generated at the point of use. Not surprisingly, given the similarity in practical limitations of the two chemicals, ozone is used for many of the same applications as chlorine dioxide including drinking water treatment, process water treatment, and occasionally as a surface sanitizer or disinfectant.

The other broad class of biocides that are used by food processors is based on surfactants (Merianos, 2001). A prime example of surfactant-based biocides is quaternary ammonium chloride (QAC). QACs are cationic surfactants that have good biocidal activity against vegetative bacteria and fungi. Most QACs are not strongly active against spores. One of the advantages that QACs have over many oxidizing biocides is that the QACs are very stable and are not as rapidly inactivated by organic matter. QACs, however, do have reduced activity in the presence of hard water. Many of the cleaners used in food processing facilities include anionic surfactants. If such a cleaner is not well rinsed before a cationic QAC is applied, the residual anionic surfactant can inactivate the QAC. There are many different types of QACs that have different structures and are used in different combinations. QACs are very commonly used as environmental biocides by many food processors.

Yet another example of surfactant-based biocides are acid anionic surfactants (Richter and Cords, 2001). These types of biocides can include several different classes of anionic surfactants in combination with acids. The antimicrobial activity of acid anionics is quite good, particularly against vegetative microorganisms. The use of anionic surfactants in these biocides can also make some of them effective cleaners. However, they are used at low pH, often below pH 3.0. This can limit the applications where they are used. Just as QACs are inactivated by anionic surfactants, so an acid anionic biocide may be inactivated by cationic detergents.

### 12.4.2 Effectiveness of biocides against biofilms

Clearly there are a lot of arrows in the biocide quiver: the question is do they work against biofilms? There has been a lot of research conducted that examined the efficacy of biocides against biofilms, and in most cases the answer to the previous question is, 'Yes, but not as well as we would hope.' A study by Kim *et al.* (2007) examined the effectiveness of several disinfectants against *Enterobacter sakazakii* in suspension and in biofilm. This study examined several different QACs, several phenolic-based disinfectants, and PAA for activity against *Enterobacter sakazakii* planktonic cells and the same strains in a biofilm on stainless steel. All of the biocides were able to significantly reduce the number of planktonic cells in a one-minute exposure time. The actual reductions varied from four to seven $\log_{10}$ in one minute, with four of the six biocides reducing the bacteria to undetectable levels. The two biocides that did not reduce the test organisms to undetectable levels in 60 seconds did achieve that reduction by five minutes. The performance of the biocides against *E. sakazakii* in biofilms, however, was much different. None of the biocides were able to reduce the microbial levels in the biofilm to undetectable levels with a 60-second exposure. Five of the six were unable to reduce the microorganisms in biofilm to undetectable levels even after 10 minutes exposure. In fact three of the four QACs tested had no significant ($P \leq 0.05$) reduction of the microbial load at any exposure time. The PAA was able to achieve a statistically significant one $\log_{10}$ reduction after five minutes and the phenolic biocides delivered a nearly one $\log_{10}$ reduction after 10 minutes. Only one of the biocides, a QAC used at pH 12, was able to reduce the biofilm to undetectable levels after five minutes. A 2002 study by Peng *et al.* also noted poor reduction of a *B. cereus* in a biofilm by QAC with the biocide only producing a one $\log_{10}$ reduction of organisms in biofilm but a five $\log_{10}$ or better reduction of planktonic cells.

Hypochlorite also performs poorly against biofilm. A study by Ryu and Beuchat (2005) compared the resistance of planktonic cells and biofilms of *E. coli* to chlorine in the form of hypochlorite. In that study, 50 ppm of chlorine was able to reduce planktonic *E. coli* to undetectable levels in one to three minutes under most conditions. However biofilms survived exposure to 50 ppm chlorine for five minutes. Some strains of *E. coli* were almost unaffected by the chlorine. Even 200 ppm for five minutes was not able to eliminate some of the biofilms tested. Peng *et al.* (2002) saw similar results when hypochlorite was used to remove a *B. cereus* biofilm. Five minutes at 50 ppm resulted in about a one-two $\log_{10}$ reduction of organisms in a biofilm compared with a five-six log reduction of planktonic cells. Ayebah *et al.* (2006) studied the ability of a solution of hypochlorous acid generated through an electrochemical reaction of NaCl in water to inactivate *L. moncytogenes* in suspension and in biofilms. The electrochemically generated hypochlorous reduced planktonic LM to undetectable levels (over 7.0 $\log_{10}$ reduction) in one minute. When the organisms were in a biofilm however the reduction was less than 4.5 $\log_{10}$. LeChevallier *et al.* (1988) also found that cells in mixed culture biofilm comprising Gram negative organisms isolated from a municipal water system were about 100 to 1000 times less sensitive to hypochlorous acid than planktonic cells.

Other oxidizing biocides do not fare much better against organisms in biofilms. Robbins *et al.* (2005) reported that 2.0 ppm ozone was able to achieve over an eight $\log_{10}$ reduction of unattached LM cells in three minutes. The same concentration and exposure time resulted in only a 4–4.5 $\log_{10}$ reduction of LM in a biofilm. Similarly, hydrogen peroxide at 3.0% could reduce planktonic cells of LM to undetectable levels in 10 minutes. However, if the LM was in a biofilm, 4% hydrogen peroxide was not able to completely eliminate the biofilm even after 15 minutes. It took 5% $H_2O_2$ to achieve complete elimination of the biofilm and even at that concentration it took 15 minutes.

Obviously, microorganisms that are in biofilms are resistant to the action of biocides compared to planktonic organisms. A good question now might be, why?

### 12.4.3 Reasons for the ineffectiveness of biocides against biofilms

One answer to this question is that the EPS produced by the biofilm is difficult for the biocides to penetrate. A report by De Beer *et al.* (1994) examined the penetration of chlorine into biofilms of *Pseudomonas aeruginosa* and *Klebsiella pneumoniae*. Experiments conducted during this study found that chlorine concentrations in biofilms may only be 20% (or less) of the concentration in the bulk liquid. Incubation for one to two hours did not allow for complete equilibration of the chlorine between the bulk phase and the biofilm. After exposure to 2.5 ppm chlorine, for one hour, only the upper 100 μm of a biofilm were penetrated by the chlorine. In another study reported by Stewart *et al.* in 2000, $H_2O_2$ was unable to penetrate a biofilm of wild-type *Pseudomonas aeruginosa*. However, when catalase deficient mutants of *P. aeruginosa* were used to generate the biofilm, hydrogen peroxide was able to penetrate the biofilm and kill the resident cells. The data presented by Stewart *et al.* (2000) suggests that biocides are inactivated during passage through the EPS. The De Beer (1994) study also found that the chlorine reacted with the EPS as well as cells as it was moving through the biofilm. Therefore a lot of the biocide's oxidizing 'muscle' was expended on the EPS in the biofilm rather than the microorganisms. A theoretical investigation carried out by Stewart (1996) suggests that for most biocides, a biofilm does not provide sufficient barrier to diffusion to explain the low activity against organisms in the biofilm. There must be some irreversible adsorption of the biocides to the EPS or a fast reaction that inactivates the biocide as it moves through the biofilm. Another possibility is that a change in the organisms growing inside the biofilm may make them more resistant to the biocide. The very existence of organisms in biofilm may cause them to become resistant to some biocides. The production of enzymes, such as catalase and superoxide dismutase, may be up regulated in microorganisms in biofilms, cells in biofilms may be growing more slowly than planktonic cells (McDonnell and Russell, 1999), or some other phenotypic change may make the organisms more resistant to biocides. However, a study by Pan *et al.* (2006) found that although *Listeria* cells that were grown in a biofilm were resistant to peroxides, QACs and chlorine, when the cells were removed from the surface, they showed no increased resistance to biocides. This suggests that if phenotypic changes do occur in cells in biofilms, the changes are not stable and removal of cells from the biofilm may cause them to become susceptible to biocides.

Not only are cells in biofilms resistant to biocides such as QACs, they may become more resistant over time. A study by Mangalappalli-Illathu and Korber (2006) showed that a *Salmonella* biofilm repeatedly exposed to sublethal levels of QAC over a 6-day period became resistant to levels of QAC that had previously been lethal (500 ppm). If biofilms are not removed by effective cleaning, and biocides alone are used to control them, they may rapidly become even more resistant to the biocides.

Whatever the mechanisms involved, the efficacy of biocides against organisms in biofilms is limited. So once again this raises a question. Why use biocides at all?

### 12.4.4 Benefits of biocide use

Although a biofilm reduces the efficacy of biocides, it does not completely eliminate most biocides' antimicrobial performance. That means that biocides can have some benefit and can help reduce microbial levels on surfaces. It is also important to remember that most biocides are applied after surfaces have been cleaned. As a result, the biocides can provide an extra level of microbial control. Somers and Wong (2004) showed that while cleaning alone could significantly reduce microorganism levels in biofilms on various surfaces, the subsequent application of a peroxygen- or chlorine-based sanitizer was able to further reduce microorganism levels in many of the biofilms tested. In many cases, the use of a biocide was able to reduce the microorganisms by an additional one $\log_{10}$ over the reduction from cleaning alone. And in some cases the sanitizers could reduce the organisms to undetectable levels, where cleaning alone had left biofilm on the test surface.

In countries where food contact sanitizers are not rinsed, it is even possible that the biocide may have some residual activity and that the chemical remaining on the surfaces after it drains could help slow biofilm regrowth.

### 12.4.5 Use of non-chemical biocides in controlling biofilms

There are, of course, biocides other than chemical. Microorganisms in biofilms may be killed by physical means such as heat. Heat treatment has some advantages over chemical biocides for biofilm control. Some surfaces that have a very complex geometry, such as a pitted or corroded metal surface, may provide niches where microorganisms could survive and possibly not be reached by chemical sanitizers. Enough heat, however, can raise the temperature even in those protected areas to a sufficient level that the organisms in a biofilm are killed. Chimelewski and Frank (2004) published a model that could predict the reduction of LM biofilms on stainless steel by heat treatment. They found that the susceptibility of LM biofilms to inactivation by heat varied from strain to strain. For example heating a *L. monocytogenes* 3990 biofilm to 80°C for 11.7 minutes gave a 75% chance to kill all the viable organisms in the biofilm. *Listeria monocygenes* Scott A however, required 12.5 minutes at 80°C for a 75% chance to eliminate the biofilm completely. Effective cleaning is still important even when heat is used to eliminate biofilms. In the Chimelewski and Frank study, the presence of food soil could significantly increase the heat resistance of some strains of LM. Of course the heat treatment does not remove the EPS and if that is not cleaned away, the EPS left on the surfaces (even though all the cells in the biofilm are dead) may make the next biofilm form more quickly.

Another non-chemical biocide is radiation. Although there have been few studies that examined the effects of radiation on biofilms, the effects of ionizing radiation on *Salmonella* biofilms was studied by Niemira and Soloman (2005). Three serovars of *Salmonella enterica* were examined for sensitivity to ionizing radiation as planktonic cells and as biofilms. Two of the serovars were more sensitive to the radiation when they were in biofilms than when they were growing as planktonic cells. The third serovar was equally sensitive to the radiation when it was grown in biofilm or as a planktonic culture. Niemira (2007) also examined the effect of ionizing radiation on *E. coli* O157:H7 strains and, as was observed for *Salmonella*, the effect varied depending on strain. Some strains showed increased sensitivity to the ionizing radiation while others decreased. It seems clear that while organisms in biofilms may respond to ionizing radiation differently than planktonic cells, the nature of that difference, increased or decreased sensitivity, will depend on the specific strain of organism encountered. The unpredictability of how a biofilm will respond to ionizing radiation makes it a questionable tool for elimination of biofilms.

However, there are fewer questions about the ability of ultraviolet (UV) radiation to control biofilm. For the most part, UV radiation is not effective against biofilm. This is not a particular surprise as UV radiation is notorious for it poor penetrating power. A study by Elasri and Miller (1999) examined *Pseudomonas* entrapped in an alginate matrix to simulate a biofilm. That study found that the alginate transmitted 1/3 or less of the UV light. This suggests that UV is not able to penetrate a biofilm and would not be very effective, particularly against organism that are deep inside a biofilm.

It seems clear that although biocides cannot eliminate biofilms on their own and may have less impact on biofilms than other control methods, there is a place for them in a biofilm control program. Although the 'heavy lifting' may have been delivered by the cleaning process, the remaining microorganisms may be further reduced or eliminated by the use of biocides, chemical or non-chemical. Unfortunately, none of the systems used in the control of biofilms are perfect. No matter what combination of cleaning process and biocides are used, few are 100% effective at eliminating biofilms. And those few that may be able to kill all the organisms in a biofilm often have other drawbacks that make them impractical for use in a food processor. Given the pervasive nature of biofilms, the quality and safety threats they pose for food processors, and the lack of any 'silver bullets' to control them, it is not surprising that there is continual research going on that explores new ways to clean or new ways to kill biofilms. Some of those technologies hold promise as ways to help control biofilms. Although none of them seem be 'silver bullets,' many of them may provide additional tools that can be used to reduce the risks posed by biofilm.

## 12.5 Future trends

One of the newer cleaning technologies that may be useful for biofilm control is enzymes. Enzymes are biological catalysts. Most of the reactions that happen in any living organism occur only because enzymes are present. There are a variety of different enzymes. Some examples include proteases (that break protein into smaller peptides), lipases (that can break apart lipids), and amylases (that can break some carbohydrates into simpler sugars.) There are also enzymes that will produce biocidal chemicals. For example, Lactoperoxidase is an enzyme that will oxidize thiocyanate in the presence of hydrogen peroxide to produce HOSCN, a potent biocide. Johansen *et al.* (1997) examined the effect of various enzymes on biofilms. A combination of lactoperoxidase and glucose oxidase enzymes was able to produce a three $\log_{10}$ reduction of *Pseudomonas* in a biofilm grown on stainless steel. The same enzyme mixture was able to reduce staphylococci in a biofilm by about 1.5–2 $\log_{10}$.

In the same study, a mixture of polysaccharide hydrolyzing enzymes was able to remove biofilm from stainless steel and polypropylene, but did not have significant biocidal activity. Another study (Whittaker *et al.* 1984) examined the ability of various cleaners including enzymes to remove biofilms from reverse osmosis membranes. Cleaning with some enzymes resulted in 90–100% removal of biofilm. In many cases the performance of the enzyme was enhanced by the inclusion of surfactants and chelating agents. One indication that enzymes are effective for cleaning biofilms is that a quick perusal of the patent literature will reveal dozens if not hundreds of patents for control of biofilms through the use of enzymes.

Another possible technology for control of biofilms is bacteriophage (phage). Phage are viruses that infect bacteria. Although the existence of phage has been known for some time and they have a long history of use as biocides, they have not been extensively explored for the ability to control biofilms in food handling settings. However, there have been a few studies that suggest that phage may be tools that can be used to control biofilms. Curtin and Donlan (2006) reported the use of phage to control biofilms on catheters. When *Staphylococcus epidermidis* biofilms were grown on catheters the bacterial level was over seven $\log_{10}$ per $cm^2$. However, pretreatment of the catheters with bacteriophage that could infect *S. epidermidis* reduced the microbial level on the catheters by nearly 4.5 $\log_{10}$. In fact, phage may be particularly well suited to controlling biofilms. Briandet *et al.* (2008) used phage that had been fluorescently labeled to track diffusion into a biofilm. The EPS did not prevent phage from entering the biofilm and diffusing through it. Some phage may be able to degrade EPS. In 1998 Hughes *et al.* reported that bacteriophage of EPS producing microorganisms could induce production of enzymes that degrade EPS polymers. Hughes *et al.* in 1998 further demonstrated that a phage could possess a polysaccharide depolymerase that was specific for the EPS produced by the bacteria that the phage could infect. The phage was found to be able to disrupt a biofilm by a combination of degradation of the EPS and lysis of the cells in the biofilm.

Another way to control biofilms is to employ the old adage, if you can't beat 'em, join 'em. Something that should be clear by now is that although biofilms can be controlled, it is not always easy to do so. A simpler way to deal with them may be to simply admit that the biofilms will be present, and rather than trying to eliminate them altogether, control the biofilm so that it comprises desirable, or at least harmless, organisms. An example of this approach was reported by Zhao *et al.* (2006). In this study, strains of *Lactococcus lactis* subsp. *lactis* and *Entrococcus durans* that produced anti-listerial compounds were applied to floor drains in a poultry plant. The drains in the plant all had detectable *Listeria* in them before application of the competitive exclusion microorganisms. After the organisms were applied to five drains, two of them no longer had detectable *Listeria* and the other three all had reduced *Listeria* levels. The reductions on all five of the drains tested ranged from about two to four $\log_{10}$ CFU/$cm^2$. Although the *Listeria* levels in the drains were reduced, the total microbial load on the drains was not changed. It seems clear that the nature of the biofilm was changed and the harmful pathogen *Listeria* had been completely or partially displaced by a harmless microorganism. What is not yet clear is how well such modified biofilms will persist in the environment and how they will respond to recontamination with *Listeria* or other pathogens. Although this approach may not work for all parts of a food processing facility, it does seem to be an elegant way to address biofilms in those areas where the problem is not the biofilm itself, but rather the constituents of that biofilm.

## 12.6 Conclusions

Controlling biofilms in a food processing setting can seem like a daunting task, especially when you consider all the new technologies and approaches that are being explored. The traditional biocides like QACs and chlorine that are usually the mainstay of microorganism control don't work very well, and then factor in the importance of cleaning, something not always seen as an antimicrobial process, it is no wonder that people can get confused or frustrated when trying to control biofilms. It is really not that complex. All that is needed is to remember is a few basic rules:

1. It is important to begin with thorough cleaning before any disinfecting or sanitizing step.
   1.1 If you optimize cleaning to remove soil you will also have optimized it for biofilm removal. You can optimize cleaning by manipulating the TACT variables.
      1.1.1 Time: Increasing contact time of the chemical will result in more biofilm removal.
      1.1.2 Action: More energy input in the form of scrubbing or impingement of the solution will remove more biofilm.
      1.1.3 Chemical: Some chemicals will be more effective at removing biofilms than others.
      1.1.4 Temperature: Higher temperatures will remove or inactivate more biofilm.
   1.2 You need to optimize the cleaning process empirically. There is no one standard process that is best for every food processor.
2. Sanitizers and disinfectants alone cannot control biofilm, but they can provide an extra measure of security if they are applied after effective cleaning.

One final useful tip for controlling biofilm is to stay informed and connected to the food safety community and seek help and advice when it is needed. There is continual research going on in the area of biofilm control and although there are no 'silver bullets' yet, there is steady stepwise improvement in ways to eliminate biofilms. Some specific groups that may have additional insight into how to control biofilms include industries that supply food processors. Equipment suppliers may know best how to eliminate biofilms from the equipment they designed; chemical suppliers can teach their customers how to get the most performance out of cleaners and biocides. Another excellent source of information on biofilm control can be found in academia. Academics are continually doing basic research into biofilms and connections with universities, especially outreach and extension programs can be useful sources of information and assistance. Many government agencies can also be valuable sources of information. Departments of health, agriculture, or commerce all around the world have research programs underway in biofilm control. Professional societies and trade associations can provide invaluable connections to others that may have encountered biofilm problems before and, hopefully, learned ways to solve the problems that they will share. Wherever you may be, it is likely that there is someone that can provide advice or assistance to help solve a biofilm problem, it is worth the effort to seek them out.

Biofilms are an issue at some point or another for almost all food processors and are a problem that sometimes may seem insurmountable. But in reality all you have to remember is to clean well, sanitize or disinfect a bit, do your best to validate that surfaces are clean (especially the hard to reach ones) and look for advice when you need it. Following these few simple steps will help ensure that you have done all you can to eliminate biofilms and minimize their formation.

## 12.7 References

ANTONIOU, KATERINA and JOSEPH F. FRANK. 2005. Removal of *Pseudomonas putida* biofilm and associated extracellular polymeric substances from stainless steel by alkali cleaning. *J. Food. Prot.*, 68:277–281.

AYEBAH, BEATRICE, YEN-CON HUNG, CHYER KIM, and JOSEPH F. FRANK. 2006. Efficacy of electrolyzed water in the inactivation of planktonic and biofilm *Listeria monocytogenes* in the presence of organic matter. *J. Food Prot.*, 69:2143–2150.

BEST, M., M. E. KENNEDY, and F. COATES. 1990. Efficacy of a variety of disinfectants against *Listeria* spp. Appl. *Environ. Microbiol.*, 56: 377–380.

BLOCK, SEYMOUR S., 2001. Peroxygen Compounds. p. 185–204 In Seymour S. Block (ed.), *Disinfection Sterilization and Preservation*. Lippincott Williams & Wilkins, Philadelphia, PA.

BRIANDET, R., P. LACROIX-GUEU, M. RENAULT, S. LECART, T. MEYLHEUC, E. BIDNENKO, K. STEENKESTE, M.-N. BELLON-FONTAINE, and M.-P. FONTAINE-AUPART. 2008. Fluorescence correlation spectroscopy to study diffusion and reaction of bacteriophages inside biofilms. *Appl. Environ. Microbiol.*, 74:2135–2143.

CHMIELEWSKI, R. A. N. and JOSEPH F. FRANK. 2004. A predictive model for heat inactivation of *Listeria monocytogenes* biofilm on stainless steel. *J. Food Prot.*, 67:2712–2718.

CURTIN, JOHN, J. and RODNEY M. DONLAN. 2006. Using bacteriophages to reduce formation of catheter-associated biofilms by *Staphylococcus epidermidis*. *Antimicrob. Agents Chemother.*, 50:1268–1275.

DE BEER, DIRK, ROHINI SRINIVASAN, and PHILIP S. STEWART. 1994. Direct measurement of chlorine penetration into biofilms during disinfection. *Appl. Environ. Microbiol.*, 60:4339–4344.

DYCHDALA, GEORGE, R. 2001. Chlorine and Chlorine compounds. p. 135–157 In Seymour S. Block (ed.), *Disinfection Sterilization and Preservation*. Lippincott Williams & Wilkins, Philadelphia, PA.

ELASRI, MOHAMED, O. and ROBERT V. MILLER. 1999. Study of the response of a biofilm bacterial community to UV radiation. *Appl. Environ. Microbiol.*, 65:2025–2031.

FLEMMING, HANS-CURT, THOMAS R. NEU, and DANIEL J. WOZNIAK. 2007. GUEST COMMENTARY The EPS matrix: the 'house of biofilm cells.' *J. Bacteriol.*, 189:7945–7947.

FOOD AND DRUG ADMINISTRATION PUBLIC HEALTH SERVICE. 2005. *Food Code.* United States Department of Health and Human Services. National Technical Information Service, Springfield, VA.

FRANK, JOSEPH F., JULIE EHLERS, and LOUISE WICKER. 2003. Removal of *Listeria monocytogenes* and poultry soil-containing biofilms using chemical cleaning and sanitizing agents under static conditions. *Food Prot. Trends*, 23:654–663.

GOTTARDI, WALDEMAR. 2001. Iodine and Iodine compounds. p. 159–184 In Seymour S. Block (ed.), *Disinfection Sterilization and Preservation*. Lippincott Williams & Wilkins, Philadelphia, PA.

HUGHES, K.A., I. W. SUTHERLAND, J. CLARK, and M. V. JONES. 1998a. Bacteriophage and associated polysaccharide depolymerase – novel tools for study of bacterial biofilms. *J. Appl. Microbio.*, 85:583–590.

HUGHES, KEVIN A., IAN W. SUTHERLAND and MARTIN V. JONES. 1998b. Biofilm susceptibility to bacteriophage attack: the role of phage-borne polysaccharide depolymerase. *Microbiol.*, 144:3039–3047.

JOHANSEN, CHARLOTTE, PER FALHOT, and LONE GRAM. 1997. Enzymatic removal and disinfection of bacterial biofilms. *Appl. Environ. Microbiol.*, 63:3724–3728.

KIM, HOIKYUNG, JEE-HOON RYU, and LARRY R. BEUCHAT. 2007. Effectiveness of disinfectants in killing *Enterobacter sakazakii* in suspension, dried on the surface of stainless steel, and in a biofilm. *Appl. Environ. Microbiol.*, 73:1256–1265.

KNAPP, JOSEPH, E. and DIANE L. BATTISTI. 2001. Chlorine Dioxide. p. 215–227 In Seymour S. Block (ed.), *Disinfection Sterilization and Preservation*. Lippincott Williams & Wilkins, Philadelphia, PA.

LAPPI, VICTORIA R., JOANNE THIMOTHE, KENDRA KERR NIGHTINGALE, KENNETH GALL, VERGINIA N. SCOTT, and MARTIN WIEDMANN. 2004. Longitudinal Studies on *Listeria* in Smoked Fish Plants: Impact of Intervention Strategies on Contamination Patterns. *J. Food. Prot.*, 67:2500–2514.

LECHEVALLIER, MARK, W., CHERYL D. CAWTHON, and RAMON G. LEE. 1988. Inactivation of biofilm bacteria. *Appl. Environ. Microbiol.*, 54:2492–2499.

MANGALAPPALLI-ILLATHU, ANIL K. and DARREN R. KORBER. 2006. Adaptive resistance and differential protein expression of *Salmonella enterica* serovar enteritidis biofilms exposed to benzalkonium chloride. *Antimicrob. Agents Chemother.*, 50:3588–3596.

MCDONNELL, GERALD and A. DENVER RUSSELL. 1999. Antiseptics and disinfectants: activity, action, and resistance. *Clin. Microbiol. Rev.*, 12:147–179.

MERIANOS, JOHN J. 2001. Surface-active agents. p. 283–320 In Seymour S. Block (ed.), *Disinfection Sterilization and Preservation*. Lippincott Williams & Wilkins, Philadelphia, PA.

NIEMIRA BRENDAN A. 2007. Irradiation sensitivity of planktonic and biofilm-associated *Escherichia coli* O157:H7 isolates is influenced by culture conditions. *Appl. Environ. Microbiol.*, 73:3239–3244.

NIEMIRA, BRENDAN A. and ETHAN B. SOLOMON. 2005. Sensitivity of planktonic and biofilm-associated *Salmonella* spp. to ionizing radiation. *Appl. Environ. Microbiol.*, 71:2732–2736.

PAN, Y., F. BREIDT, JR., and S. KATHARIOU. 2006. Resistance of *Listeria monocytogenes* biofilms to sanitizing agents in a simulated food processing environment. *Appl. Environ. Microbiol.*, 72:7711–7717.

PARKAR, S. G., S. H. FLINT, and J. D. BROOKS. 2004. Evaluation of the effect of cleaning regimes on biofilms of thermophilic bacilli on stainless steel. *J. Appl. Microbiol.*, 96:110–116.

PENG, JUI-SEN, WEI-CHONG TSAI, CHENG-CHUN CHOU. 2002. Inactivation and removal of *Bacillus cereus* by sanitizer and detergent. *International J. Food Microbio.*, 77:11–18.

RICHTER, FRANCIS, L. and BRUCE R. CORDS. 2001. Formulation of sanitizers and disinfectants. p. 473–487 In Seymour S. Block (ed.), *Disinfection Sterilization and Preservation*. Lippincott Williams & Wilkins, Philadelphia, PA.

ROBBINS, JUSTIN B, CHRISTOPHER W. FISHER, ANDREW G. MOLTZ, and SCOTT E. MARTIN. 2005. Elimination of *Listeria monocytogenes* biofilms by ozone, chlorine, and hydrogen peroxide. *J. Food Prot.*, 68:494–498.

ROUILLARD, C. 1994. *The chemistry of cleaning*. Vista Vol. 6. Diversey Corporation.

RYU, JEE-HOON and LARRY R. BEUCHAT. 2005. Biofilm formation by *Escherichia coli* O157:H7 on stainless steel: effect of exopolysaccharide and curli production on its resistance to chlorine. *Appl. Environ. Microbiol.*, 71:247–254.

SANDERS, SHERIASE, Q., JOSEPH F. FRANK, and JUDY W. ARNOLD. 2008. Temperature and nutrient effects on *Campylobacter jejuni* attachment on multispecies biofilms on stainless steel. *J. Food. Prot.*, 71:271–278.

SOMERS, EILEEN B., and AMY C. LEE WONG. 2004. Efficacy of two cleaning and sanitizing combinations on *Listeria monocytogenes* biofilms formed at low temperature on a variety of materials in the presence of ready-to-eat meat residue. *J. Food Prot.*, 67:2218–2229.

STEWART, PHILIP S. 1996. Theoretical aspects of antibiotic diffusion into microbial biofilms. *Antimicrob. Agents Chemother.*, 40:2517–2522.

STEWART, PHILIP S., FRANK ROE, JOANNA RAYNER, JAMES G. ELKINS, ZBIGNIEW LEWANDOWSKI, URS A. OCHSNER, and DANIEL J. HASSET. 2000. Effect of catalase on hydrogen peroxide penetration into *Pseudomonas aeruginosa* biofilms. *Appl. Environ. Microbiol.*, 66:836–838.

TAORMINA, P. J., and L. R. BEUCHAT. 2002. Survival of *Listeria monocytogenes* in commercial food-processing equipment cleaning solutions and subsequent sensitivity to sanitizers and heat. *J. Appl. Microbiol.*, 92:71–80.

WEAVERS, LINDA K. and G. B. WIKRAMANAYAKE. 2001. Disinfection and sterilization using ozone. p. 205–214 In Seymour S. Block (ed.), *Disinfection Sterilization and Preservation*. Lippincott Williams & Wilkins, Philadelphia, PA.

WHITTAKER, C., H. RIDGWAY, and B. H. OLSON. 1984. Evaluation of cleaning strategies for removal of biofilms from reverse-osmosis membranes. *Appl. Environ. Microbiol.*, 48:395–403.

WONG, AMY C. LEE. 1998. Biofilms in food processing environments. *J. Dairy Sci.*, 81:2765–2770.

ZHAO, TONG, TERESA C. PODTBURG, PING ZHAO, BRUCE E. SCHMIDT, DAVID A. BAKER, BRUCE CORDS, and MICHAEL P. DOYLE. 2006. Control of *Listeria* spp. By competitive-exclusion bacteria in floor drains of a poultry processing plant. *Appl. Environ. Microbiol.*, 72:3314–3320.

# 13

# Novel methods for biofilm control and removal from food processing equipment

**S. Martin and H. Feng, University of Illinois, USA**

**Abstract**: Both spoilage and pathogenic bacteria that grow in biofilms are more resistant to both physical and chemical methods to which similar planktonic cells are sensitive. In order to improve the destruction of bacterial cells in biofilms a number of new methods have been examined. Power ultrasound refers to sonic waves in frequencies ranging from 20 to 100 kHz. Power ultrasound has been shown to be effective in the removal of *Pseudomonas aeruginosa*, *Escherichia coli* and *Listeria monocytogenes* biofilms. Gas plasma discharge, magnetic fields and pulsed electric field have also been shown to reduce the biofilms of *Pseudomonas aeruginosa* and *Staphylococcus aureus*.

**Key words**: power ultrasound, gas plasma discharge, magnetic fields, pulsed electric fields.

## 13.1 Introduction

The formation of biofilms on food products or the product-contact surfaces leads to serious hygienic problems and economic losses due to food spoilage (Kumar and Anand, 1998). Mechanical blockages, biofouling, impedance of heat transfer processes and biodeterioration of the components of metallic and polymeric systems due to presence of bacterial biofilms in fluid processing operations results in billions of dollars in losses each year (Mittelman, 1998). *Listeria monocytogenes* 10403S can adhere to 17 different food-use approved materials represented by metals, rubbers and polymers. Stainless steel was found to be most suitable surface for the growth of *L. monocytogenes* biofilms (Beresford *et al.*, 2001). A *L. monocytogenes* biofilm growing on stainless steel is shown in Fig. 13.1. Cells in biofilms have shown to be significantly more resistant to sanitizers and disinfectants than

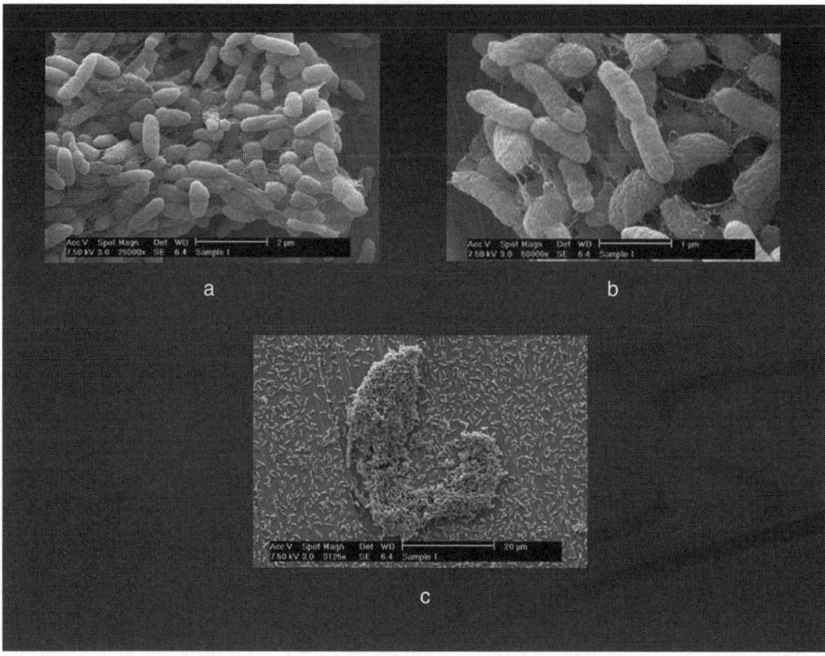


**Fig. 13.1** Scanning electron micrographs of *Listeria monocytogenes* strain Scott A biofilms on stainless steel chips. Left, top image (a) depicts dense cellular aggregation and possible microcolony formation (25000 magnification). Right, top image (b) demonstrates extracellular material and void spaces (50000 magnification). Lower image (c) demonstrates dense cellular aggregation (3125 magnification; Moltz and Martin, 2005).

planktonic cells (Ronner and Wong, 1993). *Listeria monocytogenes* has the ability to attach to various surfaces in food processing facilities. Mafu *et al.* (1990) found that *L. monocytogenes* attached to stainless steel, glass, polypropylene, and rubber which are common materials in food contact surfaces, after short contact times (20–60 min) at both ambient and refrigeration temperature. Formation of extracellular material around the attached cells was revealed by scanning electron microscope after 60 min of incubation at both temperatures. However, Moltz and Martin (2005) reported that temperature also has an effect on biofilm growth and development of *L. monocytogenes* and attachment of cells on stainless steel surfaces was less at 4° and 20°C than at 37°C. *Listeria monocytogenes* has been reported to adhere to stainless steel and polyester or polyester-polyurethane conveyor belts at levels of ~$2 \times 10^4$ cfu/cm$^2$ after incubation of these materials in a culture medium at 35°C for 24 hr (Krysinski *et al.*, 1992). Beresford *et al.* (2001) reported that *L. monocytogenes* can adhere to 17 different food-use approved materials representing metals, rubbers and polymers and found that stainless steel was the most suitable surface for the growth of *L. monocytogenes* biofilms. Blackman and Frank (1996) studied adherence of *L. monocytogenes* to surfaces of stainless steel, Teflon, nylon and polyester floor sealant after seven days of incubation in tryptic soy broth and found that attachment was markedly less at 10°C than at 21°C. It has also been reported that *L. monocytogenes* growth in minimal medium exhibited 50-fold higher attachment than cells grown in tryptic soy broth (Kim and Frank, 1994).

The other common areas involved in biofilm development are the floors, freezers, processing rooms, cases and mats utilized in the dairy industry (Moltz and Martin, 2005). Waste water pipes, bends in pipes, rubber seals, conveyor belts, Buna-n and Teflon seals have also been reported as important sites for biofilm formation (Blackman and Frank, 1996). Spurlock and Zottola (1991) investigated the attachment of *L. monocytogenes* biofilms to cast iron drains at ambient temperature. They found that *L. monocytogenes* grew and remained on these drain surfaces at $10^6$–$10^8$ cfu/cm$^2$, irrespective of drastic changes in pH from alkaline (pH 9.0) to acidic (pH 4.5).

Herald and Zottola (1988) observed the attachment of *L. monocytogenes* to stainless steel and the production of attachment fibrils. It has also been reported that hydrophobic interactions such as electrostatic and exoploymer interactions, were responsible for the attachment of *L. monocytogenes* to various surfaces (Mafu *et al.*, 1991; Blackman and Frank, 1996). In reverse osmosis process for waste water treatment, the development of microbial biofilm contributes to a significant reduction in water flux and the deterioration of overall membrane performance (Ridgeway *et al.*, 1984). In heat exchangers biofilms cause increased resistance both in liquid flow and heat transfer. In the dairy industry, improperly cleaned and sanitized equipment are usually considered to be the major sources of contamination of milk and milk products (Kumar and Anand, 1998). The persistence of accumulated microorganisms in the form of a biofilm may lead to post processing contamination, leading to lowered shelf life of the product (Zottola, 1994). Additionally, if the microorganism is pathogenic in nature such as *L. monocytogenes*, then consumption of contaminated product may pose a health risk.

The type of surface, nutrient level and type of microorganisms influence biofilm development and production of extracellular material (Ronner and Wong, 1993). Environmental flora in food processing facilities may interfere with or enhance attachment of *L. monocytogenes* to surfaces or its growth in biofilms (Lou and Yousef, 1999). It has been reported that in a flowing system, *Pseudomonas fragi*, a bacterium which strongly attaches and produces exopolysaccharride materials on surfaces enhanced attachment of *L. monocytogenes*, but *P. fragi* itself failed to attach to these surfaces in this flowing system (Sasahara and Zottola, 1993).

The presence of food components protects *L. monocytogenes* biofilms on surfaces. This may be attributed to slower loss of moisture from food layers (Ryser and Marth, 1999). When compared with phosphate buffered saline, milk residues enhanced survival of *L. monocytogenes* on stainless steel and Buna-n rubber and sometimes promoted growth (Helke *et al.*, 1993). However, albumin was found to be least favorable for the adhesion of *L. monocytogenes* to silica surfaces (Al-Makhlafi *et al.*, 1995). Cottage cheese whey as a residue did not increase survival of *L. monocytogenes* (Helke *et al.*, 1993). Temperature and moisture also affect survival of *L. monocytogenes* on surfaces. Palumbo and Williams (1990) observed enhanced survival of *L. monocytogenes* biofilms on surfaces at 5°C than at 25°C and at lower relative humidity than higher relative humidity. However, Helke and Wong (1994) reported that survival of *L. monocytogenes* was greater under high relative humidity conditions (75.5%) than under low relative humidity conditions (32.5%).

## 13.2 Power ultrasound

Power ultrasound refers to sonic waves in frequencies ranging from 20 to 100 kHz (1 Hertz = 1 cycle per second) with a sound intensity of 10 to 1000 W/cm$^2$ (Feng and Yang, 2008). The use of power ultrasound in biofilm removal is based on the surface cleaning effect of acoustic waves in a liquid, which has been known for years. Actually, ultrasonic cleaning is one of the most successful and most widely used power ultrasound applications (Povey and Mason 1997). Ultrasonic cleaners have been employed for cleaning of jewelry, lenses, coins, watches, dental and surgical instruments, fountain pens, gun parts, industrial parts, and electronic equipment.

It has been reported that biofilms form on food and food processing equipment surfaces when favorable moisture and nutritional conditions are presented. Utilizing ultrasound to remove biofilms from food contact surfaces or food processing equipment surfaces is a relatively new endeavor and not many reports are available in the literature. There are even fewer studies into the application of ultrasound for removal of biofilm from the surface of a food product. Nevertheless, surface treatment of food products with power ultrasound has been tested for poultry disinfection (Sams and Feria 1991, Lillard 1994), fresh produce decontamination (Seymour *et al.*, 2002, Ajlouni *et al.*, 2006), and vegetable seeds disinfection (Scouten and Beuchat 2002, Kim *et al.*, 2006), with some of the products having biofilms formed on the surfaces if incubation time was long enough.

## 13.3 The removal mechanism

The removal of biofilm by sonic or ultrasonic waves in a liquid medium is a process controlled by the interactions between sound waves and the biofilm via a series of physical or chemical actions taking place during sonication. When a sound wave travels through a liquid, the water molecules oscillate in longitudinal directions. For power ultrasound, the displacement of the water molecules is normally less than one micrometer, which exerts little net shear force on bacteria cells. When gas bubbles are presented in the liquid, a more disruptive effect can be observed (Pitt, 2005). In an acoustic field, the gas bubbles entrained in a flowing fluid collide with a bacteria-covered surface, resulting separation of the bacteria cells from the surface. However, it is doubtful that such gas bubbles would have a lethal effect on the bacterial cells.

Cavitation, which refers to the formation, growth, and implosive collapse of bubbles, is another phenomenon observed when ultrasound travels through a liquid. At low acoustic intensities and when the bubble size is not near the resonant size, the oscillations are at the same frequency as the sound waves. This oscillation causes micro-streaming alongside a bubble; generates shear forces sufficient to lyse red blood cells (Nyborg, 2001); and increases the permeability of cell membrane (Ananta *et al.*, 2005). The micro-streaming can also reduce the mass transfer boundary layer thickness and hence increase the transport of a sanitizer or antibiotic from exterior space to the surface of bacteria embedded in a biofilm (Qian *et al.*, 1999). At higher acoustic energy levels and when the bubble sizes approach a resonance size; the bubble's oscillation becomes non-linear, ending up with a fast implosion accompanied by very high and localized temperatures and pressures (Pitt, 2005). The implosive collapse of cavitating bubbles also generates strong shock waves, high shear forces, and free radicals which all have bactericidal functions. In addition, at the solid and liquid interface, an in-flow of liquid with a maximum speed of up to 156 m/s may occur when a bubble implodes. This phenomenon is often termed as water jet or liquid jet. The liquid jets exert an impinging force onto the solid surface and help to dislodge attached microbes and biofilm.

Maisonhaute *et al.* (2002a&b) used electrochemical approaches to explain the mechanism of cleaning by acoustic cavitation. They reported that acoustic bubbles were oscillating at a distance of only a few tens of nanometers above the surface to be cleaned. The flow resulting from the bubble collapse lead to important drag and shear forces on the surface, causing surface cleaning or eroding. The cavitation effect of ultrasound also generates a global removal stress on a particle or microorganism adhered to a surface stronger than the adhesion force of the particle resulting from Van der Waals attraction (Maisonhaute *et al.*, 2002b).

Obviously, means to maximizing the physical forces and increasing the transport of a sanitizer to bacteria will enhance the efficacy of an ultrasonic cleaning operation. This is often done by examining the effect of operational conditions, such as ultrasound power density, frequency, treatment time, temperature, etc., on the removal of biofilm, and finding optimal operation conditions to enhance the cleaning efficacy.

## 13.4 The effect of power intensity

Ultrasound power intensity is a measure of how much energy is introduced into a liquid during a sonication treatment. A higher energy input normally corresponds to a higher cavitation activity, and hence the biofilm is removed faster than at low energy levels. Both volumetric energy density ($W/cm^3$) and surface intensity ($W/cm^2$) have been used to quantify acoustic energy levels. It was reported that an average ultrasound intensity of 2 $W/cm^2$ at 70 kHz is needed to cause reduction in viability of *Pseudomonas aeruginosa* biofilm (Pitt, 2005). The tests conducted at relatively low power intensities (10–100 $mW/cm^2$) did not show a noticeable reduction of biofilm. At low power intensities, ultrasound was reported even to increase the growth rate of cells in some biofilms (Pitt and Ross, 2003). Pulsed ultrasound treatment is more effective in removing biofilms compared to a continuous ultrasound treatment at the same power intensity. In a 2 hr treatment, Qian *et al.*, (1999) applied a continuous wave with temporal average sound intensity of 10 $mW/cm^2$ and a pulsed wave with pulsed intensity of 10 $mW/cm^2$ and temporal average intensity of 1 $mW/cm^2$ to remove *P. aeruginosa* biofilm from polyethylene plugs. The $\log_{10}$ reduction in viable account of *P. aeruginosa* between the two treatments was not significantly different. This observation was confirmed by another study using ultrasound to remove *Escherichia coli* ATCC 10798 biofilm in vivo (Rediske *et al.*, 2000). When 1:3 and 1:6 duty cycles with a temporal average intensity of 100 $mW/cm^2$ was applied, the log reduction after 24 hrs sonication was comparable to that of a continuous sonication at 300 $mW/cm^2$. Since the sound intensity for the 'on' period of a 1:3 duty cycle was 300 $mW/cm^2$, it may indicate that the bacteria respond to the maximum or peak ultrasound intensity and not to the average or the total amount of energy delivered. An advantage of using pulsed ultrasound is energy savings over time. However, whether the electrical parts in an ultrasound generator and transducer ceramic disc will be durable to such frequent on-off operation is a question that remains to be answered. Noticeably, some biofilm studies did not report ultrasound power intensity, which makes it difficult to compare their results with those from other research groups.

## 13.5 The effect of frequency

When a cleaning operation was conducted at sonic frequency (80 to 120 Hz), no obvious correlation between biofilm removal and frequency was reported (Pitt, 2005). At ultrasonic frequencies, it has been demonstrated that lower frequency insonation is significantly more effective than higher frequency one in reducing bacteria viability within a biofilm (Qian *et al.*, 1997). The energy needed to induce cavitation in a liquid (cavitation energy threshold) increases when frequency increases. At high frequencies, a greater sound intensity and hence greater sound pressure amplitude has to be applied to overcome the cohesive forces of liquid and allow gas or vapor to fill the void. For this reason, 20–50 kHz frequencies are generally chosen for cleaning applications. In an ultrasound treatment, the maximized energy coupling takes place when the size of a cavity/bubble reaches its resonance size. From equation (13.1), at a lower resonance frequency $f_r$, the size of the cavitating bubble is greater.

$$R_r = \frac{1}{2\pi f_r}\left[\frac{3\gamma P_h}{\rho}\right]^{0.5} \qquad 13.1$$

where $R_r$ is the radius of the bubble at resonance frequency $f_r$, $\rho$ is the density of the liquid, $P_h$ is the hydrostatic pressure, and $\gamma$ is the ratio of the specific heats. As a result, the cavitation implosion at a lower frequency will be more powerful, which may help to more effectively remove biofilms. An increased removal of biofilms at lower frequencies has been reported by Mott *et al.* (1998) in reduction of mineralized *Proteus mirabilis* biofilms from water-filled glass tubes.

## 13.6 The synergy with other stressors

In a treatment to remove biofilm at low sound intensity (10–100 mW/cm$^2$) with ultrasound alone, the removal of biofilms is often minimal (Rediske *et al.*, 1999; Peterson and Pitt, 2000; Qian *et al.*, 1997). To enhance the efficacy, ultrasound is used in combination with other lethal factor(s), such as sanitizer or antibiotic for biofilm removal. With the aid of sonication there was a significant increase of antibiotic efficacy both *in vitro* and *in vivo* compared to a sonication alone treatment. The *in vitro* studies using ultrasound alone at 70 and 500 kHz for 2 hrs resulted in no reduction in the viability of *E. coli* biofilms at all three power intensities (2, 20, and 200 mW/cm$^2$). When ultrasound was applied with gentamicin, however, an over 2 log reduction was recorded (Peterson and Pitt, 2000). Qian *et al.*, (1997), Ensing *et al.*, (2005), and Carmen *et al.* (2005) also reported enhanced removal biofilms by sonication with the addition of antibiotics. Anguah *et al.* (2007) found that biofilm formation in florescent *E. coli* was decreased following 10 min exposure to ultrasound. In the test conducted by Berrang *et al.* (2008) to remove *L. monocytogenes* biofilm on drain pipes, addition of ultrasonication improved the performance of chlorine- and quaternary ammonium-based sanitizers, causing an additional reduction of 2.9 to 3.1 CFU/cm$^2$. It is postulated that during an ultrasound and antibiotic combined treatment, the antibiotic is transported through the outer membrane and cytoplasm membrane and then bind to biological target in the cell (Qian *et al.*, 1999). The enhanced transport may be caused by shear forces

($\tau$) on the surface of bacteria, which is given by equation (13.2), assuming Stock's law applies:

$$\tau_{\max} = \frac{3\mu}{2R} \frac{V_{of}}{(1 + A^2/\omega^2)^{0.5}} \qquad 13.2$$

where $V_{of}$ is the amplitude constant, $A = (6\pi\mu R)/m$, $\omega = 2\pi f$, and $m$ is viscosity.

In a recent study, Baumann *et al.* (2008) examined the efficacy of power ultrasound and ozonation used individually, and in tandem, for the removal of *L. monocytogenes* biofilms from stainless steel chips. In the experiments, stainless steel chips were inoculated with *L. monocytogenes*. Power ultrasound (20 kHz, 100% amplitude, 120 W) was applied for 30 or 60 sec 2.54 cm from a biofilm chip while it was submerged in 250 ml of sterile potassium phosphate buffer (PPB, pH 7.0). Ozone was cycled through the 250 ml of PPB containing the biofilm chip also for 30 or 60 sec at concentrations of 0.25, 0.5, or 1.0 ppm. Power ultrasound and ozonation were also used in tandem for testing their combined effect. Both treatments alone resulted in a significant reduction in recoverable cells, with power ultrasound being the most effective (3.8 $\log_{10}$ CFU/ml reduction after 60 sec). For the ozone in combination with power ultrasound treatment, reductions were significantly ($P < 0.05$) higher than either treatment alone. There were no recoverable cells after 60 sec of this combined treatment when an ozone concentration of 0.5 ppm was used (7.31 $\log_{10}$ CFU/ml reduction).

In an ultrasonic cleaning test, Allen *et al.* (2008) reported relatively little effect of ultrasound on biofilm on poultry transport crates when using a 4 kW ultrasonic generator alone. However, a treatment with same ultrasound source at 60°C reduced the counts of Enterobacteriaceae to below the detection limit within 1 to 3 min. The combination of ultrasound with enzymes has also been reported, especially for studies looking into improved practice to completely remove biofilm from a sample surfaces for plate counting purposes. The special requirement for this type of removal methods is to retain the viability of the cells in a biofilm. Interested readers can refer to the reports of Oulahal *et al.* (2003, 2004, and 2007) for more information.

## 13.7 Other methods

Owing to the problems associated with the presence of biofilms in the food industry, their removal and prevention are high on the list of sanitation goals. A number of methods have been examined over the years for effective removal of biofilm from various surfaces. Gibson *et al.* (1999) examined the effects of the cleaning programs in place in food production facilities. By a total plate count method, they found that the use of a high-pressure spray in conjunction with a mechanical floor scrubber had the best results. For films of *P. aeruginosa* and *Staphylococcus aureus* there was no increased removal when the spray pressure was increased from 250 psi to 1000 psi. Also, an increased time of spray from one second to ten seconds yielded no additional removal. They noted as well that all treatments generally reduced the number of cells on the surface by about 1000 CFU/ml. Electric current at low density (<100 $\mu A/cm^2$) was reported to be effective in reducing the biocides and antibiotics concentration needed to kill biofilm to those needed to kill planktonic bacteria (Costerton *et al.*, 1994). Hot water is another effective method that has been widely used by the food industry for processing equipment cleaning in an operation termed clean-in-place (CIP). In a CIP operation, a sanitizer or disinfectant is often added to enhance the cleaning efficiency (Bremer *et al.*, 2006). For example, when combined with sodium hypochlorite, povidone-iodine, or 80% v/v ethanol, a 15-sec treatment at a water temperature of >65°C eradicated *Burkholderia cepacia* biofilms (Miyano *et al.*, 2003). The effect of gamma irradiation on *P. aeruginosa*, *Listeria innocua*, and *E. coli* biofilm formed on polypropylene (PP), polyethylene (PET), and polycarbonate (PC) was examined by Byun *et al.* (2007). 3 kGy of gamma irradiation eliminated the microorganisms attached to PP, PET, and PC with minor exceptions (*P. aeruginosa* attached to PE and *E. coli* attached to PC). Of other physical methods that are currently being examined for control and removal are: super-high magnetic fields, high-pulsed electric fields with and without organic acids, and low electric fields with and without biocides (Kumar and Anand, 1998).

Chemical treatment is another effective approach for removal of biofilms. The goal of most chemical treatments is the disruption of the EPS in order for easier removal by mechanical means. Chemical use without a follow-up mechanical removal may result in a better surface for new organisms to attach and develop biofilms. Detergents have been used to bind calcium and magnesium salts as well as to destabilize the outer cellular membranes. Oxidants for EPS depolymerization have also been examined and include the use of peracetic acid, chlorine, iodine, and hydrogen peroxide (Kumar and Anand, 1998). Biofilms are much more resistant (100–1000 times) to antimicrobials and antibiotics than their planktonic counterparts (Stewart and Costerton, 2001). *Listeria monocytogenes* biofilms take about 100 times longer for the same reduction by sodium hypochlorite and heat as planktonic cells (Mittleman, 1998). They also demonstrate a greater resistance to conventional sanitizers such as quaternary ammonium compounds and acid anionic sanitizers while in biofilm form (Frank and Koffi, 1990). Several theories as to why biofilms are more resistant to these treatments have been proposed. The possible mechanisms include: a reduced diffusion throughout the film due to EPS; physiological changes of the organisms; a reduced metabolic state; and protective enzyme production (Donlan, 2002; Kumar and Anand, 1998).

Biological means of prevention have also been examined. Bacteriocins have been used to treat contact surfaces in order to deter attachment and growth of organisms. The FDA approved Nisin as GRAS in 1988 for the control of *Clostridium botulinum* spores. Nisin has also been examined in order to control *L. monocytogenes* in meat packages (Ming *et al.*, 1997).

The use of gas discharge plasmas to kill biofilm cells has been examined since the plasmas contain a mixture of charged particles, chemically reactive species, and UV radiation. *Chromobacterium violaceum*, a Gram-negative soil bacterium, biofilm was exposed to a gas discharge plasma for 10 min and the treatment was able to kill almost 100% of the cells (Abramzon *et al.*, 2006).

A magnetic field is defined as a region of space in which a magnetic body is capable of magnetizing the surroundings. The magnetic filed is measured in weber per square meter ($Wb/m^2$), also called Tesla (T). One Tesla equals 10,000 Gauss. Magnetic fields can be classified according to their field strength as low or high intensity. Low intensity magnetic fields refer to fields with intensities on the order of 10 s of Gauss, whereas the intensities of high fields are over thousands of Gauss. The inactivation of microorganisms with pulsed electric field has been conducted with limited effects reported. For a field strength of 18 T, the inactivation of *E. coli*, *L. innocua*, and *S. cerevisae* after 50 pulses was less than 1 log cycle at room temperature. Benson *et al.* (1994) reported that static magnetic fields may enhance the activity of gentamicin against a *Pseudomonas aeruginosa* biofilm.

## 13.8 Pulsed electric fields

Pulsed electric field (PEF) processing involves the application of high voltage pulses for a few microseconds to food placed or flowing between two electrodes. The concept of PEF is relatively simple. Electric energy at a low power level is collected over an extended period and stored in a capacitor. The collected energy is then discharged almost instantaneously at very high levels of power. The generation of pulsed electric fields requires two major devices: a pulsed power supply and a treatment chamber that converts the pulsed voltage into pulsed electric fields. If the electric field intensity between the electrodes reaches the range of 12–35 kV/cm, a pronounced lethal effect on microorganisms can be observed. The most common application of PEF has been focused on preservation of flowable foods, including fruit and vegetable juices, milk, and liquid eggs. The availability of equipment has increased substantially in recent years. There are a number of suppliers that build and commercialize bench-top, pilot-scale, and industrial-scale PEF machines. Perez-Roa *et al.* (2006) found that low-voltage (0.5–5 V) PEF reduced the growth of *P. aeruginosa* biofilm development. Giladi *et al.* (2008) reported that PEF could inhibit the growth of planktonic *S. aureus* and *P. aeruginosa* and that the effect was amplitude and frequency dependent, with a maximum ay 10 MHz.

## 13.9 Low electric fields (ohmic heating)

Ohmic heating refers to heating materials by passing a current through them. The current can be at very low frequencies or can be direct current. The basic principle of ohmic heating is the dissipation of electric energy into heat which is proportional to the square of the electric field strength. In principle, the ohmic heating effect is similar to that obtained with microwaves. However, the major advantage of ohmic compared to microwave heating is that the depth of penetration is not limited. The process allows food products to be heated to sterilization temperature in less than 90 sec, and hence improves product quality and gives a better retention of bioactive components. There have been some commercial applications of this technology, starting from the 1980s, in the field of liquid food pasteurization.

New and improved methods to eliminate biofilms from foods and food contact equipment are continually being developed. New and improved equipment is also being discovered. With increased desire for minimally processed foods, and the demand for greater food safety, new technologies will be required in the future. The combination of existing technologies may provide the solutions to these challenges in the future.

## 13.10 References

ABRAMZON, N., J.C. JOAQUIN, J. BRAY and G. BRELLES-MARINO, (2006), Biofilm destruction by RF high-pressure cold plasma jet, *IEEE Trans. Plasma Sci.* 34:1304–1308.

AJLOUNI, S., H. SIBRANI, R. PREMIER and B. TOMKINS, (2006), Ultrasonication and fresh produce (Cos lettuce) preservation. *J. Food Sci.* 71:62–68.

AL-MAKHLAFI, H., A. NASIR, J. MCGUIRE and M.A. DAESCHEL, (1995), Adhesion of *Listeria monocytogenes* to silica surfaces after sequential and competitive adsorption of bovine serum albumin and β-lactoglobulin, *Appl. Environ. Microbiol.* 61: 2013–2015.

ALLEN, V.M., R.T. WHYTE, C.H. BURTON, J.A. HARRIS, R.D.L. LOVELL, R.J. ATTERBURY and D.B. TINKER, (2008), Effect of ultrasonic treatment during cleaning on the microbiological condition of poultry transport crates, *Brit. Poul. Sci.* 49:423–428.

ANANTA, E., D. VOIGT, M. ZENKER and D. KNORR, (2005), Cellular injuries upon exposure of *Escherichia coli* and *Lactobacillus rhamnosus* to high-intensity ultrasound, *J. Appl. Microbiol.* 99:271–278.

ANGUAH, K., R. VELEZ, A. C. VOLLMER and E. C. EVERBACH, (2007), Effect of 810 kHz ultrasound on bacterial biofilms, *J. Acous. Sco. Am.* 122:3052–3056.

BAUMANN, A.R., S.E. MARTIN and H. FENG, (2008), Removal of biofilms by the use of power ultrasound and ozone, *J. Food Prot.* 72(6): 1306–1309.

BENSON, D.E., C.B. GRISSOM, G.L. BURNS and S.F. MOHAMMAD, (1994), Magnetic field enchancement of antibiotic activity in biofilm forming *Pseudomonas aeruginosa. ASAIO J.* 40(3):M371–376.

BERESFORD, M.R., P.W. ANDREW and G. SHAMA, (2001), *Listeria monocytogenes* adheres to many materials found in food-processing environments, *J. Appl. Microbiol.* 90:1000–1005.

BERRANG, M.E., J.F. FRANK and R.J. MEINERSMANN, (2008), Effect of chemical sanitizers with and without ultrasonication on *Listeria monocytogenes* as a biofilm within polyvinyl chloride drain pipes, *J. Food Prot.* 71:66–69.

BLACKMAN, I.C. and J.F. FRANK, (1996), Growth of *Listeria monocytogenes* as a biofilm on various food-processing surfaces, *J. Food Prot.* 59:827–831.

BREMER, P.J., S. FILLERY and A.J. MCQUILLAN, (2006), Laboratory scale clean-in-place (CIP) studies on the effectiveness of different caustic and acid wash steps on the removal of dairy biofilms, *Int. J. Food Microbiol.* 106:254–262.

BYUN, M.W., H.J. KIMB, D.H. KIM, H.J. KIM and C. JO, (2007), Effects of irradiation and sodium hypochlorite on the micro-organisms attached to a commercial food container, *Food Microbiol.* 24:544–548.

CARMEN, J.C., B.L. ROEDER, J.L. NELSON, R.L.R. OGILVIE, R.A. ROBISON, G.B. SCHAALIE and W.G. PITT, (2005), Treatment of biofilm infections on implants with low-frequency ultrasound and antibiotics, *Am. J. Infect. Control.* 33:78–82.

COSTERTON, J.W., B. ELLIS, K. LAM, F. JOHNSON and A.E. KHOURY, (1994), Mechanism of electrical enhancement of efficacy of antibiotics in killing biofilm bacteria, *Antimicro. Agents and Chemo.* 38:2803–2809.

DONLAN, R.M., (2002), Biofilms: Microbial life on surfaces, *Emerg. Infect. Dis.* 8(9):881–890.

ENSING, G.T., B.L. ROEDER, J.L. NELSON, J.R., VAN HORN, H.C. VAN DER MEI, H.J. BUSSCHER and W.G. PITT, (2005), Effect of pulsed ultrasound in combination with gentamicin on bacterial viability in biofilms on bone cements *in vivo*, *J. Appl. Microbiol.* 99:443–448.

FENG, H. and W. YANG, (2008), *Ultrasonic Process*, In Nonthermal Processing Technologies for Food. (Accepted for publication in 2009.)

FRANK, J.F. and R.A. KOFFI, (1990), Surface adherent growth of *Listeria monocytogenes* is associated with increased resistance to surfactant sanitizers and heat, *J. Food Prot.* 53:550–554.

GIBSON, J.J., H. TAYLOR, K.E. HALL and J.T. HOLAH, (1999), Effectiveness of cleaning techniques used in the food industry in terms of the removal of bacterial biofilms, *J. Appl. Microbiol.* 87:41–48.

GILADI, M., Y. PORAT, A. BLATT, Y. WASSERMAN, E.D. KIRSON, E. DEKEL and Y. PALTI, (2008), Microbial growth inhibition by alternating electric fields, *Anitmicrob. Ag. and Chemo.* 52:3517–3522.

HELKE, D.M. and A.C.L. WONG, (1994), Survival and growth characteristics of *Listeria monocytogenes* and *Salmonella typhimurium* on stainless steel and Buna-n rubber, *J. Food Prot.* 57:963–968.

HELKE, D.M., E.B. SOMERS and A.C.L. WONG, (1993), Attachment of *Listeria monocytogenes* and *Salmonella typhimurium* on stainless steel and Buna-n rubber in the presence of milk and milk components, *J. Food Prot.* 56:479–484.

HERALD, P.J. and E.A. ZOTTOLA, (1988), Attachment of *Listeria monocytogenes* to stainless steel surfaces at various temperatures and pH values, *J. Food Prot.* 53:1549–1552, 1562.

KIM, H.J., H. FENG, M.M. KUSHAD and X. FAN, (2006), Effects of ultrasound, irradiation, and acidic electrolyzed water on germination of alfalfa and broccoli seeds and *Escherichia coli* O157:H7, *J. Food Sci.* 71:168–173.

KIM, K.Y. and J.F. FRANK, (1994), Effect of growth nutrients on attachment of *Listeria monocytogenes* to stainless steel, *J. Food Prot.* 57:720–726.

KRYSINSKI, E.P., L.J. BROWN and T.J. MARCHISELLO, (1992), Effect of cleaners and sanitizers on *Listeria monocytogenes* attached to product contact surfaces, *J. Food Prot.* 55:246–251.

KUMAR, C.G. and S.K. ANAND, (1998), Significance of microbial biofilms in food industry: A review, *Int. J. Food Microbiol.* 42:9–27.

LILLARD, H.S., (1994), Decontamination of poultry skin by sonication, *Food Tech.* December: 72–73.

LOU, Y. and A.E. YOUSEF, (1999), Characteristics of *Listeria monocytogenes* important to food processors, In E.T. Ryser and E.H. Marth (ed.), *Listeria, Listeriosis and Food Safety*, 2nd ed Marcel Dekker, Inc., New York, NY. p. 131–224.

MAFU, A.A., D. ROY, J. GOULET and P. MAGNY, (1990), Attachment of *Listeria monocytogenes* to stainless steel, glass, polypropylene, and rubber surfaces after short contact times, *J. Food Prot.* 53:742–746.

MAFU, A.A., D. ROY, J. GOULET and L. SAVOIE, (1991), Characterization of physiochemical forces involved in adhesion of *Listeria monocytogenes* to surfaces, *Appl. Environ. Microbiol.* 57:1969–1973.

MAISONHAUTE, E., P.C. WHITE and R.G. COMPTON, (2002a), Surface acoustic cavitation understood via nanosecond electrochemistry. 2. The motion of acoustic bubbles, *J. Phy. Chem. B.* 106(12):3166–3172.

MAISONHAUTE, E., C. PRADO, P.C. WHITE and R.G. COMPTON, (2002b), Surface acoustic cavitation understood via nanosecond electrochemistry. Part III. Shear stress in ultrasonic cleaning, *Ultrasonics Sonochem.* 9:297–303.

MING, X.T., G.H. WEBER, J.W. AYRES and W.E. SANDINE, (1997), Bacteriocins applied to food packaging materials to inhibit *Listeria monocytogenes* on meats, *J. Food Sci.* 62:413–415.

MITTLEMAN, M.W., (1998), Structure and functional characteristics of bacterial biofilms in fluid processing operations, *J Dairy Sci.* 81:2760–2764.

MIYANO, N., S. OIE and A. KAMIYA, (2003), Efficacy of disinfectants and how water against biofilm cells of *Burkholderia cepacia*, *Biol. Pharm. Bull.* 26:671–674.

MOLTZ, A.G. and S.E. MARTIN, (2005), Formation of biofilms by *Listeria monocytogenes* under various growth conditions, *J. Food Prot.* 68:92–97.

MOTT, I.E.C., D.J. STICKLER, W.T. COAKLEY and T.R. BOTT, (1998), The removal of bacterial biofilm from water-filled tubes using axially propagated ultrasound, *J. Appl. Microbiol.* 84:509–514.

NYBORG, W.L., (2001), Biological effects of ultrasound: Development of safety guidelines. Part II: General review, *Ultrasonic Med. Biol.* 27: 301–333.

OULAHAL, N., A. MARTIAL-GROS, M. BONNEAU and L.J. BLUM, (2003), *Escherichia coli* milk biofilm removal from stainless steel surfaces: synergism between ultrasound waves and enzymes, *Biofouling*, 19:159–168.

OULAHAL, N., A. MARTIAL-GROS, M. BONNEAU and L.J. BLUM L.J., (2004), Combined effect of chelating agents and ultrasound on biofilm removal from stainless steel surfaces. Application to *Escherichia coli* milk and *Staphylococcus aureus* milk biofilms, *Biofilms*, 1:65–73.

OULAHAL, N., A. MARTIAL-GROS, M. BONNEAU and L.J. BLUM L.J., (2007), Removal of meat biofilms from surfaces by ultrasounds combined with enzymes and/or a chelating agent, *Int. J. Food Sci. Emerging Technol.* 8:192–196.

PALUMBO, S. and A.C. WILLIAMS, (1990), Effect of temperature, relative humidity, and suspending menstrua on the resistance of *Listeria monocytogenes* to drying, *J. Food Prot.* 53:377–381.

PEREZ-ROA, R.E., D.T. TOMPKINS, M. PAULOSE, C. GRIMES, M. ANDERSON and D.R. NOGUERA, (2006), Effects of localized, low-voltage pulsed electric fields on the development and inhibition of *Pseudomonas aeruginosa* biofilm, *Biofouling*, 22:383–390.

PETERSON, R.V. and W.G. PITT, (2000), The effect of frequency and power density on the ultrasonically-enhanced killing of biofilm-sequestered *Escherichia coli*, *Colloids Surf. B.* 17:219–227.

PITT, W.G. (2005), Removal of oral biofilm by sonic phenomena, *Am. J. Dent.* 18:345–352.

PITT, W.G. and S.A. ROSS, (2003), Ultrasound increases the rate of bacteria cell growth, *Biotechnol. Prog.* 19:1038–1044.

POVEY, M.J.W. and T.J. MASON, (1997), *Ultrasound in food processing.* UK: Kluwer Academic Publishers.

QIAN, Z., R.D. SAGERS and W.G. PITT, (1997), The effect of ultrasonic frequency upon enhanced killing of *Pseudomonas aeruginosa* biofilm, *Ann. Biomed. Eng.* 25:69–76.

QIAN, Z., R.D. SAGERS and W.G. PITT, (1999), Investigation of the mechanism of the bioacoustic effect, *J. Biomed. Mater. Res.* 44:198–205.

REDISKE, A.M., B.L. ROEDER, M.K. BROWN, J.L. NELSON, R.L. ROBISON, D.O. DRAPER, G.B., SCHAALJE, R.A. ROBISON and W.G. PITT, (1999), Ultrasonic Enhancement of Antibiotic Action on *Escherichia coli* Biofilms: an *in vivo* Model, *Antimicrob. Agents Chemother.* 43:1242–1251.

REDISKE, A.M., B.L. ROEDER, J.L. NELSON, R.L. ROBISON, G.B. SCHAALJE, R.A. ROBISON and W.G. PITT, (2000), Pulsed ultrasound enhance the killing of *Escherichia coli* biofilms by aminoglycoside antibiotics *in vivo*, *Antimicrob, Agents Chemother.* 44:771–772.

RIDGEWAY, H.F., M.G. RIGBY and D.G. AGRO. (1984), Adhesion of a *Mycobacterium* sp. to cellulose diacetate membranes used in reverse osmosis, *Appl. Environ. Microbiol.* 47:61–67.

RONNER, A.B. and A.C.L. WONG, (1993), Biofilm development and sanitizer inactivation of *Listeria monocytogenes* and *Salmonella typhimurium* on stainless steel and Buna-n rubber, *J. Food Prot.* 56:750–758.

RYSER, E.T. and E.H. MARTH, (1999), Listeria, *Listeriosis and Food Safety*. 2nd edn Marcel Dekker, Inc., New York, NY.

SAMS, A.R. and R. FERIA, (1991), Microbial effects of ultrasonication of broiler drumstick skin, *J. Food Sci.* 56:247–248.

SASAHARA, K., and E.A. ZOTTOLA, (1993), Biofilm formation by *Listeria monocytogenes* utilizes a primary colonizing microorganism in flowing systems, *J. Food Prot.* 56:1022–1028.

SCOUTEN, A.J. and L.R. BEUCHAT, (2002), Combined effects of chemical, heat and ultrasound treatments to kill *Salmonella* and *Escherichia coli* O157:H7 on alfalfa seeds, *J. Appl. Microbiol.* 92:668–674.

SEYMOUR, I.J., D. BURFOOT, R.L. SMITH, L.A. COX and A. LOCKWOOD, (2002), Ultrasound decontamination of minimally processed fruits and vegetables, *Int. J. Food Sci. Technol.* 37:547–557.

SPURLOCK, A.T., and E.A. ZOTTOLA, (1991), Growth and attachment of *Listeria monocytogenes* to cast iron, *J. Food Prot.* 54:925–929.

STEWART, P.S. and J.W. COSTERTON, (2001), Antibiotic resistance of bacteria in biofilms. *The Lancet.* 358:135–138.

ZOTTOLA, E.A. (1994), Microbial attachment and biofilm formation: A new problem for the food industry, *Food Technol.* 48:107–114.

# Part IV

# Biofilms in particular food industry sectors

# 14

# Biofilms in red meat processing

**B. Carpentier, Agence Française de Sécurité Sanitaire des Aliments, France**

**Abstract**: This chapter focuses on bacteria remaining on surfaces after cleaning and disinfection. The sources are many: hide, feces, the gastro-intestinal tract, lymph nodes, udder, etc. Psychrotrophic bacteria find enough nutrients and water on surfaces and grow in the cold conditions of most workplaces. They can be transferred by direct or indirect contact or by aerosols from contaminated meat to surfaces as well as from surfaces to meat. Various genera are able to persist on surfaces, among which are *Pseudomonas* and *Listeria*. In non-refrigerated areas, such as butcher shops and lairage, *Salmonella*, shigatoxin producing *Escherichia coli* can also persist. Persistence can occur where the increase due to growth is higher than the decrease due to cleaning and disinfection. Important preventative measures against bacterial persistence are a judicious use of water, cold temperature in premises, and frequent cleaning and disinfection.

**Key words**: biofilm, meat, cleaning, disinfection, persistence, bacterial transfer.

## 14.1 Introduction

The word 'biofilm' was created a long time after it had been discovered that solid surfaces have an effect upon bacterial activity (Zobell, 1943). According to a search with Scirus, a science-specific search engine on the Internet, the word biofilm appeared in the scientific literature in 1976 and a sharp increase of its usage began in 1993. In the famous paper 'How bacteria stick' (Costerton *et al.*, 1978) the word biofilm was not used at all. One of the first definitions was 'cells immobilized at a substratum and frequently embedded in a organic polymer matrix of microbial Origin' (Characklis, 1989). But progressively numerous other definitions were proposed as the one of Costerton: 'matrix-enclosed bacterial populations adherent to each other and/or to surfaces or interfaces' (Costerton *et al.*, 1995). As the polymeric matrix is recognized necessary for microcolony formation (Allison and Sutherland, 1987; Danese *et al.*, 2000; Mack *et al.*, 1996); the second definition exclude attached single cells from belonging to a biofilm. Other authors consider that a bacterium is a biofilm former only if its biofilm is visible with naked eyes on a test tube (Ghigo, 2001). Yet there is no agreement on a minimal colony size to be reached for a bacterial community to be called a biofilm. All these considerations explain the confusion among food microbiologists who often state a need to use the expression 'true biofilm'. As a consequence several authors are wondering whether there are 'true' or 'classical' biofilms or not in the food industry (Holah and Kearney, 1992; Jessen and Lammert, 2003; Whitehead *et al.*, 2008; Zottola and Sasahara, 1994).

In the food safety context, even single attached bacterial cells can be a concern, because as observed on thick biofilms, part of these single attached cells has a low susceptibility to disinfection agents when compared to susceptibility observed from suspension tests. This has been shown for adhering *Listeria monocytogenes* grown in conditions which did not allow microcolony formation. As in classical mature thick biofilms, when such bacterial communities of single cells age, the proportion of cells surviving disinfection increases (Frank and Koffi, 1990). Therefore, the present chapter will focus on the micro-organisms remaining on open surfaces of red meat processing sites following cleaning and disinfection. The chapter will not deal of bacterial communities attached to meat surface although such communities could also be considered as biofilms. Such communities have been extensively studied as well as chemical carcasses decontamination. We will describe the main bacterial flora, the factors impacting micro-organisms transfer to meat and how can persistence of undesirable micro-organisms be prevented or maintained at an acceptable level. Unfortunately very few field studies have been performed so far and laboratory experiments using environmental conditions close to these of meat processing sites are infrequent too. Indeed, classical laboratory growth media are not representative of nutriments available and growth media may have a strong impact on biofilm formation as shown for *E. coli* O157:H7 (Dewanti and Wong, 1995; Folsom *et al.*, 2006; Marouani-Gadri *et al.*, 2008) or *Listeria monocytogenes* (Folsom *et al.*, 2006). Similarly, the low temperature of meat processing sites (down to 4°C or less) is a heavy constraint to be respected in experimental studies as it considerably reduces bacterial growth rate and hence slows down results acquisition.

## 14.2 Sources of contamination

During the slaughtering process, carcasses can be contaminated by a large variety of sources including feces, hide, gastro-intestinal tract, lymph nodes, water, air, humans. Skinning and evisceration potentially cause massive dispersion of undesirable micro-organisms (Gill, 1998; Warriner *et al.*, 2002) including *Salmonella* and shigatoxin producing *E. coli* (STEC). Carcasses can be contaminated by the udders where *Staphylococcus aureus* are frequently isolated (Nickerson *et al.*, 1995) and by the hide where *Salmonella* and *E. coli* O157 are often detected (Arthur *et al.*, 2008). Food and non-food contact surfaces of processing areas can be contaminated by the carcasses, ingredients, water, air, humans, rodents and insects. Bacterial growth occurs on meat surface during chill-storage with change in the composition of the flora (Stanbridge and Davies, 1998) so that contamination of carcasses differs between fresh or chilled storage facilities.

## 14.3 Microbial load of solid surfaces after cleaning and disinfection

### 14.3.1 Reminder

Disinfection does not aim at sterilizing surfaces, but at reducing their microbial load to such an extent as to make them safe and suitable for their use (CAC, 2003), which is to produce food of satisfactory microbiological quality.

### 14.3.2 Knowledge limits due to the methods used

A few considerations on the methods used to study the open surfaces' microbiota are necessary to highlight the imperfection of our knowledge. More precisely, surface sampling and culture-dependent methods to both count and identify the detached microbial cells have serious limits. Although swabbing is well known not to remove all attached cells (Asséré *et al.*, 2008), it is still the most convenient routine method and thus is frequently used in field conditions. The removal rate of swabbing is variable depending on many factors, including the person swabbing, the bacterial species, the solid surface and the history of the microbial community. In comparing swabbing with ultrasonic methods, Asséré *et al.* (2008) found that CFU numbers recovered by swabbing were always the smallest and the highest difference of 1.7 log(CFU/cm$^2$) was observed from chlorine treated *Pseudomonas fluorescens* biofilms. CFU counts obtained from swabbing are therefore underestimations of the real load of culturable cells remaining on a surface following cleaning and disinfection.

The other point limiting our knowledge is the use of culture-dependent methods to count and identify micro-organisms. No figures can be found in the literature on the viable but non culturable cells on meat processing surfaces. Such cells can be detected by comparing culturable counts and counts obtained by quantitative PCR on ethidium monoazide treated cells. Such data were obtained from beef and pork carcasses by Guy *et al.* (2006) who found that viable coliform cells could be sometimes more numerous than culturable ones. According to studies conducted in other food sectors

(Alliot, 1999; Mettler and Carpentier, 1997) and to our laboratory study aiming at modeling the events occurring on open surfaces in meat processing sites (Peneau *et al.*, 2007), viable but non culturable cells can be far more numerous than cells culturable on classical laboratory media. In the latter study of tiles contaminated once by *Pseudomonas fluorescens* and submitted to 10 daily soiling with meat exudate and chemical treatments with agents used for cleaning and disinfection, the viable but non-culturable cells counts were greater by 2 order of magnitude than the ones of culturable cells. Furthermore, we showed that the accumulation of these cells was due to division, stressing a need to further investigate this physiological state.

### 14.3.3 Figures from fields studies

In a recent study (Marouani-Gadri *et al.*, 2009), aerobic plate counts on solid surfaces following cleaning and disinfection in a beef processing plant ranged from less than 10 CFU/cm$^2$ to 5.10$^5$ CFU/cm$^2$. Jessen and Lammert (2003) measured the aerobic count of cleaned and disinfected surfaces of processing lines for sliced cooked ham products and showed values reaching 10$^4$ CFU/cm$^2$. Bizzaro *et al.* (1990) found more than 10$^4$ CFU/cm$^2$ on cleaned and disinfected polyurethane conveyor belt for delicatessen products. Johal (1988) found 10$^2$ and 8.10$^2$ CFU/cm$^2$ following cleaning and disinfection of beef processing surfaces made of stainless steel and polypropylene respectively.

## 14.4 Micro-organisms involved

### 14.4.1 Non pathogenic bacteria

The few studies published on the microbiota found after cleaning and disinfection in red meat processing sites are not recent and reported only on culturable bacteria identified by traditional biochemical methods after culture on non-selective media (Johal, 1988; Mettler and Carpentier, 1998). The main bacterial genera found by Mettler and Carpentier on flooring materials introduced in the grinding room of a beef processing plant were: *Pseudomonas, Staphylococcus, Enterobacter, Flavobacterium, Kluyvera.* Johal (1988) isolated from beef cutting tables *Pseudomonas, Moraxella, Staphylococcus, Micrococcus and Bacillus* species and from pork cutting surfaces: *Acinetobacter, Aeromonas, Alcaligenes, Branhamella, Brochothrix, Gamella, Moraxella, Proteus, Corynebacterium, Erysipelothrix, Kurthia, Micrococcus, Staphylococcus, Streptococcus.* Recent results (Marouani-Gadri *et al.*, 2009) in a beef processing site confirmed the high prevalence of *Staphylococcus* species on surfaces following cleaning and disinfection, but very surprisingly few *Pseudomonas* were isolated within the 26 sampling performed and *Bacillus* isolates appeared to be in the majority together with *Staphylococcus*. Those two last bacterial genera, which have a high tolerance, as most Gram-positive species, against desiccation, refrigeration and reduced pH, were also found among the major Gram-positive species on fresh raw meat (Holzapfel, 1998). Regarding the low prevalence of *Pseudomonas* species in our recent study, it can be explained by two events that occurred several months before our sampling (P. Beaubois, personal communication): i) the practice of post-slaughtering washing of carcasses had been stopped, hence the high *Pseudomonas* load brought by the water was suppressed; ii) afterwards, because of an unacceptable high level of a spoiling *Pseudomonas* strain in some final products due to the persistence of the strain in the environment, the factory was subjected to drastic cleaning and disinfection operations including air filters and gully holes. Today we do not know whether such a situation is an exception or not but we must recognize that, when considering previous studies and other environments where food of animal origin are processed, *Pseudomonas* is one of the main genus isolated (Alliot, 1999; Bagge-Ravn *et al.*, 2003; Holah *et al.*, 2002). Pseudomonadaceae are also one of the major components of the Gram negative spoilage bacteria in meat. Actually all the genera isolated following cleaning and disinfection belong to the bacteria associated with meat and meat products that have been reviewed by Garcia-Lopez *et al.* (1998) for the Gram negative species and by Holzapfel (1998) for the Gram positive ones. It is recognized that contamination with Pseudomonades during processing originated from water, hands and materials (Garcia-Lopez *et al.*, 1998). Actually, there is some evidence that *Pseudomonas* from the open surfaces of the processing rooms are able to colonize the inner surfaces of the water ducts of the processing site. Microorganisms are transported by the aerosols produced during cleaning to the end of hoses or to the cleaner nozzles. Once attached, they are able to spread in the pipes. Gagnière *et al.* (2006) showed, by using a GFP tagged *Pseudomonas putida*, that the upward spread of the bacterium in a vertical supply water pipe was 4.3 cm/day. Such a phenomenon is supposed to occur in all processing facilities where water jets are used for cleaning and explain why in the study by Alliot (1999), 38% of the Pseudomonad ribotypes found in the water of cheese sites were also isolated from the open surfaces of these sites.

Although not conducted on surfaces following cleaning and disinfection, the studies of Brightwell *et al.* (2006) and the one from Takahashi *et al.* (2004) are, to our knowledge, the first ones investigating the identity of the non-culturable species present on inert surfaces from meat processing sites. Brightwell *et al.* found on a conveyor belt system in an operating lamb boning room that Sphingomonades were numerically dominant in the culture-independent clone library and were not represented among the cultured species. On the contrary, *Serratia*, *Alcaligenes* and *Microbacterium* genera were only detected after culture and Pseudomonades were detected by both methods. Takahashi *et al.* (2004) investigated a plant producing sliced ham and Vienna sausages and identified using PCR-DGGE (denaturing gradient gel electrophoresis) *Lactococcus lactis* and *Pseudomonas putida* as one of the dominant bacteria of the plant.

### 14.4.2 Pathogenic bacteria

Pathogenic bacteria from enteric origin are usually not detected in refrigerated industrial environments after cleaning and disinfection, as shown by Giovannacci *et al.* (2001) for *Salmonella* and no report on presence of shigatoxin producing *E. coli* after cleaning and disinfection in refrigerated environments can be found in the literature. However, a potential for *Salmonella* to persist has been shown in retail butcher shops or kitchen of restaurants that were poorly cleaned and disinfected. According to Berends *et al.* (1998) following a study made in the Netherlands in 1998: 'it may be that up to 10% of butcher shops or kitchens of restaurants become colonized for several weeks or months with their own endemic "house flora" of *Salmonella* sp.'. Similarly an *E. coli* O157:H7 outbreak was likely due to persistence of the pathogenic bacteria in a meat grinder of a supermarket that was cleaned and disinfected only once a week (Banatvala *et al.*, 1996). It has also been demonstrated that *E. coli* O157:H7 and *Salmonella* are able to persist in the lairage environment at beef or pork slaughterhouses (Arthur *et al.*, 2008; Small *et al.*, 2006; Swanenburg *et al.*, 2001). By contrast, there is a wealth of studies reporting persistence of *L. monocytogenes* in a variety of industrial refrigerated premises including meat processing ones (Nesbakken *et al.*, 1995; Peccio *et al.*, 2003). Persistence of *L. monocytogenes* is a serious food safety concern as it has frequently been the cause of large outbreaks (Tompkin, 2002), including the recent Canadian one (PHAC, 2008) likely due, according to press releases, to persistence in a meat slicer.

## 14.5 Possible explanations of the presence of a bacterial species after cleaning and disinfection

There are several possible explanations of the presence of a bacterial species on a surface just after well-conducted cleaning and disinfection operations. A first explaining event could be called an 'accidental' entrance of a bacterial strain. Such an entrance is considered 'accidental' either because of an 'abnormally high' cell number or/and because the bacteria reach a harborage site with the consequence that the first cleaning and disinfection operation following bacterial entrance is not be able to eliminate the bacterium. Then either the following cleaning and disinfection operation(s) will be able to eliminate it or, if growth is possible and allows an increase of the bacterial count greater than the periodical reduction caused by cleaning and disinfection, the bacterial strain will persist in the environment (Dunsmore *et al.*, 1981).

What has to be considered as a 'high' cell number is highly dependent on the bacterial species considered. As illustrated in Table 14.1 experimental persistence was much more easily achieved with a *P. fluorescens* strain than with an epidemic *E. coli* O157:H7 strain. In these experiments, only $10^2$–$10^3$ CFU/cm$^2$ before the first chemical treatment were enough for *P. fluorescens* to persist at 10°C, whereas *E. coli* O157:H7 managed to persist at 20°C after having reached $10^7$ CFU/cm$^2$ but not $10^5$ CFU/cm$^2$. Knowing that both species' biofilm reached the same population of $10^7$ CFU/cm$^2$ at the end of growth in meat exudate, our results clearly showed that biofilm formation potential expressed with respect to biofilm populations is not a good indicator of persistence potential. The *P. fluorescens* strain we used was probably one of the best candidates to achieve persistence because it was earlier isolated after having been established on materials kept at 10°C for several weeks and alternately soiled with naturally contaminated meat exudate and cleaned and disinfected with the recommended concentration of commercial products of the same family than the ones we subsequently used in the experiments presented in Table 14.1. These results suggest that conditions leading to *P. fluorescens* persistence can be reached even when applying good hygienic practices. Conditions to achieve *E. coli* O157:H7 persistence are much more unlikely to occur in meat processing environments. In our experiments, the three first daily chemical treatments managed to reduce cell counts from $10^7$ to $10^2$ CFU/cm$^2$ and this reduction occurred before growth was able to resume. Then, the increase of bacterial counts due to growth was rapidly greater than the reduction obtained by the chemical treatment which decreased over time. *P. fluorescens* and *E. coli* O157:H7 likely illustrate extreme and opposite situations.

Another explanation of the presence of a bacterial clone on environmental surfaces is its permanent ingress in the food processing premises, with for instance, raw food or ingredients. *E. coli* has been suggested to persist in chilled food processing environments (Holah *et al.*, 2004) although it is not able to grow at low temperature. However in the latter study swabbings were not all performed after cleaning and disinfection and the ratio 'number of isolates in the environment/number of isolates in the product' for the 10 assumed persistent *E. coli* ribogroups were always smaller or equal to 1. In the same study this ratio was always greater or equal to 1 for the 14 persistent *L. monocytogenes* ribogroups. Knowing that the potential for persistence of *L. monocytogenes* is now well established (ICMSF, 2002), such a difference might be explained by a permanent ingress of some *E. coli* strains with raw food.

## 14.6 Implications for food suitability and safety

Any contact between equipment, utensils, cutting implements, hands, gloves and meat allows bacterial transfer between surfaces in contact (Bell,

**Table 14.1** Experimental conditions leading or not to persistence when the bacteria suspended in meat exudate were deposited once on 4 $cm^2$ coupons made of tile (*P. fluorescens*) or of polyvinyl chloride (*E. coli* O157:H7) that were then submitted daily to a chemical treatment followed by a soiling with meat exudate

| | Initial load log suspended CFUs/ml | Day(s) before the first chemical treatment** | Log(CFU/$cm^2$) before first chemical treatment | Persistence | Ref. |
|---|---|---|---|---|---|
| *Pseudomonas fluorescens** | 3 | 1 | <2 | No | Peneau *et al.*, 2007 |
| | 5 | 1 | 2–3 | Yes | |
| *E. coli* O157:H7* | 8 | 1 | 5 | No | Marouani-Gadri, 2008 |
| | 8 | 3 | 7 | Yes | |

* The experiments were conducted at 10°C for *P. fluorescens* and at 20°C for *E. coli* O157:H7.
** The chemical treatment was a 5 min application of a chlorinated alkaline solution followed by a water rinsing and a 5 min application of a disinfectant solution containing an aldehyde and a quaternary ammonium compound.

1997; Flores and Tamplin, 2002; Flores *et al*., 2006). A comparison of two supermarket groups showed that the group applying the more stringent hygiene procedures and thus having the lowest microbial counts on equipment yielded the lowest mean bacterial counts for all meat cuts (Nortjé *et al*., 1989a, b). Such a difference was demonstrated although the carcasses received by the group applying the more stringent hygiene procedures had higher mean surface temperatures and microbial counts than those of the carcasses received by the other supermarket group.

Of course it is not necessary to wait for attached bacteria to grow and form a thick biofilm for bacterial transfer to occur. Cross-contamination during processing with micro-organisms that will not survive cleaning and disinfection deserves as much attention as contamination from transient survivors or from persistent clones but distinction between them 'is essential when investigating and correcting a contamination problem' (ICMSF, 2002). Study of cross-contamination of carcasses and equipment during pork processing using DNA fingerprinting of *E. coli* (Warriner *et al*., 2002) or of *Salmonella* (Giovannacci *et al*., 2001) showed that enteric bacteria can be transferred between pork carcasses throughout the whole processing line in spite of good manufacturing practices. An important factor impacting bacterial transfer is the presence of water, particularly when considering persistent micro-organisms. Gill *et al.* (1999) showed that when equipment used for carcass breaking was operated with wetting of the meat-contacting surfaces, bacteria from a persisting detritus were transferred to the meat-contacting surfaces and the meat.

Contacts between surfaces and food are not the only means micro-organisms can use to circulate in a meat factory: all cleaning techniques disperse viable micro-organisms in both water droplets and aerosols according to Holah *et al.* (1990).

In laboratory experiments bacterial transfer is usually studied from cells deposited and dried on a surface to facilitate attachment (Aarnisalo *et al*., 2007; Montville and Schaffner, 2003; Vorst *et al*., 2006). In our experiments we choose to study transfer from biofilms grown in immersed conditions to simulate what happen in the wet environments of meat factories. Using the principle of the method implemented by Egington *et al.* (1995) we first showed that Tryptone-soya-agar (TSA) compared to other gels, was a good experimental model to study bacterial transfer from biofilm to beef meat (Midelet, 2002). The method consisted in making 12 contacts on a biofilm, using a new slice of a TSA cylinder for each contact, and plotting the logarithm of CFU/cm$^2$ detached by each contact against the contact number. Relative attachment strength can be assessed by comparing the slopes of the transfer curves (Eginton *et al*., 1995). The slope value can also be used to calculate the total biofilm population as proposed by Veulemans *et al.* (1970) and detailed again by Midelet *et al.* (2006). We observed three types of detachment kinetics: biphasic, where the first slope may be either positive or negative, and monophasic kinetics. The biphasic curves with a first positive slope between contact number 1 to contact number 3 were observed for *L. monocytogenes* either in pure or in co-culture biofilms after a chemical treatment by a chlorinated-alkaline agent or by a quaternary-ammonium and glutaraldehyde-based agent. Among the seven factors studied, both these chemical treatments had the clearest impact on all biofilms: the CFUs which resisted the chemical treatments had greater attachment strength than these determined for untreated biofilms. This finding was corroborated by the already mentioned results of Asséré *et al.* (2008) suggesting the existence of a minority subpopulation of cells resisting the chemical treatment that could be called 'persister' cells (Lewis, 2007) and that are 'super adhesive'. Among the other factors having an impact on attachment strength was the nature of species. We showed that a *L. monocytogenes* biofilm attached more strongly to polymers for conveyor belts than did three other strains of *Staphylococcus sciuri*, *Pseudomonas putida*, and *Comamonas* sp. (Midelet and Carpentier, 2002). When *L. monocytogenes* was co-cultured with a *Kocuria varians*, the latter favored both attachment and detachment of the pathogenic bacteria (Midelet *et al.*, 2006). Attachment strength on stainless steel was lower than on polymeric materials for conveyor belts, but stainless steel, carrying a smaller population of bacteria, had a weaker contamination potential (Midelet and Carpentier, 2002; Midelet *et al.*, 2006). These results corroborate the finding of Chamberlain and Johal (1988) showing that, when both *Pseudomonas* bacterial cells and substratum surfaces were conditioned by a pork liquor prior to substratum colonization, the shear force necessary for cell detachment was smaller on stainless steel than on polypropylene.

## 14.7 Prevention of biofilm formation

The main control measure to limit biofilm formation is the refrigeration of premises so that only psychrotrophic micro-organisms are able to grow and persist in the processing environments. Refrigeration of meat during processing and storage is mandatory according to Regulation 853/2004 of the Council of the European Union: 'during cutting, boning, trimming, slicing, dicing, wrapping and packaging, the meat is maintained at not more than 3°C for offal and 7°C for other meat [...]'. Lower temperature of not more than 2°C is specified for minced meat or mechanically separated meat. The most important pathogens with respect to number of illness *i.e. Campylobacter*, *Salmonella*, STEC and *Staphylococcus aureus* are unable to persist because they are unable to grow below 5°C (Table 14.2). And when the temperature is slightly above the minimal growth temperature, the limited possible growth is not likely to overcome the reduction of the attached cells number obtained by well-conducted cleaning and disinfection operations. By contrast, *L. monocytogenes*, being able to grow at very low temperature, has an advantage over the non-psychrotrophic species. That is likely the reason why a temperature lower than 10°C was found to be one of the

**Table 14.2** Minimum growth temperature of main food borne pathogenic bacteria and estimates of the most likely annual number of cases and deaths from food borne infections in France in the 1990s (Vaillant *et al.*, 2004)

| Bacteria | Minimal growth temperature | Reservoir | Estimated annual number of cases | Estimated number of deaths |
|---|---|---|---|---|
| *Listeria monocytogenes* | –2°C | soil, ubiquitous | 304 | 78 |
| *Yersinia* spp. | 4°C | soil | 655–10909 | 4–10 |
| *Vibrio* spp. | 4°C | sea food | 14 | 1 |
| Bacillus *cereus* | 5°C | soil, ubiquitous | 219–701 | 0 |
| *Salmonella* non Typhi | 5°C | gastrointestinal tract | 30598–41139 | 92–535 |
| *Staphylococcus aureus* | 6°C | skin, nose, udders | 3257–10422 | 0 |
| *Escherichia coli* (STEC) | 6°C | gastrointestinal tract | 373–747 | 0–1 |
| *Clostridium botulinum* | proteolytic 10°C, non proteolytic 3°C | soil | 22 | 0–1 |
| *Clostridium perfringens* | 10°C | soil | 2790–8928 | 2–6 |
| *Campylobacter* spp. | 30°C | gastrointestinal tract | 12796–17322 | 13–18 |

factors associated with *L. monocytogenes* contamination of solid surfaces in poultry or pork processing plants (Chasseignaux *et al.*, 2002).

Another very efficient control measure to limit biofilm formation is dryness which is also a way to decrease bacterial circulation and, as already mentioned, transfer from solid surfaces to meat. Minimal use of water, and no water use when the product is exposed, should be the rules, in particular after a drying or thermal processing step, at slicing and packaging. In these areas water should only be applied during the cleaning and disinfection shift and drying should take place before the next work shift. Such rules have been successfully applied in plants to prevent meat product contamination with *L. monocytogenes* (M. Duke-Rohner, personal communication).

As previously explained, water ducts can be contaminated by the micro-organisms present in the processing environment. To prevent such a contamination, end of hoses and nozzles must not touch the floors and should be protected from aerosols. This can be practically done by immersing the exposed openings in a disinfecting solution. To limit biofilm formation, one of the most frequently recurring questions is about the materials that could be less prone to bacterial attachment and/or bacterial retention. To our knowledge, the first scientific paper on this subject was published by Ridenour and Armbruster (1953) . In meat processing plants food contact surfaces are open surfaces that can be cleaned by very effective mechanical action compared to the mechanical action of flow used to clean closed surfaces. Furthermore, cleaning and disinfection are done more aggressively that in other food sectors (Mettler and Carpentier, 1998). As a consequence, some materials shown to be significantly different with respect to CFU counts after the hygiene operations in a cheese-processing site or in a pastry site could not be differentiated after cleaning and disinfection in a meat processing site (Mettler and Carpentier, 1998; 1999). Accordingly, Marouani-Gadri *et al.*, (2009) observed that none of the materials of a meat processing plant, including stainless steel, polyvinyl chloride and polyethylene, appeared to be clearly more hygienic than the others when comparing their microbial load after cleaning and disinfection. The first quality a material must have is therefore to withstand the thermal, mechanical and chemical constraints linked to its usage. Such a quality seriously restrains the choice of a material. Assuming that stainless steels are materials with the highest resistance to any kind of aggression occurring in food processing premises, such materials should be chosen whenever possible.

Of course stainless steel cannot be used for all purposes. For flooring materials of food processing plants the only appropriate materials are ceramic tiles and resin-based floors (Carpentier, 2005). One obvious criterion to choose a flooring material is the absence of holes and crevices. Bubble holes are frequent in resin-based flooring materials. They are formed following gas production during polymerization and it was shown that contamination after cleaning of a flooring material containing such holes was proportional to the number of holes (Mettler and Carpentier, 1999). In Fig. 14.1 one can see such a hole full of soiling although the flooring material that had spent 8 weeks in a meat factory was observed just after very aggressive cleaning and disinfection operations. As aggregate are included in flooring materials to be anti-slippery, there may also be cracks around aggregates and holes left by removed aggregates. All these defaults are not always visible to the naked eye and the easiest way to detect them is to observe the material under a stereomicroscope (also called a binocular lens) (Carpentier, 2005). Similarly, conveyor belt polyurethane materials are preferred to polyvynil chloride because in the latter bubble holes can be found (Amgar, 2008; Midelet, 2002) as shown in Fig. 14.2.

Another material, that was classically used in meat processing and now only authorized in retail butcher shops, is wood for cutting boards. It makes no doubt that wet wood is a good porous support for bacterial cells, but it was also shown that bacteria deposited on 'dry' wood were badly recovered (Abrishami *et al.*, 1994; Ak *et al.*, 1994; Revol-Junelles *et al.*, 2005) although presence of organic matter allow a better recovery (Aantrekker *et al.* 2002). Williams *et al.* (2008) showed that wooden sawdust, frequently applied on floors in butcher shops to absorb spilled fluids, reduced *E. coli O157:H7* survival on tiles under dry conditions. These authors also showed that damp sawdust increased *E. coli O157:H7* survival. Of course, humidity and organic

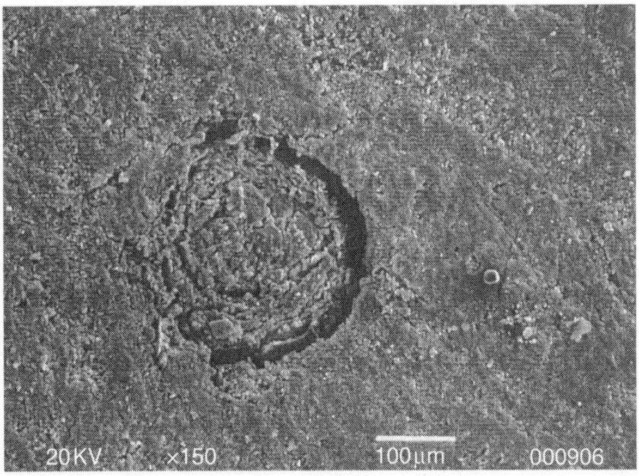


**Fig. 14.1** Top view of a polyurethane-based floor after 8 weeks in a meat processing site, observation by scanning electronic microscopy after cleaning and disinfection (Mettler, 1996).

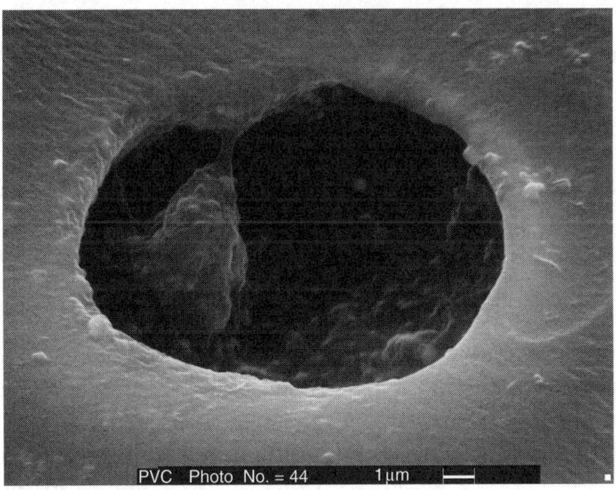


**Fig. 14.2** Observation by scanning electronic microscopy of a new conveyor belt material made of polyvinyl chloride (Midelet, 2002).

matter are present on meat cutting boards that can be highly contaminated. For this reason plastic chopping boards are generally preferred to wooden chopping ones (Braathen and Yndestad, 1996; Gerigk, 1966; Gough and Dodd, 1998) especially in industrial meat processing sites where surfaces are likely never 'dry'. However, in the comparisons of wooden and plastics cutting boards the impact of scraping as realized several times a day on wooden boards but not on plastic ones has not yet been assessed and it is plausible that, at least for butcher shops that are not so wet than industrial premises, such a scrapping gives the advantage to wood or at least allows wood to be no less hygienic than plastics boards. Because wood is far better at absorbing mechanical shocks and preventing meat and cutting implements from slipping and because it does not blunt cutting implements as much (Carpentier, 1997) it would be interesting to investigate on the impact of scrapping.

Of course, most important is the hygienic design of premises and equipment and there are many general rules to respect that have been extensively described (Lelieveld *et al.*, 2005; Wierenga and Holah, 2005). A European standard on hygienic design of machines has been recently revised (CEN, 2009). Among the most important places where hygienic design has to be respected in meat processing factories are drains, conveyors, slicers where persistence of *L. monocytogenes* has been frequently detected (Gravani, 1999; Rivera-Betancourt *et al.*, 2004).

## 14.8 Biofilm inactivation and removal

Cleaning and disinfection must be applied as soon as possible after the processing run to avoid desiccation of soils and micro-organisms. Indeed desiccation increases cells attachment strength, and many bacterial species show high survival rate to desiccation. Overall, Gram positive bacteria, but also several Enterobacteriaceae such as *Salmonella* and *E. coli* O157:H7 have good resistance to desiccation (Hiramatsu *et al.*, 2005; Janning *et al.*, 1994). It has also been reported that the sampling of dried surfaces can lead to highly underestimated CFU counts as evidenced by higher values after cleaning and disinfection than before (S. Christiaens, personal communication).

In spite of the low temperature of premises, nitrogen-rich soils from meat processing allow rapid microbial growth of undesirable micro-organisms including *L. monocytogenes* and the proteolytic Pseudomonades that are among the main spoilage bacteria of meat products. That is the reason why as already mentioned the meat sector is, among all the food sectors, the one where cleaning and disinfection operations are the most aggressive. In meat factory premises where we have performed environmental samplings (Marouani-Gadri *et al.*, 2009; Mettler and Carpentier, 1998) surfaces were submitted to a four-step cleaning and disinfection plan: pre-washing, cleaning, rinsing and disinfection. Pre-washing aims at removing the gross dirt, it was performed by scraping the surfaces followed by a pressure jet washing with 45°C water. It is of prime importance not to use warmer water, in any case not warmer than 55°C, to avoid proteins precipitation and then high difficulties for removing them. Scrapers or brushes used to perform the pre-washing step can be highly contaminated and must be thoroughly cleaned and disinfected daily. Then the surfaces should be foam-cleaned with a chlorinated–alkaline solution at room temperature. Although chlorine is a powerful disinfectant, it is applied on soiled surfaces and is, at least partially, consumed by proteins. For this reason, it is here considered to be a cleaning agent that is also appreciated for whitening polyethylene cutting boards. The foam was left on the surfaces for at least 20 minutes and surfaces were then rinsed by a pressure jet with 45°C water. The last step was disinfection with an aldehyde-quaternary-ammonium compound based product. Although surfaces may appear visually clean after the cleaning step, disinfection is really necessary and any attempt to suppress it or to decrease its frequency allows bacterial counts to reach unacceptable levels (Rahkio and Korkeala, 1996).

Knowing that very aggressive cleaning techniques including pressure water jet are used in meat factories, the already mentioned dispersion of micro-organisms must be taken into account to avoid recontamination of food contact surfaces. For example, floors and drains that are among the more contaminated surfaces (Eisel *et al.*, 1997) where *L. monocytogenes* is frequently found (Gravani, 1999; Gudbjörnsdóttir *et al.*, 2004) should be cleaned before food contact surfaces, so that the cleaning and disinfection of the latter allow removing the contamination just brought by the aerosols. Finally, according to regulation, food contact surfaces must be rinsed after disinfection.

In case undesirable bacteria persist in the environment to such an extent as to make food products unacceptably contaminated, it is necessary to first look for the harborage sites where the undesirable bacteria persist. In case persistence occurs within a machine, the latter must be dismantled and each part must be submitted to aggressive cleaning and disinfection, even better they should, after cleaning, be put in hot water (more than 70°C) if possible, and worn pieces of equipment should be changed. It is also necessary to increase the frequency of cleaning and disinfection of premises until the end of the crisis and to respect zoning. In case of *L. monocytogenes* persistence in complex equipment that cannot be completely dismantled, food producer must consider discard it and buy a new one.

Frequent cleaning and disinfection of cutting implements are needed to avoid cross-contamination and must also be performed before sharpening the blade. After cleaning with mechanical action and detergents, it is specified in European regulation (EC) No 853/2004 that slaughterhouses and meat cutting plants 'must have facilities for disinfecting tools with hot water supplied at not less than 82°C, or an alternative system having an equivalent effect.' However these operations must be done with care because when too aggressive they can damage the blade which makes the cutting difficult and thus can be the cause of musculoskeletal disorders, one of the most common occupational illness (INRS, 2004).

## 14.9 Future trends

For the meat processors to decrease meat contamination by undesirable persistent micro-organisms, improvements in the use of water should be made first in the future; it appears that it is still not well understood by all food handlers. Water allows bacteria to grow, to circulate and is frequently a source of undesirable micro-organisms. From a scientific point of view, research is needed to better understand how to avoid persistence of undesirable micro-organisms and on how to avoid them reaching the food product. By way of example, knowledge is missing on the role of bacterial interactions in persistence and both culturable and viable and non culturable cells should be taken into account in interaction studies. Knowledge is also missing on the behavior of micro-organisms in realistic conditions. However, experimenting in realistic conditions is time consuming: incubation at low temperature slows down result acquisition, working with low bacterial concentrations leads to high scatter of data so that many repetitions of experiments are needed before interpretation. For all these reasons, mathematical modeling the bacterial behavior on surfaces is a promising future tool for better understanding how to optimize hygienic practices. Finally, another important issue in the next future is how to maintain a high level of hygiene with minimal impact on the environment.

## 14.10 Sources of further information and advice

The Web discussion list Vigie-Viande (in French) http://www.vigie-viande.info/ was created by French professionals involved in animal breeding and meat production (extension centers, public institutions, associations) to promote exchanges between scientist and companies.

Book of interest: *The microbiology of meat and poultry* edited by Andrew Davies and Ron Board and published by Blackie Academic & Professional (1998). As bacteria isolated from surfaces are also these of meat, a lot of information on the species, their metabolism, physiological attributes can be found in this book.

## 14.11 References

AANTREKKER ED, BOOM RM, ZWIETERING MH and VAN SCHOTHORST M (2002), Quantifying recontamination through factory environments a review, *Int J Food Microbiol*, 80, 117–130.

AARNISALO K, SHEEN S, RAASKA L and TAMPLIN M (2007), Modelling transfer of *Listeria monocytogenes* during slicing of 'gravad' salmon, *Int J Food Microbiol*, 118, 69–78.

ABRISHAMI SH, TALL BD, BRUURSEMA TJ, EPSTEIN PS and SHAH DB (1994), Bacterial adherence and viability on cutting board surfaces, *J Food Safety*, 14, 153–172.

AK NO, CLIVER DO and KASPAR CW (1994), Cuting boards of plastic and wood contaminated experimentally with bacteria, *J Food Prot*, 57, 16–22.

ALLIOT M (1999), Ecologie microbienne des sites industriels fromagers: identification, caractérisation des flores isolées et étude des facteurs régissant leur implantation. Paris, France, Institut national agronomique de Paris-Grignon, p 145.

ALLISON DG and SUTHERLAND IW (1987), The role of exopolysaccharides in adhesion of freshwater bacteria, *J Gen Microbiol*, 133, 1319–1327.

AMGAR A (2008), Conveyor belt hygiene, *New Food*, 11, 17–19.

ARTHUR TM, BOSILEVAC JM, BRICHTA-HARHAY DM, KALCHAYANAND N, KING DA, SHACKELFORD SD, WHEELER TL and KOOHMARAIE M (2008), Source tracking of *Escherichia coli* O157:H7 and *Salmonella* contamination in the lairage environment at commercial US beef processing plants and identification of an effective intervention, *J Food Prot*, 71, 1752–1760.

ASSÉRÉ A, OULAHAL N and CARPENTIER B (2008), Comparative evaluation of methods for counting surviving biofilm cells adhering to a polyvinyl chloride surface exposed to chlorine or drying, *J Appl Microbiol*, 104, 1692–1702.

BAGGE-RAVN D, NG Y, HJELM M, CHRISTIANSEN JN, JOHANSEN C and GRAM L (2003), The microbial ecology of processing equipment in different fish industries-analysis of the microflora during processing and following cleaning and disinfection, *Int J Food Microbiol*, 87, 239–250.

BANATVALA N, MAGNANO AR, CARTTER ML, BARRETT TJ, BIBB WF, VASILE LL, MSHAR P, LAMBERT-FAIR MA, GREEN JH, BEAN NH and TAUXE RV (1996), Meat grinders and molecular epidemiology: two supermarket outbreaks of *Escherichia coli* O157:H7 infection, *J Infect Dis*, 173, 480–483.

BELL RG (1997), Distribution and sources of microbial contamination on beef carcasses, *J Appl Microbiol*, 82, 292–300.

BERENDS BR, VAN KNAPEN F, MOSSEL DA, BURT SA and SNIJDERS JM (1998), *Salmonella* spp. on pork at cutting plants and at the retail level and the influence of particular risk factors, *Int J Food Microbiol*, 44, 207–217.

BIZZARO S, DENEUVE L and VENDEUVRE, JL (1990), Etude de la contamination microbienne des surfaces en entreprise, *Viandes et Produits Carnés*, 11, 220.

BRAATHEN O and YNDESTAD M (1996), Hygienic aspects of use of wooden and plastics chopping boards, *Food Sci Technol Abstr*, 28.

BRIGHTWELL G, BOEREMA J, MILLS J, MOWAT E and PULFORD D (2006), Identifying the bacterial community on the surface of Intralox™ belting in a meat boning room by culture-dependent and culture-independent 16S rDNA sequence analysis, *Internat J Food Microbiol*, 109, 47–53.

CAC (2003), Recommended international code of practice – General principles of hygiene, including Annex on hazard analysis and critical control coint (HACCP) system and guidelines for its application – CAC/RCP 1-1969, Rev.4-2003/download/standards/23/cxp_001e.pdf. Available at http://www.codexalimentarius.net/search/advancedsearch.do (accessed 2008-03-27). Rome (Italy), Codex Alimentarius Commission.

CARPENTIER B (1997), Sanitary quality of meat chopping board surfaces: a bibliographical study, *Food Microbiol*, 14, 31–37.

CARPENTIER B (2005), Improving the design of floors, In Lelieveld HLM, Mostert MA, Holah J (eds) *Handbook of hygiene control in the food industry*, Woodhead Publishing Ltd, Cambridge, pp 168–182.

CEN (2009), European Standard EN 1672-2:2005+A1:2009 – Food processing machinery – Basic concepts – Part 2: Hygiene requirements.

CHAMBERLAIN AHL and JOHAL S (1988), Biofilms on Meat Processing Surfaces, *Biodeterioration*, 7, 57–61.

CHARACKLIS WG (1989), Microbial biofouling control, In Characklis WGaM, KC (ed) *Biofilms*, vol 15, Wiley, J. and Sons, New-York, pp 585–631.

CHASSEIGNAUX E, GERAULT P, TOQUIN MT, SALVAT G, COLIN P and ERMEL G (2002), Ecology of *Listeria monocytogenes* in the environment of raw poultry meat and raw pork meat processing plants, *FEMS Microbiol Let*, 210, 271–275.

COSTERTON J, LEWANDOWSKI Z, DOUGLAS E and KORBER R (1995), Microbial Biofilms, *Ann Rev Microbiol*, 49, 711–745.

COSTERTON JW, GEESEY GG and CHENG KJ (1978), How bacteria stick, *Scientific American*, 238, 86–95.

DANESE PN, PRATT LA and KOLTER R (2000), Exopolysaccharide production is required for development of *Escherichia coli* K-12 biofilm architecture, *J Bacteriol*, 182, 3593–3596.

DEWANTI R and WONG AC (1995), Influence of culture conditions on biofilm formation by *Escherichia coli* O157:H7, *Int J Food Microbiol*, 26, 147–164.

DUNSMORE DG, TWOMEY A, WHITTLESTONE WG and MORGAN HW (1981), Design and performance of systems for cleaning product-contact surfaces of food equipment: A review, *J Food Prot*, 44, 220–240.

EGINTON PJ, GIBSON H, HOLAH J, HANDLEY PS and GILBERT P (1995), Quantification of the ease of removal of bacteria from surfaces, *J Ind Microbiol*, 15, 305–310.

EISEL WG, LINTON RH and MURIANA PM (1997), A survey of microbial levels for incoming raw beef, environmental sources, and ground beef in a red meat processing plant, *Food Microbiol*, 14, 273–282.

FLORES RA and TAMPLIN ML (2002), Distribution patterns of *Escherichia coli* O157:H7 in ground beef produced by a laboratory-scale grinder, *J Food Prot*, 65, 1894–1902.

FLORES RA, TAMPLIN ML, MARMER BS, PHILLIPS JG and COOKE PH (2006), Transfer coefficient models for *Escherichia coli* O157:H7 on contacts between beef tissue and high-density polyethylene surfaces, *J Food Prot*, 69, 1248–1255.

FOLSOM JP, SIRAGUSA GR and FRANK JF (2006), Formation of biofilm at different nutrient levels by various genotypes of *Listeria monocytogenes*, *J Food Prot*, 69, 826–834.

FRANK JF and KOFFI RA (1990), Surface-adherent growth of *Listeria monocytogenes* is associated with increased resistance to surfactant sanitizers and heat, *J Food Prot*, 53, 550–554.

GAGNIÈRE S, AUVRAY F and CARPENTIER B (2006), Spread of a green fluorescent protein tagged *Pseudomonas putida* in a water pipe following airborne contamination, *J Food Prot*, 69, 2692–2696.

GARCIA-LOPEZ ML, PRIETO M and OTERO A (1998), The physiological attributes of Gram-negative bacteria associated with spoilage of meat and meat products, In Davies A, Board R (eds) *The microbiology of meat and poultry*, Blackie Academic & Professional, London, pp 1–34.

GERIGK K (1966), Anwendungstechnische und bakteriologische Untersuchungen von Kunstoffschneidbrettern, *Fleischwirtschaft*, 11, 1215–1216.

GHIGO JM (2001), Natural conjugative plasmids induce bacterial biofilm development, *Nature*, 412, 442–445.

GILL CO (1998), Microbial contamiantion of meat during slaughter and butchering of cattle, sheep and pigs., In Davies A, Board R (eds) *The microbiology of meat and poultry*, Blackie, Academic and Professionnal, London, pp 118–157.

GILL CO, BADONI M and MCGINNIS JC (1999), Assessment of the adequacy of cleaning of equipment used breaking beef carcasses, *Internat J Food Microbiol*, 46, 1–8.

GIOVANNACCI I, QUEGUINER S, RAGIMBEAU C, SALVAT G, VENDEUVRE JL, CARLIER V and ERMEL G (2001), Tracing of *Salmonella* spp. in two pork slaughter and cutting plants using serotyping and macrorestriction genotyping, *J Appl Microbiol*, 90, 131–147.

GOUGH NL and DODD CER (1998), The survival and disinfection of *Salmonella typhimurium* on chopping board surfaces of wood and plastic, *Food Control*, 9, 363–368.

GRAVANI R (1999), Incidence and control of *Listeria* in food-processing facilities, In Ryser ET, Marth EH (eds) *Listeria, Listeriosis, and Food Safety*, CRC Press.

GUDBJÖRNSDÓTTIR B, SUIHKO M-L, GUSTAVSSON P, THORKELSSON G, SALO S, SJÖBERG A-M, NICLASEN O and BREDHOLT S (2004), The incidence of *Listeria monocytogenes* in meat, poultry and seafood plants in the Nordic countries, *Food Microbiol*, 21, 217–225.

GUY RA, KAPOOR A, HOLICKA J, SHEPHERD D and HORGEN PA (2006), A rapid molecular-based assay for direct quantification of viable bacteria in slaughterhouses, *J Food Prot*, 69, 1265–1272.

HIRAMATSU R, MATSUMOTO M, SAKAE K and MIYAZAKI Y (2005), Ability of Shiga toxin-producing *Escherichia coli* and *Salmonella* spp. to survive in a desiccation model system and in dry foods, *Appl Environ Microbiol*, 71, 6657–6663.

HOLAH J and KEARNEY LR (1992), Introduction to biofilms in the food industry, In Melo LF, Bott TR, Fletcher M, Capdeville B (eds) *Biofilms: Science and Technology*, vol 223, Kluwer Academic Publishers, Dordrecht, Netherlands, pp 35–41.

HOLAH JT, TIMPERLEY AW and HOLDER JS (1990), Technical memorandum No. 590 – The spread of *Listeria* by cleaning systems, Chipping Campden (UK), Campden Food and Drink Research Association.

HOLAH JT, TAYLOR JH, DAWSON DJ and HALL KE (2002), Biocide use in the food industry and the disinfectant resistance of persistent strains of *Listeria monocytogenes* and *Escherichia coli*, *J Appl Microbiol*, 92 Suppl 1, 111S–120S.

HOLAH JT, BIRD J and HALL KE (2004), The microbial ecology of high-risk, chilled food factories; evidence for persistent *Listeria* spp. and *Escherichia coli* strains, *J Appl Microbiol*, 97, 68–77.

HOLZAPFEL WH (1998), The Gram-positive bacteria associated with meat and meat products, In Davies A, Board R (eds) *The microbiology of meat and poultry*, Blackie Academic & Professional, London, pp 35–84.

ICMSF (2002), Sampling to assess control of the environment, In Tompkin RB, Gram L, Roberts TA, Buchanan RL, van Schothorst M, Dahms S, Cole MB (eds) *Microorganisms in foods 7. Microbial Testing in Food Safety Management*, vol 11, Kluwer Academic/Plenum Publishers, New York, pp 199–224.

INRS (2004), Nettoyage et désinfection – Un couteau qui garde son pouvoir de coupe pour réduire les TMS – Approche participative par branche, filière viande de boucherie., Paris, Institut national de recherche et de sécurité.

JANNING B, IN 'T VELD PH, NOTERMANS S and KRAMER J (1994), Resistance of bacterial strains to dry conditions: use of anhydrous silica gel in a desiccation model system, *J Appl Bacteriol*, 77, 319–324.

JESSEN B and LAMMERT L (2003), Biofilm and disinfection in meat processing plants, *Internat Biodeter Biodegrad*, 51, 265–269.

JOHAL S (1988), Bacterial adhesion to processing surfaces in the meat indsutry. Microbiology, University of Surrey, p 409.

LELIEVELD HLM, MOSTERT MA and CURIEL GJ (2005), Hygienic equipment design, In Lelieveld HLM, Mostert MA, Holah J (eds) *Handbook of hygiene control in the food industry*, Woodhead Publishing Ltd, Cambridge, pp 122–166.

LEWIS K (2007), Persister cells, dormancy and infectious disease, *Nat Rev Microbiol*, 5, 48–56.

MACK D, FISCHER W, KROKOTSCH A, LEOPOLD K, HARTMANN R, EGGE H and LAUFS R (1996), The Intercellular adhesin involved in biofilm accumulation of *Staphylococcus epidermidis* Is a linear ß-1,6-linked glucosaminoglycan: purification and structural analysis, *J Bacteriol*, 178, 175–183.

MAROUANI-GADRI N (2008), Factors affecting biofilm formation and allowing persistence of *E. coli* O157:H7 on open surfaces of meat processing plants – Thesis, Paris (France), Institut des sciences et techniques du vivant et de l'environnement (AgroParisTech).

MAROUANI-GADRI N, CHASSAING D and CARPENTIER B (2008), Comparision of biofilm formation and susceptibility to a chemical treatment of pathogenic and nonpathogenic *E. coli* O157:H7 strains, *J Food Prot*, 72, 157–164.

MAROUANI-GADRI N, AUGIER G and CARPENTIER B (2009), Characterization of bacterial strains isolated from a beef-processing plant following cleaning and disinfection

– influence of isolated strains on biofilm formation by Sakaï and EDL 933 *E. coli* O157:H7. *Internat J Food Microbiol*, 133, 62–67.

METTLER E (1996), Etude des caractéristiques microbiologiques et physico-chimiques, après nettoyage et désinfection, de surfaces colonisées par des biofilms, dans divers ateliers de l'industrie agro-alimentaire et au laboratoire. PhD thesis. Université de Bourgogne, Dijon

METTLER E and CARPENTIER B (1997), Localisation, dénombrement et identification de la contamination microbienne après nettoyage de joints en EPDM d'un circuit de pasteurisation de I'industrie laitière, *Le Lait*, 77, 489–503.

METTLER E and CARPENTIER B (1998), Variations over time of microbial load and physicochemical properties of floor materials after cleaning in food industry premises, *J Food Prot*, 61, 57–65.

METTLER E and CARPENTIER B (1999), Hygienic quality of floors in relation to surface texture, *Trans IChemE*, 77, 90–95.

MIDELET G (2002), Etude des transferts microbiens par contact depuis des surfaces inertes vers un aliment – Application à la situation industrielle des bandes convoyeuses utilisées dans l'industrie de la viande – Thesis, University of Bourgogne, Dijon (France).

MIDELET G and CARPENTIER B (2002), Transfer of microorganisms, including *Listeria monocytogenes*, from various materials to beef, *Appl Environ Microbiol*, 68, 4015–4024.

MIDELET G, KOBILINSKY A and CARPENTIER B (2006), Construction and analysis of fractional multifactorial designs to study attachment strength and transfer of Listeria monocytogenes from pure or mixed biofilms after contact with a solid model food, *Applied and Environmental Microbiology*, 72, 2313–2321.

MONTVILLE R and SCHAFFNER DW (2003), Inoculum size influences bacterial cross contamination between surfaces, *Appl Environ Microbiol*, 69, 7188–7193.

NESBAKKEN T, KAPPERUD G and CAUGANT DA (1995), Pathways of *Listeria monocytogenes* contamination in the meat processing industry, *Internat J Food Microbiol*, 31, 161–171.

NICKERSON SC, OWENS WE and BODDIE RL (1995), Mastitis in dairy heifers: initial studies on prevalence and control, *J Dairy Sci*, 78, 1607–1618.

NORTJÉ GL, NEL L, JORDAAN E, NAUDÉ RT, HOLZAPFEL WH and GRIMBEEK RJ (1989a), A microbiological survey of fresh meat in the supermarket trade. Part 1: Carcasses and contact surfaces, *Meat Science*, 25, 81–97.

NORTJÉ GL, NEL L, JORDAAN E, NAUDÉ RT, HOLZAPFEL WH and GRIMBEEK RJ (1989b), A microbiological survey of fresh meat in the supermarket trade. Part 2: beef retail cuts, *Meat Science*, 25, 99–112.

PECCIO A, AUTIO T, KORKEALA H, ROSMINI R and TREVISANI M (2003), *Listeria monocytogenes* occurrence and characterization in meat-producing plants, *Lett Appl Microbiol*, 37, 234–238.

PENEAU S, CHASSAING D and CARPENTIER B (2007), First evidence of division and accumulation of viable but nonculturable *Pseudomonas fluorescens* cells on surfaces subjected to conditions encountered at meat processing premises, *Appl Environ Microbiol*, 73, 2839–2846.

PHAC (2008), Listeria monocytogenes outbreak http://www.phac-aspc.gc.ca/alert-alerte/listeria/listeria_2008-eng.php (accessed 25 October 2008). vol 2008, Public Health Agency of Canada.

RAHKIO M and KORKEALA H (1996), Microbiological contamination of carcasses related to hygiene practice and facilities on slaughtering lines, *A Vet Scand*, 37, 219–228.

REVOL-JUNELLES AM, MIGUINDOU-MABIALA R, ROGER-MAIGNE D and MILLIERE JB (2005), Behavior of *Escherichia coli* cells and *Bacillus cereus* spores on poplar wood crates by impedance measurements, *J Food Prot*, 68, 80–84.

RIDENOUR GM and ARMBRUSTER EH (1953), Bacterial cleanability of various types of eating surfaces, *Ame J Public Health*, 43, 138–149.

RIVERA-BETANCOURT M, SHACKELFORD SD, ARTHUR TM, WESTMORELAND KE, BELLINGER G, ROSSMAN M, REAGAN JO and KOOHMARAIE M (2004), Prevalence of *Escherichia coli* O157:H7, *Listeria monocytogenes*, and *Salmonella* in two geographically distant commercial beef processing plants in the United States, *J Food Prot*, 67, 295–302.

SMALL A, JAMES C, JAMES S, DAVIES R, LIEBANA E, HOWELL M, HUTCHISON M and BUNCIC S (2006), Presence of *Salmonella* in the red meat abattoir lairage after routine cleansing and disinfection and on carcasses, *J Food Prot*, 69, 2342–2351.

STANBRIDGE LH and DAVIES AR (1998), The microbiology of chill-stored meat, In Davies A, Board R (eds) *The microbiology of meat and poultry*, Blackie, Academic and Professionnal, London, pp 174–219.

SWANENBURG M, VAN DER WOLF PJ, URLINGS HA, SNIJDERS JM and VAN KNAPEN F (2001), Salmonella in slaughter pigs: the effect of logistic slaughter procedures of pigs on the prevalence of *Salmonella* in pork, *Int J Food Microbiol*, 70, 231–242.

TAKAHASHI H, KIMURA B, YOSHIKAWA M, GOTOU S, WATANABE I and FUJII T (2004), Direct detection and identification of lactic acid bacteria in a food processing plant and in meat products using denaturing gradient gel electrophoresis, *J Food Prot*, 67, 2515–2520.

TOMPKIN RB (2002), Control of *Listeria monocytogenes* in the food-processing environment, *J Food Prot*, 65, 709–725.

VAILLANT V, DE VALK H, BARON E *et al.* (2004), Morbidité et mortalité dues aux infections d'origine alimentaire. Institut de veille sanitaire, Saint-Maurice (France).

VEULEMANS A, JACQMAIN E and JACQMAIN D (1970), Etude d'une méthode simple pour la détermination du degré de pollution des surfaces et la comparaison du pouvoir désinfectant de divers produits d'entretien, *Revue des Fermentations et des Industries Alimentaires*, 25, 58–65.

VORST KL, TODD EC and RYSERT ET (2006), Transfer of *Listeria monocytogenes* during mechanical slicing of turkey breast, bologna, and salami, *J Food Prot*, 69, 619–626.

WARRINER K, ALDSWORTH TG, KAUR S and DODD CE (2002), Cross-contamination of carcasses and equipment during pork processing, *J Appl Microbiol*, 93, 169–177.

WHITEHEAD KA, SMITH LA and VERRAN J (2008), The detection of food soils and cells on stainless steel using industrial methods: UV illumination and ATP bioluminescence, *Internat J Food Microbiol*, 127, 121–128.

WIERENGA G and HOLAH J (2005), Hygienic plant design, In Lelieveld HLM, Mostert MA, Holah J (eds) *Handbook of hygiene control in the food industry*, Woodhead Publishing Ltd, Cambridge, pp 76–105.

WILLIAMS AP, AVERY LM, KILLHAM K and JONES DL (2008), Moisture, sawdust, and bleach regulate the persistence of *Escherichia coli* O157:H7 on floor surfaces in butcher shops, *Food Control*, 19, 1119–1125.

ZOBELL CE (1943), The effect of solid surfaces upon bacterial activity, *J Bacteriol*, 46, 39–56.

ZOTTOLA EA and SASAHARA KC (1994), Microbial biofilms in the food processing industry – should they be a concern?, *Internat J Food Microbiol*, 23, 125–148.

# 15

# Biofilms in dairy processing

**P. Bremer and B. Seale, University of Otago, New Zealand and S. Flint and J. Palmer, Massey University, New Zealand**

**Abstract**: Bacterial biofilms are believed to be a significant source of contamination to a wide range of dairy products. As dairy manufacturing plants increase in size and as product runtimes lengthen, the control of biofilms is becoming increasingly important. In this review biofilm development and the implications of contamination in the dairy industry will be discussed with particular reference to the sources of contamination, the microorganisms involved, the unique characteristics of biofilms in dairy manufacturing plants and the options for controlling biofilms in this environment.

**Key words**: dairy industry, dairy products, process biofilms, clean-in-place (CIP).

## 15.1 Introduction

The dairy industry produces a wide range of perishable (milk) and semi-perishable (cheese, butter and milk powder) foods and food ingredients (whey protein concentrates, caseinates). Bacterial contamination is of great concern to the dairy industry and strict adherence to microbiological guidelines is essential to maintain product quality, functionality and safety, and for companies to remain competitive in the international market. While milk is produced by cows as a sterile secretion, contamination can occur at all production stages from milking, through transportation, storage and processing. The manufacturing process relies on pasteurisation to eliminate pathogens and technologies such as refrigeration, pasteurisation and sanitation to control spoilage and limit contamination. Bacterial biofilms are believed to be a significant source of product contamination during processing, with biofilm control becoming increasingly important in the dairy industry. In this review the implications and control of contamination in the dairy industry will be discussed with particular reference to the sources of contamination, the microorganisms involved, the unique characteristics of dairy manufacturing plants (DMP) and the importance of biofilm development and control.

## 15.2 The microbiological flora associated with milk and dairy manufacturing

Milk is produced as a sterile secretion from healthy cows, and in theory if drawn aseptically will remain sterile. However, in reality it generally becomes contaminated prior to leaving the farm with low numbers of microorganisms ($10^2$–$10^3$ CFU $ml^{-1}$). The first potential source of contamination is from microorganisms that have invaded the teat opening such as micrococci, streptococci spp. and the diptheroid *Corynebacterium bovis* (Souto *et al.*, 2008). The second and more common source of contamination occurs during milking due to the fact that the exterior of the teats can become contaminated with a wide range of environmental isolates including thermophiles such a *Bacillus* spp, human pathogens such as *Listeria monocytogenes* or *Campylobacter* spp., psychrotrophic spoilage organisms such as *Pseudomonas* spp. and *Bacillus* spp. and coliforms including *Escherichia coli*, *Enterobacter* spp. and *Klebsiella* (Adams and Moss, 2000). As discussed later, thermoduric spore forming bacteria are of particular concern in many large milk powder manufacturing plants and their occurrence in raw milk is often correlated with the conditions on the farm, in particular the quality of silage being fed (Te Giffel *et al.*, 2002).

On farm the third and principal source of microbial contamination is believed to be milk-handling equipment such as teat cups, pipe work and storage tanks (Chambers, 2002, Gilmour and Rowe, 1990, Andersson *et al.*, 1995, Lin *et al.*, 1998, Jeong and Frank, 1994, Walker, 1971). Fouling layers formed by bacteria and milk components (protein and minerals) are believed to adversely impact on the effectiveness of on farm cleaning and sanitising regimes.

The principal mechanism to control microbial growth in milk stored on the farm is low temperature storage (<7°C), which limits growth to psychrotrophic organisms and increases the length of time the milk can be stored before processing. During transport to and storage at the manufacturing site refrigeration, while desirable, may not be required as the short times involved and the volume of the bulk milk limit temperature changes and the potential for microbial growth (Scott *et al.*, 2007, Flint *et al.*, 2001).

Generally, raw milk entering a DMP has a bacterial load of approximately $10^2$ to $10^3$ CFU $mL^{-1}$ with less than 10 thermophilic bacilli $ml^{-1}$ being present (Muir *et al.*, 1986). Poor hygiene or inadequate cooling can, however, result in numbers as high as $10^6$ CFU $mL^{-1}$ being detected in milk entering a DMP. The dairy industry relies on pasteurisation (72°C for 15 seconds) as the main process to destroy heat sensitive and pathogenic bacteria present in raw milk. However, thermoduric and thermophilic bacteria, including certain species of streptococci, micrococci, lactobacilli and spores from *Bacillus* and *Clostridium* species are able to survive this process.

## 15.3 Microorganisms in milk

Raw milk can contain a wide range of bacterial species, which may have diverse attributes and abilities to colonise, survive and grow under the different environmental niches found in DMP. In DMP is it often useful to group microorganisms on their classification (bacteria, yeasts, bacteriophages) and the temperature range over which they grow or survive.

### 15.3.1 Cold tolerant bacteria

The microorganisms which play a significant role in biological processes in low-temperature environments are referred to as being either facultative or obligative psychrophiles. Facultative psychrophiles grow over a temperature range from 0°C to approximately 40°C and are present in milk in much larger numbers than obligate psychrophiles. They are of particular significance in chilled dairy products as they can grow in refrigerated environments and cause spoilage or disease. Obligate psychrophiles have a growth temperature optimum of 15°C or lower and cannot grow beyond a maximum temperature of 20°C. These microorganisms are also often referred to as being psychrophilic or cold loving.

Species of *Pseudomonas*, *Flavovacterium*, *Alcaligenes*, *Acinetobacter*, *Listeria* and *Bacillus* can be psychrophiles. With the notable exception of *L. monocytogenes* microorganims in this group which are of concern to the dairy industry are generally nonpathogenic, but their presence can cause a variety of off-flavours, including fruity, stale, bitter, putrid and rancid flavours.

Most psychrophiles bacteria with the exception of the spore forming *Bacillus* species will be eliminated during pasteurisation. Their occurrence in and impact on dairy products is dependent upon the time and temperature of milk storage before pasteurisation and the sanitary conditions and hence recontamintion potential prevailing during subsequent production steps. *Pseudomonas* spp, for example, can produce heat stable lipases and proteases and if allowed to grow to high numbers in the raw milk, the activity of these enzymes can impact on the shelf-life of dairy products, such as soft cheeses, butter and ultra high temperature (UHT) treated milk. The influence of psychotropic bacteria on the shelf-life of pasteurised milk is dependent on the number present after packaging, their rate of growth, the storage period, and their biochemical activity.

### 15.3.2 Mesophiles

Mesophiles are the group of bacteria that grow best at temperatures between 10 and 45°C. They are common in the environment and can be found associated with soils, water, plants and animals. Most pathogenic bacteria of concern to the dairy industry are mesophiles and include *Camplyobacter* spp., *Salmonella* and *Staphylococcus aureus*. The threat of pathogen contamination is controlled by: refrigeration; pasteurisation of raw milk; and the implemtation of Good Manufacturing Practices (GMP) in the DMP. While there are a wide variety of mesophilic bacteria associated with milk and milk products (Robinson, 2002), they have not been reported as being a major compoment of biofilms in DMP plants operating under GMP. The presence of biofilms of nonstarter lactic acid bacteria, discussed later under cheese, or coliform bacteria indicate problems in the routine maintenance of dairy plant hygiene and can contribute to spoilage of products. Mesophilic sporeformers of *Bacillus* and *Clostridial* species also threaten the quality of dairy products and are a particular concern as their spores can survive harsh conditions of dairy processing and cleaning (Lindsay *et al.*, 2000).

### 15.3.3 Thermophiles

Thermophiles are a type of bacteria, which can grow in moderately high temperatures from 45° to 65°C, as they have heat stable DNA, proteins and membranes. These adaptations enable them to grow and persist in the pre-heating and evaporation sections within milk powder production plants and to act as a source of contamination to product being processed. The major thermophilic contaminants of milk powder throughout the world are *Anoxybacillus flavithermus* and *Geobacillus* spp. (Ronimus *et al.*, 2003, Rueckert *et al.*, 2004). The other bacilli found in milk powder are facultative thermophiles such as *Bacillus licheniformis*, *Bacillus coagulans*, and *Bacillus subtilis* (Ronimus *et al.*, 2003). Thermophilic bacilli are not a new problem as they have been isolated from 90-year old milk powder used in an Antarctica expedition (Ronimus *et al.*, 2006). While studies on the microorganisms initially focused on the identification and enumeration of thermophilic vegetative cells (Rueckert *et al.*, 2005a) their spores are currently receiving more interest (Rueckert *et al.*, 2005b, 2006, Seale *et al.*, 2008).

*Anoxybacillus flavithermus* appears to be the dominant thermophilic bacterium in milk powders ccworldwide and is ubiquitous in milk powder plants (Rueckert *et al.*, 2004). *Bacillus licheniformis* and *Geobacillus stearothermophilus* are also commonly isolated from various milk powders (Rueckert *et al.*, 2004).

*Anoxybacillus flavithermus*

*Anoxybacillus flavithermus* was first isolated from hot springs in New Zealand (Heinen *et al.*, 1982). The type strain *Anoxybacillus pushchinoensis* was first described as a strict anaerobe, and hence the name *Anoxybacillus*, subsequently the oxygen requirement was revised to species of the genus being aerotolerant anaerobes and facultative anaerobes (Pikuta *et al.*, 2000).

*Anoxybacillus flavithermus* is a Gram-positive, facultatively thermophilic, facultatively aerobic, motile, rod shaped spore former with a growth temperature range between 30 and 70°C and an optimal temperature for growth around 60°C at pH 6–9. Strains isolated from milk powder generally have an optimal growth temperature ranging from 50 to 65°C (Ronimus *et al.*, 2003).

*Geobacillus*

The *Geobacillus* group is composed of a closely related group of thermophilic spore formers that were previously classified in Group 5 of the *Bacillus* genus (Ash *et al.*, 1991, Nazina *et al.*, 2001). Their optimal growth temperature is above 50°C. Different strains of *Geobacillus* spp. have been isolated from temperate areas as well as areas such as hot springs, oilfields, deep sea sediments, sugar refineries and dairy factories (Ronimus *et al.*, 2003, Nazina *et al.*, 2001, Banat *et al.*, 2004, Tai *et al.*, 2004).

The most prominent *Geobacillus* species reported to be isolated from milk powder is *G. stearothermophilus* (Flint and Brooks, 2001, Ronimus *et al.*, 2003, Stadhouders, 1982), which has an optimal growth temperature in the range of 55 to 70°C (Ronimus *et al.*, 2003).

### 15.3.4 Themoduric bacteria

Thermoduric is a term given to a group of bacteria that can survive exposure to temperatures considerably above their maximum temperature for growth. In the dairy industry, the term is applied to those organisms that survive, but do not grow, at pasteurisation temperatures. This group includes species of Micrococcus, Streptococcus, Lactobacillus, Bacillus, and occasionally gram-negative rods. *Streptococcus thermophilus* can survive heat treatment used to pasteurise milk and can subsequently rapidly colonise stainless steel surfaces downstream from the pasteuriser. It has been shown that *S. thermophilus* cells attached to surfaces in DMP are more resistant to sanitisers (sodium hypochlorite or cetyltrimethylammoniumbromide) and heat than free-living cells (Flint *et al.*, 1999, Flint *et al.*, 2002).

### 15.3.5 Spores

Certain bacterial species are able to spend part of their lifecycle as dormant cellular structures known as spores. As a spore, the organism is able to survive adverse environmental conditions that would otherwise be lethal. The spore's complex structure allows it to survive extremes in heat, desiccation, pH, and exposure to UV radiation and chemicals (Setlow, 2006, Nicholson *et al.*, 2000). Its dormancy is a result of it having little to no metabolic activity (Setlow and Kornberg, 1970). When conditions become favourable spores can revert back to their vegetative state. The structure of spores (Leuschner and Lillford, 2001) and the genes involved in structural protein and regulation of sporulation, have received extensive research and review (Henriques and Moran, 2000, Driks, 1999, Aronson and Fitz-James, 1976).

Spores can survive heat treatments typically used in DMP, such as during pasteurisation, with the spores of some strains of *Geobacillus* spp. being reported to form high heat-resistant spores that can survive UHT treatment or retorting, involving heating regimes of 140°C for 3 s and 110 to 120°C for 15 to 20 min respectively (Hill and Smythe, 1994).

The number of spores present in raw milk from mesophilic and thermophilic bacteria, tested on the farm, over a period of a year has been reported to be $10^3$ and <10 spores $mL^{-1}$ milk, respectively (McGuiggan *et al.*, 2002). The development of biofilms of spore-forming bacteria in milk powder manufacturing plant is believed to ultimately originate from spores present in the raw milk surviving pasteurisation, although this has never been proven (Scott *et al.*, 2007).

### 15.3.6 Bacteriophages

A bacteriophage (from 'bacteria' and Greek *phagein*, 'to eat') is a virus that infects bacteria. The term is commonly used in its shortened form, phage. Bacteriophages may have a lytic cycle or a lysogenic cycle, with a few viruses being capable of carrying out both. Lytic bacteriophages, also described as virulent phages, replicate within an infected bacterial cell using the cell's metabolic machinery, then cause the host cells to lyse with the release of new bacteriophages. Lysogenic bacteriophages, also known as temperate phages, integrate their DNA within the genome of the host bacteria through which the phage genome is passed onto the bacterial daughter cells (Marks and Sharp, 2000).

Bacteriophage infection leads to the lysis of the starter cells used in fermentation processes such as cheese making or yoghurt production. This interrupts the fermentation of milk sugar (lactose) into lactic acid by the starter bacteria. As lactic acid production plays an important role in the creation of the food matrix and helps prevent the growth of food spoilage organisms, phage infection can delay fermentation and alter the quality or result in the loss of the product.

The control of bacteriophage in the cheese starter industry largely relies on replacing sensitive bacterial starter strains with resistant starter strains and maintaining good hygiene in the cheese manufacturing plant. It seems likely that biofilms of starter bacteria may play a role in providing a reservoir of bacteriophage in the dairy industry but this has not been proven.

### 15.3.7 Yeasts

Yeasts represent an important component of the micro-flora of many dairy products, where they play an active role in their production or spoilage. Their significance depends on the number present, the product contaminated and the type of yeast, specifically its metabolism. Owing to their diverse metabolic potential, yeasts can play an important role in the fermentation and ripening of many cheese varieties and make a positive contribution to the development of taste and aroma. They may also act as spoilage organisms causing typical defects such as yeasty off-flavour, loss of texture quality, excessive gas formation and increased acidity (Fleet, 1990).

## 15.4 Biofilms in the dairy industry

Fouling deposits in dairy manufacturing plants comprise both bacterial biofilms and milk deposits, which help to sustain microbial growth. This section will briefly discuss milk deposition before focussing on bacterial biofilms.

### 15.4.1 Milk deposition on stainless steel piping

Throughout milk processing, and especially during heating operations, the proteins and minerals in milk can form deposits on heat exchangers, evaporators and membrane filters. Two types of deposit can form depending on the process temperature. A relatively soft, bulky material, is formed at temperatures between 75°C and 115°C and is known as protein fouling due to its high protein content (50–70%, w/w) (de Jong, 1997). At the lower temperatures of this range most of the protein is β-lactoglobulin with casein predominating at higher temperatures (Changani *et al.*, 1997, Fryer *et al.*, 2006). A second type of fouling is mineral fouling which forms rapidly at temperatures above 110°C. The deposit is hard with a granular structure and a high mineral content (up to 80%, w/w) and only contains 10–20% of protein (de Jong, 1997, Changani *et al.*, 1997).

Milk deposition during dairy processing has been studied in great detail (Belmar-Beiny and Fryer, 1992, Bell and Sanders, 1944, Yoon and Lund, 1989, Fryer *et al.*, 2006, Belmar-Beiny and Fryer, 1993, Bansal and Chen, 2006). It is referred to as a transient process which follows two main stages; the induction period and the fouling period (de Jong, 1997). X-ray analysis of milk deposition on stainless steel at 73°C has shown that proteins (usually denatured) are the first to adhere to the stainless steel surface with sulphur and calcium being detected after 150 seconds contact time, but not phosphorus (Belmar-Beiny and Fryer, 1992). For process times greater than 60 mins, the deposit consisted of two layers; a protein rich outer layer and a calcium and phosphorus rich layer near the heat exchanger. This layering has been attributed to the diffusion and subsequent crystallisation of insoluble calcium phosphate (Changani *et al.*, 1997).

### 15.4.2 Bacterial biofilms

In dairy manufacturing plants biofilms can be divided into two categories:

1) 'Process biofilms' are biofilms that are unique to DMP and form on surfaces in direct contact with flowing product, and
2) 'General biofilms' that form in the general food processing environment.

While there is concern in the dairy industry about the role of biofilms associated with the general food processing environment and the presence of pathogens such as *L. monocytogenes* (Frank *et al.*, 1990), this chapter predominantly focuses on process biofilms. Process biofilms differ from general biofilms in two key ways (Flint *et al.*, 1997b). Firstly, in a process biofilm a single species often predominates because heat treatment operations reduce the competition from heat-sensitive bacteria. Secondly, process biofilms are also characterised by their rapid development, with numbers of up to $10^6$ bacteria per $cm^2$ being recorded in the regeneration section of a pasteuriser after 12 hours of operation (Bouman *et al.*, 1982). In contrast, biofilms in most other environments take several days or weeks to develop (Flint *et al.*, 1997b, Zottola and Sasahara, 1994).

In DMP many factors influence the initial attachment of bacteria to surfaces and the subsequent formation of a biofilm. Initially ions and soluble macromolecules (such as proteins) will adsorb at a solid–liquid interface (Belmar-Beiny and Fryer, 1993). These molecules affect the adhesion of bacteria and other cells, and can lead to profound alterations in the properties of the material fluid interface (Bower *et al.*, 1996). The bacterial cells that enter the interface at this stage are influenced by hydrophobic interactions, van der Waals forces, and electrostatic attraction and repulsion (Hood and Zottola, 1995). Eventually favourable conditions are reached in the microenvironment of the interface, which allow the cells to attach and grow.

In the dairy industry, surface conditioning is often referred to as fouling and takes place as soon as a dairy product is brought into contact with stainless steel (Rosmaninho *et al.*, 2007). Conflicting opinions exist on the importance of a conditioning film on the initial attachment of bacteria. The presence of skim milk on a surface of stainless steel, even at concentrations as low as 1%, has been reported to reduce the attachment of spores and vegetative cells of thermophilic bacilli (Parkar *et al.*, 2001). Skim milk was also found to reduce the attachment of *S. aureus*, *L. monocytogenes* and *Serratia marcescens* to stainless steel (Barnes *et al.*, 1999). Even individual milk components such as casein and β-lactoglobulin have been reported to reduce attachment of *L. monocytogenes* and *Salmonella typhimurium* to stainless steel (Helke *et al.*, 1993). A reason for the reported reduced attachment may be that proteins in the bulk fluid phase compete for binding sites on the surface of the stainless steel, reducing the number of sites for bacterial attachment. In contrast, however, to the studies outlined above, treating either stainless steel or rubber surfaces with either whey proteins or lactose has been reported to result in an increase in attachment of milk-associated microorganisms (Speers and Gilmour, 1985). Similarly coating stainless steel coupons with a skim milk foulant, created by denaturing skim milk on their surface by autoclaving, increased the number of *B. stearothermophilus* cells attaching (Flint *et al.*, 2001). It has been postulated that milk proteins are a good substrate for bacterial growth with the presence of milk aggregates creating a suitable supporting substrate (Yoo *et al.*, 2006).

What is not debatable is that in sections of DMP where nutrient and temperature conditions are suitable, vegetative cells or spores attach to surfaces (Hinton *et al.*, 2002) and develop into a biofilm. Product contamination occurs when either individual cells are released or when sloughing or breaking up of the biofilm occurs during processing, releasing large numbers of bacteria directly into the product passing by. This release of bacteria into the passing product is sometimes referred to as the biotransfer potential (Wirtanen *et al.*, 1996).

## 15.5 Factors affecting biofilm formation

### 15.5.1 Location

Dairy manufacturing plants consist of large surface areas all with potential to support biofilm growth; however, some specific locations are more likely to support biofilm growth than others, due to the fact that either the conditions where they are sited favour microbial growth and/or because their design makes them difficult to clean. Examples of sites where growth is favoured include plate heat exchangers, including the pasteurised milk sections of pasteurisers. Areas which are difficult to clean because they are either exposed to a wide range of contaminants, contain occluded sites or are made of porous materials include bends in pipes, rubber seals, gaskets, conveyor belts, waste water pipes and floors (Sharma and Anand, 2002a, 2002b). In addition, as the equipment ages, the surfaces will corrode, forming pits and cracks which can collect soil and bacteria (Wong, 1998).

Plate heat exchanges, including pasteurisers and evaporator pre-heaters, are among the most critical areas for biofilm formation as they contain large surface areas at temperatures suitable for bacterial growth. Growth in the regenerative section of a pasteuriser is prone to significant growth of thermo-tolerant bacteria (Bouman *et al.*, 1982) and this problem is exacerbated by long (>10 h) manufacturing runs (Knight *et al.*, 2004). In some studies, the bacteria colonising the surface of plate heat exchanger pasteurisers have been shown to be dominated by *S. thermophilus* (Te Giffel *et al.*, 1997, Bouman *et al.*, 1982, Driessen *et al.*, 1984). This is believed to be due to the fact that these bacteria are thermo-tolerant and able to survive pasteurisation, in addition, many strains produce a bio-surfactant preventing the adhesion of other bacteria (Busscher *et al.*, 1996). This area is also prone to the growth of themophilic spore forming bacteria such as *Anoxybacillus* or *Geobacillus* spp. (Scott *et al.*, 2007).

Ultra-filtration membranes are also becoming an important source of contamination as membrane separation becomes increasingly used in dairy processing for applications such as the concentration of casein micelles from skim milk or the recovery of serum proteins from whey. Biofilms growing on membrane surfaces block membrane pores resulting in a reduced flux rate and contamination of the product stream (Ridgway *et al.*, 1983, Zottola and Sasahara, 1994). Non-starter lactic acid bacteria ($10^6$ $cm^{-2}$) have been isolated from the surface of ultrafiltration plant after cleaning and sanititation (Agarwal *et al.*, 2006). *B. licheniformis* was identified as the predominant bacterium colonising a commercial dairy ultra filtration membrane plant used to concentrate milk post-pasteurisation (Lehmann, 1995). Biofilm development has been responsible for delaying the wider application of membrane filtration technology in the dairy industry (Defrise and Gekas, 1988).

### 15.5.2 Role of time/temperature

As alluded to above, temperature and time are critical factors determining the growth of biofilms in DMP. Generally the time dimension for biofilm development is short (8–12 h) compared with biofilm development in the general environment (many days), and the presence of different temperature zones create specific sites which leads to differences in the species predominating in different locations (Scott *et al.*, 2007, Langeveld *et al.*, 1995, Hinton *et al.*, 2001, Warnecke, 2001). All dairy manufacturing processes involve heat, therefore it is not surprising that surface colonisation with thermophilic and thermotolerant bacteria is a common problem. *Streptococcus thermophilus*, is a common contaminant in DMP heat exchangers. It has been reported that milk flowing through a heat exchanger inoculated with *S. thermophilus* achieved a maximum contamination level of about $10^7$ cells $mL^{-1}$ after the biofilm had been allowed to develop for 8 h (de Jong *et al.*, 2002). Interestingly, bacterial numbers on the stainless steel surface continued to increase after a steady state had been reached in the number of cells in the milk, suggesting that release of bacteria from the biofilm was not simply related to growth within the biofilm.

Biofilms can also proliferate at low temperatures and cause problems in parts of DMP where raw milk is being processed, although this does not appear to have been well studied. It has been reported that although the raw milk contained predominantly Gram-positive microorganisms, Gram-negative microorganisms, mainly *Acinetobacter* were the predominant colonisers of stainless steel and rubber surfaces exposed to the milk, (Lewis and Gilmour, 1987, Koutzayiotis, 1992). Gram-negative microorganisms predominating in milk pipelines before pasteurisation are likely to produce heat stable lipases and proteases but the role of biofilm bacteria in their production and release is unknown.

### 15.5.3 Flow rate

Turbulent flow (Re >3000) is generally achieved in DMP as flow rates of 1.5 $ms^{-1}$ are generally used. Changing the flow rate will change the nature of a biofilm with high flow rates reported to generate a thinner boundary layer (Lee, 1997) a reduced chance of bacterial adherence (de Jong *et al.*, 2002) and an increase in the rate of cell release (Lee, 1997) compared with a lower flow rates. The effect of different flow rates has not been well studied in the dairy industry but from studies in other systems it is known that the structure of biofilm is influenced by flow, with denser biofilms growing under high shear compared with those forming under low shear conditions (Stoodley *et al.*, 2002). Turbulent flow has also been suggested as enhancing bacterial attachment by bringing bacteria closer to a surface (Carpentier and Cerf, 1993). In experiments with *Bacillus cereus* spores, flow enhanced the attachment of some spores to a stainless steel surface compared with the numbers attaching under static conditions (Klavenes *et al.*, 2002). The rate of substrate consumption in a biofilm has been shown to decrease as fluid velocity increases, suggesting that increased flow reduces biofilm development (Melo and Vieira, 1999).

In pasteurisers used in cheese milk manufacture it has been reported that *S. thermophilus* biofilms reached a steady state faster at a higher flow rate than a lower flow rate and that biofilm growth rate increased with flow rate. However, the steady state biofilm density was unaffected by flow rate and the detachment rate was higher at higher flow rates (Lee, 1997). However, these experiments were carried out under laminar flow conditions and therefore differ from conditions existing in a DMP. The results do imply that once the steady state density is reached, the only way increasing flow rate will reduce microbial contamination is through the dilution effect of passing more milk over the same surface.

In summary, low flow rates have been reported to result in an open biofilm structure (Beyenal and Lewandowski, 2002), a large boundary layer and an increase in the number of attached cells (de Jong *et al.*, 2002). Higher flow rates have been reported to result in a denser biofilm (Beyenal and Lewandowski, 2002), a thinner boundary layer (Lee, 1997) and an increase in the number of cells being released (Lee, 1997).

### 15.5.4 Materials

Bacterial colonisation can occur on all surfaces used in the DMP (Kirtley and McGuire, 1989, Speers and Gilmour, 1985, Carpentier and Cerf, 1993). Stainless steel is the most common food contact material used in DMP, as it is easy to fabricate, durable, chemically and physiologically inert at a variety of food processing temperatures and pressures, generally corrosion resistant and usually easy to clean (Holah and Gibson, 2000, Tide *et al.*, 1999). However, biofilm development will occur readily on stainless steel surfaces (Langeveld *et al.*, 1995). Teflon, buna-N, plastic and electopolished stainless steel – surfaces that are often regarded as 'easy clean', have also been shown to support biofilm development (Carpentier and Cerf, 1993, Russell, 1993). Materials vary in terms of their likelihood of attracting bacteria with the level of contamination with *B. cereus* spores reported to be highest on high density polyethylene ($1.1 \times 10^7$ cells $cm^{-2}$) and lowest on 304 L stainless steel with a 2 B surface finish ($2.6 \times 10^6$ cells $cm^{-2}$) (Kives *et al.*, 2005).

Synthetic rubber seals are also prone to biofilm development in DMP. Rubber surfaces in milking equipment were found to contain 37 isolates of bacteria following cleaning (Teixeira *et al.*, 2005). Membrane surfaces are becoming increasingly common in DMP for processes such as the ultrafiltration of whey, reverse osmosis and the microfiltration of milk and milk products; however, the large scale adoption of these technologies has been limited by problems associated with membrane fouling (Brans *et al.*, 2004). Biofilms occurring in membrane systems reduce their performance and can contaminate products being processed (Ridgway *et al.*, 1983, Flemming and Schaule, 1988). The spiral wound membrane format is commonly used for applications such as reverse osmosis, due to its superior mechanical, hydraulic and economic performance. Unfortunately spiral wound membranes also provide a vast surface area for microbial colonisation. The challenge biofilms present to membrane filtration technologies is being met by advances in filter design, including the development of spiral wound membranes which incorporate a back-flushing capability, and the development of membrane materials such as polyether-urea which show some resistance to bacterial colonisation compared with polysulphone, polyethersulphone and polyamide membranes (Flemming and Schaule, 1988).

### 15.5.5 Surface roughness and micro-topography

While stainless steel is the most common food contact material used in DMP, the surface of stainless steel examined by SEM (Zoltai *et al.*, 1981) and more recently AFM (Arnold and Bailey, 2000) reveals cracks, crevices and milling lines which provide a greater area for cell attachment as well as possible protection for the bacteria from cleaning chemicals and fluid forces (Zottola and Sasahara, 1994, Mittelman, 1998, Flint *et al.*, 2000). This is especially the case for welded areas of pipe joins and corroded equipment surfaces, where pits and scratches can easily house microorganisms (Wong, 1998). It has been shown that surfaces with scratches and pits of similar size to microbial cells retained higher numbers of cells than surfaces with surface features much larger than microbial cells (Verran and Whitehead, 2006). Various groups have observed greater cell attachment on surfaces with high surface roughness and thus concluded surface roughness is an important factor in cell attachment to inert surfaces (Pederson, 1990, Lecleroq-Perlat and Lalande, 1994). However, other authors have reported that there is no correlation between surface roughness and cell attachment to inert surfaces (Mafu *et al.*, 1990, Vanhaecke *et al.*, 1990, Flint *et al.*, 2000). It has been reported that bacterial cells attachment and biofilm formation is reduced on electro-polished stainless steel compared to standard stainless steel (Arnold and Bailey, 2000, Yoon and Lund, 1989), and that biofilm formation on electro-polished stainless steel is patchy and less dense than biofilm formation on 316 stainless steel (Parkar *et al.*, 2003). In general, while there is some debate about the influence of surface roughness on attachment, there appears to be general agreement about the importance of using surfaces with minimal cracks and crevices in order to reduce bacterial adherence and biofilm growth and to enhance cleaning effectiveness.

## 15.6 Biofilm implications for process efficiency

During heat processes above 65°C whey proteins in milk begin to denature and aggregate, which can lead to faster adherence compared to proteins in their native state (Belmar-Beiny and Fryer, 1992). The adherence of proteins to surfaces can be a major problem in the dairy industry as their build-up can restrict fluid flow as well as reducing thermal transfer due to their poor thermal conductivity (Yoon and Lund, 1989). In addition, protein adherence can change the surface properties of the stainless steel (de Jong, 1997), increasing the likelihood of bacteria attaching to the surface and creating an environment that encourages bacterial attachment to an extent where in one study it was discovered that fouled surfaces attracted 10–100 times more vegetative cells and spores of *B. stearothermophilus* to the surface than the clean stainless steel (Flint *et al.*, 2001). The consequences of bacterial biofilm development and milk deposition are associated with decreased product quality due to bacterial contamination and a reduction in the functional properties of the final product, decreased production times, increased product losses and increased cleaning times as attempts are made to control the fouling problem, as well as decreased heat transfer coefficients and fluid flow due to the build-up of deposits which require increases in energy in order to maintain given flow rates (de Jong, 1997, Russell, 1993, Criado *et al.*, 1994).

## 15.7 Products

### 15.7.1 Pasteurised milk

The main source of microorganism contamination to pasteurised milk is biofilm growth on pipelines and processing surfaces (Koutzayiotis, 1992). Raw milk can contain a variety of microorganisms that form biofilms (Kirtley and McGuire, 1989). Interactions between the different species, such as *Pseudomonas* and *Lactococcus lactis* subsp. *cremoris* have been shown to increase the cell numbers of both species, resulting in an increased potential for spoilage (Kives *et al.*, 2005). Interestingly, in a study comparing the accumulation of biofilm on stainless steel surfaces exposed to either pasteurised or raw milk, the biofilm from pasteurised milk developed faster than the biofilm from raw milk, suggesting that raw milk may contain growth-inhibiting compounds that delay biofilm development (Driessen *et al.*, 1984).

Bacterial species that have been isolated from stainless steel pipelines and are commonly associated with biofilms in the dairy industry include *Pseudomonas*, *Streptococcus* and *Bacillus* species (Stone and Zottola, 1985, Flint *et al.*, 1997c, Flint *et al.*, 2001). Pseudomonads are psychrotrophic Gram-negative rods found in raw milk and are often the primary causes of milk spoilage. The bacteria are heat labile, which means they are killed by the pasteurisation process; however, colonisation of surfaces on the raw milk side of a DMP with these bacteria can still pose a risk to product quality as these bacteria produce heat-stable enzymes (lipases and proteases) that are not inactivated by pasteurisation. In addition, post pasteurisation contamination with Pseudomonads can occur, and consequently these bacteria have been isolated from biofilms on stainless steel in milk processing plants, presumably as a result of poor hygiene or due to contamination from water or aerosols (Stone and Zottola, 1985). Streptoccoci spp., in particular *S. thermophilus*, have also been associated with biofilms on the stainless steel surfaces of DMP (Carpentier and Cerf, 1993). *Streptococcus thermophilus* is a thermophilic streptococci that has been isolated from the cooling section of pasteurisers (Flint *et al.*, 1997c), indicating it has survived pasteurisation and has the potential to contaminate the product stream.

### 15.7.2 Cheese

Nonstarter lactic acid bacteria (NSLAB) can be commonly isolated from Cheddar cheese. The majority of NSLAB are *Lactobacillus* spp., although *Pediococcus* and *Leuconostoc* spp. can also be present (Peterson and Marshall, 1990). The proteolytic activity of the lactobacilli contribute to the development of desirable flavours in Cheddar cheese but can also cause accumulation of bitter peptides that result in off flavours (Arihara and Luchansky, 1995, Driessen *et al.*, 1984). In addition, NSLAB can cause defects such as gas formation and calcium lactate crystallization on the surface of the cheese, which forms a white haze. The source of NSLAB contamination is believed to be primarily from post-pasteurisation sources, including biofilms (Austin and Bergeron, 1995, Criado *et al.*, 1994, Somers *et al.*, 2001, Agarwal *et al.*, 2006, Wong, 1998).

A study by Somers *et al.* (2001), showed that biofilms can form during cheese making, and these biofilm bacteria can survive the cleaning process to contaminate a subsequent batch of cheese. In addition, biofilms present on the vat surface can dislodge during cheese making and contaminate the cheese. Therefore in the dairy environment, biofilms can serve as a reservoir for NSLAB, which may persist in the plant for months. While not all NSLAB are problematic in cheeses, modification in cleaning regimens to target biofilms may lessen undesirable contamination (Wong, 1998, Somers *et al.*, 2001, Agarwal *et al.*, 2006). Further, biofilms may provide a source of bacteriophage that threaten the viability of starter bacteria necessary for the manufacture of cheese (Sutherland *et al.*, 2004), although the significance of this compared with other potential sources of bacteriophage has not been determined.

Note that in addition to being responsible for defects in cheese, bacteria within biofilms can add desirable attributes to cheese flavour. An extreme example is seen with Ragusano cheese, a brine salted raw milk cheese produced in Eastern Sicily, to which no starters are added. The lactic acid required is produced by bacteria naturally present in the milk and bacteria released from the surface of the traditional wooden cheese vats (tinas) into the milk (Licitra *et al.*, 2007). Similarly, in a recent study Micrococci-corynebacteria, yeasts and molds were found to predominant on wooden shelves used to ripen the French raw milk smear cheese Reblochon de Savoie (Mariani *et al.*, 2007). It has been also been reported that during the ripening of raw milk smear cheeses such as Reblochon or Munster, the biofilm on the ripening shelves exert an inhibiting effect on the pathogen *L. monocytogenes*.

### 15.7.3 Infant food

*Enterobacter sakazakii* in infant formula milk has been highlighted as a significant threat to infants. The organism grows well in a biofilm on many different surfaces (Kim *et al.*, 2006). The number of *E. sakazakii* cells dried onto a stainless steel surface, from cells suspended in either water or infant formula, were shown to decrease at a faster rate as the temperature increased from 4 to 37°C. However, from an initial population of around log 7.4 CFU per coupon, cells were still recoverable after 60 days. Cells grown as biofilms on stainless steel surfaces also survived when held at 25°C at humidities ranging from 25 to 100% for in excess of 42 days (Kim *et al.*, 2008). As *E. sakazakii* is extremely sensitive to pasteurisation the threat to infants is most likely due to contamination after manufacture of the infant formula and due to growth of the bacterium on the tubing used for infant feeding (Iversen *et al.*, 2004).

### 15.7.4 Milk powder

Milk powder production involves the removal of water from milk solids. The final product is easily transportable, small in volume and has a significantly longer shelf-life than pasteurised liquid milk. The removal of water is achieved via three steps: milk treatment, evaporation and spray drying. During milk powder manufacture, heat exchangers, regeneration and evaporator sections of a plant typically operate at between 45°C to 75°C, which is a range suitable for the growth of thermophiles. The milk is concentrated approximately 10-fold to form a powder, thus if no growth occurred during the manufacturing process, the number of thermophiles in the milk powder should not exceed 100 CFU $ml^{-1}$ as the number of thermophiles in raw milk rarely exceed 10 CFU $ml^{-1}$. However, when evaporator run times exceed 16–20 hours, high numbers of thermophiles are commonly found in the final product (up to $10^5$ CFU $g^{-1}$) (Stadhouders, 1982, Murphy *et al.*, 1999). The residence time of the milk from storage tank to powder does not usually exceed 20–30 minutes, so there is obviously insufficient time for thermophilic growth in the bulk milk during processing, even with their short generation time of approximately 15–20 minutes (Scott *et al.*, 2007). Consequently, the high levels of contamination found in milk powder after 16–20 hour run times are most probably due to biofilm growth and the detachment of cells and spores from the large stainless steel surface areas found within a milk powder plant (Hinton *et al.*, 2002). Initial contamination of the powder plant probably occurs from the low number of thermophiles, predominately in spore form, in the raw milk (McGuiggan *et al.*, 2002). Spores and cells then attach to the stainless steel and foulant (Flint *et al.*, 2001, Parkar *et al.*, 2001). If conditions are favourable, spores germinate while vegetative cells divide and secrete polymers to colonise the surface and thereby initiating a biofilm. Thermophilic biofilms are typically found in the plate heat exchanger and evaporator sections of the plant (Scott *et al.*, 2007, Murphy *et al.*, 1999). Contamination of the milk subsequently occurs from spores and cells sloughing off from the biofilms. It remains unclear at what stage during biofilm development and growth that spores are formed. Scott *et al.* observed spore formation after 9 hours into an 18-hour production (Scott *et al.*, 2007). Others have observed spore formation in the basal layers in surface-air colonies of *B. subtilis* (Vlamakis *et al.*, 2008), and liquid-air biofilms of *B. cereus* (Wijman *et al.*, 2007); however, this may not be the case of surface–liquid biofilms under turbulent flow in a dairy environment.

At the end of a milk powder run, the majority of the cells within the thermophilic biofilm will be removed or killed by the cleaning regime (Parkar *et al.*, 2004); however, fouling layers may protect spores and vegetative cells (Hinton *et al.*, 2002) resulting in potential contamination of the next production run. Thermophilic bacilli are difficult to completely eliminate from processing plant surfaces due to their fast growth rate, ability to produce resistant spores, their wide temperature growth range and their ability to form biofilms (Parkar *et al.*, 2003, Flint *et al.*, 2001, Hill and Smythe, 1994). Furthermore, spores are able to survive in milk powder products for longer periods of time compared to vegetative cells (Reddy *et al.*, 1975). The spores are a particular concern to milk powder manufacturers as they are resistant to heat treatment, drying and cleaning chemicals.

### 15.7.5 Whey and lactose

Biofilm contamination of milk streams is normally associated with growth in plate heat exchangers, with contamination of whey and lactose often attributed to the presence of biofilms on plate heat exchanger and membrane surfaces (Muthukumaran *et al.*, 2005). *Bacillus subtilis* has been reported as a biofilm contaminant growing of ultra-filtration membranes and stainless steel surfaces in a whey processing plant (Flint *et al.*, 1997b).

### 15.7.6 UHT milk

The contamination of UHT milk with *Bacillus* spp. is due to the heat tolerance of some bacillus spores. These spores are present in low levels in milk and their attachment and growth in milk processing lines is believed to be a major source of spores surviving UHT treatment of milk products. Increases in heat resistance have been recorded for spores attached to stainless steel (Simmonds *et al.*, 2003). Some strains of *Geobacillus* sp. can form high heat-resistant spores that can survive UHT treatment or retorting, involving heating regimes of 140°C for 3 s and 110 to 120°C for 15 to 20 min respectively (Hill and Smythe, 1994).

## 15.8 Control of biofilms in dairy manufacturing plants

There are a number of ways to control biofilms in DMP including; prevention of cell attachment through altering the surface chemistry; treating the surface with antimicrobial agents; optimising process and equipment design; and treating biofilms through the use of cleaning regimes (Bower *et al.*, 1996).

The simplest way to control biofilms in DMP is to clean the production line using chemicals. The cleaning process revolves around the clean-in-place (CIP) system as discussed below.

### 15.8.1 Clean in place

Biofilm control in the dairy industry involves chemical cleaning of the production lines. Prior to the 1960s manual cleaning methods were used, which typically involved the complete disassembly of process piping systems and machinery, followed by brushing of product contact surfaces with detergent solutions, rinsing with water, the application of a sanitising agent, and finally, reassembling the piping system and equipment. These methods were very labour intensive and not very reliable. To improve cleaning efficiency, CIP systems were developed and have now been used by the dairy industry for more than 40 years (Stewart and Seiberling, 1996).

A CIP system has been defined as the 'cleaning of complete items of plant or pipeline circuits without dismantling or opening of the equipment and with little or no manual involvement on the part of the operator. The process involves the jetting or spraying of surfaces or circulation of cleaning solutions through the plant under conditions of increased turbulence and flow velocity' (Stewart and Seiberling, 1996). There are three primary reasons for using a CIP system. Firstly, a CIP system is far superior to any other cleaning method due to its fully automated nature, low variability and good reliability. Secondly, CIP lowers costs associated with labour and plant downtime as the time required for cleaning is reduced, thus increasing plant capacity. Also the volume of water and solvents for cleaning are significantly reduced. Thirdly, there are significant safety improvements with CIP because plant personnel no longer have to dismantle equipment, enter dangerous vessels, nor partake in such inherently hazardous activities as manual scraping, vacuuming and high-pressure water blasting. CIP also greatly reduces any possible exposure of personnel to cleaning chemicals (Stewart and Seiberling, 1996).

A typical CIP regime consists of five steps with the rinse and circulation times depending on pipe length and the equipment being cleaned. Typical steps in a CIP regime, as defined by Stewart and Seiberling (1996), are:

1. **Pre-rinse**: The purpose of this step is to remove as much 'loose' soil as possible prior to the formulated alkaline wash. Removal of most of the organic fat, carbohydrate or proteinaceous soil is generally accomplished with ambient or warm temperature water.
2. **Alkaline wash** with heated, (70–80°C) re-circulated, formulated solutions. Since relatively long contact times are required for this primary cleaning step, recirculation of cleaning solution is essential for economical operation.
3. **Post-rinse with water**, normally at ambient temperature. This step is to rinse away most of the alkaline cleaner. This solution is sometimes recovered for the pre-rinse in the next CIP cleaning programme.
4. **Acid rinse** at ambient, or heated (55–80°C) temperatures with re-circulated acid solutions. The purpose of this step is two-fold. The acid rinse will neutralise and remove residual alkaline cleaner. Alkaline cleaners form films on equipment that are not readily removed by a simple post-rinse with water. The acid rinse will also remove mineral deposits.
5. **Post-rinse** with water or re-circulated **sanitising rinse**, used to apply a bactericidal agent to all cleaned surfaces. The post-rinse is sometimes heated to permit faster drying of equipment.

A feature of CIP regimes, evident in both industrial- and laboratory-scale systems, is their variability in eliminating surface adherent bacteria

(Dufour *et al.*, 2004, Faille *et al.*, 2001, Bremer *et al.*, 2006, Austin and Bergeron, 1995). The most important factors influencing the effectiveness of the CIP are: cleaning time; cleaning agent temperature; cleaning agent concentration and chemistry; degree of turbulence of the cleaning solution; and the characteristics of the surface being cleaned (Lelievre *et al.*, 2002a, 2002b, 2001, Bremer *et al.*, 2006, Boulange-Petermann *et al.*, 2004, Changani *et al.*, 1997, Stewart and Seiberling, 1996). The cleaning time needs to be long enough to allow sufficient chemical contact with surfaces, but not so long that it becomes cost ineffective. The temperature of the cleaning agent needs to be hot enough to effectively clean, but not so hot that it bakes on milk proteins. The concentration and chemistry of the cleaning agents is dependent on the type of fouling present. The velocity of the cleaning solution is very important because it is used to lift soil from the pipelines in a manner similar to the action of manual scrubbing. A flow rate of 1.5 m $sec^{-1}$ for all cleaning and rinse solutions is often specified in process manuals; however, there appears to be no theoretical justification for this flow rate (Changani *et al.*, 1997). Finally, the surface being treated must be cleanable, with a smooth, corrosion-resistant finish (Stewart and Seiberling, 1996). During correct CIP procedure, most thermophilic bacteria are eliminated (Parkar *et al.*, 2003). However, the presence of foulant can provide protection from cleaning chemicals (Hinton *et al.*, 2002) and high numbers of thermophile spores have been reported associated with foulant in a milk powder plant (Scott *et al.*, 2007). The exact nature of the interaction between the foulant and the bacteria with a biofilm, however, remains elusive. Furthermore, there are areas within a DMP that are difficult to clean because of design limitations or the nature, age or state of the material being cleaned. For example, it has been reported that if bacterial biofilms develop on gaskets, extensive biofilms can remain present even after regular CIP procedures potentially seeding the next manufacturing run (Austin and Bergeron, 1995).

The cleaning chemicals play an important role in the CIP regime. During the cleaning process they must first be transported to the solid–liquid interface, then contact and penetrate the deposit. The cleaning solution reacts with the deposit, which is then able to be removed, due in part to the velocity of the solution through the pipes (Changani *et al.*, 1997). If a CIP is effective and used under optimum conditions for the soil, any cells that remain in the system are not viable and do not exhibit attachment fibrils (Stone and Zottola, 1985). However, under conditions of lowered temperature, and detergent and sanitiser concentrations, those cells that are present after CIP are viable and show attachment fibrils (Stone and Zottola, 1985). Therefore it is essential to maintain and control the CIP to the soil conditions so optimum cleaning is always achieved.

Cleaning and sanitising processes are designed to initially reduce the surface population of microorganism through cleaning with detergent formulations, and then to kill the residual population through the application of disinfectants (Eginton *et al.*, 1998). There are a number of factors to be considered when selecting particular cleaning agents for use in the CIP system, these include: water hardness because the combination of minerals and high fat deposits can result in foaming; cleaning temperature to be used; the time available for cleaning; availability of automatic dosing and control equipment; possible use of product in other areas such as bottle washers, fillers and cartooning machines; the availability of bulk detergent concentrate storage and, finally the materials of construction (Fryer and Bird, 1994, Bird and Fryer, 1991, Fryer *et al.*, 2006, Bird and Bartlett, 1995). There is also increased awareness of the need for disinfectants which are both highly effective and usable in concentrations which have a minimal impact on the environment. The array of cleaning chemicals available for use in CIP will now be discussed in detail.

*Caustic detergents*

The effectiveness of a CIP regime is greatly dependent on the chemical action. Of all steps in a CIP system the most effective component has been reported to be the caustic step (Dunsmore, 1981), because, as a general 'rule of thumb' most residues are easily removed with a dilute (1% or less), formulated caustic cleaner (Stewart and Seiberling, 1996). Three steps required for effective cleaning have been defined by Russell (1992) to be:

1. Separation of the soil from the substrate.
2. Dispersion of the soil in the detergent medium.
3. Prevention of soil re-depositing on the substrate.

The most common and aggressive caustic cleaner is sodium hydroxide (NaOH). It is typically used in 1–5% concentrations for plate-type and tubular heat exchangers, and other heavily soiled surfaces (Stewart and Seiberling, 1996). The active part in the caustic is the OH- ions. The hydroxyl ions, which form the bulk of the cleaning fluid, rapidly transfer to the solid–liquid interface at the surface of the biofilm. The high pH of the solution causes the deprotonation of functional groups present within proteins, leaving a net negative charge. These negative charges repel one another causing the protein to swell and become translucent (Mercadé-Prieto *et al.*, 2007). The removal of this swelled surface is by the mass transfer of material into the bulk of the solution leaving the mineral layer behind. Increasing the effectiveness of the caustic step could reduce the amount of nitric acid and the reliance on a sanitiser. The caustic wash is designed to remove organic matter such as proteins.

In order to increase the effectiveness of the caustic wash step, other compounds can be added to the basic caustic. It has been found that the addition of oxidisers and complexing agents can increase efficiency by a factor of ten compared with pure 0.25% sodium hydroxide (Changani *et al.*, 1997); for example, sodium hypochlorite is added to alkaline solutions to enhance the removal of both fat and protein soiling, and chelating agents can be used in systems where hard water is a problem (Stewart and Seiberling, 1996). These chemical solutions are termed caustic blends and have been shown in laboratory-based studies to improve the removal of dairy biofilms from stainless steel (Bremer *et al.*, 2006). One step CIP additives are also available which can be added to caustic and caustic blends to enhance CIP performance.

*Acid detergents*

An acid detergent is used to remove the mineral scale left on the plant surfaces after exposure to the caustic detergent. Acid detergents also: aid in the removal of any traces of alkaline product from equipment surfaces; enhance draining and drying; and provide bacteriostatic conditions that delay the growth of organisms that can be found in the water supply (Stewart and Seiberling, 1996). The acid step requires critical temperature-concentration combinations to be completely effective (White and Rabe, 1970). The most common acid detergent used in the dairy industry is nitric acid. A major initial concern regarding the used of acid was its corrosive effect on the stainless steel; however, it is now well established that low concentrations of nitric acid are not corrosive to stainless steel type 304 and can be used routinely for prolonged periods (White and Rabe, 1970). As with caustic chemicals, acid detergents can be formulated to contain compounds such as surfactants which improve their, surface wetting, soil penetration and cleaning properties (Bremer *et al.*, 2006).

*Sanitisers*

Sanitisers are used primarily to kill vegetative cells that remain on the surface after 'cleaning' (Bower *et al.*, 1996). In some cases sanitisers have been found to be ineffective at reducing the biofilm (Eginton *et al.*, 1998). This could be due to either increased resistance of the bacteria within the biofilm, the chemical nature of the sanitiser or the presence of milk residues. It has been reported that 48 h biofilms of *Pseudomonas fluorescens*, *L. monocytogenes*, and *B. subtilis* developed in the presence of milk were more resistance to a chlorine sanitiser at 0.012 and 0.12% sodium hypochlorite than biofilms produced on surfaces exposed to meat media (Wirtanen and Mattila-Sandholm, 1992) and biofilms of non-starter lactic acid bacteria developed in skim milk have been shown to be more resistant to chlorine and quaternary ammonium based sanitisers than biofilms developed in the presence of culture media (Somers *et al.*, 2001). Bacteria within biofilms are believed to be more resistant to the sanitiser for a variety of reasons, including phenotypic differences in the cells within a biofilm (Brown and Gilbert, 1993), the age of the biofilm (Sommer *et al.*, 1999); the thickness of the biofilm, because sanitiser cannot penetrate several layers of bacteria (Eginton *et al.*, 1998); the metabolic changes the bacteria in a biofilm undergo, because the uptake of sanitiser by the bacteria is slower (Bower *et al.*, 1996); and the production of extracellular compounds, because they can react with the sanitiser (especially if chlorine based) and prevent its diffusion in biofilm (Sommer *et al.*, 1999, Meyer, 2003, Stewart *et al.*, 2000, Dodds *et al.*, 2000). The chemical nature of the sanitiser can also impact on its effectiveness. Some sanitisers are strong oxidisers (i.e., peroxides, halogens) and their action might conceivably further cement bacteria to a surface rather than remove them or weaken their attachment (Eginton *et al.*, 1998).

It is not possible to clean frequently enough to totally prevent biofilm formation within a DMP as cell attachment can occur very rapidly (Meyer, 2003). In addition, once biofims have formed, some cells may persist after a standard CIP regime (Hood and Zottola, 1997). There is therefore considerable interest in the development of alternative control of biofilms in DMP.

## 15.9 Future trends for biofilm control in dairy manufacturing plants (DMP)

It is interesting that the CIP regimes have not significantly changed since they were first introduced in the 1950s. Reasons for this include the effectiveness of the current systems and the difficulty in validating or optimising new systems in the face of the size of many DMP and their requirement to maintain production throughput and high product safety and quality standards.

The following five factors influence the nature and rate of biofilms development in DMP, namely: the species of microorganism; the type of product being processed; the operational conditions (runtime and temperature); surface material and its condition; and the cleaning and sanitising regimes (chemicals, use and frequency) employed (Flint *et al.*, 1997b). These are areas that need to be studied in order to enhance biofilms control in DMP.

### 15.9.1 Understanding the raw material

It is well established that maintaining raw milk at temperatures ≤7°C reduces microbial spoilage. This is based on bacterial levels in the milk and sensory indicators associated with milk under different storage conditions. The source of contamination is assumed to be low levels originating in the milk from the farm; however, another potential source, biofilms forming on surfaces during raw milk handling, may be a significant source of contamination. The effect of biofilms on raw milk quality has not been well studied. The effect of organic (fat and protein) and inorganic (mineral deposits) on biofilm formation during raw milk handling is unknown. Biofilms can proliferate at low temperatures and cause problems in parts of DMP where raw milk is being processed. It is likely that changes to milk fats, proteins and carbohydrates, caused by microbial activity, will adversely affect product functionality, long-term sensory characteristics and manufacturing plant performance. The metabolic activities of microbial cells in a biofilm community differ from those bacteria not colonising surfaces. Microbial activity from biofilms may have a greater effect on sensory defects and product yields than recognised from the growth of planktonic (suspended) cells. Aged milk may promote fouling because of the activity of proteolytic enzymes from psychrotrophic bacteria. Anecdotal evidence also suggests that raw milk quality can impact final product quality. However, the link between raw milk quality and extending manufacturing run-length as well as final product quality has not been reported in the literature.

### 15.9.2 Modifying DMP operating conditions

Steps to control biofilm development by altering the runtime or operational conditions, have been given the term 'disruptive technologies' and include the introduction of temperature step changes to control *S. thermophilus* growth in plate heat exchangers (Knight *et al.*, 2004). There is also potential to expand this concept to other parts of DMP for controlling a variety of bacteria. Very little work has been done on the effect of flow rate on preventing biofilm formation; however, it has been reported that there is a greater release of bacteria from biofilm under high flow rates (Lee, 1997). Continually altering conditions, such as temperature and flow, to avoid stable conditions needed for microbial growth, is a simple concept that if succesful could be employed with minimal cost.

Many studies have also investigated methods to control biofouling by fore-warming and holding milk, especially during high temperature heat processes (de Jong, 1997). Fore-warming and holding the milk prior to the heat treatment can denature the milk proteins before the heat process step and reduces the formation of milk deposition (Bell and Sanders, 1944, Yoon and Lund, 1989) and thus bacterial attachment (Flint *et al.*, 1997a, 1997b). However, this process in itself leads to fouling of the pipes, albeit at an earlier stage, thus contamination of the process stream can still occur. A better understanding of the implications of operating condition modifications on fouling and biofilm formation throughout a DMP is likely to enhance biofilm and fouling control.

### 15.9.3 Surface modifications to reduce the attachment of microorganisms

Stainless steel is commonly used in the dairy industry since it is corrosion resistant and easily cleanable. However, several groups have reported the adhesion of bacteria to stainless steel surfaces within the dairy plant (Flint *et al.*, 2002, Parkar *et al.*, 2001, Flint *et al.*, 2001). Since microbial attachment precedes biofilm formation, surface modification to reduce the ability of bacteria to adhere, has been suggested as a possible solution to prevent biofilm formation. Several studies have examined the effects of surfaces with altered characterisitics on microbial attachment in general (Li and Logan, 2004, Cunliffe *et al.*, 1999) and specifically within the dairy industry (Faille *et al.*, 2002, Rosmaninho *et al.*, 2007, Seale *et al.*, 2008). Modifications include altering the surface characteristics of stainless steel in terms of hydrophobicity and surface charge, by impregnation of the metals with different ions or surface coatings such as Ni-Cu-P-PTFE (Santos *et al.*, 2004).

A number of publications have reported that microorganisms associated with the dairy industry, particularly spores, attach in higher numbers to surfaces with a lower surface energy (hydrophobic) and attach in lower numbers to surfaces with a higher surface energy (hydrophilic) (Faille *et al.*, 2002, Rosmaninho *et al.*, 2007). In contrast, it has been reported that some bacteria attach in lower numbers to hydrophobic surfaces (Zhao *et al.*, 2005, Zhao, 2004, Tsibouklis *et al.*, 2000). In a recent study spores from four *Geobacillus* spp. were isolated from a milk powder manufacturing line in New Zealand (Seale *et al.*, 2008). The zeta potentials (net surface charge) of the spores from the four isolates ranged from –10 to –20 mV at neutral pH, with an isoelectric point between pH 3 and 4. Through contact angle measurements, spores were found to be hydrophilic and had relative hydrophobicity values of 10 to 40%, as measured by the microbial adhesion to hexadecane assay. The most hydrophilic spore isolate with the smallest negative charge attached in the highest numbers to Thermanox and stainless steel surfaces ($1 \times 10^4$ spores $cm^{-2}$), with fewer spores attaching to glass ($3 \times 10^3$ spores $cm^{-2}$). However, spores produced by the other three strains attached in similar numbers to all three substrata ($1 \times 10^3$ spores $cm^{-2}$), indicating that there was no simple relationship between individual physicochemical interactions and spore adherence (Seale *et al.*, 2008). This data reinforce the fact that there is no simple relationship between the physicochemical interactions between an individual microorganism and its adhesion to surfaces. Rather, adhesion is dependent on complex and poorly defined relationships between a range of forces acting on both the microorganism and the substrate. The industrial significance of this is that surface treatments, which limit the attachment of one microorganism, may not be effective for all microorganisms or indeed for all strains of a single microorganism. Therefore, control regimens need to be devised with reference to the characteristics of the particular strains entering via the raw milk to the processing facility.

Surfaces that prevent non-microbial fouling are also important as the presence of milk proteins and salts can reduce heat transfer and also influence bacterial attachment depending on the organism, surface and proteins involved (Barnes *et al.*, 1999). Hydrophobic surfaces, such as Ni-Cu-P-PTFE, have been shown to reduce calcium phosphate and protein fouling (Rosmaninho *et al.*, 2007).

If a modified surface is found to reduce bacterial and non-microbial fouling, the effect on heat conductivity and the cost effectiveness also needs to be considered. For example, if a modified surface reduces heat transfer, the operation would require more energy to reach the desired temperature. Coatings need to also be either durable and persist through numerous CIPs or be easy to re-apply after cleaning. Surface modification remains an attractive approach to limit biofouling within a dairy plant (Meyer, 2003, Kumar and Anand, 1998). Molecular brush coatings using steric hindrance to block bacterial attachment, have been shown to reduce microbial attachment but the systems studied have largely used polyethylene glycol coatings on glass (Jönsson, 2004, Humphries *et al.*, 1987).

Another approach to the problem of microbial adhesion involves pre-treating the susceptible surface with an antimicrobial agent such as bacteriocins or silver ions (Meyer, 2003). This has been very successful in non-food applications such as preventing microbial accumulation on ship hulls using paints containing toxic substances that are unsuitable for the dairy industry (Srinivasan and Swain 2007). One of the most successful antimicrobial surface treatments has been the development of surfaces incorporating trichlosan (5-chloro-2-(2-4-dichlorophenoxy) phenol) (Moretro and Langsrud, 2004). This antimicrobial, commonly used in toothpastes and soaps, is now being incorporated into wall and floor materials to reduce microbial surface populations in food processing environments. Trichlosan is active against some, but not all, Gram-negative bacteria and is not suitable for surfaces in contact with food as it is a prohibited substance in foods. Antimicrobial surfaces can be prepared by coating with nisin or lysozyme to inhibit the growth of any bacteria that do attach (Bower *et al.*, 1998). However, some of these compounds may have limited use in the dairy industry because milk components may affect the stability of these proteins, and they may also reduce the necessary surface concentrations which are needed to inhibit biofilm formation (Parkar *et al.*, 2003). There are also regulatory concerns with some proposed treatments and costs that need to be considered to make these methods suitable for routine use. The approach of developing an antibacterial food processing surface is a good strategy to improve the hygiene of dairy manufacturing plants. Some of the treatments being developed for the medical industry look promising (Yakandawala *et al.*, 2007). However, to be successful, these treatments need to be cost effective. As DMP strive to reduce costs, increase production runs, enhance quality and meet increasingly stringent product specifications the immediate interest is likely to be alternative cleaning and sanitising chemicals or regimes.

### 15.9.4 New technologies to remove biofilms in DMP

Pipelines can be cleaned using a system known as 'pigging', which involves driving a piston-like object through a pipe by applying a pressure difference across the piston. As the piston moves through the pipe, it scours the pipe walls and 'cleans' them. Conventional pigging is really only effective in long straight pipes of constant diameter with no obstructions. Recently, experimental and theoretical work has been undertaken to investigate the CIP and fouling removal capability of a novel patented crushed ice pigging system (Quarini, 2002). The 'pig' consists of crushed ice in water with a freezing point depressant. The void fraction is carefully controlled so that the ice/water mix moves like a solid plug in free flow areas, but is able flow like a fluid in constricted areas. The ice pig is able to flow in pipes with sharp bends, through orifice plates, through T-junctions and even in plate heat exchangers. The experimental work evaluating the 'cleaning efficiency' of this system indicated that the ice pig could easily and efficiently remove 'soft' fouling; using a volume of ice typically less than one tenth of the volume of water. The fouling materials tested included jam and fats (food industry), toothpaste (personal hygiene products) and fine silt and sand (river water cooled exchangers).

## 15.10 Sources of further information and advice

For additional information on the diary industry the reader is directed to the *Encyclopedia of Dairy Sciences* (Anon, 2002) and the *Technology of Dairy Products* (Early, 1998). On-line access is available to the 4-volume *Encyclopedia of Dairy Sciences*, which is a complete resource on all aspects of dairy science and related food science and technology areas. Extensively cross-referenced, it contains over 400 articles covering the core theories, methods, and techniques employed by dairy scientists. The *Technology of Dairy Products* is a source of highly practical information on dairy product technology. It discusses products in which milk is either the main component or a less obvious ingredient. It explains the methods of milk product manufacture, the technology involved, and how other influences affect finished products.

Sources of additional information on dairy microbiology include *Applied Dairy Microbiology* (Marth and Steele, 2001) and *Standard Methods for the Examination of Dairy Products* (Wehr and Frank, 2004). *Applied Dairy Microbiology* provides comprehensive coverage of the latest developments and scientific advances in dairy microbiology, emphasising probiotics, fermented dairy products, disease prevention, and public health and regulatory control standards for dairy foods. *Standard Methods for the Examination of Dairy Products* (SMEDP) is designed for regulatory, industry, and commercial laboratories that need to assess the safety and quality of dairy products. It is also a compendium of microbiological, chemical, and physical methods for analysing milk and dairy products.

In this chapter reference is made to the role of surface charge and hydrophobicity on the initial attachment of microorganisms to surfaces. The mechanisms of the initial adhesive process and the methods used to study these surfaces have been reviewed and discussed by Bos *et al.* (1999).

## 15.11 References

ADAMS, M. R. & MOSS, M. O. (2000) *Food Microbiology*, Cambridge, Royal Society of Chemistry.

AGARWAL, S., SHARMA, K., SWANSON, B. G., YUKSEL, G. U. & CLARK, S. (2006) Nonstarter lactic acid bacteria biofilms and calcium lactate crystals in Cheddar cheese. *Journal of Dairy Science*, 89, 1452–1466.

ANDERSSON, A., RONNER, U. & GRANUM, P. E. (1995) What problems does the food industry have with the spore-forming pathogens *Bacillus cereus* and *Clostridium perfringens*. *International Journal of Food Microbiology*, 28, 145–155.

ANON (2002) Encyclopedia of Dairy Sciences. In Roginski, H., Fuquay, J. & Fox, P. (Eds.) *Encyclopedia of Dairy Sciences*, Academic Press.

ARIHARA, K. & LUCHANSKY, J. B. (1995) Dairy Lactobacilli. In Hui, Y. H. & Khachatourians, G. G. (Eds.) *Food Biotechnology: Microorganisms.* New York, NY, VCH Pbl.

ARNOLD, J. W. & BAILEY, G. W. (2000) Surface finishes on stainless steel reduce bacterial attachment and early biofilm formation: scanning electron and atomic force microscopy study. *Poultry Science*, 79, 1839–1845.

ARONSON, A. I. & FITZ-JAMES, P. (1976) Structure and morphogenesis of the bacterial spore coat. *Bacteriological Reviews*, 40, 360–402.

ASH, C., FARROW, J. A. E., WALLBANKS, S. & COLLINS, M. D. (1991) Phylogenetic heterogeneity of the genus *Bacillus* revealed by comparative analysis of small-subunit-ribosomal RNA sequences. *Letters in Applied Microbiology*, 13, 202–206.

AUSTIN, J. W. & BERGERON, G. (1995) Development of bacterial biofilms in dairy processing lines. *Journal of Dairy Research*, 62, 509–519.

BANAT, I. M., MARCHANT, R. & RAHMAN, T. J. (2004) *Gobacillus debilis* sp. nov., a novel obligately thermophilic bacterium isolated from a cool soil environment, and reassignment of *Bacillus pallidus* to *Geobacillus pallidus* comb. nov. *International Journal of Systematic and Evolutionary Microbiology*, 54, 2197–2201.

BANSAL, B. & CHEN, X. D. (2006) A critical review of milk fouling in heat exchangers. *Comprehensive Reviews in Food Science and Food Safety*, 5, 27–33.

BARNES, L. M., LO, M. F., ADAMS, M. R. & CHAMBERLAIN, A. H. L. (1999) Effect of milk proteins on adhesion of bacteria to stainless steel surfaces. *Applied and Environmental Microbiology*, 65, 4543–4548.

BELL, R. W. & SANDERS, C. F. (1944) Prevention of milkstone formation in a high-temperature-short-time heater by preheating milk, skim milk and whey. *Journal of Dairy Science*, XXVII, 499–504.

BELMAR-BEINY, M. T. & FRYER, P. J. (1992) Bulk and surface effects on the initial stages of whey fouling. *Food and Bioproducts Processing*, 70, 193–199.

BELMAR-BEINY, M. T. & FRYER, P. J. (1993) Preliminary stages of fouling from whey protein solutions. *Journal of Dairy Research*, 60, 467–483.

BEYENAL, H. & LEWANDOWSKI, Z. (2002) Internal and external mass transfer in biofilms grown at various flow velocities. *Biotechnology Progress*, 18, 55–61.

BIRD, M. R. & BARTLETT, M. (1995) CIP optimization for the food-industry – Relationships between detergent concentration, temperature and cleaning time. *Food and Bioproducts Processing*, 73, 63–70.

BIRD, M. R. & FRYER, P. J. (1991) An experimental study of the cleaning of surfaces fouled by whey proteins. *Transactions of the Institute of Chemical Engineers*, 69, 63–70.

BOS, R., VAN DER MEI, H. C. & BUSSCHER, H. J. (1999) Physico-chemistry of initial microbial adhesive interactions – its mechanisms and methods for study. *FEMS Microbiological Reviews*, 23, 179–230.

BOULANGE-PETERMANN, L., JULLIEN, C., DUBOIS, P. E., BENEZECH, T. & FAILLE, C. (2004) Influence of surface chemistry on the hygienic status of industrial stainless steel. *Biofouling*, 20, 25–33.

BOUMAN, S., LUND, D. B., DRIESSEN, F. M. & SCHMIDT, D. G. (1982) Growth of thermoresistant streptococci and deposition of milk constituents on plates of heat-exchangers during long operating times. *Journal of Food Protection*, 45, 806–812.

BOWER, C. K., MCGUIRE, J. & DAESCHEL, M. A. (1996) The adhesion and detachment of bacteria and spores on food-contact surfaces. *Trends in Food Science and Technology*, 7, 152–157.

BOWER, C. K., DAESCHEL, M. A. & MCGUIRE, J. (1998) Protein antimicrobial barriers to bacterial adhesion. *Journal of Dairy Science*, 81, 2771–2778.

BRANS, G., SCHROEN, C. G. P. H., VAN DER SMAN, R. G. M. & BOOM, R. M. (2004) Membrane fractionation of milk: state of the art and challenges. *Journal of Membrane Science*, 243, 263–272.

BREMER, P., FILLERY, S. & MCQUILLAN, A. J. (2006) Laboratory scale Clean-In-Place (CIP) studies on the effectiveness of different caustic and acid wash steps on the removal of dairy biofilms. *Internationl Journal of Food Microbiology*, 106, 254–262.

BROWN, M. R. W. & GILBERT, P. (1993) Sensitivity of biofilms to antimicrobial agents. *Journal of Applied Bacteriology Symposium Supplement*, 74, 87S–97S.

BUSSCHER, H. J., VAN DER KUIJL-BOOIJ, M. & VAN DER MEI, H. C. (1996) Biosurfactants from thermophilic dairy streptococci and their potential role in the fouling control of heat exchanger plates. *Journal of Industrial Microbiology and Biotechnology*, 16, 15–21.

CARPENTIER, B. & CERF, O. (1993) Biofilms and their consequences with particular reference to hygiene in the food industry. *Journal of Applied Bacteriology*, 75, 499–511.

CHAMBERS, J. V. (2002) The Microbiology of raw milk. In Robinson, R. K. (Ed.) *Dairy Microbiology Handbook*. 3rd ed., Wiley-InterScience, Inc.

CHANGANI, S. D., BELMAR-BEINY, M. T. & FRYER, P. J. (1997) Engineering and chemical factors associated with fouling and cleaning in milk processing. *Experimental Thermal and Fluid Science*, 14, 392–406.

CRIADO, M., SUAREZ, B. & FERREIROS, C. M. (1994) The importance of bacterial adhesion in the dairy industry. *Food Technology*, 2, 125–126.

CUNLIFFE, D., SMART, C. A., ALEXANDER, C. & VULFSON, E. N. (1999) Bacterial adhesion at synthetic surfaces. *Applied and Environmental Microbiology*, 65, 4995–5002.

DE JONG, P. (1997) Impact and control of fouling in milk processing. *Trends in Food Science & Technology*, 8, 401–405.

DE JONG, P., TE GIFFEL, M. C. & KIEZEBRINK, E. A. (2002) Prediction of the adherence, growth and release of microorganisms in production chains. *International Journal of Food Microbiology*, 74, 13–25.

DEFRISE, D. & GEKAS, V. (1988) Microfiltration membranes and the problem of microbial adhesion. A literature survey. *Process Biochemistry*, 23, 105–116.

DODDS, M. G., GROBE, J. & STEWART, P. S. (2000) Modelling biofilm antimicrobial resistance. *Biotechnology and Bioengineering*, 68, 456–465.

DRIESSEN, F. M., DEVRIES, J. & KINGMA, F. (1984) Adhesion and growth of thermoresistant streptococci on stainless-steel during heat-treatment of milk. *Journal of Food Protection*, 47, 848–852.

DRIKS, A. (1999) *Bacillus subtilis* spore coat. *Microbiology and Molecular Biology Reviews*, 63, 1–20.

DUFOUR, M., SIMMONDS, R. S. & BREMER, P. J. (2004) Development of a laboratory scale clean-in-place system to test the effectiveness of 'Natural' antimicrobials against dairy Biofilms. *Journal of Food Protection*, 67, 1438–1443.

DUNSMORE, D. G. (1981) Bacteriological control of food equipment surfaces by cleaning systems. 1. Detergent effects. *Journal of Food Protection*, 44, 15–20.

EARLY, R. (1998) *The Technology of Dairy Products*, Kluwer Academic/Plenum Publishers.

EGINTON, P. J., HOLAH, J., ALLISON, D. G., HANDLEY, P. S. & GILBERT, P. (1998) Changes in the strength of attachment of micro-organisms to surfaces following treatment with disinfectants and cleansing agents. *Letters in Applied Microbiology*, 27, 101–105.

FAILLE, C., FONTAINE, F. & BENEZECH, T. (2001) Potential occurrence of adhering living *Bacillus spores* in milk product processing lines. *Journal of Applied Microbiology*, 90, 892–900.

FAILLE, C., JULLIEN, C., FONTAINE, F., BELLON-FONTAINE, M. N., SLOMIANNY, C. & BENEZECH, T. (2002) Adhesion of *Bacillus* spores and *Escherichia coli* cells to inert surfaces: role of surface hydrophobicity. *Canadian Journal of Microbiology*, 48, 728–738.

FLEET, G. H. (1990) Yeasts in dairy-products. *Journal of Applied Bacteriology*, 68, 199–211.

FLEMMING, H. C. & SCHAULE, G. (1988) Biofouling on membranes – a microbiological approach. *Desalination*, 70, 95–119.

FLINT, S. H. & BROOKS, J. D. (2001) Rapid detection of *Bacillus stearothermophilus* using impedance-splitting. *Journal of Microbiological Methods*, 44, 205–208.

FLINT, S., BROOKS, J. D. & VAN DEN ELZEN, H. (1997a) Biofilms in dairy manufacturing plant – a threat to product quality. *The Food Technologist*, 27, 61–74.

FLINT, S. H., BREMER, P. J. & BROOKS, J. D. (1997b) Biofilms in dairy manufacturing plant – Description, current concerns and methods of control. *Biofouling*, 11, 81–97.

FLINT, S. H., BROOKS, J. D. & BREMER, P. J. (1997c) The influence of cell surface properties of thermophilic streptococci on attachment to stainless steel. *Journal of Applied Microbiology*, 83, 508–517.

FLINT, S. H., VAN DEN ELZEN, H., BROOKS, J. D. & BREMER, P. J. (1999) Removal and inactivation of thermo-resistant streptococci colonising stainless steel. *International Dairy Journal*, 9, 429–436.

FLINT, S. H., BROOKS, J. D. & BREMER, P. J. (2000) Properties of the stainless steel substrate influencing the adhesion of thermo-resistant streptococci. *Journal of Food Engineering*, 43, 235–242.

FLINT, S., PALMER, J., BLOEMEN, K., BROOKS, J. & CRAWFORD, R. (2001) The growth of *Bacillus stearothermophilus* on stainless steel. *Journal of Applied Microbiology*, 90, 151–157.

FLINT, S., BROOKS, J., BREMER, P., WALKER, K. & HAUSMAN, E. (2002) The resistance to heat of thermo-resistant streptococci attached to stainless steel in the presence of milk. *Journal of Industrial Microbiology and Biotechnology*, 28, 134–136.

FRANK, J. F., GILLETT, R. A. N. & WARE, G. O. (1990) Association of *Listeria* spp contamination in the dairy processing plant environment with the presence of staphylococci. *Journal of Food Protection*, 53, 928–932.

FRYER, P. J. & BIRD, M. R. (1994) Factors which eefect the kinetics of cleaning dairy soils. *Food Science and Technology Today*, 8, 36–42.

FRYER, P. J., CHRISTIAN, G. K. & LIU, W. (2006) How hygiene happens: physics and chemistry of cleaning. *International Journal of Dairy Technology*, 59, 76–84.

GILMOUR, A. & ROWE, M. T. (1990) *Micro-organisms associated with milk*, New York, USA, Elsevier.

HEINEN, W., LAUWERS, A. M. & MULDERS, J. W. (1982) *Bacillus licheniformis*, a newly isolated facultative thermophile. *Antonie van Leeuwenhoek*, 48, 265–272.

HELKE, D. M., SOMERS, E. B. & WONG, A. C. L. (1993) Attachment of *Listeria monocytogenes* and *Salmonella typhimurium*, to stainless steel and buna-N in the presence of milk and milk components. *Journal of Food Protection*, 56, 479–484.

HENRIQUES, A. O. & MORAN, C. P. (2000) Structure and assembly of the bacterial endospore coat. *Methods*, 20, 95–110.

HILL, B. M. & SMYTHE, B. W. (1994) Progress in understanding the behaviour of thermophilic bacteria during milk powder manufacture. *Milk Powders for the Future II.* Palmerston North, New Zealand, Dairy Research Institute.

HINTON, A. R., TRINH, K. T., BROOKS, D. J. & MANDERSON, G. J. (2001) Thermophile growth in the preheating section of a milk powder pilot plant. *6th World Congress of Chemical Engineering.* Melbourne, Australia.

HINTON, A. R., TRINH, K. T., BROOKS, J. D. & MANDERSON, G. J. (2002) Thermophile survival in milk fouling and on stainless steel during cleaning. *Trans IChemE*, 80, 299–304.

HOLAH, J. & GIBSON, H. (2000) Food industry biofilms. In Allison, D., Gilbert, J. P., Lappin-Scott, H. & Wilson, M. (Eds.) *Community structure and co-operation in biofilms.* Cambridge University Press.

HOOD, S. K. & ZOTTOLA, E. A. (1995) Biofilms in food processing. *Food Control*, 6, 9–18.

HOOD, S. K. & ZOTTOLA, E. A. (1997) Adherence to stainless steel by foodborne microorganisms during growth in model food systems. *International Journal of Food Microbiology*, 37, 145–153.

HUMPHRIES, M., NEMCEK, J., CANTWELL, J. B. & GERRARD, J. J. (1987) The use of graft co-polymers to inhibit the adhesion of bacteria to solid surfaces. *FEMS Microbiology Ecology*, 45, 297–304.

IVERSEN, C., LANE, M. & FORSYTHE, S. J. (2004) The growth profile, thermotolerance and biofilm formation of *Enterobacter sakazakii* grown in infant formula milk. *Letters in Applied Microbiologyl*, 38, 378–382.

JEONG, D. K. & FRANK, J. F. (1994) Growth of *Listeria monocytogenes* at 21°C in biofilms with microorganisms isolated from meat and dairy environments. *Lebensmittel-Wissenschaft und-Technologie*, 27, 415–424.

JÖNSSON, M. A. J., H.-O (2004) Effect of surface grafted olymers on the adsoprtion of different model proteins. *Colloids and Surfaces B: Biointerfaces*, 37, 71–81.

KIM, H., RYU, J. H. & BEUCHAT, L. R. (2006) Attachment and biofilm formation by *Enterobacter sakazakii* on stainless steel and enteral feeding tubes *Applied and Environmental Microbiology*, 72, 5846–5856.

KIM, H., BANG, J., BEUCHAT, L. R. & RYU, J. H. (2008) Fate of *Enterobacter sakazakii* attached to or in biofilms on stainless steel upon exposure to various temperatures or relative humidities. *Journal of Food Protection*, 71, 940–945.

KIRTLEY, S. A. & MCGUIRE, J. (1989) On differences in surface constitution of dairy product contact materials. *Journal of Dairy Science*, 72, 1748–1753.

KIVES, J., GUADARRAMA, D., ORGAZ, B., RIVERA-SEN, A., VAZQUEZ, J. & SANJOSE, C. (2005) Interactions in biofilms of *Lactococcus lactis* ssp. *cremoris* and *Pseudomonas fluorescens* cultured in cold UHT milk. *Journal of Dairy Science*, 88, 4165–4171.

KLAVENES, A., STALHEIM, T., SJOVOLD, O., JOSEFSEN, K. & GRANUM, P. E. (2002) Attachment of *Bacillus cereus* spores without and without appendages to stainless steel surfaces. *Trans IChemE*, 80, 312–318.

KNIGHT, G. C., NICOL, R. S. & MCMEEKIN, T. A. (2004) Temperature step changes: a novel approach to control biofilms of *Streptococcus thermophilus* in a pilot plant-scale cheese-milk pasteurisation plant. *International Journal of Food Microbiology*, 93, 305–318.

KOUTZAYIOTIS, C. (1992) Bacterial biofilms in milk pipelines. *South African Journal of Dairy Science*, 24, 19–22.

KUMAR, C. G. & ANAND, S. K. (1998) Significance of microbial biofilms in food industry: a review. *International Journal of Food Microbiology*, 42, 9–27.

LANGEVELD, L. P. M., VAN MONTFORT-QUASIG, R. M. G. E., WEERKAMP, A. H., WAALEWIJN, R. & WEVER, J. S. (1995) Adherence, growth and release of bacteria in a tube heat exchanger for milk. *Netherlands Milk and Dairy Journal*, 49, 207–220.

LECLEROQ-PERLAT, M.-N. & LALANDE, M. (1994) Cleanability in relation to surface chemical composition and surface finishing of some materials commonly used in food industries. *Journal of Food Engineering*, 23, 501–517.

LEE, M. W., KNIGHT, G. & NICOL R. S. (1997) Development of a biofilm of thermoduric streptococci. *25th Australian and New Zealand Chemical Engineers' Conference and Exhibition* Rotorua, New Zealand.

LEHMANN, F. L. (1995) Recontamination of industrial ultrafiltration units and pasteurisers by thermoduric-thermophilic bacteria. *Interntational Dairy Federation Symposium on Heat Treatments and Alternative Methods.* Vienna.

LELIEVRE, C., FAILLE, C. & BENEZECH, T. (2001) Removal kinetics of *Bacillus cereus* spores from stainless steel pipes under CIP procedure: influence of soiling and cleaning conditions. *Journal of Food Processing Engineering*, 24, 359–379.

LELIEVRE, C., ANTONINI, G., FAILLE, C. & BENEZECH, T. (2002a) Cleaning-in-place – Modelling of cleaning kinetics of pipes soiled by *Bacillus* spores assuming a process combining removal and deposition. *Food and Bioproducts Processing*, 80, 305–311.

LELIEVRE, C., LEGENTILHOMME, P., GAUCHER, C., LEGRAND, J., FAILLE, C. & BENEZECH, T. (2002b) Cleaning in place: effect of local wall shear stress variation on bacterial removal from stainless steel equipment. *Chemical Engineering Science*, 57, 1287–1297.

LEUSCHNER, R. G. K. & LILLFORD, P. J. (2001) Investigation of bacterial spore structure by high resolution solid-state nuclear magnetic resonance spectroscopy and transmission electron microscopy. *International Journal of Food Microbiology*, 63, 35–50.

LEWIS, S. J. & GILMOUR, A. (1987) Microflora associated with the internal surfaces of rubber and stainless steel milk transfer pipeline. *Journal of Applied Bacteriology*, 62, 327–333.

LI, B. & LOGAN, B. E. (2004) Bacterial adhesion to glass and metal-oxide surfaces. *Colloids and Surfaces B: Biointerfaces*, 36, 81–90.

LICITRA, G., OGIER, J. C., PARAYRE, S., PEDILIGGIERI, C., CARNEMOLLA, T. M., FALENTIN, H., MADEC, M. N., CARPINO, S. & LORTAL, S. (2007) Variability of bacterial biofilms of the tina wood vats used in the Ragusano cheese-making process. *Applied and Environmental Microbiology*, 73, 6980–6987.

LIN, S., SCHRAFT, H., ODUMERU, J. A. & GRIFFITHS, M. W. (1998) Identification of contamination sources of *Bacillus cereus* in pasteurized milk. *International Journal of Food Microbiology*, 43, 159–171.

LINDSAY, D., MOSUPYE, F. M., BROZEL, V. S. & VON HOLY, A. (2000) Cytotoxicity of alkaline-tolerant dairy-associated *Bacillus* spp. *Letters in Applied Microbiology*, 30, 364–369.

MAFU, A. A., ROY, D., GOULET, J. & MAGNY, P. (1990) Attachment of *Listeria monocytogenes* to stainless steel, glass, polypropylene, and rubber surfaces after short contact times. *Journal of Food Protection*, 53, 742–746.

MARIANI, C., BRIANDET, R., CHAMBA, J. F., NOTZ, E., CARNET-PANTIEZ, A., EYOUNG, R. N. & OULAHAL, N. (2007) Biofilm ecology of wooden shelves used in ripening the French raw milk smear cheese Reblochon de Savoie. *Journal of Dairy Science*, 90, 1653–1661.

MARKS, T. & SHARP, R. (2000) Bacteriophages and biotechnology: a review. *Journal of Chemical Technology and Biotechnology*, 75, 6–17.

MARTH, E. H. & STEELE, J. L. (2001) *Applied Dairy Microbiology*, Marcel Dekker.

MCGUIGGAN, J. T. M., MCCLEERY, D. R., HANNAN, A. & GILMOUR, A. (2002) Aerobic spore-forming bacteria in bulk raw milk: factors influencing the numbers of psychrotrophic, mesophilic and thermophilic *Bacillus* spores. *International Journal of Dairy Technology*, 55, 100–107.

MELO, L. F. & VIEIRA, M. J. (1999) Physical stability and biological activity of biofilms under turbulent flow and low substrate concentration. *Bioprocessing Engineering*, 20, 363–368.

MERCADÉ-PRIETO, R., SAHOO, P. K., FALCONER, R. J., PATERSON, W. R. & WILSON, D. I. (2007) Polyectrolyte screening effects on the dissolution of whey protein gels at high pH conditions. *Food Hydrocolloids*, 21, 1275–1284.

MEYER, B. (2003) Approaches to prevention, removal and killing of biofilms. *International Biodeterioration and Biodegradation*, 51, 249–253.

MITTELMAN, M. W. (1998) Structure and functional characteristics of bacterial biofilms in fluid processing operations. *Journal of Dairy Science*, 81, 2760–2764.

MORETRO, T. & LANGSRUD, S. (2004) *Listeria monocytogenes:* biofilm formation and persistance in food-processing environments. *Biofilms*, 1, 107–121.

MUIR, D. D., GRIFFITHS, M. W., PHILLIPS, J. D., SWEETSUR, A. W. M. & WEST, I. G. (1986) Effect of the bacterial quality of raw milk on the bacterial quality and some other properties of low-heat and high-heat dried milk. *Journal of the Society of Dairy Technology*, 39, 115–118.

MURPHY, P. M., LYNCH, D. & KELLY, P. M. (1999) Growth of thermophilic spore forming bacilli in milk during the manufacture of low heat powders. *International Journal of Dairy Technology*, 52, 45–50.

MUTHUKUMARAN, S., KENTISH, S., LALCHANDANI, S., ASHOKKUMAR, M., MAWSON, R., STEVENS, G. W. & GRIESER, F. (2005) The optimisation of ultrasonic cleaning procedures for dairy fouled ultrafiltration membranes. *Ultrasonics Sonochemistry*, 12, 29–35.

NAZINA, T. N., TOUROVA, T. P., POLTARAUS, A. B., NOVIKOVA, E. V., GRIGORYAN, A. A., IVANOVA, A. E., LYSENKO, A. M., PETRUNYAKA, V. V., OSIPOV, G. A., BELYAEV, S. S. & IVANOV, M. V. (2001) Taxonomic study of aerobic thermophilic bacilli: descriptions of *Geobacillus subterraneus* gen. nov., sp. nov. and *Geobacillus uzenensis* sp. nov. from petroleum reservoirs and transfer of *Bacillus stearothermophilus*, *Bacillus thermocatenulatus*, *Bacillus thermoleovorans*, *Bacillus kaustophilus*, *Bacillus thermoglucosidasius* and *Bacillus thermodenitrificans* to *Geobacillus* as the new combinations *G. stearothermophilus*, *G. thermocatenulatus*, *G. thermoleovorans*, *G. kaustophilus*, *G. thermoglucosidasius* and *G. thermodenitrificans. International Journal of Systematic and Evolutionary Microbiology*, 51, 433–446.

NICHOLSON, W. L., MUNAKATA, N., HORNECK, G., MELOSH, H. J. & SETLOW, P. (2000) Resistance of *Bacillus* endospores to extreme terrestial and extraterrestrial environments. *Microbiology and Molecular Biology Reviews*, 64, 548–572.

PARKAR, S. G., FLINT, S. H., PALMER, J. S. & BROOKS, J. D. (2001) Factors influencing attachment of thermophilic bacilli to stainless steel. *Journal of Applied Microbiology*, 90, 901–908.

PARKAR, S. G., FLINT, S. H. & BROOKS, J. D. (2003) Physiology of biofilms of thermophilic bacilli-potential consequences for cleaning. *Journal of Industrial Microbiology and Biotechnology*, 30, 553–560.

PARKAR, S. G., FLINT, S. H. & BROOKS, J. D. (2004) Evaluation of the effect of cleaning regimes on biofilms of thermophilic bacilli on stainless steel. *Journal of Applied Microbiology*, 96, 110–116.

PEDERSON, K. (1990) Biofilm development on stainless steel and PVC surfaces in drinking water. *Water Research*, 24, 239–243.

PETERSON, S. D. & MARSHALL, R. T. (1990) Nonstarter lactobacilli in Cheddar cheese: a review. *Journal of Dairy Science*, 73, 1395–1410.

PIKUTA, E., LYSENKO, A., CHUVILSKAYA, N., MENDROCK, U., HIPPE, H., SUZINA, N., NIKITIN, D., OSIPOV, G. & LAURINAVICHIUS, K. (2000) *Anoxybacillus pushchinensis* gen. nov., sp no., a novel anaerobic, alkaliphilic, moderately thermophilic bacterium from manure, and the decription of *Anoxybacillus flavithermus* comb nov. *International Journal of Systematic and Evolutionary Microbiology*, 50, 2109–2117.

QUARINI, J. (2002) Ice-pigging to reduce and remove fouling and to achieve clean-in-place. *Applied Thermal Engineering*, 22, 747–753.

REDDY, P. C., ATWAL, J. S. & SRINIVASAN, R. A. (1975) Studies on the survival of thermophilic bacteria during manufacture and storage in dried milks. *Indian Journal of Dairy Science*, 28, 289–295.

RIDGWAY, H. F., KELLY, A., JUSTICE, C. & OLSON, B. H. (1983) Microbial fouling of reverse-osmosis membranes used in advanced wastewater treatment technology: chemical, bacteriological, and ultrastructural analyses. *Applied and Environmental Microbiology*, 45, 1066–1084.

ROBINSON, R. K. (2002) *Dairy Microbiology*, Chichester, Wiley.

RONIMUS, R. S., PARKER, L. E., TURNER, N., POUDEL, S., RUCKERT, A. & MORGAN, H. W. (2003) A RAPD-based comparison of thermophilic bacilli from milk powders. *International Journal of Food Microbiology*, 85, 45–61.

RONIMUS, R. S., RUECKERT, A. & MORGAN, H. W. (2006) Survival of thermophilic spore-forming bacteria in a 90+-year-old milk powder from Ernest Shackleton's Cape Royds Hut in Antarctica. *Journal of Dairy Research*, 73, 235–243.

ROSMANINHO, R., SANTOS, O., NYLANDER, T., PAULSSON, M., BEUF, M., BENEZECH, T., YIANTSIOS, S., ANDRITSOS, N., KARABELAS, A., RIZZO, G., MÜLLER-STEINHAGEN, H. & MELO, L. F. (2007) Modified stainless steel surfaces targeted to reduce fouling – Evaluation of fouling by milk components. *Journal of Food Engineering*, 80, 1176–1187.

RUECKERT, A., RONIMUS, R. S. & MORGAN, H. W. (2004) A RAPD-based survey of thermophilic bacilli in milk powders from different countries. *International Journal of Food Microbiology*, 96, 263–272.

RUECKERT, A., RONIMUS, R. S. & MORGAN, H. W. (2005a) Development of a rapid detection and enumeration method for thermophilic bacilli in milk powders. *Journal of Microbiological Methods*, 60, 155–167.

RUECKERT, A., RONIMUS, R. S. & MORGAN, H. W. (2005b) Rapid differentiation and enumeration of the total, viable vegetative cell and spore content of thermophilic bacilli in milk powders with reference to *Anoxybacillus flavithermus*. *Journal of Applied Microbiology*, 99, 1246–1255.

RUECKERT, A., RONIMUS, R. S. & MORGAN, H. W. (2006) Development of a real-time PCR assay targeting the sporulation gene, spo0A, for the enumeration of thermophilic bacilli in milk powder. *Food Microbiology*, 23, 220–230.

RUSSELL, M. J. (1992) Live long and prosper. *Food Engineering*, Dec, 77–80.

RUSSELL, P. (1993) *The formation of biofilms*. Milk Industry. London.

SANTOS, O., NYLANDER, T., ROSMANINHO, R., RIZZO, G., YIANTSIOS, S., ANDRITSOS, N., KARABELAS, A., MÜLLER-STEINHAGEN, H., MELO, L., BOULANGÉ-PETERMANN, L., GABET, C., BRAEM, A., TRÄGÅRDH, C. & PAULSSON, M. (2004) Modified stainless steel surfaces targeted to reduce fouling-surface characterization. *Journal of Food Engineering*, 64, 63–79.

SCOTT, S. A., BROOKS, J. D., RAKONJAC, J., WALKER, K. M. R. & FLINT, S. H. (2007) The formation of thermophilic spores during the manufacture of whole milk powder. *International Journal of Dairy Technology*, 60, 109–117.

SEALE, R. B., FLINT, S. H., MCQUILLAN, A. J. & BREMER, P. J. (2008) Recovery of spores from thermophilic dairy bacilli and effects of their surface characteristics on attachment to different surfaces. *Applied and Environmental Microbiology*, 74, 731–737.

SETLOW, P. (2006) Spores of *Bacillus subtilis:* their resistance to and killing by radiation, heat and chemicals. *Journal of Applied Microbiology*, 101, 514–525.

SETLOW, P. & KORNBERG, A. (1970) Biochemical studies of bacterial sporulation and germination: XXII energy metabolism in early stages of germination of *Bacillus megaterium* spores. *Journal of Biological Chemistry*, 245, 3637–3644.

SHARMA, M. & ANAND, S. K. (2002a) Bacterial biofilm on food contact surfaces: A review. *Journal of Food Science and Technology-Mysore*, 39, 573–593.

SHARMA, M. & ANAND, S. K. (2002b) Characterization of constitutive microflora of biofilms in dairy processing lines. *Food Microbiology*, 19, 627–636.

SIMMONDS, P., MOSSEL, B. L., INTARAPHAN, T. & DEETH, H. C. (2003) Heat resistance of *Bacillus* spores when adhered to stainless steel and its relationship to spore hydrophobicity. *Journal of Food Protection*, 66, 2070–2075.

SOMERS, E. B., JOHNSON, M. E. & WONG, A. C. L. (2001) Biofilm formation and contamination of cheese by nonstarter lactic acid bacteria in the dairy environment. *Journal of Dairy Science*, 84, 1926–1936.

SOMMER, P., MARTIN-ROUAS, C. & METTLER, E. (1999) Influence of the adherent population level on biofilm population, structure and resistance to chlorination. *Food Microbiology*, 16, 503–515.

SOUTO, L. I. M., MINAGAWAY, C. Y., TELES, E. O., GABUGLIO, M. A., AMAKU, M., DAIS, R. A., SAKATA, S. T. & BENITES, N. R. (2008) Relationship between occurance of mastitis pathogens in the dairy cattle herds and raw milk indicators of hgyienic-sanitary quality. *Journal of Dairy Research*, 75, 121–127.

SPEERS, J. G. & GILMOUR, A. (1985) The influence of milk and milk components on the attachment of bacteria to farm dairy equipment surfaces. *Journal of Applied Bacteriology*, 59, 325–332.

SRINVASAN, M. & SWAIN, G. (2007) Managing the use of copper-based antifouling paints. *Environment Management*, 39(3), 423–441.

STADHOUDERS, J., HUP, G. & HASSING, F. (1982) The conceptions index and indicator organisms discussed on the basis of the bacteriology of spray-dried milk powder. *Netherlands Milk and Dairy Journal*, 36, 231–260.

STEWART, J. C. & SEIBERLING, D. A. (1996) The secrets out: Clean in place. *Chemical Engineering (New York)*, 103, 72–79.

STEWART, P. S., ROE, F., RAYNER, J., ELKINS, J. G., LEWANDOWSKI, Z., OCHSNER, U. A. & HASSETT, D. J. (2000) Effect of catalase on hydrogen peroxide penetration into *Pseudomonas aeruginosa* biofilms. *Applied and Environmental Microbiology*, 66, 836–838.

STONE, L. S. & ZOTTOLA, E. A. (1985) Effect of cleaning and sanitizing on the attachment of *Pseudomonas fragi* to stainless steel. *Journal of Food Science*, 50, 951–956.

STOODLEY, P., SAUER, K., DAVIES, D. G. & COSTERTON, J. W. (2002) Biofilms as complex differentiated communities. *Annual Reviews in Microbiology*, 56, 187–209.

SUTHERLAND, I. W., HUGHES, K. A., SKILLMAN, L. C. & TAIT, K. (2004) The interaction of phage and biofilms. *FEMS Microbiology Letters*, 232, 1–6.

TAI, S. K., LIN, H. P., KUO, J. & LIU, J. K. (2004) Isolation and characterization of a cellulolytic *Geobacillus thermoleovorans* T4 strain from a sugar refinery wastewater. *Extremophiles*, 8, 345–349.

TE GIFFEL, M. C., BEUMER, R. R., LANGEVELD, L. P. M. & ROMBOUTS, M. (1997) The role of heat exchangers in the contamination of milk with *Bacillus cereus* in dairy processing plants. *International Journal of Dairy Technology*, 50, 43–47.

TE GIFFEL, M. C., WAGENDORP, A., HERREWEGH, A. & DRIEHUIS, F. (2002) Bacterial spores in silage and raw milk. *Antonie van Leeuwenhoek*, 81, 625–630.

TEIXEIRA, P., LOPES, Z., AZEREDO, J., OLIVEIRA, R. & VIEIRA, M. J. (2005) Physico-chemical surface characterization of a bacterial population isolated from a milking machine. *Food Microbiology*, 22, 247–251.

TIDE, C., HARKIN, S. R., GEESEY, G. G., BREMER, P. J. & SCHOLZ, W. (1999) The influence of welding procedures on bacterial colonization of stainless steel weldments. *Journal of Food Engineering*, 42, 85–96.

TSIBOUKLIS, J., STONE, M., THORPE, A. A., GRAHAM, P., NEVELL, T. G. & EWEN, R. J. (2000) Inhibiting bacterial adhesion onto surfaces: the non-stick coating approach. *International Journal of Adhesion and Adhesives*, 20, 91–96.

VANHAECKE, E., REMON, J. P., MOORS, M., RAES, F., DE RUDDER, D. & VAN PETEGHEM, A. (1990) Kinetics of *Pseudomonas aeruginosa* adhesion to 304 and 316-L stainless steel: role of cell surface hydrophobicity. *Applied and Environmental Microbiology*, 56, 788–795.

VERRAN, J. & WHITEHEAD, K. A. (2006) Assessment of organic materials and microbial components on hygienic surfaces. *Food and Bioproducts Processing*, 84, 253–259.

VLAMAKIS, H., AGUILAR, C., LOSICK, R. & KOLTER, R. (2008) Control of cell fate by the formation of an architecturally complex bacterial community. *Genes & Development*, 22, 945–953.

WALKER, S. J. (1971) Major spoilage micro-organisms in milk and dairy products. *Journal of the Society of Dairy Technology*, 41, 91–92.

WARNECKE, F. (2001) The ecology of thermophilic bacilli of milk powder processing plants. Hamilton, NZ, M. Phil., Thesis, University of Waikato.

WEHR, H. M. & FRANK, J. F. (2004) *Standard Methods for the Examination of Dairy Products*. American Public Health Association.

WHITE, J. C. & RABE, G. O. (1970) Evaluating the use of nitric acid as a detergent in dairy cleaned-in-place systems. *Journal of Milk and Food Technology*, 33, 25–28.

WIJMAN, J. G. E., DE LEEUW, P. P. L., MOEZELAAR, R., ZWIETERING, M. H. & ABEE, T. (2007) Air-liquid interface biofilms of *Bacillus cereus*: formation, sporulation, and dispersion. *Applied and Environmental Microbiology*, 73, 1481–1488.

WIRTANEN, G. & MATTILA-SANDHOLM, T. (1992) Effect of the growth phase of food-borne biofilms on their resistance to a chlorine sanitizer. Part II. *Lebensm. -Wiss. u. -Technol*, 25, 50–54.

WIRTANEN, G., HUSMARK, U. & MATTILA-SANDHOLM, T. (1996) Microbial evaluation of the biotransfer potential from surfaces with *Bacillus* biofilms after rinsing and cleaning procedures in closed food-processing systems. *Journal of Food Protection*, 59, 727–733.

WONG, A. C. (1998) Biofilms in food processing environments. *Journal of Dairy Science*, 81, 2765–2770.

YAKANDAWALA, N., GAWANDE, P. V., LOVERTRI, K. & MADHYASTHA, S. (2007) Effect of ovotransferrin, protamine sulphate and EDTA combination on biofilm formation by catheter-associated bacteria. *Journal of Applied Microbiology*, 102, 722–727.

YOO, J.-A., HARDIN, M. T. & CHEN, X. D. (2006) The influence of milk composition on the growth of *Bacillus stearothermophilus*. *Journal of Food Engineering*, 77, 96–102.

YOON, J. & LUND, D. B. (1989) Effect of operating conditions, surface coatings and pretreatment on milk fouling in a plate heat exchanger. In Kessler, H. G. (Ed) *Fouling and Cleaning in Food Processing*, pp 59–80, University of Munich, Germany.

ZHAO, Q. (2004) Effect of surface free energy of graded Ni-P-PTFE coatings on bacterial adhesion. *Surface and Coatings Technology*, 185, 199–204.

ZHAO, Q., LIU, Y., WANG, C., WANG, S. & MULLER-STEINHAGEN, H. (2005) Effect of surface free energy on the adhesion of biofouling and crystalline fouling. *Chemical Engineering Science*, 60, 4858–4865.

ZOLTAI, P. T., ZOLTOLA, E. A. & MCKAY, L. (1981) Scanning electron microscopy of microbial attachment to milk contact surface. *Journal of Food Protection*, 44, 204–208.

ZOTTOLA, E. A. & SASAHARA, K. C. (1994) Microbial biofilms in the food-processing industry – should they be a concern. *International Journal of Food Microbiology*, 23, 125–148.

# 16

# Biofilms and brewing

**E. Storgårds and O. Priha, VTT Technical Research Centre of Finland, Finland**

**Abstract**: Beer is generally a hostile environment for most microorganisms rendering it safe in terms of food-borne illnesses. However, species adapted to the brewery environment can cause product spoilage. In this chapter, microorganisms associated with brewing and the production of beer as well as with beer dispensing are briefly described. Keeping the beer free from contaminating organisms is important for image reasons and, additionally, production plant hygiene has both economic and environmental motivation. In the effort of improving brewery hygiene, studies regarding brewery biofilms have lately been carried out. This chapter aims at reviewing the current knowledge of biofilms occurring in the brewery. It also summarises the means available for prevention of biofilm formation with special emphasis on emerging technologies and future trends in biofilm combat and their application in breweries.

**Key words**: beer, biofilms, brewing, dispense, hygiene, functional materials, quorum sensing.

## 16.1 Introduction

Beer is generally regarded as safe in terms of food-borne illnesses, due to the belief that pathogens are not able to grow in beer. The biological stability of modern brewery products is also very good, with best before dates ranging from 6 to 12 months or more from production. Why then is hygiene still considered so important in the brewing industry? The brewing process itself is prone to growth of microorganisms because of the nutrient-rich environment of wort and the additional growth factors produced by the brewing yeast. The comparatively long production run in traditional brewing, with batch fermentations of up to several weeks, gives plenty of time for unwanted microorganisms to develop, when given the opportunity. The microbiological sensitivity of rapid continuous fermentation systems using immobilised yeast is well documented (Kronlöf and Haikara 1991, Haikara and Kronlöf 1995, Haikara *et al.* 1997, Virkajärvi *et al.* 2001). However, work carried out for over one century in the field of brewery microbiology, ever since the pioneering studies of Louis Pasteur (1876) and E.C. Hansen (1896), has resulted in the high hygienic standard of modern breweries. Both economic and environmental reasons oblige the breweries to avoid the risk that the image of a beer brand would suffer because of quality losses due to microbiological problems in the process.

Beer production and dispensing takes place mainly in closed systems, where cleaning-in-place procedures without the need for dismantling are applied. Long runs between cleaning are also typical for these systems. Such systems are susceptible to bacterial attachment and accumulation on surfaces. Areas in which biofilms mainly develop are those that are the most difficult to rinse, clean and disinfect and also those most difficult to sample. Contaminations in the brewery can be divided into primary contaminations originating from the yeast, wort, fermentation, maturation or pressure tanks, and secondary contaminations originating from bottling, canning or kegging. Secondary contaminations are responsible for at least half of the incidents of microbiological spoilage in breweries not using tunnel pasteurisation (Storgårds 2000, Back 2003). Thus, all points with direct or indirect contact with cleaned or filled unsealed bottles are possible sources of contamination. Microbiological problems seldom arise suddenly but result from the gradual build-up of microbial communities on critical sites.

## 16.2 Biofilms in beer production and dispensing

### 16.2.1 Microorganisms associated with beer production

Owing to the fact that beer is a hostile environment for most microorganisms, only a few species can actually cause spoilage by growing in beer (e.g., Back 1994, Storgårds *et al.* 2006a). These include mainly bacteria representing the genera *Lactobacillus*, *Pediococcus*, *Pectinatus* or *Megasphaera* and yeast representing *Saccharomyces* or *Dekkera*. The beer spoilage organisms are microaerophiles or anaerobes and they prefer or tolerate acidic conditions. The presence of inhibitors such as hop compounds, alcohol, carbon dioxide and sulphur dioxide as well as the shortage of nutrients and oxygen and the low pH all make beer resistant to microbial contamination. Moreover, processes such as filtration, storage at low temperatures and possible pasteurisation reduce contamination. While the special environment in the brewing process restricts the range of microorganisms likely to cause spoilage to relatively few species, on the other hand, species adapted to the brewery environment have often not been isolated elsewhere (Back 1994, Haikara and Helander 2006). These organisms may survive for a long time in niches of the process, probably outside the direct product stream, without causing signs of contamination. Then suddenly, they may contaminate the entire process as a consequence of technological faults or insufficient cleaning.

New species adapted to the brewery environment are constantly being described as modern methods are implemented also into the field of brewing. Such 'newcomers' among beer spoilage bacteria are *Pediococcus claussenii* (Dobson *et al.* 2002), *Lactobacillus paracollinoides* (Suzuki *et al.* 2004), *Megasphaera paucivorans*, *Megasphaera sueciensis* and *Pectinatus haikarae* (Juvonen and Suihko 2006). Despite of this, the most common beer spoilage bacterium continues to be *Lactobacillus brevis* (Back 1994, Priest 2003).

A wide range of Gram-negative bacteria, mainly members of the family Enterobacteriaceae and *Pseudomonas* spp. are also frequently encountered in breweries even if they are unable to grow in beer. Some of these may grow in early stages of the process causing off flavours in the final product if present in sufficient numbers (Back 1994). Although the contaminants found may cause quality defects, pathogens have not been reported to grow in standard beer products (e.g. Back 1994, L'Anthoën and Ingledew 1996, Baxter and Hughes 2001).

### 16.2.2 Biofilms in beer production and bottling

Beer bottling plants consist of complex machines with plenty of niches for biofilm formation. At least half of the incidents of microbiological spoilage in breweries can be traced to the bottling department (Back 1994, Storgårds 2000). To reduce the oxygen content in the bottle head space, jets causing over foaming of the bottle are used, thereby supplying a constant flow of nutrients on the site. Thus conveyors are in constant contact with beer, which might favour adaptation of microorganisms to beer. Conveyors are even thought to distribute microorganisms within the bottling plant. In the brewery filling department, biofilms have been found on conveyor systems, on pasteurisers, on points close to the fillers and can or bottle warmers, as well as on side rails, wear strips, interior and exterior surfaces of conveyor carriages, drip plates, struts linking the chains, and on the bottom of, and between chain links (Czechowski and Banner 1992, Banner 1994). Primary biofilm colonisers were found to accumulate on sterile stainless steel surfaces on the filler within 2–12 hours after the start of production, and effective daily cleaning reduced the number of microorganisms on the surfaces only momentarily (Storgårds *et al.* 2006b). In a study analysing first attached biofilm organisms on cleaned brewery surfaces, *Acinetobacter* spp. accounted for up to 71% of all detected cells (Lipski 2005). Experiments also showed that *Acinetobacter* spp. adhered more readily and firmly to stainless steel than most other microorganisms, thus probably providing a foundation on which other microorganisms can settle and proliferate to form a fully developed biofilm (Lipski 2005). Additionally, four new *Chryseobacterium* species were recently described that are thought to be involved in the rapid recolonisation of cleaned surfaces and biofilm formation in beer bottling plants (Herzog *et al.* 2008). In a follow-up study of 8 weeks, sites on the filling machines were identified that were 100% covered by biofilm already in one week (Storgårds *et al.* 2006b). Biofilms 1–2 mm high were reported to have developed 7 months after last major extensive cleaning on sites exposed to regular cleaning stress (Timke *et al.* 2005a).

Bacteria were shown to form the dominating community in brewery biofilms with members of *Eukarya* and *Archaea* representing below 3% of total cell numbers (Timke *et al.* 2005a). It is assumed that the beer-spoilage bacteria themselves are not the primary colonisers in biofilms; instead they are thought to benefit from the multiple interactions within complex communities. Slime-producing acetic acid bacteria were considered to be primary colonisers in brewery biofilms where they contribute to favourable conditions for beer spoilage organisms by reducing the oxygen level and producing an acidic environment (Back 2003). However, this hypothesis was later proved wrong (Timke 2004, Timke *et al.* 2005a, 2005c, Storgårds *et al.* 2006b). Although acetic acid bacteria are present in some brewery biofilms, they don't seem to represent a significant group. Cultivation-based methods give an incomplete picture of the community as they seem to favour detection of gammaproteobacteria (Timke *et al.* 2005a). Thus Enterobacteriaceae and *Pseudomonas* spp. are often isolated from brewery surfaces (Back 1994, Timke *et al.* 2005c, Storgårds *et al.* 2006b).

As often seen in practice, primary colonisers enable secondary organisms, such as product spoiling organisms, to become part of the biofilm. Thus, brewery biofilms may harbour beer spoiling microorganisms. About 25% of brewery biofilms were indeed found to contain microorganisms which were able to multiply in beer with 4.8% of ethanol (Timke *et al.* 2005b). Contradictory to this result, spoilage bacteria were seldom found in microbial diversity studies of brewery biofilms (Timke 2004, Timke *et al.* 2005a, Storgårds *et al.* 2006b). Although the oxygen-requiring metabolism of the biofilm community is establishing anaerobic micro-environments, anaerobic beer spoilage bacteria such as *Pectinatus* spp. normally only represent an extremely low proportion of the brewery biofilm community. However, they may multiply on heavily contaminated sites becoming a threat to the microbial stability of the beer (unpublished observations, Timke *et al.* 2005a).

Although yeasts are outnumbered by bacteria in brewery biofilms, they are frequently isolated from brewery sites. The most common yeasts isolated from brewery samples have been classified to belong to *Candida* spp., *Debaryomyces* spp. and *Saccharomyces* spp. (Van der Aa Kühle and Jespersen 1998, Storgårds *et al.* 2006b, Timke *et al.* 2008). All

*Saccharomyces cerevisiae* strains isolated from brewery biofilms were shown to grow in beer (Timke *et al.* 2008). Thus, the spoilage potential of brewery biofilms described by Timke and co-workers (2005b), was probably due to the yeast population present. One of the most common yeast species in brewery biofilms *Candida pelliculosa* (teleomorph *Pichia anomala*) is presumably involved in attachment and primary biofilm formation in beer bottling plants, while *S. cerevisiae* colonises preformed biofilm increasing its beer spoiling potential (Timke *et al.* 2008).

An unexpected high diversity of microorganisms has been found in brewery biofilms despite intensive cleaning practices (Timke 2004, Timke *et al.* 2005a, b, c, Storgårds *et al.* 2006b, Timke *et al.* 2008). Moreover, no spatial concentration was observed in these biofilms and no typical brewery biofilm community could be identified even on distinct sites of the bottling plant. Instead, these communities seemed to be subjected to shifts of populations (Timke *et al.* 2005b). Thus a large gene pool present enables the biofilm community to react to changing cleaning and disinfection practices. The common presence of *Methylobacterium* spp. in brewery biofilms may also contribute to resistance of biofilms against disinfection as they are potentially degrading formaldehyde-based sanitisers (Timke *et al.* 2005a). Considering the present knowledge, the solution to biofilm-related problems should rather be prevention of microbial adhesion and biofilm formation than attempts to remove established biofilms by extensive cleaning.

The use of immobilised yeast in continuous beer fermentations represents an application of beneficial biofilms in beer production (Kronlöf 1994). Immobilised yeast reactors are used in industrial scale both for primary fermentation and for maturation (Kronlöf and Virkajärvi 1999, Pajunen *et al.* 1991) and for non-alcoholic beer production (Lommi *et al.* 1997, Van Iersel *et al.* 1999). Although this technology speeds up beer fermentations considerably compared to the traditional batch process, the fear of contamination has hampered its implementation. The microbiological stability of an immobilised yeast system is dependent on the nutrients available. Thus, primary fermentation systems are much more sensitive to contamination than secondary fermentation systems and consequently, the contaminants found in primary and secondary fermentation reactors differ distinctively (Haikara *et al.* 1997). Gram-negative bacteria, such as enterobacteria and acetic acid bacteria, are the most common contaminants in immobilised yeast reactors used for primary fermentation. In reactors used for secondary fermentation, lactic acid bacteria are the most hazardous contaminants. Typically lactic acid bacteria showing only poor attachment to stainless steel, such as *Lactobacillus lindneri* strain VTT E-95589, attached firmly to yeast and carrier material when inoculated to a secondary fermentation immobilised yeast reactor (Fig. 16.1). Besides bacteria, wild yeasts may contaminate both primary and secondary fermentation

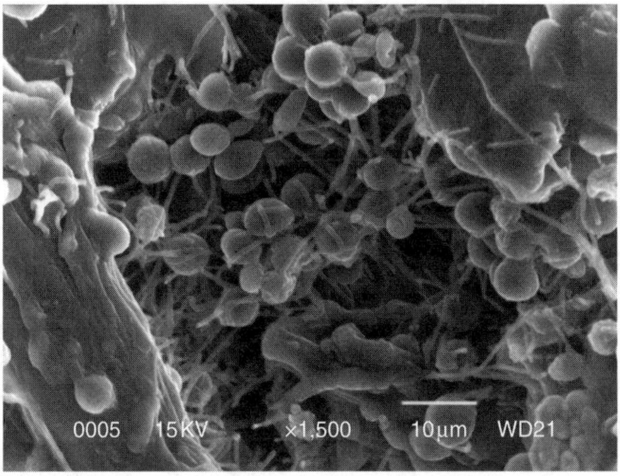


**Fig. 16.1** Scanning electron micrograph of *Lactobacillus lindneri* VTT E-95589 cells tightly attached to yeast cells and carrier material in an immobilised yeast reactor for secondary beer fermentation (Storgårds *et al.* 1997).

reactors (Kronlöf and Haikara 1991, Haikara and Kronlöf 1995, Haikara *et al.* 1997).

### 16.2.3 Microorganisms associated with beer dispensing systems

A wider range of microorganisms can cause problems in beer dispensing equipment than in the brewing process or in packaged beer. This is mainly due to the higher oxygen levels in the dispensing tubes and to less strict temperature control throughout the system. Aerobic conditions prevail at the dispensing tap and at the keg tapping head, and the pipe lines may also be comparatively oxygen permeable. The dispensing lines are not always cooled at full length – at least close to the tap there may be a non-cooled area. These conditions favour contamination by microorganisms such as acetic acid bacteria, moderate levels of coliforms and aerobic wild yeast in addition to the oxygen-tolerant beer spoilage organisms found in the brewery environment (Harper 1981, Ilberg *et al.* 1995, Schwill-Miedaner *et al.* 1996, Taschan 1996, Storgårds 1997, Walker *et al.* 2007). Bacteria and yeasts from the following genera have been isolated during surveys of beer dispensing systems: *Acetobacter*, *Gluconobacter*, *Obesumbacterium*, *Lactobacillus* (among them *L. brevis*), *Pediococcus*, *Zymomonas*, *Brettanomyces*/*Dekkera*, *Debaryomyces*, *Kloeckera*, *Pichia*, *Rhodotorula*, *Saccharomyces* (brewing and wild yeast strains), *Torulopsis* (Harper 1981, Casson 1985, Thomas and Whitham 1997, Storgårds 2000). Harper (1981) also reported that the acetic acid bacteria isolated from dispensing systems were able to grow in a microaerophilic environment, in contrast to corresponding laboratory strains. However, the microbiology of draught beer has received very little attention in the professional or scientific literature over the decades (Quain 2007).

### 16.2.4 Biofilms in beer dispensing system lines

As a result of poor or inadequate cleaning, biofilms have repeatedly been observed in beer dispensing system lines (Harper 1981, Casson 1985, Thomas and Whitham 1997, Walker *et al.* 2007). Recent audits still showed that 25% of accounts surveyed had dirty lines (Quain 2007). *Pediococcus* spp. and acetic acid bacteria as well as brewing and wild yeasts have been found to adhere to dispensing lines (Thomas and Whitham 1997). Biofilms have been observed on dispensing system lines made of polyvinyl chloride (PVC), polythene and nylon (Harper 1981). Casson (1985) studied the colonisation of dispensing systems and found that an organic conditioning film adsorbed onto the PVC pipe after 24 h exposure to beer. He concluded that the adsorbed organic material consisting of polysaccharides or glycoproteins may arise from the original wort or yeast cell wall material. Contaminants introduced into the dispensing system are attracted to the pipe surface by electrostatic interactions, but cannot actually adhere on the conditioning film due to close range charge repulsion. The yeasts overcome this charge barrier by extending surface fimbriae, which anchor them to the conditioning film (Fig. 16.2). Subsequently more fimbriae are produced and

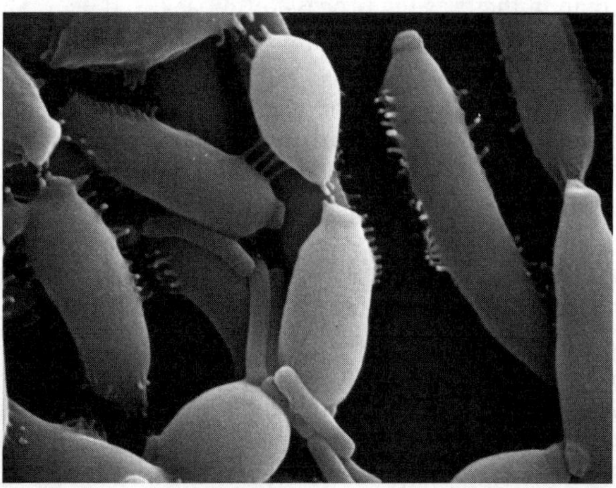

**Fig. 16.2** Scanning electron micrograph of a biofilm comprising yeast and bacteria on the inner surface of a dispensing line. Note protruding fimbriae of the yeast cells. (Adapted from Storgårds 2000).

finally the cells produce EPS to consolidate their position and protect the cells. According to Casson (1985), this polymeric matrix may then harden and become rigid, making the removal of these deposits very difficult. Even if the cells in the film are killed during cleaning, the remaining deposit provides perfect sites for recolonisation when new viable cells are introduced into the dispensing system. Thomas and Whitham (1997) found that PVC tubing inserted into trade dispensing lines carrying cask ale contained adhering microorganisms after two weeks at levels comparable to control samples of dispensing lines used for more than 18 months. Average levels of adhesion in these samples after washing ranged from 10 to $3.5 \cdot 10^4$ cells per $cm^2$. Approximately comparable numbers of bacteria and yeast were found to be adhered. *Pediococcus* spp. and acetic acid bacteria were common contaminants in many lines, along with brewing and wild yeast (Thomas and Whitham 1997).

### 16.2.5 Implications for food quality and safety

Pathogenic bacteria able to cause food poisoning are generally unable to grow in normal beer (Back 1994, L'Anthoën and Ingledew 1996, Baxter and Hughes 2001). However, L'Anthoën and Ingledew (1996) reported that some pathogenic bacteria including *Escherichia coli* O157:H7, *Salmonella typhimurium*, *Pseudomonas aeruginosa*, *Shigella flexneri*, *Klebsiella pneumoniae*, *Yersinia enterocolitica* and *Enterococcus faecium* were able to grow in degassed nutrient-rich alcohol-free (<0.5% v/v) beer with an abnormally high pH of 5.25. Members of Enterobacteriaceae isolated from brewery biofilms were not able to grow in standard beer, but they were multiplying in non-alcoholic beer under access of oxygen (Timke 2004). Although it is unlikely that growth would occur in the presence of carbon dioxide, safety issues should be carefully examined especially in the production of low-alcohol cereal-based drinks.

Some bacteria contaminating the brewing process may indirectly be a safety concern. *Bacillus* spp. occurring in the mash or sweet wort can reduce nitrate to nitrite whereas members of Enterobacteriaceae can reduce nitrate in early stages of fermentation. The nitrite may then react with the amines of the wort producing non-volatile nitrosoamines (Baxter and Hughes 2001). In addition, biogenic amines can be formed during beer production as a result of bacterial contamination (Donhauser *et al.* 1993, Izquierdo-Pulido *et al.* 1996).

Post-process recontamination is considered to be a significant cause of outbreaks of food pathogens (Reij and Aantrekker 2004). In breweries, the risk for recontamination is low due to the minimal contact with the process stream after wort boiling. Although coliform bacteria such as *Obesumbacterium proteus* can be present in poorly handled yeast, pathogens have not been reported. However, the occurrence of coliforms in beer dispensing systems is a cause of concern. As it is common that pubs and restaurants serve both beer and food, a cross-contamination from the food to the beer is possible. Thus the possible survival in beer of acid-tolerant pathogens such as *E. coli* O157:H7 should not be overlooked.

## 16.3 Prevention of biofilm formation in breweries and in beer dispensing

### 16.3.1 Means for reduction of microorganisms

Prevention of microbial spoilage of beer can best be achieved by eliminating the sources of contamination. However, the brewing process is not aseptic, and contaminating microorganisms will often be encountered. Traditionally, various means are used for reduction of the numbers of microorganisms in beer production (Table 16.1). Of these, hygienic equipment design and cleaning and disinfection can be seen as means aiming at prevention and reduction of biofilms in breweries.

### 16.3.2 Hygienic equipment design

Hygienic design is essential in controlling biofilm formation and/or minimising the biotransfer potential in food processing equipment such as tanks, pipelines, joints and accessories. The requirements for hygienic design are well documented and they state in detail how equipment should be constructed so that all surfaces in contact with the food or beverage are easy to clean (http://www.ehedg.org/guidelines.htm). These include suitable choice of equipment, materials and accessories, correct construction, process layout and process automation such as outlined by the European Hygienic Engineering and Design Group (EHEDG) or

**Table 16.1** Means used for reduction of microorganisms in beer production

| Process | Purpose |
|---|---|
| Acid washing of pitching yeast | Reduction of contaminating microorganisms in pitching yeast |
| Cooling | Retardation of the growth of contaminating microorganisms during fermentation and maturation |
| Filtration | Removal of pitching yeast, reduction of contaminating microorganisms |
| Pasteurisation | Elimination of vegetative cells in final beer |
| Hygienic equipment design | Prevention of contamination and facilitating the removal of contaminants from equipment |
| Cleaning and disinfection | Inactivation and removal of microorganisms from equipment and facilities |

corresponding US-based organisations. Generally, all product-contact surfaces should be smooth (preferably $R_a$ < 0.8 μm), pits, crevices, sharp edges and dead ends should be avoided and all equipment and pipelines should be self-draining.

In the filling hall, constructions should be open to facilitate cleaning and should not allow any liquid to remain on surfaces. Drains must be correctly sized and placed in order to avoid any water and organic residues and floor coverings must be chosen so that they can be effectively cleaned and maintained. Drop plates should be avoided when possible since they collect dirt. Cable installations should be avoided in the wet area whenever possible or they should be in closed pipes with access from below (Paier and Ringhofer 1997). Nowadays the importance of hygienic design of filling machines and the whole filling department is well understood and advances in construction design have been made (Fischer 2003, Ricker 2006, Pahl and Folz 2007). Strict ensuring of hygiene during filling or aseptic packaging is applied in breweries that do not tunnel-pasteurise their products. Clean room technology can be used to prevent secondary contaminations at bottling, canning and kegging (Takagi 1993, Ryder *et al.* 1994, Oliver-Daumen 2004). Aseptic cold filling (ACF) technology is used in the bottling of microbiologically susceptible beverages such as non-alcoholic drinks that contain little or no carbon dioxide (Oliver-Daumen 2007). Machinery constructions are modified to make them more easily cleanable (Takagi 1993, Ryder *et al.* 1994, Oliver-Daumen 2004, Keller 2007).

In the construction of beer dispensing systems, hygienic design is equally important as in the construction of beer production equipment. However, many weak points have been identified in these systems, including the dispensing tap and tap armature, fittings and joints (Hauser 2003). The dispense lines are generally more prone to microbial colonisation and biofilm formation than the parts of the system made of stainless steel (Walker *et al.* 2007). The dispensing systems should be constructed so that pipes, pumps and refrigeration equipment are self draining and no gas pockets or dead ends are left in the system (Hauser 2003, Nieroda 2006).

### 16.3.3 Cleaning and disinfection

Cleaning-in-place (CIP) procedures are employed in closed processing lines of the brewing process (Table 16.2). The mechanical input in cleaning has been shown to be critical in removing biofilms (Carpentier and Cerf 1993, Salo *et al.* 2006). Mechanical force can be achieved by turbulence flow in the pipelines and spray nozzles in the cylindrical tanks, but in practice there are places in the process where the mechanical action is low, such as, for example, the large cylindroconical fermentation tanks (Salo *et al.* 2006). Low cleaning temperatures have been found to be ineffective in the removal of biofilms (Holah and Gibson 1999), but unfortunately, high temperatures can only partly be employed in cleaning of

**Table 16.2** Typical CIP programmes used in the brewery. The programmes are adapted to the part of the process to be cleaned, and some of the steps: alkalic, acidic, or disinfection, can be left out

| Action | Temperature | Duration |
|---|---|---|
| Prerinsing | cold or hot | 5–10 min |
| Alkali cleaning; sodium hydroxide (1.5–4%) | cold or hot (60–85°C) | 10–60 min |
| Intermediate rinsing | cold or hot | 10–30 min |
| Acidic cleaning; phosphoric, nitric or sulphuric acid (1–2%) | cold | 10–30 min |
| Intermediate rinsing | cold | 10–30 min |
| Disinfection | | |
| – by disinfectant solution | cold | 10–30 min |
| – by hot water | 85–90°C | 45–60 min |
| Final rinsing if necessary | cold | 5–10 min |
| – may contain a disinfectant at low concentration | | |

**Table 16.3** Foam cleaning and disinfection programme for open surfaces

| Action | Agent |
|---|---|
| Prerinsing | Water |
| Foaming | Foam cleaner |
| Soak time | Foam cleaner |
| Intermediate rinsing | Water |
| Spraying | Disinfectant solution |
| Final rinsing | Water |

brewery vessels. Chlorinated alkaline detergents were found to be the most effective in removing biofilms of brewery-related species in CIP (Czechowski and Banner 1992).

The cleaning of open surfaces in the brewery, such as filling machines, bottle inspectors and conveyor belts in the bottling hall, is usually performed using low-pressure foam systems or thin film cleaning (Table 16.3). The use of hot solutions or strong chemicals is limited for safety reasons, but disinfectants also effective in cold conditions can be used to ensure the hygiene. Foam cleaning and subsequent spraying with a disinfectant after every production shift – sometimes also during pauses – is commonly recommended (Paludan-Muller 2003, Wershofen 2003). However, care must be taken to avoid transmission of spoilage organisms resulting from aerosols produced during pressure-cleaning. In addition to daily cleaning and disinfection, regular major equipment overhauls including dismantling of components that are difficult to inspect visually when intact, are important.

Mechanical or chemical breakage of the polysaccharide matrix is essential for successful biofilm control, as the matrix protects the microorganisms from the effects of detergents and disinfectants. Enzymes have been demonstrated to cause effective breakage of the EPS matrix, thus helping in the removal of biofilms, and oxidoreductases have bactericidal activity against biofilm bacteria (Carpentier and Cerf 1993, Johansen *et al.* 1997, Kumar and Anand 1998). Multicomponent enzymes could provide a supplement to the present cleaning and disinfection agents. A broad spectrum of enzymes encompassing proteolytic and non-proteolytic enzymes was found to be most effective for removal of biofilms from stainless steel (Walker *et al.* 2007). However, the efficacy of the enzyme treatment was dependent on both microbial species and the surface material to which the biofilm was attached. Thus, it might be difficult to obtain an enzyme mix which can universally treat all biofilms found in a specific environment (Walker *et al.* 2007).

The aim of disinfection is to reduce the surface population of viable microorganisms after cleaning and to prevent microbial growth on surfaces during the interproduction time. The borderline between cleaning and disinfection is somewhat diffuse because microorganisms are to a great extent eliminated already during the cleaning stage. Some detergents are bactericidal and some disinfectants depolymerise EPS, causing detachment of biofilms from surfaces, e.g. oxidants such as chlorine and hydrogen peroxide (Carpentier and Cerf 1993). Sodium hydroxide, the most common cleaning agent in CIP, was shown to have microbicidic activity against organisms encountered in the brewery. In a suspension of 0.5% sodium hydroxide at 20°C, a 5 log reduction of brewer's yeast was achieved in 2 min, of *Lactobacillus brevis* in 3 min and of *Pediococcus. damnosus* in 5 min (Donhauser *et al.* 1991). Generally, chemical cleaners are more effective in eliminating attached bacteria from surfaces than disinfectants. Furthermore, disinfectants are generally most effective in the absence of organic material (Donhauser *et al.* 1991, Czechowski and Banner 1992). Thus the control and inactivation of adherent microbes or biofilms requires detergent cleaning of the surface followed by treatment with a disinfectant.

In choosing disinfectants for use in the brewery, the following characteristics are of importance (Donhauser *et al.* 1991):

- Effective against Gram-positive and Gram-negative bacteria and against yeasts (preferably also against moulds).
- Effective in the presence of proteins.
- Effective at low temperatures (often contradictory with efficiency against proteins).
- Wetting ability (contradictory with rinsability).

- CIP-suitability (low foam formation, compatibility with carbon dioxide, concentration measurable by conductibility, reusable/not easily decomposable).
- Environmental aspects (easily rinsable, readily biodegradable).
- Economy (effective at low concentrations, reusable, easily rinsable).
- Health aspects – safe to use.
- Product compatibility – no adverse effects on the product.

Some disinfectants commonly recommended for brewery use are listed in Table 16.4. Formulations based on peracetic acid (PAA) and hydrogen peroxide are frequently used for post-cleaning disinfection. Because of its acidic and non-foaming properties, PAA is suitable for CIP disinfection under a carbon dioxide atmosphere such as in fermentation tanks and lines (Banner 1995). The PAA agents quickly lose their activity in a basic environment, making careful rinsing after alkaline cleaning essential. PAA and hydrogen peroxide-based disinfectants also perform well in

**Table 16.4** Disinfectants and their use in the brewery

| Disinfectant type | Use |
|---|---|
| Hydrogen peroxide – peracetic acid<br>a) peracetic acid (2.5–15%)<br>b) with organic or inorganic acids and surfactants | Brewhouse<br>Fermentation, storage, pressure tanks<br>Pipelines<br>Filler<br>Bottle washing<br>Dialysis/reverse osmosis |
| Halogenes<br>a) alkaline chlorine<br>b) acid iodophores | Malthouse<br>Brewhouse<br>Tanks and pipelines<br>Hoses, fittings, filtration<br>Bottle washing (rinse water) |
| Surface active agents<br>a) quaternary ammonium compounds, pH 4–9<br>b) amphoterics | General plant cleaning<br>Malthouse, hoses, fittings<br>Mixing machines, filler |
| Halogenated carbonic acids<br>– chlorine/iodine/bromine with inorganic acids | Tanks and pipelines |
| Alkylamines (foam disinfection) | Filler<br>Environmental hygiene |
| Biguanides | Soaking of small utensils and instruments |
| Aldehydes<br>– formaldehyde/glutaraldehyde | Air sanitation by fogging (bottling hall)<br>Water treatment systems (glutaraldehyde): cooling, pasteuriser, can/bottle warmer |
| Chlorine dioxide | Bottle washing (rinse water)<br>Conveyor systems<br>Filler surface disinfection |

the presence of organic soil, but they are markedly less effective when the temperature is decreased from ambient (20°C) to 4°C (Donhauser *et al.* 1991). At low temperatures, such as in the fermentation cellar, higher concentrations are needed to obtain a good result. Chlorine- and iodine-based disinfectants also perform well at low temperature. However, these disinfectants are inactivated by proteins, which reduce their efficacy in the presence of wort or beer residues (Donhauser *et al.* 1991, Banner 1995). Chlorine is often used as hypochlorite solutions under alkaline conditions, whereas iodine disinfectants are most active around pH 2–3, making the latter more suitable for use in brewery CIP (Banner 1995). Chlorine dioxide, like chlorine, is a powerful oxidising agent which has lately found more applications in the brewery (Homann and Möller-Hergt 2004). Quaternary ammonium compounds (QAC) have low surface tension giving them a good penetrating ability but are also difficult to rinse, which may have an adverse effect on the products (Donhauser *et al.* 1991, Gingell and Bruce 1998). Acidic QACs are effective against a wide spectrum of microorganisms, especially yeasts. (Gingell and Bruce 1998). Residual films of QACs may reduce the foam level of beer and can also interfere with the growth and metabolism of brewery yeast. Because of this and also due to foaming and rinsing problems, QAC products are not used in CIP operations but mainly for soaking purposes (Banner 1995, Gingell and Bruce 1998). When heat is used for disinfection, moist heat is far more effective than dry heat. *L. brevis* was shown to withstand more than 60 min at 80°C in dry conditions (Donhauser *et al.* 1991). In the process, such dry conditions may occur if microorganisms are trapped between metal surfaces, between a seal and a stainless steel surface or in microscopically small cracks in the process materials.

## 16.4 Future trends

### 16.4.1 Functional surface materials

Although extensive cleaning and major equipment overhauls within short intervals would probably be effective, excessive cleaning means cost for the industry and also has a negative impact on the environment due to increased chemical waste. The development of materials science has introduced new products aimed at antimicrobial properties and improved cleanability of surface materials. Application of functional materials hindering microbial attachment to process surfaces in microbiologically susceptible sites in the brewing industry could be a novel means to improve cleanability. Along with more easily cleanable surfaces less aggressive cleaning methods could be used, thus reducing the environmental impact and saving costs.

Published results concerning microbial adhesion and low surface energy are partly controversial (Liu and Zhao 2005). In a number of studies, however, low surface energy has been shown to reduce bacterial adhesion

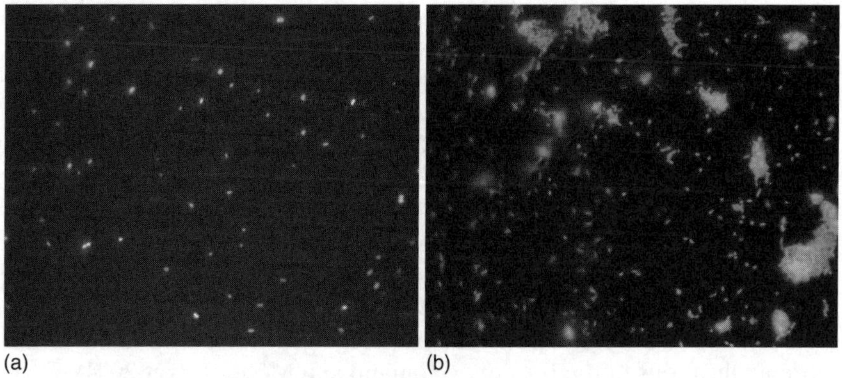

**Fig. 16.3** Microbial adhesion on fluoropolymer coated (a) and non-coated (b) cold-rolled stainless steel (AISI304) surfaces of a mixture of bacteria and yeasts. The bacterial (*Lactobacillus paracasei* VTT E-032064, *Pseudomonas fluorescens* VTT E-031889, *Serratia marcescens* VTT E-031888) and yeast (*Pichia anomala* VTT C-02470) strains were originally isolated from brewery filler surfaces. The contact time in weak (1%) wort sucrose broth was 1 hour at 25°C, after which the surfaces were rinsed and the attached cells stained with acridine orange and analysed by epifluorescence microscopy (Storgårds *et al.* 2007).

(Liu and Zhao 2005, Tsibouklis *et al.* 1999, 2000). A stabile low surface energy leads to low adhesion of deposits with low roll-off angle for droplets, which in turn collects weakly adhered deposits and microbes leading to a clean surface. In laboratory experiments with microbes isolated from breweries, microbial adhesion on fluoropolymer coated cold-rolled stainless steel was substantially weaker than on the non-coated reference surface (Fig. 16.3).

Antimicrobial properties of certain metals have been known and taken advantage of for a long time. In ancient times, silver items were kept in water bowls to keep the water fresh. Antimicrobial steels containing silver (0.042%) were found to reduce the number of adhering bacteria on brewery filling machine surfaces in operation with even 2 log units compared to normal steel meaning a reduction of 99% in the microbial load (Storgårds *et al.* 2007).

One of the most widely used functional materials nowadays is photocatalytic titanium dioxide ($TiO_2$). This material has a dual effect: it generates reactive oxygen upon illumination with UVA light in normal sunlight and thus decomposes organic matter and kills bacteria (Kikuchi *et al.* 1997, Cho *et al.* 2004). In addition, cleanability of the $TiO_2$ coating is enhanced by reduction in the hydrophobicity of the surface in UVA light (Li and Logan 2005). $TiO_2$ coatings have been shown to have potential for reducing biofilms (Raulio *et al.* 2006). An obstacle in the use of photocatalytic materials is the requirement of UVA light to activate the surfaces. However, by combining photocatalytic surfaces with antimicrobial metals, the effect of

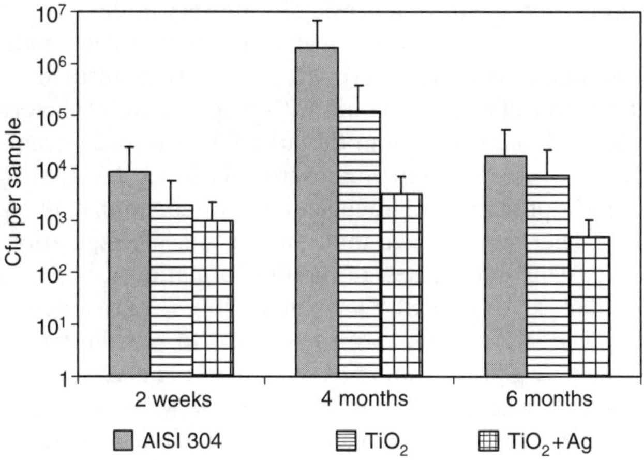


**Fig. 16.4** Number of aerobic bacteria on non-coated stainless steel (AISI 304), on titanium dioxide coated steel ($TiO_2$) and on titanium dioxide + silver coated steel ($TiO_2$ + Ag) on brewery filling machines after exposure to normal production conditions for various times, average colony forming units (cfu) per coupon (75 × 25 mm) (Adapted from Storgårds *et al.* 2007).

the material can be improved even under very weak UV light illumination such as in the brewery filling department. The action mechanism of $TiO_2$ combined with Cu ions was shown to work in two steps in a copper resistant strain of *E. coli* (Sunada *et al.* 2003). First the outer membrane was partially decomposed by photocatalysis, and then the Cu ions caused disorder of the cytoplasmic membrane. Thus, the change in survival rate on Cu/$TiO_2$ under weak UV light was similar to the effect of photocatalysis under strong UV light. When photocatalytic $TiO_2$ surfaces were improved by a low concentration of silver, they were found to further strengthen the effect of $TiO_2$ (Fig. 16.4). This way the total microbial load could be reduced in production environment where the illumination basically is day light and strip lights (Storgårds *et al.* 2007).

When introducing new materials into beverage production areas, it is important to ensure not only that the total microbial load is reduced but also that the population does not consist of more harmful microbes than before. By analysing the bacterial populations adhering onto the antimicrobial steels and to the photocatalytic coatings by PCR-DGGE, the distribution of species was shown to be more dependent on the location than on the material (Storgårds *et al.* 2007, unpublished results). In conclusion, the variation in bacterial populations was generally more dependent on the site than on the material. Thus antimicrobial steels did not seem to increase harmful bacteria compared to ordinary stainless steel. This should, however, be verified after longer contact times including also yeasts and moulds.

### 16.4.2 Disturbance of quorum sensing in brewery biofilms

Bacteria are no longer regarded as undifferentiated cells with a single-minded programme focused on growth and division. Instead it has been shown that cell-to-cell signalling both within and between bacterial species is a widely spread phenomenon in the microbial world. In general, Gram-negative bacteria use acylated homoserine lactones (AHLs) and gram-positive bacteria processed oligonucleotides to communicate (Miller and Bassler 2001). There is evidence that quorum sensing regulation also participates, at least in some cases, in biofilm formation (Shrout *et al.* 2006, Sarkar and Chakrabotry 2008). Quorum sensing is most often studied by using bacterial monitor strains, which respond to signalling molecules by physical changes (e.g., McClean *et al.* 1997, Ravn *et al.* 2001). In the study of Timke (2004), where 106 brewery isolates were investigated using reporter strain *Agrobacterium tumefaciens* NTL4, nine isolates were AHL positive. Five of the positive isolates were cultivated from the shims which had only a thin biofilm, hence Timke (2004) proposed that the AHL positive isolates could be surface colonisers. When Gram-negative bacterial strains isolated from breweries were screened for the production of signalling in our laboratory, many strains produced long chain (C10-C14) AHL-compounds and some strains produced AHL-type molecules detected with *Agrobacterium tumefaciens* NTL4 (unpublished results).

Several natural and synthetic compounds inhibiting either the synthesis or use of quorum sensing molecules have been detected, and a possibility to regulate unwanted bacterial actions like virulence has been proposed (Rasmussen and Givskov 2006). Although it has proven not to be a simple task, it might in the future be possible to interfere with the biofilm formation of bacteria by controlling their quorum sensing signalling.

### 16.4.3 Synergy of different means to improve brewery hygiene

Killing microorganisms within the biofilm is often understood as cleaning. Nevertheless, disinfection leads to dead biomass, which also serves as a nutrient for build-up of new biofilm. The main aim should be on prevention of the formation of the biofilm rather than removal of the biofilm already formed. This is best achieved by not only one means, but by combining multiple methods giving synergistic advantages. Entry of contaminants to the process should be kept low by controlling the quality of raw materials, process air and traffic at the process area, and by hygienic working practices of the personnel. The build-up of biofilms is best hindered by hygienic design of processing equipment and by regular and efficient cleaning, and could be further reduced by implementing functional surface coatings reducing microbial attachment and/or facilitating their removal on critical places of the process. When choosing cleaning chemicals, they should be evaluated for compatibility with the functional materials in use. In addition, physical means like UVC disinfection could serve as a preventive means of biofilm formation at critical points (Hoyer 1998, Sastry *et al.* 2000). In the future, when the activity of microbial communities are better understood, it could be possible to disturb the signalling of microbial communities, thus again aiding management of process hygiene. Ultimately, implementation of scientifically based hygiene concepts will lead to reducing hazards and avoiding recalls caused by poor microbiological quality of the products.

## 16.5 Sources of further information and advice

For detailed information regarding biofilms in brewery bottling plants, see the excellent PhD thesis of Timke (2004, http://elib.ub.uni-osnabrueck.de/publications/diss/E-Diss385_thesis.pdf). Comprehensive information regarding practically all relevant microbiological aspects involved in brewing is given in the book *Brewing Microbiology* (Priest and Campbell 2003). For a more detailed review on various aspects of process hygiene control in beer production and dispensing, see Storgårds (2000, http://www.vtt.fi/inf/pdf/publications/2000/P410.pdf). A summary of different possibilities to detect and identify the microorganisms likely to be encountered in breweries is given in Storgårds *et al.* (2006a).

## 16.6 References

BACK W (1994), *Farbatlas und Handbuch der Getränkebiologie. Teil 1. Kultivierung/Methoden, Brauerei, Winzerei*, Nürnberg, Verlag Hans Carl.

BACK W (2003), 'Biofilme in der Brauerei und Getränkeindustrie', *Brauwelt*, 143, 766–777.

BANNER M J (1994), 'Perspectives on conveyor track treatment', *Master Brewers Association of the Americas, Technical Quarterly*, 31, 142–148.

BANNER M J (1995), 'The selection of disinfectants for use in food hygiene', In Rossmore H W, *Handbook of biocide and preservative use*, London, Chapman & Hall, 315–329.

BAXTER E D and HUGHES P S (2001), *Beer: Quality, safety and nutritional aspects*, Cambridge, The Royal Society of Chemistry.

CARPENTIER B and CERF O (1993), 'A review: biofilms and their consequences, with particular reference to hygiene in the food industry', *J Appl Bacteriol*, 75, 499–511.

CASSON D (1985), 'Microbiological problems of beer dispense', *The Brewer*, 71, 417–421.

CHO M, CHUNG H, CHOI W and YOON J (2004), 'Linear correlation between inactivation of *E. coli* and OH radical concentration in $TiO_2$ photocatalytic disinfection', *Water Research*, 38, 1069–1077.

CZECHOWSKI M H, BANNER M (1992), 'Control of biofilms in breweries through cleaning and sanitizing', *Master Brewers Association of the Americas, Technical Quarterly*, 29, 86–88.

DOBSON C M, DENEER H, LEE S, HEMMINGSEN S, GLAZE S and ZIOLA B (2002), 'Phylogenetic analysis of the genus *Pediococcus*, including *Pediococcus claussenii* sp. nov., a novel lactic acid bacterium isolated from beer, *International Journal of Systematic and Evolutionary Microbiology*, 52, 2003–2010.

DONHAUSER S, WAGNER D and GEIGER E (1991), 'Zur Wirkung von Desinfektionsmitteln in der Brauerei', *Brauwelt*, 131, 604–616.

DONHAUSER S, WAGNER D and GEIGER E (1993), 'Biogenic amines. Significance, occurrence and assessment', *Brauwelt International*, 11, 100–107.

FISCHER S (2003), 'Hygienic design of filling lines', *European Brewery Convention Monograph 32, Symposium on sanitary engineering and HACCP, Bonn December 2003*, Nürnberg, Verlag Hans Carl (CD-ROM) 15, 8 pp.

GINGELL K and BRUCE P (1998), 'Latest developments in acid cleaning and related sanitation in breweries', *Proceedings of the 25th Convention of the Institute of Brewing – Asia Pacific Section, Perth 22–27 March*, 134–137.

HAIKARA A and HELANDER I (2006), '*Pectinatus, Megasphaera* and *Zymophilus*', In Dworkin M, *The Prokaryotes: A handbook on the biology of bacteria*, 3rd ed. Singapore, Springer, 4, 965–981.

HAIKARA A and KRONLÖF J (1995), 'Hygiene and microbiological control requirements of immobilized systems, *European Brewery Convention Monograph 24, Symposium Immobilized yeast applications in the brewing industry, Espoo, Finland*, Nürnberg, Verlag Hans Carl, 194–206.

HAIKARA A, VIRKAJÄRVI I, KRONLÖF J and PAJUNEN E (1997), ' Microbiological contaminations in immobilized yeast bioreactors for primary fermentation', *Proceedings of the 26th European Brewery Convention Congress, Maastricht 24–29 May 1997*, Nürnberg, Verlag Hans Carl, 439–446.

HANSEN E C (1896), 'Practical studies in fermentation', Cited in Young T W 1996, 'The biochemistry and physiology of yeast growth', In Priest F G and Campbell I, *Brewing Microbiology*, 2nd ed. London, Chapman & Hall, 13–42.

HARPER D R (1981), 'Microbial contamination of draught beer in public houses', *Process Biochem*, 16, 2–7, 21.

HAUSER G (2003), 'Hygienic design and cleaning of systems for draught beer', *European Brewery Convention Monograph 32, Symposium on sanitary engineering and HACCP, Bonn December 2003*, Nürnberg, Verlag Hans Carl (CD-ROM) 16, 8 pp.

HERZOG P, WINKLER I, WOLKING D, KÄMPFER P and LIPSKI A (2008), '*Chryseobacterium ureilyticum* sp. nov., *Chryseobacterium gambrini* sp. nov., *Chryseobacterium pallidum* sp. nov. and *Chryseobacterium molle* sp. nov., isolated from beer-bottling plants, *International Journal of Systematic and Evolutionary Microbiology*, 58, 26–33.

HOLAH J and GIBSON H (1999), 'Biofouling: cleanability problems in food processing?' *Proceedings of the 30th R³-Nordic Contamination Control Symposium*, Espoo, VTT Symposium 193, 177–186.

HOMANN F and MÖLLER-HERGT G (2004), 'Cleaning and disinfection of conveyor systems for bottling and canning lines', *Brauwelt*, 144 (45), 1466–1471 (in German).

HOYER O (1998), 'Testing performance and monitoring of UV systems for drinking water disinfection', *Water Supply*, 16, 419–442.

ILBERG V, SCHWILL-MIEDANER A and SOMMER K (1995), 'Mikrobiologische Probleme an Schankanlagen', *Brauindustrie*, 80, 992–995.

IZQUIERDO-PULIDO M, HERNÁNDEZ-JOVER T, MARINÉ-FONT A and VIDAL-CAROU M C (1996), 'Biogenic amines in European beers', *J Agric Food Chem*, 44, 3159–3163.

JOHANSEN C, FALHOLT P and GRAM L (1997), 'Enzymatic removal and disinfection of bacterial biofilms', *Appl Environ Microbiol*, 63, 3724–3728.

JUVONEN R and SUIHKO M L (2006), '*Megasphaera paucivorans* sp. nov., *Megasphaera sueciensis* sp. nov. and *Pectinatus haikarae* sp. nov., isolated from brewery samples, and emended description of the genus *Pectinatus*', *International Journal of Systematic and Evolutionary Microbiology*, 56, 695–702.

KELLER M (2007), 'Cold sterile filling and much more', *Brauwelt*, 147 (40), 1092–1095 (in German).

KIKUCHI Y, SUNADA K, IYODA T, HASHIMOTO K and FUJISHAMA A (1997), 'Photocatalytic bactericidal effect of $TiO_2$ thin films: dynamic view of the active oxygen species responsible for the effect', *Journal of Photochemistry and Photobiology A: Chemistry*, 106, 51–56.

KRONLÖF J (1994), *Immobilized yeast in continuous fermentation of beer*, PhD Thesis, Espoo, VTT Technical Research Centre of Finland, VTT Publications 167.

KRONLÖF J and HAIKARA A (1991), 'Contamination of immobilized yeast bioreactors', *J Inst Brew*, 97, 375–380.

KRONLÖF J and VIRKAJÄRVI I (1999), 'Primary fermentation with immobilized yeast', *Proceedings of the 27th European Brewery Convention Congress, Cannes*, Zoeterwoude, European Brewery Convention, 761–770.

KUMAR C G and ANAND S K (1998), 'Significance of microbial biofilms in food industry: a review', *Int J Food Microbiol*, 42, 9–27.

L'ANTHOËN N C and INGLEDEW W M (1996), 'Heat resistance of bacteria in alcohol-free beer', *Journal of the American Society of Brewing Chemists*, 54, 32–36.

LI B and LOGAN B E (2005), 'The impact of ultraviolet light on bacterial adhesion to glass and metal oxide coated surface', *Colloids and Surfaces B, Biointerfaces*, 41, 153–161.

LIPSKI A (2005), 'New focus in combating biofilms: The pioneer organisms on cleaned surfaces', *Brau Forum*, 20(7), 178–179 (in German).

LIU Y and ZHAO Q (2005), 'Influence of surface energy of modified surfaces on bacterial adhesion', *Biophysical Chemistry*, 117, 39–45.

LOMMI H, SWINKELS W and VAN DIEREN B (1997), 'Process for the production of non-alcoholic or low alcohol malt beverage', *Biotechnol Adv*, 15, 789.

MCCLEAN K H, WINSON M K, FISH L, TAYLOR A, CHHABRA S R, CAMARA M, DAYKIN M, LAMB J H, SWIFT S, BYCROFT B W, STEWART G S and WILLIAMS P (1997), 'Quorum sensing and *Chromobacterium violaceum*: exploitation of violacein production and inhibition for the detection of acylhomoserine lactones', *Microbiology*, 143, 3703–3711.

MILLER M B and BASSLER B L (2001), 'Quorum sensing in bacteria', *Annual Review of Microbiology*, 55, 165–199.

NIERODA A (2006), 'New aspects of hygiene legislation concerning the operation of dispensing systems', *Brauwelt*, 146(6), 140–141 (in German).

OLIVER-DAUMEN B (2004), 'Cold aseptic filling plants. Part 2: Line configurations', *Brauwelt*, 144 (29), 877–881 (in German).

OLIVER-DAUMEN B P (2007), 'Aseptic cold filling plants – Hygiene concepts', *Brauwelt Int*, 25 (4), 232–238.

PAHL R and FOLZ R (2007), 'The devils in the details. Hygienic design in filling lines versus actual filling practice', *Brauindustrie*, 92 (8), 30–33 (in German).

PAIER H J, RINGHOFER R (1997), 'Biologische Betriebskontrolle in der Praxis. Teil 3: Schwachstellen mit ihren Gegenmassnahmen', *Brauindustrie*, 82, 628–634.

PAJUNEN E, GRÖNQVIST A and RANTA B (1991), 'Immobilized yeast reactor application in continuous secondary fermentation in industrial scale operation', *Proceedings of the 23rd European Brewery Convention Congress, Lisboa, Portugal*. Oxford, IRL Press, 361–368.

PALUDAN-MULLER T (2003), 'Selection of the best system for filling lines for soft drinks of varying pH, viscosity and hygienic requirements is not easy', *Scand Brew Rev*, 60 (6), 180–20, 22–23.

PASTEUR L (1876), 'Etudes sur la bière, ses maladies, causes qui les provoquent. Procédés pour la rendre inaltérable, avec une théorie nouvelle de la fermentation', Masson, Paris. Cited in Van Vuuren H J J 1996, 'Gram-negative spoilage bacteria', In Priest F G and Campbell I, *Brewing Microbiology*, 2nd ed. London, Chapman & Hall, 163–191.

PRIEST F G (2003), 'Gram-positive bacteria', in Priest F G and Campbell I, *Brewing Microbiology*, 3rd ed. New York, Kluwer Academic/Plenum Publishers, 181–217.

PRIEST F G and CAMPBELL I (2003), *Brewing Microbiology*, 3rd ed. New York, Kluwer Academic/Plenum Publishers.

QUAIN D (2007), 'Draught beer quality – challenges and opportunities', *Proceedings of the 31st European Brewery Convention Congress, Venice, Italy*, Nürnberg, Fachverlag Hans Carl, (CD-ROM) 86, 791–801.

RASMUSSEN T B and GIVSKOV M (2006), 'Quorum sensing inhibitors: a bargain of effects', *Microbiology*, 152, 895–904.

RAULIO M, PORE V, AREVA S, RITALA M, LESKELÄ M, LINDÉN M, ROSENHOLM J B, LOUNATMAA K and SALKINOJA-SALONEN M (2006), 'Destruction of *Deinococcus geothermalis* biofilm by photocatalytic ALD and sol-gel $TiO_2$ surfaces', *Journal of Industrial Microbiology and Biotechnology*, 33, 261–268.

RAVN L, BECK CHRISTENSEN A, MOLIN S, GIVSKOV M and GRAM L (2001), 'Methods for detecting acylated homoserine lactones produced by gram-negative bacteria and their application in studies of AHL-production kinetics', *Journal of Microbiological Methods*, 44, 239–251.

REIJ M W, DEN AANTREKKER E D (2004), 'Recontamination as a source of pathogens in processed foods', *International Journal of Food Microbiology*, 91, 1–11.

RICKER T (2006), 'New filler generation: The front table was yesterday', *Fluessiges Obst*, 73 (4), 205–206 (in German).

RYDER D, BARNEY M, DANIELS D, BORSH S and CHRISTIANSEN K (1994), 'The philosophy and practice of process hygiene in sterile filtration and aseptic filling in modern brewery operations', *European Brewery Convention Monograph 21, Symposium Process Hygiene, Nutfield March 1994*, Nürnberg, Verlag Hans Carl, 65–80.

SALO S, JOHNSEN S, JENSEN B B B, STORGÅRDS E and FRIIS A (2006), 'The mechanical effect of rinsing in clening of fermentation tank bottoms', in Wilson D I, Chew J Y M, Fryer P J and Hasting A P M, *Proceedings of the Conference Fouling, Cleaning & Disinfection in Food Processing*, University of Cambridge, 162–171.

SARKAR S and CHAKRABOTRY R (2008), 'Quorum sensing in metal tolerance of *Acinetobacter junii* BB1A is associated with biofilm production', *FEMS Microbiology Letters*, 282, 160–165.

SASTRY S K, DATTA A K and WOROBO R W (2000), 'Ultraviolet light', *J Food Sci Suppl*, 65, 90–92.

SCHWILL-MIEDANER A, ILBERG V and SOMMER K (1996), 'Der Hygienestatus von Schankanlagen', *Brauwelt*, 136, 395–400.

SHROUT J D, CHOPP D L, JUST C L, HENTZER M, GIVSKOV M and PARSEK M R (2006), 'The impact of quorum sensing and swarming motility of *Pseudomonas aeruginosa* biofilm formation is nutritionally conditional', *Molecular Microbiology*, 62, 1264–1277.

STORGÅRDS E (1997), 'Microbiological quality of draught beer – is there reason for concern?', *European Brewery Convention Monograph 25, Symposium Draught beer, packaging & dispense, Edinburgh September 1996*, Nürnberg, Verlag Hans Carl, 92–103.

STORGÅRDS E (2000), *Process hygiene control in beer production and dispensing*, PhD thesis, University of Helsinki, Espoo, VTT Technical Research Centre of Finland, VTT Publications 410, http://www.vtt.fi/inf/pdf/publications/2000/P410.pdf.

STORGÅRDS E, PIHLAJAMÄKI O and HAIKARA A (1997), 'Biofilms in the brewing process – a new approach to hygiene management', *Proceedings of the 26th European Brewery Convention Congress, Maastricht 24–29 May 1997*, Oxford, IRL Press, 717–724.

STORGÅRDS E, HAIKARA A and JUVONEN R (2006a), 'Brewing control systems: microbiological analysis', In Bamforth C W *Brewing. New technologies*, Cambridge, Woodhead Publishing Limited, 391–426.

STORGÅRDS E, TAPANI K, HARTWALL P, SALEVA R and SUIHKO M L (2006b), 'Microbial attachment and biofilm formation in brewery bottling plants', *Journal of the American Society of Brewing Chemists*, 64, 8–15.

STORGÅRDS E, JUVONEN R and MÄNTYLÄ T (2007), 'Functional materials – a possibility to improve brewery hygiene?', *Proceedings of the 31st European Brewery Convention Congress, Venice, 6–10 May 2007*, Nürnberg, Fachverlag Hans Carl, (CD-ROM) 158, 1385–1392.

SUNADA K, WATANABE T and HASHIMOTO K (2003), 'Bactericidal activity of copper-deposited $TiO_2$ thin film under weak UV light illumination', *Environ Sci Technol*, 37, 4785–4789.

SUZUKI K, FUNAHASHI W, KOYANAGI M and YAMASHITA H (2004), '*Lactobacillus paracollinoides* sp. nov., isolated from brewery environments', *International Journal of Systematic and Evolutionary Microbiology*, 54, 115–117.

TAKAGI S (1993), 'Sterile packaging of beer', *Ferment* 6, 185–189.

TASCHAN H (1996), 'Mikrobiologische Untersuchung von Bieren aus Schankanlagen in der Gastronomie', *Brauwelt*, 136, 1014–1016.

THOMAS K and WHITHAM H (1997), 'Improvement in beer line technology', *European Brewery Convention Monograph 25, Symposium Draught beer, packaging & dispense, Edinburgh September 1996*, Nürnberg, Verlag Hans Carl, 124–137.

TIMKE M (2004), *Analysis of biofilm communities in breweries*, PhD thesis, University of Osnabrück, http://elib.ub.uni-osnabrueck.de/publications/diss/E-Diss385_thesis.pdf.

TIMKE M, WANG-LIEU N Q, ALTENDORF K and LIPSKI A (2005a), 'Community structure and diversity of biofilms from a beer bottling plant as revealed using 16S rRNA gene clone libraries', *Appl Environ Microbiol*, 71, 6446–6452.

TIMKE M, WANG-LIEU N Q, ALTENDORF K and LIPSKI A (2005b), 'Fatty acid analysis and spoilage potential of biofilms from two breweries', *J Appl Microbiol*, 99, 1108–1122.

TIMKE M, WOLKING D, WANG-LIEU N Q, ALTENDORF K and LIPSKI A (2005c), 'Microbial composition of biofilms in a brewery investigated by fatty acid analysis, fluorescence in situ hybridisation and isolation techniques', *Appl Microbial Biotechnol*, 66, 100–107.

TIMKE M, WANG-LIEU N Q, ALTENDORF K and LIPSKI A (2008), 'Identity, beer spoling and biofilm forming potential of yeasts from beer bottling plant associated biofilms', *Antonie van Leeuwenhoek*, 93, 151–161.

TSIBOUKLIS J, STONE M, THORPE A A, GRAHAM P, PETERS V, HEERLIEN R, SMITH J R, GREEN K L, NEVELL T G and EWEN R J (1999), 'Preventing bacterial adhesion onto surfaces: the low-surface-energy approach', *Biomaterials*, 20, 1229–1235.

TSIBOUKLIS J, STONE M, THORPE A A, GRAHAM P, NEVELL T G and EWEN R J (2000), 'Inhibiting bacterial adhesion onto surfaces: the non-stick coating approach', *International Journal of Adhesion & Adhesives*, 20, 91–96.

VAN DER AA KÜHLE A and JESPERSEN L (1998), 'Detection and identification of wild yeasts in lager breweries', *Int J Food Microbiol*, 43, 205–213.

VAN IERSEL M F M, VAN DIEREN B, ROMBOUTS F M and ABEE T (1999), 'Flavor formation and cell physiology during the production of alcohol-free beer with immobilized *Saccharomyces cerevisiae*', *Enzyme Microbial Technol*, 24, 407–411.

VIRKAJÄRVI I, VAUHKONEN T and STORGÅRDS E (2001), 'Control of microbial contamination in continuous primary fermentation by immobilized yeast', *Journal of the American Society of Brewing Chemists*, 59, 63–68.

WALKER S L, FOURGIALAKIS M, CEREZO B and LIVENS S (2007), 'Removal of microbial biofilms from dispense equipment: The effect of enzymatic pre-digestion and detergent treatment', *J Inst Brew*, 113, 61–66.

WERSHOFEN T (2003), 'Practical experiences in the handling of cold aseptic filling lines from the viewpoint of detergent and disinfectant manufacturers', *Brauwelt*, 143 (44), 1434–1440 (in German).

# 17

# Biofilms in poultry processing

**Judy W. Arnold, United States Department of Agriculture, Agricultural Research Service, USA**

**Abstract**: The chapter conveys the importance of biofilm study in the environment of the poultry production and processing industries. Implications for food safety and security are established for sites of occurrence and causes of biofilm formation in poultry environments. Regulations and testing methods that are incorporated into industrial sanitation systems are related to the pathogens involved in recalls of poultry products. The physical, chemical, and electrochemical methods used to treat microbial contamination in poultry facilities are compared, and a comparison of efficacy is tabulated. The types of sanitation products used in the processing environment are compared, and alternative interventions are reported. Future trends, including regulatory controls, predictive microbiology, and outsourcing are suggested.

**Key words**: biofilm, poultry, sanitation, disinfectant, *Listeria*, *Campylobacter*, *Salmonella*.

## 17.1 Introduction

### 17.1.1 Why biofilm study is important for poultry processing

Biofilms in food processing plants are a risk to food safety and a potential source of spoilage bacteria and equipment deterioration. Pathogens released from biofilms by the shear forces of processing can contaminate carcasses that follow in the processing line (Arnold, 1998). Thus, the proliferation of pathogens within the protective covering of biofilms can result in product recalls. Comparative studies between attached bacteria and planktonic bacteria (those not attached) showed that when microorganisms attached to surfaces and formed biofilms, they became more resistant to the chemicals used for plant sanitation (Krysinski *et al*., 1992; Mosteller and Bishop, 1993). Bacterial biofilms can create mechanical blockages, impedance of heat transfer processes, and biodeterioration of equipment components resulting in billions of dollars in losses each year for industrial processing operations.

### 17.1.2 Implications for food safety and security

Reduction of bacterial contamination of poultry products during processing is of major concern among processors and those concerned with food safety, because of the frequent incrimination of products such as chicken, turkey and eggs in outbreaks of food-borne illness (Gibbs *et al.*, 1978; Adams and Mead, 1983; Franco *et al.*, 1995; Smith and Fratamico, 1995). In response to consumer demand, the number of carcasses processed by a single plant each day has vastly increased in recent years. The addition of equipment to increase automation has resulted in the presentation of new surface areas for carcasses to contact repeatedly, and thus, new opportunities for bacterial attachment and cross-contamination (McEldowney and Fletcher, 1988; Arnold, 2007). Research on the physical and chemical characteristics of bacterial attachment that prevent the formation or buildup of biofilms presents the opportunity for reduction of pathogens and spoilage organisms (Arnold, 1998). To establish scientifically based regulatory guidelines, agencies need facts on the growth and survival of specific pathogens in situ. Understanding the conditions conducive to bacterial attachment and biofilm formation will provide important information for successful development of Hazard Analysis Critical Control Point (HACCP) plans for poultry production and processing. Food safety could be enhanced by increasing the use of materials that prevent attachment of microorganisms prone to develop into biofilms while decreasing the use of materials that encourage bacterial attachment. An increased understanding of bacterial attachment and biofilm formation will assist in the development of interventions to counteract these processes, and thereby enhance plant sanitation and pathogen control. Prevention of biofilm development can also reduce the need for chemicals in food plant sanitation, thereby lowering the costs to produce a safe, wholesome product.

### 17.1.3 Main themes and content of sections

This chapter reports on research and its implications for biofilms in the poultry production and processing environment. The sites where biofilms are found, which poultry products are affected, and the source of the biofilms are discussed in relation to the pathogens which they can sequester. Neither all products nor all pathogens can be discussed in one chapter. Pathogens that are most frequent, such as *Campylobacter jejuni*, and have the most deadly consequences for food safety, such as *Listeria monocytogenes*, are the focus in this work. Sanitation products and methods that kill bacteria or reduce contamination in response to governmental regulations are contrasted with alternative interventions for control and prevention of contamination to enhance sanitation practices.

## 17.2 Biofilm formation in poultry processing

### 17.2.1 Sites of occurrence and causes

Substantial bacterial contamination of the poultry processing environment, (e.g., carcasses and plant surfaces) depends on the attachment over time of many different species of microbes to other microbes, debris, and inert surfaces. Bacteria are attracted to the interfaces between surfaces (Arnold and Shimkets, 1988; Hermanowicz *et al.*, 1995), and biofilms occur on solid surfaces in contact with liquid (Costerton *et al.*, 1987; Zottola, 1994). Organic and inorganic material in the liquid can sediment to the solid surface. Subsequently, biologically active microorganisms will be attracted to these conditioned surfaces and adhere. When bacterial cells adhere or attach to a surface, they produce extracellular polymers that anchor the cells and provide a favorable environment for growth and subsequent attachment of more microorganisms and sediment (Steinberg and Poole, 1981; Arnold and Shimkets, 1988). This complex community forms a biofilm that is resistant to sanitation practices.

Bacteria from the carcasses can attach to wet equipment surfaces, form biofilms, and provide a source of cross-contamination for subsequent carcasses (Arnold, 1998). The attachment phenomena and biofilm formation have been examined by multiple laboratories. Stainless steel is the most common material found in the processing plant, and bacterial attachment to stainless steel typifies the attachment process for most other materials (Arnold and Suzuki, 2003). Cells of *Pseudomonas fragi* attached and extended appendages to surfaces of stainless steel and glass (Zoltai *et al.*, 1981), and cells of *L. monocytogenes* attached to stainless steel and other surfaces within 20 min of contact (Mafu *et al.*, 1990). *Listeria* grew on stainless steel, Teflon®, nylon, and polyester for 7 to 18 days, and its biofilm formation was supported at 21°C, but was reduced at 10°C (Blackman and Frank, 1996). Gram-negative bacteria and gram-positive bacteria, including *Pseudomonas* and *Klebsiella*, were isolated from samples of steel, rubber, and cast iron chips positioned on floor drains and food contact surfaces in several food processing environments (Helke *et al.*, 1993; Zottola, 1994). The nature of bacterial attachment to solid surfaces and the attachment rate depends upon the bacterial species, cell density, and surface properties as well as environmental conditions. The tolerance for chlorine and heat treatments of *Listeria* spp. and other microorganisms in biofilms was increased after attachment (Mead and Adams, 1986; Frank and Koffi, 1990; Wirtanen and Mattila-Sandholm, 1992; Oh and Marshall, 1995). *Salmonella* Enteritidis cells that were attached were more than twice as resistant to heat treatment as planktonic cells (Dhir and Dodd, 1995). Attached cells of *C. jejuni*, *Escherichia coli* O157:H7, *L. monocytogenes*, and *Salmonella* Typhimurium were more resistant to trisodium phosphate treatment than their corresponding planktonic cells (Somers *et al.*, 1994).

Commercial poultry processing plants mechanically defeather carcasses after feathers are loosened by scalding. Rubber fingers on mechanical pickers that remove feathers from broiler carcasses have been considered a contributor to cross-contamination in poultry processing plants (Lillard, 1986). The contaminating bacteria may come from the skin of birds, where microorganisms may be attached or trapped (Kim and Doores, 1993). During the defeathering process, which includes scalding and picking, the intact skin may be damaged, allowing bacteria to become lodged underneath superficial layers and subsequently proliferate (Thomas and McMeekin, 1980). Water sprays that may enhance microbial dispersal are used during defeathering and chilling, and the process creates an aerosol of bacteria and other solids which may contaminate other carcasses and equipment (Keel and Parmelee, 1968; Mead *et al.*, 1996).

Raw poultry products can be contaminated by bacteria in biofilms, which are usually mixed microbial communities on processing plant surfaces (Boothe and Arnold, 2003). In order to successfully control biofilms in these environments, including poultry processing facilities, knowledge of the sources and properties of the microorganisms commonly present is needed. Substrate utilization profiles (Boothe *et al.*, 1999; Boothe and Arnold, 2002) and volatile products (Arnold and Senter, 1998; Senter *et al.*, 2000) of individual bacterial species and mixed microbial communities (biofilm) associated with chicken meat samples are relevant to the poultry processing industry. Differences occur in the rapidity and frequency of utilization of individual and groups (i.e., polymers, carbohydrates, etc.) of substrates by the microflora present on poultry meat stored at 4°C and 13°C. Senter *et al.* (2000) also reported differences in the volatile compounds produced by the microbial consortia associated with the samples stored at these two temperatures.

Storage quality of cut-up chicken meat is related to the carry over of bacteria present on the carcass prior to slaughter, efficiency of processing, and the conditions and time of storage that allow proliferation prior to consumption (Senter *et al.*, 2000). Because of the seasonal and 'flock-to-flock' variations of bacteria on the carcasses, it is difficult to determine storage quality other than by enumeration of microbial load and/or characterization of odors. It has been shown that poultry products are considered unfit for consumption when aerobic plate counts (APC) reach $\log_{10}$ 7–8 colony forming units (CFU) per gram. At this stage of contamination, proteolysis and slime production begins under aerobic conditions (Senter *et al.*, 2000). Surface odors are noted when the bacterial counts reach about $\log_{10}$ 6–7 CFU per gram, and at storage temperatures of 10°C or lower. The odors were caused primarily by species of *Pseudomonas* and *Acinetobacter* (Cox *et al.*, 1975). Understanding and controlling the metabolic processes of microorganisms associated with chicken meat can lead to safer poultry products with a longer shelf life (Boothe *et al.*, 1999).

### 17.2.2 Pathogen survival

*Campylobacter*

To establish scientifically based regulatory guidelines, information is needed on growth and survival of specific pathogens *in situ*, such as *C. jejuni*. Although *C. jejuni* is a commensal organism in poultry (Pearson *et al.*, 1993), it is also a major food-borne pathogen associated with poultry products. In humans, *C. jejuni* is the etiologic agent of acute inflammatory enteritis which may be associated with Guillain-Barré syndrome, a disease of the peripheral nervous system. Consumption of undercooked or improperly handled poultry products is considered to be a major route of infection (Pearson *et al.*, 1993). Reduction of the incidence of this bacterium on raw poultry products will lead to a decrease in related food-borne illness.

Pathogens can survive in the presence of various temperature and nutrient conditions. One source of *Campylobacter* in poultry processing facilities may be equipment surfaces which have come in contact with contaminated broiler carcasses (Trachoo and Frank, 2002). When bacterial cells attach to a surface, their extracellular fibrils form a complex matrix with other microbes and debris in a biofilm composite (Arnold, 1998). Release of bacteria, such as *C. jejuni*, from the biofilm may subsequently contaminate product passing over the surface of processing equipment. Trachoo and Frank (2002) demonstrated that *C. jejuni* survives in biofilm formed on polyvinyl chloride used in poultry house water systems. Research on the attachment and viability on chicken skin of *C. jejuni* constitutively expressing green fluorescent protein (gfp) indicated that the bacterium can attach to skin, survive, and multiply during product storage at refrigeration or room temperature (Chantarapanont *et al.*, 2003; Sanders *et al.*, 2007).

Pathogens can survive for extended periods in the processing environment. When biofilms containing natural populations of bacteria and *C. jejuni* were produced on stainless steel, biofilm coverage increased during seven days when grown in Bolton's broth at 37°C (Sanders *et al.*, 2007). In addition, viability of *C. jejuni* with gfp was maintained for seven days and the fluorescence of gfp for 14 days while in nutrient depleted environments. Although other bacteria are not necessary for attachment of *C. jejuni* to stainless steel, the presence of natural populations of bacteria collected from a poultry processing facility was associated with biofilm of *C. jejuni* (Fig. 17.1). Additional studies are needed to determine why this occurs. Identifying the role that pathogens play in bacterial attachment and biofilm formation will help to determine the relative importance to food safety of pathogens found in the poultry processing plant. The risk of food-borne disease could be reduced by determining how the pathogen interacts with other bacteria in biofilms in food processing environments and using this information to develop effective intervention strategies that minimize contamination of products.

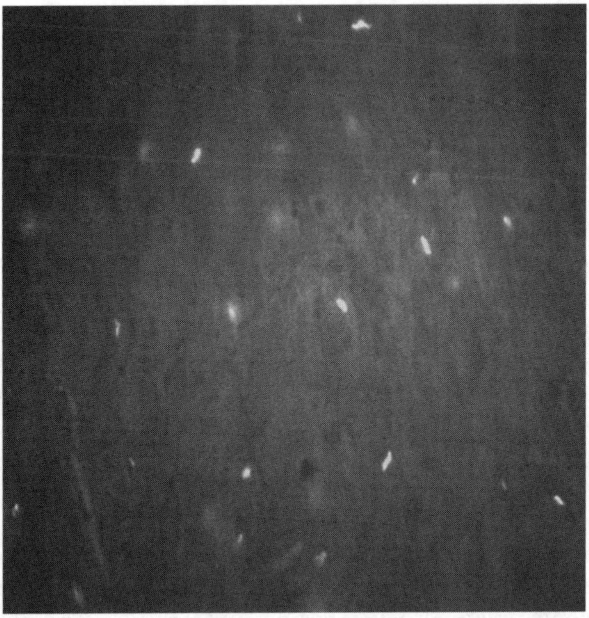

**Fig. 17.1** Light micrography of *C. jejuni* cells within a biofilm cultured from a broiler carcass rinse, Hoescht stain, magnification = 1000×.

*Listeria*

Control of *L. monocytogenes* is a major concern for regulatory, research, and industrial associations, because the bacterial pathogen causes death in 20% of its victims and has been involved in most food product recalls in the last few years (Arnold *et al.*, 2006). A modified crystal violet assay was used to quantify biofilm production of *L. monocytogenes* that mimicked wet food processing and production areas. The crystal violet dye stained bacterial cells and biofilm matrix well, and the assay was fast and economical (Arnold, 2008). Because of the urgent need for control of this pathogen, the assay was used to evaluate an improved inhibitory compound (described below) that was targeted against *L monocytogenes* grown in biofilms among other bacteria from the food processing and production environment (Arnold, 2008). An effective compound could be used for sanitation in food and health areas to help reduce cases of listeriosis.

## 17.3 Biofilm inactivation and removal

Biofilm formation is a multi-step process whereby bacteria adhere to equipment surfaces and surround themselves with a protective layer of proteins and polysaccharides. Spoilage bacteria and pathogens can be trapped inside the biofilms and contaminate product and compromise food safety during processing (Arnold, 1998). Chlorine, or some chemical form of chlorine, is the most commonly used antimicrobial agent in poultry processing in the United States. Although it is economical, the potential health risk to workers mandates the search for safer products.

### 17.3.1 Development of sanitation products used in poultry processing plants

Conditions from plant to plant and at specific locations within each plant can vary greatly. Sanitation programs must be tailored according to the needs of the facility (Arnold, 2007). Most disinfectant products do not have an adequate kill rate to completely eliminate contamination at specific points in the processing line (Arnold, 2003). Control measures to minimize cross contamination are most effective if applied in the early stages of the process near the pickers for some products (Mafu *et al.*, 1990), or in the latter stages near the chillers for other products (Arnold *et al.*, 2006).

*Silver-based compound*

Silver-based technologies were assessed by standard disinfectant testing for application in the later stages of poultry processing. Compounds with different formulations were screened for effectiveness on attachment and growth of bacterial pathogens to enhance sanitation practices and control pathogens that are responsible for food-borne illness. The activity, stability, and residual disinfection time of designated compounds were determined in planktonic and sessile conditions in the laboratory. One formulation was selected for further tests, and certification of the concentrations of each component in the formulation was acquired from a contracted laboratory. Changes in the stability of stock formulations were monitored for two weeks without degradation of efficacy.

In separate experiments the reduction of four pathogens after treatment with 1, 5, and 10 ppm of the selected silver-based formulation was measured (Arnold and Yates, 2009). Treatment at each concentration was maintained for 1, 5, 10 min for each of four pathogens. Two gram-positive and two gram-negative pathogens were selected to identify possible differences in the efficacy of the silver-based compound against bacteria with different cell wall types. Gram-positive test organisms were *Staphylococcus aureus* and *L. monocytogenes*, and gram-negative bacteria were *E. coli* and *S.* Enteritidis (Table 17.1). Solutions used in testing were prepared as follows. Sterile Butterfield's phosphate buffer (BPB) was made according to procedures in the *Bacteriological Analytical Manual* (FDA, 1998). Disinfectant was prepared in 50 ppm BPB, with final concentrations of 1, 5, or 10 ppm (10-mL total volume). Neutralization broth (D/E, Becton Dickinson) was prepared according to the manufacturer's directions.

**Table 17.1** Efficacy of silver-based compound against four pathogens

| Organism | Exposure time (minutes) | Inoculum size (cfu/ml) | Compound (ppm) | Log reduction | Percentage reduction |
|---|---|---|---|---|---|
| *S.* Enteritidis | 1 | $3.87 \times 10^5$ | 1 | 0.33 | 52.2 |
| | | | 5 | 0.43 | 61.6 |
| | | | 10 | 1.31 | 95.0 |
| | 5 | $7.74 \times 10^5$ | 1 | 0.79 | 82.5 |
| | | | 5 | 5.89 | 100 |
| | | | 10 | 5.89 | 100 |
| | 10 | $1.11 \times 10^6$ | 1 | 4 | 99.6 |
| | | | 5 | 6.04 | 100 |
| | | | 10 | 6.04 | 100 |
| *E. coli* | 1 | $1.59 \times 10^6$ | 1 | 0.12 | 24.1 |
| | | | 5 | 0.3 | 49.5 |
| | | | 10 | 1.01 | 89.8 |
| | 5 | $1.77 \times 10^6$ | 1 | 0.88 | 85.7 |
| | | | 5 | 6.24 | 100 |
| | | | 10 | 6.24 | 100 |
| | 10 | $3.98 \times 10^5$ | 1 | 2.81 | 96.8 |
| | | | 5 | 5.16 | 100 |
| | | | 10 | 5.16 | 100 |
| *L. monocytogenes* | 1 | $1.33 \times 10^6$ | 1 | 1.96 | 98.9 |
| | | | 5 | 6.08 | 100 |
| | | | 10 | 6.08 | 100 |
| | 5 | $0.96 \times 10^6$ | 1 | 5.98 | 100 |
| | | | 5 | 5.98 | 100 |
| | | | 10 | 5.98 | 100 |
| | 10 | $3.39 \times 10^5$ | 1 | 5.53 | 100 |
| | | | 5 | 5.53 | 100 |
| | | | 10 | 5.53 | 100 |
| *S. aureus* | 1 | $2.46 \times 10^6$ | 1 | 0 | 0 |
| | | | 5 | 0.04 | 7.50 |
| | | | 10 | 0.05 | 10.8 |
| | 5 | $2.82 \times 10^6$ | 1 | 1.49 | 50.0 |
| | | | 5 | 1.26 | 54.8 |
| | | | 10 | 1.56 | 69.0 |
| | 10 | $2.97 \times 10^5$ | 1 | 3.03 | 88.7 |
| | | | 5 | 5.47 | 100 |
| | | | 10 | 5.47 | 100 |

Testing was conducted as follows. Each bacterial culture was incubated for 21 hours in tryptic soy broth (TSB) at 37°C and centrifuged for 10 minutes (10000 × g). The supernatant was removed, and the pellet was resuspended in BPB. This culture was used to prepare a test culture in BPB of $10^6$ (calculated) bacteria per milliliter for a final volume of 10 mL per test tube. The requisite volume of 50 ppm formulation was added to each test tube with $10^6$ bacteria per milliliter to yield the desired final concentration. Following exposure of the bacteria to the solution for 0, 1, 5, or 10 minutes, a subsample (1 mL) was removed from each test tube and mixed with neutralization broth (9 mL) to terminate the reaction. The aerobic plate count (APC) confirmed the number of CFU per milliliter of culture in the test and original cultures. Ten-fold serial dilutions of the culture of interest were performed in BPB, with cultures in neutralization broth providing the initial dilution in the series for the treated cultures. Duplicate volumes (100 μL) of each dilution in a series were spread onto APC plates, which were incubated for 24 hours at 37°C prior to determination of the number of CFU.

Table 17.1 shows the test results for each of the pathogens at each concentration for each time period. Results from the experiments indicate that the selected formulation will be a viable treatment in practice. As a general 'rule of thumb,' the processing industry is interested in products with a four-log kill or better. The efficacy of the silver-based product was: *S.* Enteritidis, 5 ppm for 5 minutes, *E. coli*, 5 ppm for 5 minutes, *L. monocytogenes*, 1 ppm for 5 minutes, and *S. aureus*, 5 ppm for 10 minutes. Table 17.1 also indicates the percentage reduction for each of the trial pairs, and lists other parameters of the experiments. Although effective, silver-based compounds are not the most economical.

*Iodine-based compound*

Assays that measured biofilm formation, bacterial growth, and its inhibition on processing surfaces were used to assess a non-toxic iodine-based compound. The selected formulation was effective against bacterial and fungal cultures (Arnold and Yates, 2009). At 5 to 250 ppm, the compound was effective against bacterial cultures of *Enterobacter cloacae*, *Pseudomonas aeruginosa*, *L. monocytogenes*, *S. aureus*, *S.* Enteritidis, and *E. coli*. *S.* Enteritidis, *E. coli*, *L. monocytogenes*, and *S. aureus* are important food safety pathogens. *P. aeruginosa* is a major player in food spoilage (reduced shelf-life) and *E. cloacae* has been implicated for food spoilage and pathogenicity. Samples of bacteria that were collected from the processing plant were also assessed. Washing with the compound was found to enhance the resistance of stainless steel and rubber to bacterial contamination.

Quality control assays of the compound were performed for stock solutions to assure activity and stability of the compound. The subsequent goal was to target the early phases of poultry processing, stainless steel equipment, and defeathering machinery. The iodine-based formulation (25 ppm) achieved total kill against bacterial cultures of *E. cloacae*, *P. aeruginosa*, *L. monocytogenes*, *S. aureus*, *S.* Enteritidis, and *E. coli* at bacterial concentrations up to $10^9$ CFU. Here, the species are listed in relative order, from the most susceptible to the most resistant. APC confirmed the number of CFU per milliliter culture in the test and original cultures. For each test, ten-fold serial dilutions of the culture of interest were performed in TSB, and triplicate volumes (100 μl) of each dilution in the series of ten were

**Table 17.2** Efficacy of an iodine-based compound against bacterial species

| | 5 ppm | | 2.5 ppm | | 1.25 ppm | |
|---|---|---|---|---|---|---|
| | % kill | log reduction | % kill | log reduction | % kill | log reduction |
| *S.* Enteritidis | 100 | 9.82 | 99.9 | 4.02 | 18.7 | 0.09 |
| *E. coli* | 100 | 10.0 | 100 | 8.89 | 88.0 | 0.92 |
| *L. monocytogenes* | 100 | 9.06 | 84.4 | 0.81 | 71.2 | 0.54 |
| *S. aureus* | 100 | 8.87 | 87.5 | 0.90 | 0 | 0 |
| *P. aeruginosa* | 100 | 8.71 | 100 | 9.13 | 71.7 | 0.55 |
| *E. cloacae* | 100 | 9.04 | 99.9 | 3.08 | 38.4 | 0.21 |

spread onto plates, which were incubated for 24 hours at 37°C prior to determination of the number of surviving bacteria. Table 17.2 shows that, at concentrations of 5 ppm and above, all samples had 100 per cent reduction (complete kill) when calculated against bacterial samples without disinfectant. Results from the experiments indicate that the selected formulation will be a viable treatment in practice.

For many years, the use of rubber fingers on mechanical pickers to remove feathers from broilers after scalding was considered a major contributor to cross-contamination of product and equipment. Antimicrobial compounds that can be applied to the rubber surfaces or incorporated into the rubber to inhibit the attachment and growth of bacteria are of interest to the industry. The iodine-based formulation was tested for efficacy on rubber picker fingers used in defeathering machines in commercial poultry processing operations. Rubber samples were cut from the main body of each picker finger (14.5 cm in length): each sample was approximately 10 mm in length and 14 mm in diameter and contained 3 rib sections. Initial studies, to develop the methodology, indicated the presence of bacteria on these pieces. Accordingly, the samples were exposed to ultraviolet (UV) light for 24 hours prior to use in experiments in order to eliminate these endogenous microflora. Effective treatment by this UV exposure was confirmed by the absence of bacterial growth following introduction of the samples into sterile TSB and incubation at 37°C overnight.

The iodine-based formulation (50 ppm), was further tested on rubber surfaces with attached bacteria. Experimental procedures mimicked the sanitation program in the processing plant, which used water rinses to clean the rubber fingers in the defeathering machinery (Arnold and Yates, 2009). A reduction (from one to seven logs) of the number of attached gram-negative bacteria tested, *S.* Enteritidis and *E. coli*, was noted following exposure of rubber fingers to the iodine-based formulation (50 ppm). The compound was more effective against *S.* Enteritidis than *E. coli*, although initial culture densities ($\log_{10}$ 8–9 CFU) were approximately equal. For the gram-positive test bacteria, *S. aureus* and *L. monocytogenes*, a reduction (from six to seven logs) of the number of attached bacteria was also evident after exposure of picker fingers to 50 ppm. Initial culture densities of these gram-positive bacteria were approximately the same ($\log_{10}$ 7 CFU), and the test compound was equally effective against them. The usual culture growth of the gram-positive test bacteria was one to two orders of magnitude lower than the gram-negative test bacteria.

Three trials in each of three processing plants were undertaken to assess the bacterial populations present and to determine the efficacy of the iodine-based compound in situ for equipment components. Rubber picker fingers that had been in use in commercial processing equipment were evaluated as described above. The numbers of bacteria present on fingers from the plant were lower than on fingers exposed in the laboratory to bacterial cultures. The results indicated that the iodine-based compound, at 50 ppm, could eliminate *S.* Enteritidis on rubber picker fingers in the processing plant (Arnold, 2007). Scanning electron microscopy (SEM) was used to visualize the surface before and after testing to determine whether bacteria were sequestered by the rubber, and whether there were detrimental changes to the rubber surface or the bacterial cells caused by the test compound (Arnold and Yates, 2009). During examination of the micrographs, no detrimental changes were observed on the rubber surfaces that were exposed to the compound, and bacteria and beginning biofilms attached to the rubber surfaces. Figure 17.2 shows *S.* Enteritidis on the rubber surface before treatment (Fig. 17.2A) and after treatment (Fig. 17.2B) with the iodine-based compound. Further, the results supported the previous efficacy data, indicating that bacterial cells and biofilms were removed by treatment with the test compound.

*Quaternary ammonium-based compound*

The reduction of microbial pathogens in meat and poultry products, particularly *L. monocytogenes*, is a pressing food safety issue. *L. monocytogenes* is responsible for the most product recalls relating to pathogen contamination, and causes a 20 per cent death rate among its human victims. To establish baseline disinfectant technology for attached bacteria, the effectiveness of a quaternary ammonium-based compound was measured (Arnold *et al.*, 2006). Improved methodology for the standard use-dilution test (modification of the AOAC Official method 955.11) for *L. monocytogenes* was developed to measure the lowest effective dilution of the compound on attached cells as compared to phenol. Stainless steel penicylinders were contaminated with test bacteria and added separately to a series of test tubes. Aliquots of the test compound were placed into the tubes for five, ten, and 15 minutes (Arnold *et al.*, 2006). Negative controls of penicylinders and media tubes, positive controls of inoculated penicylinders and neutralization buffer, and controls for variability of bacterial numbers were included. The concentration of bacteria on each penicylinder was $10^5$ CFU.

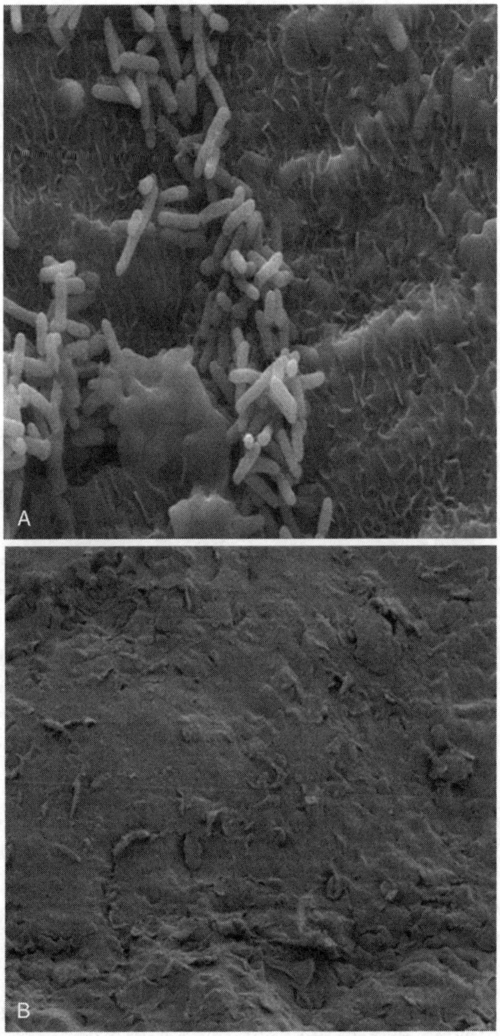


**Fig. 17.2** SEM of *S.* Enteritidis biofilm on picker finger rubber before treatment (A) and after treatment (B) with an iodine-based compound, magnification = 3000×.

After exposure, the carriers were removed and placed into two series of neutralization media, incubated at 37°C for 48 hr and observed for growth. The quantitative assessment of disinfectant activity was used against 15 field isolates and ten laboratory strains. *S. aureus*, *S.* Enteritidis, *E. coli*, and *P. aeruginosa*, as representatives of bacterial pathogenicity and spoilage activity, were compared with *Listeria* species. The minimum inhibitory concentration to clear each pathogen was determined for isolates and controls. The product was most effective against *L. monocytogenes*, and was more effective against field isolates than on laboratory strains. Growth of the gram-positive *Listeria* sp. and *S. aureus* was prevented by 5–25 ppm of the disinfectant, but 250 ppm was required to prevent growth of the gram-negative bacteria.

In further testing of the quaternary ammonium-based compound against *Listeria*, test methods for biofilm removal were employed (Arnold, 2008). Four strains of *Listeria* (two laboratory and two field strains) were grown and evaluated by a colorimetric assay of crystal violet staining, with the field strains again showing greater susceptibility to the chemical. A bioreactor was used to grow *Listeria* biofilms on stainless steel coupons under flow conditions similar to poultry processing areas. An APC was performed to assess the initial presence of the bacteria in the reactor as well as on the coupons after consecutive 24-hour periods. An equal portion of the coupons were stained with crystal violet dye, extracted with ethanol, and measured for biofilm production. Figure 17.3 shows *L. monocytogenes* biofilm before (Fig. 17.3A) and after (Fig. 17.3B) treatment with the quaternary ammonium-based compound. Live cells could not be cultured from the treated samples. The biofilm mass, before and after treatment, was confirmed by confocal microscopy (not shown). An optimized disinfectant formulation that is highly effective against *L. monocytogenes* biofilm will be of great use for sanitation in food and healthcare setting to help reduce cases of listeriosis.

Laboratory studies indicate that the three test compounds, with activity based on silver, iodine, or quaternary ammonium, merit investigation as commercial techniques of application for use as controls for pathogens and spoilage organisms. Applications of these products are most effective in food processing areas where the surfaces are susceptible to bacterial attachment and biofilm formation, and the humidity and temperature are favorable. Combining the surface material most resistant to bacterial attachment with cleaning by the most effective antimicrobial concentration at such locations will maximize the potential for a useful product.

## 17.4 Prevention of biofilm formation

Sanitation programs in processing facilities control bacterial contamination by physical and chemical methods. Physical methods include the use of heat, ultra-low temperatures, desiccation, osmotic pressure, filtration, and radiation (Arnold, 1998). The common steps in a processing plant sanitation program begin by pre-rinsing with a water spray followed by washing or scrubbing with a chemical application. However, each of these methods can cause degradation of the surface.

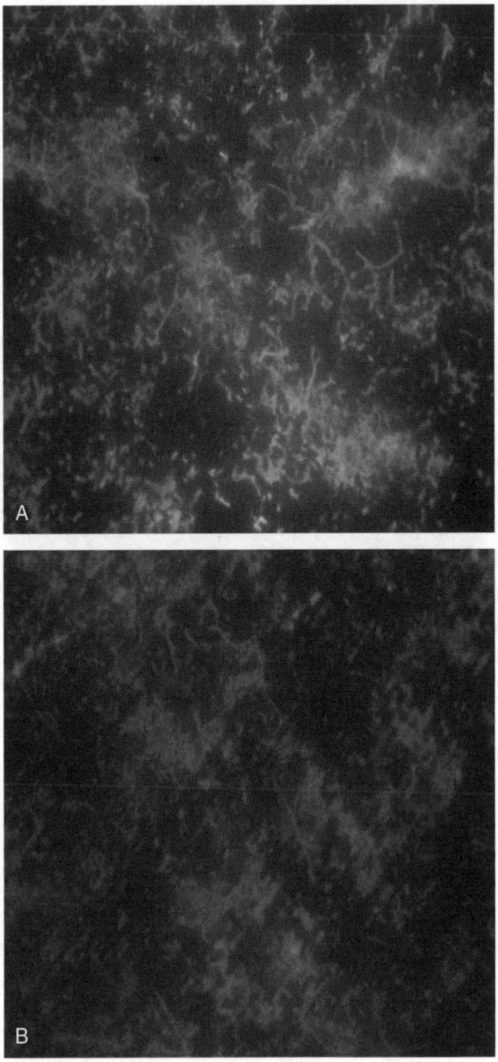


**Fig. 17.3** Epifluorescence micrograph of *L. monocytogenes* biofilm on stainless steel before treatment (A) and after treatment (B) with a quat-based compound, DAPI stain, magnification = 1000×.

### 17.4.1 Rubber picker fingers

Picker finger rubber resists bacterial attachment better than stainless steel and other processing equipment surface materials. Previous studies suggested that new picker-finger rubber can inhibit the growth of natural microbial populations from broiler carcasses (Arnold and Silvers, 2000). As discussed above, the minimal concentration of chemicals necessary to completely eliminate bacterial contamination on the fingers can be measured and applied.

### 17.4.2 Electrochemical

Surfaces that have been modified to resist bacterial attachment provide an alternative process for intervention (Arnold *et al.*, 2001). Prevention of attachment would allow bacteria dislodged from carcasses to be washed away during processing. Methods were developed to measure attached bacteria and to identify factors that make surface finishes susceptible or resistant to bacterial attachment and biofilm formation (Arnold and Bailey, 2000). Samples of the treated surfaces: sandblasted, sanded, and electropolished, were exposed to natural bacterial populations from chicken carcass rinses to allow growth of bacteria and development of biofilms on the surfaces. The kinetics of bacterial growth during surface exposure was followed by UV and visible spectrophotometry, and counts of bacteria and early biofilm formation were measured along with assessment by SEM. The surface morphology of the samples was analyzed by atomic force microscopy, using samples from each of the batches of treatments used in the SEM studies. The surface types varied in affinity for bacteria, and both physical and electrochemical treatments improved resistance of stainless steel to bacterial attachment (Arnold and Silvers, 2000). Electropolished stainless steel was the least rough surface and showed significantly fewer bacterial cells and beginning biofilm formations than the other treated surfaces. Food safety could be improved if bacterial populations could be reduced during processing by increasing the use of materials that are resistant to bacterial contamination.

### 17.4.3 Alternative interventions

The use of chemicals in food plant sanitation for killing microorganisms could be reduced by the use of alternative nonchemical interventions. Air-cleaning devices installed in ductwork, UV rays, and high efficiency particulate air filtration are products intended to remove microorganisms and particles from the air. Negative air ionization has been shown to be a promising technology that is a safe, non-toxic, and non-chemical means of eliminating dust and pathogens from the air in poultry environments (Arnold and Mitchell, 2002). The electrostatic space charge system (ESCS) has been shown to reduce airborne particulates and bacteria by charging the particles and collecting them on special grounded collectors or on walls and floors. The ESCS was tested against five pathogens: *C. jejuni*, *E. coli*, *S.* Enteritidis, *L. monocytogenes*, and *S. aureus*; mixed species of bacteria from chicken carcass rinses, and spores of *Bacillus stearothermophilus*. The system reduced bacterial contamination and bacterial spores on stainless steel surfaces up to 99.9% (Arnold *et al.*, 2004). These results with biofilms are comparable to the reductions obtained with aerosolized *S.* Enteritidis and negative air ionization treatment in a caged layer room to show reduction efficiency for airborne levels of *S.* Enteritidis. When exposed to negative air ionization, levels of bacteria on XLT 4 plates were reduced by 95% (Mitchell *et al.*, 2002). Negative air ionization could potentially have a significant impact on the microbial load in a poultry processing facility. The economic feasibility of using this technology in a poultry processing facility remains to be determined; but, if shown to be effective on a large scale, the basic equipment cost would probably be comparable to existing disinfection equipment and chemicals.

## 17.5 Future trends

The results reported in this chapter suggest that a combination of appropriate surface material and efficacious sanitation procedures are necessary to prevent significant contamination. Combining the surface material most resistant to bacterial attachment with cleaning by the most effective sanitation system will maximize the potential for reduction of bacterial contamination and chemical impact on the environment. Consumers demand safer products for health and the environment. Improvements are needed in the food industry for sampling biofilms, and for identification, enumeration, and detection of pathogens within a biofilm. In the future food processors will frequently rely on outsourcing to contract labs for this information. Regulatory controls must be founded on science-based data with the help of predictive microbiology. A combination of surface modification, chemical control, or alternative interventions that are tailored to site-specific problem areas will be needed to eliminate bacterial contamination by the end of the processing line.

## 17.6 Sources of further information and advice

### 17.6.1 References to consult

Sanitation steps for processing plant surfaces, table of chemicals for control of microbial growth, biofilm formation and prevention (Arnold, 2003).

History, generation methods, hazards, and uses of chlorine, ozone, bromine, UV radiation, and other alternative disinfectants (White, 1999).

### 17.6.2 Industry and interest groups, web sites

Biofilms Online – news, features, trade information, and applied research about microbial biofilms. http://www.biofilmsonline.com

Poultry Science Association – Association news, journal contents, jobs, meeting information, and related items. http://www.poultryscience.org

Food Safety and Inspection Service – the regulatory agency in the US Department of Agriculture responsible for ensuring that the nation's commercial supply of meat, poultry, and egg products is safe, wholesome, and correctly labeled and packaged, as required by the Federal Meat Inspection Act, the Poultry Products Inspection Act, and the Egg Products Inspection Act. http://www.fsis.usda.gov

US Poultry & Egg Association – industry organization, sponsor of largest poultry trade show on earth. http://www.poultryegg.org

## 17.7 References


ADAMS B W, and MEAD G C (1983), 'Incidence and properties of *Staphylococcus aureus* associated with turkeys during processing and further-processing operations', *J Hyg Camb*, 91, 479–490.

ARNOLD J W (1998), 'Development of bacterial biofilms during poultry processing', *Av Poult Biol Rev*, 9 (1), 1–9.

ARNOLD J W (2003), 'Advances in food sanitation: Use of intervention strategies', in Schmidt R H, and Rodrick G E, *Food Safety Handbook*, Hoboken, NJ, John Wiley & Sons, Inc., 337–352.

ARNOLD J W (2007), 'Bacterial contamination on rubber picker fingers before, during, and after processing', *Poult Sci*, 86 (12), 2671–2675.

ARNOLD J W (2008), 'Colorimetric assay for biofilms in wet processing conditions', *J Ind Microbiol Biotechnol*, 35, 1475–1480.

ARNOLD J W, and BAILEY G W (2000), 'Surface finishes on stainless steel reduce bacterial attachment and early biofilm formation: scanning electron and atomic force microscopy study', *Poult Sci*, 79 (12), 1839–1845.

ARNOLD J W, and MITCHELL B W (2002), 'Use of negative air ionization for reducing microbial contamination on stainless steel surfaces', *J Appl Poult Res*, 11 (2), 179–186.

ARNOLD J W, and SENTER S D (1998), 'Use of digital aroma technology and SPME GC-MS to compare volatile compounds produced by bacteria isolated from processed poultry', *J Sci Food Ag*, 78 (3), 343–348.

ARNOLD J W, and SHIMKETS L J (1988), 'Inhibition of cell-cell interactions in *Myxococcus xanthus* by congo red', *J Bacteriol*, 170 (12), 5765–5770.

ARNOLD J W, and SILVERS S (2000), 'Comparison of poultry processing equipment surfaces for susceptibility to bacterial attachment and biofilm formation', *Poult Sci*, 79 (8), 1215–1221.

ARNOLD J W, and SUZUKI O (2003), 'Effects of corrosive treatment on stainless steel surface finishes and bacterial attachment', *Trans ASAE*, 46 (6), 1595–1602.

ARNOLD J W, and YATES I E (2009), 'Interventions for control of *Salmonella*: clearance of microbial growth from rubber picker fingers', *Poult Sci*, 88, 575–581.

ARNOLD J W, BOOTHE D H, and BAILEY G W (2001), 'Parameters of treated stainless steel surfaces important for resistance to bacterial contamination', *Trans ASAE*, 44, 347–356.

ARNOLD J W, BOOTHE D H, and MITCHELL B W (2004), 'Use of negative air ionization for reducing bacterial pathogens and spores on stainless steel surfaces', *J Appl Poult Res*, 13 (2), 200–206.

ARNOLD J W, DELAUBENFELS E, and ZAMBELLI-WEINER A (2006), 'Quantitative assessment of hard surface disinfectant activity against the foodborne pathogen *Listeria monocytogenes*', *J AOAC Int*, 89 (6), 1617–1621.

BLACKMAN I C, and FRANK J F (1996), 'Growth of *Listeria monocytogenes* as a biofilm on various food-processing surfaces', *J Food Prot*, 59, 827–831.

BOOTHE D D, and ARNOLD J W (2002), 'Nutrient substrates used by bacterial isolates from the poultry processing environment', *Poult Sci*, 81 (9), 1392–1405.

BOOTHE D D, ARNOLD J W, and CHEW V (1999), 'Utilization of substrates by bacterial communities (biofilm) as they develop on stored chicken meat samples', *Poult Sci*, 78 (12), 1801–1809.

BOOTHE D H, and ARNOLD J W (2003), 'Resistance of bacterial isolates from poultry products to therapeutic veterinary antibiotics', *J Food Prot*, 66 (1), 94–102.

CHANTARAPANONT W, BERRANG M, and FRANK J F (2003), 'Direct microscopic observation and viability determination of *Campylobacter jejuni* on chicken skin', *J Food Prot*, 66 (12), 2222–2230.

COSTERTON J W, CHENG K J, GEESEY G G, LADD T I, NICKEL J C, DASGUPTA M, and MARRIE T J (1987), 'Bacterial biofilms in nature and disease', *Ann Rev Microbiol*, 41, 435–464.

COX N A, JUVEN B J, THOMSON J E, MERCURI A J, and CHEW V (1975), 'Spoilage odors in poultry meat produced by pigmented and nonpigmented *Pseudomonas*', *Poult Sci*, S4, 2001–2006.

DHIR V K, and DODD C E R (1995), 'Susceptibility of suspended and surface-attached *Salmonella* Enteritidis to biocides and elevated temperatures', *Appl Environ Microbiol*, 61 (5), 1731–1738.

FDA (1998), *Bacteriological Analytical Manual*, Gaithersburg, MD, AOAC International.

FRANCO C M, QUINTO E J, FENTE C, RODRIGUEZ-OTERO J L, DOMINGUEZ L, and CEPEDA A (1995), 'Determination of the principal sources of *Listeria* spp. contamination in poultry meat and a poultry processing plant', *J Food Prot*, 58 (12), 1320–1325.

FRANK J F, and KOFFI R A (1990), 'Surface-adherent growth of *Listeria monocytogenes* is associated with increased resistance to surfactant sanitizers and heat', *J Food Prot*, 53 (7), 550–554.

GIBBS P A, PATTERSON J T, and THOMPSON J K (1978), 'The distribution of *Staphylococcus aureus* in a poultry processing plant', *J Appl Bacteriol*, 44, 401–410.

HELKE D M, SOMERS E B, and WONG A C L (1993), 'Attachment of *Listeria monocytogenes* and *Salmonella typhimurium* to stainless steel and Buna-N in the presence of milk and individual milk components', *J Food Prot*, 56 (6), 479–480.

HERMANOWICZ S W, SCHINDLER U, and WILDERER P (1995), 'Fractal structure of biofilms: new tools for investigation of morphology', *Wat Sci Tech*, 32 (8), 99–105.

KEEL J E, and PARMELEE C E (1968), 'Improving the bacteriological quality of chicken fryers', *J Milk Food Tech*, 31, 377–381.

KIM J-W, and DOORES S (1993), 'Influence of three defeathering systems on microtopography of turkey skin and adhesion of *Salmonella typhimurium*', *J Food Prot*, 56 (4), 286–291.

KRYSINSKI E P, BROWN L J, and MARCHISELLO T J (1992), 'Effect of cleaners and sanitizers on *Listeria monocytogenes* attached to product contact surfaces', *J Food Prot*, 55 (4), 246–251.

LILLARD H S (1986), 'Role of fimbriae and flagella in the attachment of *Salmonella typhimurium* to poultry skin', *J Food Sci*, 51 (1), 54–56.

MAFU A A, ROY D, GOULET J, and MAGNY P (1990), 'Attachment of *Listeria monocytogenes* to stainless steel, glass, polypropylene, and rubber surfaces after short contact times', *J Food Prot*, 53 (9), 742–746.

MCELDOWNEY S, and FLETCHER M (1988), 'Bacterial desorption from food container and food processing surfaces', *Microb Ecol*, 15, 229–237.

MEAD G C, and ADAMS B W (1986), 'Chlorine resistance of *Staphylococcus aureus* isolated from turkeys and turkey products', *Lett Appl Microbiol*, 3, 131–133.

MEAD G C, SCOTT M J, HUMPHREY T J, and MCALPINE K (1996), 'Observations on the control of *Campylobacter jejuni* infection of poultry by 'competitive exclusion', *Av Path*, 25 (1), 69–79.

MITCHELL B W, BUHR R J, BERRANG M E, BAILEY J S, and COX N A (2002), 'Reducing airborne pathogens, dust and *Salmonella* transmission in experimental hatching cabinets using an electrostatic space charge system', *Poult Sci*, 81 (1), 49–55.

MOSTELLER T M, and BISHOP J R (1993), 'Sanitizer efficacy against attached bacteria in a milk biofilm', *J Food Prot*, 56 (1), 34–41.

OH D-H, and MARSHALL D L (1995), 'Destruction of *Listeria monocytogenes* biofilms on stainless steel using monolaurin and heat', *J Food Prot*, 57 (3), 251–255.

PEARSON A D, GREENWOOD M, HEALING T D, ROLLINS D, SHAHAMAT M, DONALDSON J, and COLWELL R R (1993), 'Colonization of broiler chickens by waterborne *Campylobacter jejuni*', *Appl Environ Microbiol*, 59 (4), 987–996.

SANDERS S Q, BOOTHE D H, FRANK J F, and ARNOLD J W (2007), 'Culture and detection of *Campylobacter jejuni* within mixed microbial populations of biofilms on stainless steel', *J Food Prot*, 70 (6), 1379–1385.

SENTER S D, ARNOLD J W, and CHEW V (2000), 'APC values and volatile compounds formed in commercially processed, raw chicken parts during storage at 4 and 13°C and under simulated temperature abuse conditions', *J Sci Food Ag*, 80 (10), 1559–1564.

SMITH J L, and FRATAMICO P M (1995), 'Factors involved in the emergence and persistence of food-borne diseases', *J Food Prot*, 58 (6), 696–708.

SOMERS E B, SCHOENI J L, and WONG A C (1994), 'Effect of trisodium phosphate on biofilm and planktonic cells of *Campylobacter jejuni*, *Escherichia coli* O157:H7, *Listeria monocytogenes* and *Salmonella typhimurium*', *Int J Food Microbiol*, 22, 269–276.

STEINBERG M S, and POOLE T J (1981), 'Strategies for specifying form and pattern: adhesion-guided multicellular assembly', *Phil Trans R Soc Lond*, B295, 451–460.

THOMAS C J, and MCMEEKIN T A (1980), 'Contamination of broiler carcass skin during commercial processing procedures: an electron microscopic study', *Appl Environ Microbiol*, 40 (1), 133–144.

TRACHOO N, and FRANK J F (2002), 'Effectiveness of chemical sanitizers against Campylobacter jejuni-containing biofilms', *J Food Prot*, 65 (7), 1117–1121.

WHITE G C (1999), *Handbook of chlorination and alternative disinfectants*, New York, John Wiley & Sons, Inc.

WIRTANEN G, and MATTILA-SANDHOLM T (1992), 'Effect of the growth phase of foodborne biofilms on their resistance to a chlorine sanitizer Part II', *Lebensm Wiss U Technol*, 25, 50–54.

ZOLTAI P T, ZOTTOLA E A, and MCKAY L L (1981), 'Scanning electron microscopy of microbial attachment to milk contact surfaces', *J Food Prot*, 44, 204–208.

ZOTTOLA E A (1994), 'Scientific status summary "Microbial attachment and biofilm formation: a new problem for the food industry?"', *Food Technol*, 48 (7), 107–114.

# 18

# Beneficial biofilms: wastewater and other industrial applications

**Nasib Qureshi, United States Department of Agriculture, USA**

**Abstract**: This chapter describes the use of beneficial biofilms for the production of industrial chemicals such as ethanol, butanol, lactic acid, acetic acid/vinegar, succinic acid, and fumaric acid. It also emphasizes application of biofilm reactors for treatment of dairy industry wastewater, oily sea water, and wastewater in general. The bioreactor designs where these biofilms are used can be batch, continuous stirred tank (CSTR), packed bed (PBR), fluidized bed (FBR), airlift (ALR), or any other suitable reactor configuration. Biofilm reactors are simple, offer higher productivities, and biofilm supports such as bone char, and clay brick particles are economically available.

**Key words**: biofilms, industrial chemicals, ethanol, butanol, lactic acid, acetic acid/ vinegar, biofilm reactors, biofilm supports.

## 18.1 Introduction

Biofilm formation is a natural process, where microbial cells attach to solid surfaces and grow in the form of uniform layers of cells. Common examples of biofilms include slimy and slippery films found on stones and rocks submerged in water streams and deep under the ocean. Other examples of biofilms include cell growth on hulls of ships (water submerged surfaces), inside heat exchange equipment, inside surfaces of pipes carrying liquid foods and even on teeth and medical equipment. The amount of cell growth and formation of biofilm layers depend upon availability of nutrients and favorable environmental conditions. Since biofilm formation on ship surfaces and heat exchange equipment is not desirable, removing

Mention of trade names of commercial products in this article is solely for the purpose of providing scientific information and does not imply recommendation or endorsement by the United States Department of Agriculture.

them is an involved process which costs the shipping and chemical industry a significant amount of money. However, this natural process of biofilm formation either by multi or single microbial species can be exploited for obtaining economic gains. The reader is advised that biofilm formation requires no exogenous chemical or physical process as bacteria (or other microbial cells) can grow on nearly all surfaces. A question arises as why cells grow on surfaces? Attempts to answer this question have been, at least partly, reported in literature (Costerton *et al.*, 1987; O'Toole *et al.*, 2000; Qureshi *et al.*, 2005) and hence are considered beyond the scope of this article.

Bioconversion and microbial biotechnology is an environmentally friendly and rapidly growing area where chemicals and biochemicals can be produced using microbial cultures. The production of such chemicals and biochemicals involves various costs including substrate, process, and recovery. In order to reduce the overall cost of production of a chemical or biochemical, the processing cost should be reduced. One of the goals in microbial bioconversion is to increase cell concentration inside the reactors by cell immobilization or cell recycle for the purpose of obtaining high reaction rates thus reducing reactor size and process time. Reduction in reactor size and process time reduces both capital and process costs thus producing chemicals more economically. Cell immobilization generally requires expensive chemicals thus offsetting economic gains obtained by increasing cell concentration inside the reactor. Similarly cell recycle technology is cost intensive. Since increasing cell concentration inside the reactor either by immobilization or cell recycle techniques is costly, various authors have employed the natural process of adsorption and biofilm formation on solid surfaces (outlined in the first paragraph of this section) and used them in reactors for the production of chemicals cost effectively (Qureshi and Maddox, 1987; Krug and Daugulis, 1983; Tyagi and Ghose, 1982; Crueger and Crueger, 1989).

Microbial cells can adhere and multiply onto solid surfaces of natural origin and this property can be employed for increasing cell concentration inside the reactors and produce chemicals at faster rates thus reducing capital and process costs that are listed in the above paragraph. It has been observed that cell concentration in adsorbed cell reactors can be increased to 50–60 $gL^{-1}$ (Qureshi *et al.*, 2005) as compared to 3–4 $gL^{-1}$ in free cell batch reactors. Such an increase in cell concentration results in enhanced reaction rates. It should be noted that the process of cell adsorption results in a number of advantages including:

i) requirements for no exogenous chemicals
ii) the process of adsorption is simple and can be performed in place
iii) reactors where adsorbed cell biofilms are used are simple in concept and construction
iv) reaction rates are high

v) adsorption supports are inexpensive and are of natural origin and environmentally friendly
vi) high cell concentrations are achieved inside the reactors.

The details of the reactors that can be used for biofilms have been given in Section 18.4. The purpose of this chapter is to inform the reader that biofilms can be successfully employed for production of various chemicals and hence providing economic benefits to mankind.

## 18.2 Various types of biofilms

Biofilms can be categorized into two groups, namely: i) multi-species biofilms and ii) single species biofilms.

**i)** In multi-species bacterial communities each species fills an ecological niche within the biofilm depending on its metabolism and morphology (Stoodley *et al.*, 2002). The nature of mixed culture biofilms is dependent on which species are present and what role each species plays. For example, a single species may utilize anaerobic fermentation deep within one biofilm in one environment, but may utilize an aerobic metabolism in another environment in the presence of different neighboring biofilm species. It has been observed that multi-species biofilms are important both clinically as well as industrially. Clinically, biofilms are undesirable and important as a source of persistent infections. They are responsible for dental caries and nosocomial infections, as well as a variety of other infections and diseases (Costerton *et al.*, 1999). Industrially, biofilms are detrimental in many cases and beneficial in many others. For example, natural biofilms can reduce heat transfer in heat exchangers and cooling towers (Mortensen and Conley, 1994), foul ultrafiltration and reverse osmosis membranes (McDonogh *et al.*, 1994), and contaminate food processing equipment (Carpentier and Cerf, 1993). Beneficial uses of multi-species biofilms include the treatment of wastewater for removal of organics (Taras *et al.*, 2005; Hall, 1987) and heavy metals (Meyer and Wallis, 1997).

**ii)** Single species biofilms are used to produce commercially important chemicals (Qureshi and Maddox, 1990; Maddox *et al.*, 1988). Such biofilms can exist in some situations and are important industrially, although in nature they are not the norm. For chemical production, single species biofilms are important because they allow for control and maximization of desired products. In this case, a single species is inoculated into a sterile environment and allowed to form a biofilm before being used to produce a particular chemical product.

Both multi and single species biofilms can further be divided into two groups, namely: a) biofilms that grow onto supports including bonechar, charcoal, resin, concrete, clay brick, or sand particles and, b) biofilms that are formed as a result of flocs and aggregate formation. On the above mentioned supports, biomass grows all around the particles and the size of the biofilm particles grows over time usually to several mm in diameter. The density of the support particles is usually higher than the fermentation broth and for this reason bioparticles tend to remain in the lower section of the reactor. The second type of biofilm that is mentioned above is where no support is used and cells form biomass granules and flocs that also grow in size over time. This type of biofilm is called granular biofilm and the reactor where this biofilm is used is called a granular biofilm reactor (GBR). Granule formation may take from several weeks to several months. During biofilm formation, the cells produce extracellular polymeric substances (EPS) that bind the cells firmly in the form of flocs and aggregates.

## 18.3 Factors enhancing biofilm formation

Several parameters determine how quickly biofilms form and mature, including the surfaces of adsorbents and bacterial species, the cellular makeup of the culture, and environmental factors (such as nutrients present in the medium, temperature, and pH). Rough surfaces and pores inside porous supports tend to enhance biofilm formation (Characklis *et al.*, 1990) as shear forces are lower near a rough surface, and there is a larger surface area to which cells can adhere and multiply. Pores provide a protected environment for cells to attach. Biofilm formation also tends to increase with the hydrophobicity of the surface material (Donlan and Costerton, 2002). Biofilms tend to form more readily in the presence of ample nutrients (Cowan *et al.*, 1991). One function of the biofilm is to anchor cells in a friendly, nutrient-rich environment. Phosphorus is a particularly important nutrient. Cells saturated with phosphate have a greater tendency to flocculate and adhere due to their increased hydrophobicity, while those cells depleted in phosphate are more hydrophilic and less likely to adhere (Bücks *et al.*, 1988). A hydrophobic cell will be more able to overcome the initial electrostatic repulsion with the solid surface and adhere more readily. The presence of fimbriae, proteinaceous bacterial appendages high in hydrophobic amino acids, can increase cell surface hydrophobicity (Donlan and Costerton, 2002; Donlan, 1992). Flagellated cells, where flagellar motility may serve to overcome initial electrostatic surface repulsion, also show increased ability to attach to surfaces.

## 18.4 Biofilm reactors

Biofilm reactors can be assembled in a number of configurations including batch, continuous stirred tank (CSTR; including agitating continuous reactors, and rotary continuous reactors), packed bed (PBR), trickling bed

(TBR), fluidized bed (FBR), airlift (ALR), upflow anaerobic sludge blanket (UASB), expanded granular sludge bed (EGSB), expanded bed (EBR), granular biofilm (GBR), moving bed biofilm (MBBR), and vertical moving biofilm (VMBR) reactors. The details of some of these reactors including schematic diagrams are published elsewhere (Qureshi *et al.*, 2005).

Airlift reactors contain two concentric tubes, a riser (an inner tube) and a downcomer (an outer tube). In these reactors, mixing is achieved by circulating or injecting air at the bottom of the reactor. As a result of force applied by the air (at the bottom of the inner tube), the liquid in the inner tube moves up which then overflows (the inner tube) downward thus creating eddies to mix the liquid. In some airlift reactors the downcomer is replaced with an external loop to circulate fermentation broth or medium. Such reactors where air is replaced by anaerobic gases ($N_2$ or $CO_2$ and $H_2$) are called gaslift reactors. UASB reactors (contain granular biofilm particles; developed by Lettinga *et al.*, 1980) are used for anaerobic treatment of wastewater/industrial effluents. As the name suggests, the flow in these reactors is in an upward direction. At the top of the reactor, provisions are made for gas(es) to escape and sludge particles to settle to the bottom part of the reactor. Reactor effluent is removed from the top of the reactor.

## 18.5 Biofilms in biological wastewater treatment

Many chemical, biological, and food processing industries produce large amounts of effluents that cannot be discharged into water streams as they are environmental pollutants and are toxic to marine life and humans. Hence, their treatment is sought prior to such a discharge. Use of biofilm technology is one area which has been successfully employed to treat such wastewater streams. The application of biofilm technology in wastewater treatment originated from the industrial operation of trickling filters in the early 1880s in Wales, Great Britain (Lazarova and Manem, 2000). Biofilm processes in wastewater treatment can be divided into two groups, namely: i) the fixed-medium systems where the biofilm media are static in the reactors and the biological reactions take place in the biofilm developed on the static media, and ii) the moving-medium systems where the biofilm media are kept continually moving by means of mechanical, hydraulic, or air forces (Rodgers and Zhan, 2003). The fixed-medium systems include trickling filters and biological aerated filters, while the moving-medium systems include rotating biological contactors (RBC), MBBRs, VMBRs, and FBRs. Rotating biological contactors treat wastewater streams using a thin biofilm of aerobic microorganisms on rotating cylinders or biodiscs. The rate of rotation is selected to provide optimum contact of the waste stream with the biofilm for efficient oxygen transfer and bioactivity (reviewed in Qureshi *et al.*, 2005).

Before a biofilm-based treatment system is to be considered for the treatment of wastewater, it is necessary to determine whether the naturally occurring microorganisms are able to produce biofilms, while simultaneously reducing the COD (chemical oxygen demand) of the wastewater, or if there is a need to inoculate the reactor with external bacterial strains. The most commonly used RBC is the rotating biodisc and its various modifications (Kargi and Eker, 2001). In this study, Kargi and Eker (2001) have shown that a rotating-perforated-tube biofilm reactor is effective in COD removal from synthetic wastewater composed of diluted molasses, urea, $KH_2PO_4$, and $MgSO_4$. The liquid phase in the tank was not aerated, (the total biofilm surface area was 1.34 $m^2$), and the rotation speed of the tubes was 5 rpm (Kargi and Eker, 2001).

The fluidized bed biofilm reactors (FBBR; also called FBR), in which particles move up and down within the expanded bed in the well defined zone of the reactor, have been used for more than two decades for treating industrial wastewater (Jesis *et al.*, 1977; Rabah and Dahab, 2004). Adsorbed cell bacterial systems used in FBBRs offer some technical advantages. Since chemical wastes are injected into the recycle stream (of FBBR), toxic chemicals present in feed are immediately diluted, which make the microorganisms more adapting to chemical toxicity than in many other conventional reactor systems. In addition, since FBBRs are usually oxygenated by supplying air into the recycle loop, a high level of microbial activity may be supported with minimal air stripping of volatile chemicals.

Jesis *et al.* (1977) studied FBBR, and they found that the use of small, fluidized media enabled the FBBR to retain high biomass concentrations and, thereby, operate at significantly reduced hydraulic retention times. In pilot scale operations these authors reported that when the volatile solids concentrations were between 30000 to 40000 $mgL^{-1}$ during denitrification operation, 99% of influent nitrates could be removed under hydraulic retention times as low as 6 min. In a similar study, Rabah and Dahab (2004) (during their evaluation of the use of fluidized-bed biofilm reactors for nitrate removal) concluded that the FBBR system is capable of handling an exceptionally high nitrate nitrogen concentration of 1000 $mg(N)L^{-1}$ with very high removal efficiency (up to 99.8%). The efficiency of the FBBR can be up to 10 times greater than that of the activated sludge system and typically occupies only 10% of the space required by stirred tank reactors of similar capacities (Rabah and Dahab, 2004). Higher biomass concentration in the FBBRs (40000 $mgL^{-1}$) compared to 3000 $mgL^{-1}$ in the activated sludge have been shown to be the reason for this greater efficiency (Shieh and Keenan, 1986). Anaerobic treatment of wastewater in FBBRs is another area that has been studied extensively (Iza, 1991). In this article, Iza (1991) presented the theoretical basis for design and operation of a FBR for anaerobic treatment of wastewater.

Use of UASB is another process that has been successfully applied in wastewater treatment (Seghezzo *et al.*, 1998). The development of UASB saw a significant increase in anaerobic removal of various chemicals from the wastewater using these reactors. The examples include anaerobic removal of pentachlorophenol (Ye *et al.*, 2004), nitrogen removal (Schmidt *et al.*, 2004), dechlorination using *Dehalospirillum multivorns* (Horber *et al.*, 1998), anaerobic treatment of municipal solid leachate (Iza *et al.*, 1992), and starch degradation (Fang and Hui, 1994). More studies on this subject have been reviewed in Veeresh *et al.* (2005) (anaerobic treatment of phenol and cresols in UASB reactors). The success of the UASB concept relies on the establishment of a dense sludge bed in the bottom of the reactor which is usually a result of microbial growth and incoming sludge.

Seghezzo *et al.* (1998) reported that in a pilot plant UASB reactor, internal mixing was not optimal for sewage treatment (4–20°C) thus producing dead space and reduced process efficiency. In order to improve the process efficiency, an adequate influent distribution was sought. The use of effluent recirculation in combination with a taller reactor (a larger height to diameter ratio) resulted in the development of EGSB. Usually EBRs, as opposed to EGSB, have biofilm that is adsorbed onto support particles. An example of EBR is that of Tsuno *et al.* (1996) who employed them for the degradation of pentachlorophenol (PCP) anaerobically where the granular activated carbon was used as a support.

## 18.6 Biofilms employed to treat dairy industry wastewater

Application of biofilms in treating dairy industry wash water is another important area. In a recent report, dairy industry wash water was treated using a horizontal flow biofilm reactor system (Rodgers *et al.*, 2008). In this study, a horizontal flow biofilm reactor (HFBR) with step-feed was tested in the laboratory to remove organic carbon and nitrogen from an agricultural strength synthetic wash water. This study was conducted over a 212-day period and under steady state 99.7% carbon was removed. The nitrogen removal percentage was 98.3%. The authors reported that no clogging of media occurred during the study. The reactor system was simple to construct and operate.

## 18.7 Biofilms employed to treat oily sea water

Often there are oil leaks into sea or water streams that are toxic to marine life or may ultimately reach ground water. To avoid toxicity of such oil leaks, contaminated water or sea water should be treated to degrade toxic chemicals. In a number of studies, removal of oil from sea water has been reported (Al-Awadhi *et al.*, 2003; Radwan *et al.*, 2002). Al-Awadhi *et al.* (2003) established biofilms rich in hydrocarbon degrading bacteria. The biofilms were established on gravel particles and glass plates. The microbial consortia in the biofilms included filamentous cynobacteria, picoplankton, and diatoms. Hydrocarbon utilizing bacteria *Acinetobactor calcoaceticus* and nocardioforms were, in part, attached to filaments of cyanobacteria. These studies were performed in batch cultures and it was demonstrated that the attached biofilms were able to remove (or perhaps utilize) crude oil from contaminated sea water. Radwan *et al.* (2002) employed a mineral medium containing crude oil as the sole carbon and energy source to identify 10 different macroalgae and bacteria capable of using oil. It was concluded that macroalgae found in Arabian Gulf sea water were coated with biofilms rich in oil utilizing/degrading bacteria. In another report bacterial biofilms were found associated with corroded oil pipelines (Neria-Gonzalez *et al.*, 2006).

## 18.8 Biofilms for production of industrial chemicals

### 18.8.1 Biofilms for ethanol production

Biofuels such as ethanol and butanol are large volume low value products. For this reason they should be produced economically using the most efficient reactor systems (Table 18.1). In order to increase the rate of ethanol

**Table 18.1** A comparison of ethanol productivities in adsorbed cell biofilm reactors using various biofilm supports and cultures

| System/Support | Reactor type | Culture | Productivity [$gL^{-1}$ $h^{-1}$] | Reference |
|---|---|---|---|---|
| **Biofilm reactors** | | | | |
| Resin | Packed bed | *Z. mobilis* | 377.4 ($P_{dv}$) | Krug and Daugulis, 1983 |
| Vermiculite | Packed bed[d] | *Z. mobilis* | 210 ($P_{dv}$) | Bland *et al.*, 1982 |
| Sugarcane bagasse[a] | Packed bed | *S. cerevisiae* | 28.6[b] | Tyagi and Ghose, 1982 |
| Polypropylene | Packed bed[e] | *Z. mobilis* | 536[c] | Kunduru and Pometto, 1996 |
| Plastic composite | Packed bed | *S. cerevisiae* | 76[c] | Kunduru and Pometto, 1996 |
| **Batch/continuous suspended cell (control)** | | | | |
| Continuous | CSTR | *Z. mobilis* | 5.0[b] | Kunduru and Pometto, 1996 |
| | | *S. cerevisiae* | 5.0[b] | Kunduru and Pometto, 1996 |
| Continuous | CSTR | *S. cerevisiae* | 3.35[b] | Tyagi and Ghose, 1982 |

a – The support was reported as an adsorbent of natural origin (perhaps sugarcane bagasse); b – Not reported whether based on total reactor volume or void volume; c – Not reported (possibly based on void volume); d – Cone shaped reactor; e – Trickling packed bed reactor; $P_{dv}$ – Productivity based on reactor void volume.

production Bland *et al.* (1982) employed a *Zymomonas mobilis* biofilm reactor to produce ethanol in an expanded bed bioreactor. The cells of *Z. mobilis* were adsorbed onto vermiculite and the culture formed an active biofilm. In this reactor, a productivity of 105 $gL^{-1} h^{-1}$ (based on total volume) was obtained at a dilution rate of 3.6 $h^{-1}$ while in a control batch free cell reactor a productivity of $<4 gL^{-1} h^{-1}$ was achieved. The enhanced (>25 times) productivity reported here is due to the formation of active biofilm onto the adsorbent.

In another report on ethanol production, adsorbed cells of *Saccharomyces cerevisiae* were used in a packed bed continuous bioreactor (Tyagi and Ghose, 1982; Table 18.1). The cells were immobilized onto a support of natural origin, possibly sugarcane bagasse and the amount of cells that was adsorbed was 0.13 $gg^{-1}$ support. In this biofilm reactor, the authors reported a productivity of 28.6 $gL^{-1} h^{-1}$ as compared to 3.35 $gL^{-1} h^{-1}$ in a free cell continuous process. The dilution rates in the biofilm reactor and free cell continuous system were 0.47 $h^{-1}$ and 0.65 $h^{-1}$, respectively. Although such biofilm reactors (also known as immobilized cell reactors) are typically operated at higher dilution rates than the free cell continuous reactors, it is not clear why the authors used a lower dilution rate in the biofilm reactor. It should be noted that product yield was also improved as compared to that achieved in a batch reactor.

Ion exchange resins are good supports for forming biofilms. Since resins have charge on them, bacterial cells can be adsorbed onto them thus forming uniform biofilm layers. This concept was employed by Krug and Daugulis (1983) to produce ethanol in high productivity reactors using *Z. mobilis*. These authors showed that a cationic macroreticular resin was the most efficient adsorbent to immobilize cells of *Z. mobilis*. The adsorbed (immobilized) cells were used in a continuous column and 100 $gL^{-1}$ glucose solution was fed to the reactor. As a result of formation of biofilm, the reactor productivity was measured at 135.8 $gL^{-1} h^{-1}$ (void volume based productivity, $P_{dv}$ = 377.4 $gL^{-1} h^{-1}$). Unfortunately, the reactor stopped working due to excessive cell growth after a period of 200 h.

Kunduru and Pometto (1996) and Demirci *et al.* (1997) are among other authors that used biofilm reactors for ethanol production. Kunduru and Pometto (1996) studied ethanol production in continuous reactors using biofilm supports of polypropylene or plastic composite. Employing *Z. mobilis* and a bacterial support of polypropylene, an overwhelming high productivity of 536 $gL^{-1} h^{-1}$ was obtained at a dilution rate of 15.36 $h^{-1}$. In a control free cell fermentation, a productivity of 5 $gL^{-1} h^{-1}$ was obtained at a dilution rate of 0.5 $h^{-1}$. The direction of feed flow in the biofilm reactor was from top to bottom. Kunduru and Pometto (1996) used another biofilm reactor of *S. cerevisiae* (support material: plastic composite) and reported a productivity of 76 $gL^{-1} h^{-1}$ at a dilution rate of 2.88 $h^{-1}$. In a control experiment a reactor productivity of 5 $gL^{-1} h^{-1}$ was obtained at a dilution rate of 0.5 $h^{-1}$. Unlike the above biofilm reactor, this biofilm reactor was fed at the bottom, and the product was obtained from the top. It is suggested that for an appropriate comparison, direction of liquid flow in both the reactors should have been the same.

In a report on enhancing biofilm formation, Demirci *et al.* (1997) developed a new support for the growth of *S. cerevisiae*. A mixture of ground soybean hulls (or oat hulls), complex nutrients, and polypropylene was extruded at high temperature into disks and rings. The purpose of including nutrients into the support was to encourage cell growth. However, it is possible that heat sensitive nutrients may have been inactivated during extrusion at high temperature. Also, polypropylene film may have covered the nutrients, thus making them unavailable to the culture for cell growth. Since no data have been provided on the time period of formation of biofilm or thickness of biofilm, it is difficult to compare this support with others.

In a more recent study, Qureshi *et al.* (2004a) produced ethanol in a biofilm reactor of genetically engineered *Escherichia coli* using xylose, a five carbon sugar as substrate. The biofilm was formed on clay brick particles and the bioreactor was operated in a continuous mode for 103 days. The reactor was operated at various dilution rates, and reactor productivity was found to be improved compared to a free cell batch process.

### 18.8.2 Butanol production in biofilm reactors

Butanol is a superior biofuel and has 30% more energy content than ethanol on a mass basis. It can be used in internal combustion engines and can be mixed with gasoline in any proportion. This chemical and biofuel can be produced from various substrates including cellulosic biomass (these cultures can utilize hexoses and pentoses), whey permeate, and starch using *Clostridium acetobutylicum* or *C. beijerinckii*. Butanol production has been investigated in batch, fed-batch, free cell continuous, immobilized (including biofilm) cell continuous, and cell recycle continuous reactors (Maddox, 1989).

Continuous immobilized cell and cell recycle reactors offer higher productivities than batch and free cell continuous reactors. In addition to achieving a high productivity, a major advantage of immobilized cell technology (and cell recycle) is that cell mass is not washed out at high flow rates.

An early report on adsorption (biofilm formation) of cells of *C. acetobutylicum* for the production of butanol was that of Forberg and Haggstrom (1985). They used beechwood shavings as an adsorption support for cells. Over a period of time, an active biofilm was formed on the wood shavings, and a reactor productivity as high as 1.53 $gL^{-1} h^{-1}$ was observed (compared to <0.1–0.38 $gL^{-1} h^{-1}$ in control batch fermentation). This work was followed by examining production of butanol in an adsorbed cell biofilm reactor of *C. acetobutylicum* P262 from whey permeate (Qureshi and Maddox, 1987). It should be noted that biofilm formation on this support was rapid, and a reactor productivity of 4.5 $gL^{-1}$ $h^{-1}$ was achieved, which was superior to that obtained in any previously reported butanol production reactor system.

Following these reports, Welsh *et al.* (1987) investigated the use of a number of adsorption supports for butanol production by *C. acetobutylicum* in batch and continuous systems. The adsorbents that were used included coke, kaolinite, and gel white (a montmorillonite clay). Coke was found to be superior to the other supports for adsorption. A maximum concentration of acetone butanol ethanol (ABE) in the effluent of the reactor was reported to be 12 $gL^{-1}$ at a dilution rate of 0.1 $h^{-1}$, thus resulting in a productivity of 1.2 $gL^{-1}$ $h^{-1}$.

Following the above reports, an intensive study was performed on biofilm formation using *C. acetobutylicum* P262 onto numerous supports (Qureshi *et al.*, 2005). It has been observed that *C. acetobutylicum* and *C. beijerinckii* form visual biofilm layers in 2–4 days (in packed bed reactors), and reactors become productive after the 4th day of continuous operation. The techniques of adsorption and reactor operation have been reported elsewhere (Qureshi and Maddox, 1987; Qureshi *et al.*, 2000; Welsh *et al.*, 1987). It has been observed that not all the supports are suitable for adsorption. It has also been observed that during biofilm formation onto bonechar, the culture produces higher concentration of polysaccharide between days 2 and 4 (Qureshi *et al.*, 2005).

It is interesting that some supports accumulated greater cell concentrations (*C. acetobutylicum* P262) and were more solventogenic than the others (Qureshi *et al.*, 2005). At this stage, we are not aware what makes some supports better than others for biofilm formation and cell accumulation. From some supports it was easier to wash away the cells while from others, such as bonechar and clay brick, it was more difficult. During our studies on butanol production, it was observed that approximately 0.9–1.0 $gL^{-1}$ cells were present in the effluent (Qureshi *et al.*, 1988; Qureshi *et al.*, 2004b) of the reactor. We have demonstrated that the cells that were present in the effluent of the reactor were those that grew on the surface of the support, rather than those that grew in liquid medium inside the reactor (Qureshi *et al.*, 2004b), suggesting that a tremendous amount of cell growth activity occurred on the surface of the biofilm in these *C. beijerinckii*/*C. acetobutylicum* cultures. The thickness of biofilm that is formed in *C. acetobutylicum* cultures can range from few cell layers to as many as 35 or more. Figure 18.1 shows adsorbed cells and biofilm formed by *C. acetobutylicum* P262 onto bonechar. Similar observations on biofilm formation were observed for *C. beijerinckii* (Qureshi *et al.*, 2000).

Butanol production in various types of reactor systems has been intensively studied (Maddox, 1989; Qureshi and Maddox, 1988; Ezeji *et al.*, 2005). The biofilm reactor systems for butanol production can be varied from vertical PBR, horizontal PBR, compartmentalized reactor, double series reactors, and FBR. The PBRs and FBRs are different in the sense that FBR

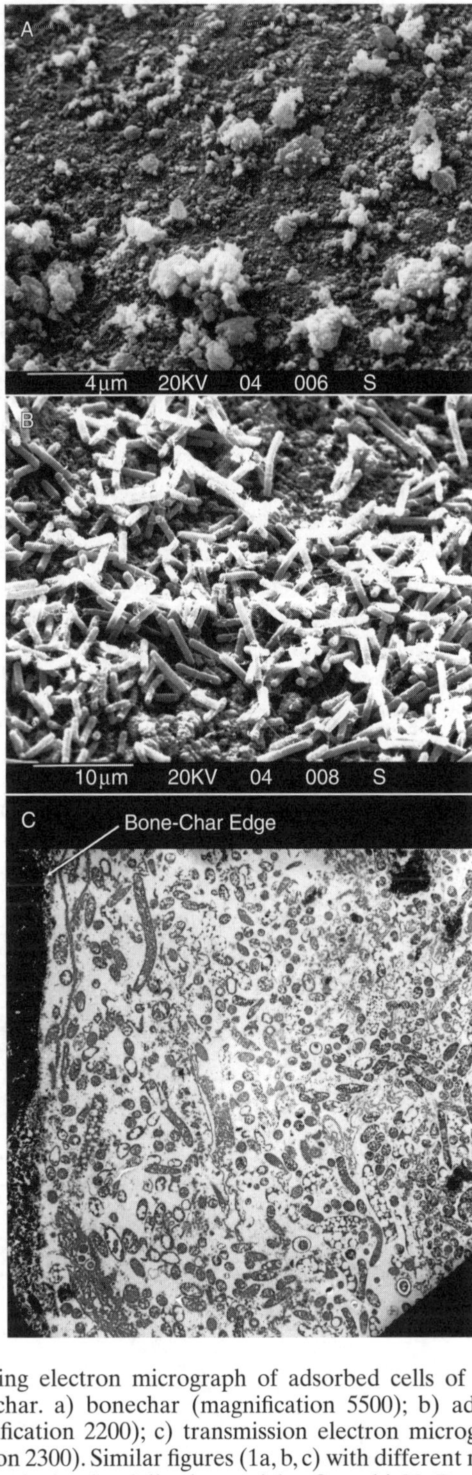


**Fig. 18.1** Scanning electron micrograph of adsorbed cells of *C. acetobutylicum* P262 onto bonechar. a) bonechar (magnification 5500); b) adsorbed cells onto bonechar (magnification 2200); c) transmission electron micrograph of adsorbed cells (magnification 2300). Similar figures (1a, b, c) with different magnification were published previously in the following article: Qureshi N, Paterson A H J and Maddox I S (1988), 'Model for continuous production of solvents from whey permeate in a packed bed reactor using cells of *Clostridium acetobutylicum* immobilized by adsorption onto bonechar', *Appl Microbiol Biotechnol*, 29, 323–328. Reprinted 'with kind permission from Springer Science+Business Media' (see above article).

is started with support <10% of the reactor's volume while packed beds are filled up to 90% of their volume. PBRs often clog due to excessive cell growth while FBRs do not clog. In FBRs, the bed is fluidized either by recycling fermentation broth (mixture of fermentation medium containing cell mass, residual sugar and products), using anaerobic gases ($N_2$ or $CO_2$ and $H_2$ in case butanol fermentation) or air (for other aerobic systems). Cell growth occurs all around the support particles and over a period of time the volume of biofilm particles becomes many fold greater than the support particle (Fig. 18.2). It should be noted that in FBRs cell growth occurs on the particles in spite of the broth's high flow rates (Qureshi and Maddox, 1991). In this type of fluidized bed reactor liquid flow velocity ranges from 40 to 60 ms$^{-1}$. The reader is advised that despite such a high flow velocity, the culture maintains its growth as a biofilm. This biofilm reactor was used for the production of butanol from whey permeate in continuous operation for >4 months (unpublished results – Qureshi and Maddox). Newly adsorbed *C. acetobutylicum* cells onto bonechar grow in an exponential manner and accumulation of biomass continues with time. Figure 18.3 shows a picture of a fluidized bed reactor for butanol production from whey permeate using cells of *C. acetobutylicum* P262.

Among the various types of reactors used for butanol production, adsorbed cell biofilm reactors (cells adsorbed onto bonechar and clay brick) offered the highest reactor productivities ranging from 6.5 (Qureshi and Maddox, 1988) to 15.8 (Qureshi *et al.*, 2000) gL$^{-1}$ h$^{-1}$ (as compared to 0.10–0.38 gL$^{-1}$ h$^{-1}$ in batch reactors) (Table 18.2). Membrane cell reactors

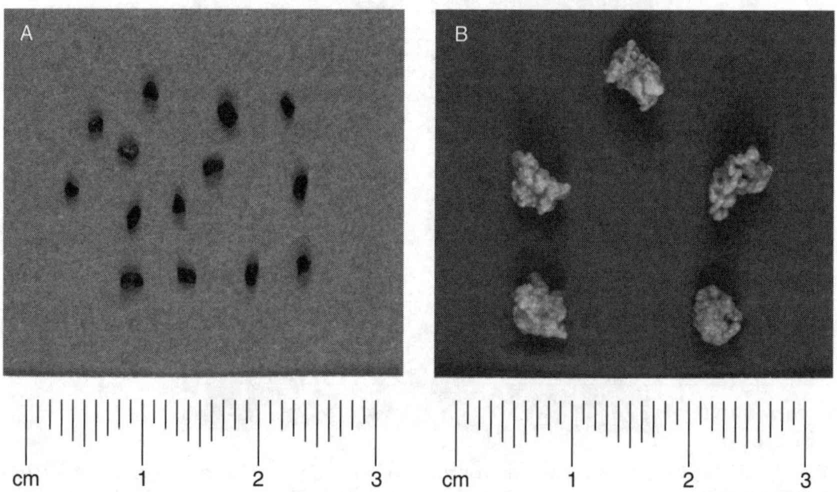


**Fig. 18.2** A picture of biofilm particles of *C. acetobutylicum* P262 used in a fluidized bed reactor for butanol production from whey permeate (Adapted from Qureshi *et al.*, 2005; Open Access Journal). a) bonechar support particles before growth; b) bonechar particles covered after cell growth and formation of biofilm.

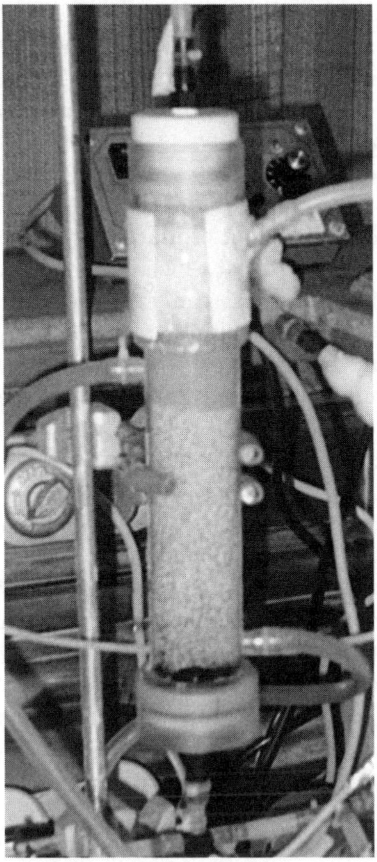

**Fig. 18.3** A picture of a fluidized bed biofilm reactor used for butanol production from whey permeate (Adapted from Qureshi *et al.*, 2005; Open Access Journal).

**Table 18.2** Production of solvents (acetone butanol ethanol; ABE) in packed bed biofilm reactors of *C. acetobutylicum/C. beijerinckii*

| Culture/ support | Maximum solvent [$gL^{-1}$] | Maximum productivity [$gL^{-1}$ $h^{-1}$] | Accumulated biomass [$gL^{-1}$ reactor vol] | Reference |
|---|---|---|---|---|
| ***C. acetobutylicum*** | | | | |
| Bonechar | 9.3 (0.30) | 6.50 (1.5) | 74.0 | Qureshi and Maddox, 1988 |
| ***C. beijerinckii*** | | | | |
| Clay brick | 7.9 (2.00) | 15.8 (2.00) | 73.7 | Qureshi *et al.*, 2000 |

Numbers in bracket are dilution rates ($h^{-1}$) at which solvent and productivity were obtained.

also offer high productivities (6.5 $gL^{-1}$ $h^{-1}$) (Pierrot *et al.*, 1986; Afschar *et al.*, 1985); however, biofilm reactors have proved to be superior to these reactors. It should be noted that of the various supports tested, bonechar and clay brick were found to be most suitable, and strong biofilms were formed on these supports.

### 18.8.3 Biofilms in 2,3-butanediol production

2,3-butanediol is another chemical which has numerous applications in the chemical industry. This chemical can be produced from glucose, whey permeate, and hydrolyzed agricultural residues using biofilm reactors of *Klebsiella pneumonia* or *K. oxytoca*. In an attempt to improve reactor productivity in 2,3-butanediol fermentation, Maddox *et al.* (1988) adsorbed cells of *Klebsiella pneumoniae* on to bonechar in a similar fashion as *C. acetobutylicum* P262 (Qureshi and Maddox, 1987). During the 2,3-butanediol fermentation, a productivity of 11.7 $gL^{-1}$ $h^{-1}$ was obtained, which was the highest reported productivity for this system. Prior to this work Shazer and Speckman (1984) reported a productivity of 1.04 $gL^{-1}$ $h^{-1}$ in 2,3-butanediol fermentation using *Bacillus polymyxa* in a membrane cell reactor. This work clearly demonstrated that bonechar adsorbed cells of *K. pneumoniae* resulted in superior productivities. Table 18.3 compares 2,3-butanediol reactor productivity achieved in a biofilm reactor as compared to various other reactor types. It should be noted that although high reactor productivity was obtained in the adsorbed cell reactor, *Klebsiella pneumonia* cells did not adsorb onto bonechar surfaces as strongly as *C. acetobutylicum* P262. Rather, cells were entrapped in between bonechar particles. However, it is anticipated that there were significant amounts of cells sitting on the surface of bonechar as the bonechar surface area was large. At the end of fermentation, it was observed that unlike cells of *C. acetobutylicum*, *K. pneumoniae* cells were easily washed away. It is not known whether *K. pneumoniae* cells produce

**Table 18.3** A comparison of 2,3-butanediol productivity in a packed bed biofilm reactor with productivities reported in other reactor systems

| Reactor type | Culture | Substrate | Productivity [$gL^{-1}$ $h^{-1}$] | Reference |
|---|---|---|---|---|
| Biofilm, continuous | *K. pneumoniae* | Whey permeate | 11.70 | Maddox *et al.*, 1988 |
| Batch (control) | *A. aerogenes* | Glucose | 1.10 | Syblayrolles and Goma, 1984 |
| Continuous reactor (free cells) | *K. pneumoniae* | Glucose | 4.25 | Ramachandran and Goma, 1987 |

polysaccharide which can facilitate adsorption of cells to the adsorbent's surface. Even though *K. pneumoniae* cells do not form firm layers of cells, these reactors were still highly productive.

### 18.8.4 Production of acetic acid or vinegar using biofilms

Acetic acid or vinegar has numerous applications in food, feed, and the chemical industry and can be produced employing a biological process. Acetic acid production in trickling bed biofilm reactors is a mature technology and is exercised at the commercial scale (Table 18.4) (Crueger and Crueger, 1989). In addition to the use of trickling bed biofilm reactors, a submerged process was also developed in the late 1940s. Acetic acid is produced by one of the bacteria grouped in the two genera, *Gluconobacter* and *Acetobacter*. The species that are used commercially include *Acetobacter aceti*, *A. pasteurianus*, and *Gluconobacter oxydans*. In the trickling bed biofilm reactor (volume 60 $m^3$), beechwood shavings are packed and the starting material (alcohol solution) is sprayed over the surface. Initially, nutrients and bacteria are added to this solution to promote growth and biofilm formation on the beechwood shavings. The liquid trickles to the bottom of the reactor containing acetic acid. In order to increase concentration of acetic acid, the liquid is cooled and recycled back to the top of the reactor. Of the alcohol added, approximately 90% is converted to acetic acid during the trickling process. Approximately 120 $gL^{-1}$ acetic acid is obtained in 72 h, thus resulting in a productivity of 1.67 $gL^{-1}$ $h^{-1}$.

**Table 18.4** Production of various other chemicals in biofilm reactors (Adapted from Qureshi *et al.*, 2005)

| Product/reactor type | Adsorption support | Productivity [$gL^{-1}$ $h^{-1}$] | Reference |
|---|---|---|---|
| **Acetic acid** | | | |
| Trickling bed biofilm reactor | Beechwood shavings | 1.67 (120) | Crueger and Crueger, 1989 |
| **Lactic acid** | | | |
| Agitating continuous reactor | Fibrous bed (cloth) | 2.5 (126) | Tay and Yang, 2002 |
| **Fumaric acid** | | | |
| Rotary continuous reactor | Plastic discs | 4.25 (85) | Cao *et al.*, 1996 |
| Strirred-tank (control) | None | 0.91 | Cao *et al.*, 1996 |
| **Succinic acid** | | | |
| Repeated batch fermentations | Plastic discs | – | Urbance *et al.*, 2003 |

Numbers in bracket – product concentration in $gL^{-1}$.

### 18.8.5 Production of lactic acid employing biofilms

Production of lactic acid in biofilm reactors is another example of industrial chemical production using such reactors. Demirci *et al.* (1993a) evaluated a number of supports for biofilm formation using lactic acid producing cultures. It has been reported that the best biofilms were obtained with *Pseudomonas fragi*, *Streptomyces viridosporus*, and *Thermoactinomyces vulgaris* when used in combination with polypropylene composite chips. The polypropylene composite chips were obtained by extruding polypropylene and 25% (w/w) agricultural material to the desired size and shape. Subsequently, a number of reports appeared from the same group on synthesizing, evaluating, and using various supports for biofilm formation and lactic acid production (Demirci *et al.*, 1993b; Ho *et al.*, 1997a, b, c). In one of the reports (Ho *et al.*, 1997a), lactic acid was produced in repeated batch cultures using a biofilm reactor. The reactor productivity was improved from 2.78 to 4.26 $gL^{-1}$ $h^{-1}$. A maximum lactic acid concentration of 60 $gL^{-1}$ was produced in biofilm reactors where a plastic composite support was used for cell adsorption.

In a study on the production of lactic acid by adsorbed cells of *Rhizopus oryzae*, the culture was immobilized on a fibrous-bed and used in a bioreactor (Tay and Yang, 2002). In this reactor (fed-batch), a productivity of 2.5 $gL^{-1}$ $h^{-1}$ was obtained with a high yield of 90% and a high product concentration of 127 $gL^{-1}$. When glucose (substrate) was replaced with cornstarch, yield improved to 100% and productivity decreased to 1.65 $gL^{-1}$ $h^{-1}$. Using starch as a substrate, a product concentration of 126 $gL^{-1}$ was achieved.

Other reports on using cell support for cell growth and lactic acid production are those of Park *et al.* (1998) and Sun *et al.* (1999). Park *et al.* (1998) used 3 $gL^{-1}$ mineral support (Aid-Plus; ML-50D, Mizusawa Chemical Co., Niigata, Japan) and 5 ppm polyethylene oxide to flocculate the culture and change mycelial morphology from a large pellet to mycelial flocs. Sun *et al.* (1999) immobilized cells of *R. oryzae* in polyurethane foam cubes. In a recent report on production of lactic acid using lactic acid bacteria (LAB), it was reported that the investigated bacteria (*Lactobacillus plantarum*, *L. brevis*, and *L. fructivorans*) adhered, accumulated, and formed biofilms on glass cover slips (Kubota *et al.*, 2008).

### 18.8.6 Biofilms for fumaric acid production

Biofilm reactors have also been used successfully for the production of fumaric acid (Cao *et al.*, 1996) and mineral ore treatment (Crueger and Crueger, 1989). In a study, Cao *et al.* (1996) used plastic discs to adsorb cells of *Rhizopus oryzae* to produce fumaric acid from glucose. The use of the biofilm reactor resulted in an increase in reactor productivity from 0.9 $gL^{-1}$ $h^{-1}$ (in a free cell stirred-tank reactor) to 4.25 $gL^{-1}$ $h^{-1}$ in the biofilm reactor. In the latter reactor, fumaric acid concentration up to 85 $gL^{-1}$ was obtained from 100 $gL^{-1}$ glucose. The fermentation time was shorter and took 20 h as compared to 72 h in the free cell reactor.

### 18.8.7 Biofilms for succinic acid production

The industrial potential of succinic acid fermentation was recognized as early as the late 1970s (Zeikus *et al.*, 1999). Succinic acid ($HOOCCH_2CH_2COOH$) is a dicarboxylic acid, and can be used as a feedstock chemical for the production of high value products including 1,4-butanediol, tetrahydrofuran, adipic acid, γ-butyrolactone, and n-methylpyrrolidone (Zeikus *et al.*, 1999) for numerous applications in agriculture, food, medicine, plastics, cosmetics, and textiles. Succinic acid can be produced in biofilm reactors. In a recent study on succinic acid production using *Actinobacillus succinogenes*, Urbance *et al.* (2003) employed customized plastic composite support (PCS) (Cotton *et al.*, 2001) and 20 other different PCS blends. The customized plastic composite support (PCS) blends were screened for biofilm formation and succinic acid production. These customized PCS blends demonstrated 70% yields for succinic acid production as compared to 64% in suspended cell bioreactor using *A. succinogenes* (Urbance *et al.*, 2003). Table 18.4 shows production of various industrial chemicals (in particular acids) in biofilm reactors.

## 18.9 Length of operation of biofilm reactors

While producing chemicals in biofilm reactors, it should be one of the objectives to operate the reactor for a long period of time, thus enhancing the economics of the process. Packed bed biofilm reactors often clog due to excessive cell growth. It should be noted that reactor blockage depends on a number of factors including cell growth rate, packing density of the support, and supply of nutrients. Packed bed biofilm reactors have been typically operated ranging from 2 weeks to 4 months. Tyagi and Ghose (1982) operated a packed bed biofilm reactor of *S. cerevisiae* for a period of 35 days, while Qureshi *et al.* (2004a) used a packed bed biofilm reactor of *E. coli* for a period of 103 days for ethanol production in continuous operation. However, it was observed that packed bed biofilm reactors of *C. acetobutylicum* P262/*C. beijerinckii* blocked sooner than 103 days due to excessive cell growth of these cultures. In order to prolong life of the reactor, feed media that limit cell growth (perhaps deficient in nutrients) and still produce chemicals at a rapid rate should be attempted as successfully applied by Qureshi and Maddox (1987) and Qureshi *et al.* (2004a). In addition to the reactor blockage due to excessive growth, influent to the reactor plays an important role in prolonging the life of the reactor. Reactor feed containing suspended and particulate solids, in particular wastewater influents, may

**Table 18.5** Length of operation of various biofilm reactors used for the production of industrial chemicals (Adapted from Qureshi *et al.*, 2005)

| Chemical produced | Reactor type | Length of operation [days] | Dilution rate [$h^{-1}$] (productivity $gL^{-1} h^{-1}$) | Reference |
|---|---|---|---|---|
| Butanol | Packed bed | 61 days | 0.30–1.00 (0.98–4.10) | Qureshi and Maddox, 1987 |
| | Fluidized bed | >4 months | 0.33–1.37 (1.65–5.10) | Unpublished data[1] |
| Lactic acid | Various reactors | 30 days | – – | Tay and Yang, 2002 |
| Ethanol | Packed bed | 35 days | 0.12–0.48 (7.80–28.60) | Tyagi and Ghose, 1982 |
| | Packed bed | 103 days | 0.04–0.12 (1.10–2.58) | Qureshi *et al.*, 2004a |
| | Packed bed | 60 days | 0.50–5.76 (5.00–74.88) | Kunduru and Pometto, 1996 |

[1] Qureshi and Maddox.
– Not reported.

also clog the reactor. Table 18.5 shows the lengths of operation of various biofilm bioreactors.

## 18.10 Industrial/pilot-plant level biofilm reactors

Biofilm reactors have successfully been used for wastewater treatment (Lettinga *et al.*, 1980; 1983; Nicolella *et al.*, 2000; Seghezzo *et al.*, 1998; Tsuno *et al.*, 1996; Iza, 1991; Schmidt *et al.*, 2004) (Fig. 18.4). In these industrial biofilm reactors, cell mass concentration as high as 30–40 $gL^{-1}$ could be maintained (Nicolella *et al.*, 2000 [Fig. 18.4]; Shieh and Keenan, 1986). As a result of superior efficiency, biofilm reactors are being used throughout the world with a number of full-scale applications for industrial chemical production and wastewater treatment. Examples of these reactors operating in The Netherlands and Brazil have been reported elsewhere (Nicolella *et al.*, 2000; reviewed in Qureshi *et al.*, 2005). Similarly, commercial production of acetic acid or vinegar using biofilm reactors has been exercised for many years. Production of these chemicals has been reported by Crueger and Crueger (1989). Large biofilm fermentors up to 60 000 L have been used. Often beechwood shavings are used as a support for biofilm formation. For this system, trickling bed reactors have been used with an exit product concentration up to 120 $gL^{-1}$ and a productivity of 1.67 $gL^{-1} h^{-1}$.

**Fig. 18.4** Full-scale biofilm reactors: (a) biothane biofilm airlift suspension and expanded granular sludge blanket (Biobed) reactors at Gist Brocades, Delft (The Netherlands); (b) Pagues CIRCOX (foreground; 140 $m^3$) and internal circulation (background; 385 $m^3$) reactors at a brewery in Brazil. Reprinted from 'Nicolella C, van Loosdrecht M C M, Heijnen S J (2000), "Particle-based biofilm reactor technology", *Trends in Biotechnology*, 18, 312–320', with permission from Copyright Clearance Center's Rightslink service. Elsevier has partnered with Rightslink to license its content online.

## 18.11 Future trends and conclusions

A comparison of biofilm reactors with other reactor systems suggests that biofilm reactors are simple, offer higher productivity, and biofilm supports are economically available. In biofilm reactors, cells can be adsorbed within the reactor without the use of any chemicals, and the reactors can be operated for prolonged periods, thus helping to reduce the process cost. Some examples of their use at a large scale have been given in the above section. It is clear that use of biofilm reactors (in laboratory scale) has been consistently increasing for the production of various industrial chemicals. As productivity in these simple biofilm reactors is high, their full potential should be employed for biotechnological/biological conversion processes. Biofilm reactors have potential to be operated for long periods (Table 18.5) thus offering economic benefits over other reactor types. For further reading on biofilms and their formation, the reader is referred to the comprehensive articles published by Costerton *et al.* (1987) and O'Toole *et al.* (2000).

## 18.12 Acknowledgements

N. Qureshi would like to thank Michael A Cotta (United States Department of Agriculture, National Center for Agricultural Utilization Research [USDA/NCAUR], Peoria, IL) for reading this manuscript critically and making useful comments. NQ is grateful to Mr. Doug Hopcroft (AgResearch Ltd., Palmerston North, New Zealand; formerly known as Department of Scientific and Industrial Research [D.S.I.R; Biotechnology Division]) for performing the electron microscopy reprinted in Fig. 18.1. It is acknowledged that some sections of this chapter have been taken from our previous publication on biofilm reactors (Qureshi *et al.*, 2005; Open Access Journal) and updated.

## 18.13 References

AFSCHAR A S, BIEBL H, SCHALLER K and SCHUGERL K (1985), 'Production of acetone and butanol by *Clostridium acetobutylicum* in continuous culture with cell recycle', *Appl Microbiol Biotechnol*, 22, 394–398.

AL-AWADHI H, AL-HASAN R H, SORKHOH N A, SALAMAH S and RADWAN S S (2003), 'Establishing oil-degrading biofilms on gravel particles and glass plates', *Int Bioremediation & Biodegradation*, 51, 181–185.

BLAND R R, CHEN H C, JEWELL W J, BELLAMY W D and ZALL R R (1982), 'Continuous high rate production of ethanol by *Zymomonas mobilis* in an attached film expanded bed fermentor', *Biotechnol Lett*, 4, 323–328.

BÜCKS J, MOZES N, WANDREY C and ROUXHET P G (1988), 'Cell adsorption control by culture conditions', *Appl Microbiol Biotechnol*, 29, 119–128.

CAO N, DU J, GONG C S and TSAO G T (1996), 'Simultaneous production and recovery of fumaric acid from immobilized *Rhizopus oryzae* with a rotary biofilm contactor and an adsorption column', *Appl Env Microbiol*, 62, 2926–2931.

CARPENTIER B and CERF O (1993), 'Biofilm and their consequences with particular reference to hygiene in the food industry', *J Appl Bacteriol*, 75, 499–511.

CHARACKLIS W G, MCFETERS G A and MARSHALL K C (1990), 'Physiological ecology in biofilm systems', in Characklis W G and Marshall K C, *Biofilms*, New York: John Wiley and Sons, 37, 67–72.

COSTERTON J W, CHENG K J, GEESEY G G, LADD T I, NICKEL J C, DASGUPTA M and MARRIE T J (1987), 'Bacterial biofilms in nature and disease', *Annu Rev Microbiol*, 41, 435–464.

COSTERTON J W, STEWART P S and GREENBURG E P (1999), 'Bacterial biofilms: a common cause of persistent infections', *Science*, 284(5418), 1318–1322.

COTTON J C, POMETTO III A L and GVOZDENOVIC-JEREMIC J (2001), 'Continuous lactic acid fermentation using a plastic composite support biofilm reactor', *Appl Microbiol Biotechnol*, 57, 626–630.

COWAN M M, WARREN T M and FLETCHER M (1991), 'Mixed species colonization of solid surfaces in laboratory biofilms', *Biofouling* 1991, 3, 23–34.

CRUEGER W and CRUEGER C (1989) (Eds), 'Organic acids.' In *Biotechnology A textbook of industrial microbiology*. Sinauer Associates, Inc., Sunderland. MA.

DEMIRCI A, POMETTO III AL and JOHNSON K E (1993a), 'Evaluation of biofilm reactor solid support for mixed-culture lactic acid production', *Appl Microbiol Biotechnol*, 38, 728–733.

DEMIRCI A, POMETTO A L III and JOHNSON K E (1993b), 'Lactic acid production in a mixed-culture biofilm reactor', *Appl Environ Microbiol*, 59, 203–207.

DEMIRCI A, POMETTO III A L and HO K L G (1997), 'Ethanol production by *Saccharomyces cerevisiae* in biofilm reactors', *J Ind Microbiol Biotechnol*, 19, 299–304.

DONLAN R M (1992), 'Biofilms: microbial life on surfaces', *Emerg Inf Dis*, 8(9), 881–890.

DONLAN R M and COSTERTON J W (2002), 'Biofilms: survival mechanisms of clinically relevant microorganisms', *Clin Microbiol Rev*, 15(2), 167–193.

EZEJI T C, QURESHI N, BLASCHEK H P (2005), 'Industrially relevant fermentations', in Durrie P, *Handbook on Clostridia*, New York, NY: CRC Press, 797–812.

FANG H H P and HUI HH (1994), 'Effect of heavy metals on the methanogenic activity of starch-degrading granules', *Biotechnol Lett*, 16(10), 1091–1096.

FORBERG C and HAGGSTROM L (1985), 'Control of cell adhesion and activity during continuous production of acetone and butanol with adsorbed cells', *Enz Microbial Technol*, 7, 230–234.

HALL E R (1987), 'Biofilm reactors in anaerobic wastewater treatment', *Biotech Adv*, 5, 257–269.

HO K L G, POMETTO III A L and HINZ P N (1997a), 'Optimization of L-(+)-lactic acid production by ring and disc plastic composite supports through repeated-batch biofilm fermentation', *Appl Environ Microbiol*, 63, 2533–2542.

HO K L G, POMETTO III A L, HINZ P N and DEMIRCI A (1997b), 'Nutrient leaching and end product accumulation in plastic composite supports for L-(+)-lactic acid biofilm fermentation', *Appl Environ Microbiol*, 63, 2524–2532.

HO K L G, POMETTO III A L, HINZ P N, DICKSON J S and DEMIRCI A (1997c), 'Ingredient selection for plastic composite supports for L-(+)-lactic acid biofilm fermentation by *Lactobacillus casei* subsp. *Rhmnosus*', *Appl Environ Microbiol*, 63, 2516–2523.

HORBER C, CHRISTIANSEN N, ARVIN E and AHRING B K (1998), 'Improved dechlorinating performance of upflow anaerobic sludge blanket reactors by incorporation of *Dehalospirillum multivorans* into granular sludge', *Appl Env Microbiol*, 64(5), 1860–1863.

IZA J (1991), 'Fluidized bed reactors for anaerobic wastewater treatment', *Water Sci Technol*, 24(8), 109–132.

IZA J, KEENAN P J and SWITZENBAUM M S (1992), 'Anaerobic treatment of municipal solid waste landfill leachate: operation of a pilot scale hybrid UASB/AF reactor', *Water Sci Technol*, 25(7), 255–264.

JESIS J S, OWEN R W, HICKEY R and FLOOD F (1977), 'Biological fluidized-bed treatment for BOD and nitrogen removal', *J Water Pollut Control Fed*, 49, 816–831.

KARGI F and EKER S (2001), 'Rotating-perforated-tubes biofilm reactor for high-strength wastewater treatment', *J Environ Eng*, 127, 959–963.

KRUG T A and DAUGULIS A J (1983), 'Ethanol production using *Zymomonas mobilis* immobilized on an ion exchange resin', *Biotechnol Lett*, 5, 159–164.

KUBOTA H, SENDA S, NOMURA N, TOKUDA H and UCHIYAMA H (2008), 'Biofilm formation by lactic acid bacteria and resistance to environmental stress', *J Biosci Bioeng*, 106, 381–386.

KUNDURU M R and POMETTO III A L (1996), 'Continuous ethanol production by *Zymomonas mobilis* and *Saccharomyces cerevisiae* in biofilm reactors', *J Ind Microbiol Biotechnol*, 16, 249–256.

LAZAROVA V and MANEM J (2000), 'Innovative biofilm treatment technologies for water and wastewater treatment', in Bryers J D, *Biofilm II: Process Analysis and Applications*, New York: Wiley-Liss Press, 159–206.

LETTINGA G, VAN NELSEN A F M, HOBMA S W, DE ZEEUW W and KLAPWIJK A (1980), 'Use of the upflow sludge blanket (USB) reactor concept for biological wastewater treatment, especially for anaerobic treatment', *Biotechnol Bioeng*, 22, 699–734.

LETTINGA G, ROERSMA R and GRIN P (1983), 'Anaerobic treatment of raw domestic sewage at ambient temperature using a granular bed UASB reactor', *Biotechnol Bioeng*, 25, 1701–1723.

MADDOX I S (1989), 'The acetone butanol ethanol fermentation: recent progress in technology', *Biotechnol Genetic Eng Reviews*, 7, 190–220.

MADDOX I S, QURESHI N and MCQUEEN J (1988), 'Continuous production of 2,3-butanediol from whey permeate using cells of *Klebsiella pneumoniae* immobilized on to bonechar', *New Zealand J Dairy Sci Technol*, 23, 127–132.

MCDONOGH R, SCHAULE G and FLEMMING H C (1994), 'The permeability of biofouling layers on membranes', *J Membr Sci*, 87, 199–217.

MEYER A and WALLIS F M (1997), 'Development of microbial biofilms on various surfaces for the treatment of heavy metal containing effluents', *Biotechnol Tech*, 11(12), 859–863.

MORTENSEN K P and CONLEY S N (1994), 'Film fill fouling in counterflow cooling towers: mechanisms and design', *CTI J*, 15, 10–25.

NERIA-GONZALEZ I, WANG E T, RAMIREZ F, ROMERO J M and HERNANDEZ-RODRIGUEZ C (2006), 'Characterization of bacterial community associated to biofilms of corroded oil pipelines from the southeast of Mexico', *Anaerobe*, 12, 122–133.

NICOLELLA C, VAN LOOSDRECHT M C M and HEIJNEN S J (2000), 'Particle-based biofilm reactor technology', *Trends in Biotechnol (TIBTECH)*, 18, 312–320.

O'TOOLE G, KAPLAN H B and KOLTER R (2000), 'Biofilm formation as microbial development', *Annu Rev Microbiol*, 54, 49–79.

PARK E Y, KOSAKAI Y and OKABE M (1998), 'Efficient production of L-(+)-lactic acid using mycelial cotton-like flocs of *Rhizopus oryzae* in an air-lift bioreactor', *Biotechnol Prog*, 14, 699–704.

PIERROT P, FICK M and ENGASSER J M (1986), 'Continuous acetone butanol fermentation with high productivity by cell ultrafiltration and recycling', *Biotechnol Lett*, 8, 253–256.

QURESHI N and MADDOX I S (1987), 'Continuous solvent production from whey permeate using cells of *Clostridium acetobutylicum* immobilized by adsorption onto bonechar', *Enz Microbial Technol*, 9, 668–671.

QURESHI N and MADDOX I S (1988), 'Reactor design for the ABE fermentation using cells of *Clostridium acetobutylicum* immobilized by adsorption onto bonechar', *Bioprocess Eng*, 3, 69–72.

QURESHI N and MADDOX I S (1990), 'Novel bioreactors for the ABE fermentation using cells of *Clostridium acetobutylicum* immobilized by adsorption onto bonechar', in Yu P L, *Fermentation Technologies: Industrial Applications*, London: Elsevier Appl Sci Publishers, 343–348.

QURESHI N and MADDOX I S (1991), 'Integration of continuous production and recovery of solvents from whey permeate: use of immobilized cells of *Clostridium acetobutylicum* in fluidized bed bioreactor coupled with gas stripping', *Bioprocess Eng*, 6, 63–69.

QURESHI N, PATERSON A H J and MADDOX I S (1988), 'Model for continuous production of solvents from whey permeate in a packed bed reactor using cells of *Clostridium acetobutylicum* immobilized by adsorption onto bonechar', *Appl Microbiol Biotechnol*, 29, 323–328.

QURESHI N, SCHRIPSEMA J, LIENHARDT J and BLASCHEK H P (2000), 'Continuous solvent production by *Clostridium beijerinckii* BA101 immobilized by adsorption onto brick', *World J Microbiol Biotechnol*, 16, 377–382.

QURESHI N, BRINING H, ITEN L, DIEN B, NICHOLS N, SAHA B and COTTA M A (2004a), 'Adsorbed cell dynamic biofilm reactor for ethanol production from xylose and corn fiber hydrolysate', The 36th Great Lakes Regional Meeting of the American Chemical Society, Peoria, IL, October 17–20.

QURESHI N, LAI L L and BLASCHEK H P (2004b), 'Scale-up of a high productivity continuous biofilm reactor to produce butanol by adsorbed cells of *Clostridium beijerinckii* BA101', *Transactions Institution Chemical Engineers (Trans IChemE) (Chemical Engineering Research and Design)*, 82(C2), 164–173.

QURESHI N, ANNOUS B, EZEJI T C, KARCHER P and MADDOX I S (2005), 'Biofilm reactors for industrial bioconversion processes: employing potential of enhanced reaction rates', *Microbial Cell Factories*, 4, 24.

RABAH F K J and DAHAB M F (2004), 'Nitrate removal characteristics of high performance fluidized-bed biofilm reactors', *Water Res*, 38, 3719–3728.

RADWAN S S, AL-HASAN R H, SALAMAH S and AL-DABBOUS S (2002), 'Bioremediation of oily sea water by bacteria immobilized in biofilms coating macroalgae', *Int Bioremediation & Biodegradation*, 50, 55–59.

RAMACHANDRAN K B and GOMA G (1987), 'Effect of oxygen supply and dilution rate on the production of 2,3-butanediol in continuous bioreactor by *Klebsiella pneumoniae*', *Enz Microbial Technol*, 9, 107–111.

RODGERS M and ZHAN X M (2003), 'Moving-medium biofilm reactors', *Rev Env Sci Bio/Technol*, 2, 213–224.

RODGERS M, PAOR D DE and CLIFFORD E (2008), 'Dairy washwater treatment using a horizontal flow biofilm system', *J Environ Manag*, 86, 114–120.

SCHMIDT J E, BATSTONE D J and ANGELIDAKI I (2004), 'Improved nitrogen removal in upflow anaerobic sludge blanket (UASB) reactors by incorporation of Annammox bacteria into the granular sludge', *Water Sci Technol*, 49(11–12), 69–76.

SEGHEZZO L, ZEEMAN G, VAN LIER J B, HAMELERS H V M and LETTINGA G (1998), 'A review: the anaerobic treatment of sewage in UASB and EGSB reactors', *Bioresource Technol*, 65, 175–190.

SHAZER W H and SPECKMAN R A (1984), 'Continuous fermentation of whey permeate by *Bacillus polymyxa*', *J Dairy Sci*, 67 (Suppl 1, D5G), 1216–1218.

SHIEH W K and KEENAN J D (1986), 'Fluidized bed biofilm reactor for wastewater treatment', *Adv Biochem Eng Biotechnol*, 33, 131–169.

STOODLEY P, SAUER K, DAVIES D G and COSTERTON J W (2002), 'Biofilms as complex differentiated communities', *Annu Rev Microbiol*, 56,187–209.

SUN Y, LI Y L and BAI S (1999), 'Modeling of continuous L(+)-lactic acid production with immobilized *R. oryzae* in an airlift bioreactor', *Biochem Eng J*, 3, 87–90.

SYBLAYROLLES J M and GOMA G (1984), 'Butanediol production by *Aerobactor aerogenes* NRRL B199: effect of initial substrate concentration and aeration agitation', *Biotechnol Bioeng*, 26, 148–155.

TARAS M, HAKANSSON K and GUIEYSSE B (2005), 'Continuous acetonitrile degradation in packed-bed bioreactor', *Appl Microbiol Biotechnol*, 66, 567–574.

TAY A and YANG S T (2002), 'Production of L-(+)-lactic acid from glucose and starch by immobilized cells of *Rhizopus oryzae* in a rotating fibrous bed bioreactor', *Biotechnol Bioeng*, 80, 1–12.

TSUNO H, KAWAMURA M and SOMIYA I (1996), 'Anaerobic degradation of pentachlorophenol (PCP) in biological expanded-bed reactor', *Water Sci Technol*, 34(5–6), 335–344.

TYAGI R D and GHOSE T K (1982), 'Studies on immobilized *Saccharomyces cerevisiae*. I. Analysis of continuous rapid ethanol fermentation in immobilized cell reactor', *Biotechnol Bioeng*, 24, 781–795.

URBANCE S E, POMETTO III A L, DISPIRITO A A and DEMIRCI A (2003), 'Medium evaluation and plastic composite support ingredient selection for biofilm formation and succinic acid production by *Actinobacillus succinogenes*', *Food Biotechnol*, 17, 53–65.

VEERESH G S, KUMAR P and MEHROTRA (2005), 'Treatment of phenol and cresols in upflow anaerobic sludge blanket (UASB) process: a review', *Water Research*, 39, 154–170.

WELSH F W, WILLIAMS R E and VELIKY I A (1987), 'Solid carriers for a *Clostridium acetobutylicum* that produced acetone and butanol', *Enz Microbial Technol*, 9, 500–502.

YE F X, SHEN D S and FENG X S (2004), 'Anaerobic granule development for removal of pentachlorophenol in an upflow anaerobic sludge blanket (UASB) reactor', *Process Biochem*, 39, 1249–1256.

ZEIKUS J G, JAIN M K and ELANKOVAN P (1999), 'Biotechnology of succinic acid production and markets for industrial products', *Appl Microbiol Biotechnol*, 51, 545–552.

# 19

# Biofilms in fish processing

**K. T. Rajkowski, United States Department of Agriculture, Agricultural Research Service, (USDA-ARS) USA**

**Abstract**: The seafood industry is expanding to include traditional wild-caught species, as well as farm-raised species (aquaculture). With this expansion comes different growing and processing technologies. Biofilm formation has been problematic in the aquaculture industry from farm (preharvest) through the processing plant (post harvest). This chapter discusses the three product groups, mollusk, crustacean and finfish, with regard to possible biofilm formation. Of concern is the possibility that the biofilm may contain pathogenic microorganisms, i.e. *Listeria monocytogenes*, *Salmonella* spp., *Vibrio* spp., *Bacillus* spp., *Aeromonas* and *Pseudomonas* spp. These microbes are known to form biofilms and information of their involvement in biofilm formation in the seafood industry is discussed. In the aquaculture industry water is a major source of carrying these microbes from one operation to another. Methods of preventing and cleaning/disinfecting to remove the biofilm, which possibly harbors pathogens, are discussed. Future areas of investigation include the improving of the effluent water system for ponds, filtration/disinfection steps for the recirculating systems and processing plant sanitization.

**Key words**: biofilm, aquaculture, human pathogens, biofilm removal techniques.

## 19.1 Introduction

In 2004, an estimated 75% of the world fish production (105.5 million tons) was for human consumption, and the remaining 25% (34.8 million tones) went into non-food production, ie, production of finfish-derived products and oil (FAO, 2006). Fish consumption, which provides a major protein source in the diet, can vary from 1–100 kg/capita depending on the geographical area and individual country. The amount of wild caught fish has become a concern in that natural replacement may not keep up with the demand. Aquaculture, the farming of fish, is helping to meet the demand, and much agricultural land has been converted to aquaculture farming, particularly in southeast Asia. In 2004, aquaculture contributed an estimated 43% of the available fish consumed, and for some countries this can result in an increase of exported and imported fish. The exporting and importing of fish and fishery products has major trade implications, and the quality and safety of the products can be assessed for detention or rejection in international fish trade.

In the United States, fish is defined in Section 21 of the United States' Code of Federal Register part 123.3 as 'fresh or saltwater finfish, crustaceans, other forms of aquatic animal life (including, but not limited to, alligator, frog, aquatic turtle, jellyfish, sea cucumber, and sea urchin and the roe of such animals) other than birds or mammals, and all mollusks, where such animal life is intended for human consumption' (FDA, 2006). Another term used to describe fish is seafood, which is divided into three categories: finfish (carp, tilapia, catfish), crustacean (shrimp and lobster), and mollusk shellfish (oysters and clams) and will be the category terms used in the discussion of biofilm formation in seafood processing.

In North America, consumption of fish has remained at about 23 kg/capita (1996–2006). A report issued by United States General Accounting Office (GAO, 2004) on food safety related to the seafood program in 2004, stated that 80% of the consumed seafood was imported and that eating contaminated seafood (crustaceans and finfish) resulted in about 15% of the reported food borne outbreaks in the US, which 'is a greater percentage than either meat or poultry even though meat and poultry are consumed at 8 and 6 times the rate of seafood, respectively.' Bacterial pathogens were listed as the cause of the seafood-related illnesses (GAO, 2004). The United States Food and Drug Administration collected and tested 11 312 imported and 768 domestic seafood samples from 1990 to 1998 and reported that overall 7.2% of imported and 1.3% of domestic seafood were positive for *Salmonella*. Similar survey results for seafood products were reported from other countries: Norway (Guerin *et al.*, 2004), Italy (Ripabelli, 2004); Spain (Herrera *et al.*, 2006); Austria (Suppin and Smulders, 2005); and Ghana (Ampofo and Clerk, 2002). These survey results indicated an increased awareness of the role of aquaculture in the spread of pathogens. In a review of food borne microbial pathogens of aquaculture-grown seafood, Greenlees *et al.* (1998) reported on the food borne microbial pathogens associated with these products and identified *Escherichia coli* and *Salmonella* as bacterial inhabitants of pond water. *Salmonella* was isolated from both fish and shellfish, whereas *E. coli* was isolated only from finfish. Andrews *et al.* (1977) surveyed retail fresh and frozen channel catfish (*Ictalurus punctatus*) and reported that the number of samples positive for *Salmonella* in farm-raised catfish varied with the seasons with a 0.9% incidence for January to March versus 5.7% for July to September. Huss *et al.* (2000) reviewed the hazards of consuming seafood and identified both *Salmonella* and *E. coli* O157:H7 as pathogens contaminating seafood. Atanassova *et al.* (2008) surveyed sushi and reported the isolation of both

*Salmonella* and *E. coli* from fresh fish with the prevalence of *E. coli* higher than *Salmonella.* As more reports become available, there will be an increased emphasis on food safety of seafood products.

## 19.2 The water environment

With the awareness that seafood products can become contaminated with both human pathogens and spoilage bacteria, the attachment and survival of these microbes are now being studied and reported. The water environment is not particularly rich in nutrients. With a biological oxygen demand of <2 mg/liter, water has a nutrient content sufficient for growth and survival of microorganisms such as *Escherichia coli* O157:H7 (Rajkowski and Rice, 1999), *Salmonella* spp., and *Vibrio cholerae* (Rajkowski *et al.*, 1996), *Pseudomonas* and *Shigella* (Rajkowski and Rice, 2001), and *Aeromonas* (Palumbo *et al.*, 1996). Biofilms developing from estuary or marine water sources were documented by Zobell (1943) over 60 years ago. Bacterial biofilms colonize the tissue of fish, including mollusks and crustaceans and can reflect the health of the animal (Geesey *et al.*, 1992). One area of study is how the bacteria from the nutrient-limited water environment attach and form biofilms. Kjelleberg and Hermansson (1984) studied the attachment of *Vibrio* and *Pseudomonas* during starvation and concluded that there were marked changes in the cells physiochemically. The cells irreversibly bound to the solid surfaces and remained stable during starvation. With microscopic examination, it was reported that the cell surface became rough, which seemed to aid in attachment. Environmental stress, including changes in temperature and pH, was reported to affect the ability of *Listeria monocytogenes* Scott A to attach to food contact surfaces when grown in phosphate buffer, a nutrient deprived medium (Smoot and Pierson, 1998a; 1998b).

## 19.3 Microorganisms of concern in biofilm formation

Biofilms on fish surfaces and bacteria from marine water source can contaminate seafood processing facilities. *Pseudomonas*, *Vibrio*, *Staphylococcus*, *Bacillus*, and *Aeromonas* are among the genera that can cause food borne illness and form biofilms (Cahill, 1990). Other genera that were isolated from seafood included *Listeria*, *Salmonella*, and *Shigella* biofilm formers (Somers *et al.*, 1994).

Listeria monocytogenes

*Listeria monocytogenes* is considered a pathogen of seafood. In flowing systems Sashara and Zottola (1993) demonstrated that *Listeria* attaches and forms biofilms. This microbe is water borne and was isolated from fresh channel catfish, *Ictalurus punctatus* (Erdenlig *et al.*, 2000), from fresh water fish in India (Jallewar *et al.*, 2007), and was reported to survive and grow on refrigerated fish fillets (Jørgensen and Huss, 1998; Augustin *et al.*, 2005). *L. monocytogenes* was identified as a contaminant of crab meat (Brackett and Beuchat, 1990) and in vacuum-packaged cold-smoked salmon fillet (Heinitz *et al.*, 2000). The control of *L. monocytogenes* biofilms is covered in Chapter 7.

#### Salmonella *spp.*

*Salmonella* spp. were identified as contaminants of farm raised shrimp (Zhao *et al.*, 2006). This water borne contaminant can develop as a biofilm by attaching to the crustacean's chitin during growth and can be internalized during feeding. More recently, *Salmonella* was identified as the cause of an outbreak following the consumption of smoked eel in Germany (Fell *et al.*, 2000; Heinitz *et al.*, 2000). *Salmonella* was isolated from fresh fish in Iran (Basti *et al.*, 2006). Shin *et al.* (2004) reported that the use of antimicrobial ice containing chlorine dioxide effectively lowered the levels of *Salmonella* and *L. monocytogenes* on the surface of fish.

#### Vibrio *spp.*

*Vibrio* spp. are recognized as water borne pathogens. In the estuary environment, vibrios are associated with pollution (Miller *et al.*, 2006). When polluted estuaries are found, these areas are closed for harvesting fish, particularly mollusks and wild caught finfish (Chen *et al.*, 2006). Removal or reduction of vibrios in the marine environment is temperature dependent, and the bacteria are found in higher levels when the water temperature ranges from 14 to 19°C (Kaneko and Colwell, 1973). Harvesting is not recommended during those times. Biofilms containing vibrios can occur on the zooplankton in fresh water (Kaneko and Colwell, 1975). When the zooplankton are ingested by mollusks, which are filter feeders, so is the pathogen. Mollusk can accumulate this pathogen internally and on the shell surface as biofilms. Removal from the surface can be done through a brushing process; however, the internalized bacteria are more difficult to remove. Chilling immediately after harvesting is one procedure, and placing the mollusks in sterile water for a set time for reduction in the number of vibrios is another. High pressure processing has been suggested as a means to destroy the pathogen in internalized biofilms (Lopez-Cabellero *et al.*, 2000; He *et al.*, 2002). More recently vibrios were identified as bacterial contaminants on finfish, which can occur when the fish are raised in contaminated water or through cross contamination during the filleting process (Fouz *et al.*, 2006).

*Vibrio parahaemolyticus* is recognized as a human pathogen of concern and is associated with seafood borne illness. This microbe has been isolated from estuary water in Oregon and Washington (Chiu *et al.*, 2007), Texas and New York (DePaola *et al.*, 2000), China (Yano *et al.*, 2006), Brazil

(Riberiro, 2006), northern Mexico (Gomez-Gil, 2007), and Southern Thailand (Vuddhakul *et al.*, 2006). This pathogen is mainly associated with contaminated mollusks, which are eaten raw but has also been isolated from other seafood products, such as shrimp and finfish. Dawson *et al.* (1981) suggested that adhesion (biofilm formation) is a survival strategy of marine vibrios during starvation (nutrient limiting environments). In his review of *V. parahaemolyticus*, Levin (2006) describes the genes responsible for the formation of the lateral flagella, which results in swarmer cells when grown on solid media – the first step in biofilm formation. Kim and McCarter (2007) studied the gene that controlled the expression of lateral flagella and reported that ScrG, a GGDEF-EAL protein, regulated swarming and attachment of *V. parahaemolyticus* (biofilm formation). Shime-Hattori *et al.* (2006) reported that two type IV pili are involved in the formation of biofilms as a strategy for survival of *V parahaemolyticus* in the environment.

*Vibrio vulnificus* is now recognized as an emerging fish pathogen in the aquaculture environment, as well as an important human pathogen. In the aquaculture of both eel and tilapia, this microbe causes infection of the fish and results in economic loss (Fouz *et al.*, 2006).

#### Bacillus *spp.*

*Bacillus* spp. can become part of biofilms either as bacterial spores or as vegetative cells. However, since most biofilms form during the starvation state, most likely the *Bacillus* cells are in the spore form. Rahmati and Labbe (2008) examined fresh and processed retail seafood samples for the presence of *Bacillus cereus*. The pathogen was isolated from seafood samples at levels ranging from 3.6 to greater than 1000 CFU/g. The strains possessed both the hemolysin BL and nonhemolytic enterotoxin or only one of two toxins. Therefore, seafood is a vehicle for food borne illness caused by *B. cereus*.

#### Aeromonas

*Aeromonas.* Within the estuary and aquaculture environments, *Aeromonas hydrophila* was identified as a finfish pathogen and as a possible human pathogen. This microbe infects the finfish resulting in economic loss. If the finfish contaminated with this pathogen are harvested and improperly handled or prepared, *Aeromonas* can grow during refrigerate storage (Leung *et al.*, 1992; Fapohunda *et al.*, 1994; Fernandes *et al.*, 1998), and this may result in food borne illness.

#### Pseudomonas *spp.*

*Pseudomonas* spp. are opportunistic pathogens and are the major seafood spoilage organisms. They are slimy biofilm formers and histamine producers causing both quality and safety issues (Geesey *et al.*, 1992). *Pseudomonas* species were the dominant bacteria on fish skin regardless of time of the year (Tryfinopoulou *et al.*, 2002).

*Other bacteria*

Other bacteria: *Escherichia coli* and *Shigella* spp. were isolated from seafood (Greenlees *et al.*, 1998; King *et al.*, 2004) and have been shown to be associated with biofilms.

## 19.4 Control or removal of biofilms during seafood processing

### 19.4.1 Preharvest

*Open air ponds*

Control of bacteria in the pond environment is difficult. Figure 19.1 shows an aerial view of a catfish farm in Stoneville, MS. The pond water is fed and drained from a natural source, and the depth of the pond varies depending on the catfish growth. Since the bacteria in ground water can attach and develop biofilms on the indigenous zooplankton, removal is almost impossible. In geographic areas where human pathogens cannot be removed from the effluent, the ponds are drained, lined with heavy plastic, and the water disinfected before flowing into the pond. When ground water is chlorinated with a residual chlorine content of >0.1 mg/l, no *L. monocytogenes* or *Salmonella* were recovered (Kampelmacher and van Noorle Jansen, 1974). These plastic lined ponds can now be cleaned with similar protocols using Good Aquaculture Practices (Koonse, 2005) in order to reduce bacterial attachment to the zooplankton and thus also on the finfish or shrimp.

During harvesting, baskets or buckets are used to move the seafood from the pond to the transporting vehicle, and these can be made of either metal

**Fig. 19.1** Aerial view of catfish farm in Stoneville, Mississippi. Pond water is fed through a natural source. The pond depth varies depending on the size of the catfish.

**Fig. 19.2** Harvesting of catfish using plastic basket. The arrow at 'A' indicates biofilm formation in transportation tank at the water air interface. The arrow at 'B' indicates spaces where water plants and dirt collect.

or plastic. This equipment, shown in Fig. 19.2, can have trapped particulate matter and floating vegetation, and bacteria can attach onto the basket or bucket with possible biofilm formation. Regarding transport of seafood, vehicles should be sanitized between uses with potable water. During transport, the fish are tightly packed and contamination on the surface of the fish transfers to the surfaces of the transport vehicle. Good Aquaculture Practices are needed to prevent the spread of bacteria.

*Recirculating systems*

Another area of pre-harvest seafood processing in which biofilms containing human pathogens and spoilage microbes can occur is in recirculating aquaculture system tanks shown in Fig. 19.3. Aquaculture farming in these tanks can be done using either fresh or salt water. Biofilm formation can occur on the surfaces of all materials used: buna-N rubber, polyvinyl chloride (PVC), chlorinated PVC, glass, fiberglass, and stainless steel. Notice in Fig. 19.3, the biofilm on the tank surface in contact with the water, which is a reservoir for bacterial recontamination. Water quality is a major concern, and it is recirculated through a filtration and disfection system (Fig. 19.4 – catch tank of system) to remove uneaten food and the animal waste. Notice the biofilms formed on the pipe water-air interface indicated in the photograph. In some systems the water is passed through a sand filtration apparatus (Fig. 19.5) to remove particulate matter, but the bacteria are not always trapped in the sand filter and further disinfection steps are applied. The most commonly used intervention is ultraviolet light (UV). When the bacteria attached to particulate matter (excess food particles or

**Fig. 19.3** Recirculating tank. A) Filter where the water is removed from the recirculating tank, arrow indicates where biofilms form on the filter; B) Air water interface, arrow indicates where biofilms form on tank surface.

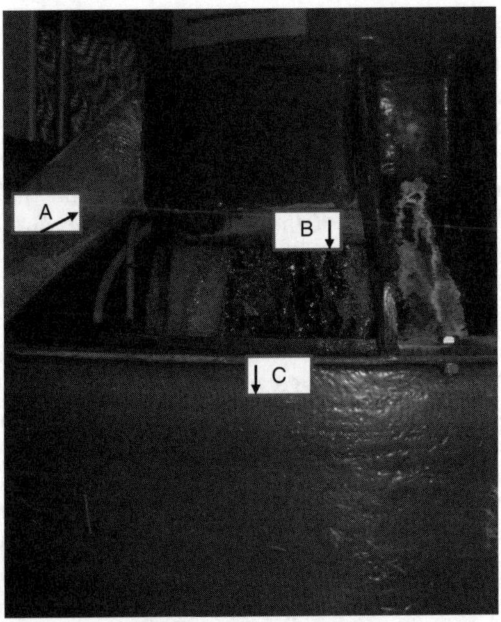


**Fig. 19.4** Water exiting the sand filtration system. A) Water pumped to top of sand filter; B) Water exiting sand filter; C) Water overflowing the catch tank for reintroduction into recirculation fish tank.

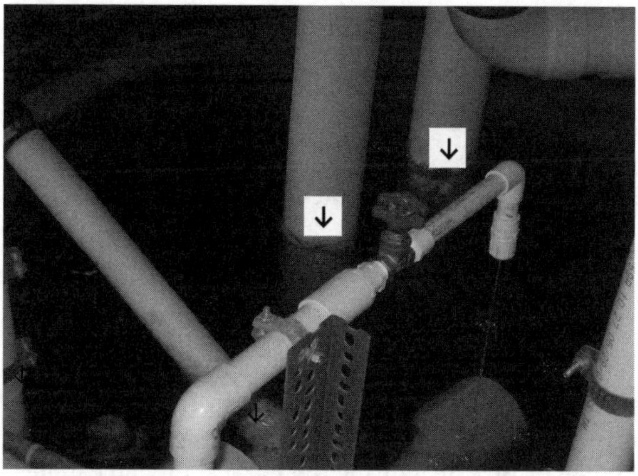

**Fig. 19.5** Water holding tank of filtered water before re-entry into growth tank. Arrows indicate places where biofilms form.

fish waste) and are not removed by filtration, UV light was not 100% effective to inactivate the bacteria (Sharrer *et al.*, 2005). The effectiveness of dissolved ozone on water borne microflora has been examined (Bullock *et al.*, 1997; Summerfelt *et al.*, 1997), and there is concern that the residual dissolved ozone would be toxic to the finfish. Summerfelt *et al.* (2004) showed that dissolved ozone can be inactivated by using UV light. Further studies showed that using ozone followed by UV light was an effective process to inactive the bacteria (Summerfelt, 2003; Sharrer and Summerfelt, 2007).

Pipes are used in recirculation systems. The flow characteristics of the liquid in these pipes influence biofilm development. Stoodley *et al.* (1999) studied biofilm formation under laminar and turbulent flow and after increasing the liquid's nutrient content. They reported that the structure of the biofilms was different based on the type of flow. Under laminar flow, the biofilms were patchy and consisted of circular cell clusters, whereas under turbulent flow, the biofilms were patchy and consisted of ripples and streamer type formation. When the nutrient content of the fluid was increased under both laminar and turbulent flow, the cell cluster grew rapidly (laminar) and the biofilm thickened and the ripples disappeared (turbulent).

In addition to the water flow characteristics, the pipe material affects biofilm formation. Kerr *et al.* (1999) reported that medium density polyethylene and unplasticised polyvinyl chloride supported the lowest numbers of bacteria in a steady state biofilm. Further research is needed in this area to determine which piping material would inhibit the formation of biofilms.

After the seafood is harvested, the recirculating system is cleaned and sanitized. King *et al.* (2004) showed that proper cleaning and sanitizing with sodium hypochlorite or peroxacetic acid is effective in removing biofilms from the surfaces of materials. Proper cleaning/sanitizing of the recirculating system before start-up and maintaining the filtering/disinfection of the water would reduce the bacterial populations involved in biofilm formation.

### 19.4.2 Postharvest

Postharvest seafood processing methods differ somewhat depending on where the seafood is grown, i.e. wild caught or aquaculture grown. Wild catches from the open seas are processed by more traditional methods, either commercial or private. During the harvest of wild catches, preventing bacterial growth and biofilm formation on processing or cutting surfaces and knives for any of the above mentioned pathogens is the first step. For wild catches, the finfish are deheaded and the guts removed. The ship's deck then becomes contaminated with particulate matter, and bacteria attach to surfaces. The seafood is iced during transport. Control of bacterial growth is temperature dependent, requiring a quick decrease in temperature immediately after harvesting and continuance the cool temperature during transport (Bao *et al.*, 2007). Proper cleaning and sanitizing of the harvest vessel and equipment used is essential to prevent biofilm formation and spread of these pathogens.

Aquaculture harvesting processes entail removal of fish from ponds and placing the fish into transportation vehicles. The fish arrive alive at the processing plant. Depending on the variety, finfish are processed by descaling and/or deskinning and/or deboning. The end product can be fillets or cut portions either shipped unfrozen on ice or frozen. Shrimp are collected, placed on ice, and brought to processing plants. The shrimp may then be deheaded, shelled, deveined, or cooked. If the heads are not removed, the shrimp may be soaked in an antioxidant/preservative agent to reduce shell oxidation.

### 19.4.3 Mollusk processing

Biofilms can be found attached to the surface of the mollusk shell and can be removed by proper cleaning. However, if not removed, the biofilms can become a source of cross contamination, particularly of the ice used for cooling. Mollusks are filter feeders, and the bacteria can accumulate within the flesh (Miller *et al.*, 2006). Physical removal of bacteria from the flesh would cause a quality issue. The US Food and Drug Administration is now recommending that upon harvest, oysters should be refrigerated immediately to prevent the growth of *V. parahaemolyticus*, a major problem with raw oysters. Such chilling can also inactivate other temperature-sensitive vibrios. Another intervention step involves placing the mollusk in clean water for a period of time to dilute out any pathogenic microbes, but this step would also require recirculation of the water and filtration and disinfection processing steps.

Some mollusks are sold un-cooked after the meat is removed from the shell and packaged. Some packaged mollusks are given a mild heat treatment, which inactivates the microbial population, but does not remove the biofilm. Other means of inactivating microbes from the raw mollusks are gamma irradiation, which was approved for use on mollusks in the United States (FDA, 2005) and high pressure treatment (Lopez-Cabellero *et al.*, 2000; He *et al.*, 2002). In areas where harvesting occurred from known polluted estuaries or aquaculture environments, it is recommended that mollusks be cooked before eating.

### 19.4.4 Crustacean processing

Aquaculture-raised shrimp have been found to be contaminated with *Salmonella* spp., *L. monocyogenes*, *Vibrio* spp., and *Shigella* spp. During shrimp processing, under conditions where freezing or reduced storage temperatures are used, the vibrios become less problematic. However, other microorganisms were reported to survive the freeze/thaw cycle (Rajkowski, 2007). All of the microorganisms are biofilm formers and survive in the water environment. As biofilms, they are known to attach to the crustacean chitin (skin) and form biofilms, as well as becoming internalized since the crustaceans are also filter feeders. Physical removal of bacteria from the outer skin would cause quality deterioration. During processing, crustaceans are deheaded, deveined (gut removed) and individually quick frozen (considered raw) or cooked with or without the outer skin. If cooked, a glaze usually containing a polyphosphate is applied. Some researchers consider this chemical application as an antimicrobial step, but polyphosphates are also water binding agents and can reduce drip when the product is thawed, reducing the possibility of cross contamination. It is recommended that these frozen products be cooked or blanched before eating.

### 19.4.5 Seafood finishing processes

Seafood arrives at the processing plant with biofilm on their outer skin, which may transfer to equipment during processing. Trisodium phosphate was shown to inactivate attached *Salmonella* and *L. monocytogenes* on catfish; however, the skin mucus slightly decreased the antimicrobial effect (Kim and Marshall, 2002). Some finfish varieties lend themselves to further processing by heat, i.e. canning of salmon and tuna. Although this thermal processing step does not physically remove biofilms, the bacteria are inactivated.

Canned salted anchovies and sardines are traditional foods in the Mediterranean area. During harvesting, the biofilms that formed in the fish boxes were reported to contain *Salmonella* spp. and *Staphylococcus aureus*, and cross contamination of the fish occurred. In order to control these pathogens, a salt ripening period of 90 days was shown to provide a safe fish product (Arkoudelos *et al.*, 2003). Another process that is used particularly for salmon is cold or hot smoking (Heinitz and Johnson 1998). *Listeria monocytogenes* is a problem in these smoked packaged products, and the packaging process was identified as an area in the plant where contamination for the hot or cold smoked product occurs due to biofilm formation on the fresh salmon. Controls for this pathogen during processing are similar to other food plants. HACCP requires thorough cleaning and sanitizing in the finfish industry (Lupin, 2003).

### 19.4.6 Removal of biofilms at the processing plant

The equipment and materials used for fish processing are similar to those used in other food plants and include buna-N rubber, polyvinyl chloride (PVC), chlorinated PVC, glass, fiberglass and stainless steel (Ronner and Wong, 1993; Kerr *et al.*, 1999; Joseph *et al.*, 2001). Biofilms were shown to form on these materials, and various sanitizers were tested. Sodium hypochlorite and peracetic acid containing sanitizers were found to be more effective than quaternary ammonium or ozone in the reduction of microflora from processing materials where seafood biofilms developed (King *et al.*, 2004). Bagge-Ravn *et al.* (2003) reported that *Pseudomonas* spp., were the dominant microflora remaining following cleaning and disinfection. *Pseudomonas* is a spoilage organism in seafood. It is recommended that a rinse type cleaning plan be replaced by a more vigorous method followed by use of an appropriate sanitizer. Sharma *et al.* (2005) showed that vigorous methods applying microbubbles to the cleaning liquid flow can stimulate microbial detachment from surfaces. For biofilm inactivation, a residual chlorine level was shown to be the most effective sanitizer (LeChevallier *et al.*, 1988).

## 19.5 Prevention

In his review of potential hazards in aquaculture fish, Lupin (2003) proposes strategies to reduce the risk, which includes identification of the critical control points in aquaculture. Proper aquaculture techniques using good practices for farmers as outlined by the FDA have shown that Good Aquaculture Practices are essential. Good water quality is the first step in prevention. The water should be disinfected before use in ponds and in recirculating systems. A HACCP program in any processing plant would require proper clean-up of particulate matter, washing, and sanitizing. Plant personnel should be trained to give them an understanding that removal of adhering microflora is essential for Good Hygienic Practices.

### 19.5.1 Biofilm inactivation

Various reports have been published describing methods to inhibit the growth of microorganism on raw seafood. Gamma irradiation is proposed for finfish and inactivates pathogens at low levels of irradiation. Rajkowski (2008) reported that seafood isolates had irradiation destruction values similar to *L. monocytogenes* strains isolated from other sources. Doke (1990) reported that low dose irradiation was effective for disinfecting sun-dried fish. Shelf life extension was achieved by pretreatment with ozone (Gelman *et al.*, 2005) or nisin/lactoperoxidase (Elotmani and Assobhei, 2003; Neetoo and Chen, 2008) before cold storage. Ultraviolet light was also used to reduce pathogens on raw finfish fillets giving about a one log reduction (Ozer and Demirci, 2006) and increased the shelf life of iced vacuum packaged fish by 4 days (Huang and Toledo, 1982). UV-light was shown to be an effective water and smooth surface disinfectant (Liltved and Landfald, 2000).

### 19.5.2 Research needs

Aquaculture farming is a relatively new practice. Much is still unknown about the process and where biofilm formation can become problematic. Research is needed to improve the quality of the effluent water system for ponds, filtration/disinfection steps for the recirculating systems, and the processing plant.

## 19.6 References

AMPOFO J A and CLERK G C (2002), 'Infestation of fish-culturing communities with fish-borne bacteria: the Ghanaian case', *Int J Environ Health Res*, 3, 277–282.

ANDREWS W H, WILSON C R, POELMA P L and ROMERO A (1977), 'Bacteriological survey of the channel catfish (*Ictalurus punctatus*) at the retail level', *J Food Sci*, 42, 359–363.

ARKOUDELOS J S, SAMARAS F J and TASSOU C C (2003), 'Survival of *Staphylococcus aureus* and *Salmonella* Enteritidis on salted sardines (*Sardina pilchardus*) during ripening', *J Food Prot*, 66(8), 1479–1481.

ATANASSOVA V, REICH F and KLEIN G (2008), 'Microbiological quality of sushi from sushi bars and retailers', *J Food Prot*, 71(4), 860–864.

AUGUSTIN J-C, ZULIANI V, CORNU M and GUILLIER L (2005), 'Growth rate and growth probability of *Listeria monocytogenes* in dairy, meat and seafood products in suboptimal conditions', *J Appl Microbiol*, 99, 1019–1042.

BAGGE-RAVN D, NG Y, HJELM M, CHRISTIANSEN J N, JOHANSEN C and GRAM L (2003), 'The microbial ecology of processing equipment in different fish industries-analysis of the microflora during processing and following cleaning and disinfection', *Int J Food Microbiol*, 87, 239–250.

BAO H N D, ARASON S and IÓRARINSDÓTTIR K A (2007), 'Effects of dry ice and superchilling on quality and shelf life of Arctic Char (*Salvelinus alpinus*) fillets', *Int J Food Eng*, 3(3), 1–27.

BASTI A A, MISAGHI A, SALEHI T Z and KAMKAR A (2006), 'Bacterial pathogens in fresh, smoked and salted Iranian fish', *Food Control*, 17, 193–188.

BRACKETT R E and BEUCHAT, L R (1990), 'Pathogenicity of *Listeria monocytogenes* grown on crabmeat', *Appl Environ Microbiol*, 56(5), 1216–1220.

BULLOCK G L, SUMMERFELT S T, NOBLE A C, WEBER A L, DURANT M D and HANKINS J A (1997), 'Ozonation of a recirculating rainbow trout culture system I. Effects on bacterial gill disease and heterotrophic bacteria', *Aquaculture*, 158, 43–55.

CAHILL M M (1990), 'Bacterial flora of fishes: a review', *Microb Ecol*, 19, 21–41.

CHEN C-Y, CHAO C-B and BOWSER P R (2006), 'Infection of tilapia *Oreochromis* sp. by *Vibrio vulnificus* in freshwater and low-salinity environments', *J World Aquaculture Society*, 37(1), 82–88.

CHIU T-H, DUAN J and SU Y-C (2007), 'Characteristics of virulent *Vibrio parahaemolyticus* isolated from Oregon and Washington', *J Food Prot*, 70(4), 1011–1016.

DAWSON M P, HUMPHREY B A and MARSHALL K C (1981), 'Adhesion: a tactic in the survival strategy of a marine vibrio during starvation', *Curr Microbiol*, 6, 195–199.

DEPAOLA A, KAYSNER C A, BOWERS J and COOK D W (2000), 'Environmental investigations of *Vibrio parahaemolyticus* in oysters after outbreaks in Washington, Texas, and New York (1997 and 1998)', *Appl Environ Microbiol*, 66(11), 4649–4654.

DOKE S N (1990), 'The packaging of sun-dried fish disinfection by irradiation', *Fleischwirtsch*, 70(11), 1303–1304.

ELOTMANI F and ASSOBHEI O (2003), '*In vitro* inhibition of microbial flora of fish by nisin and lactoperoxidase system', *Letts Appl Microbiol*, 38, 60–65.

ERDENLIG S, AINSWORTH A J and AUSTIN F W (2000), 'Pathogenicity and production of virulence factors by *Listeria monocytogenes* isolates from channel fish', *J Food Prot*, 63, 613–619.

FAO (2006), *The state of world fisheries and aquaculture*, Rome, Italy.

FAPOHUNDA A O, MCMILLIN K W, MARSHALL D L and WAITES W M (1994), 'Growth of selected cross-contaminating bacterial pathogens on beef and fish at 15 and 35°C[+]', *J Food Prot*, 57(4), 337–340.

FDA, FEDERAL REGISTER (2005), *Irradiation in the production, processing, and handling of food – Final Rule, 21 CFR Part 179*, 70(157), 48057–48073. http://www.cfsan.fda.gov/~lrd/fr080822.html.

FDA, CODE OF FEDERAL REGUATIONS (2006), *Definitions, 21 CFR Part 123.3*, 2, 269–270. http://www.cfsan.fda.gov/~lrd/FCF123.html.

FELL G, HAMOUDA O, LINDNER R, REHMET S, LIESEGANG A, PRAGER R, GERICKE B and PETERSEN L (2000), 'An outbreak of *Salmonella blockley* infections following smoked eel consumption in Germany', *Epidemiol Infect*, 125, 9–12.

FERNANDES C F, FLICK G J and THOMAS T B (1998), 'Growth of inoculated psychrotrophic pathogens on refrigerated fillets of aquacultured rainbow trout and channel fish', *J Food Prot*, 61, 313–317.

FOUZ B, LARSEN J L and AMARO C (2006), '*Vibrio vulnificus* serovar A: an emerging pathogen in European anguilliculture', *J Fish Diseases*, 29, 285–291.

GAO (2004), *Food Safety FDA's imported seafood safety program shows some progress, but further improvements are needed: GAO-04-246 Federal oversight of imported seafood*, .gao.gov/new.items/d04246.pdf, accessed 7/16/08.

GEESEY G G, STUPY M W and BREMER P J (1992), 'The dynamics of biofilms', *Int Biodeterioration and Biodegradation*, 30, 135–154.

GELMAN A, SACHS O, KHANN Y, DRABKIN V and GLATMAN L (2005), 'Effect of ozone pretreatment on fish storage life at low temperatures', *J Food Prot*, 68(4), 778–784.

GOMEZ-GIL B, FAJER-AVILA E and GARCÍA-VARGAS F (2007), 'Vibrios of the spotted rose snapper *Lutjanus guttatus* Steindachner, 1869 from northwestern Mexico', *J Appl Microbiol*, 102, 1518–1526.

GREENLEES K J, MACHADO J, BELL T and SUNDLIF S F (1998), 'Food borne microbial pathogens of cultured aquatic species', *Vet Clin North Am Food Anim Pract*, 14, 101–112.

GUERIN P J, DE JONG B, HEIR E, HASSELTVEDT V, KAPPERUD G, STYRMO K, GONDROSEN B, LASSEN J, ANDERSSON Y and AAVITSLAND P (2004), 'Outbreak of Salmonella Livingstone infection in Norway and Sweden due to contaminated processed fish products', *Epidemiol Infect*, 132, 889–895.

HE H, ADAMS R M, FARKAS D F and MORRISSEY M T (2002), 'Use of high-pressure processing for oyster shucking and shelf-life extension', *J Food Sci*, 67(2), 640–645.

HEINITZ M L and JOHNSON J M (1998), 'The incidence of *Listeria* spp., *Salmonella* spp., and *Clostridium botulinum* in smoked fish and shellfish', *J Food Prot*, 61(3), 318–323.

HEINITZ M L, RUBLE R D, WAGNER D E and TATINI S R (2000), 'Incidence of *Salmonella* in fish and seafood', *J Food Prot*, 63(5), 579–592.

HERRERA F C, SANTOS J A, OTERO A and GARCÍA-LÓPEZ M-L (2006), 'Occurrence of foodborne pathogenic bacteria in retail prepackaged portions of marine fish in Spain', *J Appl Microbiol*, 100, 527–536.

HUANG Y-W and TOLEDO R (1982), 'Effect of doses of high and low intensity UV irradiation on surface microbiological counts and storage-life of fish', *J Food Sci*, 47, 1667–1669, 1731.

HUSS H H, REILLY A, BEN EMBAREK P K (2000), 'Prevention and control of hazards in seafood', *Food Control*, 11, 149–156.

JALLEWAR P K, KALOREY D R, KURKURE N V, PANDE V V and BARBUDDHE S B (2007), 'Genotypic characterization of *Listeria* spp. Isolated from fresh water fish', *Int J Food Microbiol*, 114, 120–123.

JØRGENSEN L V and HUSS H H (1998), 'Prevalence and growth of *Listeria monocytogenes* in naturally contaminated seafood', *Int J Food Microbiol*, 42, 127–131.

JOSEPH B, OTTA S K, KARUNASAGAR INDRANI and KARUNASAGAR I (2001), 'Biofilm formation by *Salmonella* spp. on food contact surfaces and their sensitivity to sanitizers', *Int J Food Microbiol*, 64(3), 367–372.

KAMPELMACHER E H and VAN NOORLE JANSEN ML (1974), '*Occurrence of L. monocytogenes in effluents*', In VI International Symposium of Listeriosis, vol. 23, Sutton Bonington, England.

KANEKO T and COLWELL R R (1973), 'Ecology of *Vibrio parahaemolyticus* in Chesapeake Bay', *J Bacteriol*, 113(1), 24–32.

KANEKO T and COLWELL R R (1975), 'Adsorption of *Vibrio parahaemolyticus* onto chitin and copepods', *Appl Environ Microbiol*, 29(2), 269–274.

KERR C J, OSBORN K S, ROBSON G D and HANDLEY P S (1999), 'The relationship between pipe material in biofilm formation in a laboratory model system', *J Appl Microbiol*, 85, 29S–38S.

KIM J and MARSHALL D L (2002), 'Influence of catfish skin mucus on trisodium phosphate inactivation of attached *Salmonella* Typhimurium, *Edwardsiella tarda*, and *Listeria monocytogenes*', *J Food Prot*, 65(7), 1146–1151.

KIM Y-K and MCCARTER L L (2007), 'ScrG, a GGDEF-EAL protein, participates in regulating swarming and sticking in *Vibrio parahaemolyticus*', *J Bacteriol*, 189(11), 4094–4107.

KING R K, FLICK, JR. G J, PIERSON M D, SMITH S A, BOARDMAN G D and COALE, JR. C W (2004), 'Identification of bacterial pathogens in biofilms of recirculating aquaculture systems', *J Aquatic Food Prod Technol*, 13, 125–133.

KJELLEBERG S and HERMANSSON M (1984), 'Starvation-induced effects on bacterial surface characteristics', *Appl Environ Microbiol*, 48, 497–503.

KOONSE B (2005), 'Good aquaculture practices for farmers – an update', *FAO Fisheries Proceedings*, 1, 76–77.

LECHEVALLIER M W, CAWTHON C D, LEE R G (1988), 'Inactivation of biofilm bacteria', *Appl Environ Microbiol*, 54, 2492–2499.

LEUNG C-K, HUANG Y-W and HARRISON M A (1992), 'Fate of *Listeria monocytogenes* and Aeromonas hydrophila on packaged channel catfish fillets stored at 4°C', *J Food Prot*, 55(9), 728–730.

LEVIN R E (2006), '*Vibrio parahaemolyticus*, a notably lethal human pathogen derived from seafood: a review of its pathogenicity, characteristics, subspecies characterization, and molecular methods of detection', *Food Biotechnol*, 20, 93–128.

LILTVED H and LANDFALD B (2000), 'Effects of high intensity light on ultraviolet-irradiated and non-irradiated fish pathogenic bacteria', *Wat Res*, 34(2), 481–486.

LÓPEZ-CABALLERO M E, PÉREZ-MATEOS M, MONTERO P and BORDERÍAS A J (2000), 'Oyster preservation by high-pressure treatment', *J Food Prot*, 63(2), 196–201.

LUPIN H M (2003), 'Potential hazards in aquaculture fish', *FAO Fisheries Proceedings*, 65–75.

MILLER W A, MILLER M A, GARDNER I A, ATWILL E R, BYRNE B A, JANG S, HARRIS M, AMES J, JESSUP D, PARADIES D, WORCESTER K, MELLI A and CONRAD P A (2006), '*Salmonella* spp., *Vibrio* spp., *Clostridium perfringens*, and *Plesiomonas shigelloides* in marine and freshwater invertebrates from coastal California ecosystems', *Microbial Ecol*, 52, 198–206.

NEETOO H, YE M, and CHEN H (2008), 'Potential antimicrobials to control *Listeria monocytogenes* in vacuum-packaged cold-smoked salmon pâté and fillets' *Int J Food Microbiol*, 123, 220–227.

OZER N P and DEMIRCI A (2006), 'Inactivation of Escherichia coli O157:H7 and *Listeria monocytogenes* inoculated on raw salmon fillets by pulsed UV-light treatment', *Int J Food Sci Technol*, 41, 354–360.

PALUMBO S A, CALL J, HUYNH B and FANELLI J (1996), 'Survival and growth potential of *Aeromonas hydrophila* in reconditioned pork-processing-plant water', *J Food Prot*, 59, 881–885.

RAHMATI T and LABBE R (2008), 'Levels and toxigenicity of *Bacillus cereus* and *Clostridium perfringens* from retail seafood'. *J Food Prot*, 71(6), 1178–1185.

RAJKOWSKI K T (2007), 'The survival of *Shigella sonnei* in frozen media after ultraviolet treatment of ice', *Ice World J*, 14–23.

RAJKOWSKI K T (2008), 'Radiation $D_{10}$ – values on thawed and frozen catfish and tilapia for finfish isolates of *Listeria monocytogenes*', *J Food Prot*, 71(11), 2278–2282.

RAJKOWSKI K T and RICE E W (1999), 'Recovery and survival of *Escherichia coli* O157:H7 in reconditioned pork', *J Food Prot*, 62(7), 731–734.

RAJKOWSKI K T and RICE E W (2001), 'Growth and recovery of selected gram-negative bacteria in reconditioned wastewater', *J Food Prot*, 64(11), 1761–1767.

RAJKOWSKI K T, RICE E W, HUYNH B and PATSY J (1996), 'Growth of *Salmonella* spp. and *Vibrio cholerae* in reconditioned wastewater', *J Food Prot*, 59(6), 577–581.

RIBEIRO R V, HOFER E, DOS SANTOS LAZARO N and DOS PRAZERES RODRIGUES D (2006), 'Enzymatic factors associated with virulence in *Vibrio* species from environment in Rio De Janeiro, Brazil', *Rev Bras med Vet*, 28(1), 20–25.

RIPABELLI G, SAMMARCO M L, FANELLI I and GRASSO G M (2004), 'Ricerca di *Salmonella*, *Listeria* spp., *Vibrio* spp. e *Yersinia enterocolitica* in prodotti ittici congelati e surgelati del commercio e correlazione con gli indici di contaminazione fecals: implicaziono in Sanità Pubblica', *Ann Ig*, 16, 531–539.

RONNER A B and WONG A C L (1993), 'Biofilm development and sanitizer inactivation of *Listeria monocytogenes* and *Salmonella typhimurium* on stainless steel and Buna-n rubber', *J Food Prot*, 56, 750–758.

SASAHARA K C and ZOTTOLA E A (1993), 'Biofilm formation by *Listeria monocytogenes* utilizes a primary colonizing microorganism in flowing systems', *J Food Prot*, 56, 1022–1028.

SHARMA P K, GIBCUS M J, VAN DER MEI H C and BUSSCHER H J (2005), 'Influence of fluid shear and microbubbles on bacterial detachment from a surface', *Appl and Envrion Microbiol*, 71(7), 3668–3673.

SHARRER M J and SUMMERFELT S T (2007), 'Ozonation followed by ultraviolet irradiation provides effective bacterial inactivation in a freshwater recirculating system', *Aquacultural Engineering*, 37, 180–191.

SHARRER M J, SUMMERFELT S T, BULLOCK G L, GLEASON L E and TAEUBER J (2005), 'Inactivation of bacteria using ultraviolet irradiation in a recirculating salmonid culture system', *Aquaculturual Engineering*, 33, 135–149.

SHIME-HATTORI A, IIDA T, ARITA M, PARK K-S, KODAMA T and HONDA T (2006), 'Two type IV pili of *Vibrio parahaemolyticus* play different roles in biofilm formation', *FEMS Microbiol Lett*, 264, 89–97.

SHIN J-H, CHANG S and KANG D-H (2004), 'Application of antimicrobial ice for reduction of foodborne pathogens (*Escherichia coli* O157:H7, *Salmonella* Typhimurium, *Listeria monocytogenes*) on the surface of fish', J Appl Microbiol, 97, 916–922.

SMOOT L M and PIERSON M D (1998a), 'Effect of environmental stress on the ability of *Listeria monocytogenes* Scott A to attach to food contact surfaces, *J Food Prot*, 61(10), 1293–1298.

SMOOT L M, PIERSON M D, (1998b), 'Influence of environmental stress on the kinetics and strength of attachment of *Listeria monocytogenes* Scott A to buna-n rubber and stainless steel', *J Food Prot*, 61(10), 1286–1292.

SOMERS E B, SCHOENI J L, WONG A C L (1994), 'Effect of trisodium phosphate on biofilm and planktonic cells of *Campylobacter jejuni*, *Eschericha coli* O157:H7, *Listeria monocytogenes* and *Salmonella typhimurium*', *Food Microbiol*, 22, 269–276.

STOODLEY P, DODDS I, BOYLE J D and LAPPIN-SCOTT H M (1999), 'Influence of hydrodynamics and nutrients on biofilm structure', *J Appl Microbiol Symp Suppl*, 85, 19S–28S.

SUMMERFELT S T (2003), 'Ozonation and UV irradiation – an introduction and examples of current applications', *Aquacultural Engineering*, 28, 21–36.

SUMMERFELT S T, HANKINS J A, WEBER A L and DURANT M D (1997), 'Ozonation of a recirulating rainbow trout culture system II. Effects on microscreen filtration and water quality', *Aquaculture*, 158, 57–67.

SUMMERFELT S T, SHARRER M J, HOLLIS J, GLEASON L E and SUMMERFELT S R (2004), 'Dissolved ozone destruction using ultraviolet irradiation in a recirculating salmonid culture system', *Aquacultural Engineering*, 32, 209–223.

SUPPIN D and SMULDERS F J M (2005), 'Zum hygienestatus von frischen Fischfilets auf Handelsebene', *Vet Med Austria*, 92, 259–264.

TRYFINOPOULOU P, TSAKALIDOU E and NYCHAS G-J (2002), 'Characterization of *Pseudomonas* spp. associated with spoilage of gilt-head sea bream stored under various conditions', *Appl Environ Microbiol*, 68(1), 65–72.

VUDDHAKUL V, SOBOON S, SUNGHIRAN W, KAEWPIBOON S, CHOWDHURY A, ISHIBASHI M, NAKAGUCHI Y and NISHIBUCHI M (2006), 'Distribution of virulent and pandemic strains of *Vibrio parahaemolyticus* in three molluscan shellfish species (*Meretrix meretrix*, *Perna viridis*, and *Anadara granosa*) and their association with food-borne disease in southern Thailand', *J Food Prot*, 69(11), 2615–2620.

YANO Y, KANENIWA M, SATOMI M, OIKAWA H and CHEN S-S (2006), 'Occurrence and density of Vibrio parahaemolyticus in live edible crustaceans from markets in China', *J Food Prot*, 69(11), 2742–2746.

ZHAO S, MCDERMOTT P F, FREIDMAN S, QAIYUMI S, ABBOTT J, KEISSLING C, AYERS A, SINGH R, HUBERT S, SOFOS J, and WHITE D G (2006), 'Characterization of antimicrobial-resistant *Salmonella* isolated from imported foods', *J Food Prot*, 69, 500–507.

ZOBELL C E (1943), 'The effect of solid surfaces upon bacterial activity', *J Bacteriol*, 46, 39–56.

# 20

# Biofilms in fresh fruit and vegetables

**B. A. Annous, J. L. Smith, and P. M. Fratamico, USDA, Agricultural Research Service, USA and E. B. Solomon, DuPont Chemical Solutions Enterprise, USA**

**Abstract**: Bacteria can attach to surfaces and form biofilms, which have a characteristic structure consisting of microcolonies enclosed in a hydrated matrix of microbially produced proteins and polysaccharides. In this complex biofilm network, the cells act less as individual entities and more as a collective living system, often with channels to deliver water and nutrients to the cells in the inner portion of the biofilm. Human pathogens form biofilms on food and food contact surfaces, which enhances their ability to survive harsh environments, to resist antimicrobial treatments, and to spread and persist in the food processing environment. The formation of biofilms is a food safety concern since persistent low-level contamination of foods can occur.

**Key words**: *Escherichia coli* O157:H7, *Listeria monocytogenes*, *Salmonella enterica*, *Sigella*, biofilm, pathogens, planktonic cells, produce.

## 20.1 Introduction

The great majority of microorganisms present in the natural environment exist within biofilms. Whereas pure cultures of bacteria have been studied in detail for over a century, research focused on understanding the ecology of mixed populations began only a short while ago. Biofilms are defined as 'microbially derived sessile communities characterized by cells that are irreversibly attached to a substratum or interface or to each other, are embedded in a matrix of extracellular polymeric substances that they have produced, and exhibit an altered phenotype with respect to growth rate and gene transcription' (Donlan and Costerton, 2002). Cells present within biofilms are distinctly different from their planktonic counterparts in terms of physiology and response to environmental stresses. Most studies on biofilms have described work with bacteria inhabiting abiotic surfaces; however, recent studies have also focused on the interaction between biofilms and biotic surfaces (including animal and plant tissues) in order to discern the interactions between microbe and host.

The aerial tissues of plants are colonized by a wide variety of bacteria, fungi, and yeasts. These colonizing microorganisms are known as epiphytes. Bacteria are the primary colonizers of leaf surfaces, with $10^6$–$10^8$ bacteria often found per gram of leaf tissue (Lindow and Brandl, 2003). Most of the research on bacterial biofilms on plants has focused on discerning the spatial relationships of epiphytic microorganisms on the leaf surface (Morris *et al.*, 1997), the role they play in pathogenesis (Rigano *et al.*, 2007), the ability of cells entrapped within biofilms to survive environmental stresses compared to singular cells (Monier and Lindow, 2003), and the relationship between bacteria and their plant hosts (Brandl, 2006; Morris and Monier, 2003). Results from this research demonstrate that native biofilms on plant surfaces are composed of a wide variety of bacterial species, many from unknown genera. Furthermore, the population dynamics of biofilms vary greatly corresponding to environmental conditions such as temperature, relative humidity, and the availability of nutrients. Specific bacterial phenotypes are more successful colonizers of the leaf surface because of their ability to modify their habitat. Successful leaf colonizers generally have the ability to produce surfactants, plant growth regulators, and copious amounts of exopolysaccharide to shield them from dessication, sunlight, wind and other environmental stresses. (Further discussion of these topics can be found in reviews Andrews and Harris, 2000; Lindow and Brandl, 2003; Morris and Monier, 2003; Ramey *et al.*, 2004.)

The number of outbreaks of foodborne illness arising from consumption of fresh produce has increased dramatically over the last two decades (Sivapalasingam *et al.*, 2004). From 1990–2005, fresh produce was associated with 713 outbreaks, resulting in 34 049 cases of illness (DeWaal and Bhuiya, 2007). Several reasons have been cited for this increase. First, the per capita consumption of fresh produce has increased significantly. US consumption, between 1982 and 1997, of raw fruits and vegetables increased by 18 and 29%, respectively (Garrett *et al.*, 2003). Second, the produce industry has become increasingly global, with large volumes of produce transported around the globe. Third, consumption of convenience foods such as fresh-cut fruits and bagged salads has also grown exponentially, yet these products are more conducive to microbial growth and spoilage than the whole produce from which they are prepared (Brandl, 2008).

After several large outbreaks in the 1990s (apple cider, sprouts, and cantaloupes), regulatory agencies as well as individual researchers began to focus on quantifying the presence of human pathogens on raw fruits and vegetables. A large number of studies were published indicating that extremely low levels of pathogens could be isolated when sampling produce from various parts of the production chain (field, transport, storage, and consumer). This research gave way to an even greater number of studies undertaken to find postharvest treatment regimes capable of inactivating pathogenic microorganisms attached to fresh produce (Buck *et al.*, 2003). A wide variety of chemical rinses (e.g., chlorine, peroxide, surfactants, organic acids, etc.) and physical treatments (e.g., UV, irradiation) were evaluated as disinfection steps in produce production, yet none have been proven to successfully eliminate pathogens attached to fresh produce without negatively impacting quality.

Because of the extremely low levels of pathogens found during surveys of fresh produce, it was believed that the majority of illnesses linked to fresh produce were a result of post-harvest contamination. However, recent evidence points to the potential for pre-harvest interactions between pathogenic microorganisms and growing plants to result in contaminated crops at harvest (Erickson *et al.*, 2004; Islam *et al.*, 2004a, 2004b, 2004c, 2004d; Jiang *et al.*, 2004). As a result of these findings, the ecology of human pathogens on plant surfaces has begun to be scrutinized (Brandl, 2006). Distinct molecular mechanisms of attachment have not yet been elucidated, although mutants of both *Escherichia coli* and *Salmonella* deficient in cell surface moieties (and other genes) exhibit diminished attachment to plant tissues (Matthysse *et al.*, 2008; Shaw *et al.*, 2008; Torres *et al.*, 2005). Additionally, there are indications that specific strains are better equipped than others for survival on plant tissue (Shi *et al.*, 2007). While not as fit as traditional epiphytic bacteria, human pathogens have demonstrated an ability to compete well with native microflora on plant surfaces and to occupy microenvironments that are conducive to their survival (Brandl and Mandrell, 2002). In addition, human pathogens such as *Campylobacter jejuni* and *Salmonella enterica* have been shown to multiply on plant surfaces under specific environmental conditions (Brandl and Mandrell, 2002; Brandl *et al.*, 2004). A number of field-level studies have shown the ability of pathogens to survive for long periods of time in both the plant phyllosphere and rhizosphere.

The goal of this chapter is to briefly review the literature describing the ability of human pathogens to attach, proliferate, and survive on the surface of fresh produce, with special attention paid to the specific phenomenon of biofilm formation. Although spoilage microorganisms are a significant concern, the majority of research has centered on the bacterial pathogens *E. coli* O157:H7, *Salmonella* spp., *Shigella* spp., *Listeria monocytogenes*, *Clostridium botulinum*, *Campylobacter*, and *Bacillus cereus*.

## 20.2 Several pathogens of concern in produce

### 20.2.1 *Escherichia coli* O157:H7

During the period of 1992–2002, *E. coli* O157:H7 was responsible for 21% of the produce-linked foodborne outbreaks in the United States (Rangel *et al.*, 2005). A multistate outbreak involving 26 states occurred in 2006 and was linked to *E. coli* O157:H7 contamination of bagged spinach (Wendel *et al.*, 2009). Uhlich *et al.* (2008) found that the presence of a strong biofilm-forming *E. coli* strain in mixed species biofilms with a weak-biofilm forming *E. coli* O157:H7 spinach outbreak strain resulted in the increase in numbers of the outbreak strain within the biofilm. Thus, the presence of non/less-virulent strong biofilm-forming strains could play an important role in the persistence of *E. coli* O157:H7 on produce surfaces.

### 20.2.2 *Salmonella* spp.

*Salmonella* have been associated with outbreaks arising from commodities as diverse as sprouts, melons, tomatoes, and mixed greens (Sivapalasingam *et al.*, 2004). In 2008, a large outbreak involving 43 states, the District of Columbia, and Canada was linked to jalapẽno and serrano peppers contaminated with *Salmonella* Saintpaul (CDC, 2008). The ability of this organism to survive in a variety of environments may be related to its ability to colonize plant surfaces and produce biofilms. Brandl and Mandrell (2002) investigated the ability of *Salmonella enterica* serovar Thompson to form biofilms in the phyllosphere of cilantro plants. Osmoregulated periplasmic glucans (OPG) of *S.* Typhimurium has shown to play a role in biofilm formation (Liu *et al.*, 2009). Colonization of mouse organs and biofilm formation in wash water from leafy greens was greater by wild-type *S.* Typhimurium in the presence of mutant strains that lacked OPG.

### 20.2.3 *Shigella* spp.

Infection with *Shigella* spp. is usually transmitted person-to-person, but may also occur through consumption of contaminated food and water. Fruit and vegetables have been linked to *Shigella* infections, and several large outbreaks were associated with contaminated raw vegetables (Agle and Blaschek, 2007; Davis *et al.*, 1988; Kapperud *et al.*, 1995; Martin *et al.*, 1986). Additionally, *Shigella* has been shown to form biofilms and become entrapped within an exopolysaccharide matrix on parsley leaf surfaces (Agle, 2003).

### 20.2.4 Listeria monocytogenes

*L. monocytogenes* is a ubiquitous organism and is found on raw fruits and vegetables. It can grow on fresh produce stored at refrigeration temperature. Outbreaks associated with *L. monocytogenes* have been associated with ready-to-eat foods such as hot dogs, luncheon meats, cold cuts, and soft cheeses; however, outbreaks involving produce have also occurred (Aureli *et al.*, 2000; Ho *et al.*, 1986; Schlech *et al.*, 1983). For example, an outbreak involving 20 cases of illness and 5 deaths occurred in the Boston area in 1979. *L. monocytogenes* type 4b was the cause of the illness, and suspected food items were raw tomatoes, lettuce, and celery (Ho *et al.*,

1986). Another large outbreak that occurred in Italy in 1997 linked corn contaminated with *L. monocytogenes* 4b as the source of infection (Aureli *et al.*, 2000). The pathogen can persist for long periods in soil and food processing facilities and is capable of persisting and growing on produce. These investigators found that the strain grew on corn kept for at least 10 h at 25°C. Ells and Hansen (2006) showed that *L. monocytogenes* was able to attach and colonize cabbage tissue. The development of cell clusters and biofilm structures resulted in a decreased ability to remove the pathogen from cabbage leaf surfaces. *L. monocytogenes* attached preferentially to cut cabbage tissues compared to intact surfaces.

### 20.2.5 *Campylobacter* spp.

*Campylobacter* is a leading cause of bacterial foodborne illness worldwide, and food of animal origin, particularly poultry, is responsible for a majority of infections. However, *Campylobacter* spp. have been isolated from a variety of produce items (Park and Sanders, 1992), and outbreaks associated with produce have been reported (Allen, 1985; CDC 1998). *Campylobacter* spp. have been shown to form biofilms in single species and mixed species biofilms (Gunther and Chen, 2009; Sanders *et al.*, 2007).

### 20.2.6 Spore-forming bacteria

*Bacillus cereus*, *C. botulinum*, and *C. perfringens* are spore-forming bacteria that can contaminate fruits and vegetables, since these organisms are found in soil. Human infections can occur if produce is subjected to treatments that allow germination of spores, for example modified atmosphere packaging. Lindsay *et al.* (2006) showed that *B. cereus* and *B. subtilis* spores attached to surfaces germinated under different conditions and biofilms that were formed contained both spores and vegetative cells. Outbreaks linked to garlic in oil and coleslaw contaminated with *C. botulinum* have been reported (Solomon and Kautter, 1988; Solomon *et al.*, 1990). *B. cereus* was associated with an outbreak linked to the consumption of mixed seed sprouts (Portnoy *et al.*, 1976).

Two factors have spurred interest in biofilm formation by enteric pathogens on fresh produce. First, the ability of chemical sanitizers to remove or inactivate bacteria attached to fresh produce is reduced significantly with increased contact time (the time elapsed between application of bacteria and introduction of the sanitizer). Early investigators speculated that this reduction in efficacy was a result of biofilm formation (Ukuku and Sapers, 2001). Second, a number of microscopic investigations into the microflora of fresh produce identified the presence of large aggregates of bacteria at various sites (Fett, 2000; Fett and Cooke, 2003). The potential of pathogens becoming incorporated into mixed-species biofilms would render them far less susceptible to antimicrobial treatments as well as increase their tolerance to other stresses such as desiccation and UV.

## 20.3 Biofilms on produce

Foodborne illness linked to fresh fruits and vegetables has increased significantly during the past thirty years (Sivapalasingam *et al.*, 2004). The rise in the number of produce-related outbreaks, coupled with the lack of an effective intervention, has given rise to an intense research effort into the ecology of human pathogens in the produce production environment. In contrast to earlier ideas, human pathogens have been found to survive in water, animal manure, and a variety of agricultural soils for long periods. In addition, human pathogens are capable of attaching to and colonizing the surfaces of growing plants. Once attached, human pathogens are capable of forming biofilms on plant tissues leading to the failure of washing treatments to remove or inactivate human pathogens on produce surfaces (Annous *et al.*, 2001, 2004, 2005a).

Microscopic studies indicate that plant-associated epiphytic bacteria form biofilms on surfaces of a wide variety of plants (Fett and Cooke, 2003; Morris *et al.*, 1997). Between 30 and 80 percent of bacteria on plant surfaces exist within biofilms (Lindow and Brandl, 2003). The formation of biofilms on plant surfaces is probably a survival strategy for the bacterial cells to withstand the harsh environment of the plant surface including wide temperature changes, desiccation, UV, and oxidative stress. Biofilm-associated bacteria embedded in a matrix of extracellular polysaccharides (EPS) are more difficult to remove from surfaces of produce or food-processing areas than their solitary counterparts (Annous *et al.*, 2004, 2005a; Fett and Cooke, 2003). The production of EPS shields bacterial cells within biofilms from desiccation and aids in the resistance of the bacteria to antimicrobial compounds. Three commodities that have been repetively linked to outbreaks are cantaloupe melons, apples (unpasteurized juice or cider), and leafy greens, each associated with a different human pathogen. Research into the interactions between *S. enterica*, *E. coli* O157:H7, and *Shigella* and these three commodities are discussed below.

### 20.3.1 Leafy greens

Minimally processed leafy greens and pre-packaged salads have emerged as important vehicles of transmission for foodborne pathogens. Since 1995, USFDA (2005) has documented eighteen *E. coli* O157:H7 outbreaks associated with lettuce and one outbreak associated with spinach. Parsley, green onions, basil, cilantro, and cabbage have also been associated with recent outbreaks of foodborne illness (DeWaal and Barlow, 2002; CDC, 2005). Hundreds of illness and some deaths were attributed to these outbreaks.

Generally, the source(s) of food contamination are unknown, but often, pathogens are found in environmental samples obtained from surrounding and immediate areas of the facilities responsible for specific outbreaks (Cummings *et al.*, 2001; CDC, 2005).

Many commercial post-harvest handling operations marketing leafy vegetables, including the majority of packaged salad processors, clean the surfaces of leafy greens by triple wash treatments combining physical removal with various types of disinfection. However, while planktonic bacterial pathogens are easily controlled by relatively low levels of wash water disinfectant, the degree to which leafy products may be sanitized is highly limited and is dependent on diverse intrinsic factors and on the effectiveness of how the commercial unit is operated. It is well known that conventional methods of washing with hypochlorite or other sanitizing agents can achieve only limited pathogen reduction levels (1–2 logs) and therefore cannot assure microbiological safety due to the fact that many enteric pathogens, such as *E. coli* O157:H7, cause illness at very low infectious doses (Sapers, 2005; USFDA, 2001a). Greater reductions of the pathogen are not possible due to strong microbial attachment and internalization of the organisms (Seo and Frank, 1999; Takeuchi and Frank, 2000) as well as to aggregate or biofilm formation (Solomon *et al.*, 2002). This prompted an investigation into survival, uptake, and biofilm formation by *E. coli* O157:H7 on freshly harvest lettuce leaves (Keskinen *et al.*, 2009). Stem ends of freshly harvested salad bowl lettuce leaves were submerged into water containing *E. coli* O157:H7, and were allowed to incubate for one week at 4°C. SEM confirmed that *E. coli* O157:H7 cells survived, colonized, and formed biofilms on lettuce leaves (Fig. 20.1).

In August 1998, eight outbreaks of *Shigella sonnei* occurred in four states and two Canadian provinces (Naimi *et al.*, 2003). The outbreaks, while geographically dispersed, were all linked to fresh parsley from a supplier in Mexico. In 1999, an outbreak of *Shigella boydii* associated with bean salad prompted an investigation into the ability of *Shigella* to persist and form biofilms on the surface of parsley (Agle, 2003). *S. boydii* survived on parsley for more than 20 days when stored at refrigeration temperatures. SEM confirmed that the organism become entrapped within a matrix of extracellular polymeric material on the surface of a parsley leaf (Fig. 20.2).

### 20.3.2 Apples

It is interesting that although seven outbreaks of *E. coli* O157:H7 were associated with apple juice or cider between 1982 and 2002 (Rangel *et al.*, 2005), none were associated with the consumption of fresh apples. However, the presence of pathogens on the surface of apples has implications for the safety of the fresh juice and fresh-cut fruit markets. Therefore, the FDA imposed a mandatory HACCP program for juice processors requiring a 5-log reduction in the level of the target pathogen. Laboratory washing

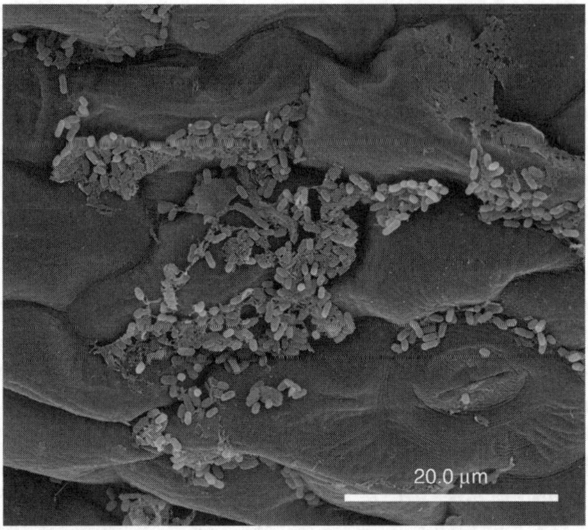


**Fig. 20.1** Scanning electron microscopy image (SEM) showing attachment and biofilm formation by *Escherichia coli* cells on lettuce leaf. Three week old lettuce plants were inoculated and allowed to grow for another week prior to imaging.

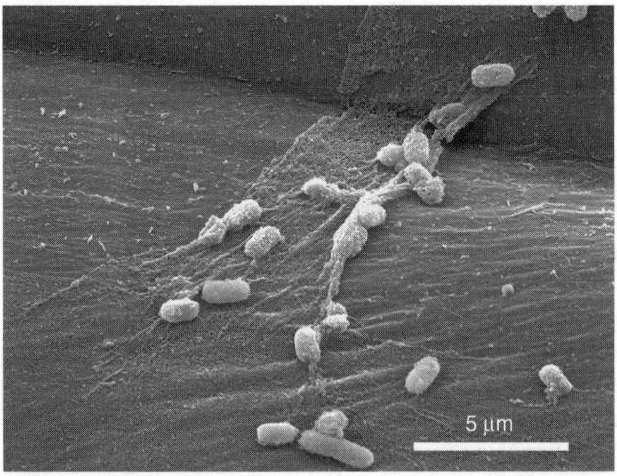


**Fig. 20.2** SEM of showing attachment and biofilm formation by *Shigella boydii* on parsley leaf. (Image is courtesy of Scott Robinson and Meredith Agle, University of Illinois.)

studies of apples, using water, detergents or sanitizing agents, produced a maximum of 3-log reduction in the levels of *E. coli* (Sapers *et al.*, 2000, 2002). However, these same washing treatments using a commercial flatbed brush washer gave less than a 1-log reduction in *E. coli* populations (Annous *et al.*, 2001). Survival of bacteria during washing was probably due to the attachment of *E. coli* cells to inaccessible regions of the apple (stem and calyx). Infiltration into the calyx channel or the core of the apple (Annous *et al.*, 2001; Fatemi *et al.*, 2006; Sapers *et al.*, 2002), and incorporation of bacterial cells within biofilms in inaccessible sites or on the surface of the apple (Annous *et al.*, 2001). Burnett *et al.* (2000) demonstrated that cells attached to subsurface structures were not inactivated by chlorine. Annous *et al.* (2001) reported that the majority of cells remaining in apples following washing were located in the stem and calyx areas. An SEM study of these areas demonstrated that *E. coli* cells were able to penetrate into the core of the apple and were able to form biofilms within the calyx region (Fig. 20.3) and on the stem (Fig. 20.4).

### 20.3.3 Cantaloupe melons

Six multistate outbreaks of salmonellosis have been traced to the consumption of cantaloupe since 1990. A survey conducted in response to a 1997 outbreak of *S. enterica* serovar Saphra indicated that approximately 5 percent of imported cantaloupes tested positive for *S. enterica* (USFDA, 2001b). Consumption of melons imported from Mexico was responsible for three successive outbreaks (2000–2002) of salmonellosis and prompted the USFDA to issue an import alert detaining all cantaloupes from Mexico offered for entry at US ports (USFDA, 2002).

A variety of sanitizers did not remove or inactivate *S. enterica* on cantaloupes (Annous *et al.*, 2005a; Ukuku and Sapers 2001). Furthermore, the

**Fig. 20.3** SEM showing attachment and biofilm formation by *Escherichia coli* O157:H7 cells in the calyx area of an inoculated Golden Delicious apple. Apples were inoculated and allowed to dry at 4°C for 72 hours prior to imaging. Note the extracellular matrix encapsulating cells.

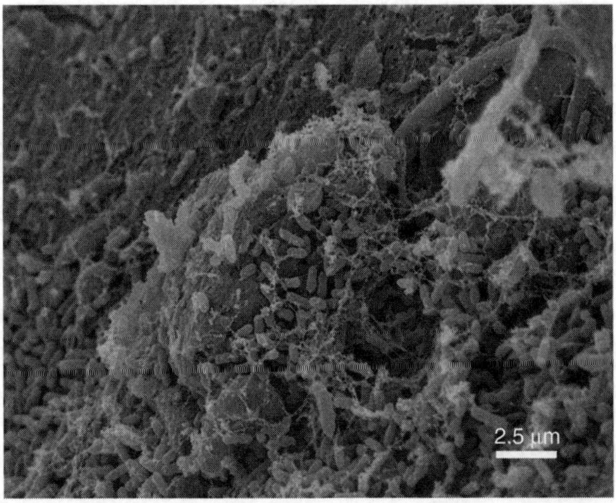


**Fig. 20.4** SEM showing attachment and biofilm formation by *Escherichia coli* O157:H7 cells on the stem of an inoculated Golden Delicious apple. Apples were inoculated and allowed to dry at 4°C for 72 hours prior to imaging. Note the extracellular matrix encapsulating cells.

efficacy of sanitizers applied to cantaloupes decreased significantly if the organism was allowed to reside on the rind surface for more than 24 hours suggesting that increased residence time led to the formation of a biofilm prior to application of the sanitizer (Annous *et al.*, 2004). Therefore, an investigation of the ability of *S. enterica* to form biofilms on cantaloupe surfaces was undertaken (Annous *et al.*, 2004, 2005b). Scanning electron microscopy (SEM) indicated that biofilm formation by *S. enterica* occurred rapidly following inoculation of cells onto the rind (Annous *et al.*, 2004, 2005b). Fibrillar material could be seen on the cantaloupe rind after just two hours following inoculation and drying at 20°C (Fig. 20.5). Once attached, *S. enterica* cells developed biofilms by growth and excretion of extracellular material following 24 hours of storage at either 10 or 20°C (Annous *et al.*, 2004, 2005b). Attachment and biofilm formation by *S. enterica* serovar Poona and serovar Michigan attached and formed biofilms inside the netting of an inoculated cantaloupe (Figs 20.6 and 20.7, respectively). Cell attachment to inaccessible sites such as netting of the rind along with biofilm formation is probably responsible for bacterial resistance to aqueous sanitizers (Annous *et al.*, 2004, 2005a).

These investigations into the interactions between enteric pathogens and plant tissues show that these pathogens form biofilms. It is probable that this phenomenon is primarily responsible for the consistent finding that aqueous sanitizers are not effective at inactivating human pathogens on plant tissues. Therefore, new strategies are needed to improve sanitizers

**Fig. 20.5** SEM showing attachment and initiation of biofilm formation by *Salmonella enterica* serovar Michigan cells inside the netting of inoculated cantaloupe. Cantaloupes were inoculated and allowed to dry at 20°C for two hours prior to imaging.

**Fig. 20.6** SEM showing attachment and biofilm formation by *Salmonella enterica* serovar Poona cells inside the netting of inoculated cantaloupe. Cantaloupes were inoculated and allowed to dry at 20°C for 24 hours prior to imaging. Note the extracellular matrix encapsulating cells.

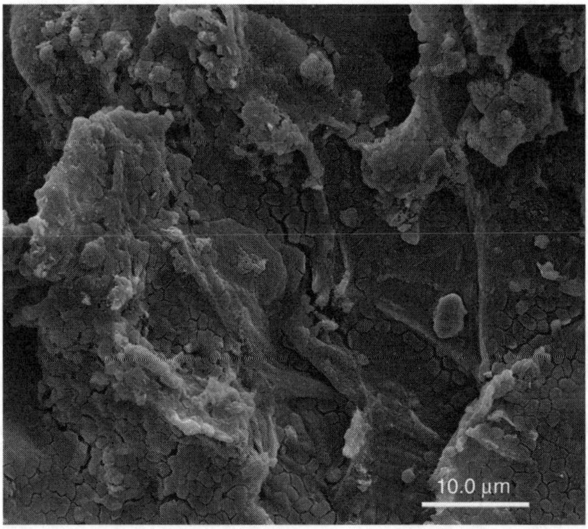


**Fig. 20.7** SEM showing attachment and biofilm formation by *Salmonella enterica* serovar Michigan cells inside the netting of inoculated cantaloupes following storage at 20°C for: 24 hours. Cells are visible arranged within the cracks on the netting; note the extracellular matrix encapsulating cells.

and/or physical processes in order to overcome the protective ability of the biofilm habitat.

## 20.4 Control of biofilms on produce, and research needs

Production and consumption of fruits and vegetables has increased in the United States, as well as the number of documented outbreaks of gastroenteritis associated with the consumption of contaminated fruits and vegetables. Brandl (2006) describes various factors associated with the increase in produce-related outbreaks. The formation of biofilms/aggregates, which may harbor different bacterial species, as well as pathogens, may protect the bacteria within the extracellular matrix of the aggregate. Investigations into the interactions between enteric pathogens and plant tissues have begun to document the ability of these pathogens to form biofilms (Annous *et al.*, 2005b; Brandl and Mandrell, 2002). Biofilm formation can protect bacteria from environmental stress, increase their resistance to antimicrobials, and enhance persistence on foods and solid surfaces. It is likely that this phenomenon is responsible for the consistent finding that aqueous sanitizers are ineffective in inactivating bacteria in plant tissues. Furthermore, pathogens are often protected against exposure to antimicrobials because they are found in subsurface areas of fruits and vegetables, for example in the inner spaces of leaves and inside apple calyces (Fatemi and Knabel, 2006). Therefore, new treatments that reach and inactivate pathogens in biofilms and in subsurface areas on produce must be developed.

New technologies that are being evaluated to kill pathogens within biofilms and/or internalized in the tissues of produce include the use of irradiation, cold plasma, hot water surface pasteurization, and biological controls (Niemira *et al.*, 2009). Cold plasma is a promising and relatively new nonthermal sanitizing technology in the field of food processing that can effectively reduce human pathogens on fresh produce. Cold plasma systems when applied to foods and inert surfaces have been reported to achieve 5-log reductions in populations of *E. coli*, *Staphylococcus aureus*, *B. subtillis, and Saccharomyces cerevisiae* (Niemira *et al.*, 2009).

Irradiation, a nonthermal intervention technology, has been approved in 2008 for use on fresh lettuce and fresh spinach to improve food safety and shelf life (USFDA, 2008). While an extensive body of research has demonstrated that this technology is safe and effective, research has recently begun to assess the ability of irradiation to inactivate pathogens in biofilm and/or internalized within the produce tissues. Penetrating power of irradiation has been reported to eliminate pathogens within biofilms and/or internalized within the tissues of fresh produce (Niemira, 2008) which are otherwise protected from the action of the conventional hypochloride wash treatment. However, irradiation has been reported to cause tissue softening and loss of some nutritional qualty (Fan *et al.*, 2008b). These quality losses due to irradiation treatments are reported to be minimized by combination with other sanitizers or techniques such as MAP, heat treatment, calcium infiltration and antibrowning agents (Niemira *et al.*, 2009).

Although, hot water or steam surface pasteurization has been shown to be an effective treatment in reducing levels of human pathogens on the surface of meat and poultry (Purnell *et al.*, 2004, Whyte *et al.*, 2003), intact eggs (Schuman *et al.*, 1997), and cantaloupes (Annous *et al.*, 2004; Sites *et al.*, 2008), it is not usually used in the fresh and fresh-cut produce industry. Traditionally, hot water surface pasteurization of fresh produce has been used to control insects and is the most effective method for destroying microorganisms, including postharvest plant pathogens that cause spoilage. Fresh fruits and vegetables investigated for surface pasteurization include apples, melons, mango, lemon, orange, cucumber, pear, tomato, and alfalfa seeds. Resistance of *S. enterica* to washing treatments with aqueous sanitizers was reported to be due to the attachment of *S. enterica* cells to inaccessible sites (net-like structure) on the rind of the cantaloupe and development of biofilm (Figs 20.5–20.7), thus avoiding contact with the washing solution (Annous *et al.*, 2004; 2005b). Hot water surface pasteurization treatment was reported to thermally reduce *S. enterica* populations in biofilm on the surface of inoculated cantaloupes by more than 5 logs (Annous *et al.*, 2004; Sites *et al.*, 2008; Solomen *et al.*, 2006). Therefore, hot water surface pasteurization of cantaloupes can decrease the risk of foodborne illnesses associated with this commodity as previously reported (Annous *et al.*, 2004; Solomon *et al.*, 2006). Fan *et al.* (2006, 2008a) reported that hot water surface pasteurization was able to significantly reduce total aerobic microorganisms on the surface of the rind and on the fresh-cut fruit prepared from the treated cantaloupes. Also, these researchers reported that the total aerobic microorganisms on the packaged fresh-cut fruit were significantly lower than the controls following seven days of storage at 4°C. These results demonstrate the utility of hot water for the inactivation of *Salmonella* on cantaloupes, extending the shelf life of the fresh and fresh-cut cantaloupes, and provide a framework to producers of fresh-cut melon for the potential use of hot water as an intervention treatment for enhancing the microbiological safety of this commodity.

Hot water treatments of a wide variety of fruits and vegetables can greatly improve microbiological qualities and increase shelf life, while maintaining sensory qualities of the produce. However, over processing the produce can result in thermal injury to apples and juice extracted from treated oranges, and significantly reduces seed germination. These adverse effects can be controlled by the process temperature and time. Since individual commodities have different thermal tolerance, the hot water immersion treatment should be tailored to each commodity. While the rind of a cantaloupe (Annous *et al.*, 2004) and the peel of an orange (Pao and Davis, 1999) effectively insulated the flesh from thermal damage at temperatures above 70°C, the peel of an apple did not protect the flesh from thermal damage at temperatures above 60°C (Sapers *et al.*, 2002). Thus, the tolerance to hot water immersion over a range of temperatures must be determined for individual commodities at different maturity stages (Lurie *et al.*, 1998).

New strategies which apply sanitizers, and physical processes, singly and in combination, are being investigated for their ability to overcome the protective ability of the biofilm habitat. Preventing contamination in the field and in processing plants and preventing bacterial growth, as well as the use of effective methods for removal or killing of pathogens on produce will lower the risk of human infections. Research to understand whether resistance of pathogens to specific antimicrobial treatments is enhanced when found in biofilms on produce is lacking. It is also important to understand the ecology of pathogens and microorganisms naturally found on produce to develop better intervention strategies.

## 20.5 Sources of further information and advice

http://www.ugacfs.org/research/BiofilmsResearch.html
http://www.biofilmsonline.com/cgi-bin/biofilmsonline/index.html
http://www.cdc.gov

## 20.6 References

AGLE ME (2003), '*Shigella boydii* 18: characterization and biofilm formation', PhD Thesis, Urbana Illinois, University of Illinois.

AGLE ME and BLASCHEK HP (2007), '*Shigella*: survival on produce and biofilm formation', In Blaschek HP, Wang HH and Agle ME, *Biofilms in the food environment*, Blackwell Publishing, Ames, Iowa, 19–45.

ALLEN AB (1985), 'Outbreak of campylobacteriosis in a large educational institution–British Columbia', *Can Dis Weekly Rep.* 2, 28–30.

ANDREWS JH and HARRIS RF (2000), 'The ecology and biogeography of microorganisms on plant surfaces', *Annu Rev Phytopathol.* 38, 145–180.

ANNOUS BA, SAPERS GM, MATTRAZZO AM and RIORDAN DCR (2001), Efficacy of washing with a commercial flatbed brush washer, using conventional and experimental washing agents, in reducing populations of *Escherichia coli* on artificially inoculated apples. *J Food Prot.* 64, 159–163.

ANNOUS BA, BURKE A and SITES JE (2004), Surface pasteurization of whole fresh cantaloupes inoculated with *Salmonella* Poona or *Escherichia coli*. *J Food Prot.* 67, 1876–1885.

ANNOUS BA, SAPERS GM, JONES DM and BURKE A (2005a), Improved recovery procedure for evaluation of sanitizer efficacy in disinfecting contaminated cantaloupes. *J Food Sci.* 70, M242–M247.

ANNOUS BA, SOLOMON EB, COOKE PH and BURKE A (2005b), Biofilm formation by *Salmonella* spp. on cantaloupe melons. *J Food Safety.* 25, 276–287.

AURELI P, FIORUCCI GC, CAROLI D, MARCHIARO G, NOVARA O, LEONE L and SALMASO S (2000), An outbreak of febrile gatroenteritis associated with corn contaminated by *Listeria monocytogenes*. *N Engl J Med.* 342(17), 1236–1241.

BRANDL MT (2006), Fitness of human enteric pathogens on plants and implications for food safety. *Annu Rev Phytopathol.* 44, 367–392.

BRANDL MT (2008), Plant lesions promote the rapid multiplication of *Escherichia coli* O157:H7 on postharvest lettuce. *Appl Environ Microbiol.* 74(17), 5285–5289.

BRANDL MT and MANDRELL RE (2002), Fitness of *Salmonella enterica* serovar Thompson in the cilantro phyllosphere. *Appl Environ Microbiol.* 68, 3614–3621.

BRANDL MT, HAXO AF, BATES AH and MANDRELL RE (2004), Comparison of survival of *Campylobacter jejuni* in the phyollosphere with that in the rhizosphere of spinach and radish plants. *Appl Environ Microbiol.* 70, 1182–1189.

BUCK JW, WALCOTT RR and BEUCHAT LR (2003), Recent trends in microbiological safety of fruits and vegetables. Plant Management Network. Available at: http://www.apsnet.org/online/feature/safety/Buck.pdf. Accessed March 10, 2009.

BURNETT SL, CHEN J and BEUCHAT LR (2000), Attachment of *Escherichia coli* O157:H7 to surfaces and internal structures of apples as detected by confocal scanning laser microscopy. *Appl Environ Microbiol.* 66, 4679–4687.

[CDC] CENTER FOR DISEASE CONTROL AND PREVENTION (1998), Outbreak of *Campylobacter* enteritis associated with cross-contamination of food Oklahoma, 1996. *Morb Mortal Wkly Rep.* 47, 129–131.

[CDC] CENTER FOR DISEASE CONTROL AND PREVENTION (2005), Outbreaks of *Salmonella* Infections Associated with eating roma tomatoes – United States and Canada, 2004. *Morb Mortal Wkly Rep.* 54(13), 325–328.

[CDC] CENTER FOR DISEASE CONTROL AND PREVENTION (2008), Outbreak of *Salmonella* serotype Saintpaul infections associated with multiple raw produce items – United States, 2008. *Morb Mortal Wkly Rep.* 57(34), 929–934.

CUMMINGS K, BARRETT E, MOHLE-BOETANI JC, BROOKS JT, FARRAR J, HUNT T, FIORE A, KOMATSU K, WERNER SB and SLUTSKER L (2001), A multistate outbreak of *Salmonella enterica* serotype Baildon associated with domestic raw tomatoes. *Emerg Infect Dis.* 7, 1046–1048.

DAVIS H, TALYOR JP, PERDUE JN, STELMA JR GN, HUMPHREYS JR JM, ROWNTREE III R and GREENE KD (1988), 'A shigellosis outbreak traced to commercially distributed shredded lettuce', *Am J Epidemiol.* 128(6), 1312–1321.

DEWAAL CS and BARLOW K (2002), 'Outbreak alert! Closing the gaps in our federal food-safety net', 5th ed. Center for Science in the Public Interest. Washington, DC.

DEWAAL CS and BHUIYA F (2007), 'Outbreak alert! Closing the gaps in our federal food-safety net', 9th ed. Center for Science in the Public Interest, Washington, DC, http://www.cspinet.org/foodsafety/outbreak_alert.pdf.

DONLAN RM and COSTERTON JW (2002), 'Biofilms: survival mechanisms of clinically relevant microorganisms', *Clin Microbiol Rev.* 15(2),167–193.

ELLS TC and HANSEN LT (2006), 'Strain and growth temperature influence *Listeria* spp. attachment to intact and cut cabbage', *Int J Food Microbiol.* 111, 34–42.

ERICKSON MC, ISLAM M, SHEPPARD C, LIAO J and DOYLE MP (2004), 'Reduction of *Escherichia coli* O157:H7 and *Salmonella* enterica serovar Enteritidis in chicken manure by larvae of the black soldier fly', *J Food Prot.* 67(4), 685–690.

FAN X, ANNOUS BA, SOKORAI KJB, BURKE A and MATTHEIS JP (2006), Combination of hot-water surface pasteurization of whole fruit and low-dose gamma irradiation of fresh-cut cantaloupe. *J Food Prot.* 69, 912–919.

FAN X, ANNOUS BA, BEAULIEU J and SITES J (2008a), Effect of hot water surface pasteurization of whole fruit on shelf life and quality of fresh-cut cantaloupes. *Journal Food Science.* 73(3), M91–98.

FAN X, NIEMIRA BA and PRAKASH A (2008b), Irradiation of fresh and fresh-cut fruits and vegetables. *Food Technol.* 3, 36–43.

FATEMI P and KNABEL SJ (2006), 'Evaluation of sanitizer penetration and its effect on destruction of *Escherichia coli* O157:H7 in Golden Delicious apples', *J Food Prot.* 69(3), 548–555

FATEMI P, LABORDE LF, PATTON J, SAPERS GM, ANNOUS BA and KNABEL SJ (2006), Influence of punctures, cuts, and surface morphologies of golden delicious apples on penetration and growth of *Escherichia coli* O157:H7. *J Food Prot.* 69, 267–275.

FETT WF (2000), 'Naturally occurring biofilms on alfalfa and other types of sprouts', *J Food Prot.* 63(5), 625–632.

FETT WF and COOKE PH (2003), Scanning electron microscopy of native biofilms on mung bean sprouts. *Can J Microbiol.* 49, 45–50.

GARRETT EH, GORNY JR, BEUCHAT LR, FARBER JN, HARRIS LJ, PARISH ME, SUSLOW TV and BUSTA EF (2003), Microbiological safety of fresh and fresh-cut produce: description of the situation and economic impact. *Comprehensive Reviews in Food Science and Food Safety*, Volume 2 (Supplement), 13–19.

GUNTHER, NW 4th and CHEN CY (2009), 'The biofilm forming potential of bacterial species in the genus *Campylobacter*', *Food Microbiol.* 26(1), 44–51.

HO JL, SHANDS KN, FRIEDLAND G, ECKIND P and FRASER DW (1986), 'An outbreak of type 4b *Listeria monocytogenes* infection involving patients from eight Boston hospitals', *Arch Intern Med.* 146(3), 520–524.

ISLAM M, DOYLE MP, PHATAK SC, MILLNER P and JIANG X (2004a), 'Persistence of enterohemorrhagic *Escherichia coli* O157:H7 in soil and on leaf lettuce and parsley grown in fields treated with contaminated manure composts or irrigation water', *J Food Prot.* 67(7), 1365–1370.

ISLAM M, MORGAN J, DOYLE MP and JIANG X (2004b), 'Fate of *Escherichia coli* O157:H7 in manure compost-amended soil and on carrots and onions grown in an environmentally controlled growth chamber', *J Food Prot.* 67(3), 574–578.

ISLAM M, MORGAN J, DOYLE MP, PHATAK SC, MILLNER P and JIANG X (2004c), 'Fate of *Salmonella* enterica serovar Typhimurium on carrots and radishes grown in fields treated with contaminated manure composts or irrigation water', *Appl Environ Microbiol.* 70(4), 2497–2502.

ISLAM M, MORGAN J, DOYLE MP, PHATAK SC, MILLNER P and JIANG X (2004d), 'Persistence of Salmonella enterica serovar typhimurium on lettuce and parsley and in soils on which they were grown in fields treated with contaminated manure composts or irrigation water', *Foodborne Pathog Dis.* 1(1), 27–35.

JIANG X, ISLAM M, MORGAN J and DOYLE MP (2004), 'Fate of *Listeria monocytogenes* in bovie manure-amended soil', *J Food Prot.* 67(8), 1676–1681.

KAPPERUD G, RORVIK LM, HASSELTVEDT V, HOIBY EA, IVERSEN BG, STAVELAND K, JOHNSEN G, LEITAO J, HRIKSTAD H, ANDERSSON Y, LANGELAND G, GONDROSEN B and LASSEN J (1995), 'Outbreak of *Shigella sonnei* infection traced to imported iceberg lettuce. *J Clin Microbiol.* 33(3), 609–614.

KESKINEN LA and ANNOUS BA (2009), Efficacy of chlorine, acidic electrolyzed water and aqueous chlorine dioxide solutions to decontaminate *Escherichia coli* O157:H7 from lettuce leaves. *Int J Food Microbiol.* (In press).

LINDOW SE and BRANDL MT (2003), 'Microbiology of the phyllosphere', *Appl Environ Microbiol.* 69 (4), 1875–1883.

LINDSAY D, BRÖZEL VS and VON HOLY A (2006), 'Biofilm-spore response in *Bacillus cereus* and *Bacillus subtilus* during nutrient limitation', *J Food Prot.* 69(5) 1168–1172.

LIU L, TAN S, JUN W, SMITH A, MENG J and BHAGWAT AA (2009), 'Osmoregulated periplasmic glucans are needed for competitive growth and biofilm formation by *Salmonella* enterica serovar Typhimurium in leafy-green vegetable wash waters and colonization in mice', *FEMS Microbiol Lett.* 292(1), 13–20.

LURIE S, FALLIK E, KLEIN JD, KOZAR F and KOVACS K (1998), Postharvest heat treatment of apples to control San Jose Scale (Quadraspidiotus perniciosus Comstock) and blue mold (Penicillium expansum link) and maintain fruit firmness, *J Am Soc Hort Sci.* 123, 110.

MARTIN DL, GUSTAFSON TL, PELOSI JW, SUAREZ L and PIERCE GV (1986), 'Contaminated produce: a common source for two outbreaks of *Shigella* gastroenteritis', *Am J Epidemiol.* 124(2), 299–305.

MATTHYSSE AG, DEORA R, MISHRA M and TORRES AG (2008), 'Polysaccharides cellulose, poly-B-1,6-*N*-acetyl-D-glucosamine, and colanic acid are required for optimal binding of *Escherichia coli* O157:H7 strains to alfalfa sprouts and K-12 strains to plastic but not for binding to epithelial cells', *Appl Environ Microbiol.* 74, 2384–2390.

MONIER JM and LINDOW SE (2003), 'Differential survival of solitary and aggregated bacterial cells promotes aggregate formation on leaf surfaces', *Proc Natl Acad Sci USA.* 100(26), 15977–15982.

MORRIS CE and MONIER J-M (2003), 'The ecological significance of biofilm formation by plant-associated bacteria', *Annu Rev Phytopathol.* 41, 429–453.

MORRIS CE, MONIER J-M and JACQUES M-A (1997), Methods for observing microbial biofilms directly on leaf surfaces and recovering them for isolation of culturable microorganisms. *Appl Environ Microbiol.* 63, 1570–1576.

NAIMI TS, WICKLUND JH, OLSEN SJ, KRAUSE G, WELLS JG, BARTKUS JM, BOXRUD DJ, SULLIVAN M, KASSENBORG H, BESSER JM, MINTZ ED, OSTERHOLM MT and HEDBERG CW (2003), Concurrent outbreaks of *Shigella sonnei* and enterotoxigenic *Escherichia coli* infections associated with parsley: implications for surveillance and control of foodborne illness. *J Food Prot.* 66, 535–541.

NIEMIRA BA (2008), 'Irradiation compared with chlorination for elimination of *Escherichia coli* O157:H7 internalized in lettuce leaves: influence of lettuce variety', *J Food Sci.* 73(5), M208–M213.

NIEMIRA BA, ANNOUS BA, FAN X, LIAO CH and SITES JE (2009), Inactivation of Microbial Contaminants in Fresh Produce. In Al-Taher F, Jackson L and DeVries J, *Intentional and Unintentional Food Contaminants*. ACS Press, Washington DC (In press).

PAO S and DAVIS CL (1999), Enhancing microbiological safety of fresh orange juice by fruit immersion in hot water and chemical sanitizers, *J Food Prot.* 62, 756.

PARK CE and SANDERS GW (1992), 'Occurrence of thermotolerant campylobacters in fresh vegetables sold at farmers' markets and supermarkets', *Can J Microbiol.* 38(4), 313–316.

PORTNOY BL, GOEPFERT JM and HARMON BM (1976), 'An outbreak of *Bacillus cereus* food poisoning resulting from contaminated vegetable sprouts', *Am J Epidemiol.* 103(6), 589–594.

PURNELL G, MATTICK K and HUMPHREY T (2004), The use of 'hot wash' treatments to reduce the number of pathogenic and spoilage bacteria on raw retail poultry, *J Food Eng.* 62, 29.

RAMEY BE, KOUTSOUDIS M, VON BODMAN SB and FUQUA C (2004), 'Biofilm formation in plant-microbe associations', *Curr Opin Microbiol.* 7, 602–609.

RANGEL JM, SPARLING PH, CROWE C, GRIFFIN PM and SWERDLOW DL (2005), Epidemiology of *Escherichia coli* outbreaks, United States, 1982–2002. *Emerg Infect Dis.* 11, 603–609.

RIGANO LA, SICILIANO F, ENRIQUE R, SENDIN L, FILLIPPONE P, TORRES PS, QÜESTA J, DOW JM, CASTAGNARO AP, VOJNOV AA and MARANO MR (2007), 'Biofilm formation, epiphytic fitness, and canker development in *Xanthomonas axonopodis* pv. *citri*', *Mol Plant Microbe Interact.* 20(10), 1222–1230.

SANDERS SQ, BOOTHE DH, FRANK JF and ARNOLD JW (2007), 'Culture and detection of *Campylobacter jejuni* within mixed microbial populations of biofilms on stainless steel', *J Food Prot.* 70(6), 1379–1385.

SAPERS GM (2005), Washing and sanitizing treatments for fruits and vegetables. In Sapers GM, Gorny JR and Yousef AE, *Microbiology of fruits and vegetables.* Boca Raton, FL, CRC Taylor & Francis, 376–387.

SAPERS GM, MILLER RL, JANTSCHKE M and MATTRAZZO AM (2000), Factors limiting the efficacy of hydrogen peroxide washes for decontamination of apples containing *Escherichia coli. J Food Sci.* 65, 529–532.

SAPERS GM, MILLER RL, ANNOUS BA and BURKE AM (2002), Improved antimicrobial wash treatments for decontamination of apples. *J Food Sci.* 67(5), 1886–1891.

SCHLECH WF 3rd, LAVINGNE PM, BORTOLUSSI RA, ALLEN AC, HALDANE EV, WORT AJ, HIGHTOWER AW, JOHNSON SE, KING SH, NICHOLLS ES and BROOME CV (1983), 'Epidemic listeriosis-evidence for transmission by food', *N Engl J Med.* 308(4), 203–206.

SCHUMAN JD, SHELDON BW, VANDEPOPULIERE JM and BALL JR HR (1997), Immersion heat treatments for inactivation of Salmonella enteritidis with intact eggs, *J Appl Microbiol.* 83, 438.

SEO KH and FRANK JF (1999), Attachment of *Escherichia coli* O157:H7 to lettuce leaf surface and bacterial viability in response to chlorine treatment as demonstrated by using confocal scanning laser microscopy. *J Food Prot.* 62, 3–9.

SHAW RK, BERGER, CN, FEYS B, KNUTTON S, PALLEN MJ and FRANKEL G (2008), 'Enterohemorrhagic *Escherichia coli* exploits EspA filaments for attachment to salad leaves', *Appl Environ Microbiol.* 74(9), 2908–2914.

SHI X, NAMVAR A and KOSTRZYNSKA M (2007), 'Persistence and growth of different *Salmonella* serovars on pre- and postharvest tomatoes', *J Food Prot.* 70, 2725–2731.

SITES JE, ANNOUS BA, WALKER PN and BURKE AM (2008), Development and validation of a pilot scale enhanced biosafety level two containment for performance evaluation of produce disinfection technologies. *Appl Biosafety.* 13(1), 30–44.

SIVAPALASINGAM S, FRIEDMAN CR, COHEN L and TAUXE RV (2004), 'Fresh produce: a growing cause of outbreaks of foodborne illness in the United States, 1973 through 1997' *J Food Prot.* 67, 2342–2353.

SOLOMON EB, YARON S and MATTHEWS KR (2002), Transmission of *Escherichia coli* O157:H7 from contaminated manure and irrigation water to lettuce plant tissue and its subsequent internalization. *Appl Envion Microbiol.* 68, 397–400.

SOLOMON EB, HUANG L, SITES JE and ANNOUS BA (2006), Thermal inactivation of salmonella on cantaloupes using hot water. *J Food Sci.* 71(2), M25–M30.

SOLOMON HM and KAUTTER DA (1988), 'Outgrowth and toxin production by *Clostridium botulinum* in bottled chopped garlic', *J Food Prot.* 51(11), 862–865.

SOLOMON HM, KAUTTER DA, LILLY T and RHODEHAMEL EJ (1990), 'Outgrowth of *Clostridium botulinum* in shredded cabbage at room temperature under modified atmosphere', *J Food Prot.* 53(10), 831–833.

TAKEUCHI K and FRANK JF (2000), Penetration of *Escherichia coli* O157:H7 into lettuce tissues as affected by inoculum size and temperature and the effect of chlorine treatment on cell viability. *J Food Prot.* 63, 434–440.

TORRES AG, JETER C, LANGLEY W and MATTHYSSE AG (2005), 'Differential binding of *Escherichia coli* O157:H7 to alfalfa, human epithelial cells, and plastic is mediated by a variety of surface structures', *Appl Environ Microbiol.* (12), 8008–8015.

UHLICH GA, SINCLAIR JR, WARREN NG, CHMIELECKI WA and FRATAMICO P (2008), 'Characterizaion of Shiga toxin-producing *Escherichia coli* isolates associated with two multistate foodborne outbreaks that occurred in 2006'. *Appl Environ Microbiol.* 74(4), 1268–1272.

UKUKU DO and SAPERS GM (2001), Effect of sanitizer treatments on *Salmonella* Stanley attached to the surface of cantaloupe and cell transfer to fresh-cut tissues during cutting practices. *J Food Prot.* 64, 1286–1291.

[USFDA] US FOOD AND DRUG ADMINISTRATION (2001a), *Escherichia coli* O157:H7. Bad Bug Book. Foodborne Pathogenic Microorganisms and Natural Toxins Handbook. CFSAN, US Food and Drug Administration. Available at: http://www.cfsan.fda.gov/~mow/chap15.html. Accessed May 11, 2009.

[USFDA] US FOOD AND DRUG ADMINISTRATION (2001b), FDA survey of imported fresh produce. US Food and Drug Administration. Center for Food Safety and Applied Nutrition; Office of Plant and Dairy Foods and Beverages. Available at: http://www.cfsan.fda.gov/~dms/prodsur6.html. Accessed June 18, 2008.

[USFDA] US FOOD AND DRUG ADMINISTRATION (2002), FDA issue import alert on cantaloupes from Mexico. US Food and Drug Administration. Office of Public Affairs. Available at: http://www.fda.gov/bbs/topics/ANSWERS/2002/ANS01167.html. Accessed June 18, 2008.

[USFDA] US FOOD AND DRUG ADMINISTRATION (2005), Letter to California Firms that Grow, Pack, Process, or Ship Fresh and Fresh-cut Lettuce. Available at: http://www.cfsan.fda.gov/~dms/prodltr2.html. Accessed March 10, 2009.

[USFDA] US FOOD AND DRUG ADMINISTRATION (2008), Irradiation in the Production, Processing and Handling of Food (Final Rule). Available at: http://www.cfsan.fda.gov/~lrd/fr080822.html. Accessed March 11, 2009.

WENDEL AM, HOANG JOHNSON DH, SHARAPOV U, GRANT J, ARCHER JR, MONSON T, KOSCHMANN C and DAVIS JP (2009), 'Multistate outbreak of *Escherichia coli* O157:H7 Infection Associated with Consumption of Packaged Spinach, August-September 2006: The Wisconsin investigation', *Clin Infect Dis.* 48(8), 1079–1086.

WHYTE P, MCGILL K and COLLINS JD (2003), An assessment of steam pasteurization and hot water immersion treatments for the microbiological decontamination of broiler carcasses, *Food Microbiol.* 20, 111.

# Part V

# Appendix

# 21

# Sampling and quantification of biofilms in food processing and other environments

**D. E. Nivens, and B. M. Co, Purdue University, USA and**
**M. J. Franklin, The Center for Biofilm Engineering, USA**

**Abstract**: In the food industry, assessment of food contact surfaces is necessary to determine whether equipment is properly cleaned and/or sanitized and whether living problematic microorganisms are present. Existing quantitative detection technologies are limited by the inability to directly detect living cells in sporadically dispersed biofilms on large surface areas. Thus, precise and accurate sampling strategies must be coupled with detection technology. This chapter discusses sampling methods and standard (e.g., plating and ATP-bioluminescence) and emerging (e.g., spectrometry, immunosensor, and nucleic acid-based) quantitative techniques to detect biofilms on food contact surfaces with a survey of function, analytical performance, and limitations.

**Key words**: biofilm, sampling methods, spectrometry, immunosensors, nucleic acid-based.

## 21.1 Introduction

Definitions of biofilms vary but most investigators agree that biofilms usually consist of four main components: (i) water, (ii) the living microbial population, (iii) the surface to which the cells are adhered, and (iv) the associated extracellular matrix, which can consist of secreted polymers (2, 17, 20, 107). These polymers, usually polysaccharides and/or protein, allow the cells to form three-dimensional structures but are not required for biofilms to proliferate on a surface (79). Biofilms can be contiguous, but are more likely to form networks of micro-colonies and yield patchy distributions (Fig. 21.1). Microorganisms are thought to contaminate 'clean' surfaces by the following sequential steps: transfer to the surface, attachment

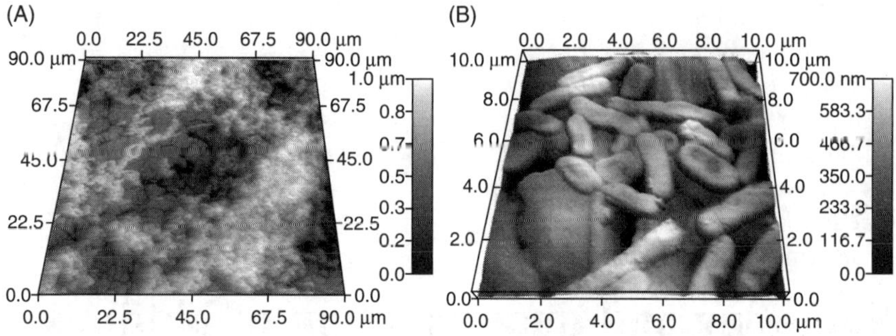


**Fig. 21.1** 3D z-sensor AFM image of air dried *Escherichia coli* O157:H7 biofilm (1-hour inoculation in $10^7$ cells/ml and 8-hour biofilm development in minimal salts media) on stainless steel surface. Imaged in air with AC240 cantilever in AC mode. A) 90 µm × 90 µm and B) 10 µm × 10 µm.

or entrapment on the surface, and biofilm development. Sources of the initial colonizing microorganisms are air, the food product, human contact, rinse and cleaning solutions, or an up-stream surface in sequential step food processes.

Attachment occurs when the forces associated with surfaces (electrostatic, van der Waals, hydrogen bonding, etc.) overcome forces associated with transport (e.g., convection). This process can also occur in protective microenvironments such as scratches, grooves, or interfaces between different materials. Development of biofilms occurs when the adhered cell(s) have proper environmental conditions (e.g., temperature and humidity) and suitable source of nutrients to grow. Once colonized on surfaces, biofilms can live in single or multi species microenvironments and through detachment become a source of contamination for other sites.

On food processing equipment, detection assays are necessary to provide evidence mainly for (i) determining that living problematic microorganisms (e.g., pathogens) are not present and/or (ii) determining whether equipment is properly cleaned and/or sanitized. Idealized monitoring technologies should provide automated, continuous, rapid, inexpensive, and rugged operation. Other desirable attributes for the idealized detection system include the ability to scan large surface areas to accurately quantify and to selectively identify all potential problematic microorganisms without sample preparation. Obviously, this idealized technology does not exist; however, progress has been made towards the development of technology that can achieve some on these goals. One of the major limitations of existing methodology is the lack of techniques that can be used to directly detect living cells in sporadically dispersed biofilms on large surface areas associated with food processing equipment. To circumvent this problem, many food industries have incorporated sampling techniques, such as rinsing and/or swabbing, as a step prior to detection and analysis.

Precise and accurate sampling of biofilms requires cells to be efficiently removed, collected, and prepared for the analytical detection method of choice. To quantitatively remove biofilms from microenvironments, the sampling technique should provide enough force to overcome the intermolecular forces that secure the biofilm matrix or cells within the biofilm to the surface. In sampling, biofilm removal is usually accomplished using shear forces. Removal can be achieved by providing convectional forces from rinsing and/or contact shear forces from methods such as swabs or sponges. For many assays, the goal of sampling is to remove living cells under conditions that avoid cell death in order to obtain an accurate assessment of the number and type of living microorganism present. Chemicals such as detergents can be added to sampling solutions to facilitate removal; however, in many cases, these chemical treatments can kill cells, thereby systematically biasing data. After removal from the surface, the cells or fragments from the biofilm must be removed from any sampling materials and quantitatively transferred to a sterile/clean and sometimes enzyme free container for analysis. Some assays require enrichment in a growth medium to increase the number of cells to a concentration above the minimum detectable quantity and reduce the interferences from sample matrices. Although necessary, enrichments can bias quantitative results since all growth media has a certain degree of selectivity, especially in multiplexed analysis schemes; and accurate models of growth are rarely determined or applied.

Recently, advances in bio-analytical chemistry and molecular microbiology are providing a new understanding of quantitative microbiology; and new techniques are being developed for the food industry. These efforts have led to the development of novel immunosensor, immunoassay, and nucleic acid detection technologies that allow more rapid and selective data to be obtained. This chapter will discuss sampling methods as well as standard (widely utilized in the food industry) and emerging/non-routine bioanalytical techniques (novel and not widely utilized) that can be coupled with these sampling technologies and detect the presence of problematic species within biofilms on food contact materials (Fig. 21.2). The chapter is not intended to be an exhaustive review of quantitative techniques applied to sampled biofilms, but instead a survey of techniques with highlights of key recent work and reviews in the hopes of providing an overall sense of the function, analytical performance, limitations, and scope of standard and state-of-the-art detection/identification platforms.

## 21.2 Sampling surfaces

On food contact surfaces, the US Public Health Service recommends no more than 100 colony forming units (CFUs) per 50 cm$^2$ sampled (28). However, in most cases the type of microorganism is more important than

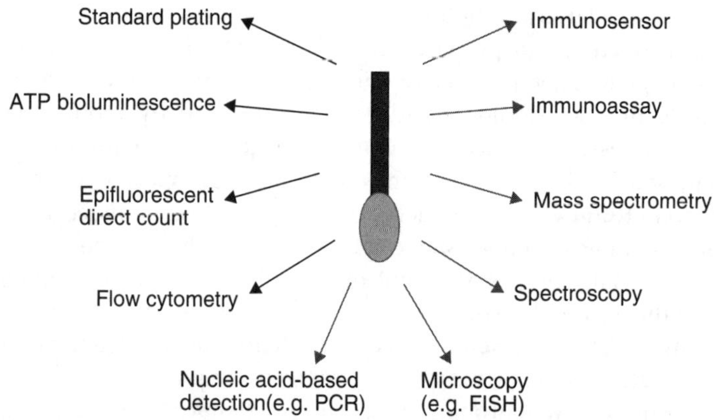


**Fig. 21.2** Standard (black) and emerging detection technologies (gray).

the number. Sampling and testing are the best means to monitor microbial activity on process surfaces; and for meaningful results, the appropriate sampling strategy for examining surfaces needs to be selected. Which sampling strategy used is dependent on the sampling surface material, surface structure, location, and expected microbial contamination level (109). Some sampling strategies include random, representative, selective, and convenience sampling (56). Food processing companies should develop a sampling program to selectively test identified hazards and risks at critical control points (CCPs) within their Hazard Analysis Critical Control Point (HACCP) system (28). Assessment should be performed at all sites with the potential to harbor microorganisms that may directly or indirectly contaminate the product. For instance, sampling should occur at points where there is higher risk of residual moisture and also at points that are representative of the production equipment. Thus, focus should not only be on direct food contact surfaces because microorganisms can be transferred from indirect contact surfaces to direct contact surfaces. Direct contact surfaces include the inside of pipes, conveyors, storage vessels, fillers, and mixers; and indirect contact surfaces include the outside of equipment, walls, floors, and tools. Sampling should always be performed sterilely to avoid contamination of the sample. Moreover, the sample should be properly labeled, stored, and transported and quickly analyzed after being collected (13).

Sampling methods can be qualitative or quantitative. Qualitative sampling detects the presence or absence of microorganisms at numbers above a given threshold; whereas, quantitative sampling determines the number of microorganisms on the sampled surface. Sampling methods inherently increase random error which decreases precision and systematic errors which cause deviations from true values (56). Errors in sampling are associated with the removal, recovery, and dispersal of living microorganisms within a sampled biofilm isolated from a processing surface. In quantitative sampling, approximately 95% of the replicate results should lie within plus or minus two standard deviations of the mean (59).

There are two main sampling methods that have been found acceptable: indirect and direct sampling (13, 28). In indirect surface sampling, the most common method is the rinse test. The rinse method can be used to sample containers and processing equipment systems such as tanks, pipelines, and fillers or areas inaccessible to direct sampling. The test involves rinsing a large area of the surface and collecting the rinse water. Water used for rinsing should be heat sterilized, filter sterilized, or chlorinated followed by neutralization (28). The rinse water is usually added upstream of the process and collected downstream at various points in the process for sampling. The number of CFUs is then obtained from the rinse samples. Indirect sampling has the advantage of sampling a larger surface area and areas inaccessible without disassembling equipment. The disadvantages of indirect sampling are that it may not possibly provide enough shear force to remove biofilms physically attached to the surface and that large rinsing volumes yield high material and disposal costs (13).

Microbial assessment of contact surfaces is generally tested by direct contact methods. In direct surface sampling, the most common method is the swab test. The swab contact method involves using a sterile nonabsorbent swab (e.g., cotton, calcium alginate, Dacron, or rayon) to sample the surface (13, 27, 28). The swab is first moistened with rinse solution and then rubbed slowly and thoroughly over the surface to be sampled with rinsing after each swab. Generally, swabs are used to wipe a defined area (e.g., 25 $cm^2$) of the surface and can be used for surfaces with cracks, corners, or crevices. Flexible stencils made of inert materials like Teflon® or silicon can be used to mark the area to be sampled. Since sampling is mechanical, attached biofilms can be sampled. After sampling, the swab is broken or cut and collected in small vials containing buffered rinse solution with appropriate neutralizers. The sample can be used for both qualitative and quantitative analyses. For quantitative analysis, accurate enumeration depends greatly on the ability of the swab to remove microorganisms from the surface, the release of the microorganisms from the swab into an enrichment media. The type of swab and media used for sampling greatly affects the number of microorganisms detected (74). Underestimation may be due to microorganisms becoming injured during swabbing (73). Other disadvantages of swabbing include a small sample area and low reproducibility and repeatability due to variation among sampling technique such as amount of pressure applied when rubbing (86). For instance, applying too little pressure may cause microorganisms to remain adhered to the surface and applying too much pressure can damage the swab (98). Self-contained media-based swabs include a semi-solid selective agar culture tube for microbial detection (73). These swabs have been found to be more sensitive than traditional swabbing techniques but only give qualitative data. Sampled areas should be carefully cleaned after use with 70% alcohol to prevent agar residue from promoting growth (13).

An alternative to swabs are sponges, which can sample large surface areas and identify areas that harbor pathogens for better control. The sponge contact method involves using sterile cellulose or polyurethane sponges free of antimicrobial preservatives (28). When sampling, sterile tongs or gloves should be used to hold the sponge. The sponge is moistened with rinse solution and rubbed over the surface (up to several meters). After sampling, the sponge is placed in a sterile plastic bag. The sponge can be introduced directly into enrichment broth or diluent can be added to the bag. Before analysis the sponge should be massaged in the diluent to release the microorganisms.

Another commonly used method involves using solid culture media in the form of a RODAC plate (Replicate Organism Detection and Counting), which consists of a plastic dish containing convex culture medium (6). The RODAC plate (agar contact) method is a simple method for sampling flat surfaces that have been previously cleaned or sanitized. It should not be used for crevices, curved and irregular surfaces, difficult to access areas, or heavily contaminated areas. During sampling, a representative and random number of sites should be obtained (101). The RODAC plate should be free of condensation and pressed against the surface so that the entire agar meniscus contacts the surface for 5–10 seconds (28). Applied pressure and contact time need to be optimized. This method is more sensitive than swabbing when the surface density of attached cells on surfaces is low with a recovery rate of 80% versus 1%, respectively (86). When surface density is high, the RODAC plate method will underestimate the number.

Commercially available alternatives to RODAC plates are PetriFilm™ (2M medical-Surgical Division, St. Paul, Minn.), Con-Tact-It (Birko Chemical Corp., Dever, Colo.), and dip slides (BIOSAN Laboratories, Inc., Warren, Mich.) (28, 73). The Petrifilm™ aerobic count or direct-contact method is a simple way to sample flat or curved surfaces but should not be used for surfaces with crevices or cracks. Plates come sealed in foiled pouches and need to be prehydrated before use with sterile water. When sampling, the top film with the gel should be lifted and pressed against the sample surface. Then the top film is placed back onto the bottom film and incubated. The Con-Tact-it® method is also a simple method to sample surfaces. When sampling, a fresh piece of tape is removed from the dispenser and pressed onto the surface to be tested. Then it is pressed onto the Con-Tact-it Petri plate of selective media, leaving one zone as a control, and incubated. Dipslides, which consist of nutrient agar on a sampling paddle, are another method to sample surfaces. When sampling, the contact slide is removed from its tube and pressed against the surface to be tested. Then the slide is turned over and pressed again against the surface. The slide is returned to the tube and incubated for 24–48 hours. Contamination levels are assessed by using a comparison chart provided in the commercially available kits.

Given all sampling techniques have limitations that bias results, it is recommended to compare results from more than one method to obtain representative data (98). Good hygienic design of equipment should minimize problematic sites (e.g., joints, pipe corners, gaskets, and o-rings, and fasteners) where contamination may occur. Furthermore, materials used in food processing should withstand a wide range of temperatures and be durable so that the surface is free of cracks, scratches, and pits, and most importantly microorganisms (66).

## 21.3 Quantitative detection technologies for sampling

### 21.3.1 Standard detection technologies

*Plating assay*

Culture-based methods are traditional and still useful approaches for identification of bacteria isolated from food-processing surfaces due to the ease of use and low cost. These methods typically involve three steps: (i) growth of microorganisms in nonselective and selective enrichment to greater than detectable limits (detection), (ii) obtaining a pure culture of the targeted microorganism (isolation), and (iii) identifying the targeted microorganism and applying a confirmatory assay (identification and confirmation) (1). Given their reliability, they generally are considered the 'gold standard' for microbiological analysis (8). For nonselective enrichment, methods such as the aerobic plate count (APC) or standard plate count (SPC) are used to detect viable microorganisms (42). Selective enrichment and differentiating media are then used to test for desired species of bacteria and to prevent overestimating by accounting for harmless bacteria. Isolated colonies are often further characterized and/or identified using bacterial identification based-systems such as test kits (e.g., API 20E and API Listeria test strips) or nucleic acid-based assays to confirm the microorganism detected is correctly identified. Recently, emerging light-scattering technology using a laser (BARDOT) has been shown to nondestructively detect and distinguish bacterial species growing on a Petri dish with high specificity in seconds (8). The technique, however, is limited by the inability to detect microorganisms on membrane filters and the lack of defined scatter signatures for transparent colonies. Other promising emerging confirmatory assays are: mass spectrometry (18, 106), vibrational spectrometry (49, 50, 61, 77), and hyperspectral imaging (39, 71), which are discussed later.

A major drawback of culture-based methods is that only a small percentage of viable bacteria obtained from any environmental sample are culturable on a given medium. Conventional techniques recover less than 1% of the total species in a sample (70). Compounding the problem, injured or stressed bacteria may need time to recover and increase to detectable limits; and the lack of cultivability may be due to a variety of factors such as inappropriate growth medium or growth conditions, including concentrations of oxygen or carbon dioxide that may not be optimal for the bacteria. Many bacteria from natural environments are in what has been termed a 'viable but nonculturable' or more recently referred to as an 'active but not culturable' state (68). These bacteria, although still active in the environment and potentially problematic in terms of food spoilage or sources of food poisoning, will not grow immediately on a selected culture medium. Another issue affecting accuracy is cell aggregation or clumping, which results in an underestimation of the number of bacteria present. Other disadvantages of conventional cultivation are that it is time-consuming, labor intensive, and may take days before results are obtained, particularly when the sample is a complex of microbial species.

*ATP-bioluminescence assay*

Adenosine triphosphate (ATP) bioluminescence measurement is a commonly used method for monitoring the presence of surface contamination (43). The technique determines the amount of total organic soil on surfaces based on ATP content (86). Luciferase and luciferin are added to the sample in a magnesium-containing buffer. The enzyme luciferase uses the chemical energy in ATP to drive the oxidation of luciferin to oxyluciferin. For each ATP, one photon of light is emitted and detected with a luminometer within 15 minutes (42, 47). However, since ATP is within all actively metabolizing cells, the method does not differentiate between microorganisms on the surface and ATP present in food residues. Additional steps need to be performed to segregate the microbial ATP, which are laborious and time consuming. Drawbacks of the ATP-bioluminescence assay are the lack of selectivity and dependence of the results on cell physiology. The limit of detection is $10^3$–$10^4$ CFU/ml (86).

*Epifluorescent direct count*

Direct count (DC) methods using standard optical microscopy offer an excellent method to visualize and enumerate extracted remnants of biofilms. The methods detect all types of cells including: living, 'viable but nonculturable', and intact dead cells. Furthermore, visualization of the sample can identify systematic errors associated with cell aggregates or particles. To improve accuracy, the sample can be treated with a detergent or other dispersing agents to separate cells prior to visualization.

Although microorganisms can be directly visualized, stains such as acridine orange (12, 90) or 4′,6-diamindino-2-phenylindole (DAPI) (68, 90) are commonly used to enhance visualization. For example, acridine orange binds to nucleic acids and when exposed to ultraviolet (UV) light, causes the cells to fluoresce in the green and orange wavelengths. To improve the limit of detection, the sample can be filtered onto commercially available stained black 0.2 micrometer pore size filter membranes. Another epifluorescence method involves binding an antibody with a fluorescent label to targeted microorganisms for visualization. Direct counting methods have excellent limit of detection of approximately $10^3$ microorganism/ml and provide great insight into sample consistency and make-up of biofilms. However, the approach is labor intensive, has difficulties differentiating live and dead cells, and cannot accurately determine the number of counts from clumped or particle associated biofilms.

*Flow and chemical cytometry*

Cytometry is a cell counting method that measures physicochemical properties of single cells. In flow cytometry, cells from a liquid suspension are hydrodynamically focused onto a small column with internal dimensions similar to the size of a single cell and are passed one by one through a detection zone that measures changes in optical or electrical properties. Cytometers can also be used to nondestructively sort cells to be used in further studies (19). Optical parameters that are measured include absorbance, light scattering, and fluorescence techniques. For instance, light scattering techniques can be used to determine the size of the particle passing through the detection zone; however, bacteria can be difficult to discern from some culture debris since they are similar in size. Thus, better analytical performance is obtained with larger cells such as mammalian cells. To increase selectivity, microorganisms can be stained or tagged with fluorescence antibodies for counting (110). Multiple tags can be used to detect and quantify multiple targeted microorganisms in a single sample suspension. The technique for homogeneous microbial populations has a limit of detection of $<10^3$ CFU/ml (24). The technique is useful for on-line monitoring to obtain a population distribution of a heterogeneous sample; but it also may require enrichment and use of sample preparation steps to ensure that only single cells are present in the analytical sample. In addition, cytometry use has been limited due to its cost and complexity with the need for trained operators to properly handle samples and waste (24, 68, 100).

Chemical cytometry detects chemical biomarkers obtained from a lysed cell. Biomarkers that can be analyzed are nucleic acids, proteins, and metabolites (19). The technique is currently slow and destroys the cells, but offers the capability of quantitative analysis and confirmatory assays in a single detection system. The technique has many configurations but can be coupled with other automated analytical separation techniques including one and two-dimensional capillary electrophoresis and amplification detection techniques (e.g., quantitative polymerase chain reaction technology). Using microfluidics and other advances in nanotechnology, investigators are working to miniaturize the technology for use in microchip format (7, 112).

*Enzyme-linked immunosorbent assay*

The enzyme-linked immunosorbent assay (ELISA) based on immunological techniques couples the specificity of antibody and antigen interactions with the sensitivity of enzyme assays to detect analytes (in this case problematic microorganisms or their biomarkers) in a 96-well microtiter plate format (64). Common enzymes linked to specific antibodies in these assays are alkaline phosphatase, horseradish peroxidase, and beta-galactosidase. Enzyme-labeled antibodies bound to antigens then catalyze the conversion of a substrate into a measurable product. For example, alkaline phosphatase will dephosphorylate *p*-nitrophenyl phosphate (substrate) to produce *p*-nitrophenol, which is a detectable yellow-colored product with absorbance at 405 nm.

Different types of ELISA include direct, indirect, sandwich, and competitive. In the direct method, the antigen is bound to the surface of the plate and an enzyme-labeled antibody binds the antigen forming a surface bound complex. After washing to remove excess antibodies, the appropriate substrate for the enzyme is added and reacts to give a signal. If the signaling molecule accumulates in the well, the signal will be amplified. The amount of analyte containing the antigen (problematic cells or biomarkers) is quantified using a calibration curve. The indirect method is similar except it uses an unlabeled primary antibody and an enzyme labeled second antibody. Direct ELISA eliminates the risk of cross-reactivity of the second antibody but lacks signal amplification; and indirect ELISA has increased sensitivity but may result in additional nonspecific binding.

Sandwich ELISA involves sandwiching the antigen between two layers of antibodies. In this assay, a capture antibody is used to bind the antigen to the surface of the plate. After washing, a second enzyme-labeled antibody is added and also washed away. An appropriate substrate is used to quantify the amount of bound antigen; thus, the measured signal is proportional to the concentration. Competitive ELISA on the other hand uses labeled and unlabeled antigens to compete for surface-bound antibodies. Since the labeled antigens are used to quantify the concentration, the measured signal is inversely proportional to the concentration. The limit of detection for commercial ELISA kits is $10^5$–$10^6$ CFUs/ml (95). Thus, a pre-enrichment step is needed for detection. ELISA assays can be used to quantify problematic microorganisms; however, the assay is prone to problems associated with non-specific adsorption (leading to false positives) and is time consuming with multiple reagent addition and sample washing steps.

### 21.3.2 Emerging or non-routine quantitative detection technologies

In biofilm research, various analytical technologies have been used to characterize biofilms and biofilm components (26, 105), study interspecies and intraspecies interactions, identify various types of 3-dimensional microbial architecture (52, 93), and study real-time dynamic responses and processes occurring within biofilms (80). Capital cost and the complexity of instrumentation currently may preclude their routine use in the food industry; however, several of these bio-analytical fields of study may potentially provide quantitative information and/or confirmatory results. Furthermore, companies are in the process of developing application specific miniaturized equipment that may prove to be useful in the future.

### 21.3.3 Vibrational and spectrometry detection technologies

*Vibrational spectrometry*

Spectroscopic techniques, which measure quantized energy states of matter, have been used to detect absorption peaks or bands that indicate the presence of biofilms or components of biofilms, such as proteins and polysaccharides in wavelengths ranging from UV-visible to mid-IR. Mid-IR (10 to 2.5 μm) absorption spectrometry (typically Fourier transform infrared (FT-IR)) measures quantized vibration modes that indicate the presence of specific functional groups within molecules of a sample. With IR spectrometry, the absorption bands with the greatest intensity are associated with vibrational modes which produce changes in dipole moments. Fourier transform infrared spectrometry analysis allows the rapid determination of relative amounts of biomolecules (e.g., proteins and polysaccharides) from isolated dried or hydrated biofilms. For example, biofilms have been removed from surfaces, suspended in water, spotted directly onto IR-cards, dried, and analyzed by transmission methods (99). In real-time analyses, specialized substrata that totally reflect (attenuated total reflection) IR radiation are used to analyze biofilm formation (79). In addition, Fourier transform infrared spectrometry, which uses an interferometer, has been used to tentatively identify bacteria from isolated colonies by comparing unknown spectra with library data (49, 50, 61, 77). Limitations in FT-IR analysis of microorganisms include: selectivity, high limit of detection (approximately $10^8$ cells), and interferences from water. Selectivity is a potential problem because infrared bands are rarely unique to biofilms and relative intensities of the bands within a given spectrum are dependent on: (i) the physiology of the cells within the biofilm, (ii) the expression of phenotypic makers, and (iii) the species and serotype of microorganism present. Thus, with spectral fingerprinting analysis, cells within biofilms should be isolated and/or enriched under the same growth conditions as those used for library development. In addition, water strongly absorbs mid-IR radiation and must either be removed from the sample (e.g., analysis using dried biofilms on IR-cards) or subtracted from the sample spectra (analysis of aqueous samples using attenuated total reflectance). In response to the need for inexpensive technology, many investigators and some analytical instrument companies are developing small handheld and less expensive spectrometers that could potentially be used for routine monitoring and/or confirming the presence of problematic microorganisms (21–23).

Near-IR absorption spectrometry (770 to 2500 nm) uses similar instrumentation to visible spectrometry, but provides more spectral information. The technology measures spectral bands that are associated with overtones and vibrational modes, which yield excellent fingerprint data for identification of unknown materials with less interferences due to water. The technology provides similar spectral information to mid-IR spectrometry, with the advantage of being very inexpensive. The technology has been used in conjunction with multivariate data analysis to successfully identify and classify *Lactococcus lactis, Listeria innocua, Pseudomonas fluorescens, Pseudomonas putida,* and *Pseudomonas mendocina* suspended in saline solutions (4).

Raman, like IR spectrometry, can be used to detect vibrational modes in molecules and to characterize and help identify chemical structure. With Raman spectrometry, light interacts with the sample, and some of the radiation is scattered at wavelengths shifted from the incident radiation (referred to as inelastic scattering). Chemical shifts are caused by vibrational modes of bonds in the sample molecule. The vibrational energy can be added to or subtracted from the incident radiation. Raman spectral data is considered to be complementary to IR data. In general, Raman active vibrational modes associated with changes in polarizability can detect totally symmetrical vibrations, which are not detected with IR spectrometry. Raman is considered to have a poor limit of detection because the inelastic scattering of light has low efficiency. The efficiency of inelastic scattering can be greatly improved when the energy of incident light approaches the energy required for an electronic transition. This phenomenon is referred to as resonance Raman spectrometry and can lead to signal enhancement up to $10^6$. Resonance Raman has been used to fingerprint bacteria. For example, the technique has been used to identify eight different strains of lactic acid bacteria from yogurt (34).

Surface enhanced Raman spectrometry (SERS) is another technique that dramatically increases the intensity of the Raman scattered light. However in this case, molecules are absorbed onto specialized surfaces. The surface enhancement occurs when light of specific wavelengths interacts with free electrons in the conducting band (surface plasmon) producing an intensified electromagnetic field on thin roughened metal surfaces or metal colloids (51). Metals are usually silver or gold with silver having a potential toxic effect on microorganisms. The stimulation of the plasmon creates an intense Raman signal with enhancements ranging from $10^4$ to $10^{15}$. SERS has been used to detect a number of bacteria including *Bacillus anthracis, Escherichia coli* K12, and *Salmonella typhimurium* (92). Micro-Raman technique has also been used to detect single bacteria and offers the potential to tentatively identify problematic microorganisms without enrichment (45, 46, 96, 102).

### *Mass spectrometry*

Mass spectrometry is a vacuum technique that creates, separates, and detects mass ions (mass to charge ratio) from molecules of interest and has many types of configurations and platforms for various types of applications (37).

The technique is conceptually simple: the samples are ionized in a source and the ions are separated by a mass analyzer that in turn focuses the ions onto a mass detector. However, selection of the proper mass spectrometry platform requires knowledge of the molecule(s) of interest, potential interferences that may also be contained within the sample matrix, and the type of data required. Platforms vary by the sample preparation technique (e.g., chromatography), the type of ionization source, and the types and number of mass analyzers. For example, mass spectrometry is routinely interfaced with gas chromatography (electron ionization/chemical ionization source) and liquid chromatography (electro spray or atmospheric pressure chemical ionization) to analyze cellular components (biomarkers) that are specific to a particular microorganism. Matrix assisted laser desorption/ionization time of flight (MALDI-TOF) mass spectrometry is used to identify proteins in proteomic research and to create spectral fingerprint patterns of bacteria isolated from agar plates for tentative identification (18, 106). As with mid-IR and near-IR, researchers and companies are working toward miniaturization of the technology while maintaining enough application specific analytical performance. Recently, a 10 kg tandem mass spectrometer (33) with a mass to charge limit of 2000 has been developed. The mass spectrometer uses an electrospray ionization source and a resonant ion ejection technique to analyze ions from proteins that were as large as 17 000 Daltons (57). With further development, miniature mass spectrometers may also play a role in detection of biofilm contamination on surfaces or also be used as a detection system for a confirmatory assay.

*Camera/imaging spectroscopy*

Owing to the sporadic nature of biofilms on food processing equipment, no analytical technologies offer the ideal attributes that will allow inexpensive, noninvasive, rapid, and direct detection of biofilms on the surfaces of food processing equipment. Camera technology offers the ability to provide images that directly detect problematic regions on food processing equipment without the need for sampling. However, microcolonies can be translucent and small in size and thus are not readily detectable with standard visible light cameras. Hyperspectral or imaging spectroscopy is an emerging technology that is being used for detecting deleterious effects in food and agriculture systems (41). The technology, originally designed for use in remote sensing (38), collects simultaneous spatial and spectral information to help locate regions of interest on surfaces with proper illumination. For example, the spatial data is collected using a visible NIR camera system that is sensitive between 400 and 1000 nm, while the spectral data is generated with either acousto-optic or liquid crystal tunable filters. The results are typically in the form of a data cube, where the x and y axes represent the 2-D images, and the z-axis plots the spectral information collected from each pixel. The system can be operated in reflectance, transmission, or fluorescence imaging modes. Once a given system is well characterized with hyperspectral data, multispectral imaging is useful in reducing the amount of data for storage. Multispectral imaging collects relevant 2-D images showing significant differences at selected wavelengths. The technology has been used to detect defects in foods such as apples (71) to allow more efficient sorting and fungal infections on citrus fruits before they are detected visually (39).

### 21.3.4 Antibody antigen-based detection technologies

A biosensor uses a biological element (e.g., proteins) to selectively recognize a targeted analyte (e.g., chemicals, biomolecules, viruses, pathogens, etc.) and create a signal, which in turn can be detected with an analytical transducer. The transducer converts the signal to an electrical form for further processing and potential quantification. Biosensors are usually inexpensive, potentially easy to use, and at least in part reusable or disposable. Typically, biosensors are classified by the biological recognition element and type of transducer used for detection.

*Immunosensors*

Biosensors using antibody (monoclonal or polyclonal) and antigen interactions to mediate detection are referred to as immunosensors; and use of the technology in the food industry has recently been reviewed (108). Common analytical transducers for immunosensors use: electrochemistry (3, 113), surface plasmon resonance (SPR) (9, 54), and optical (11), or piezoelectricity (44, 63) for detection. Measurements are made directly by capturing analytes (the antigen directly or the cell with the antigen) with attached, unlabeled antibodies using piezoelectric, SPR, or impedance immunosensors. Indirect measurements detect the label (e.g., enzymes, fluorescent, chemiluminescent, and visible) linked to the antibodies which usually lowers the limit of detection but adds cost to the system. As with most immumo-technologies, the antibody antigen interaction and non-specific adsorption play a key role in determining the selectivity of the device.

Piezoelectricity yields an electrical potential response on anisotropic materials (e.g., quartz or $LiTaO_3$) when a mechanical stress is applied to its surface. Piezoelectric immunosensors attach analyte-specific antibodies to the active surface of either quartz crystal microbalances (104) or surface acoustic wave (SAW) devices (44, 63). These immunosensors typically use the inverse piezoelectric effect, where an alternating potential is applied to or across the material to establish a surface bound wave. Quartz crystal microbalances typically create oscillations by coating an electrode on both sides of thin AT-cut quartz crystals to measure mass changes at its surface and have been used to monitor biofilm formation directly in liquids (78). The frequency of the oscillation is dependent on a number of factors including material and medium properties, temperature, and most importantly for sensors, adsorbed mass. Surface acoustic wave devices use two sets of interdigitated electrode patterns, one to transmit the wave and the other to receive it, to create surface waves across surfaces (44). Depending on the cut and type of anisotropic material, different types of surface waves can propagate across the surface. These devices are smaller, more rugged, and most importantly operate at higher frequencies and thus are more sensitive to mass changes. Applications usually involve attaching the antibodies onto the surface of the device and exposing the surface to a liquid medium containing the pathogen or biomarker of interest to generate a change in frequency and/or phase. For example, *Escherichia coli* (*E. coli*) O157:H7 cells were detected using a pure shear horizontal SAW immunosensor (10); however, best results were obtained when the sensor was removed from solution and dried. In another study, approximately 2000 *Bacillus thuringiensis* spores were detected in aqueous solutions using antibodies attached to polystyrene or polyimide coated shear horizontal SAW immunosensor on $LiTaO_3$ (15). The SAW devices offer promise, but the technology has not been fully developed for wide spread utilization in the food industry.

Surface plasmon resonance (SPR) is another phenomenon that has been used to quantify antibody and antigen interactions in solution. SPR occurs when the free electron (conducting band) at interfaces of thin metal films are excited with light (9, 54). At particular angles of incidence and under conditions for total reflection (light traveling through a transparent material with a higher refractive index), light impinging off the metal interface excites free electrons in a metal film (plasmon), which in turn causes measurable changes in the properties of the light at photodetectors. Although the underlying physics is conceptually complex, the instrumentation and resultant signals are relatively straightforward to understand. For example, *Listeria monocytogenes* cells were detected using the single-chain variable-fragment (scFv) antibody physically adsorbed onto a SPR sensor with a limit of detection of $2 \times 10^6$ cells (76). Laboratory-based systems have been commercialized and used to study the kinetics and thermodynamics of antibody and antigen interactions. A number of advancements in optoelectronics, integrated optics, microfluidic and nano-fabrication techniques have made the technology more attractive for pathogen detection in the food industry (53).

Impedance measurements involve applying an oscillating voltage perturbation onto a system of electrodes and monitoring the alternating current response. In impedance microbiology, commercially available instrumentation has been used for a number of years to quantify microbial growth or activity in fermentation by directly measuring changes in conductance of a liquid medium. This conventional technology usually has a limit of detection ranging from $10^6$ to $10^7$ cells/ml and by measuring increases in biomass, can be used to distinguish living from dead cells (103). Initial application of the technology for pathogen detection involved the development of growth medium that (i) optimizes the sensitivity of the response and (ii) increases the medium specificity for a particular pathogen or group of pathogens (e.g., *Salmonella* in foods) (35). More recently, impedance spectroscopy, which stimulates the electrode system at a wide range of frequencies, has been used to develop immunosensor technology. The presence of pathogens can be determined using double interdigitated electrode arrays in which a pair of addressable array of finger-like parallel electrodes is fabricated onto surfaces in an interlacing pattern (67, 113, 114). These interdigitated electrode systems have been used to follow the growth of *Salmonella* and *E. coli* (62). The technology was further improved with the miniaturization of the electrode systems onto a microchip device and the integration of a concentration technique using dielectrophoresis (DEP) (40). DEP is the translational motion of particles as a nonuniform electric field acts on dipoles or induced dipoles (88, 89). The concentration technique used microfluidics, a cell deviation electrode, and a set of interdigitated microelectrodes to concentrate two species of *Listeria* by a factor calculated to be between $10^4$ or $10^5$. Once concentrated, the cells are transferred to a smaller chamber using impedance and a second interdigitated microelectrode system for detection. This concentration step decreased the response time by several hours to approximately 1 hour for a culture with a final concentration of $8 \times 10^5$ *L. monocytogenes* cells. In a different microchip set-up, $9 \times 10^5$ *E. coli* cells/ml were detected using a microfluidic chip, antibodies, and impedance base detection system (14). In that study, the antibodies were coated onto a glass coverslip on a surface opposite of the electrode system. As the cells become trapped by the antibodies on the glass coverslip, the number of cells in solution decreases, thereby attenuating the conductance signal, which in turn is used to quantify the number of cells.

Magnetoelastic immunosensors have been developed to improve the sensitivity of direct immunosensors (97). The transient response to a magnetic impulse applied to the sensor is captured and converted using a fast Fourier transform to obtain the resonant frequency, which provides chemical and environmental information. In the food industry, magnetoelastic immunosensors have been used to detect pathogens (e.g., *E. coli* O157:H7) by monitoring antibody-antigen reactions (97). An enzyme catalyzes a reaction to form an insoluble product that deposits on the sensor surface, which changes the mass and in turn the resonance frequency. Changes in resonance frequency can then be correlated to cell concentration (CFU/ml) by a calibration curve. Recently the technology has also been used to detect in real-time pathogens in liquid medium (55). The microorganisms' consumption of nutrients and growth change the viscosity, which changes the resonance frequency. The technology is highly sensitive with a limit of detection of $10^2$ to $10^6$ CFU/ml (55, 97). Other advantages of magnetoelastic sensors include being cost effective and small, while yielding comparable results to optical, electrochemical, and mass-sensing methods (97).

Although many antibody antigen-based configurations show promise, no platform is able to detect pathogens directly from a food contact surface, thus enrichment steps such as growth in a liquid medium are required prior to detection. Enrichment is advantageous in some ways since it: (i) provides a mechanism to increase the concentration of cells, (ii) minimizes the effect of interferences, and (iii) may decrease the amount cell aggregations or biofilm fragments in swabbed or rinsed biofilm samples. Drawbacks associated with enrichment are the dramatic increase in analysis time, the capability to enrich only a subset of the population, and the introduction of systematic errors associated with variables with lag or inhibitors when modeling the growth data. In addition, some of these detection systems may also require washing steps to remove materials that non-specifically adsorb to the antibody. Improvements in the analytical performance of these techniques are being made with new microchip technologies that integrate sample preparation steps.

*Lateral flow immunoassays*

Similar to an immunosensor, an immunoassay involves antibody and antigen interactions in an assay to detect analytes but without an on-device transducer. Therefore, in the strictest sense immunoassays should not be classified as a biosensor. Immunoassays have many configurations, with some of the simplest to use being lateral flow immunoassays (LFIAs) (Fig. 21.3), which are also sometimes referred to as immune-chromatographic assays (91). LFIAs involve applying a test liquid onto a sample pad and allowing capillary forces to draw the sample through the device. Liquid applied to the sample pad is wicked over a conjugate pad (containing reagents) and onto a chromatographic strip membrane (e.g., nitrocellulose)

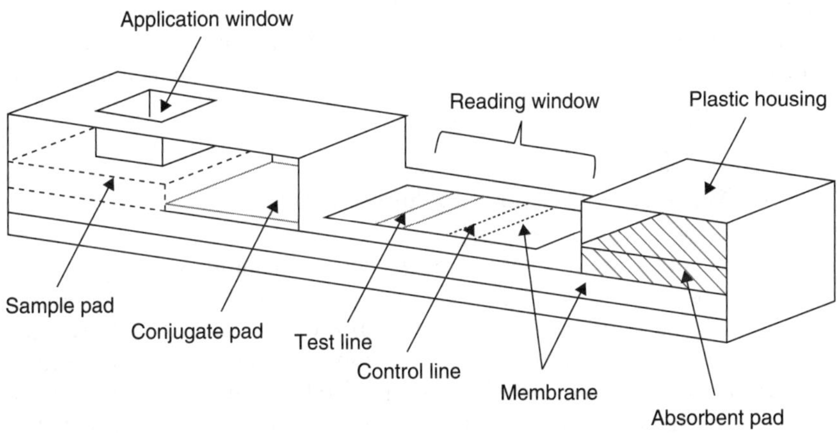


**Fig. 21.3** Lateral flow immunoassay (LFIA) adapted from Posthuma-Trumpie, Korf *et al.* 2009 (91).

with zones (e.g., lines) of dried and/or immobilized reagents. The zones are used to provide selectivity and/or a means to visually detect the analyte and control. Excess sample is trapped by an absorbent pad.

As in ELISA, sandwich and competitive assay formats exist for LFIAs. The sandwich assay typically has labeled antibodies in the conjugate pad and a second antibody at the test line. When analyte is present in the sample and passes across the conjugate pad, it binds to a labeled antibody. This antibody-labeled analyte complex will then form a sandwich complex with the antibody at the test line and yield a positive signal. In contrast, a competitive assay has labeled analytes in the conjugate pad and an antibody at the test line. Thus, when an analyte is present in the sample, it will compete with the labeled analytes in the conjugate pad for the antibody at the test line. Generally, labels are colored or fluorescent nanoparticles. Color signals can lead to ambiguous results in cases where the analyte concentration is near the limit of detection resulting in false positive and false negatives, whereas fluorescent signals are quantitative and have a better limit of detection but require devices to readout the signals. In order to eliminate ambiguity and obtain more quantitative results, manufacturers are looking to develop and launch handheld and portable LFIA readers (29).

Presently, immunoassays are most often used in medical or home diagnostic kits (e.g., pregnancy test) but are also being used to detect pathogens in food samples due to rapid detection (10–20 minutes) (91). Typically, pathogens from enrichments are detected directly with sandwich LFIAs. For the food industry, commercially available kits are available for detection of many pathogens, including *Escherichia coli* O157, *Listeria*, and *Salmonella* to allow screening of food samples after enrichment. This pathogen LFIA technology is similar to ELISA commonly used in a 96-well format, where in sequential steps a targeted pathogen is captured with a surface-bound antibody, tagged with second enzyme labeled antibody, and quantitatively detected when the enzyme interacts with a substrate to produce a detectable signal. LFIAs can also be used to detect RNA, DNA and protein from specific pathogens (16).

### 21.3.5 Nucleic acid-based detection technologies

*Polymerase chain reaction-based approaches for bacterial detection*

The polymerase chain reaction (PCR) is a powerful tool for the identification of bacterial pathogens on food preparation surfaces. Real-time PCR can be used to quantify target sequences, and it has a broad dynamic range. PCR primers may be developed to be specific for a certain organism/gene or can be designed to detect certain genera or classes of organisms. For example, studies have been designed to characterize the presence of *E. coli* O157:H7 by targeting the PCR primers to be specific for Shiga toxin genes or to the genes encoding the O157 antigen. In contrast, more general primers may target the *E. coli* ribosomal RNA genes (particularly the 16S rRNA gene). Ribosomal RNAs contain regions that are highly conserved, for identifying general classes of organisms, including the different domains, Bacteria and Archaea, as well as regions that are more divergent. The diverged regions may be used to identify organisms to the genus and often the species level.

Polymerase chain reaction analysis of 16S rDNA may be used to study community structure. Denaturing gradient gel electrophoresis (DGEE) is used to separate the PCR products based on the annealing properties of the double-stranded DNA molecule, rather than the size of the PCR product (32, 58, 75). For these studies, a 'clamp' is added to the end of the PCR product by using PCR primers with a 5′ end that is rich in guanine/cytosine bases. This gives the PCR product a region that does not denature. Using a denaturing gradient gel, the PCR products migrate on the gel based upon their denaturing properties (i.e., when the remainder of the PCR product melts to single-stranded form), allowing separation of the products based on their sequence, rather than their size. For community structural analysis, each community will give a different banding pattern on a DGEE gel. For each community, the bands on the gel may be extracted and sequenced for species identification.

Two potential drawbacks for using DGEE for biofilm community profiling are PCR amplification bias – some sequences may amplify more efficiently than others, and the resolving power of the DGEE gels – the community may have many more species than may be resolved on the gel. The development of high-throughput sequencing approaches has allowed greater resolution of microbial community structures. For these experiments rather than using electrophoresis to resolve PCR products, the PCR products (for example, targeting 16S rDNA of a microbial community) are cloned *en masse* into a plasmid vector. Each PCR insert may then be sequenced, providing a qualitative assessment of community structure, generally displayed using a phylogenetic tree, as well as a semi-quantitative analysis, providing information on the number of different types of organisms obtained. The PCR amplification and sequencing strategy also suffers from potential amplification bias.

Another potential drawback of PCR amplification strategies (either DGEE or cloning) is that the PCR reaction cannot distinguish between viable cells, dead cells, or simply DNA released into the environment by cell lysis. Recently, Nocker *et al.* used a combination of PCR with propidium monoazide to enrich for PCR products from only viable cells (81). DNA-intercalating dyes are used to differentiate live versus dead cells, since they often only penetrate into cells with compromised membranes (dead cells). Once intercalated into the DNA of dead cells or extracellular DNA, propidium monoazide inhibits the PCR reaction. Using this dye, only DNA from cells with intact membranes is amplified, providing an indication of not only the community structure, but also the viable cells in that community. This approach has been tested on pathogens that contaminate food, including *Listeria monocytogenes*, *Salmonella typhimurium*, and *Escherichia coli* O157:H7 (82).

*Quantitative polymerase chain reaction and quantitative reverse transcriptase polymerase chain reaction*

Gel-based approaches for PCR analysis are generally not quantitative since a PCR amplification step is required. However, quantitative PCR (qPCR) and quantitative reverse transcriptase PCR (qRT-PCR) now provide excellent approaches for determining the number of specific DNA or RNA molecules in a sample, respectively (36, 48, 60, 83). qPCR and qRT-PCR are very sensitive and may be used on samples with small amounts of biological material, such as swab samples from food processing surfaces. The techniques also have a large and generally linear dynamic range. Commonly used qPCR methods are the SYBR green labeling approach and the dual-labeled probe approach. In the SYBR green approach, a fluorescent dye is incorporated into the PCR product. As the amount of PCR product increases during the PCR cycles, the amount of fluorescence that is detected also increases exponentially. A threshold cycle (Ct value) is obtained during the exponential increase of the PCR product/fluorescence and compared to a linear curve of values for standards with known amounts of starting DNA. The amount of starting DNA for a particular gene may then be determined for each unknown sample. In the dual-labeled probe approach, a DNA probe that is specific for the PCR product of interest is labeled with both a fluorescent molecule and a fluorescence quenching molecule. While both molecules are contained on the DNA probe, fluorescence is quenched. As the probe is degraded during the exponential increase of the PCR product by *Taq* polymerase, the quenching molecule is released from the fluorescence molecule, resulting in increased fluorescence. Fluorescence is detected, and a Ct value is obtained, and compared to standard curves. qPCR may be applied directly to the DNA isolated from samples. For analysis of gene expression, mRNA is isolated from cells and converted to cDNA by reverse transcriptase prior to performing the qPCR reaction.

The advantage of the SYBR-green approach is that it is the most cost-effective. In addition, primer specificity may be determined by a melt-curve analysis. Following the PCR reaction a melt curve analysis may be performed. If only one peak is observed that generally indicates that the PCR primers were specific for a unique PCR product. While the cost of the dual-labeled probes is higher than that of the SYBR-green dye, an advantage of dual-labeled probe analysis is that it allows multiplexing. Different probes may be labeled with different color fluorescent dyes. Therefore, multiple PCR products (for example, designed for different genes or for different organisms) may be produced in one reaction, allowing more rapid analysis of community structure or multiple gene expressions.

For detailed protocols on isolation of DNA and RNA from samples with small amounts of biological material, please refer to protocols by Perez-Osorio and Franklin (84, 85).

### 21.3.6 Microscopy detection technologies

*Fluorescence in situ hybridization (FISH)*

While PCR techniques allow amplification of DNA or cDNA from low biomass samples, approaches are available that allow direct probing of the samples without an amplification step. Fluorescence *in situ* hybridization (FISH) allows direct probing of samples using nucleic acid probes that are specific for a target sequence. As with PCR-based techniques, the probes may be specific for a particular genus or more general, to detect different domains of organisms. Generally, rRNA is targeted for hybridization, since the rRNA is the most abundant RNA in the cells, and therefore provides the greatest signal (5, 25, 72). Targeting mRNA is also possible, but generally requires an amplification step, since individual mRNAs are in low abundance. *In situ* reverse transcription polymerase chain reaction (IS-RT-PCR) was used to amplify mRNA present in low copy number from intact cells (69). However, this technique is not quantitative for the gene expression since the amplification step is required.

Using FISH, cells are treated to allow penetration of the probes into the cells. The fluorescently labeled probes then hybridize to the target molecule, usually the rRNA target, and cells from a given group are quantified using epifluorescence microscopy. Recently, difficulties with FISH, such as lack of penetration of the probes through extracellular matrix material and low fluorescence intensities, have been addressed by modifying the approach. For example, PNA-FISH uses peptide nucleic acid probes (65). These probes while still specific for a certain target sequence use a peptide backbone that has increased penetration into cells and through biofilm matrix material. CARD-FISH (catalyzed reporter deposition) is another advantage, in that the probes provide increased fluorescence intensity and therefore enhanced detection of target sequences (30, 31, 87).

Most of the PCR and probing techniques described above are specific for target nucleic acids. However, the techniques generally do not provide an indication of cell viability or activity. FISH is often combined with microautoradiography (MAR) to determine cell activity. For MAR, cells are exposed to a radiolabeled substrate, allowing the cells to incorporate the label into cellular material. The active cells may be detected by microautoradiography, using a photographic emulsion. This approach, when combined with FISH allows cell identification as well as an indication of cell activity. The FISH-MAR technique can be tedious and is not easily amenable to the rapid study of many samples (111).

*Live-dead staining*
Alternative approaches for examining cell viability involve differential staining with fluorescent stains followed by epifluorescence or confocal scanning laser microscopy. Among the stains commonly used are the 'live-dead' stains. The 'dead' stain, such as propidium iodide primarily enters cells with compromised membranes, and therefore does not generally stain viable cells. Counterstains that stain all cells are then used to distinguish the viable cells from the dead cells. This approach must be tested on each system to ensure that the individual stains are specific for the live and/or the dead cells. Tetrazolium dyes such as 5-cyano-2,3-ditolyl tetrazolium chloride (CTC) can be useful for studying actively respiring cells within a population. CTC is a soluble, non-fluorescent substrate that forms a red fluorescent precipitate when it is reduced by respiring bacteria. Therefore, actively respiring cells may be detected using CTC. As with the live-dead stain, the specificity of CTC dye must be tested for each system. Recently, Rani *et al.* used pulse labeling with the thymidine analog 5-bromo-2-deoxyuridine to label cells that are actively synthesizing new DNA (replicating cells) (94). Following DNA labeling, the brominated DNA was probed with fluorescently labeled antibodies. This approach represents an alternative to examining cell viability by characterizing cells that are actively incorporating nucleotides into their DNA.

## 21.4 Conclusions

In general, successful monitoring of food contact surfaces with respect to microbial contamination is ultimately dependent on the attributes (e.g., limit of detection, specificity, precision, accuracy, etc.) of the analysis methodology and the statistical sampling plan. However, the implementation of a particular analysis strategy is not solely dependent on the merits of the method but also on economic considerations. Thus, evaluation of analysis strategies are based on the ability of the assay to: (i) provide precise and accurate data in a timely and cost-effective manner, (ii) provide the proper degree of sensitivity and specificity, and (iii) provide a statistically significant result. Furthermore, although presence/absence assays may suffice, accurate quantification of the number of viable microorganisms is desirable since it provides a measure of the level of risk associated with the contaminated product. Once quantified, a confirmatory assay that has a high degree of selectivity should be applied to verify results.

Ideally, quantification of microorganisms on food contact surfaces could be performed directly to eliminate the sampling step, which introduces sampling error. However, currently no such technique exists, so sampling strategies must be considered when developing a sampling plan in order to minimize error and obtain precise and accurate results. Accurate assessment of food contact surfaces integrates both representative and critical control points. These critical control points should include areas identified as problematic such as corners, cracks, and interfaces, which can harbor microorganisms and are difficult to clean. Since all sampling methods are limited, using more than one sampling method is recommended to obtain representative data.

Plate count assays quantify the number of bacteria growing at a given set of nutrients, thereby providing an estimate of the number of viable bacteria in the sample. The methodology is sensitive but provides only limited specificity and does not provide rapid results. Delays in obtaining results are a source of lost revenue and are a major reason why companies are seeking new methods. Plate count assays can underestimate populations since no plating medium supports growth of all types of microorganisms. The lack of specificity can be a crucial flaw if the assay is compromised by the presence of viable but nonculturable problematic microorganisms (underestimates risk) or by culturing harmless bacteria (overestimates risk). Other sources of systematic errors are biofilm fragments causing cell aggregates or contaminates detected on growth plates. Plating on selective medium should be considered presumptive evidence, and should be verified by a confirmatory test. Bioanalytical techniques including infrared, Raman, and mass spectrometry can be used for confirmatory testing if more inexpensive instrumentation that allows fingerprinting and library searches from bacterial isolates becomes available.

Immunosensors and immunoassays like impedance-based immunosensors which can detect living cells may prove to offer more rapid analysis and be more selective than plating assays. However, these detection systems currently have problems associated with selectivity because of nonselective binding of samples. Therefore, a second confirmatory assay should also be used with these sensors and assays to verify results. In addition, investigators should recognize that samples may contain food ingredients and biofilm fragments that will complicate detection and/or quantification. Additional sample preparation steps may be necessary for more accurate and precise results.

Molecular microbiological detection strategies have variable specificity allowing the detection of all bacteria, as well as particular groups of bacteria (e.g., *Salmonella* species), or even a distinct strain or serotype (e.g., *E. coli* O157:H7). Thus, nucleic acid based analysis strategies can be used as indicator and confirmatory assays. For example, determination of the total bacterial concentration of the product can indicate a problem, while analyzing for specific harmful bacteria confirms the existence of the problem. Instruments capable of automated and real-time detection of PCR products are presently available and provide results in hours rather than days.

Each of the quantitative detection technologies for sampling of food contact surfaces surveyed in this chapter have limitations and fall short of the attributes for an ideal detection method: cost effective, rapid, selective, sensitive, and rugged. Advancements and continued developments in the various technologies are making progress toward these goals. The majority of these technologies have the potential to be used for sampling of food when fully developed.

## 21.5 Acknowledgements

The authors would like to acknowledge the Agricultural Research Service of the United States Department of Agriculture, the Center for Food Safety and Engineering at Purdue University, and the National Alliance for Food Safety and Security.

## 21.6 References

1. ANON. 2008. *Detection, Isolation and Identification of Escherichia coli O157:H7 from Meat Products, Microbiology Laboratory Guidebook*. Food Safety and Inspection Service.
2. AGLE, M. E. 2007. *Biofilms in the Food Industry*. In H. P. Blaschek, H. H. Wang, and M. E. Agle (ed.), Hans P. Blaschek, Hua H. Wang, Meredith E. Agle. Blackwell Publishing and the Institute of Food Technologists.
3. AHMED, M. U., M. M. HOSSAIN, and E. TAMIYA. 2008. Electrochemical biosensors for medical and food applications. *Electroanalysis* **20:**616–626.
4. ALEXANDRAKIS, D., G. DOWNEY, and A. G. M. SCANNELL. 2008. Detection and identification of bacteria in an isolated system with near-infrared spectroscopy and multivariate analysis. *Journal of Agricultural and Food Chemistry* **56:**3431–3437.
5. AMANN, R. I. 1995. Fluorescently labelled, rRNA-targeted oligonucleotide probes in the study of microbial ecology. *Molecular Ecology* **4:**543–553.
6. ANGELOTTI, R., J. WILSON, W. LITSKY, and W. WALTER. 1964. Comparative evaluation of the cotton swab and Rodac methods for the recovery of *Bacillus subtilis* spore contamination from stainless steel surfaces. *Health Laboratory Science* **1:**289–296.
7. ATEYA, D. A., J. S. ERICKSON, P. B. HOWELL, L. R. HILLIARD, J. P. GOLDEN, and F. S. LIGLER. 2008. The good, the bad, and the tiny: a review of microflow cytometry. *Analytical and Bioanalytical Chemistry* **391:**1485–1498.
8. BANADA, P. P., K. HUFF, E. BAE, B. RAJWAD, A. AROONNUALA, B. BAYRAKTARE, A. ADIL, J. PAUL ROBINSOND, E. D. HIRLEMAN, and A. K. BHUNIA. 2009. Label-free detection of multiple bacterial pathogens using light-scattering sensor. *Biosensors and Bioelectronics* **24:**1685–1692.
9. BERGWERFF, A. A., and F. VAN KNAPEN. 2006. Surface plasmon resonance biosensors for detection of pathogenic microorganisms: Strategies to secure food and environmental safety. *Journal of Aoac International* **89:**826–831.
10. BERKENPAS, E., P. MILLARD, and M. P. DA CUNHA. 2006. Detection of *Escherichia coli* O157:H7 with langasite pure shear horizontal surface acoustic wave sensors. *Biosensors & Bioelectronics* **21:**2255–2262.
11. BHUNIA, A. K., P. BANADA, P. BANERJEE, A. VALADEZ, and E. D. HIRLEMAN. 2007. Light scattering, fiber optic- and cell-based sensors for sensitive detection of foodborne pathogens. *Journal of Rapid Methods and Automation in Microbiology* **15:**121–145.

12. BITTON, G., B. KOOPMAN, K. H. JUNG, G. VOILAND, and M. KOTOB. 1993. Modification of the standard epifluorescence microscopic method for total bacterial counts in environmental samples. *Water Research* **27:**1109–1112.
13. BOCHERT, D. 2004. *Cleaning Validation*, GMP Manual (Up04) Maas & Peither AG – GMP Publishing.
14. BOEHM, D. A., P. A. GOTTLIEB, and S. Z. HUA. 2007. On-chip microfluidic biosensor for bacterial detection and identification. *Sensors and Actuators B-Chemical* **126:**508–514.
15. BRANCH, D. W., and S. M. BROZIK. 2004. Low-level detection of a Bacillus anthracis simulant using Love-wave biosensors on 36 degrees YX LiTaO3. *Biosensors & Bioelectronics* **19:**849–859.
16. CARTER, D. J., and R. B. CARY. 2007. Lateral flow microarrays: a novel platform for rapid nucleic acid detection based on miniaturized lateral flow chromatography. *Nucleic Acids Research* **35:**1–11.
17. CHARACKLIS, W. G. 1990. *Biofilms*, 1 ed. John Wiley & Sons.
18. CLAYDON, M. A., S. N. DAVEY, V. EDWARDSJONES, and D. B. GORDON. 1996. The rapid identification of intact microorganisms using mass spectrometry. *Nature Biotechnology* **14:**1584–1586.
19. COHEN, D., J. A. DICKERSON, C. D. WHITMORE, E. H. TURNER, M. M. PALCIC, O. HINDSGAUL, and N. J. DOVICHI. 2008. Chemical Cytometry: Fluorescence-Based Single-Cell Analysis. *Annual Review of Analytical Chemistry* **1:**165–190.
20. COSTERTON, J. W., Z. LEWANDOWSKI, D. E. CALDWELL, D. R. KORBER, and H. M. LAPPINSCOTT. 1995. Microbial Biofilms. *Annual Review of Microbiology* **49:** 711–745.
21. CROCOMBE, R. A. 2008. Miniature optical spectrometers, Part III: Conventional and laboratory near-infrared spectrometers. *Spectroscopy* **23:**40.
22. CROCOMBE, R. A. 2008. Miniature optical spectrometers: Follow the money – Part II: The telecommunications boom. *Spectroscopy* **23:**56.
23. CROCOMBE, R. A. 2008. Miniature optical spectrometers: The art of the possible, Part IV: New near-infrared technologies and spectrometers. *Spectroscopy* **23:**26.
24. DAVEY, H. M., and D. B. KELL. 1996. Flow cytometry and cell sorting of heterogeneous microbial populations: The importance of single-cell analyses. *Microbiological Reviews* **60:**641.
25. DELONG, E. F., G. S. WICKHAM, and N. R. PACE. 1989. Phylogenetic stains: ribosomal RNA-based probes for the identification of single cells. *Science* **243:**1360–1363.
26. DENKHAUS, E., S. MEISEN, U. TELGHEDER, and J. WINGENDER. 2007. Chemical and physical methods for characterisation of biofilms. *Microchimica Acta* **158:**1–27.
27. DEVENISH, J. A., B. W. CIEBIN, and M. H. BRODSKY. 1985. Evaluation of Millipore swab-membrane filter kits. *Journal of Food Protection* **48:**870.
28. EVANCHO, G. M., W. H. SVEUM, L. J. MOBERG, and J. F. FRANK. 2001. Microbiological monitoring of the food processing environment. In F. P. Downes and K. Ito (ed.), *Compendium of Methods for the Microbiological Examination of Foods*, 4 ed. American Public Health Association.
29. FAULSTICH, K., R. GRULER, M. EBERHARD, D. LENTZSCH, and K. HABERSTROH. 2008. *Lateral Flow Immunoassay.* In R. Wong and H. Tse (ed.), illustrated ed. Springer.
30. FAZI, S., S. AMALFITANO, J. PERNTHALER, and A. PUDDU. 2005. Bacterial communities associated with benthic organic matter in headwater stream microhabitats. *Environmental Microbiology* **7:**1633–1640.
31. FERRARI, B. C., N. TUJULA, K. STONER, and S. KJELLEBERG. 2006. Catalyzed reporter deposition-fluorescence in situ hybridization allows for enrichment-

independent detection of microcolony-forming soil bacteria. *Applied and Environmental Microbiology* **72:**918–922.

32. FERRIS, M. J., G. MUYZER, and D. M. WARD. 1996. Denaturing gradient gel electrophoresis profiles of 16S rRNA-defined populations inhabiting a hot spring microbial mat community. *Applied and Environmental Microbiology* **62:**340–346.
33. GAO, L., Q. Y. SONG, G. E. PATTERSON, R. G. COOKS, and Z. OUYANG. 2006. Handheld rectilinear ion trap mass spectrometer. *Analytical Chemistry* **78:**5994–6002.
34. GAUS, K., P. ROSCH, R. PETRY, K. D. PESCHKE, O. RONNEBERGER, H. BURKHARDT, K. BAUMANN, and J. POPP. 2005. Presented at the 11th European Conference on Spectroscopy of Biological Molecules (ECSBM 2005), Aschaffenburg, Germany, Sep 03–08.
35. GIBSON, D. M. 1987. Some modification to the media for rapid automated detection of salmonellas by conductance measurement. *Journal of Applied Bacteriology* **63:**299–304.
36. GIBSON, U. E. M., C. A. HEID, and P. M. WILLIAMS. 1996. A novel method for real time quantitative RT PCR. *Genome Research* **6:**995–1001.
37. GLISH, G. L., and R. W. VACHET. 2003. The basics of mass spectrometry in the twenty-first century. *Nature Reviews Drug Discovery* **2:**140–150.
38. GOETZ, A. F. H. 1985. Imaging spectrometry – A great leap in remote-sensing. *Journal of the Optical Society of America a-Optics Image Science and Vision* **2:**P83.
39. GOMEZ-SANCHIS, J., E. MOLTO, L. GOMEZ-CHOVA, N. ALEIXOS, G. CAMPS-VALLS, and F. JUSTE. 2004. Hyperspectral computer vision system for the detection of Penicillium digitatum in citrus packing lines, 2004 CIGR International Conference, Beijing, China.
40. GOMEZ-SJOBERG, R., D. T. MORISETTE, and R. BASHIR. 2005. Impedance microbiology-on-a-chip: Microfluidic bioprocessor for rapid detection of bacterial metabolism. *Journal of Microelectromechanical Systems* **14:**829–838.
41. GOWEN, A. A., C. P. O'DONNELL, P. J. CULLEN, G. DOWNEY, and J. M. FRIAS. 2007. Hyperspectral imaging – an emerging process analytical tool for food quality and safety control. *Trends in Food Science & Technology* **18:**590–598.
42. GRACIAS, K. S., and J. L. MCKILLIP. 2004. A review of conventional detection and enumeration methods for pathogenic bacteria in food. *Canadian Journal of Microbiology* **50:**883–890.
43. GRIFFITHS, M. W. 1996. The role of ATP bioluminescence in the food industry: New light on old problems. *Food Technology* **50:**62.
44. GRONEWOLD, T. M. A. 2007. Surface acoustic wave sensors in the bioanalytical field: Recent trends and challenges. *Analytica Chimica Acta* **603:**119–128.
45. HARZ, A., P. ROSCH, and J. POPP. 2009. Vibrational Spectroscopy – A Powerful Tool for the Rapid Identification of Microbial Cells at the Single-Cell Level. *Cytometry Part A* **75A:**104–113.
46. HARZ, M., P. ROSCH, K. D. PESCHKE, O. RONNEBERGER, H. BURKHARDT, and J. POPP. 2005. Micro-Raman spectroscopic identification of bacterial cells of the genus Staphylococcus and dependence on their cultivation conditions. *Analyst* **130:** 1543–1550.
47. HAWRONSKYJ, J. M., and J. HOLAH. 1997. ATP: A universal hygiene monitor. *Trends in Food Science & Technology* **8:**79–84.
48. HEID, C. A., J. STEVENS, K. J. LIVAK, and P. M. WILLIAMS. 1996. Real time quantitative PCR. *Genome Research* **6:**986–994.
49. HELM, D., H. LABISCHINSKI, and D. NAUMANN. 1991. Elaboration of a procedure for identification of bacteria using Fourier-Transform IR spectral libraries: a stepwise correlation approach. *Journal of Microbiological Methods* **14:**127–142.

50. HELM, D., H. LABISCHINSKI, G. SCHALLEHN, and D. NAUMANN. 1991. Classification and identification of bacteria by Fourier-transform infrared spectroscopy. *Journal of General Microbiology* **137:**69–79.
51. HERING, K., D. CIALLA, K. ACKERMANN, T. DORFER, R. MOLLER, H. SCHNEIDEWIND, R. MATTHEIS, W. FRITZSCHE, P. ROSCH, and J. POPP. 2008. SERS: a versatile tool in chemical and biochemical diagnostics. *Analytical and Bioanalytical Chemistry* **390:**113–124.
52. HEYDORN, A., A. T. NIELSEN, M. HENTZER, C. STERNBERG, M. GIVSKOV, B. K. ERSBOLL, and S. MOLIN. 2000. Quantification of biofilm structures by the novel computer program COMSTAT. *Microbiology-UK* **146:**2395–2407.
53. HOA, X. D., A. G. KIRK, and M. TABRIZIAN. 2007. Towards integrated and sensitive surface plasmon resonance biosensors: A review of recent progress. *Biosensors & Bioelectronics* **23:**151–160.
54. HOMOLA, J., S. S. YEE, and G. GAUGLITZ. 1999. Surface plasmon resonance sensors: review. *Sensors and Actuators B-Chemical* **54:**3–15.
55. HUANG, S. J., P. F. PANG, X. L. MAO, L. W. HE, Q. Y. CAI, and C. A. GRIMES. 2008. A wireless, remote-query sensor for real-time detection of *Escherichia coli* O157:H7 concentrations. *Sensors and Actuators B-Chemical* **131:**489–495.
56. INCZEDY, J., T. LENGYEL, and A. M. URE. 1998. *Compendium of Analytical Nomenclature*, 3 edn Blackwell Science.
57. JANFELT, C., N. TALATY, C. C. MULLIGAN, A. KEIL, Z. OUYANG, and R. G. COOKS. 2008. Mass spectra of proteins and other biomolecules recorded using a handheld instrument. *International Journal of Mass Spectrometry* **278:**166–169.
58. JANY, J. L., and G. BARBIER. 2008. Culture-independent methods for identifying microbial communities in cheese. *Food Microbiology* **25:**839–848.
59. JARVIS, B. 2000. Sampling for microbiological analysis. In B. M. Lund, A. C. T. Baird-Parker, and G. W. Gould (ed.), *The Microbiological Safety and Quality of Food*, vol. 1–2. Springer.
60. KAWASAKI, S., P. M. FRATAMICO, N. HORIKOSHI, Y. OKADA, K. TAKESHITA, T. SAMESHIMA, and S. KAWAMOTO. 2009. Evaluation of a Multiplex PCR System for Simultaneous Detection of *Salmonella* spp., *Listeria monocytogenes*, and *Escherichia coli* O157:H7 in Foods and in Food Subjected to Freezing. *Foodborne Pathogens and Disease* **6:**81–89.
61. KIRSCHNER, C., K. MAQUELIN, P. PINA, N. A. N. THI, L. P. CHOO-SMITH, G. D. SOCKALINGUM, C. SANDT, D. AMI, F. ORSINI, S. M. DOGLIA, P. ALLOUCH, M. MAINFAIT, G. J. PUPPELS, and D. NAUMANN. 2001. Classification and identification of enterococci: a comparative phenotypic, genotypic, and vibrational spectroscopic study. *Journal of Clinical Microbiology* **39:**1763–1770.
62. LACZKA, O., E. BALDRICH, F. X. MUNOZ, and F. J. DEL CAMPO. 2008. Detection of *Escherichia coli* and *Salmonella typhimurium* using interdigitated microelectrode capacitive immunosensors: The importance of transducer geometry. *Analytical Chemistry* **80:**7239–7247.
63. LANGE, K., B. E. RAPP, and M. RAPP. 2008. Surface acoustic wave biosensors: a review. *Analytical and Bioanalytical Chemistry* **391:**1509–1519.
64. LAZCKA, O., F. J. DEL CAMPO, and F. X. MUNOZ. 2007. Pathogen detection: A perspective of traditional methods and biosensors. *Biosensors & Bioelectronics* **22:**1205–1217.
65. LEHTOLA, M. J., E. TORVINEN, L. T. MIETTINEN, and C. W. KEEVIL. 2006. Fluorescence in situ hybridization using peptide nucleic acid probes for rapid detection of *Mycobacterium avium* subsp avium and *Mycobacterium avium* subsp paratuberculosis in potable-water biofilms. *Applied and Environmental Microbiology* **72:**848–853.
66. LELIEVELD, H. L. M., M. A. MOSTERT, and G. J. CURIEL. 2003. Hygienic equipment design. In H. L. M. Lelieveld, M. A. Mostert, J. Holah, and B. White (ed.),

*Hygiene in Food Processing: Principles and Practice*, Illustrated ed. Woodhead Publishing.

67. LISDAT, F., and D. SCHAFER. 2008. The use of electrochemical impedance spectroscopy for biosensing. *Analytical and Bioanalytical Chemistry* **391:**1555–1567.
68. MADRID, R. F., and C. J. FELICE. 2005. Microbial biomass estimation. *Critical Reviews in Biotechnology* **25:**97–112.
69. MAGNUSON, T. S., A. L. NEAL, and G. G. GEESEY. 2004. Combining in situ reverse transcriptase polymerase chain reaction, optical microscopy, and X-ray photoelectron spectroscopy to investigate mineral surface-associated microbial activities. *Microbial Ecology* **48:**578–588.
70. MAUKONEN, J., J. MATTO, G. WIRTANEN, L. RAASKA, T. MATTILA-SANDHOLM, and M. SAARELA. 2003. Methodologies for the characterization of microbes in industrial environments: a review. *Journal of Industrial Microbiology & Biotechnology* **30:**327–356.
71. MEHL, P. M., Y. R. CHEN, M. S. KIM, and D. E. CHAN. 2004. Development of hyperspectral imaging technique for the detection of apple surface defects and contaminations. *Journal of Food Engineering* **61:**67–81.
72. MOLLER, S., A. R. PEDERSEN, L. K. POULSEN, E. ARVIN, and S. MOLIN. 1996. Activity and three-dimensional distribution of toluene-degrading *Pseudomonas putida* in a multispecies biofilm assessed by quantitative in situ hybridization and scanning confocal laser microscopy. *Applied and Environmental Microbiology* **62:**4632–4640.
73. MOORE, G., and C. GRIFFITH. 2002. A comparison of surface sampling methods for detecting coliforms on food contact surfaces. *Food Microbiology* **19:** 65–73.
74. MOORE, G., and C. GRIFFITH. 2007. Problems associated with traditional hygiene swabbing: the need for in-house standardization. *Journal of Applied Microbiology* **103:**1090–1103.
75. MUYZER, G., E. C. DEWAAL, and A. G. UITTERLINDEN. 1993. Profiling of complex microbial populations by denaturing gradient gel electrophoresis analysis of polymerase chain reaction-amplified genes coding for 16S rRNA. *Applied and Environmental Microbiology* **59:**695–700.
76. NANDURI, V., A. K. BHUNIA, S. I. TU, G. C. PAOLI, and J. D. BREWSTER. 2007. SPR biosensor for the detection of L-monocytogenes using phage-displayed antibody. *Biosensors & Bioelectronics* **23:**248–252.
77. NGO-THI, N. A., C. KIRSCHNER, and D. NAUMANN. 2003. Characterization and identification of microorganisms by FIF-IR microspectrometry. *Journal of Molecular Structure* **661:**371–380.
78. NIVENS, D. E., J. Q. CHAMBERS, T. R. ANDERSON, and D. C. WHITE. 1993. Long-term, on-line monitoring of microbial biofilms using a quartz crystal microbalance. *Analytical Chemistry* **65:**65–69.
79. NIVENS, D. E., D. E. OHMAN, J. WILLIAMS, and M. J. FRANKLIN. 2001. Role of alginate and its O acetylation in formation of *Pseudomonas aeruginosa* microcolonies and biofilms. *Journal of Bacteriology* **183:**1047–1057.
80. NIVENS, D. E., R. J. PALMER, and D. C. WHITE. 1995. Continuous nondestructive monitoring of microbial biofilms: a review of analytical techniques. *Journal of Industrial Microbiology* **15:**263–276.
81. NOCKER, A., P. SOSSA-FERNANDEZ, M. D. BURR, and A. K. CAMPER. 2007. Use of propidium monoazide for live/dead distinction in microbial ecology. *Applied and Environmental Microbiology* **73:**5111–5117.
82. NOCKER, A., K. E. SOSSA, and A. K. CAMPER. 2007. Molecular monitoring of disinfection efficacy using propidium monoazide in combination with quantitative PCR. *Journal of Microbiological Methods* **70:**252–260.

83. NOLAN, T., R. E. HANDS, and S. A. BUSTIN. 2006. Quantification of mRNA using real-time RT-PCR. *Nature Protocols* **1:**1559–1582.

84. PEREZ-OSORIO, A. C., and M. J. FRANKLIN. 2008. *Isolation of RNA and DNA from biofilm samples obtained by Laser Capture Microdissection Microscopy*. Cold Spring Harbor Protocols.

85. PEREZ-OSORIO, A. C., and M. J. FRANKLIN. 2008. *qRT-PCR of Microbial Biofilms*. Cold Spring Harbor Protocols.

86. PEREZ-RODRIGUEZ, F., A. VALERO, E. CARRASCO, R. M. GARCIA, and G. ZURERA. 2008. Understanding and modelling bacterial transfer to foods: a review. *Trends in Food Science & Technology* **19:**131–144.

87. PERNTHALER, A., J. PERNTHALER, and R. AMANN. 2002. Fluorescence in situ hybridization and catalyzed reporter deposition for the identification of marine bacteria. *Applied and Environmental Microbiology* **68:**3094–3101.

88. POHL, H. A., and J. S. CRANE. 1971. Dielectrophoresis of Cells. *Biophysical Journal* **11:**711.

89. POHL, H. A., and I. HAWK. 1966. Separation of Living and Dead Cells by Dielectrophoresis. *Science* **152:**647.

90. PORTER, K. G., and Y. S. FEIG. 1980. The use of DAPI for identifying and counting aquatic microfloral. *Limnology and Oceanography* **25:**943–948.

91. POSTHUMA-TRUMPIE, G. A., J. KORF, and A. VAN AMERONGEN. 2009. Lateral flow (immuno) assay: its strengths, weaknesses, opportunities and threats. A literature survey. *Analytical and Bioanalytical Chemistry* **393:**569–582.

92. PREMASIRI, W. R., D. T. MOIR, M. S. KLEMPNER, N. KRIEGER, G. JONES, and L. D. ZIEGLER. 2005. Characterization of the Surface Enhanced Raman Scattering (SERS) of bacteria. *Journal of Physical Chemistry B* **109:**312–320.

93. RANI, S. A., B. PITTS, H. BEYENAL, R. A. VELUCHAMY, Z. LEWANDOWSKI, W. M. DAVISON, K. BUCKINGHAM-MEYER, and P. S. STEWART. 2007. Spatial patterns of DNA replication, protein synthesis, and oxygen concentration within bacterial biofilms reveal diverse physiological states. *Journal of Bacteriology* **189:**4223–4233.

94. RANI, S. A., B. PITTS, H. BEYENAL, R. A. VELUCHAMY, Z. LEWANDOWSKI, W. M. DAVISON, K. BUCKINGHAM-MEYER, and P. S. STEWART. 2007. Spatial patterns of DNA replication, protein synthesis, and oxygen concentration within bacterial biofilms reveal diverse physiological states. *J Bacteriol* **189:**4223–4233.

95. ROBISON, B. J. 1995. Detection of Foodborne Pathogens In J. Howard and D. M. Whitcombe (ed.), *Methods in Molecular Biology: Diagnostic Bacteriology Protocols*, vol. 46. Humana Press.

96. ROSCH, P., M. HARZ, M. SCHMITT, K. D. PESCHKE, O. RONNEBERGER, H. BURKHARDT, H. W. MOTZKUS, M. LANKERS, S. HOFER, H. THIELE, and J. POPP. 2005. Chemotaxonomic identification of single bacteria by micro-Raman spectroscopy: Application to clean-room-relevant biological contaminations. *Applied and Environmental Microbiology* **71:**1626–1637.

97. RUAN, C. M., K. F. ZENG, O. K. VARGHESE, and C. A. GRIMES. 2003. Magnetoelastic immunosensors: Amplified mass immunosorbent assay for detection of *Escherichia coli* O157:H7. *Analytical Chemistry* **75:**6494–6498.

98. SALO, S., A. FRIIS, and G. WIRTANEN. 2008. Cleaning validation of fermentation tanks. *Food and Bioproducts Processing* **86:**204–210.

99. SARKISOVA, S., M. A. PATRAUCHAN, D. BERGLUND, D. E. NIVENS, and M. J. FRANKLIN. 2005. Calcium-induced virulence factors associated with the extracellular matrix of mucoid *Pseudomonas aeruginosa* biofilms. *Journal of Bacteriology* **187:**4327–4337.

100. SCHMID, I., C. LAMBERT, D. AMBROZAK, G. E. MARTI, D. M. MOSS, and S. P. PERFETTO. 2007. International society for analytical cytology biosafety standard for sorting of unfixed cells. *Cytometry Part A* **71A:**414–437.

101. SCHULZE, G., and G. HILDEBRANDT. 2007. The representativeness of the RODAC-technique. *Fleischwirtschaft* **87:**213–219.
102. SCHUSTER, K. C., E. URLAUB, and J. R. GAPES. 1999. Presented at the Conference on Analysis of Microbial Cells at the Single Cell Level – Why, How, When, Como, Italy, Mar 25–27.
103. SILLEY, P., and S. FORSYTHE. 1996. Impedance microbiology – A rapid change for microbiologists. *Journal of Applied Bacteriology* **80:**233–243.
104. SKOTTRUP, P. D., M. NICOLAISEN, and A. F. JUSTESEN. 2008. Towards on-site pathogen detection using antibody-based sensors. *Biosensors & Bioelectronics* **24:**339–348.
105. STEWART, P. S., and M. J. FRANKLIN. 2008. Physiological heterogeneity in biofilms. *Nature Reviews Microbiology* **6:**199–210.
106. STINGU, C. S., A. C. RODLOFF, H. JENTSCH, R. SCHAUMANN, and K. ESCHRICH. 2008. Rapid identification of oral anaerobic bacteria cultivated from subgingival biofilm by MALDI-TOF-MS. *Oral Microbiology and Immunology* **23:**372–376.
107. SUTHERLAND, I. W. 2001. The biofilm matrix – an immobilized but dynamic microbial environment. *Trends in Microbiology* **9:**222–227.
108. TOKARSKYY, O., and D. L. MARSHALL. 2008. Immunosensors for rapid detection of *Escherichia coli* O157:H7 – Perspectives for use in the meat processing industry. *Food Microbiology* **25:**1–12.
109. UPMANN, M., and C. BONAPARTE. 1999. Rapid methods food hygiene inspection. In R. K. Robinson, C. A. Batt, and P. Patel (ed.), *Encylopedia Microbiology*, Illustrated ed, vol. 1-3. Academic Press.
110. VEAL, D. A., D. DEERE, B. FERRARI, J. PIPER, and P. V. ATTFIELD. 2000. Fluorescence staining and flow cytometry for monitoring microbial cells. *Journal of Immunological Methods* **243:**191–210.
111. WAGNER, M., P. H. NIELSEN, A. LOY, J. L. NIELSEN, and H. DAIMS. 2006. Linking microbial community structure with function: fluorescence in situ hybridization – microautoradiography and isotope arrays. *Current Opinion in Biotechnology* **17:**83–91.
112. YAN, H., B. Y. ZHANG, and H. K. WU. 2008. Chemical cytometry on microfluidic chips. *Electrophoresis* **29:**1775–1786.
113. YANG, L., and R. BASHIR. 2008. Electrical/electrochemical impedance for rapid detection of foodborne pathogenic bacteria. *Biotechnology Advances* **26:** 135–150.
114. YANG, L. J. 2008. Electrical impedance spectroscopy for detection of bacterial cells in suspensions using interdigitated microelectrodes. *Talanta* **74:**1621–1629.

# Index